Ergonomics

PRENTICE HALL INTERNATIONAL SERIES
IN INDUSTRIAL AND SYSTEMS ENGINEERING

W. J. Fabrycky and J. H. Mize, Editors

AMRINE/RITCHEY/MOODIE/KMEC—*Manufacturing Organization and Management,* 7/E
ASFAHL—*Industrial Safety and Health Management,* 4/E
BABCOCK—*Managing Engineering and Technology,* 2/E
BADIRU—*Expert Systems Application in Engineering and Manufacturing*
BANKS/CARSON/NELSON—*Discrete Event System Simulation,* 3/E
BLANCHARD—*Logistics Engineering and Management,* 5/E
BLANCHARD/FABRYCKY—*Systems Engineering and Analysis,* 3/E
CANADA/SULLIVAN/WHITE—*Capital Investment Analysis for Engineering and Management,* 2/E
CHANG/WYSK/WANG—*Computer Aided Manufacturing,* 2/E
ELSAYED/BOUCHER—*Analysis and Control of Production Systems,* 2/E
FABRYCKY/THUESEN/VERMA—*Economic Decision Analysis,* 3/E
FISHWICK—*Simulation Model Design and Execution: Building Digital Worlds*
FRANCIS/MCGINNIS/WHITE—*Facility Layout and Location; An Analytical Approach,* 2/E
GRAEDEL/ALLENBY—*Industrial Ecology*
GROOVER–*Automation, Production Systems, and Computer- Integrated Manufacturing,* 2/E
HAMMER/PRICE—*Occupational Safety Management and Engineering,* 5/E
HAZELRIGG—*Systems Engineering: An Approach to Information—Based Design*
IGNIZIO/CAVALIER—*Linear Programming*
KROEMER/KROEMER/KROEMER-ELBERT–*Ergonomics: How to Design for Ease and Efficiency,* 2/E
LEEMIS—*Reliability; Probabilistic Models and Statistical Methods*
MUNDEL/DANNER—*Motion and Time Study: Improving Productivity,* 7/E
OSTWALD—*Engineering Cost Estimating,* 3/E
PINEDO—*Scheduling: Theory, Algorithms, and Systems,* 2/E
SHTUB/BARB/GLOBERSON—*Project Management: Engineering, Technology, and Implementation*
THUESEN/FABRYCKY—*Engineering Economy,* 9/E
TURNER/MIZE/CASE/NAZEMETZ—*Introduction to Industrial and Systems Engineering,* 3/E
WOLFF—*Stochastic Modeling and the Theory of Queues*

Ergonomics

How to Design for Ease and Efficiency

Second Edition

Karl H.E. Kroemer, Dr.Ing., CPE
Professor Emeritus, Virginia Tech, Blacksburg, VA

Henrike B. Kroemer, Ph.D.
Licensed Psychologist, Main Street Psychologists and Associates, Petoskey, MI

Katrin E. Kroemer-Elbert, Ph.D.
Principal Engineer, Gynecare Ethicon, Johnson & Johnson, Somerville, NJ

Prentice Hall International Series in Industrial and Systems Engineering
W.J. Fabrycky and J.H. Mize, Editors

Prentice Hall
Upper Saddle River, NJ 07458

Library of Congress Cataloging-in-Publication Data

Kroemer, K.H.E.,
Ergonomics: how to design for ease and efficiency / by Karl Kroemer, Henrike Kroemer, Katrin Kroemer-Elbert. —2nd ed.
p. cm.
Includes bibliographical references and index.
ISBN 0-13-752478-1 (case)
Human Engineering. 2. Engineering design. I. Kroemer, H.B. II. Kroemer-Elbert, K.E. (Katrin E.) III. Title.

TA166 .K77 2000
620.8'2-dc21 00-035677

Associate editor: *Laura Curless*
Editorial assistant: *Laurie Friedman*
Production editor: *Audri Anna Bazlen*
Vice-president and editorial director of ECS: *Marcia Horton*
Executive managing editor: *Vince O'Brien*
Managing editor: *David A. George*
Vice-president of production and manufacturing: *David W. Riccardi*
Art director: *Gus Vibal*
Art editor: *Adam Velthaus*
Cover design: *Bruce Kenselaar*
Manufacturing buyer: *Pat Brown*

Reprinted with corrections August, 2003.

Printed in the United States of America
20 19 18 17 16 15 14 13 12

ISBN 0-13-752478-1

Prentice-Hall International (UK) Limited, *London*
Prentice-Hall of Australia Pty. Limited, *Sydney*
Prentice-Hall Canada Inc., *Toronto*
Prentice-Hall Hispanoamericana, S.A., *Mexico*
Prentice-Hall of India Private Limited, *New Delhi*
Prentice-Hall of Japan, Inc., *Tokyo*
Pearson Education Asia Pte. Ltd., *Singapore*
Editora Prentice-Hall do Brasil, Ltda., *Rio de Janeiro*

For Hiltrud and Anne

Contents

Preface to the Second Edition

"What do we know about the human body and mind at work? Given what we know, how then should we design the work task, tools, the interface with the machine, and work procedures so that the human can perform safely, efficiently, and with satisfaction—perhaps even enjoy working?

These challenges are the main themes of this book. The solutions are the WHY and HOW of ergonomics/human engineering."

These first lines in the 1994 edition of *Ergonomics—How to Design for Ease and Efficiency* are still true in the year 2000. Thus, our book still has the same goals. With this second edition, getting there is a bit more convenient, because we have updated the content, improved the layout, and made the text easier to read.

We have left out many of the references that, in the early 1990s, were still needed to support ergonomic findings and recommendations which are now well established and internationally accepted. For ease of reading, we have abbreviated the author citations in the text: If there are two or more names, we give only the first, followed by "et al." Of course, we do provide the complete reference citations for all newly included material, so the third part of the book remains a substantive source of current ergonomic information.

Major additions are in the "basic knowledge," found in Chapters 1 through 6. Chapter 7 "Ergonomic Models, Methods, and Measurements" is essentially new. The material on application and design, in Chapters 8 through 12, has been thoroughly updated. The 1994 postcript has been expanded into Chapter 13 (no, we are not superstitious!), which includes discussions of macroergonomics and of the "economics of ergonomics."

We are grateful to all those who gave us suggestions for improvements. Special thanks go to M. Susan Hallbeck, William F. Moroney, Jeremy Rickards, and, especially, Jon A. Sangeorzean, who helped in many respects, particularly in the discussion of diving.

We encourage our readers to tell us of new developments to be considered in the future and remind us of past knowledge not to be forgotten.

- *Send mail* to Prof. Emeritus Karl Kroemer, Grado Department of Industrial and Systems Engineering, Virginia Tech., 205 Durham Hall, Blacksburg, VA 24061-0118, USA.
- *Telephone:* 540–231–6656.

- *Fax:* 540–231–3322
- E-mail: kroemer@vt.edu

We would like to hear from you!

Karl H. E. Kroemer
Henrike B. Kroemer
Katrin E. Kroemer-Elbert

Our special thanks are due to the outstanding professionals at Prentice Hall/Pearson Education: Brian and Edith Baker, copy editors par excellence; Audri Anna Bazlen, production editor; Laura Curless, acquisitions editor; Wolter J. Fabrycky, series editor; Marcia Horton, editorial director and Dolores Mars, the editorial supervisor who mended all that broke.

INTRODUCTION

Goals of Ergonomics

WHAT ERGONOMICS IS

Ergonomics is the application of scientific principles, methods, and data drawn from a variety of disciplines to the development of engineering systems in which people play a significant role. Among the basic disciplines are psychology, cognitive science, physiology, biomechanics, applied physical anthropometry, and industrial systems engineering. The engineering systems to be developed range from the use of a simple tool by a consumer to a multiperson, sociotechnical system.

Ergonomic specialists involved in the system design process, according to the National Research Council [1983, pages 2–3], "are united by a singular perspective on [that] process: . . . [D]esign begins with an understanding of the user's role in overall system performance and that systems exist to serve their users, whether they are consumers, system operators, production workers, or maintenance crews. This user-oriented design philosophy acknowledges human variability as a design parameter. The resultant designs incorporate features that take advantage of unique human capabilities as well as build in safeguards to avoid or reduce the impact of unpredictable human error."

Success is measured by improved productivity, efficiency, safety, acceptance of the resultant system design and, last, but truly not least, improved quality of human life.

WHAT ERGONOMICS DOES

There is a hierarchy of goals in ergonomics. The fundamental task is to generate "tolerable" working conditions that do not pose known dangers to human life or health. When this basic requirement is assured, the next goal is to generate "acceptable" conditions upon which the people involved can voluntarily agree, according to current scientific knowledge and under given sociological, technological, and organizational circumstances. Of course, the final goal is to generate "optimal" conditions which are so well adapted to human characteristics, capabilities, and desires, that physical, mental, and social well-being is achieved.

To be more specific, we may define ergonomics (also called human factors or human engineering in the United States) as the study of "human characteristics for the appropriate design of the living and work environment." Its fundamental aim is that all human-made tools, devices, equipment, machines, and environments should advance, directly or indirectly, the safety, well-being, and performance of human beings.

Thus, ergonomics has two distinct aspects:

1. Study, research, and experimentation, in which we determine specific human traits and characteristics that we need to know for engineering design.
2. Application and engineering, in which we design tools, machines, shelter, environments, work tasks, and job procedures to fit and accommodate the human. This aspect includes, of course, the observation of the actual performance of humans and equipment in the environment, to assess the suitability of the designed human–machine system and to determine possible improvements.

Ergonomics adapts the human-made world to the people involved because it focuses on the human as the most important component of our technological systems. Thus, the utmost goal of ergonomics is the "humanization" of work. This goal may be symbolized by the "E & E" of ease and efficiency, for which all technological systems and their elements should be designed. Such design requires knowledge of the characteristics of the people involved, particularly of their dimensions, their capabilities, and their limitations.

Ergonomics is neutral: It takes no sides—neither employers' nor employees'. It is not for or against progress. It is not a philosophy, but a scientific discipline and technology.

GOALS OF THIS BOOK

In this book, we discuss the human interaction with work task and technology. Our intention in exploring this interaction is to build a knowledge-based understanding so that we can

- amplify human capabilities,
- utilize human abilities,
- facilitate human efficiency, and
- avoid overloading or underloading ourselves in our tasks on the job.

This understanding will benefit a variety of specialists and generalists who are concerned with people's performance and well-being at work. Among these professionals are the designer, engineer, architect, ergonomist, human factors specialist, industrial hygienist, industrial physician, occupational nurse, manager, student in any of these areas, and, of course, everybody interested in "humanizing work"—i.e., making work safe, efficient, and satisfying.

HOW THIS BOOK IS ORGANIZED

The book has three major parts:

Part One, "The Ergonomics Knowledge Base," consists of Chapters 1 through 6. Here, we explore the properties of the human body and mind, as manifested in people's interactions with the environment. The focus is on human dimensions, capabilities, and limitations—the human factors to be considered in designing for ease and efficiency. For convenience, we use anatomy, physiology, and psychology as traditional disciplinary divisions, whereas, in fact, the human functions holistically and synergistically.

Part Two, "Design Applications," contains Chapters 7 through 13. Here, we discuss the design of tasks, equipment and environment in the light of our knowledge about human size, strengths, and weaknesses—the knowledge base developed in Part One. This information is brought to light with the aim of matching human capabilities with demands.

Part Three, "Further Information," includes a listing of all references mentioned in the text, an extensive glossary with concise descriptions and definitions, and a detailed index that refers the reader to specific pages.

HOW TO USE THE BOOK

You may use this book in three ways:

1. Read it straight through from beginning to end, as in a university course, and work on the "challenges" listed at the end of each chapter.
2. Read a chapter of interest, absorb the background information, and proceed to design applications that make use of that information.
3. Start with the index, pick a topic of interest, and look up the information in the book sections that the index cites.

THE DEVELOPMENT OF ERGONOMICS

Using objects found in the environment as tools or weapons is an ancient and fundamental activity that distinguishes humans from many other primates. Pieces of stone, bone, and wood were at first not shaped but rather selected for their fit to the human hand and their suitability as scrapers, pounders, and missiles. Purposeful shaping of these tools was the next step. Creating finished products from raw materials and manufacturing in quantity followed. Fitting clothes and making shelters also were early and fundamental "ergonomic" activities.

As human society grew more complex, organizational and managerial challenges developed. Training workers and soldiers, for example, became necessary, together with forming and controlling their behavior. For major projects, such as building the pyramids of ancient Egypt, assembling armies for warfare, sheltering the inhabitants of cities, and supplying them with food and water, sophisticated knowledge of human needs and desires was required, and careful planning and complex logistics had to be mastered. The aims and means of training became sophisticated as well. Roman soldiers, for example, underwent well-organized training and conditioning until they could perform military exercises without sweat accumulating uselessly on their skin.

☛☛☛ *"Drying the legions" of the Roman Empire relied, consciously or by experience, on the principle of training and adapting the physiological capabilities of the recruits to their physical requirements—until they did not show sweat anymore on their skin; they were "dry," meaning fit.* ☚☚☚

Evolution of Disciplines

Artists, military officers, employers, and sports enthusiasts were always interested in body build and physical performance. Specialized "medicine men" and "herb women" treated illnesses and injuries. The disciplines of anatomy and anthropology began to develop. About 400 B.C., Hippocrates described a scheme of four body types, which were supposedly determined by their fluids. The "moist" type was believed to be dominated by black gall, the "dry" type by yellow gall, the "cold" type by slime, and the "warm" type by blood.

Over the centuries, more exact information accumulated into specialized disciplines. In the 15th to 17th centuries, gifted persons such as Leonardo da Vinci and Alfonso Giovanni Borrelli could still master all the existing knowledge of anatomy, physiology, and equipment design; these individuals were artist, scientist, and engineer in one.

In the 18th century, the sciences of anatomy and physiology diversified and accumulated specific detailed knowledge. Psychology began to develop as a separate science. Well into the 19th century, these sciences tended to be oriented toward theories, trying to understand the complex human being: the stereotype is the scientist in a white coat leading a life devoted to research in the laboratory. But increasing industrialization with its employment of human workers, together with the old interest in military deployment of the humans to military needs, brought forward "applied" aspects of the formerly "pure" sciences. In the early 1800s, in France, Lavoisier, Duchenne, Amar, and Dunod researched energy capabilities of the working human body, Marey developed methods to describe human motions at work, and Bedaux made studies to determine work payment systems, before Taylor and the Gilbreths did similar work in the United States in the early 1900s. In England, the Industrial Fatigue Research Board considered theoretical and practical aspects of the human at work. In Italy, Mosso constructed dynamometers and ergometers to research fatigue. In Scandinavia, Johannsson and Tigerstedt developed the scientific discipline of work physiology. A Work Physiology Institute was founded in Germany by Rubner in 1913. In the United States, Benedict and Cathcard (1913) described the efficiencies of muscular work. The Harvard Fatigue Laboratory was established in the 1920s. (For more details, check the books by Lehmann 1962, and Brouha 1967.)

In the first half of the 20th century, industrial physiology and psychology were well advanced and widely recognized, both in their theoretical research "to study human characteristics" and in the application of this knowledge "for the appropriate design of the living and work environment." Two distinct approaches to studying human characteristics had developed, one concerned chiefly with physiological and physical properties of the human, the other interested mainly in psychological and social traits. Although there was much overlap between these approaches, the physical and physiological aspects were studied mainly in Europe and the psychological and social aspects in North America.

Directions in Europe

Based on a broad fundament of anatomical, anthropological, and physiological research, applied or "work" physiology assumed great importance in Europe, particularly during the hunger years associated with the First World War. Marginal living conditions stimulated research on the minimal nutrition required to perform certain activities, the consumption of energy while carrying out agricultural, industrial, military, and household tasks, the relationships between energy consumption and heart rate, the use of muscular capabilities, suitable body postures at work, the design of equipment and workplaces to fit the human body, and related topics. Another development in the 1920s was "psychotechnology," which involved testing persons for their ability to perform physical and mental work, their vigilance and attention, their ability to carry mental workload, their behavior as drivers of vehicles, their ability to read road signs, and related topics.

Directions in North America

Most psychologists around 1900 were strictly scientific and deliberately avoided studying problems that strayed outside the boundaries of pure research. Some investigators, however, had practical concerns, such as sending and receiving Morse code, measuring perception and attention at work, using psychology in advertising, and promoting industrial efficiency.

A particularly important step was the development of "intelligence testing," used to screen military recruits during the First World War and, later, to screen industrial workers for jobs appropriate to their mental capabilities. The terms "intelligence testing" and "industrial psychology" won acceptance. Gould (1981) provides a partly amusing, partly disturbing account of the early years of intelligence testing.

The term "industrial psychology" first appeared as a typographical error, actually meant to read "individual" psychology.

Among the best known, most puzzling findings in industrial psychology are those yielded by the experiments at the Hawthorne Works near Chicago in the mid-1920s. This study was designed to assess relationships between lighting and efficiency in workrooms where electrical equipment was produced. The bizarre finding was that the workers' productivity increased whether or not the illumination was changed, apparently in response to the attention paid to the workers by the researchers (Roethlisberger and Dickson 1943; Jones 1990; Parsons 1974, 1990). This phenomenom became known as the Hawthorne Effect.

Industrial psychology divided into special branches, including personnel psychology, organizational behavior, industrial relations, and engineering psychology. Under the pressures of the Second World War, the "human factor" as part of a "man–machine system" became of major concern. Technological development led to machines and systems that put higher demands on the attention, strength, and endurance of individuals and teams than many could muster. For example, operators had to observe radar screens over periods of many hours, with the intent of detecting and distinguishing some blips from others. In high-performance aircraft the pilot was subjected to forceful accelerations, such as those experienced during sharp turns. In these instances, the pilot might be unable to operate hand controls properly and could even black out. Crew members had to fit into tank and aircraft cockpits that were narrow and low, requiring that small persons be selected. Stressful conditions made it difficult to maintain combat morale and performance. Thus, military activities and related efforts on the "home front" generated the need to consider human physique and psychology, purposefully and knowingly, in the design of tasks, equipment, and environments.

Names for the Discipline: "Ergonomics" and "Human Factors"

In Europe and North America, various terms were used to describe the activities of anthropologists, physiologists, psychologists, sociologists, statisticians, and engineers who studied the human and used the information obtained thereby in design, selection, and training.

On January 13 and 14, 1950, British researchers met in Cambridge, England, to discuss the name of a new society to represent their activities. Among others, the term "ergonomics" was proposed. This word had been coined in late 1949 by K. F. H. Murrell, who derived it from the Greek terms *ergon,* indicating work and effort, and *nomos,* meaning law or usage, apparently reinventing a word already used in Poland a hundred years earlier (Monod and Valentin 1979). The term was neutral, implying no priority of contributing disciplines, such as physiology or psychology or functional anatomy or engineering. It was easily remembered and recognized and could be used in any language. "Ergonomics" was formally accepted as the name of the new society at its council meeting on February 16, 1950 (Edholm and Murrell 1974).

Two sidelights are worthy of a brief note. The original proposal for the name included two alternative suggestions. One was the Ergonomic Society. Note that there is no "s" at the end of the word "Ergonomic"; apparently, a final "s" somehow slipped onto the ballot and since that time has made deriving an adjective or adverb difficult. The alternative name, "Human Research Society," bears some similarity to the term "human engineering."

In the United States, a group of persons convened in 1956 to establish a formal society. The name "ergonomics" was rejected, and instead "human factors" was selected. Often, the word "engineering" is added or substituted to indicate applications, as in "human (factors) engineering" (Christensen et al., 1988). In 1992, the Human Factors Society renamed itself the Human Factors and Ergonomics Society.

☛☛☛ *There has been some discussion of whether human factors differs from ergonomics—whether one relies more heavily on psychology or physiology or is more theoretical or practical than the other. Today, the two terms are usually considered synonymous, as exemplified by the naming practice of the Canadian Society, which uses "human factors" in its English name, and "ergonomie" in its French version.* ☚☚☚

THE ERGONOMIC KNOWLEDGE BASE

Ergonomics today is growing and changing. The discipline's development stems from increasing and improving knowledge about the human and is driven by new applications and new technological developments.

As mentioned, several classic sciences provide fundamental knowledge about human beings. The anthropological basis of such knowledge consists of anatomy, describing the build of the human body; orthopedics, concerned with the skeletal system; physiology, dealing with the functions and activities of the living body, including the physical and chemical processes involved; medicine, concerned with illnesses and their prevention and healing; psychology, the science of mind and behavior; and sociology, concerned with the development, structure, interaction, and behavior of individuals or groups. Of course, physics, chemistry, mathematics, and statistics also supply knowledge, approaches, and techniques.

From these basic sciences, a group of more applied disciplines developed into the core of ergonomics. These disciplines consist primarily of anthropometry, the measuring and description of the physical dimensions of the human body; biomechanics, describing the physical behavior of the body in mechanical terms; industrial hygiene, concerned with the control of occupational health hazards that arise as a result of doing work; industrial psychology, discussing people's attitude and behavior at work; management, dealing with and coordinating the intentions of the employer and the employees; and work physiology, applying physiological knowledge and measuring techniques to the body at work. Of course, many other disciplinary areas such as labor relations, have developed that also are part of, or contribute to, or partly overlap with, ergonomics.

Several distinct application areas use ergonomics as components of their knowledge base and work procedures. Among these are industrial engineering, concerned with the interactions among people, machinery, and energies; bioengineering, working to replace worn or damaged body parts; systems engineering, in which the human is an important component of the overall work unit; safety engineering and industrial hygiene, which focus on the well-being of humans; and military engineering, which relies on the human being as a soldier or an operator. Naturally, other application areas are in urgent need of ergonomic information and data, such as computer-aided design, in which information about the human must be provided in computerized form. Oceanographic, aeronautical, and astronautical engineering also rely on ergonomic knowledge.

The development from basic sciences to applied disciplines in ergonomics and the use of ergonomic knowledge in specific areas are depicted schematically in Figure 1. As more knowledge about humans becomes available, as new opportunities develop to make use of human capabilities in modern systems, and as needs arise for protecting the person from outside events, ergonomics changes and develops.

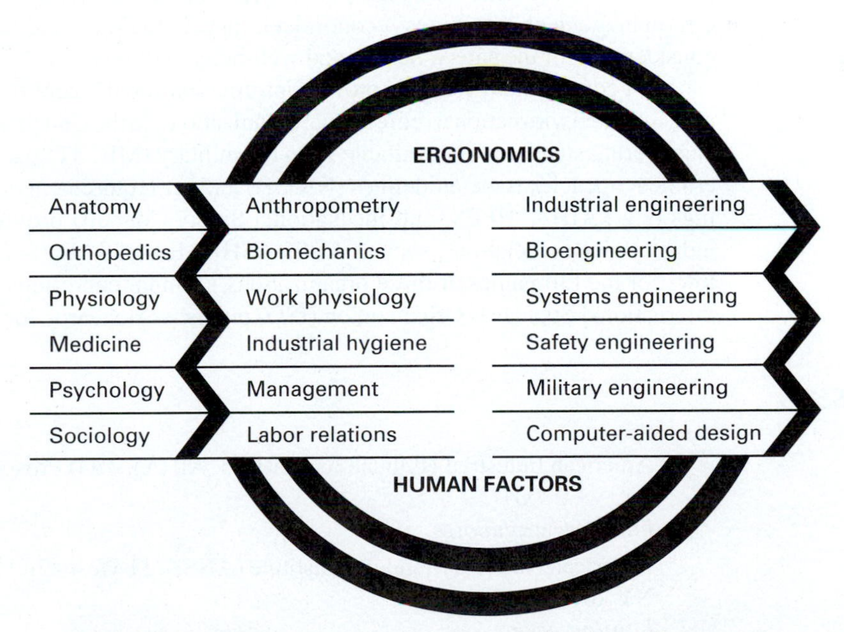

FIGURE 1: Origins, developments, and applications of ergonomics.

Professional Organizations

The **Ergonomics Society** is located in the United Kingdom. (See the section on addresses at the end of this introduction.) Founded in 1950, it is the oldest ergonomics organization. The Ergonomics Society has some 800 members, mostly in the United Kingdom. The Society supports two journals, *Ergonomics* and *Applied Ergonomics,* and organizes an annual professional congress.

The largest single (national) professional organization in ergonomics/human factors is the **Human Factors and Ergonomics Society,** in the United States, with about 5,000 members and various special technical interest groups. The society publishes two journals, *Human Factors* and *Ergonomics in Design,* and organizes an annual professional congress.

More than three dozen national and regional societies exist, many carrying the term "Ergonomics" in their name. Most are members of the **International Ergonomics Association,** founded in 1959. Over 16,000 ergonomists are working worldwide.

SOURCES OF ERGONOMIC INFORMATION

After the initial development from basic "theoretical sciences" to applications of ergonomic knowledge, several major consumers and generators of human-factors information emerged. First among them is the military (like it or not), which must "integrate" people (soldiers) and equipment (weapons). Thus, military regulations and standards (such as MIL STD 1472 in the United States) are still major sources of ergonomic information. Industries that supply the military are immediately affected, because they must comply with human engineering requirements.

Industries and occupations (such as mining, forestry, agriculture, construction, steel, tire manufacturing, and nurses and their aides) that require much physical work are in need of physiological and biomechanical information. Newer, high-tech industries (such as manufacturing of automobiles, aircraft, spacecraft, and computers) need similar knowledge, but with a shift in emphasis toward sensorimotor, perception, decision-making, and sociopsychological aspects. Modern management combines economical, psychological, and engineering aspects with a consideration of the safety, health, and well-being of humans.

Of course, many journals provide information about recent findings. Most of these journals are associated with national professional organizations. In the United States, well-established human engineering standards are available from the military (MIL STD) and NASA. Furthermore, government agencies issue guidelines (NIOSH) and set requirements (OSHA). Professional groupings (e.g., AIHA, HFES, and the National Safety Council) provide specialized guides, as do industrywide associations, such as ANSI, ASHRAE, and SAE. (See the list of addresses that follows for the full names of these organizations.) Similar conditions exist in many countries. The International Standards Organization (ISO) publishes standards for worldwide use.

ADDRESSES

American Industrial Hygiene Association (AIHA), 2700 Prosperity Avenue, Suite 250, Fairfax, VA 22031
http://www.aiha.org

American National Standards Institute (ANSI), 11 West 42nd Street, 13th Floor, New York, NY 10036
http://www.ansi.org

American Society of Heating, Refrigerating, and Air-conditioning Engineers (ASHRAE), 1791 Tullie Circle, Atlanta, GA 30329
http://www.ASHRAE.org

Crew System Ergonomics Information Analysis Center (CSERIAC) AFRL/HEF/CSERIAC, Bldg. 196, 2261 Monahan Way, Wright-Patterson AFB, OH 45433-7022
http://cseriac.wpafb.af.mil

Ergonomics Society, Devonshire House, Devonshire Square, Loughborough, Leics., LE 11 3DW, UK
http://www.ergonomics.org.uk

Human Factors and Ergonomics Society, P.O. Box 1369, Santa Monica, CA 90406-1369, USA
http://www.hfes.org

Military and federal standards, handbooks, and specifications are available from the following sources:

National Technical Information Service (NTIS), 5285 Port Royal Road, Springfield, VA 22161
http://www.NTIS.org

NASA National Aeronautics and Space Administration, SP 34-MSIS, LBJ Space Center, Houston, TX 77058
http://www.nasa.gov

National Academy of Sciences, 2101 Constitution Avenue, Washington, DC 20418
http:www.nas.edu

National Safety Council, 1121 Spring Lake Drive, Itasca, IL 60143-3201
http://www.nsc.org

National Institute of Occupational Safety and Health (NIOSH), 4676 Columbia Parkway, Cincinnati, OH 45226
http://www.cdc.gov.niosh

Occupational Safety and Health Agency (OSHA), 200 Constitution Avenue, NW, N3651, Washington, DC 20210
http://www.osha.gov

Society of Automotive Engineers (SAE), 400 Commonwealth Drive, Warrendale, PA 15096-0001
http://www.sae.org

ISO International Organization for Standardization (ISO), 1 Rue Varembe, Case Postale 56, CH-1211 Genève 20, Switzerland
http://www.ISO.org

part I

THE ERGONOMIC KNOWLEDGE BASE

Chapter 1

The Anatomical and Mechanical Structure of the Human Body

OVERVIEW

To design things to "fit" the human body, we must know the dimensions of people. The development of the human race has led to much variability in body build among humans. Body sizes of some populations have been measured, but have only been estimated for most people on earth. For North Americans, the dimensions of adult civilians must be derived from data taken on soldiers. Percentile values and relationships among measurements allow us to calculate estimated body dimensions.

Biomechanically, one can describe the human body as a basic skeleton whose parts are linked in joints; the members have volumes and mass properties and are moved by muscles. Understanding the properties, capabilities, and limitations of the body allows us to design equipment and tools that use and enhance human strengths.

DEVELOPMENT OF THE HUMAN RACE

Development of the human being

The human species grew like a bush: Some branches developed and died, while others grew more and more twigs, some of which vanished while others flourished. In this development, variations occurred in body dimensions of groups of people. Ergonomic design must take these variations into account.

The development of the human race can be traced from fossils and from reconstruction of mitochondrial DNA over several million years in Africa, for hundreds of thousands of years in Europe and Asia, and for some 20 thousand years in the Americas.

Homo in Africa

Apparently, *Australopithecus* was a predecessor of the genus *Homo* about 3 million years ago in Africa, where *Homo Erectus* subsequently developed. One humanoid branch started about 250 thousand years ago and remained in Africa. Another branch developed 60 or 70 thousand years later. Some of its members stayed in Africa and others spread all over the earth.

Neanderthals and Cro-Magnons

Remains of anatomically modern humans who lived about 130 to 180 thousand years ago have been found in South Africa and in the Levant. There and in central Europe, about 150 thousand years ago, the Neanderthals emerged. Stocky, heavy set, and well adapted to cold, they had a brain as big as modern humans. Neanderthals lived tens of thousands of years side by side with Cro-Magnons, but vanished about 30 thousand years ago. Cro-Magnons lived in the area of what today is Israel around 90 thousand years ago; they probably developed into *Homo sapiens*. Popular notions about the different appearances of Cro-Magnons and Neanderthals are mostly based on conjecture, often in the style of a Hollywood movie. For example, there is no indication that the Cro-Magnons were dark or the Neanderthals light skinned. Furthermore, there is no evidence of violent struggles for superiority between the two races.

First humans in Africa

Bushmen and Pygmies probably occupied most of subequatorial Africa until about 2 thousand years ago. Bantu-speaking people living in the Congo area learned to use iron, developed agriculture, and domesticated animals. The flourishing Bantu then drove Bushmen and Pygmies into areas unsuitable for agriculture. Subsequently, 60 million Bantu occupied half the African continent.

Humans spreading from Africa

From Africa, genus *Homo* spread over the earth. About 50 thousand years ago, Australia was settled by early humans who arrived from eastern Indonesia. Their descendants became the Aboriginal population. Most of the current inhabitants of Indonesia, the Philippines, and parts of Southeast Asia are descendants of a population that emigrated from Taiwan about 6 thousand years ago.

Emigrants from Asia settled the Americas. Waves of peoples, the earliest probably about 20 thousand years ago, crossed what was then the Bering land bridge to Alaska and went south along the Pacific coast. It is believed that other groups later moved into the areas that are today Canada and the United States. The descendants of these waves of peoples became the ancestors of North and South American Indians.

Europe, after its long history of pre-Neanderthals, Neanderthals, and Cro-Magnons, has been reconstituted twice fairly recently: about 8 thousand years ago by farmers from the Near East and about 2 thousand years later by Indo-Europeans from southern Russia.

In 1776, the German anthropologist Johann Friedrich Blumenbach (1752–1840) divided groups of humans into "races": Caucasians (Europeans, "white"), Mongolians (East Asians, "yellow"), Malayans (Southeast Asians and Pacific Islanders, "brown"), Ethiopians (sub-Saharan Africans, "black"), and (Native) Americans ("red"). These divisions, leaving out Native Australians entirely, relied on superficial differences in skin color, hair form, and eye lid shape.

Population growth

Thus, the human stock, with its many current branches, appears African in origin and about a quarter-million years old. Today, the number of people is growing fast; "population explosions" are occurring in some parts of the earth. The total number of humans was about 10^9 (a thousand million, ie, 1 billion) around 1800. In 1900, about 1.7 billion people lived on earth. The second billion was reached by 1930. The third billion was attained in 1960, the fifth in 1987. In 1998, about 5.8 billion were alive, 4.6 billion of which living in developing countries.

If current birth and death rates continue, 80 billion people will live on earth in 2100, and 150 billion will crowd it in 2125. For 2050, between 9 and 11 billion people are probable, with the current 1.2 billion or so in developed countries still about the same then. Most children will be born in third-world countries, where food supplies are insufficient even now; of the 5 billion people on earth in 1987, a relatively small proportion of about 1.2 billion lived in industrialized countries.

Local population changes

Emigration from certain areas and immigration to others are on a much smaller scale than population growth, but can be of great importance locally. In North America, for example, during the last few centuries, waves of immigrants from certain geographical areas have been changing the composition of the population, replacing most Native Americans by Europeans. Today, in the United States, the influx of Cubans and Haitians is strongly felt in Florida, the arrival of South Americans affects southwestern states, and Asians are quite in evidence along the Pacific coast.

Marco Polo (1254–1324) traveled to the Far East and stayed in China for 20 years. There he found a nation far in advance of Europe in population, wealth, technology, and the civilized amenities. He returned in 1292 to Italy, where he was taken prisoner in the war between Venice and Genoa. While detained, he began to dictate his reminiscences of China, published in 1298. Although largely disbelieved, his book was immensely popular and stimulated much interest in the study of other countries (Asimov 1989).

ANTHROPOLOGY AND ANTHROPOMETRY

Measurement of the human body

Anthropology, the study of mankind, was primarily philosophical and esthetical in nature until about the middle of the nineteenth century. Yet, the size and proportions of the human body have always been of interest to artists, warriors, and physicians. Physical anthropology is that scientific subdiscipline of anthropology in which the parts of the body—particularly its bones—are measured and compared. In the middle of the 19th century, the Belgian statistician Adolphe Quetelet first applied statistics to anthropological data. This was the beginning of modern anthropometry, the measurement of the human body. By the end of the 19th century, anthropometry was a widely applied scientific discipline, used both in measuring the bones of early people and in assessing the body sizes and proportions of contemporaries. A new offspring, biomechanics, had already developed. Today, engineers have become highly interested in the application of anthropometric and biomechanical information, especially to the design of equipment and the arrangement of workstations.

In 1316, Mondino D. Luzzi, professor at the medical school of Bologna, Italy, published the first book devoted entirely to anatomy. The Flemish anatomist Andreas Vesalius, in the early 16th century, upset many traditional, but false, Greek and Egyptian notions about the anatomy of the human body in his book De Corporis Humani Fabrica, *concerning the structure of the human body. The book contained careful illustrations of anatomical facts, drawn by a student of Titian (Asimov 1989).*

Standardization of measurement

The unification of measuring methods became necessary and was achieved primarily by anthropologists who convened 1906 in Monaco and 1912 in Geneva. They established bony landmarks on the body, to and from which to take measurements. In 1914 an authoritative textbook was published, Martin's *Lehrbuch der Anthropologie,* editions of which shaped the discipline for several decades. Beginning in the 1960s, new engineering needs for anthropometric information, newly developing measuring techniques, and advanced statistical considerations stimulated the need for updated standardization. In the 1980s, the International Standardization Organization (ISO) began efforts to standardize anthropometric measures and measuring techniques worldwide.

Measurement Techniques

Body measurements are usually defined by the two endpoints of the distance measured. For example, forearm length is often measured as elbow-to-fingertip distance; stature (height) starts at the floor on which the subject stands and extends to the highest point on the skull.

> For the measurement of stature, the subject assumes one of four customary positions: (1) standing naturally upright, (2) standing stretched to maximum height, (3) leaning against a wall with the back flattened and the buttocks, shoulders, and back of the head touching the wall, and (4) lying on the back. The difference between measurements when the standing subject either stretches or just stands upright can easily be 2 cm or more. Lying supine results in the greatest measure. This example shows that standardization is needed to assure uniform postures and comparable results.

Specific terminology and measuring conventions have been described in English by Garrett and Kennedy (1971), Gordon et al. (1989), Kroemer (1999a), Kroemer et al. (1997), Lohman et al. (1988), NASA/Webb (1978), and Roebuck (1995). These publications provide complete information about traditional measurement procedure and techniques.

Height is a straight-line, point-to-point vertical measurement. **Breadth** is a straight-line, point-to-point horizontal measurement running across the entire body or a body segment. **Depth** is a straight-line, point-to-point horizontal measurement running fore and aft the body. **Distance** is a straight-line, point-to-point measurement between landmarks on the body. **Curvature** is a point-to-point measurement following a contour; this measurement is usually neither closed nor circular. **Circumference** is a closed measurement that follows a body contour; hence, this measurement is not circular. Reach is a point-to-point measurement following the long axis of the arm or leg.

The standard reference planes are the medial (mid-sagittal), the frontal (or coronal), and the transverse planes, usually thought to meet in the center of mass of the whole body.

Figure 1–1 shows reference planes and descriptive terms. Figures 1–2 and 1–3 illustrate anatomical landmarks on the human body.

For most measurements, the subject's body is placed in a defined upright straight posture, with body segments either in line with each other or at a right angle to each other. For example, the subject may be required to "stand erect; heels together; buttocks, shoulder blades and back of head touching a wall; arms and fingers straight and vertical." This is similar to the so-called anatomical position.

The head is positioned to be "straight": the pupils are on the same horizontal level, and the Ear–Eye Line is angled 15 degrees against the horizon, as shown in Figure 1–4. (This was formerly called putting the head into the "Frankfurt Plane.") The EE line is shown in Figure 1–4.

When measurements are taken on a seated person, the flat and horizontal surfaces of the seat and the foot support are so arranged that the thighs are horizontal, the lower legs vertical, and the feet flat on their horizontal support. The subject is nude, or nearly so, and does not wear shoes.

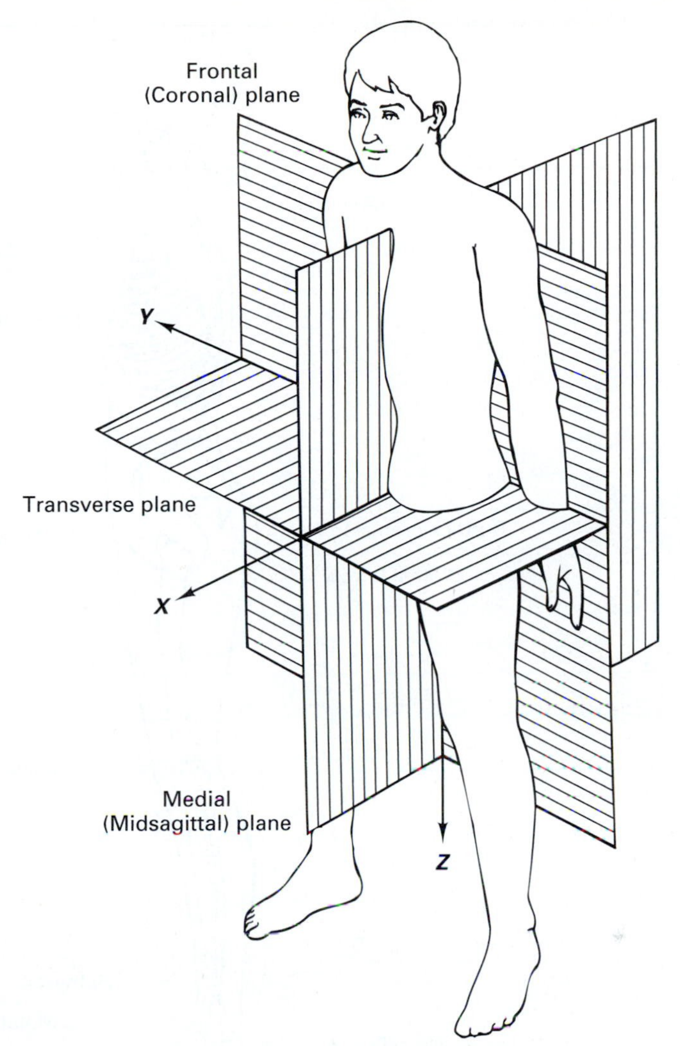

Figure 1–1. Terms and measuring planes used in anthropometry.

The grid technique

Classical measuring techniques. The conventional measurement devices are quite simple. In the *Morant technique,* one uses a set of grids, usually attached to the inside corner of two vertical walls meeting at right angles. The subject is placed in front of the grids, and projections of the body onto the grids are used to measure anthropometric variables. Boxlike jigs with grids provide references for the measurement of head and foot dimensions.

Measuring instruments

Many bony landmarks cannot be projected easily onto grids. In this case, special instruments are used. The most important is the *anthropometer*, a graduated rod with a sliding edge at a right angle. The rod can be taken apart for transport and storage, but, put together, it is 2 meters long. (Anthropometric data are traditionally recorded in metric units.) The *spreading caliper* consists of two curved branches joined in a hinge. The distance between the tips of the branches is read from a scale. A small *sliding caliper* can be used for short measurements, such as finger

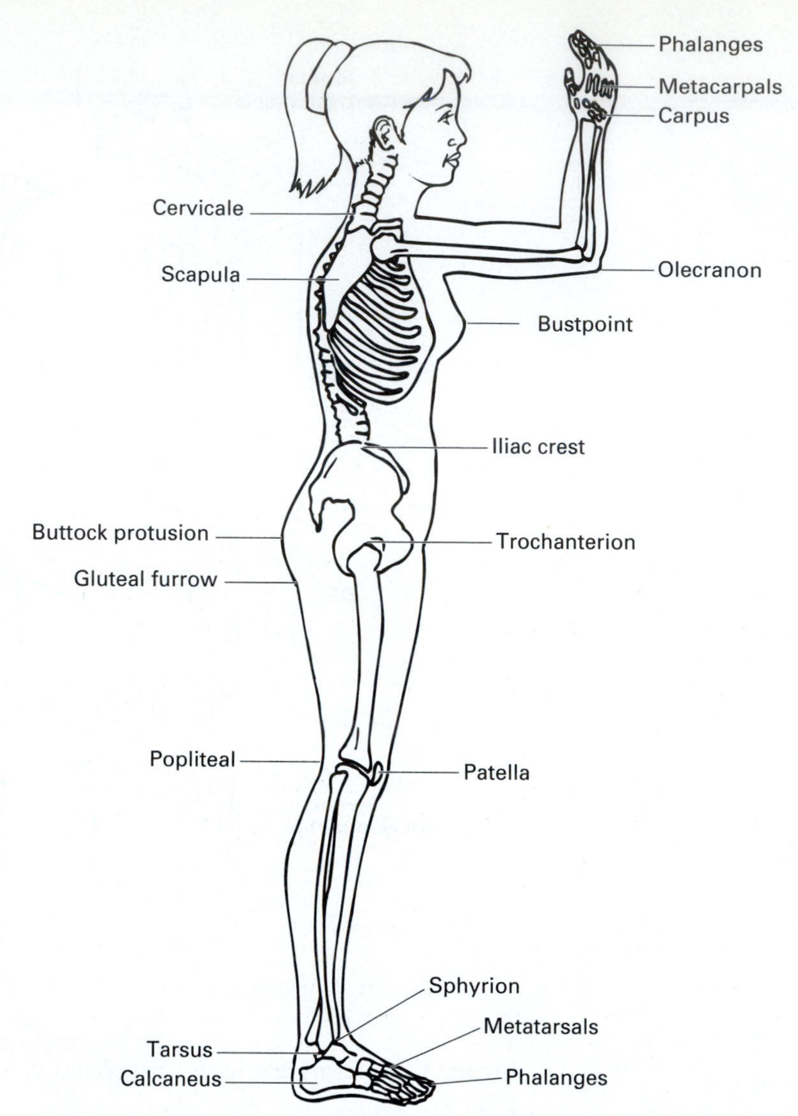

Figure 1–2. Anatomical landmarks in the sagittal view. (From Kroemer et al. 1990, *Engineering Physiology: Bases of Human Factors/Ergonomics,* 2d ed. With permission by the publisher, Van Nostrand Reinhold. All rights reserved.)

thickness or finger length. A special caliper is used to measure the thickness of skinfolds. A *cone* is employed to measure the diameter around which fingers can close. Circular holes of increasing size drilled in a thin plate serve to measure the external diameter of the finger. Circumferences and curvatures are measured with *tapes*. A scale is used to measure the weight of the body. Many other measuring methods can be applied in special cases, such as the shadow technique, the use of templates, or casting; these methods are explained in the reference books listed earlier.

Simple but clumsy

Most traditional measuring instruments are applied by the hand of the measurer to the body of the subject. This approach is simple, but time consuming; also, it requires that each measure-

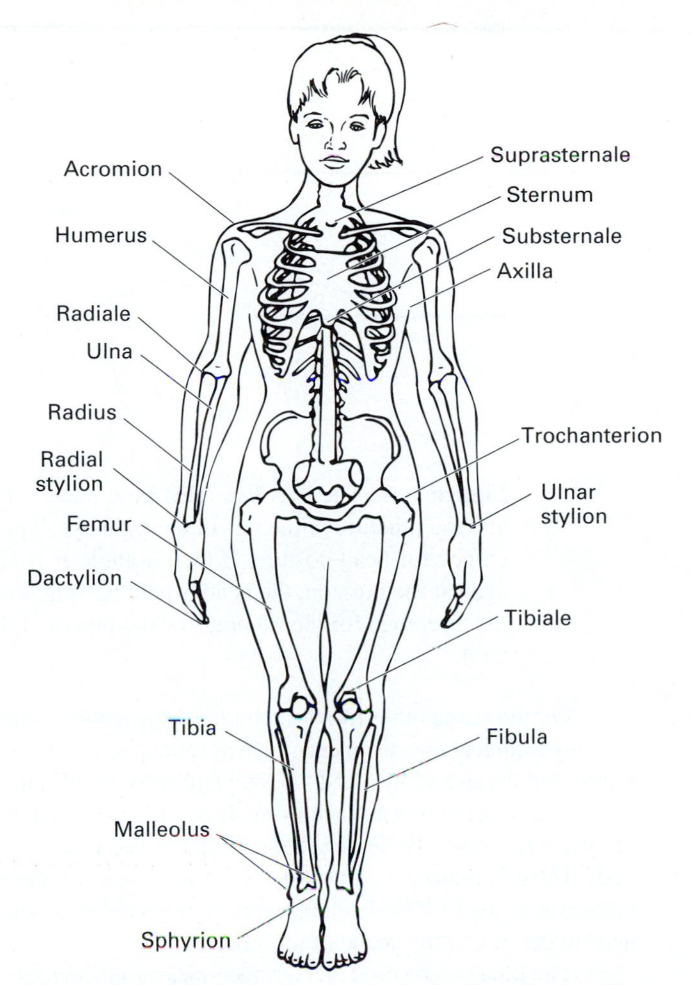

Figure 1–3. Anatomical landmarks in the frontal view. (From Kroemer et al. 1990, *Engineering Physiology: Bases of Human Factors/Ergonomics,* 2d ed. With permission by the publisher, Van Nostrand Reinhold. All rights reserved.)

ment and tool be selected in advance and implies that what was not measured in the test session remains unknown.

Not 3-D

A major shortcoming of the classical measurement techniques is that they leave many of the body dimensions unrelated to each other in space. For example, as one looks at a subject from the side, the person's stature, eye height, and shoulder height are located in different undefined frontal planes. Another shortcoming is that contact measurements cannot be made on certain parts of the body, such as the eyes, which are sensitive.

Photo to video

New measurement techniques. Photographs can record all three-dimensional aspects of the human body. Photos allow the recording of practically infinite numbers of measurements, which can be taken from the record at one's convenience. Photos also have drawbacks, however: The body is depicted in two dimensions, a scale may be difficult to establish, distortions occur due to parallax, and bony landmarks under the skin cannot be palpated on the photograph.

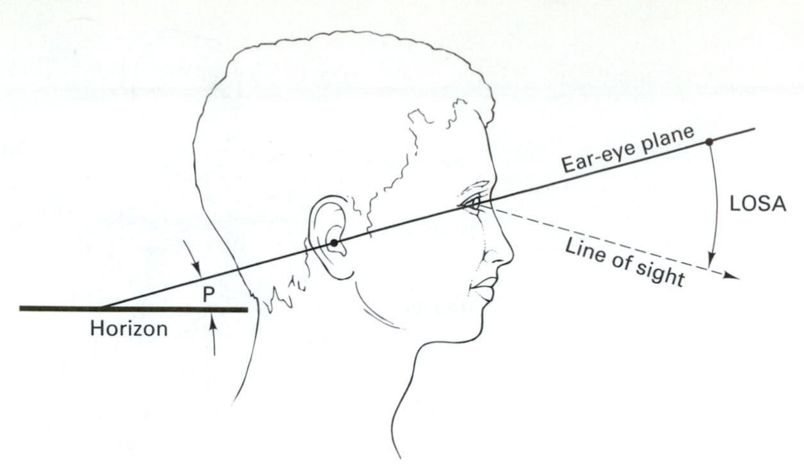

Figure 1–4. The Ear–Eye (EE) Line passes through the right ear hole and the juncture of the lids of the right eye. The EE Line serves as a reference for head posture: If the tilt angle *P* is about 15 degrees inclined against the horizon, the head is held upright (erect). The EE Line is also the reference for the tilt angle of the line of sight, as discussed in Chapter 4.

For these and other reasons, photographic anthropometry has not been widely used, in spite of many technical improvements, such as stereophotometry with several cameras or mirrors, holography, and the use of film and videotape instead of still photography.

Many techniques for acquiring three-dimensional anthropometric data have been proposed in the past. Some rely on projecting a regular geometric grid onto the irregularly shaped human body. The grid remains regular when viewed along its axis of projection, but appears distorted when viewed at an angle. The displacements of projected grid points from their regular positions can be used to determine the shape of the surface.

Laser use

The laser can be used as a distance-measuring device to determine the shape of irregular bodies. Either the body to be measured is rotated, or the sending and receiving units of the laser device rotate around the body. Markers may be placed on points of the surface so that the laser can recognize them, for example, as indicating the location of an important bone landmark. A computer is used to store and organize the data, whereby the body surface can be described in minute, 3-D detail, with mathematical techniques derived from topography. Laser-based anthropometry should allow the exact description of the shape of the body not only in the traditional static position, but also when it changes with motion, training, or aging.

☛☛☛ *In 1998, the Civilian American and European Surface Anthropometry (CAESAR) project was started. This is a laser-based 3-D survey of some 10 thousand persons, males and females 18 to 65 years of age. Most of the subjects are from the United States, but several thousand persons from the Netherlands and Italy are included because they represent especially tall and short populations. For the measurements, about 70 markers are attached to the skin to indicate the location of bony landmarks, such as the acromia.* ☚☚☚

With the increased use of computer models of the human body, requirements for anthropometric data have become much more complex now than they were just a few decades ago. Most such models represent the long bones of the human body as links, connected in joints with varying degrees of freedom, and powered by muscles spanning one or two joints. Adding external contours, mass, and motion properties yields the "human body model" for which we design in engineering anthropometry.

The "stick person" is the 17th-century concept of Giovanni Alfonso Borelli, taken up two centuries later by the Webers (1836) in their discussion of the mechanics of the legs, by Harless (1860) and von Meyer (1863) in their considerations of body mass properties, and by Braune and Fischer (1889) in their analysis of the biomechanics of a infantryman firing a gun. In 1873, von Meyer modeled body segments as ellipsoids and spheres. This biomechanical model was refined and expanded by Dempster in the 1950s. The Simons and Gardner model of 1960 still depicted body segments as uniform geometric shapes: cylinders for the appendages, neck, and torso, and a sphere for the head. With the use of equations developed by Barter in 1957, inertial parameters were computed for the geometric forms and the moment of inertia for the total body. Barter's elementary work still is the basis for much current biodynamic modeling.

Available Anthropometric Information

Anthropometric sourcebooks

In the past, interest in the body build of populations other than one's own group was based mostly on curiosity and a general "wish to know." More recently, as industry and marketing reach around the globe, body size has become a matter of practical interest to designers and engineers. In the early 1970s, a conference on ethnic variables in human-factors engineering first attempted a compilation of worldwide ergonomic information (Chapanis, 1975). A thorough collection of data, available in the mid-1970s, was published in the NASA/Webb (1978) anthropometric sourcebook. Since then, an increasing number of publications describing national populations has appeared in the literature. For example, body sizes of southeast Asians are becoming well known, reflecting both scientific interest and economic concern. In 1990, Li et al. showed how the proper use of modern technology and statistics allows one to carry out an anthropometric assessment (in this case, of Taiwan) rapidly and exactly. Gordon et al. (1989) demonstrated, in exemplary fashion, how to prepare, perform, and report the results of a large-scale survey (of the U.S. Army).

Earth's population

Juergens et al. (1990) attempted to classify the total population of the earth into 20 groups living in various areas and to estimate 19 of their main anthropometric dimensions. Because of gaps in the existing data, many had to be guesstimated, and certain subgroups (eg, pygmies) are not represented in this global survey. An excerpt from their global estimates is given in Table 1–1. More exact data on specific population samples, measured in recent years, but often only on small groups, are compiled in Table 1–2.

Variability

Anthropometric data show considerable variability, stemming from four sources:

Poor data

Variability in measurements. More or less care can be exercised in selecting population samples, using measuring instruments, storing the measured data, and applying statistical treatments. Depending on the care applied, the resulting information may be quite variable.

Changes with time

Intraindividual variability. The size of the same body segment of a given person changes from youth to old age, depending also on nutrition, physical exercise, and health. Such changes become apparent in *longitudinal* studies, in which an individual is observed over years and decades. Most (but not all) of these changes with age follow the scheme shown in Figure 1–5. During childhood and adolescence, body dimensions such as stature change rapidly. From the early twenties into the fifties, little change occurs in general, with stature remaining almost steady. From the

TABLE 1–1. Average Anthropometric Data (in cm) Estimated For 20 Regions of the Earth

	Stature		*Sitting height*		*Knee height, sitting*	
	Females	*Males*	*Females*	*Males*	*Females*	*Males*
NORTH AMERICA	165.0	179.0	88.0	93.0	50.0	55.0
LATIN AMERICA						
Indian population	148.0	162.0	80.0	85.0	44.5	49.5
European and Negroid population	162.0	175.0	86.0	93.0	48.0	54.0
EUROPE						
Northern	169.1	181.0	90.0	95.0	50.0	55.0
Central	166.0	177.0	88.0	94.0	50.0	55.0
Eastern	163.0	175.0	87.0	91.0	51.0	55.0
Southeastern	162.0	173.0	86.0	90.0	46.0	53.5
France	163.0	177.0	86.0	93.0	49.0	54.0
Iberia	160.0	171.0	85.0	89.0	48.0	52.0
AFRICA						
North	161.0	169.0	84.0	87.0	50.5	53.5
West	153.0	167.0	79.0	82.0	48.0	53.0
Southeast	157.0	168.0	82.0	86.0	49.5	54.0
NEAR EAST	161.0	171.0	85.0	89.0	49.0	52.0
INDIA						
North	154.0	167.0	82.0	87.0	49.0	53.0
South	150.0	162.0	80.0	82.0	47.0	51.0
ASIA						
North	159.0	169.0	85.0	90.0	47.5	51.5
Southeast	153.0	163.0	80.0	84.0	46.0	49.5
SOUTH CHINA	152.0	166.0	79.0	84.0	46.0	50.5
JAPAN	159.0	172.0	86.0	92.0	39.5	51.5
AUSTRALIA (European population)	167.0	177.0	88.0	93.0	52.5	57.0

SOURCE: Juergens, Aune, and Pieper 1990.

sixties on, many dimensions decline, while others—for example, weight or bone circumference—often increase (Annis and McConville 1996).

People differ

Interindividual variability. Individuals differ from each other in arm length, weight, height, and, in fact, nearly all measurements. Data describing a population sample are usually collected in a *cross-sectional* study, in which every subject is measured at the same moment in time. This means that people of different ages, nutrition, fitness, and so on, are included in the sample set. The anthropometric data found in most textbooks, including this one, are gathered in cross-sectional studies.

Today's soldiers are too big to fit into medieval body armor.

Long-term trends: taller and fatter

Secular variations. There is some factual and much anecdotal evidence that people today are larger, on average, than their ancestors. Reliable anthropometric information on this development has been available only for about the last hundred years. During the last six or so decades, stature has increased in North America and Europe by about 1 cm per decade, on average, while body weight has increased about 2 kg per decade. The reason is probably that improved nutrition and hygiene have allowed persons to achieve more of their genetically determined body-size potential. (If this explanation is correct, then the rate of increase should slowly taper off until a final body size is reached. This seems to be happening. (See later.) Data from Japan initially indicated that present-day Japanese were growing much faster than average compared to Caucasians, but the rate seems to be slowing down now (Roebuck et al. 1988). The most reliable data for observing such secular trends are from military surveys.

TABLE 1–2. Recent Anthropometric Data on International Population Samples, Averages and (in Parentheses) Standard Deviations

	Sample Size	*Stature, mm*	*Sitting Height, mm*	*Knee Height, Sitting, mm*	*Weight, kg*
Algerian females (Mebarki and Davies 1990)	666	1,576 (56)	795 (50)	487 (36)	61.3 (12.9)
Brazilian males (Ferreira 1988; cited by Al-Haboubi 1991)	3,076	1,699 (67)	—	—	—
Chinese females (Singapore) (Ong, Koh, Phoon, and Low 1988)	46	1,598 (58)	855 (31)	—	—
Chinese females (Taiwan) (Huang and You 1994)	300	1,582 (49)	—	—	51.2 (6.9)
Cantonese males (Evans 1990)	41	1,720 (63)	—	—	60.0 (6.2)
Egyptian females (Moustafa, Davies, Darwich, and Ibraheem 1987)	4,960	1,606 (72)	838 (43)	499 (25)	62.6 (4.4)
German (East) females	123	1,608 (59)	854 (31)	497 (24)	—
German (East) males (Fluegel, Greil, and Sommer 1986)	30	1,715 (66)	903 (34)	531 (27)	—
Indian females	251	1,523 (66)	775 (39)	483 (28)	49.5 (9.9)
Indian males (Chakarbarti 1997)	710	1,650 (70)	937 (45)	520 (30)	57 (11)
Central Indian male farmworkers (Gite and Yadav 1989)	39	1,620 (50)	739 (26)	509 (30)	49.3 (6.0)
South Indian males (workers) (Fernandez and Uppugonduri 1992)	128	1,607 (60)	791 (40)	542 (38)	56.6 (5.1)
East Indian male farmworkers (Yadav, Tewari, and Prasad 1997)	134	1,621 (58)	809 (22)	515 (29)	53.6 (67)
Indonesian females	468	1,516 (54)	719 (34)	—	—
Indonesian males (Sama'mur 1985, cited by Intaranont 1991)	949	1,613 (56)	872 (37)	—	—
Irish males (Gallwey and Fitzgibbon 1991)	164	1,731 (58)	911 (30)	508 (28)	73.9 (8.7)
Italian females	753	1,610 (64)	850 (34)	495 (30)	58 (8.3)
Italian males (Coniglio, Fubini, Masali, Masiero, Pierlorenzi, and Sagone 1991)	913	1,733 (71)	896 (36)	541 (30)	75 (9.6)
Jamaican females	123	1,648	832	—	61.4
Jamaican males (Lamey, Aghazadeh, and Nye 1991)	30	1,749	856	—	67.6
Japanese females	240	1,584 (50)	855 (28)	475 (20)	54 (6)

TABLE 1–2. *(Continued)*

	Sample Size	*Stature, mm*	*Sitting Height, mm*	*Knee Height, Sitting, mm*	*Weight, kg*
Japanese males (Kagimoto 1990)	248	1,688 (55)	910 (30)	509 (22)	66 (8)
Korean female workers (Fernandez, Malzahn, Eyada, and Kim 1989)	101	1,580 (57)	833 (32)	460 (22)	53.9 (6.9)
Saudi-Arabia males (Dairi 1986, cited by Al-Haboubi 1991)	1,440	1,675 (61)	—	—	—
Singapore males (pilot trainees) (Singh, Peng, Lim, and Ong 1995)	832	1,685 (53)	894 (32)	—	—
Sri Lanka females	287	1,523 (59)	774 (22)	—	—
Sri Lanka males (Abeysekera 1985, cited by Intaranont 1991)	435	1,639 (63)	833 (27)	—	—
Sudan males					
Villagers	37*	1,687 (63)	—	—	57.1 (7.6)
City dwellers	16*	1,704 (72)	—	—	62.3 (13.1)
	48**	1,668	—	—	51.3
Soldiers	21*	1,735 (71)	—	—	71.1 (8.4)
	104**	1,728	—	—	60.0
*(ElKarim, Sukkar, Collins, and Doré 1981) **(Ballal et al. 1982, cited by Intaranont 1991)					
Taiwan females (Huang and You 1994)	300	1,582 (49)	—	—	51.2 (6.9)
Thai females	250*	1,512 (48)	—	—	—
	711*	1,540 (50)	817 (27)	—	—
Thai males*	250*	1,607 (20)	—	—	—
	1,478**	1,654 (59)	872 (32)	—	—
*(Intaranont 1991) **(NICE, cited by Intaranont 1991)					
Turkey females					
Villagers	47	1,567 (52)	792 (38)	486 (27)	69.1 (13.8)
City dwellers (Goenen, Kalinkara, and Oezgen 1991)	53	1,563 (55)	786 (05)	471 (05)	65.9 (13.0)
Turkey males (soldiers) (Kayis and Oezok 1991)	5,108	1,702 (60)	888 (34)	513 (28)	63.3 (7.3)
American Vietnamese					
Females	30	1,559 (61)	—	—	48.6
Males (Imrhan, Nguyen, and Nguyen 1993)	41	1,646 (54)	—	—	58.9
U.S. Midwest workers, with shoes and light clothes					
females	125	1,637 (62)	—	—	64.7 (11.8)
males (Marras and Kim 1993)	384	1,778 (73)	—	—	84.2 (15.5)
U.S. male miners (Kuenzi and Kennedy 1993)	105	1,803 (65)	—	—	89.4 (15.1)

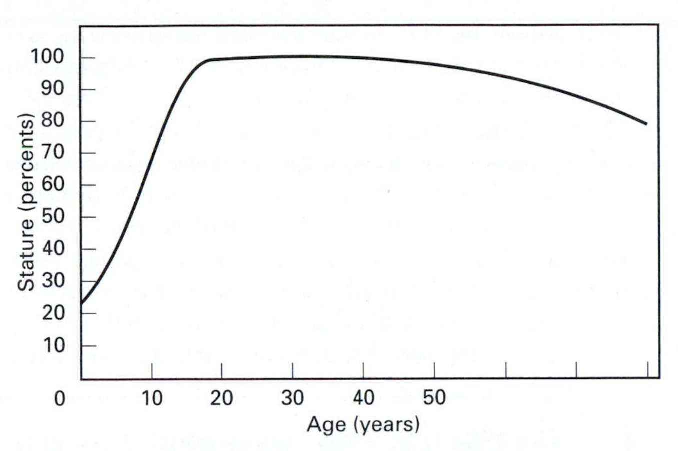

Figure 1–5. Approximate changes in stature with age.

Measurements of U.S. soldiers have been minutely recorded since the Civil War. The military, however, is a selected sample of the general population, excluding, for example, people older than about 50 years and people who are unusual in their body dimensions, such as extremely short or tall individuals; also, only fairly healthy persons are in the military.

Body sizes of U.S. soldiers

Anthropometric secular trends in 22 body dimensions of Caucasian, Afro-American, Hispanic, and Asian female and male U.S. Army soldiers were investigated in 1990 by Greiner and Gordon. They found that some dimensions changed very little while others showed fairly clear trends. The fast increases in stature and in sitting height seem to be slowing down: It now takes about 20 years before the gains are measurable with current techniques, by which time they are approximately another centimeter. Leg length (measured as crotch height), in contrast, is not changing appreciably; thus, it seems that mainly the trunk is still getting taller, though more slowly. Yet, body weight continues its rapid increase, by 2 to 3 kg per decade. Shoulder breadth and chest circumference are increasing at rates of about 1 cm or more per decade. Altogether, Caucasian, Afro-American, and Hispanic U.S. Army soldiers show similar changes, while U.S. soldiers of Asian extraction exhibit quite different trends, which can be explained by recent immigration of Asians to the United States.

Population Samples

Body dimensions of soldiers have long been of interest for a variety of reasons, among which is the necessity for the armed forces to provide uniforms, armor, and equipment. Armies have medical personnel capable of performing body measurements on large samples on command. Hence, anthropometric information about soldiers has a long history and is rather complete. For example, the anthropometric data bank of the U.S. Air Force (CSERIAC) contains the data of approximately 100 surveys from many nations, though mostly on U.S. military personnel. Similarly, the Human Biometry Data Bank Ergodata at the Université René Descartes in Paris contains anthropometric information on Europeans.

U.S. Civilians' Body Sizes

In earlier publications (eg, Kroemer 1981, Kroemer et al. 1994), we had to rely on estimates for U.S. civilians' body sizes based on data measured in the 1960s and '70s. In 1988, a thorough anthropometric survey of U.S. Army personnel was conducted (Gordon et al. 1989) in which 2,208 female and 1,774 male soldiers were measured. These subsamples were carefully selected to match the proportions of ages and racial or ethnic groups in the active-duty army of June 1988.

Among the U.S. military services, the Army is the largest and anthropometrically least biased sample of the total U.S. adult population. The measured sample in the 1988 survey is a mix of older and younger subjects: among the men, 30 percent were aged 31 and over, 25 percent were between 25 and 30, and the others were younger. Sixty-six percent were white, 26 percent Afro-American, and 4 percent Hispanic, with the remaining 4 percent from other racial or ethnic groups. Among the women, 22 percent were aged 31 and over, 32 percent were between 25 and 30 years, and the others were younger. Fifty-two percent of the females were white, 42 percent black, 3 percent Hispanic, and 4 percent from other racial or ethnic groups. Altogether, this survey is a reasonably good mix of ages and persons of various origins, and, in any event, it provides better information about the anthropometry of the civilian U.S. adult population than decades-old estimates. Therefore, the 1988 Army anthropometric data are used in this book to represent the U.S. adult population.

The 1988 U.S. Army survey constitutes, at present, the best estimate of body sizes of the U.S. adult population. The 1988 data set contains 180 measurements (including 48 head and face dimensions) and 60 derived dimensions calculated from the measured data. The data are correlated in various ways. Thus, for information on data not reported here, the publication by Gordon et al. (1989) and the associated reports listed there should be consulted. Table 1–3 is an excerpt of anthropometric data, which should describe the adult U.S. civilian population well enough until better information becomes available.

REACHES

For standardization purposes, anthropometric measurements are done on persons standing or sitting erect with body joints at 0, 90, or 180 degrees—body postures not usually maintained at work. For the design of workstations and equipment, "functional" data are needed. Such data are often reported and used in engineering design guidelines (see, e.g., the sections on body posture, controls, and office design in this book), but they are dependent on stated or implied assumptions. A typical example is that of reach contours, shown in Figure 1–6. Here, an upright trunk posture is presumed, with the shoulder blades remaining in contact with the seat back and the Seat Reference Point (SRP, the theoretical central juncture of the planes of seat back and pan) providing the origin of all reach measures.

Anthropometric Statistics

Parametrics

Fortunately, anthropometric data are usually dispersed in a reasonably normal (Gaussian) distribution (with the occasional exception, especially of muscle strength data). Hence, regular parametric statistics apply in most cases. The data cluster in the center of the set at the 50th percentile, which coincides with the mean *m* (the average). The peakedness or flatness of the data cluster is measured by the standard deviation (SD). Table 1–4 lists commonly used formulae to calculate the most often needed statistical descriptors of normal distributions.

There are two ways to determine given percentile values. One is simply to use a graph of a data distribution and find the critical percentile values from the graph (by measuring, counting, or estimating). This technique works well whether the distribution is normal, skewed, binomial, or in any other form. Fortunately, most anthropometric data are normally distributed, which allows the second, even easier (and usually more exact), approach: to calculate the percentile values, as shown in Table 1–4.

TABLE 1–3. Body Dimensions of U.S. Civilian Adults, Female/Male, in mm

	Percentiles			
	5th	*50th*	*95th*	*SD*
HEIGHTS				
(*f* above floor, *s* above seat)				
Stature ("height")[f]	527.8/1,646.9	1,629.4/1,755.8	1,737.3/1,866.5	63.6/66.8
Eye height[f]	1,415.2/1,528.2	1,516.1/1,633.9	1,621.3/1,742.9	62.5/65.7
Shoulder (acromial) height[f]	1,240.9/1,341.6	1,333.6/1,442.5	1,432.0/1,545.6	57.9/62.0
Elbow height[f]	926.3/995.2	997.9/1,072.5	1,074.0/1,152.8	44.8/48.1
Wrist height[f]	727.9/777.9	790.3/846.5	855.1/915.2	38.6/41.5
Crotch height[f]	700.2/764.4	771.4/837.2	845.8/916.4	44.1/46.2
Height (sitting)[f]	795.3/854.5	852.0/913.9	910.2/971.9	34.9/35.6
Eye height (sitting)[s]	684.6/735.0	738.7/792.0	794.3/848.0	33.2/34.2
Shoulder (acromial) height (sitting)[s]	509.1/548.5	555.5/597.8	603.6/646.3	28.6/29.6
Elbow height (sitting)[s]	175.7/184.1	220.5/230.6	264.4/273.7	26.8/27.2
Thigh height (sitting)[s]	140.4/148.6	158.9/168.2	180.2/189.9	12.1/12.6
Knee height (sitting)[f]	474.0/514.4	515.4/558.8	560.2/605.7	26.3/27.9
Popliteal height (sitting)[f]	351.3/394.6	389.4/434.1	429.4/476.3	23.7/24.9
DEPTHS				
Forward reach (to tip of thumb)	676.7/739.2	734.6/800.8	796.7/867.0	36.4/39.2
Buttock–knee distance (sitting)	542.1/569.0	588.9/616.4	639.8/667.4	29.6/29.9
Buttock–popliteal distance (sitting)	440.0/458.1	481.7/500.4	527.7/545.5	26.6/26.6
Elbow–fingertip distance	406.2/447.9	442.9/484.0	482.5/524.2	23.4/23.3
Chest depth	208.6/209.6	239.4/243.2	277.8/280.4	21.1/21.5
BREADTHS				
Forearm–forearm breadth	414.7/477.4	468.5/546.1	528.4/620.6	34.7/43.6
Hip breadth (sitting)	342.5/328.7	384.5/366.8	432.2/411.6	27.2/25.2
HEAD DIMENSIONS				
Head circumference	522.5/542.7	546.2/567.7	570.5/593.5	14.6/15.4
Head breadth	136.6/143.1	144.4/151.7	152.7/160.8	4.9/5.4
Interpupillary breadth	56.6/58.8	62.3/64.7	68.5/71.0	3.6/3.7
FOOT DIMENSIONS				
Foot length	224.4/248.8	244.4/269.7	264.6/292.0	12.2/13.1
Foot breadth	81.6/92.3	89.7/100.6	97.8/109.5	4.9/5.3
Lateral malleolus height[f]	52.3/58.4	60.6/67.1	69.7/76.4	5.3/5.5
HAND DIMENSIONS				
Circumference, metacarpal	172.5/198.5	186.2/213.8	200.3/230.3	8.5/9.7
Hand length	165.0/178.7	180.5/193.8	196.9/210.6	9.7/9.8
Hand breadth, metacarpal	73.4/83.6	79.4/90.4	85.6/97.6	3.8/4.2
Thumb breadth, interphalangeal	18.6/21.9	20.7/24.1	22.9/26.5	1.3/1.4
WEIGHT (in kg)	39.2*/57.7*	62.0/78.5	84.8*/99.3*	13.8*/12.6*

*Estimated (from Kroemer 1981).

Source: Adapted from U.S. Army data reported by Gordon et al. (1989).

Note: In this table, the entries in the 50th-percentile column are actually mean (average) values. The 5th- and 95th-percentile values are from measured, not calculated, data (except for weight). Thus, the values given may be slightly different from those obtained by subtracting 1.65 SD from the mean (50th percentile) or by adding 1.65 SD to it.

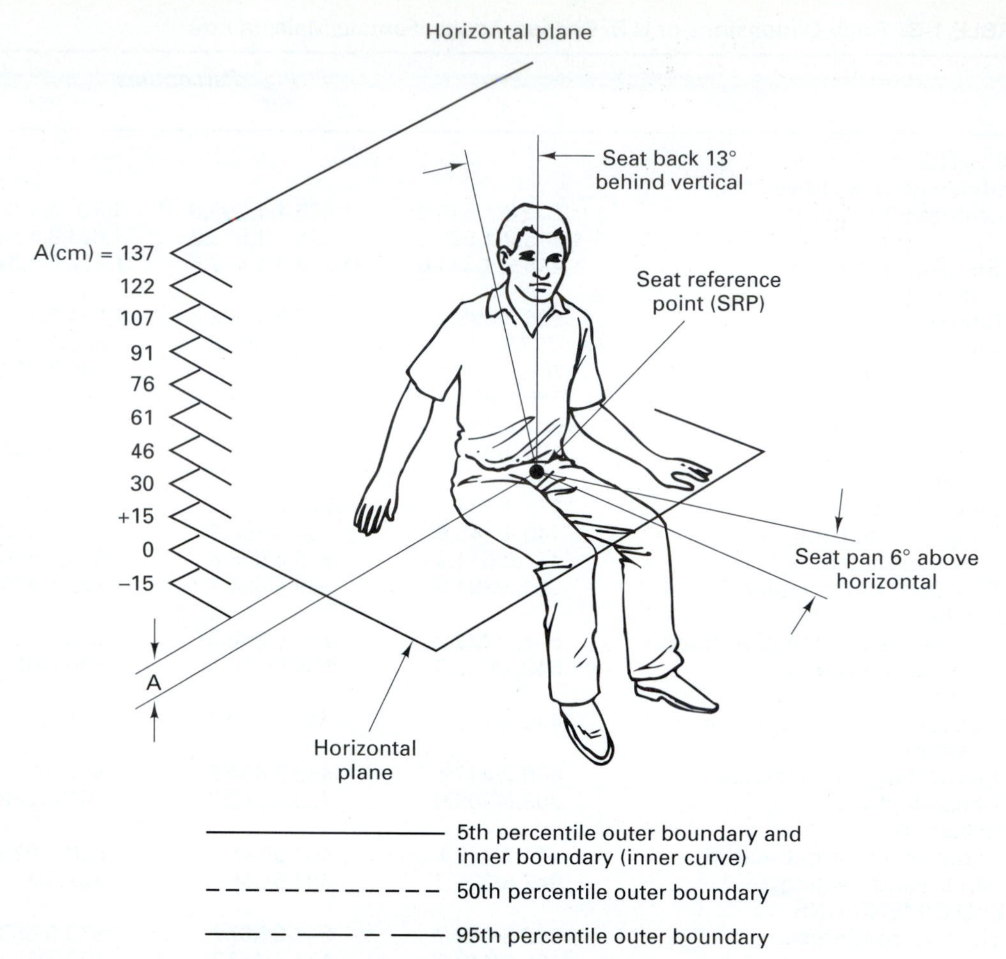

Figure 1–6a. Definition of reach contours of U.S. Air Force males and females (adapted from NASA 1989).

The distribution of anthropometric information is, for practical purposes, well described by the mean *m* (often called the average), standard deviation SD, and sample size *n*. The range indicates the smallest and largest values.

Check the CV One easy way to check on how diverse the data are is to divide the standard deviation of the data in question by their mean to get the coefficient of variation (CV). For most body dimensions, the CV is in the neighborhood of 3 to 10 percent; larger values are suspect and should prompt a thorough examination of the data. However, in most strength data, the CV is between 10 and 85 percent.

Use percentiles Anthropometric data often are best presented in percentiles. They provide a convenient means of describing the range of body dimensions to be accommodated, making it easy to locate the percentile equivalent of a measured body dimension. Also, the use of percentiles avoids the misuse of the average in design (as is discussed later).

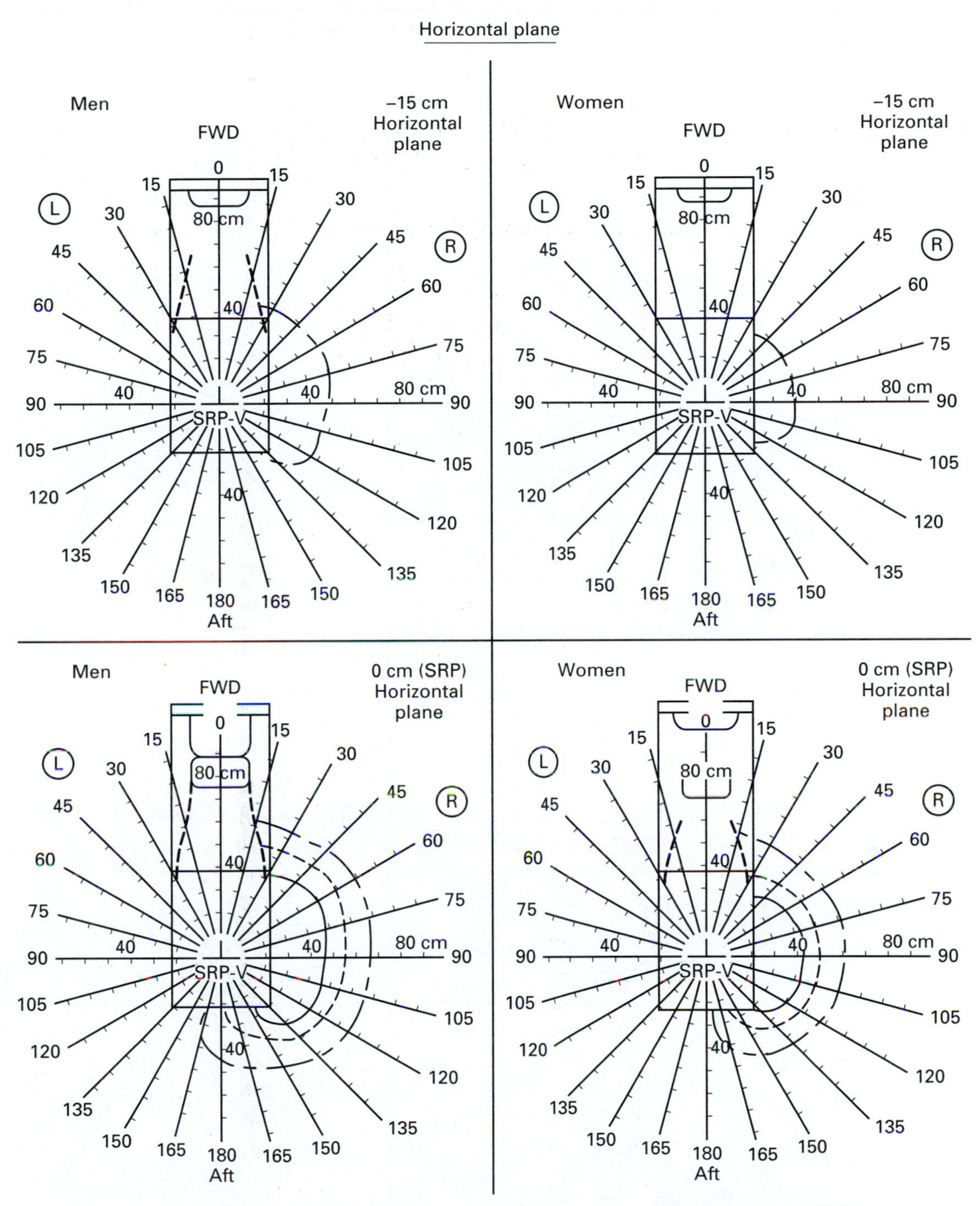

Figure 1–6b. Reach contours in planes 15 cm below and at seat (SRP) height.

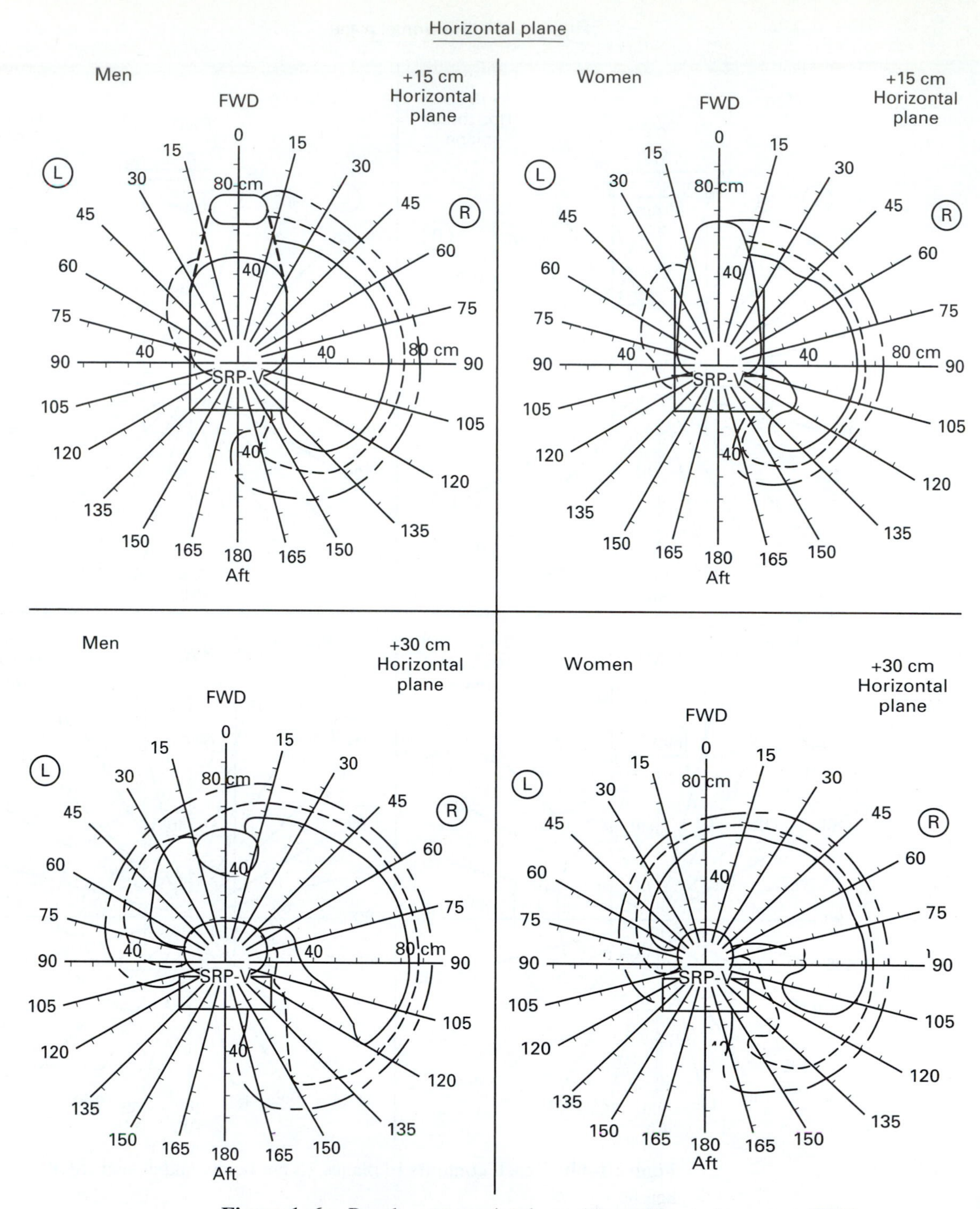

Figure 1–6c. Reach contours in planes 15 and 30 cm above seat (SRP) height.

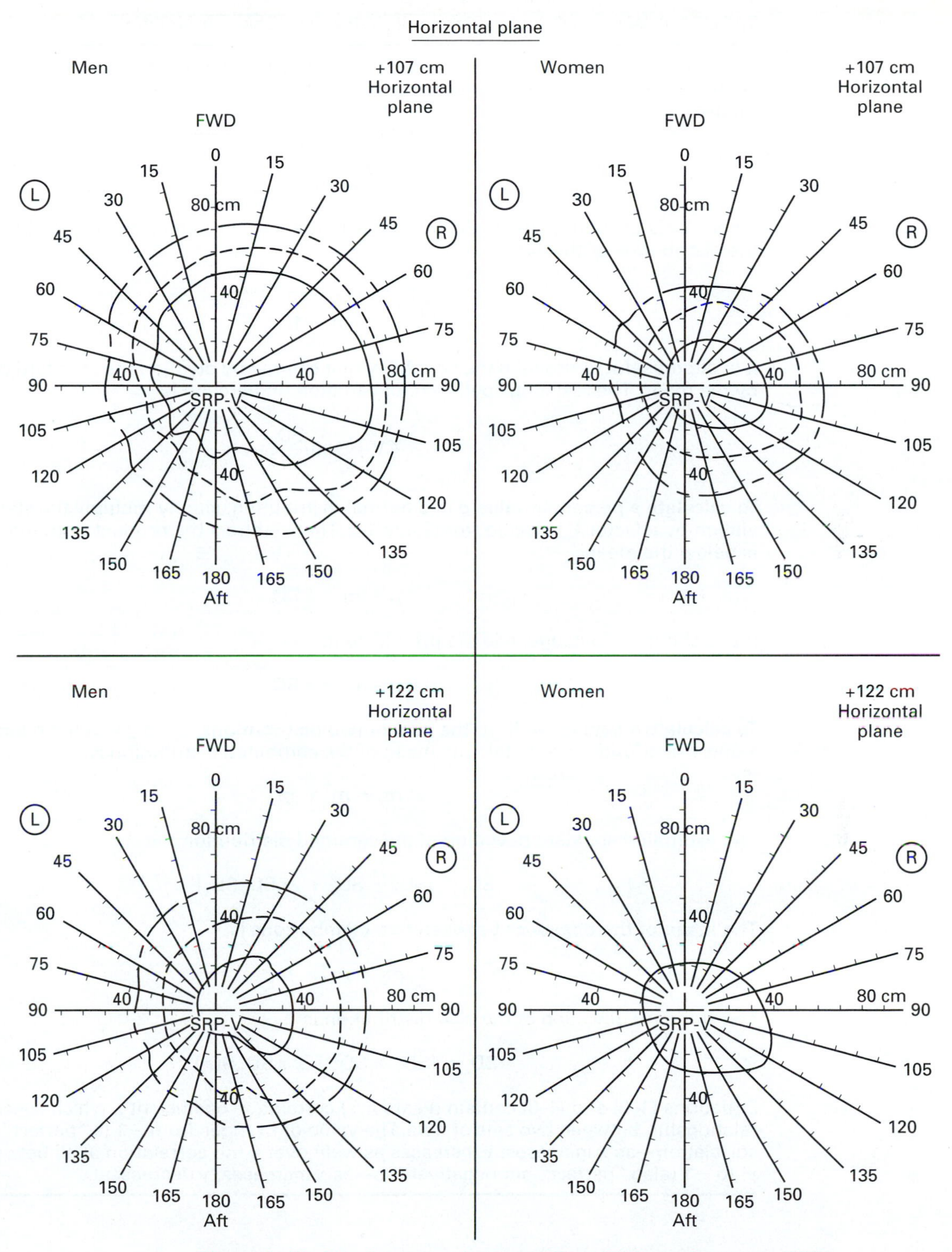

Figure 1–6d. Reach contours in planes 107 and 122 cm above seat (SRP) height.

TABLE 1–4 Ways to Calculate Percentile Values and Their Variabilities

A normally distributed set of n data is described by two simple statistics: The 50th percentile is, by definition, the same as the mean m (also commonly called the average). Mathematically,

$$m = \frac{\Sigma x}{n}, \tag{1–1}$$

where Σx is the sum of the individual measurements. The standard deviation SD describes the distribution of the data:

$$SD = \left[\frac{\Sigma(x - m)^2}{n - 1}\right]^{1/2}. \tag{1–2}$$

It is often useful to describe the variability of a sample by dividing the standard deviation by the mean. The resulting coefficient of variation (in percent) is

$$CV = 100\frac{SD}{m}. \tag{1–3}$$

To calculate a percentile value p of a normal distribution, simply multiply the standard deviation by a factor k, selected from Table 1–5. Then subtract the product from the mean if p is below the mean:

$$p = m - k\,SD. \tag{1–4a}$$

If p is above the average, add the product to the mean:

$$p = m + k\,SD. \tag{1–4b}$$

To calculate a new mean from the *sum* of two distributions, simply add the means of the x and y distributions to obtain the mean of the combined distribution z:

$$m_z = m_x + m_y. \tag{1–5}$$

The estimated standard deviation of the summed distribution z is

$$SD_z = [SD_x^2 + SD_y^2 + 2r\,SD_x SD_y]^{1/2}. \tag{1–6}$$

The mean of the *difference* between two distributions is

$$m_Z = m_X - m_y. \tag{1–7}$$

The standard deviation of the new distribution is

$$SD_z = [SD_x^2 + SD_y^2 - 2\,r\,SD_x SD_y]^{1/2}. \tag{1–8}$$

Equations (1–6) and (1–8) contain (Pearson's) correlation coefficient r, which describes the relationship between two sets of data. The value of r ranges from $+1$ (a "perfect" positive correlation—as x increases, y increases as well) over 0 (no correlation at all between x and y) to -1 (also "perfect," but negatively so—as x increases, y decreases).

For a normal distribution, percentiles are easily calculated from the mean and standard deviation, as shown in Table 1–4. We simply multiply the standard deviation by a factor *k*, selected from Table 1–5, and then deduct the result from the average to arrive at the desired percentile value below the 50th, or else add the result to the average (which coincides with the 50th percentile) to arrive at a value above the average.

More details about use of percentiles are discussed by Kroemer et al. (1997) and Kroemer (1999a). Several examples are contained in the "box of samples" that follows.

TABLE 1–5. Factor *k* For Computing Percentiles From Mean $\bar{X}$ and Standard Deviation *S*

k	*Percentile p* associated with X	
	$X = \bar{X} - kS$	$X = \bar{X} + kS$
2.576	00.5	99.5
2.326	1	99
2.06	2	98
1.96	2.5	97.5
1.88	3	97
1.65	5	95
1.28	10	90
1.04	15	85
1.00	16.5	83.5
0.84	20	80
0.67	25	75
0	50	50

Examples:

To determine the 95th percentile, use $k = 1.65$.

To determine the 20th percentile, use $k = 0.84$.

BOX OF SAMPLES

KEYBOARD HEIGHT ABOVE THE SEAT

The Task: Establish the surface height of a keyboard so that the sitting operator has the forearms and wrists horizontal.

The Solution: Assume that having tops of the keys at elbow height will allow the operator to keep the wrists straight, forearms horizontal, and upper arms effortless hanging down. In all likelihood, adjusting the key height properly for the 10th percentile female elbow clearance to the 90th percentile male clearance will be appropriate. The elbow height above the seat pan is listed for Americans in Table 1–3: females have a mean of 220.5 mm with a SD of 26.8 mm; the corresponding numbers for males are 230.6 mm and 27.2 mm.

The multiplication factor used to calculate the 10th and 90th percentiles is given as 1.28 in Table 1–5. According to Equation (1–4) in Table 1–4, the 10p values are 186 mm for females and 196 mm for males; the

90p values are 265 mm for males and 255 mm for females. Thus, the height of the key tops should be adjustable from 186 to 265 mm, so the adjustment range is 79 mm under the given assumptions.

ARM LENGTH

The Task: Calculate the 95p shoulder-to-fingertip length.
The Solution: You know that the mean lower arm (LA) link length (with the hand) is 442.9 mm with a standard deviation of 23.4 mm. The mean upper arm (UA) link length is 335.8 mm, and its standard deviation is 17.4 mm.
The Solution: The multiplication factor of $k = 1.65$ (from Table 1–5) leads to the 95th percentile. But using the sum of the two 95-p lengths would be mistaken, because that would disregard their correlation; instead, calculate the sum of the mean values first:

$$m_A = m_{LA} + m_{UA} = 442.9 + 335.8 = 778.7 \text{ mm}$$

[See Equation (1–5).]

Next, calculate its standard deviation, using an assumed correlation coefficient of 0.4:

$$SD_A = [23.4^2 + 17.4^2 + 2 \times 0.4 \times 23.4 \times 17.4]^{1/2} = 34.3 \text{ mm}.$$

[See Equation (1–6).]

The 95-p total arm length can now be calculated: $A_{95} = 778.7$ mm $+ 1.65 \times 34.3$ mm $= 835.3$ mm. [See Equation (1–4b).]

OPERATING FORCE

The Task: Design the critical, hand-operated shutoff lever in a chemical plant so that the necessary torque to close it can be achieved in an emergency even by "weak" operators; at the other extreme, the valve must be structurally so solid that even brute strength applied does not break it.
The Solution: Measuring the pull and push force capabilities of a random sample of 100 plant employees on a simulated valve handle provided the information that the average pull force of the weaker hand was 116 N with a standard deviation of 37 N, while the mean push force of both hands combined on the lever was 331 N with a standard deviation of 173 N. The human-factors engineer decided to select a force near the 100th percentile (331 N $+ 3 \times 173$ N; see Table 1–5)—that is, a force of 850 N—to design for structural strength of the valve assembly, while the operating force was selected to be near the 5th percentile—that is, at 55 N, (116 N $- 1.65 \times 37$ N; see Table 1–5).

TORSO MASS

The Task: Determine the mass of the torso of a 75-p female.
The Solution: You know that the estimated mass of the torso and head combined has an average of 35.8 kg and a standard deviation of 5.2 kg.

The estimated mass of the head, measured separately, has a mean of 5.8 kg with a standard deviation of 1.2 kg. You assume the correlation between head and torso to be 0.1.

The mean torso mass is the difference between the average values of the torso and head masses:

$$mean_{torso} = 35.8 \text{ kg} - 5.8 \text{ kg} = 30.0 \text{ kg}.$$ [See Equation (1–7).]

The standard deviation of the mean torso mass is calculated from $SD_{torso} = [5.2^2 + 1.2^2 - 2 \times 0.1 \times 5.2 \times 1.2]^{1/2}$ kg = 5.2 kg. [See Equation (1–8).]

The mass of a 75th percentile torso is (with $k = 0.67$ taken from Table 1–5) $mass_{torso\ 75p} = 30.0$ kg + 0.67 × 5.2 kg = 33.5 kg. [See Equation (1–4b).]

Percentiles serve the designer in several ways. First, they help to establish the portion of a user population that will be included in (or excluded from) a specific design solution. For example, a certain product may need to fit everybody who is taller than the 5th percentile or smaller than the 95th percentile in a specified dimension, such as grip size or arm reach. Thus, only the 5 percent having values smaller than 5th percentile and the 5 percent having values larger than 5th percentile, will not be fitted. The central 90 percent of all users will be accommodated.

Second, percentiles are easily used to select subjects for fit tests. For example, if a product needs to be tested, persons having 5th- or 95th-percentile values in the critical dimensions can be employed for use tests.

Third, any body dimension, design value, or score of a subject can be located exactly. For instance, a certain foot length can be described as a given percentile value of that dimension, a certain seat height can be described as fitting a certain percentile value of popliteal height (a measure of lower leg length), or a test score can be described as being a certain percentile value.

Finally, the use of percentiles helps in the selection of persons who can use a given product. For example, if the cockpit of an airplane is designed to fit 5th to 95th percentiles, one can select cockpit crews whose body measures are between those percentiles in the critical design dimensions.

The body beautiful

Body proportions. We often judge the human body by how its components "fit" together; our images of the beautiful body are affected by esthetic codes, canons, and rules often founded on ancient (eg, Egyptian, Greek, or Roman) beauty concepts of the human body. Leonardo da Vinci's drawing of the body within a frame of graduated circles and squares has been adopted, in simplified form, as the emblem of the U.S. Human Factors and Ergonomics Society.

Types of body build

The categorization of body builds into different types is called *somatotyping* (from the Greek *soma,* for body). About 400 B.C., Hippocrates developed a scheme that included four body types, supposedly determined by their fluids. (The "moist" type was dominated by black gall and the "dry" by yellow gall; the "cold" type was governed by slime, the "warm" by blood.) In 1921, the psychiatrist Ernst Kretschmer described a system of three body types that was intended to relate body build to personality traits: the asthenic, pyknic, and athletic body builds. (The "athletic" type referred to character traits, not sports performance capabilities.) In the 1940s, the anthropologist W. H. Sheldon also established a system of three body types, intended to describe (male) body proportions. Sheldon rated each person's appearance in terms of ecto-, endo-, and mesomorphic components (stocky/round, strong/sturdy, and lean/fragile body builds, respectively). This typology was originally based on intuitive assessment, not on actual body measurements; the latter were introduced into the system later by Sheldon's disciples. In 1967, Heath and Carter standardized the somatotyping procedure, and their body typology using Sheldon's terms has been widely employed since. Unfortunately, these and other attempts at somatotyping have not provided reliable predictors of human performance with regard to technological systems. Hence, somatotyping is of little value for engineers or managers.

How we see ourselves

Body image. We have a mental picture of the physical appearance of our body. This body image may affect our behavior and even our lifestyle—in particular, when it is disturbed. Body image disturbance is defined as any form of cognitive, affective, behavioral, or perceptual disturbance that is directly concerned with an aspect of our physical appearance (Thompson 1995). Quite a few people, especially in North America and Europe, are dissatisfied with their own body image. Distortion of, and dissatisfaction with, one's body image are associated with some weight and eating problems.

Several generalizations can be made about the effect of body image on variables concerning body shape and size:

- In general, men with body image disturbance tend to underestimate their size, while women with that affliction tend to overestimate their size. Interestingly, men also tend to overestimate the body size they think is most attractive to women, while women underestimate the body size they think is most attractive to men (Fallon and Rozin 1985).
- In Western societies that scorn the obese, women tend to exhibit greater body image dissatisfaction and body image distortion than do men. This "normative discontent" (Rodin 1993) has been implicated in higher rates of eating disorders in women compared with men.
- Even following much reduction of body weight, some individuals perceive themselves as having lost almost no weight: They still overestimate their body size. On the other hand, some obese individuals who have lost weight tend to underestimate their body size. In general, however, even small reductions in weight may enhance body image satisfaction in obese individuals (Foster et al. 1997, Wadden et al. 1994). There is probably little effect of body image distortion on the success of treatment for weight loss, but body image satisfaction may play a role in preventing recidivism after an initially successful weight-loss program.
- The extent of body image disturbance can be severe enough to warrant psychiatric diagnosis and treatment, such as in the case of body dysmorphic disorder (DSM IV, American Psychiatric Association, 1995). People with this disorder are excessively concerned that there is something wrong with the shape or appearance of a body part, usually a portion of the face, torso, or skin. They often request medical procedures or plastic surgery to correct their imagined defects, and their preoccupation causes significant distress and social and/or occupational dysfunction.
- Individuals with certain eating disorders such as anorexia nervosa exhibit often striking body image disturbances with respect to overestimation of body size and distortion of body proportions. A person with anorexia nervosa may be objectively emaciated yet still complain of looking and feeling "fat".

- Individuals who engage intensely in athletics requiring rigid weight or body shape expectations are also likely to exhibit body image disturbance. Studies of ballet dancers, wrestlers, runners, and gymnasts have shown higher rates of body image disturbance among these athletes than in appropriate controls. Underlying body image disturbance may also play a role in maintaining compulsive ("obligatory") exercise in some individuals (Thompson 1990).

Several measurement devices are available to assess body image, such as figural stimuli, questionnaires, and interviews. Thompson (1995) reviewed more than 30 assessment methods to determine various cognitive, affective, perceptual and behavioral aspects of body image. The origins of body image disturbances have also been widely researched, and many explanations have been proposed. These include theories involving cortical factors, developmental factors, sociocultural influences, and cognitive variables.

Asking about, instead of measuring, body size? Self-reporting is a notoriously inaccurate method: As a rule, both women and men tend to underreport their weight (Bowman and Delucia 1992), but some individuals may overestimate their weight and body size due in part to body image disturbance. Anthropometric surveys in the United States and Europe have shown that short people tend to overestimate their stature, while heavy people often underestimate their weight. If one applies appropriate correction factors to counteract these tendencies, simply asking people (instead of measuring them, which takes more effort) for their height and weight can lead to fairly reliable information.

Healthy body weight?

"Desirable" body weight. People who are severely overweight have a higher risk of health problems and of early death than their slimmer contemporaries do. Indeed, the more a person is overweight, the higher is the risk. Adipose (fat-containing) tissue is a normal part of the human body; in most individuals it contains about 85% fat. Adipose tissue stores fat for use as energy under high metabolic demands. Obesity has been defined in various terms, but usually its definition entails a specific excess of such fatty tissue. The reasons for obesity may be both behavioral and genetic and include too much caloric intake, too little physical activity, and metabolic and endocrine malfunctions. The investigator measuring body composition or fatness has at least two things to assess: total fat and body fat distribution. However, the determination of a healthy (vs. obese) body weight and healthy (vs. undesirable) body fat distribution is a complicated enterprise, in both methodology and quantification.

Methodologies for Determining Body Composition. All methods for determining body composition are indirect and are based on measuring various properties, such as body density or the gamma-ray decay of certain isotopes. Mathematical functions are employed to relate the known or measured property or component to the unknown component. Typically, these mathematical procedures utilize regression analysis or known biological–physiological ratios. Methods of determining body composition are varied, utilizing knowledge of total body water (TBW), dual-energy X-ray absorptiometry (DEXA), ultrasound and other imaging techniques (eg, CT and MRI), weight/stature indices (such as Quetelet's body mass index (discussed later), bioimpedance analysis (BIA), total body electrical conductivity (TOBEC), hydrodensitometry, anthropometry (eg, skinfold thicknesses and circumferences), and infrared interactance. These methods range from the very inexpensive and fairly valid (such as in the measurement of body mass index discussed later) to the very expensive and highly valid (such as DEXA and MRI), and investigators must balance a host of issues in their selection, including precision, bias, expense, and safety (Heymsfield et al. 1995).

Quantifying obesity

Since there are no given cutoff points, any quantitative definition of normality or obesity is arbitrary. In 1985, a specially assembled committee of experts agreed that 20% or more above "desirable" body weight should be called obese; so one must first establish a desirable reference weight. Several methods are in use in the United States. "Relative weight" is the measured body weight divided by the midpoint of the weight recommended for a "medium frame" in the Metropolitan Life Insurance tables (Metropolitan Life Foundation 1983). The 1990 USDA Weight Table (U.S. Dept. of Agriculture 1990) uses the "body mass index" (BMI), calculated by dividing the body weight (in kilograms) by the square of the body height (stature, in meters). While the BMI calculation yields a better estimate of body composition than simply the weight of an individual, it has weak discriminant validity, because it taps into both lean and fat mass. Body composition varies among individuals of the same height and weight (Andres 1985, National Institutes of Health 1985).

☛☛☛ *In the general U.S. population, body weight relates to stature only moderately; that is, the coefficient of correlation is smaller than 0.5.* ☚☚☚

Besides general interindividual variability, age is an important moderator of the BMI–fatness relationship; at equal BMIs, older adults are fatter than younger adults (Cronk and Roche 1982). BMI also changes in a markedly nonlinear fashion with age in children; therefore, age-standardized BMIs should be utilized when assessing youngsters (Heymsfield et al. 1995, Siervogel et al. 1991). Finally, because women generally have smaller bones and less muscle tissue than men, one might expect that women's BMIs would be less than those of men for any given percentile of the BMI distribution. However, this generalization actually applies only below the 75th percentile: In the upper quarter of the BMI distribution, women's BMIs are generally higher than men's (Williamson 1995).

☛☛☛ ***"Overweight" or "obese"?*** *Obesity probably should be defined as an excess of body fat, not of body weight. It is possible for an individual to have a high weight-for-height ratio but not be "too fat" because he or she is very muscular. In such a case, it would be more appropriate to use the term "overweight" rather than "obese" when employing the BMI measure. The National Center for Health Statistics has defined overweight as a BMI of greater than or equal to 27.8 in men and 27.3 in women (Williamson 1995). "Severe overweight" is defined as a BMI of 31.3 or more in men and 32.3 or greater in women, and "morbid obesity" is 39.0 or more in both men and women.* ☚☚☚

Degree of obesity is only one aspect of obesity that is of interest to researchers. Numerous systems of typing obesity have been proposed since an initial distinction was made in the early 1900s between "endogenous" and "exogenous" obesity. Current systems of typology include those based on cellularity (i.e. hyperplastic vs. hypertrophic), body fat distribution (upper vs. lower body; subcutaneous fat vs. visceral fat), and age at onset (Brownell 1995). Obesity is a heterogeneous condition, and the ergonomist may need to consider it from a variety of perspectives.

☛☛☛ *People commonly say they weigh too much for their height. Others jokingly say they are too short for their weight. But there is little correlation between stature and weight.* ☚☚☚

Dealing with Statistics

Relations among body dimensions

Some body dimensions, such as stature and eye height, are closely related to each other. But stature is not well correlated with head length, waist circumference, or weight. Table 1–6 shows selected correlation coefficients among body dimensions of U.S. Army personnel, male and female. (More detailed tables are contained in publications by NASA/Webb 1978, Cheverud et al. 1990, Kroemer et al. 1997, and Kroemer 1999.)

Nonsense design template Given the varying correlations among body measures, any attempt to express all body dimensions as a portion of stature is futile. Unfortunately, some designers still believe in an outdated scheme that expressed body heights, body breadths, and lengths of segments in terms of fixed percentages of stature. For instance, hip breadth was said to be 19.1 percent of height—misleading nonsense, of course, because hip breadth varies widely among individuals and between males and females as groups. Furthermore, nothing can be designed for a fixed "average" hip breadth.

Covariation

In the human body, some groups of anthropometric data vary with each other in such a way that as one increases, another increases (or several others increase) as well; this is true among many body heights, weights, and circumferences. Conversely, as one dimension increases, others may decrease; for example, as one advances into old age, many body heights decrease. In statistics, this relationship is called covariation. The simple correlation coefficient r (also called Pearson product–moment correlation) is a measure of the strength of the linear relationship between two variables.

Determination

The coefficient of determination, R^2, measures the proportion of variation in the dependent variable y associated with the independent variable x; that is, R^2 measures the strength of association represented by the regression. R^2 is the square of the correlation coefficient between the two variables used in a bivariate regression equation, or among more variables in multiple regression equations.

"0.7 for design decisions" It is common practice in engineering anthropometry (in fact, in ergonomics altogether) to require a correlation coefficient of at least 0.7 as a basis for design decisions. The reason for this "0.7 convention" is that one should be able to explain at least 50 percent of the variance of the predicted value from the predictor variable: This requires r to be at least 0.5, so r is at least 0.7075. (Note that r depends on the sample size n.)

Regression

A bivariate regression expresses the linear relationship between a dependent variable y and an independent variable x according to the equation $y = a + bx$, where a is the intercept and b is the slope. Note that a linear relationship between x and y is often assumed, but not verified.

☛☛☛ *Clothing tariffs are examples of the use, misuse, and nonuse of correlations. In the United States, the sizing of clothes for men is a fairly well organized and standardized procedure. Most men's jacket sizes run from "38" to "56," meaning that they should fit men with chest circumferences between 38 and 56 inches, in increments of 1 or 2 inches. Chest circumference, then,*

TABLE 1–6. Simple Correlation Coefficients for Anthropometric Data on U.S. Army Personnel, Women above the Diagonal, Men below the Diagonal. Values larger than 0.7 are indicated by an asterisk

		1 Age	*2* W	*3* St	*4* OFR	*5* WH	*6* CH	*7* SH	*8* PH	*9* SC	*10* CC
1	Age [302]		0.219	0.041	0.017	0.044	−0.055	0.066	−0.07	0.155	0.193
2	Weight [125]	0.195		.529	.493	.491	.370	.422	.242	.845*	.806*
3	Stature [100]	−0.021	.546		.928*	.848*	.840*	.755*	.808*	.377	.222
4	Overhead Fingertip Reach [84]	−0.013	.525	.937*		.704*	.905*	.554	.868*	.384	.199
5	Wrist Height, Standing [128]	0.028	.527	.856*	.749*		.625	.754*	.587	.300	.255
6	Crotch Height [39]	0.090	.351	.852*	.890*	.673		.330	.915*	.267	.093
7	Sitting Height [94]	0.026	.447	.741*	.578	.692	.347		.343	.285	.202
8	Popliteal Height, Sitting [87]	−0.094	.341	.852*	.883*	.673	.924*	.383		.188	.023
9	Shoulder Circumference [91]	0.122	.861*	.399	.413	.334	.250	.326	.256		.808*
10	Chest Circumference [34]	0.279	.873*	.312	.308	.357	.135	.287	.137	.859*	
11	Waist Circumference [115]	0.364	.849*	.276	.251	.343	.060	.298	.074	.703*	.839*
12	Buttock Circumference [24]	0.190	.935*	.401	.380	.412	.204	.373	.191	.781*	.815*
13	Span [99]	−0.016	.497	.815*	.908*	.535	.840*	.398	.844*	.445	.281
14	Biacromial Breadth [11]	0.034	.496	.487	.506	.295	.370	.407	.394	.633	.419
15	Hip Breadth, Standing [66]	0.209	.831*	.453	.416	.457	.239	.464	.224	.672	.727
16	Head Circumference [62]	0.125	.508	.342	.312	.302	.224	.303	.240	.433	.421
17	Head Length [63]	−0.002	.371	.346	.315	.295	.260	.302	.268	.295	.271
18	Head Breadth [61]	0.198	.320	.114	.098	.112	.034	.128	.035	.303	.311
19	Hand Length [60]	0.032	.453	.650	.724*	.464	.676	.300	.679	.372	.242
20	Foot Length [52]	−0.012	.512	.700	.734*	.537	.687	.383	.697	.409	.299

(continued)

TABLE 1–6. (concluded)

		*11*WC	*12*BC	*13*Sp	*14*BB	*15*HiB	*16* HC	*17*HeL	*18*HeB	*19*HaL	*20*FL
1	Age [302]	0.299	.0258	0.011	0.025	0.283	0.073	0.027	0.044	0.044	0.026
2	Weight [125]	.767*	.897*	.438	.440	.778*	.428	.329	.420	.430	.493
3	Stature [100]	.167	.361	.787*	.505	.372	.348	.354	.124	.637	.673
4	Overhead Fingertip Reach [84]	.132	.313	.907*	.535	.294	.337	.345	.095	.737	.732*
5	Wrist Height, Standing [128]	.217	.363	.453	.303	.397	.250	.261	.403	.403	.468
6	Crotch Height [39]	.061	.185	.870*	.418	.146	.287	.302	.043	.706*	.703*
7	Sitting Height [94]	.142	.351	.336	.384	.438	.246	.255	.159	.256	.330
8	Popliteal Height, Sitting [87]	−0.031	.063	.840*	.420	.051	.241	.271	.020	.685	.671
9	Shoulder Circumference [91]	.697	.726*	.395	.574	.601	.353	.264	.261	.355	.379
10	Chest Circumference [34]	.781*	.707*	.167	.304	.603	.393	.191	.246	.186	.288
11	Waist Circumference [115]		.738*	.109	.214	.673	.223	.117	.229	.127	.170
12	Buttock Circumference [24]	.859*		.258	.327	.915*	.313	.226	.220	.258	.323
13	Span [99]	.201	.352		.565	.203	.345	.338	.083	.827*	.775*
14	Biacromial Breadth [11]	.311	.411	.575		.294	.287	.259	.152	.441	.456
15	Hip Breadth, Standing [66]	.799	.902*	.355	.404		.232	.160	.196	.180	.250
16	Head Circumference [62]	.376	.427	.320	.301	.364		.824*	.497	.342	.360
17	Head Length [63]	.222	.301	.304	.235	.259	.820*		.131	.337	.339
18	Head Breadth [61]	.277	.268	.131	.180	.235	.541	.120		.082	.113
19	Hand Length [60]	.166	.320	.810*	.433	.298	.330	.306	.137		.825*
20	Foot Length [52]	.220	.390	.766*	.445	.377	.333	.304	.161	.806*	

Source: Cheverud, Gordon, Walker, et al. (1990), with their numbering given in brackets. Note that all pairs of data that correlate above 0.700 appear in both the male and female groups; note also that the correlations below 0.300 are similar for both genders.

is used as the primary "predictor variable" for other design variables, such as coat length, shoulder width, and sleeve length. Similarly, slacks are ordered by waist circumference and shirts by neck circumference.

In men's shirts, a given neck circumference is associated with a given chest circumference, while sleeve length may vary by 1- or 2-inch increments. This is an attempt to cover various body dimensions with a few shirt sizes, but it has obvious shortcomings: If a person needs a large neck size (eg, size 17) such a shirt also usually comes with ballooning chest and waist circumferences, which the buyer may not need. Worse, there is a trend to further consolidation of size ranges, providing shirts only in three neck sizes, "small," "medium," and "large," having only one sleeve length associated with each. Production variability is significantly reduced in this simplified tariff, but fewer customers are fitted.

By contrast, in the United States, women's clothes are not well standardized. There appears to exist only one ill-defined prototype "size 12" (based mostly on 1941 measurements), from which larger and smaller sizes are derived in nonstandard manners, as deemed suitable by each manufacturer. Hence, a woman well fitted by clothes of size 10 made by one manufacturer may need a size 12 or 8 in clothing tailored by another company. This situation has allowed several manufacturers to specialize in catering to "petite" or "mature" and other special customers.

HOW TO GET MISSING DATA

Anthropometrically, Europe and North America have the best-known populations of the earth. Yet even on those continents, the civilian populations are not assessed exactly, and current data on subgroups are sparse. Perhaps the ergonomist is interested in body sizes of Italians visiting swimming beaches (Coviglio et al. 1991), Turkish schoolchildren (Vayis and Oezok 1991), Irish workers (Gallery and Fitzgibbon 1991), American farmers (Casey 1989), pregnant American women (Culver and Viano 1990), or North Americans' hand sizes (Greiner 1991). But in many cases, the exact body dimensions needed for a design are not available in the literature.

Measure or calculate?

Several routes exist to obtain the needed information. One is to actually measure a sufficiently large and well-selected sample of the population to be fitted. This is a complex and time-consuming task that should be done by anthropometrists or other specialists, but occasionally one can simply measure a few coworkers to get a rough estimate for the missing data. Another approach is to take the data of a population of known dimensions if one has good reason to believe that population to be similar to the one on which data are missing. Yet, are Taiwanese similar in size to all Chinese? Do Japanese males have body dimensions that are similar to those of their American contemporaries? In fact, they don't, as Nakanishi and Nethery showed in 1999. In such cases, one should seek help from an anthropologist or other expert. The literature provides some help in discussing important assessment aspects, such as sample selection, sample size, and composite populations (Chapanis 1975, Kroemer 1999, Kroemer et al. 1997, Lohman et al. 1988, Pheasant 1996, Roebuck 1995).

A rather interesting task is the prediction of future body dimensions, needed when equipment must be designed for use in decades to come. In the 1960s, for example, NASA was concerned about the body sizes of astronauts in the 1980s and 1990s. In 1988, Roebuck et al. used a large variety of sources, military and civilian, U.S. and foreign, together with regression equations, to forecast the body dimensions of astronauts in the year 2000. Figure 1–7 shows the stature values predicted by these researchers for male U.S. Air Force and NASA flight crews. Figure 1–8 shows their predictions for American and Asian women.

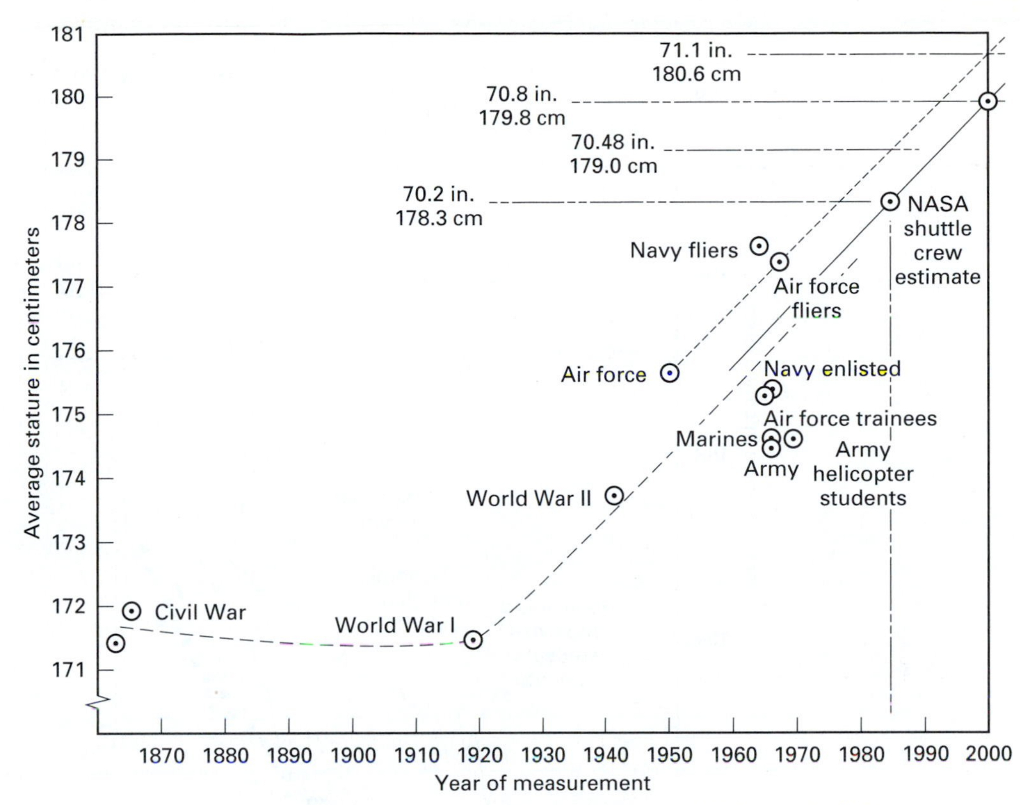

Figure 1–7. Predicted average stature of USAF and NASA male flying personnel (with permission from Roebuck et al. 1988).

Phantoms, Ghosts, and the "Average Person" *Several misleadingly simple body proportion templates have been used in the past (Drillis and Contini 1966). In fact, all "fixed" design templates fall into that category if they assume that all body dimensions, such as lengths, breadths, and circumferences, can be represented as given fixed proportions (percentages) of one body dimension—for example, stature. Obviously, such a simplistic assumption contradicts reality: The relationships among body dimensions are neither necessarily linear nor the same for all persons. In spite of the obvious fallacy of the model, "single-percentile constructs" have been generated, assuming that persons exist whose body segments are all of the same percentile value. Not only the 50th-percentile phantom (the "average person") has been used as a design template, but other ghostly figures have been created that have, for example, all 5th- or 95th-percentile values. Of course, designs for these figments do not fit actual users.*

Fitting what to what?

"Fitting" Design Procedures. Information about body size is needed when an object must fit the human body. Examples of such objects are tool handles to hold, protective equipment and clothing to wear, chairs to sit on, windows to look through, and workstations in general. Several of these applications are discussed elsewhere in this book (see, for example, the sections on hand tools in Chapter 8 and computer workstations in Chapter 9). Different fitting methods are used, such as choosing exact percentiles on the continuum of the measuring scale to determine ranges (say, from the 5th to the 67th percentile), or to assure that the largest persons will fit through an

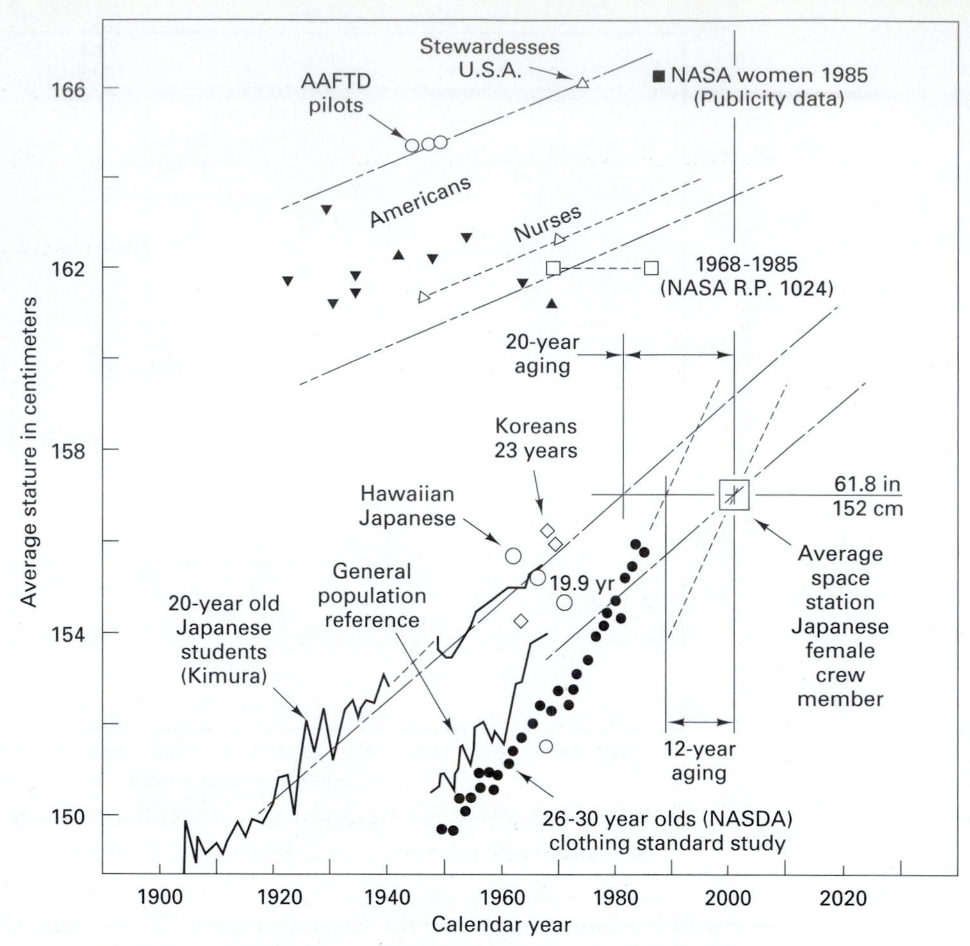

Figure 1–8. Predicted average stature of American and Asian women space crews (with permission from Roebuck et al. 1988).

opening, or to ensure that even the smallest can use the equipment. In this context, one often speaks of "functional" (or dynamic) anthropometry, meaning body data that depend on the coordinated efforts of several body segments to achieve a desired posture or perform an activity. These data may define zones of convenience, of expediency, or the minimally required or the largest covered space.

Workspaces of hands and feet

Zones of convenience or expediency are difficult to define, because the criterion is not absolute (in the sense of minimal or maximal), but depends on the situation, the subject, and the task. The various "normal working areas" first shown in the 1940s, usually in the form of partial spheres around the elbow or shoulder, are examples of plausible, yet ill-defined, convenience contours. It is difficult to accept that a man should have a working area within a "radius of 394 mm from the shoulder," while a woman should have a working area limited by a "radius of 356 mm" (Nicholson 1991). Why those exact dimensions? Of course, it makes sense that work should be done within easy reach—see Figure 1–9—one just has to define what "easy" means.

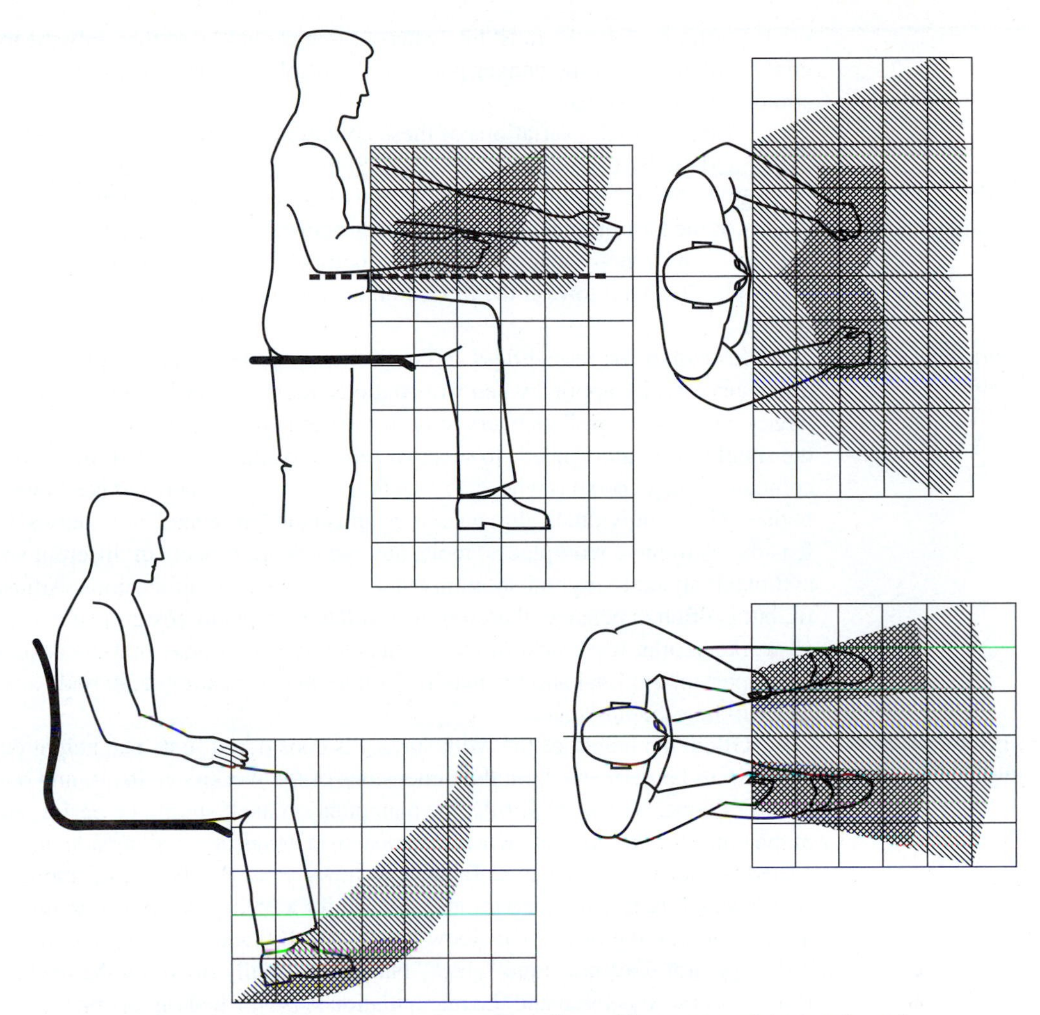

Figure 1–9. The concept of "preferred" working zones of the hands and feet.

An example of a clear and defined procedure is the determination of a "safe distance" from a danger point. The danger point is that edge of a hazardous gadget (such as of a press mold, cutting edge, or pinch point) closest to the operator from which the operator's body (usually the finger or toe) must be kept away. The safe distance is the straight-line distance between the danger point and the barrier (wall, safety guard, enclosure of an opening) beyond which the operator's body cannot proceed toward the hazard. That distance should be increased by a safety margin.

Finger and toe safety distances

For finger safety, the distance may be determined either beyond an opening that allows only the finger to penetrate, or pass an opening or a barrier that can be overreached by the arm. In the first case, the safe distance would be the length of the longest possible finger, with a safety margin; in the second case, the longest arm and finger reach would determine the distance.

For foot safety, the most likely barrier is at the ankle, so that only the toes and foot can penetrate further toward the danger point; or the whole leg may have to be considered, probably restrained in the hip area.

There are many variations of these conditions, such as those in the German Standard DIN 31001 and the British Standard BS 5304. Some are shown in Figure 1–10.

Safety margins

In each case, the longest possible body segment should be considered, under the given conditions of the barrier and mobility. A predetermined safety margin (of, say, 10 percent) should be applied to those body lengths. Certain conditions, such as holding an object that, if entrapped, might pull the hand toward the hazard point, could be good reason to extend the safety distance further.

"Min–max strategy" for fitting small and tall persons

Instead of the thoughtless and useless average-person concept, the "min–max strategy" is often successfully applied when workstations, tools, and tasks must be designed to fit small and large operators, as well as everybody in-between. The "minimal" dimensions are derived from the smallest operators' needs to see, to reach, or to apply force; yet these dimensions may not accommodate large persons who can reach farther, can see higher, and need more open space for their bodies. These individuals' dimensions establish the "maximal" boundaries. One good solution, if feasible, is to have workplaces, tools, and other work objects in different sizes between minima and maxima; another good solution is to have adjustable dimensions. Adjustability allows good fit, but is often expensive; therefore, one often attempts to design just one workstation that suits all workers, or at least most of them. This requires, of course, that the designer first decide what the important minimal and the maximal values are (for example, in body size, reach, or strength) that will be accommodated.

Work at elbow height

Often, the height of the work surface is considered first. The height depends on the physical work to be performed, on the dimensions of the workpiece itself, and on the need to observe the work done. As a general rule, the manipulation itself should be performed at about the height of the elbow of the operator when the upper arm hangs down alongside the trunk or is slightly elevated forward and sideways. Table 1–3 shows an elbow height of about 93 cm for a 5th-percentile standing female operator and 107 cm for a 95th-percentile standing female. For standing male operators, the respective elbow heights are 100 and 115 cm. One may reduce these heights if the operator does not stand "erect," but that is usually offset by the heel height of shoes worn. If the workpiece is large and the manipulation is performed on its upper part, the support surface (bench height) must be lowered to allow the hands to be at elbow level. Some work demands close visual observation, requiring an appropriate viewing distance. In that case, particularly if the manipulation needs fairly little force and energy, the work area might be elevated well above elbow height. (But that, in turn, might require support for the elevated hands and forearms.) These conditions are illustrated in Figure 1–11.

To determine actual design values for a workstation, the relevant body dimensions (often, especially elbow height and eye height) of the expected operator population must be selected and adjusted according to body postures and specific work requirements. For a sitting operator, the elbow height is referenced not to the floor, but to the seat surface. For example, the 5th- and 95th-percentile elbow heights of a sitting female are given in Table 1–3 as 18 and 26 cm, respectively, and the corresponding values are 18 and 27 cm for male operators (note how similar the values are). However, the support surface can be lowered only until it nearly touches the upper side of the thighs; the thigh height shown in the table ranges from about 14 cm to about 19 cm above the seat height for the 5th- to 95th-percentile operator, whether

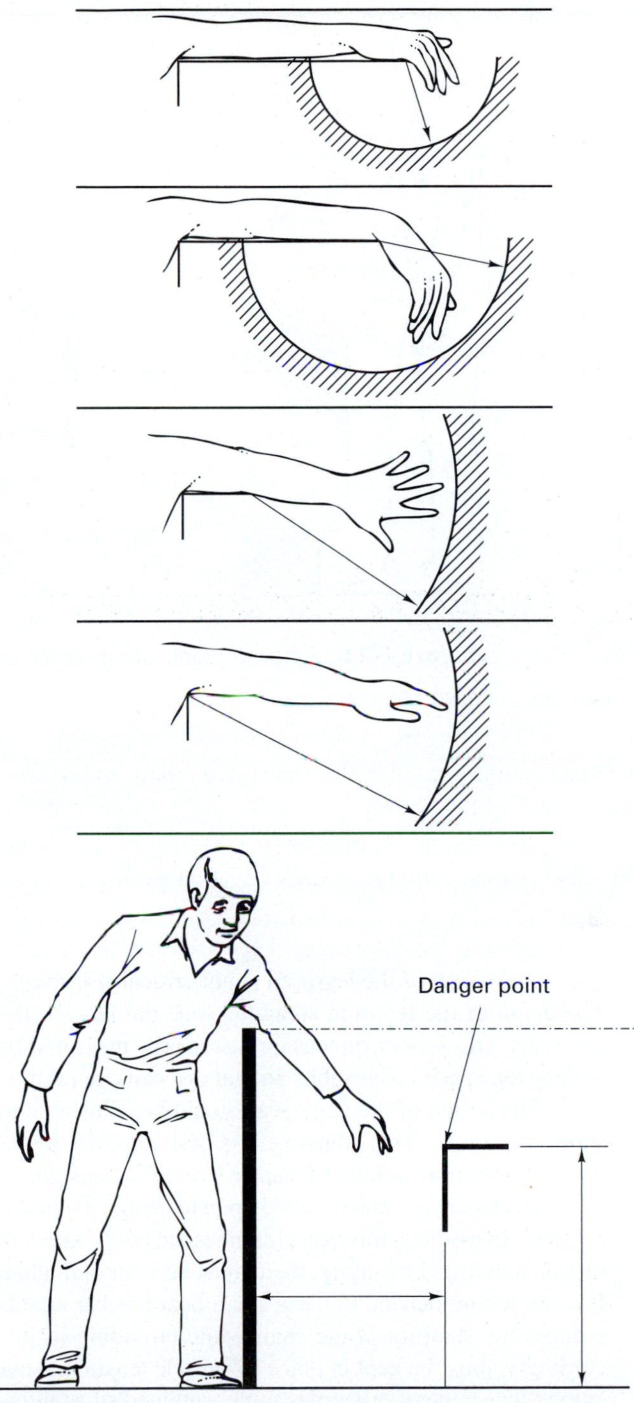

Figure 1–10. Examples of "safe distances" (modified from DIN 31001 and BS 5304).

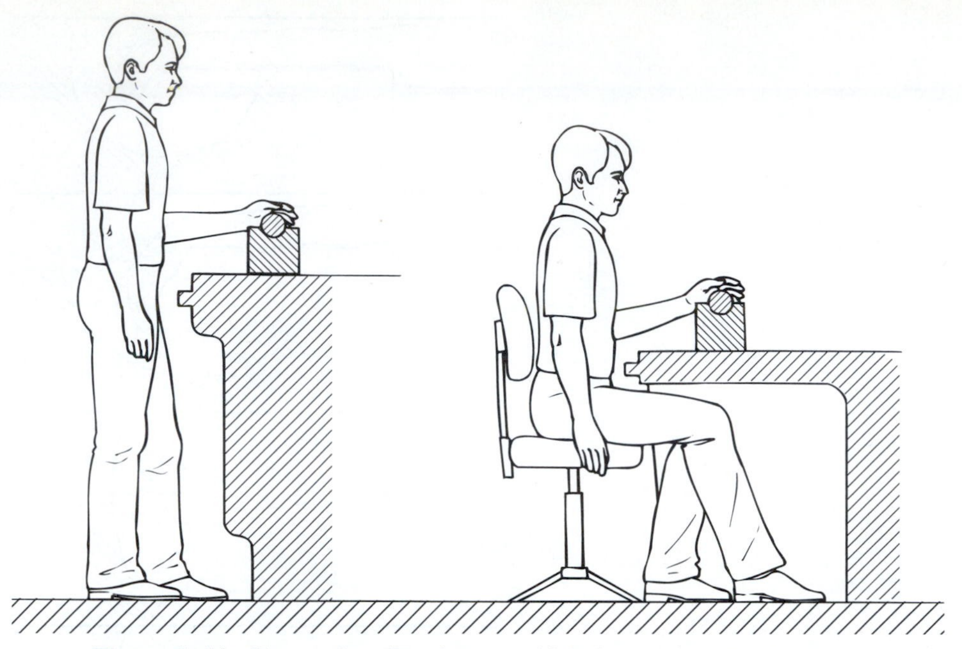

Figure 1–11. Shape of workstations at which the operator stands or sits.

female or male. These values establish the necessary height of the space underneath the working surface to accommodate the legs of the sitting operator—here, at least 19 cm, which conflicts with the 18-cm minimal elbow height just determined. Another way to determine the needed height of the leg space is to use the "knee height," also given in the table, plus some allowance for shoe heels.

The *width* of the legroom is not critical if it exceeds the hip width of the widest operator. The *depth* of the legroom should exceed the largest distance from the front of the belly to the kneecaps. This is not a dimension customarily measured by anthropometrists; it has to be estimated. A deep leg space is desirable, so that one can extend the lower legs and push the feet forward.

The height of the work seat should be adjustable to fit persons with long and persons with short lower legs. This adjustment is best achieved by varying the height of the seat surface, as discussed in more detail in Chapter 9 on office design.

Sit–Stand station

Occasionally, one is called upon to design a workstation at which the operator can either sit or stand. In essence, this task combines the "min–max" requirements of the stations for either sitting or standing. For sitting, there must be a very tall chair and a high support surface for the feet. It is not recommended to use a small board or bar attached to the chair as a foot rest, because it reduces the stability of the chair while providing little support surface for the feet, which, accordingly, must be kept in place by muscle tension instead of being able to move to different positions. The general principles for a combined sit–stand workstation are sketched in Figure 1–12. Some "toe space" at the bottom of the workstation allows a standing operator to step in closely to the work. This space should be high enough to accommodate persons wearing thick soles, but shallow enough so that one does not hit the edge of the foot-space cutout with the instep of the foot. Thus, a depth not to exceed 10 centimeters and a height of not less than about 10 cm should be appropriate—as sketched in Figures 1–11 and 1–12.

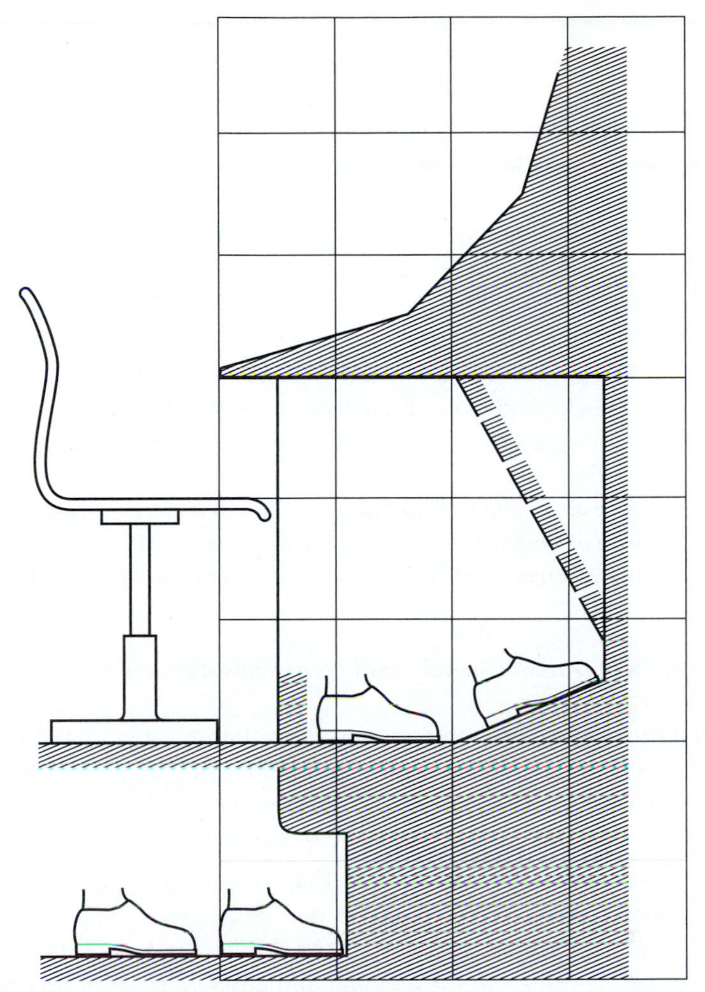

Figure 1–12. Workstation for standing or sitting, or for alternately sitting and standing.

DESIGN PROCEDURES

"GOOD"

Several practical procedures lead to proper consideration of the different sizes and proportions in which the human body comes. Basically, one identifies the critical dimensions and ensures that they are fitted. This means that one has to decide which are the smallest percentile values (not necessarily the 5th) and the largest values (not always the 95th) to be accommodated. If both the smallest and the largest percentiles are taken into account, one is fairly sure that the intermediate range will be accommodated. The calculation can be relatively simple, particularly if the problem is one dimensional or if there are no statistical relationships among several relevant dimensions.

"CAUTION"

Stature and weight are often used to scale an imaginary *n*th-percentile person; but many other dimensions are not highly correlated with either stature or weight, and hence, these two variables are often not good predictors. Check Table 1–6.

"BAD"

A blatant, but unfortunately rather common, example of misuse is combining the two variables of stature and weight into one index, in order, say, to calculate "undesirable" weight–height ratios. The correlation between the two values is low—in the neighborhood of 0.3 for women and 0.4 for men in the general population. Yet, there are still people who try to predict hip breadth from stature (Hamill and Hardin 1997) or construct single-percentile phantoms (such as the 5th-, 50th-, "average", or 95th-percentile "person"), even after Daniels tried to eradicate this nonsense in 1952.

☛☛☛ *If one tries to "stack" values of a given percentile, such as the 5th, one does not end up with a composite figure that, in its sum, is also 5th percentile. For example: 5p hip height, plus 5p trunk height, plus 5p head height does* not *add up to 5p stature.* ☚☚☚

One useful and correct general design procedure consists of the following four steps:

Step 1: Select those anthropometric measures that directly relate to defined design dimensions. Examples are hand length related to handle size, shoulder and hip breadth related to escape-hatch diameter, head length and breadth related to helmet size, eye height related to the heights of windows and displays, knee height and hip breadth related to the legroom in a console.

Step 2: For each of the pairings chosen in Step 1, determine whether the design must fit only one given percentile of the body dimension or a range along that body dimension. Examples are as follows: The escape hatch must be fitted to the largest extreme values of shoulder breadth and hip breadth, considering the clothing and equipment worn; the handle size of pliers is probably selected to fit a smallish hand; the legroom of a console must accommodate the tallest knee heights; the height of a seat should be adjustable to fit persons with short and those with long lower legs. (Table 1–4 shows how to calculate percentile values.)

Step 3: Combine all selected design values in a careful drawing, mock-up, or computer model to ascertain that they are compatible. For example, the required legroom clearance height, needed for sitting persons with long lower legs, may be very close to the height of the working surface, determined from elbow height.

Step 4: Determine whether one design will fit all users. If not, several sizes or an adjustment must be provided to fit all. Examples are as follows: one large bed size fits all sleepers; gloves and shoes must come in different sizes; seat heights are adjustable.

Fit checking

The ultimate test of a design is through its actual use by real persons. This can be simulated with test subjects who represent the min–max values selected by the designer. Suppose those subjects are women who had 10th-percentile values in the critical design-related dimensions and men with 99th-percentile values. A reasonable way to present a 10p woman (or a 99p man) is to measure a group of women (or men) who have that stature and weight and to calculate, from their measures the median values of other dimensions of interest in the group. If one then checks each resulting value, it is likely not to be exactly at the 10th (99th) percentile, but close to it.

The following paraphrased story appeared in the Washington Post *of May 25, 1984: The Navy has adopted new flight training standards that will require its aviators, as a whole, to have longer arms and shorter legs. The standards will exclude 73 percent of all college-age women and 13 percent of college-age men, according to a military spokesman who said that the new standards were devised because some aviation candidates could not reach rudder pedals or see over instrument panels. Some taller pilots were so tightly wedged that their helmets bumped the aircrafts' canopies. "We found out that manufacturers are still building airplanes the way they want, but God is not making people to fit them" Previously, 39 percent of the women applicants and 7 percent of the men were ineligible to become aviation candidates because of their size.*

Six years later, Buckle and coworkers (1990) checked the cockpit dimensions of aircraft used throughout the world (the Boeing 737-200, 747, and 757 and the Lockheed TriStar) with respect to eight critical body dimensions of pilots, including eye height, hand and leg sizes, and reaches. In many cases, the fit was marginal at best. For example, on the basis of eye height, 13 percent of British male and 73 percent of British female pilot candidates would have to be excluded from being crew members.

HUMAN BIOMECHANICS

Biomechanics is the study of characteristics of the body in mechanical terms. The biomechanical approach is not new: Biomechanics has been applied to the statics and dynamics of the human body, to explain effects of vibrations and impacts, to explore characteristics of the spinal column, and to examine the use of prosthetic devices, to mention just a few examples.

Leonardo da Vinci (1452–1519) and Giovanni Alfonso Borelli (1608–1679) combined mechanical with anatomical and physiological explanations to describe the functioning of the biological body. Since Borelli, the human body has often been modeled as consisting of long bones (links) that are connected in the articulations (joints), powered by muscles that bridge the articulations. The physical laws developed by Isaac Newton (1642–1727) explained the effects of external impulses applied to the human body.

Today, we still use the biomechanical model that Borelli suggested more than 300 years ago: The human body is built on a structure of solid links connecting to each other in joints of various degrees of freedom; the body segments are powered by muscles that cross joints, volumes and mass properties are attributed to the body as well.

Treating the human body as a mechanical system entails gross simplifications, such as disregarding mental functions. Still, many components of the body may be well considered in terms of analogies such as the following:

- bones—lever arms, central axes, structural members
- articulations—joints and bearing surfaces

- tendons—cables transmitting muscle forces
- tendon sheaths—pulleys and sliding surfaces
- flesh—volumes and masses
- body contours—surfaces of geometric bodies
- nerves—control and feedback circuits
- muscles—motors, dampers, or locks
- organs—generators or consumers of energy

THE SKELETAL SYSTEM

Bones

The human skeleton is normally composed of 206 bones, together with associated connective tissue and articulations.

Bone framework

The main function of human skeletal bone is to provide an internal framework for the whole body, see Figure 1–13. The long, more or less cylindrical bones that connect body joints are of particular interest to the biomechanist. They are the lever arms at which muscles pull.

Young and old bone

While bone is firm and hard, and thus can resist high strain, it has certain elastic properties. In childhood, when mineralization is relatively low, bone is rather flexible. In contrast, the bones of the elderly are highly mineralized and therefore more brittle. Also, they change their geometry, similar to pipes getting wider in diameter, but thinner in their walls. Thus, osteoporosis in the elderly means, mechanically speaking, a hollowing of bones, a decrease in bone mass, and the thinned walls becoming brittle (Ostlere and Gold 1991). Yet, the moment of inertia, $I = \pi/4\,(R^4_{outer} - R^4_{inner})$, remains about the same. Bone cells are nourished through blood vessels.

Use it or lose it

Bone material is continuously resorbed and rebuilt throughout one's life. Local strain encourages growth and disuse encourages resorption, as long as a suitable threshold is not exceeded (Wolff's law): Overstrain can cause structural damage through micro- or macrofractures.

☛☛☛ *The French physician Marie Francois Xavier Bichat (1771–1802) performed many postmortems during which he observed that the various organs were built of a mixture of simpler structures. Basically, these structures are flat and delicately thin, and in his 1800 book* Treatise on Membranes, *Bichat called them tissues. (Asimov 1989).* ☚☚☚

Connective Tissues

Several types of tissues connect parts of the body:

- Muscles are the organs that generate force and movement between bone linkages.
- Tendons are strong yet elastic elongations of muscle, tissues connecting muscle to bones.
- Ligaments connect bones and provide capsules around joints.
- Cartilage is a translucent, viscoelastic, flexible material capable of rapid growth, located at the ends of the ribs, as discs between the vertebrae, and, in general, as articulation surfaces at the joints.

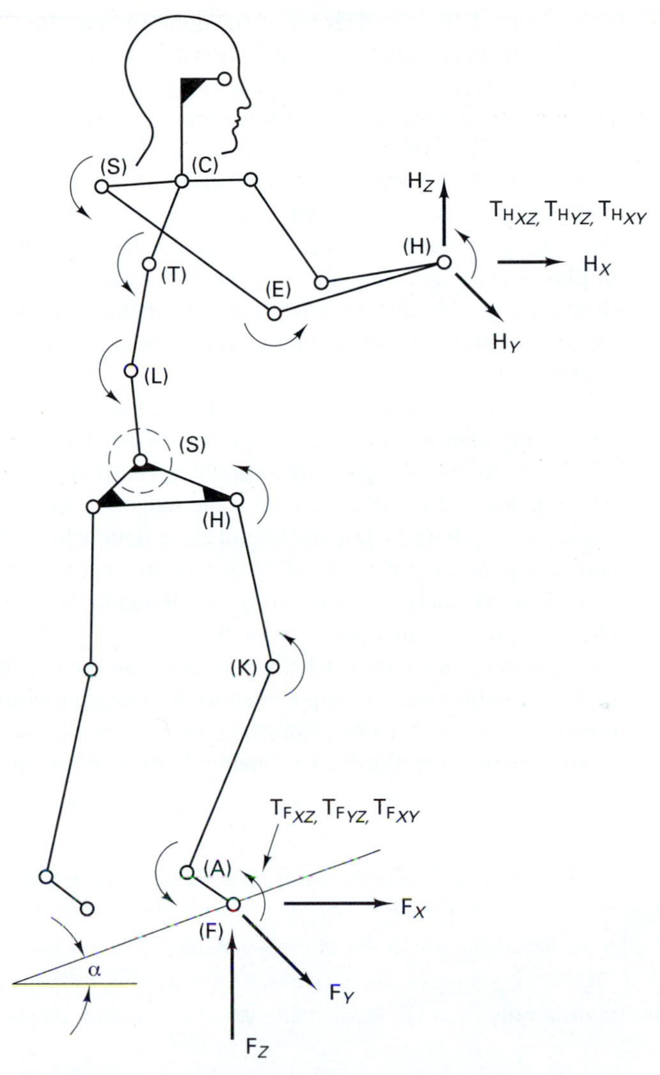

Figure 1–13. Human skeleton simplified as "links connected in joints" (*H* for hand, *E* for elbow, *S* for shoulder, etc.). (From Kroemer et al. 1990, *Engineering Physiology: Bases of Human Factors/Ergonomics,* 2d ed. With permission by the publisher, Van Nostrand Reinhold. All rights reserved.)

Joints

Joint design

The shape of the bones at their articulations, the encapsulation of the joint by ligaments, the supply of cartilaginous membranes, and the provision of discs or volar plates determine, together with the action of muscles, the mobility of body joints.

Joint mobility

Some bony joints, such as the seams in the skull of an adult, have no mobility left in them; others, such as the connections of the ribs to the sternum, have very limited mobility. Joints with "one degree of freedom" are simple hinge joints, like the elbow or the distal joints of the fingers.

Other joints have two degrees of freedom, such as the ill-defined wrist joint (discussed later), wherein the hand may be bent in flexion and extension and may pivot laterally (deviate). (The capability to twist, pronate and supinate, is located in the forearm, not in the wrist.) Shoulder and hip joints have three degrees of freedom.

Joint lubrication

Synovial fluid in a joint facilitates movement of the adjoining bones by providing lubrication. For example, while a person is running, the cartilage in the knee joint can show an increase in thickness of about 10 percent, brought about in a short time by synovial fluid seeping into it from the underlying bone marrow cavity. Similarly, fluid seeps into the spinal discs (which are composed of fibrous cartilage) when they are not compressed (eg, when one lies down to sleep). This makes them more pliable directly after getting up than during the day, when they are "squeezed out" by the load of body masses and their accelerations. Thus, immediately after getting up, one stands taller than after a day's effort.

Ranges of motion

The term "mobility," or "flexibility," indicates the range of motion that can be achieved in a body articulation. The actual range, which depends on training (use), age, and gender, is properly measured by an angle from a known reference position, such as the so-called neutral position, which is located somewhere within the range. Or the range may be measured as the enclosed angle between the smallest and largest excursions achieved by adjacent body segments about their common joint. Figure 1–14 indicates common mobility measurements.

"Neutral" position?

Unfortunately, many mobility data found in the literature are questionable or unexplained. Quite often, the actual point of rotation moves with the motion, the arms of the angles are ill-defined, and it is not clear whether the positions were achieved by internal muscular force alone (active mobility) or with help from outside (passive mobility). The "neutral" position is often intuitively assumed to mean "straight" (such as when hand and forearm are aligned), but, for most joints, remains undefined. (See Chapter 8 for a discussion of postures at work.)

A study by Staff (1983) provided reliable information about active (maximal voluntary, unforced) mobility in the major body joints of physical education students. The study, done on 100 female subjects, was carefully controlled to resemble an earlier study by Houy (1982) on 100 male subjects. The results are compiled in Table 1–7. Of the 32 measurements taken, 24 showed significantly more mobility by women, while men were more flexible only in ankle flexion and wrist abduction. The differences were small throughout, however.

"Flexibility" in body joints should be of practical importance. Intuitively, it appears that more flexibility should indicate better physical performance and reduced risk of injury. Yet, in comparison with untrained persons, athletes in the following categories have been found less flexible: soccer players, runners, persons participating in sports for five years or longer, and even ballet dancers in some hip movements (Burton 1991).

Artificial Joints

Natural joints may fail as a result of disease, trauma, or long-term wear and tear. If conservative medical treatment fails, joints may be replaced with artificial, manufactured devices, a procedure that is routinely done in fingers, but mostly in hips and knees. In the United States, about half a million joints are implanted annually, predominantly in elderly persons.

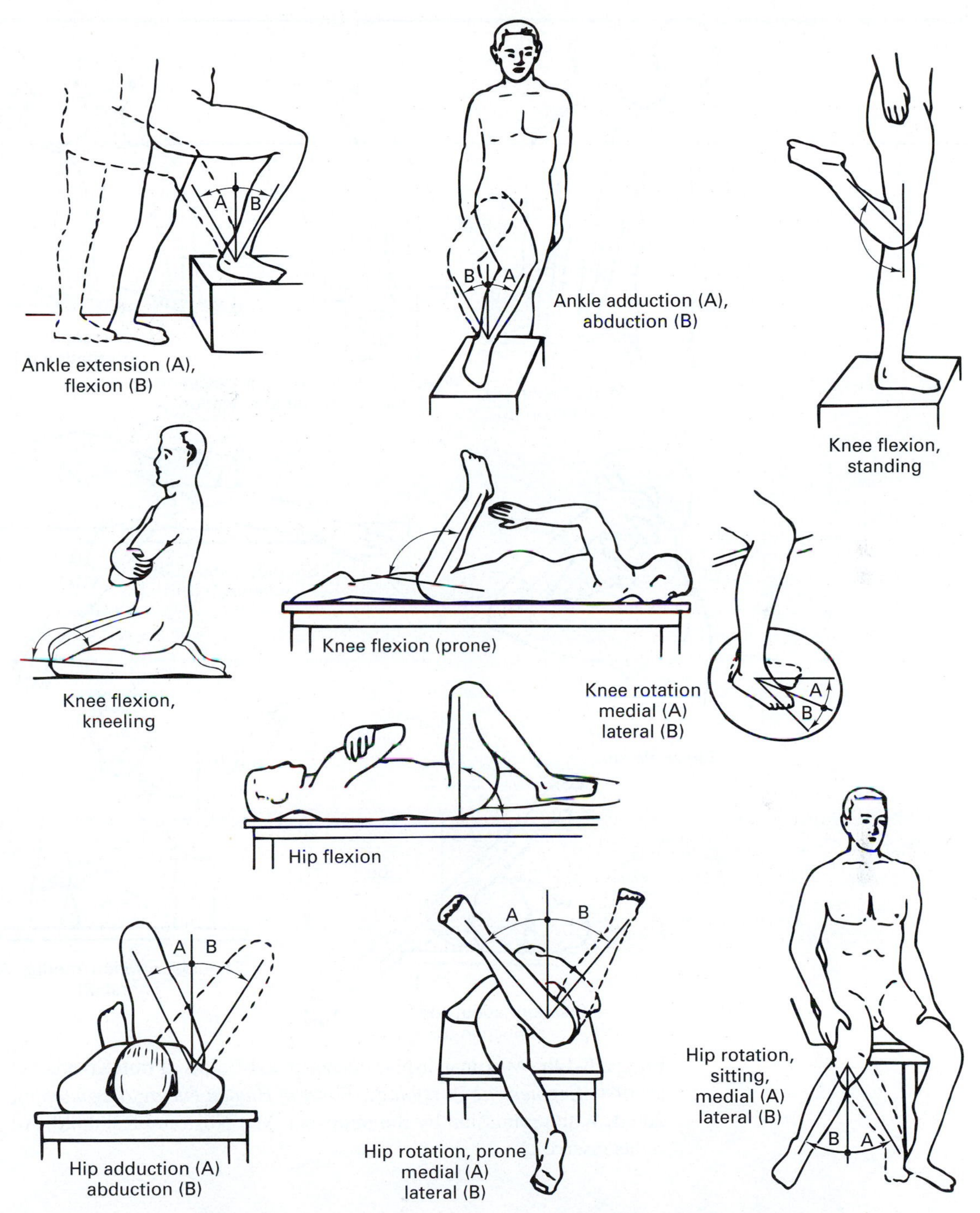

Figure 1–14a. Maximal displacements in body joints. (From Kroemer et al. 1990, *Engineering Physiology: Bases of Human Factors/Ergonomics*, 2d ed. With permission by the publisher, Van Nostrand Reinhold. All rights reserved.)

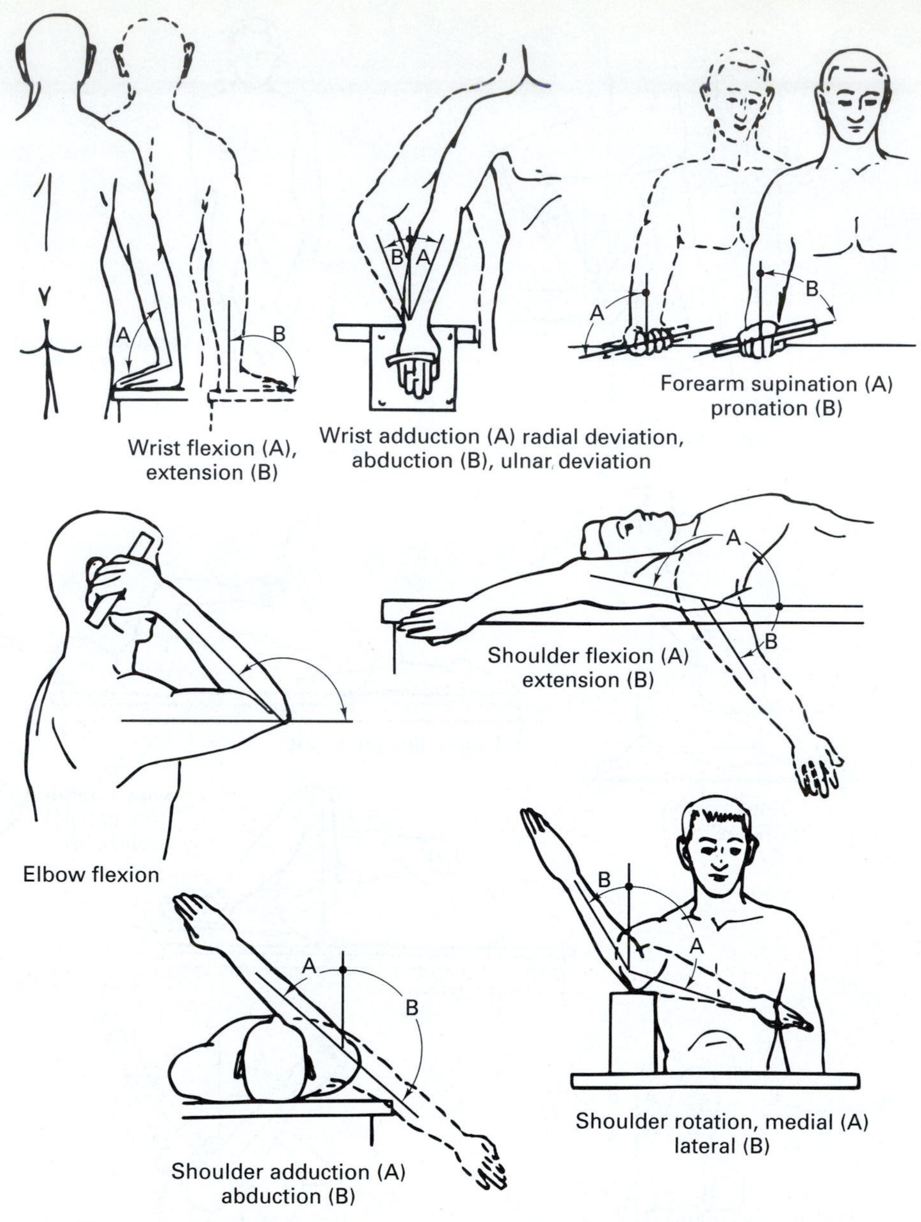

Figure 1–14b. Maximal displacements in body joints. (From Kroemer et al. 1990, *Engineering Physiology: Bases of Human Factors/Ergonomics,* 2d ed. With permission by the publisher, Van Nostrand Reinhold. All rights reserved.)

The degeneration of cartilage in the major joints (such as the hip and knee) due to trauma or arthritis may lead to the replacement of the articulating surfaces with artificial joints. Although total joint replacement typically restores function and mobility to the patient, its primary and most appreciated purpose is to relieve pain.

TABLE 1–7. Comparison of Mobility Data for Females and Males (in degrees)

Joint	Movement	5th Percentile Female	5th Percentile Male	50th Percentile Female	50th Percentile Male	95th Percentile Female	95th Percentile Male	Difference* Female − Male
Neck	Ventral flexion	34.0	25.0	51.5	43.0	69.0	60.0	+8.5
	Dorsal flexion	47.5	38.0	70.5	56.5	93.5	74.0	+14.0
	Right rotation	67.0	56.0	81.0	74.0	95.0	85.0	+7.0
	Left rotation	64.0	67.5	77.0	77.0	90.0	85.0	NS
Shoulder	Flexion	169.5	161.0	184.5	178.0	199.5	193.5	+6.5
	Extension	47.0	41.5	66.0	57.5	85.0	76.0	+8.5
	Adduction	37.5	36.0	52.5	50.5	67.5	63.0	NS
	Abduction	106.0	106.0	122.5	123.5	139.0	140.0	NS
	Medial rotation	94.0	68.5	110.5	95.0	127.0	114.0	+15.5
	Lateral rotation	19.5	16.0	37.0	31.5	54.5	46.0	+5.5
Elbow–forearm	Flexion	135.5	122.5	148.0	138.0	160.5	150.0	+10.0
	Supination	87.0	86.0	108.5	107.5	130.0	135.0	NS
	Pronation	63.0	42.5	81.0	65.0	99.0	86.5	+16.0
Wrist	Extension	56.5	47.0	72.0	62.0	87.5	76.0	+10.0
	Flexion	53.5	50.5	71.5	67.5	89.5	85.0	+4.0
	Adduction	16.5	14.0	26.5	22.0	36.5	30.0	+4.5
	Abduction	19.0	22.0	28.0	30.5	37.0	40.0	−2.5
Hip	Flexion	103.0	95.0	125.0	109.5	147.0	130.0	+15.5
	Adduction	27.0	15.5	38.5	26.0	50.0	39.0	+12.5
	Abduction	47.0	38.0	66.0	59.0	85.0	81.0	+7.0
	Medial rotation (prone)	30.5	30.0	44.5	46.0	58.5	62.5	NS
	Lateral rotation (prone)	29.0	21.5	45.5	33.0	62.0	46.0	+12.5
	Medial rotation (sitting)	20.5	18.0	32.0	28.0	43.5	43.0	+4.0
	Lateral rotation (sitting)	20.5	18.0	33.0	26.5	45.5	37.0	+6.5
Knee	Flexion (standing)	99.5	87.0	113.5	103.5	127.5	122.0	+10.0
	Flexion (prone)	116.0	99.5	130.0	117.0	144.0	130.0	+13.0
	Medial rotation	18.5	14.5	31.5	23.0	44.5	35.0	+8.5
	Lateral rotation	28.5	21.0	43.5	33.5	58.5	48.0	+10.0
Ankle	Flexion	13.0	18.0	23.0	29.0	33.0	34.0	−6.0
	Extension	30.5	21.0	41.0	35.5	51.5	51.5	+5.5
	Adduction	13.0	15.0	23.5	25.0	34.0	38.0	NS
	Abduction	11.5	11.0	24.0	19.0	36.5	30.0	+5.0

Source: From Kroemer, Kroemer, and Kroemer-Elbert 1990, *Engineering Physiology: Bases of Human Factors/Ergonomics,* 2d ed. With permission of the publisher, Van Nostrand Reinhold.

*Only significant ($\alpha < .05$) differences at the 50th percentile are listed. NS = Not significant.

Joint replacement

Joint degeneration is associated with pain and with progressive and severe limitations of motion. If needed, in the hip, typically, the head of the femur (the thigh bone) is removed and replaced by a spherical metal ball on a stem, and the acetabular socket is resurfaced with a plastic liner. In the knee, the articulating surfaces on the bottom of the femur are replaced with metal, and articulating surfaces at the top of the tibia (the shinbone) and on the patella (the kneecap) are resurfaced with plastic. To date, the metals used have been stainless steel, a cobalt–chromium alloy, and a titanium alloy. These joints all have the same type of design for the major load-bearing components: The metallic component is convex and the plastic component is concave. The plastic now used (after an early, disastrous attempt with Teflon) is apolyethylene of ultrahigh molecular weight.

☛☛☛ *Replacements for the ball-and-socket joint at the hip have been attempted for about a century, but routinely successful total hip replacement started in England in the 1960s with Charnley. He pioneered the use of the metal-on-plastic articulations and the use of polymethyl methacrylate (PMMA) as bone cement. Today, at least 90 percent of patients with hip and knee replacements report that they are pleased with their new joints and continue to function well 10 years and longer after surgery.* ☚☚☚

Attachment to bone

A polymeric bone cement is often used to fill the space between the metal implant surfaces and the reamed bone cavity. This cement has no adhesive properties of its own, but serves as a grout to link the prosthesis and the bone mechanically. Recently, devices have been designed that need no bone cement. They have rough surfaces with a pore size of less than 1 mm, into which the bone is intended to grow. To encourage bony ingrowth, an osteoinductive or conductive chemical coating may be sprayed on the porous surface of the metal.

Linings

A new development in total joint replacement has been the use of ceramics. The acetabular cup liner is made from polyethylene, and the head of the femur replacement is ceramic, placed on a traditional metal stem. Ceramics have theoretical and practical advantages in reducing wear of the artificial joint, but their high cost, low material toughness, and difficulties being manufactured, in terms of quality control, continue to raise problems. To date, ceramic joint designs for hip and knee replacements have not performed better than metal-on-plastic joint designs, which may have success rates of 95 to 90 percent at 5- to 10-year follow-up tests.

Finger joints

Finger joints are usually replaced by a one-component, molded plastic integral hinge. This simple artificial joint is successful for several reasons, not least among which is the fact that the loads carried by the joints are low, and that wear generates only little debris.

Design requirements

The design of joint replacements is governed by biologic and mechanical considerations. The device must be biologically compatible with the body, both when intact and as wear debris. In the body's corrosive and warm environment, the material from which the device is made must be nontoxic, have low reactivity, exhibit minimal wear, and maintain its structural strength. Of course, the artificial joint must be implantable (in terms of both complexity and size) and should yield near-normal range of motion. Finally, the design of the device should consider the possibility of salvage: Sufficient bone and soft tissue should remain to allow for replacement of the device or fusion of the joint if needed.

Failures

When joint replacements fail, the patient usually presents with severe pain. Upon examination, infection is commonly found, and the device is often no longer firmly attached to the surrounding bone. This is mostly a mechanical problem, frequently associated with wear particles from the metallic component, the lining, or the cement. The debris may trigger a biologic response that leads to resorption of the bone and loss of support for the implant, as well as inflammation, reduced range of motion, and pain (Elbert 1991).

Testing

Physical testing of artificial joints, especially to evaluate novel designs or materials, may involve animal models. The testing may also be *ex vivo,* such as on isolated bones and joint segments. Implantation techniques or the range of motion of a device are often evaluated using these isolated segments. Strain gages, brittle stress coatings, and photoelasticity may be used to determine the extent of the strains (and infer that of the stresses) on the surfaces of the devices and cortical bone, but it is difficult to measure the strain response inside the material.

Numerical testing may rely on elasticity analysis, such as composite beam theory, plate/shell theory, torsion theory, and beam-on-elastic foundation theory. More recently, linear and nonlinear finite-element analyses have been used to calculate the stresses inside and on devices and idealized bone models. The accuracy of the numerical models is limited by the assumptions made in the formulation of the model (such as assuming certain boundary conditions) and by the sophistication of the material modeled (in terms of homogeneity, elasticity, and continuity, for example).

To design better and longer-lasting artificial joints, one must understand the loads to which the joints are subjected. Indirect and some limited direct methods to determine joint loads have been used. Loading of the joint may be estimated indirectly by using the classical technique of correlating limb position, velocity, acceleration, and force-plate readings or motion to determine the balance of forces across the joint. Some sophisticated studies incorporate electrical activities of muscles to model the distribution of forces at the joint; recent numerical work has focused on nonlinear optimization of the forces in the tendons and ligaments at a joint, along with electromyographic information regarding muscle activity, to predict force distribution during various activities (Elbert 1991).

Recently, the joint loads in the human body have been directly measured. For example, a special total hip replacement with a three-axis load cell in the neck of the metal component has been implanted in the thighbone. Then the forces occurring during various activities were telemetered to recorders in the laboratory. The actual *in vivo* loads were found to be somewhat smaller than those calculated previously using less invasive techniques.

The Spinal Column

Structure

The spine is a complex structure consisting of 24 movable vertebrae (seven cervical, 12 thoracic, and five lumbar), together with the sacrum and the coccyx, which are a fused group of rudimentary bones. These sections are held together in cartilaginous joints of two different kinds. First, there are fibro-cartilage discs between the main bodies of the vertebrae. Second, each vertebra has two protuberances extending posterior–superiorly—the superior articulation processes—which end in rounded surfaces fitting into cavities on the underside of the next-higher vertebra. These synovial facet joints are covered with sensitive tissue, whereas the discs between the main bodies of the vertebra have no pain sensors.

Functions

The spine transfers forces and both bending and twisting moments from head and shoulder bones to the pelvis. It also protects the spinal cord, which runs through posterior openings (the spinal canal), carrying signals between the brain and all sections of the body. This complex rod, transversing the trunk and keeping the shoulders separated from the pelvis, is held in delicate balance by ligaments connecting the vertebrae and by muscles that pull along the posterior and the sides of the spinal column. Longitudinal muscles located along the sides and the front of the trunk also both balance and load the spine.

Bends

Figure 1–15 shows schematically the stack of the vertebrae, indicating that—in the side view—the column is not straight, but has two forward bends (lordoses) in the cervical and lumbar sections and one backward bend (kyphosis) in the chest area. Only in the frontal view is the spinal column straight; distortion in this view is called scoliosis.

Configuration

Figure 1–16 is a schematic of the lumbar section of the spinal column, showing particularly the bearing surfaces at the main bodies and at the facets. Figure 1–17 shows a top view of a vertebra, indicating its main body, the structure surrounding the canal for the spinal cord (the vertebral foramen), and the five major protuberances, two to the side, two upward–backward, and one extending straight to the rear, to which ligaments and muscles attach. (For more exact geometric details of the vertebrae, see Panjabi et al. 1992.)

Flexibility under load

The spine is capable of withstanding considerable loads, yet is flexible enough to allow a large range of motions. There is, however, a trade-off between the load carried and flexibility. If no external load is pressing on the spine, only its anatomical structures (joints, ligaments, and muscles) restrict its mobility. Applying load to the spinal column reduces its mobility until, under heavy load, the range of possible postures becomes limited.

Stability under load

The traditional model of the spine has been that of a straight column, as depicted in Figure 1–18A. This simplification allows a unique description of its geometry and strain under the applied load (Aspden 1988, Yettram and Jackman 1981). If one considers the spinal column as an arch (Figure 1–18B), its load-bearing mechanism depends on its curvature. Since the spinal arching is not fixed, there is no unique solution that describes its strain. Force along the arch is thought to be transmitted along a straight line, called the thrust line. The theorem of plasticity assumes that if the arch is to be stable, then the thrust line must lie within the cross section of the arch components throughout the entire length of the arch. If the thrust line lies outside the arch at any point, the tensile force must keep the arch within its possible range of positions, or it will buckle—see Figure 1–18B.

Compression

A major load on the spine is that due to compression. Figure 1–19 illustrates that the compressive force C results from the pull force M of the trunk muscles and the weight due to segment masses and any external load. Spinal compression may be somewhat relieved by the upward-directed force P due to intraabdominal pressure (IAP; see Chapter 11).

Other strains

Owing mostly to the slanted arrangement of load-bearing surfaces at discs and facet joints, the spine is also subjected to shear S. Furthermore, the spine must withstand both bending and twisting torques T. Aspden (1988) calculated spinal strain according to his model and obtained three interesting results:

1. The calculated compression loads in a stable arched spine are considerably lower than those computed using a straight-spine model.
2. The loads depend on the adopted posture—that is, on the geometry of the spine.
3. Intraabdominal pressure (IAP) can stiffen the lumbar spine.

The effect of IAP in stiffening the lumbar spine is shown in Figure 1–18C. The thrust line would be outside the spinal column if kyphotic flattening of the lumbar area were maintained. Yet, if lumbar lordosis is introduced, the thrust line can be kept within the spinal components, and the arch is stable. The larger the IAP, the better lordosis can be maintained, even under heavy axial loading (compression) of the spinal column. Such lordotic curvature of the spinal column is, supposedly, used in competitions by weight lifters so that they can lift large weights with relatively small compressive force in the spine. In contrast, the straight spinal column (as presumed in traditional spinal modeling) generates large compression forces for the same external load (Aspden 1988).

Such observations have led to recommendations to use abdominal belts while lifting loads, but both theoretical considerations and practical experiences explain the limited success of these devices. (See Chapter 11 on "handling loads".)

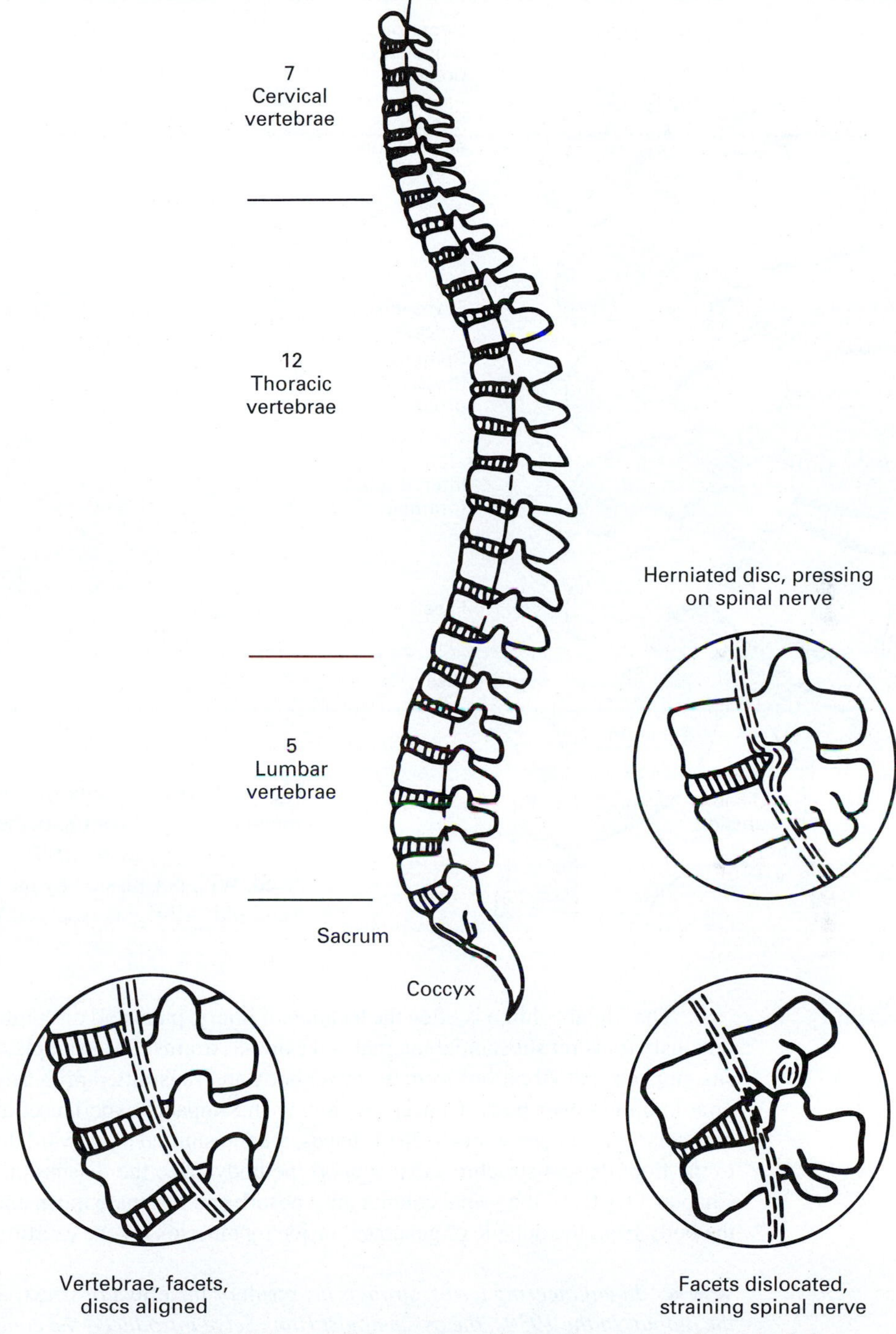

Figure 1–15. Scheme of the human spinal column, seen from the left side. (From Kroemer et al. 1990, *Engineering Physiology: Bases of Human Factors/Ergonomics,* 2d ed. With permission by the publisher, Van Nostrand Reinhold. All rights reserved.)

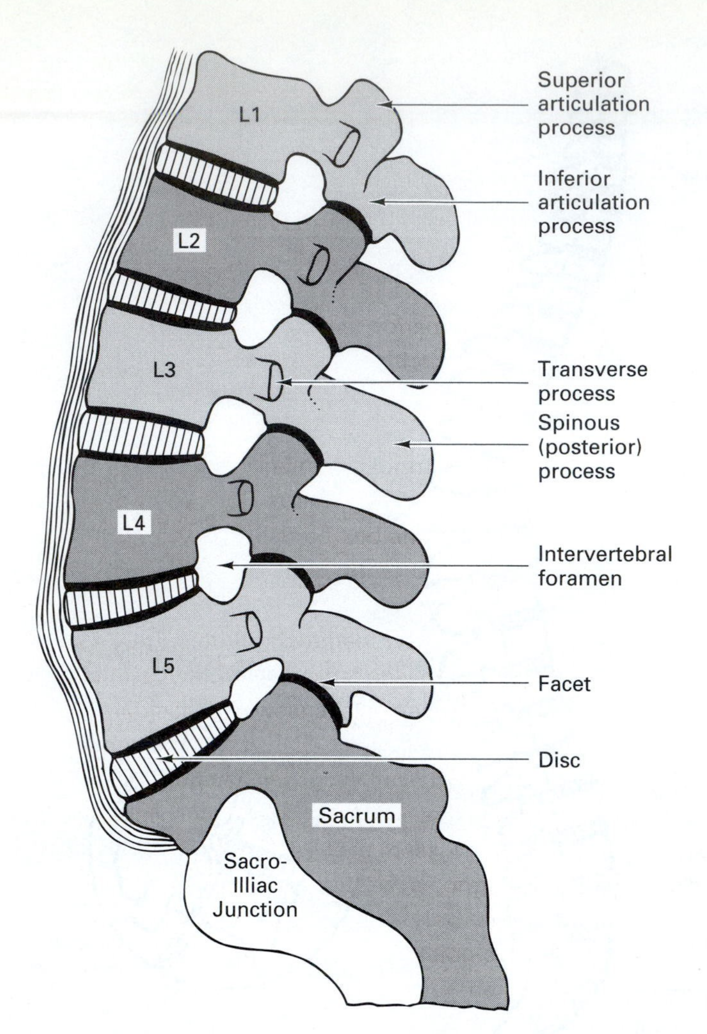

Figure 1–16. Scheme of the lumbar section of the spinal column. (From Kroemer et al. 1990, *Engineering Physiology: Bases of Human Factors/Ergonomics,* 2d ed. With permission by the publisher, Van Nostrand Reinhold. All rights reserved.)

Back pain

The spinal column is often the location of injury, pain, and discomfort, because it must continuously transmit substantial internal and external strains. For example, when a person stands or sits, impacts and vibrations from the lower body are transmitted primarily through the spinal column into the upper body. Conversely, forces and impacts experienced through the upper body, particularly when one works with the hands, are transmitted downward through the spinal column to the floor or seat structures that support the body. (See the discussion of material handling in Chapter 11.) Thus, the spinal column must absorb and dissipate much energy, be it transmitted to the body from the outside or generated inside by muscles for the exertion of work to the outside.

☛☛☛ *In engineering terms, strain is the* result *or the* effect *of stress: Stress is the input, strain the output. In the 1930s, the psychologist Hans Selye introduced the concept of stressors* causing *stress (or distress if excessive).*

It is confusing to use the term "stress" with two different meanings: either the cause or the result. (What is "job stress"?) To avoid confusion, in this text the engineering terminology will be used: Stress produces strain. ☚☚☚

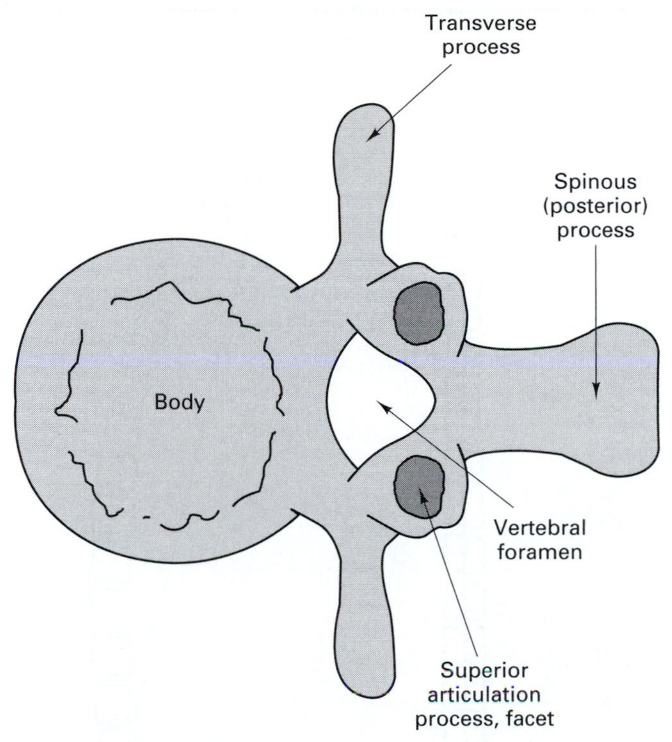

Figure 1–17. Scheme of a vertebra. (From Kroemer et al. 1990, *Engineering Physiology: Bases of Human Factors/Ergonomics,* 2d ed. With permission by the publisher, Van Nostrand Reinhold. All rights reserved.)

The spinal disc

The gel-like round core of the intervertebral disc, the nucleus pulposus, is the main load-bearing and load-transmitting element of the spine. It is kept in place by surrounding layers of elastic material called the annulus fibrosus. The disc and its surrounding ligaments allow the relative movements of adjacent vertebrae and together form a physiological shock absorber that, when not functioning properly because of injury or deterioration, transmits unsuitable strain to the cartilaginous end plates of the vertebrae and possibly also to the facet joints. Displacement of the cartilaginous components of the disc or displacement of the bony structures of the spinal column may reduce the opening of the neural canal and can impinge on the nerve roots emanating from the spinal cord through the intervertebral foramen (See Figure 1–16) or can impinge directly on the cord.

Activity helps maintenance

The nucleus pulposus has no blood supply of its own, but is nourished through the exchange of tissue fluid, which circulates through the disc as a result of osmotic forces, gravitation, and the pumping effects of body movements on the spinal column. Thus, disc nourishment improves with activity and is adversely affected by immobilization. Tissue fluid circulation is needed to provide a proper balance of water, solutes, glycosaminoglycans, protein, and collagen. If this proper balance is not achieved, the annulus may degenerate. Then fissures may develop in the annulus

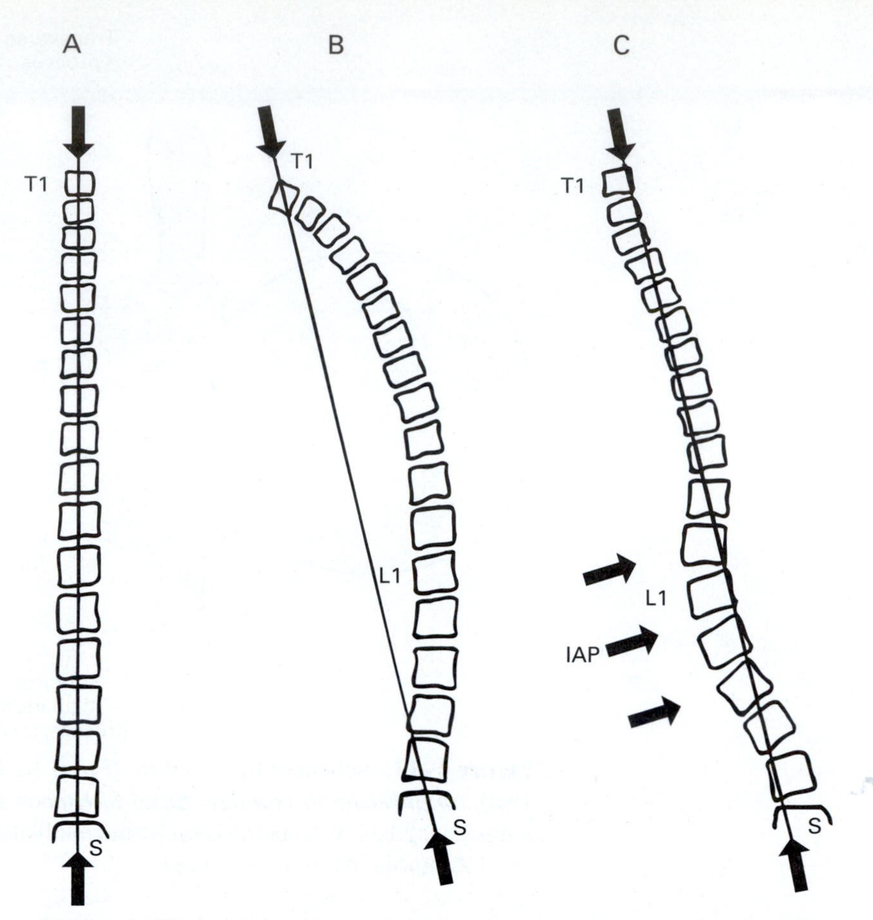

Figure 1–18. Models of the spinal column: (A) as a straight column, (B) arched, (C) supported by intraabdominal pressure. (Modified from Aspden 1988.)

through which material of the nucleus pulposus can penetrate and herniate peripheral areas. These are sensitive to mechanical and chemical stimulation, and we feel that "something is wrong."

"8-in-10 chance" of back pain

Low back pain (LBP) is the result of disorders that have been with humans since ancient times. It was diagnosed among Egyptians 5,000 years ago and was discussed in 1690 by Bernadino Ramazzini. The problem is not confined to mankind, since quadrupeds suffer from LBP as well. Everyone has an "8-in-10 chance" of suffering from back pain sometime during his or her life (Snook 1988b).

Why low back pain?

Low back pain is just that: a painful sensation of disorder apparently existing in the low back area. LBP may stem from a large number of sources, many believed to be basically associated with changes in the spinal column and its supporting ligaments and muscles due to aging, starting in the teen years and usually increasing as one gets older. These changes result from a

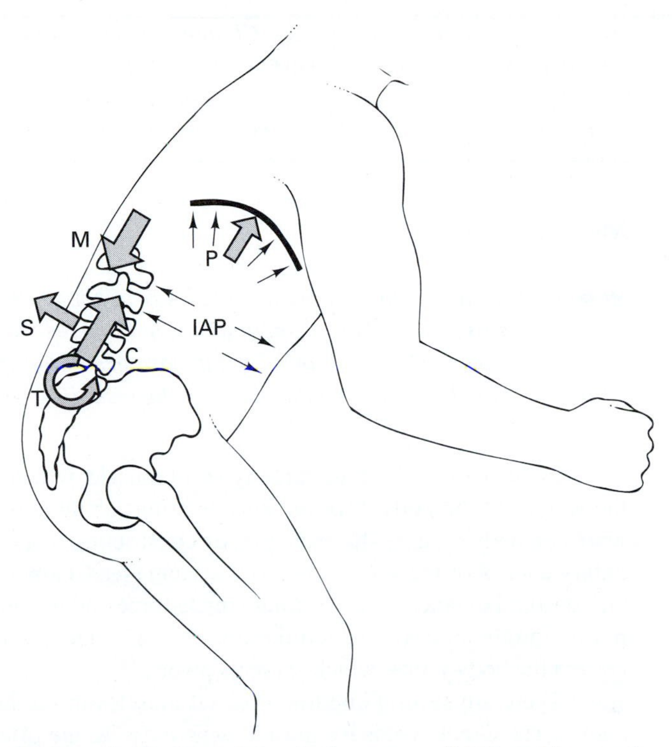

Figure 1–19. Intraabdominal pressure (IAP) and its resulting force vector (*P*) reduce the compressive force (*C*), which is produced by trunk muscle pull (*M*). Shear force (*S*) and torque (*T*) also load the spine.

combination of repetitive trauma and the normal aging process. Strong activity demands may trigger the occurrence of various LBP symptoms. However, except in cases of acute injuries, the causes of or reasons for LBP usually remain unclear. Rowe (1983) found that only 4 percent in a large sample of industrial LBP cases were related to traumatic injuries during work. The classification of LBP is difficult, and different clinicians frequently diagnose it differently. Furthermore, many persons who do have objective signs of spinal degeneration (as diagnosed by spinal imaging, via techniques such as X rays, CT scans, MRI scans, myelograms, and discograms) do not feel any pain.

Changes in discs

Among the changes commonly found in the spinal column are damage to the cartilaginous end plates at the main body of the vertebrae, degeneration of the annulus fibrosus and of the nucleus pulposus of the disc, and "drying out" of the disc structure. All make the disc behave less like a hydrostatic device and change the spine's biomechanical motion characteristics.

☛☛☛ *A theory that has become popular in recent decades is that many overexertion injuries of the spinal column can be traced to, or explained by, compression of spinal discs. Such excessive compression is thought to damage, temporarily or permanently, the fibro-cartilage disc. In turn, the damage may lead to the intrusion of disc structures into their surroundings, particularly toward spinal nerve. Experimental measurements of the compression within the disc in the living human body, however, are difficult to perform. Thus, many assessments rely on calculations using biomechanical models (see Chapter 7), which often simply assume static strains. Yet the spinal joints (discs and facets) are subject not only to compression, but also to shear, bending, and twisting. To consider such combined strains dynamically, while taking into account the actions of trunk muscles and ligaments, is a major task of biomechanical research and modeling in the near future.*

Muscle

The Greek physician Galen (129 A.D. to about 199 A.D.) studied human physiology, first at a gladiator school in Pergamon and then in Rome. He identified muscles and showed that they worked in groups. He also showed the importance of the spinal cord by cutting it in various positions in animals and noting the extent of the resulting paralysis (Asimov 1989).

Three types of muscle

In the human, there are three types of muscle, which together make up about 40 percent of the weight of the body. *Cardiac muscle* brings about contractions of the heart. *Smooth muscle* works on body organs—for example, by constricting blood vessels. *Skeletal muscle* is under voluntary control of the somatic nervous system (see Chapter 3) and serves two purposes: to maintain postural balance by generating tensile force and to cause local motion of body segments by pulling on the bones to which the muscles are attached, thereby creating torques or moments around the body joints, which serve as pivots.

There are several hundred skeletal muscles in the human body, identified by their Latin names. The Greek words for muscle, *mys* and *myo,* are often used as prefixes. The proximal end of the muscle (pointing toward the center of the body) is called the origin; the other, distal end moving with the moving body segment is called the insertion.

Muscle tension

Muscles actively perform their functions by contracting—that is, by quickly and temporarily developing internal lengthwise tension. Contraction is often, but not always, accompanied by shortening. Muscle may also be lengthened passively beyond its resting length by a force external to it; in response, muscle tissue develops internal tension both by elastic resistance and by attempted active contraction.

Architecture of Muscle

Water and Proteins. The main components of muscle besides water (about 75 percent by weight) are proteins (20 percent). Collagen, an abundant protein in the body, constitutes the insoluble fiber of the binding and supportive substance in muscle tissue. The proteins actin and myosin form rod-shaped polymerized molecules that attach end to end, creating thin strands.

Contracting structures

Filaments. The actin and myosin filaments are the contracting microstructure of muscle. As seen in a crosscut through the muscle, several actin molecules surround each myosin molecule. Lengthwise, the thin actin strands are wound around the thicker myosin in the form of a spiral (double helix). The actin filaments project from the Z-discs (see shortly) like bristles of a brush. When a muscle is at rest, troponin and tropomysin proteins separate the actins from the myosin. During a contraction, these are pulled away, and the actin strands temporarily connect with the myosin via so-called cross bridges (which resemble tiny golf-club heads) serving as temporary ratching attachments by means of which the actins pull themselves along the myosin rod. Figure 1–20 illustrates the relative locations of actin and myosin rods within contracted, relaxed, and stretched muscle.

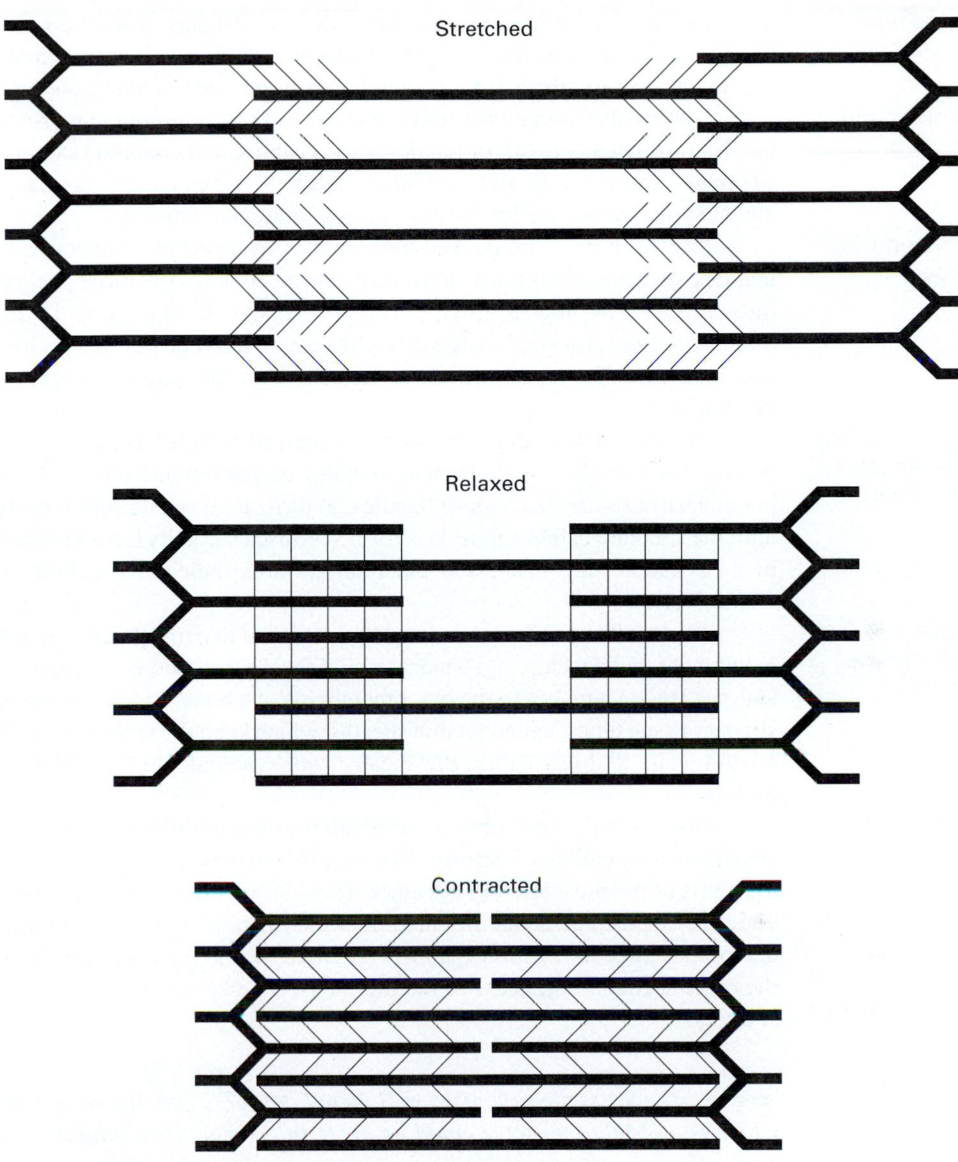

Figure 1–20. Schemes of the location of actin and myosin filaments and of the cross bridges within stretched, relaxed, and contracted muscle. (From Kroemer et al. 1990, *Engineering Physiology: Bases of Human Factors/Ergonomics,* 2d ed. With permission by the publisher, Van Nostrand Reinhold. All rights reserved.)

Striation

Fibrils and striation. Between 10 and 500 muscle filaments are bundled tightly into what is known as a fibril. (Since contraction of myofibrils can be easily observed, they are often called "contractile elements.") Along the length of each fibril, lighter and darker stripes appear in the electron microscope, mostly depending on the density of actin and myosin molecules. This banding or striping has led to the name "striated" muscle. The striations show a repeating pattern about every 250 angstroms, which makes the fibril appear to have a series of disclike partitions.

Z-discs anchors actins

One distinct stripe indicates the Z-disc (or Z-band; from the German *zwischen,* meaning between). The distance between two adjacent Z-discs is called the sarcomere. Z-discs are dense membranes across fibrils from which the actin rods extend along the myosins.

Plumbing and fueling

Z-discs also contain the transverse parts of the "plumbing and fueling" system of the muscle, from which a network of tubules, sacs, cisterns, and channels extends along and between the filaments. This network allows fluid to transport oxygen and energy carriers, as well as chemical and electrical messages for the execution of muscular activities.

Mitochondria in fibers

Fibrils, in turn, are packed into bundles wrapped by connective tissue called fascia. The bundles are bathed in a fluid called sarcoplasm. Bundles of fibrils packed together constitute the muscle fiber. The fiber is a cylindrical unit, 10^5 to 10^6 angstroms in diameter and 1 to 50 mm long. Each fiber is a single large cell with several hundred cell nuclei located at regular intervals near its surface. The nuclei contain the mitochondria, the "energy factories of the muscle" discussed in Chapter 2.

Origin—muscle belly—insertion

Muscles and tendons. Between 10 and 150 muscle fibers constitute the so-called primary bundle (fasciculus). Several primary bundles are packed into a secondary bundle, again wrapped in connective tissue. Secondary bundles, in turn, are grouped into tertiary bundles, and so forth, until the structure called a muscle is formed. Muscles usually have several hundred thousand fibers in their middle, the belly, and taper off towards their ends, called the origin and insertion, respectively.

Tendon connects muscle with bone

The bundles and the total muscle are wrapped in perimysium—tough and dense collagenous membranes that develop into bendable, slightly elastic tendons at origin and insertion. On its other end, the tendon attaches to the outer membrane of a bone. Many (but not all) tendons run through slippery tissue tubes, called tendon sheaths, which keep the tendon in position and guide it around a bend while it slides within the sheath, with synovial fluid serving as a lubricant for smooth gliding.

Strength of muscle

Often, fibrils (and fibers) are parallel to the middle axis of the muscle. Such a parallel arrangement is called a fusiform. However, fibers may also be arranged at various angles to the long axis of the muscle, as in a feather. These arrangements are called penniforms, and there are several of them. The actual orientation of the fibers with respect to the centerline of the muscle determines the contraction capability (strength) of the muscle. Muscle strength also depends on the cross-sectional thickness of the muscle (its number of fibers) and on the types of fibers (see shortly).

☛☛☛ *The Swiss physiologist von Haller (1708–1777) published in 1766 his experimental work which showed that muscles could be made to contract by a stimulus transmitted to them by a nerve. He demonstrated that all nerves connect with the brain or spinal cord, the centers of sense perception and of responsive actions. Von Haller is considered the founder of modern neurology. In 1780, the Italian anatomist Luigi Galvani (1737–1798) observed that muscles of dissected frog's legs twitched when an electrical spark struck them. While his explanation of the phenomenon was wrong, the finding that electricity was involved with nerve and muscle action was correct (Asimov 1989).* ☚☚☚

Innervation

The motor unit. Several muscle fibrils share a common connection area with the axon end of a motor nerve (called an alpha-motoneuron). This junction between nerve and muscle is called the motor end plate. Each nerve innervates a number, usually hundreds or thousands, of muscle fibers. These fibers, which are under a common control are called a motor unit, all stimulated by the same signal. (One alpha-motoneuron may innervate more than one motor unit, however.)

Nervous control

Motor units can be classified by the innervation ratio, which describes the number of fibers innervated by one neuron. Muscles used for finely controlled actions (eg, rotation of the eyeball) have a ratio such as 1:7, while muscles for gross activities may have ratios of 1:1,000 or more.

Slow, enduring fibril

Another classification of motor units describes their types of muscle fibrils. *Type I* is usually associated with muscles exerting fine and enduring control (eg, over the fingers or eyes) or the muscles of the back. The Type I fibril is short and appears red, because it is penetrated by many capillaries, which provide good blood supply and oxygen storage; hence, it resists fatigue. A contraction is produced by a relatively low action potential, but takes a relatively long time, 60 to 120 ms, to peak. Therefore, Type I is called a slow fibril.

Fast, fatiguing fibril

Type II fibrils appear light in comparison to Type I, because they are not profuse with capillaries. Type II fibrils are less resistant to fatigue, but perform better under anaerobic conditions. They require high action potentials, but produce fast twitches, 15 to 50 ms to peak. (Type II fibrils are further subdivided according to their supply of capillaries.)

Note that the times just given for slow or fast reactions have been measured in isolated muscle preparations, usually taken from cats, and with artificial stimulation. In the human muscle, distinct groupings of muscle fibrils are not as prevalent, and the actual behavior of the stimulated muscle depends on many factors, such as fatigue and external resistance.

Activation of the Motor Unit. As described earlier, a semipermeable membrane called the sarcolemma covers the muscle fiber consisting of thousands of fibrils. When a muscle is at rest, sodium (positive) and some potassium (positive) ions accumulate on the outside of the membrane, while (negative) chlorine ions are on the inside. This separation of positive from negative ions establishes a polarized sarcolemma, with a transmural electrical potential of nearly 100 mV.

Axon action impulse

An action impulse arriving from an alpha-motoneuron at the motor end plate must be strong enough to depolarize the membrane potential by at least 40 mV. If this threshold is not achieved, the motor unit does not react; if the threshold is achieved, the motor unit contracts completely. This all-or-none principle governs muscle contraction.

Potential propagates

Given sufficient depolarization, the permeability of the sarcolemma is increased so that (the positive) sodium ions can penetrate and neutralize (the negative) chlorine ions. This local depolarization, called end-plate potential, propagates at a speed of about 5 m s^{-1} along the membrane. The depolarization wave, acting like an electric current, causes hydrolysis of water molecules, releasing hydrogen and hydroxal ions that, in turn, split off a phosphate group from adenosine triphosphate (ATP) with activated myosin (ATPase) as the catalyst.

Splitting ATP frees energy

The decomposition of ATP results in the formation of ADP and phosphoric acid, a reaction that liberates energy of about 11 kcal per mole of ATP.* This reaction is the primary source of energy for muscular contraction. (See the discussion of energy release in Chapter 2.)

In death, the ADP complex disintegrates and firmly bonds the myosin cross bridges to the actin molecules, leading to rigor mortis.

Actin–myosin cross bridges allow contraction

The action potential propagates into the fibrils through the "plumbing" network of the sarcoplasmic reticulum, where calcium ions are trapped by membranes that do not allow them to escape when the muscle is at rest. However, the arriving action potential makes the cistern membranes permeable to the calcium ions, which, discharged into the myofibrillar fluid, raise its calcium concentration about a thousandfold. This, in turn, allows some calcium ions to combine with myosin

*One mole is the quantity of a chemical compound whose weight in grams equals its molecular weight.

molecules to form the activated myosin catalyst (ATPase). In addition, the released calcium binds to the troponin and tropomysin molecules, pulling these protein strands away from the binding sites between the actin and myosin molecules. This separation allows cross bridges to be established between actin and myosin, and the actual contraction process of the muscle fibril can commence.

Twitch. After an action impulse arrives at the motor unit, there is an initial latency (of about 10 ms) during which no perceptible muscle tension or change in muscle length takes place. Within this time span, the alpha-motoneuron signal generates an end-plate potential, releases calcium ions from the cisterns, activates the ATP fuel element, and establishes cross bridges between actin and myosin.

Then follows the period of contraction, which takes up to 40 ms in isolated fast-twitch fibrils and up to 110 ms in isolated slow-twitch fibrils. In living muscle, those times are considerably longer. During the period of contraction, energy from the decomposition of ATP allows actins to pull along the myosin filaments. This reduces the length of sarcomere and hence shortens the length of the entire muscle.

Then follow 30 to 50 milliseconds of relaxation, during which the muscle returns to its resting length. If no new alpha-motoneuron impulse arrives, actin and myosin filaments separate, and the cross bridges disengage. During this time of relaxation, ADP picks up a phosphate group, using the so-called Lohman reaction, to resynthesize ATP.

The final recovery phase lasts 30 ms or longer if no contraction signal arrives. During this time, leftover phosphoric acid combines with glucose to form diphosphate, which can be completely oxidized to carbon dioxide and water in the mitochondria if oxygen is available. This reaction yields 675 kcal per mole of glucose, enough energy to regenerate 38 moles of ATP.

Twitches build upon each other

Still looking at a single motor unit, one sees summation (also called superposition) of twitches when they are initiated frequently after each other, so that a contraction is not yet completely released by the time the next stimulus arrives. In that case, the new contraction builds on a level higher than if the fiber were completely relaxed, and accordingly, higher contractile tension is generated in the muscle. This staircase effect takes place when excitation impulses arrive at frequencies of 10 or more per second.

When a muscle is stimulated above frequencies of 30 to 40 stimuli per second, successive contractions fuse together, resulting in a maintained contraction called tetanus. In superposition of twitches, the muscle tension generated may be two or three times as large as that produced by a single twitch, and a full tetanus may build up to five times the single-twitch tension.

Rate Coding. As just shown, in a single motor unit, the frequency of contractions and the strength of a contraction are controlled by the frequency of the excitatory nervous signals. This is called rate coding. Yet, the fibers belonging to one motor unit are generally in various locations within the muscle. Therefore, the activation of one motor unit brings about a weak contraction throughout the muscle.

Recruitment Coding. If, at the same time, more than one alpha-motoneuron excite their motor units to contract, one speaks of recruitment coding. The larger the number of motor units contracting simultaneously, the higher is the contractile strength exerted by the total muscle. Thus, nervous control of muscle strength (see later) follows a complex pattern of rate and recruitment coding.

Fatigue relieved by rest

Fatigue. If, after contraction, not enough time is provided for relaxation and recovery before the next contraction, then the muscle experiences an "oxygen deficit", lactic acid is not removed sufficiently, potassium ions accumulate while sodium is depleted in extracellular fluid, phosphate accumulates in intracellular fluid, and ATP rebuilding is hindered. (See Chapter 2.) This combination of events leads to muscle fatigue, which prevents the continuation of strong contractions. Muscular fatigue is overcome by rest, during which time the accumulated metabolic by-products are removed.

Fatigue of a single motor unit or of a whole muscle depends on the frequency and intensity of muscular contraction and on the period of time over which it is maintained. The more strength exertion required of a given muscle, the shorter the period is through which the strength can be maintained. Figure 1–21 schematically shows this relationship between strength exertion and endurance. (The relationship applies strictly to static, or isometric, efforts—see later in the chapter.)

As a practical rule, maximal muscle strength can be maintained for only a few seconds; 50 percent of strength is available for about 1 minute; less than 20 percent can be applied continuously for long times.

Length–strength Relationships. In engineering terms, skeletal muscles exhibit viscoelastic behavior. These muscles are viscous because their behavior depends both on the amount by which they are deformed and on the rate of deformation. They are elastic in that, after deformation, they return to their original length and shape. These behaviors, however, are not pure in the muscle, because it is nonhomogeneous, anisotropic, and discontinuous in its mass. Nevertheless, nonlinear elastic theory and viscoelastic descriptors can be used to describe major features of muscular performance (Schneck 1992).

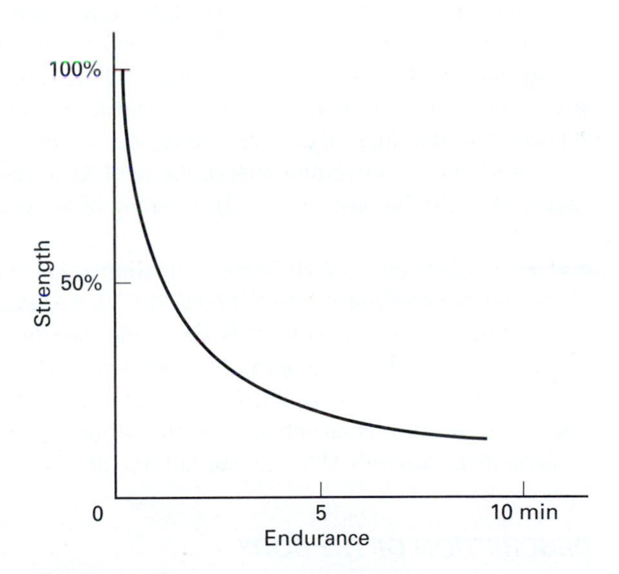

Kroemer 1.20

Figure 1–21. Strength and endurance of muscular exertion. (From Kroemer et al. 1990, *Engineering Physiology: Bases of Human Factors/Ergonomics,* 2d ed. With permission by the publisher, Van Nostrand Reinhold. All rights reserved.)

Tension when shortened

Under a "no-load" condition, in which no external force applies and no internal contraction occurs, the muscle is at its resting length. Without external load, nervous stimulation causes the muscle to contract to its smallest possible length, which is about 60 percent of resting length. In this condition, the actin proteins are completely curled around the myosin rods, so that the Z-lines are as close as possible. This is the shortest possible length of the involved sarcomeres, and the muscle cannot develop any further active contraction force.

Stretching the muscle beyond resting length by applying external forces between insertion and origin lengthens the muscle passively; that is, it increases the length of the involved sarcomeres. The external stretch force slides the actin and myosin fibrils along each other. When the muscle is stretched to 120 or 130 percent of its resting length, the cross bridges between the actin and myosin rods are in an optimal position to generate a contractile force. But if the fibrils are elongated further, the overlap in the cross bridges between the protein rods is reduced. At about 160 percent of resting length, so little overlap remains that no contraction force can be developed internally. Thus, the curve of active contractile force developed within a muscle is zero at approximately 60 percent of resting length, becomes about 0.9 at resting length, goes to unit value at about 120 to 130 percent of resting length, and then falls back to zero at about 160 percent resting length. (These values apply to an isometric, or static, contraction; see later in the chapter.)

Strong tension when stretched

Passive reaction to external stretch also occurs, as in a rubber band. This passive stretch resistance increases strongly from the muscle's resting length to the point at which the muscle or tendon (attachment) will break. Thus, when the length of the muscle is greater than its resting length, the tension in the muscle is the summation of active and passive strain. This explains why we stretch ("preload") muscles when we wish to exert a strong force, as in bringing the arm behind the shoulder before throwing.

The viscoelastic theory also explains why the tension that we can develop statically by holding a stretch is the highest possible, while in active shortening (a dynamic concentric movement), muscle tension is decidedly lower. The higher the velocity of muscle contraction, the faster actin and myosin filaments slide by each other, and the less time is available for the cross bridges to develop and hold. This reduction in force capability of the muscle holds true for both concentric and eccentric activities. In eccentric activities, however, where the muscle becomes increasingly lengthened beyond its resting length, the total force resisting the stretch increases with greater length, owing to the (just discussed) summing of active and passive tension within the muscle.

☛☛☛ *In 1680, the book* De Motu Animalium (On the Motion of Animals), *by the Italian physiologist Giovanni Alfonso Borelli (1608–1679), was posthumously published. Borelli explained muscular action on a mechanical basis, describing the actions of bones and muscles in terms of a system of levers. This kind of explanation showed that natural laws govern life and nonlife, animals and humans, alike. In 1721, Jean Bernoulli wrote his* Physiomechanicae Dissertatio de Motu Musculorum (Biomechanics of the Motion of Muscles), *followed in 1799 by Charles-Augustin de Coulomb's* Mémoire sur la force des hommes (On Human Strength). ☚☚☚

BIOMECHANICAL DESCRIPTION OF THE BODY

What do muscles optimize?

The question "How does the body perform activities if it is free to choose different ways?" is of basic interest. It becomes quite important in biomechanical modeling (see Chapter 7), in which the human body is treated as a mechanical system of linkages and masses, activated by muscles that span joints. Strain tolerance limits in muscles and tendons, joints, and joint-enclosing ligaments are of particular biomechanical interest in maximal efforts. However, even in submaximal exertions, the body apparently uses some kind of optimization in the shared efforts of muscles.

Minimum or maximum?

In routine exertions, the body's overall goal may be the minimization of total energy spent, of contraction effort in a muscle (eg, in the number of muscles employed or the forces endured in joints and ligaments), or of muscle fatigue, perceived pain, and bone loading. In emergencies, one can presume maximization of the overall output—of strength exerted to the outside, of effort. What then does the body optimize in everyday work, and how is that determined and achieved?

Mathematical modeling requires the identification of the elements that are of interest to the modeler—for example, all muscles participating in an activity or all joint loadings. This can result in a large number of unknown forces and torques, often much in excess of the available number of equations of equilibrium. To solve such an indeterminate problem, optimization assumptions and techniques may be employed. Many of these use linear or nonlinear methods to find a unique solution.

Mathematical optimization techniques need to be bound by limits and assumptions that realistically describe the functions of the human body. They must be based on sound physiological and biomechanical principles, rather than being simply conceptually or mathematically convenient. Much research remains to be done in this area.

Links, Joints, and Masses

"Stick man"

The human skeletal system is often simplified into a relatively small number of straight-line links (representing long bones) and joints (representing major articulations). Figure 1–22 shows a typical link–joint system with identifying numbers. Note that the feet are not subdivided into their components, and the spinal column is represented by only three links. If more details are needed, they can be generated, as is done, for example, for the hand shown in Figure 1–23.

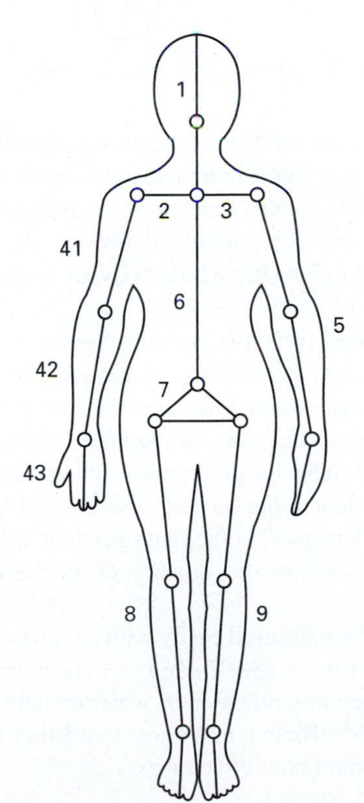

Figure 1–22. Numbering scheme for the body-link system.

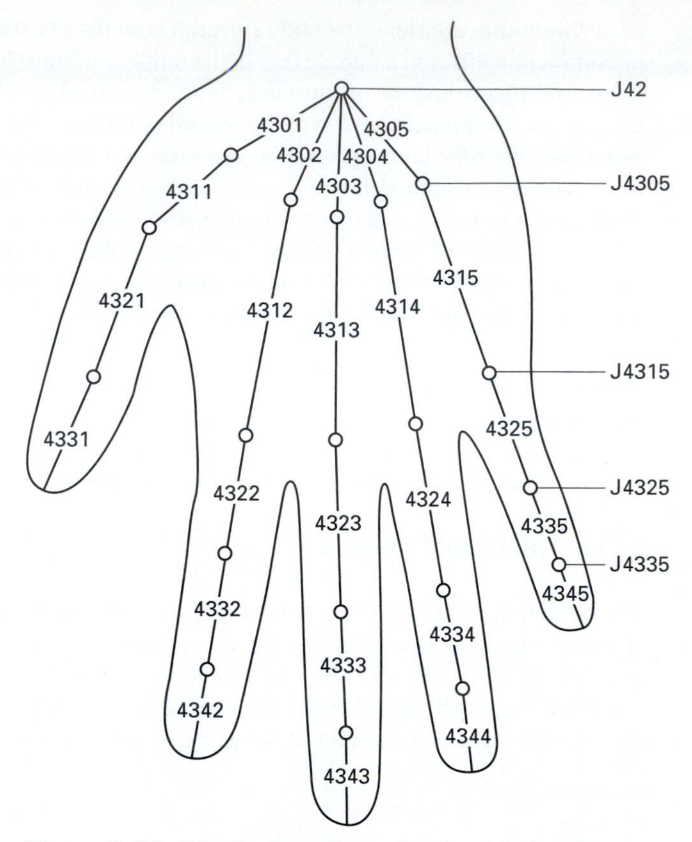

Figure 1–23. Numbering scheme for the right hand.

Joint locations and link lengths

The determination of the location of the joint center of rotation is relatively easy for simple articulations, such as the hinge joints in fingers and elbows. However, it is much more difficult for complex joints with several degrees of freedom, such as those in the hip or shoulder. Once the joints are established, a straight-line link length is defined as the distance between adjacent joint centers. Knowledge of the volume of the whole body, or of the segment of interest, is necessary to calculate inertial properties.

Volume

The use of Archimedes' principle provides the body volume: The subject is immersed in a container filled with water, and the displaced water yields the volume. Other data can be obtained by dissecting cadavers or calculating with models (Kroemer et al. 1988). Kroemer et al. (1997) extracted definitions of the joint centers, of the links between them, and of mass properties from the literature and compiled these for use by biomechanicists and engineers.

Weight and density

Weight W is a force depending on body mass m and gravitational acceleration g according to the formula $W = mg$. Density D is the mass per unit volume: $D = mV^{-1} = Wg^{-1}V^{-1}$. Mass is, of course, $m = DV$. The specific density D_s is the ratio of D to the density of water $D_w: D_s = DD_w^{-1}$.

Mass properties

Body weight W can be measured easily with a variety of scales and then can be used to predict body segment weight or mass. (See Table 1–8.) The human body is not homogeneous throughout; its density varies, depending on cavities, water content, fat tissue, bone components, and so on. Still, in many cases it is sufficient to assume that either the body segment in question or even the whole body is of constant (average) density.

TABLE 1–8. Equations for Estimating Segment Mass (in kg) from Total Body Weight W (kg)

Segment	*Empirical Equation*	*Standard Error of Estimate*
Head	$0.0306W + 2.46$	0.43
Head and neck	$0.0534W + 2.33$	0.60
Neck	$0.0146W + 0.60$	0.21
Head, neck, and torso	$0.5940W - 2.20$	2.01
Neck and torso	$0.5582W - 4.26$	1.72
Total arm	$0.0505W + 0.01$	0.35
Upper arm	$0.0274W - 0.01$	0.19
Forearm and head	$0.0233W - 0.01$	0.20
Forearm	$0.0189W - 0.16$	0.15
Hand	$0.0055W + 0.07$	0.07
Total leg	$0.1582W + 0.05$	1.02
Thigh	$0.1159W - 1.02$	0.71
Shank and foot	$0.0452W + 0.82$	0.41
Shank	$0.0375W + 0.38$	0.33
Foot	$0.0069W + 0.47$	0.11

Source: Adapted from Kroemer, Kroemer, and Kroemer-Elbert 1990.

Lean body

Lean body mass (or lean body weight) is often used to distinguish between body compositions. Basic structural components such as skin, muscle, and bone are relatively constant in percentage from individual to individual. Fat, however, varies in percentage of total mass or weight throughout the body and for different persons. This allows body weight to be expressed as lean body weight plus fat weight. There are several techniques for determining body fat. Many rely on skinfold measures: In selected areas of the body, the thickness of a fold of skin with its underlying fatty tissue is measured with a special caliper.

Center of mass

Body mass is often considered to be concentrated at one point in the body, where its physical characteristics respond in the same way as if it were distributed throughout the body. The location of the center of mass (also called, less appropriately, center of gravity) shifts with body posture, muscular contractions, the ingestion or excretion of food and fluid, and respiration. Table 1–9 lists relative locations of centers of mass. (For further information about biomechanical properties of the human body, see eg, Chandler et al. (1975), Hay (1973), Kroemer et al. (1988, 1997), Kroemer (1999a, b), NASA/Webb (1978), Kaleps et al. (1984), McConville et al. (1980), NASA (1989), and Roebuck (1995).

> Newton's first law states that a mass remains in uniform motion (which includes being at rest) until acted upon by unbalanced external forces. The second law, derived from the first, indicates that force is proportional to the acceleration of a mass. The third law states that any given action is counterbalanced by an equal and opposite reaction.

Force

According to Newton's second law, force is not a basic unit, but a derived one. Thus no device exists that measures force directly. Instead, all measuring devices for force (or torque) rely on other physical phenomena, which are then transformed and calibrated in units of force (or torque). The quantities used to assess force are usually either displacement (such as the bending of a metal beam) or acceleration.

TABLE 1–9. Locations of the Centers of Mass of Body Segments Measured on the Straight Body, in Percent from their Proximal Ends

	Harless (1860)	*Braune and Fischer (1889)*	*Fischer (1906)*	*Dempster (1955)*	*Clauser, McConville, and Young (1969)*
Sample Size	*2*	*3*	*1*	*8*	*13*
Head	36.2%	—	—	43.3%	46.6%
Trunk*	44.8	—	—	—	38.0
Total arm	—	—	44.6%	—	41.3
Upper arm	—	47.0%	45.0	43.6	51.3
Forearm and hand*	—	47.2	46.2	67.7	62.6
Forearm*	42.0	42.1	—	43.0	39.0
Hand*	39.7	—	—	49.4	18.0
Total leg*	—	—	41.2	43.3	38.2
Thigh*	48.9	44.0	43.6	43.3	37.2
Calf and foot	—	52.4	53.7	43.7	47.5
Calf	43.3	42.0	43.3	43.3	37.1
Foot	44.4	44.4	—	42.9	44.9
Total body	58.6**	—	—	—	58.8**

Source: Adapted from Kroemer, Kroemer, and Kroemer-Elbert 1990.

*The values in these rows are not directly comparable, since the different investigators used differing definitions for segment lengths.

**Percent of stature, measured from the floor up.

FORCE UNITS

In the International System of Measurement (Système International, or SI), the correct unit for force measurement is the newton; one pound-force unit is approximately 4.45 newtons, and 1 kg_f (kilogram-force, also occasionally called 1 kilopond, kp) equals 9.81 newtons. The pound (lb), ounce (oz), and gram (g) are not units of force, but rather, are mass units.

Torque or moment

Torque (also called moment) is the product of force and its lever arm (distance) with respect to the articulation about which it acts; the direction of the force must be at a right angle to its lever arm. In kinesiology, the lever arm is often called the mechanical advantage.

Vectors

Force and torque are vectors, which means that they must be described not only as a magnitude, but also by direction, namely, their line of application. For practical reasons, the point of application must also be known.

Body Kinetics

The "stick-person" concept, consisting of links and joints embellished with volumes and masses and driven by muscles, can be used to understand and model human motion and strength capabilities. (See also Chapter 7.)

Force transmission in the body

Recall Figure 1–13, which shows a model of the human body exerting forces or torques with the hand that act on an outside object. Within the body, the forces or torques are transmitted along the links. First, the force *H* exerted with the right hand, modified by the existing mechanical advantages, must be transmitted across the right elbow *E*. (Also, at the elbow, additional force or torque must be generated to support the mass of the hand and forearm; however, for the moment, the model will be considered massless.) Similarly, the shoulder *S* must transport the same effort, again modified by existing mechanical conditions. In this manner, all subsequent joints convey the effort exerted with the hands along the trunk, hips, and legs, and finally relay it from the foot to the floor. Here, force and torque vectors can be separated again into their component directions, similar to the vector analysis at the hands. Assuming that energy is not stored in the body, the same sum of vectors must exist at the feet as at the hands.

Of course, the assumption of no body mass is unrealistic and can be remedied by incorporating information about mass properties of the human body, as mentioned earlier. Body motions, instead of the static position assumed in the figure, complicate the model.

Describing Human Motion (Kinematics)

In the past, human movement was described with anatomically derived terms used in the medical profession: the nouns "flexion," "tension," "duction," "rotation," and "nation," together with the prefixes "ex," "hyper," "ad," "ab," "in," and "out." Unfortunately, the same terms were applied indiscriminately to rotational movements about joints and to translational displacements of limbs; thus, certain motions said to occur in a given plane could also occur in others. A scheme such as this makes it very difficult to describe relative locations of body parts, and a "zero position" is often not specified.

New taxonomy

Based on Roebuck's 1968 attempt to generate a more appropriate terminology (described in detail by Roebuck et al. 1975), a new taxonomy uses as a coordinate system the common frontal, medial, and transverse planes that may be centered at each body articulation considered. (See Figure 1–24.)

To describe movement, the verb "twist" indicates rotation about a long axis. The familiar "flexion" and "extension" are maintained. "Pivot" indicates rotation about an axis perpendicular to the flexion–extension axis (Kroemer et al. 1989). Roebuck's prefixes are kept: "e-" signifies out or up from the standard position; "in-" is the opposite of "e-" (ie, in or down), and "posi-" and "negi-" are self-explanatory. Table 1–10 lists the terms. Any motion can be identified by a descriptive term composed of the designations for the plane, the direction, and the type of motion. For example, "sag-in-flexion" indicates that flexion occurs in the sagittal plane and is inward (eg, toward the origin of the reference system).

The consistent use of reference points, planes, and motion terms should reduce the uncertainty that exists regarding human motion capabilities and allow their measurement and reporting, utilizing computer models and systems.

☛☛☛ *About 500* B.C.*, the Greek physician Alcmaeon deliberately and carefully dissected human cadavers. He noted the difference between arteries and veins and found that the sense organs were connected to the brain by nerves. (The word "anatomy" stems from the Greek "to cut up.") The Greek physician Galen (129–ca.199* A.D.*) worked first at the gladiator school at Pergamon, where he gained rough-and-ready hints on human anatomy, especially on muscles and how they worked in groups. Afterward, in Rome from 161 on, he could dissect only animals, which occasionally misled him when he transferred his observations to the human body. In medieval times,*

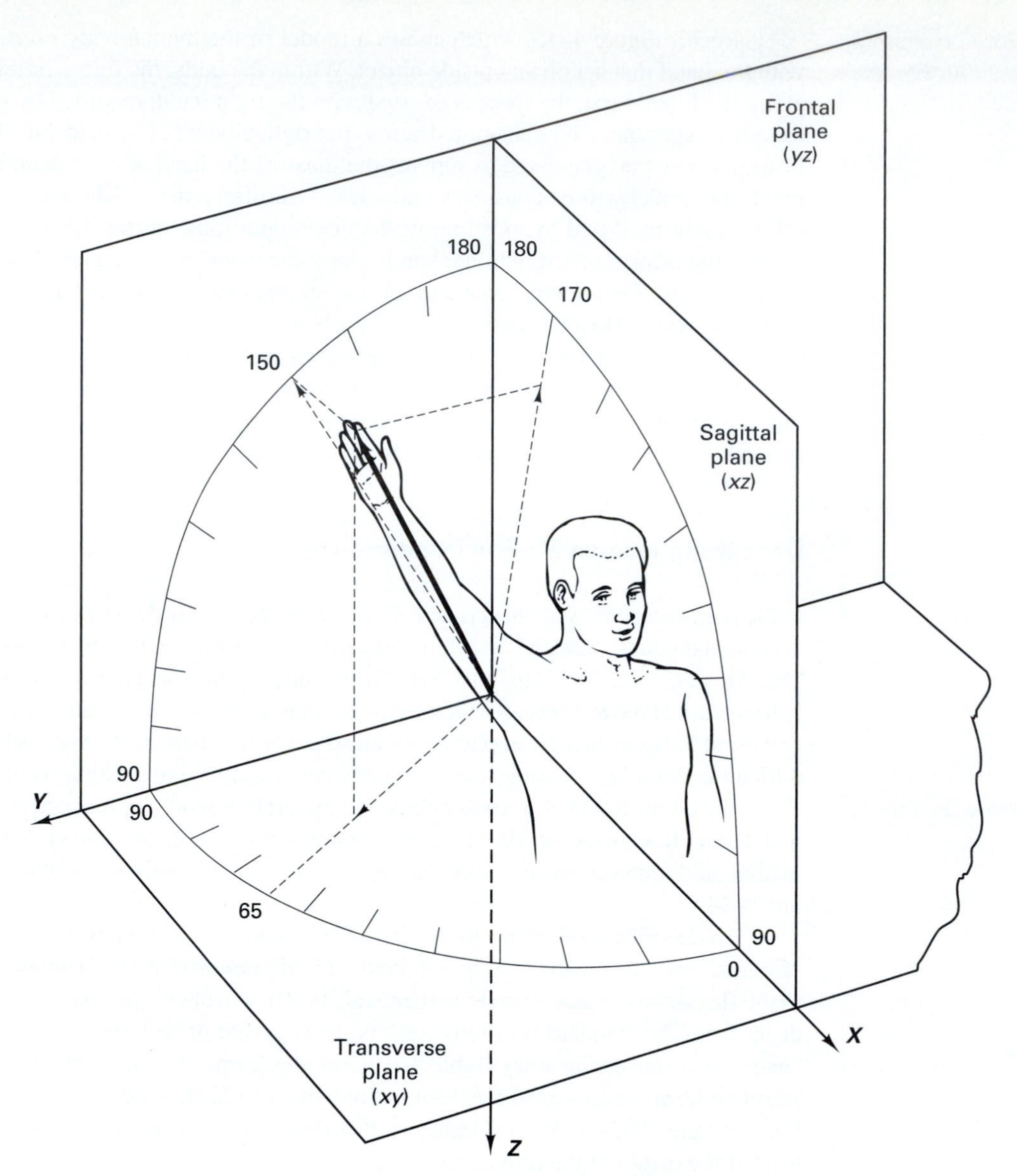

Figure 1–24. Descriptor of the angular position of the right arm in the new notation system. The position of the hand point on the arm vector is $F = 150$ degrees, $S = 170$ degrees, $T = 65$ degrees. Note that the directions for *x, y,* and *z* are commonly used in biomechanics to describe accelerations of the "vehicle" that carries a human occupant. (From Kroemer et al. 1990, *Engineering Physiology: Bases of Human Factors/Ergonomics,* 2d ed. With permission by the publisher, Van Nostrand Reinhold. All rights reserved.)

physicians looked upon surgery as inferior work and left it to those who also cut hair—that is, barbers. In 1543, however, the Flemish anatomist Andreas Vesalius (1514–1564) published his splendidly illustrated book De Corporis Humani Fabrica (On the Structure of the Human Body), *in which he corrected more than 200 errors that Galen had made (Asimov 1989).*

TABLE 1–10. New Terminology to Describe Body Motions

New Terms	*Meaning*	*Replacing*
Pivot and flexion, extension	Rotation of a body segment about its proximal joint	duction vection rotation
Twist	Rotation of a body segment about its internal axis	pronation supination rotation
Ex-	Away (up, out) from zero	ab- e-
In-	Toward (in, down) zero	ad-
Clock(wise) (or none)	In clockwise direction, as seen on own body	supi- or pro- (depending on body segment)
Counter- (clockwise)	In counterclockwise direction, as seen on own body	pro- or supi- (depending on body segment)
Front-, trans-, sag-	Reference to the plane in which motion is described	uncertainty and confusion

Examples:
Front-ex-pivot = pivoting movement in the frontal plane, away from zero.
Trans-in-twist = twisting movement in the transverse plane, clockwise.

Source: From Kroemer, Kroemer, Kroemer-Elbert 1990, *Engineering Physiology: Bases of Human Factors/Ergonomics,* 2d ed. With permission of the publisher, Van Nostrand Reinhold.

HUMAN STRENGTH

"Two-thirds" activation

As previously discussed, muscular contraction occurs by the activation, concurrently or sequentially, of many motor units. The cooperative effort of the participating units, controlled by both rate and recruitment coding, determines the contraction of the whole muscle. In general, one cannot voluntarily contract more than about two-thirds of all fibers of a muscle at once. Apparently, this limitation ensures that structural tensile capacity is not exceeded, although that can occur as a result of a reflex, which might damage or even tear a muscle or tendon.

Paired muscle groups

In the human body, muscles are arranged in pairs or groups that act in opposite directions around their common joint. For example, the triceps muscle opens the elbow angle, while the biceps, brachialis, and brachioradilis muscles reduce the elbow angle. Simultaneous activation (coactivation) of these agonistic (also called protagonistic) and antagonistic muscles allows the body to control strength and motion.

Muscle strength definition

Voluntary muscle strength is the torque that a given muscle (or group of muscles) can maximally develop voluntarily around a skeletal articulation which is spanned by the muscle(s). This statement acknowledges the fact that, with current technology, it is impossible to measure the force or tension developed within a muscle of a living human. If it becomes feasible to measure this internal force directly, voluntary muscle strength can be redefined as "the maximal voluntary force that a muscle can exert along its length."

Engrams and motivation

The model shown in Figure 1–25 helps to understand the events involved in the exertion of muscular strength. The control initiatives generated in the central nervous system start by calling up an "executive program" (called an engram) that exists for all normal muscular activities, such as walking or pushing and pulling objects. The engram is modified by "subroutines" appropriate for the specific case, such as walking quickly upstairs, or pulling hard, or pushing carefully (Schmidt 1988). The "subroutines" in turn are modified by "motivation," which determines how (and how much of the structurally possible) strength will be exerted under the given conditions.

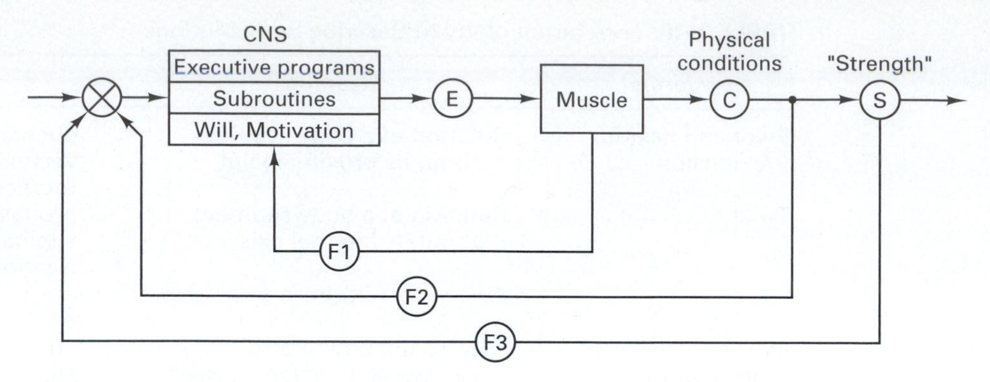

E: Efferent excitation impulses generated in the CNS
F: Afferent feedback loops

Figure 1–25. Model of the regulation of muscle strength exertion. (From Kroemer et al. 1990, *Engineering Physiology: Bases of Human Factors/Ergonomics,* 2d ed. With permission by the publisher, Van Nostrand Reinhold. All rights reserved.)

A qualitative listing of circumstances that may increase or decrease one's willingness to exert strength is given in Table 1–11.

Muscle tension excitation

The complex interactions just described result in excitation signals *E* transmitted along the efferent nervous pathways to motor units, which are triggered to contract. The contraction tension developed in the muscle depends on the motor units involved, on the rate and frequency of signals received, on muscle length and thickness, and possibly on fatigue existing in the muscle as a consequence of previous contractions. It also depends on whether the muscle length changes or remains constant when the muscle contracts. (See the discussion of static and dynamic exertions later in the chapter.)

Internal transmission

The output of the muscular effort is modified by the existing mechanical conditions, such as the lever arm at which a muscle tendon pulls about the bridged articulation, and the pull angle with

TABLE 1–11. Factors Likely to Increase (+) or Decrease (−) Maximal Muscular Performance

	Likely Effect
Feedback of results to subject	+
Instructions of how to exert strength	+
Arousal of ego involvement, aspiration	+
Pharmaceutical agents (drugs)	+
Startling noise, subject's outcry	+
Hypnosis	+
Setting of goals, incentives	+ or −
Competition, contest	+ or −
Verbal encouragement	+ or −
Fear of injuries	—
Spectators	?
Deception	?

Source: From Kroemer, Kroemer, Kroemer-Elbert 1990, *Engineering Physiology: Bases of Human Factors/Ergonomics,* 2d ed. With permission of the publisher, Van Nostrand Reinhold. All rights reserved.

respect to the lever arm. If the torque is transmitted across another articulation, and even further articulations, the new lever arm(s) and pull angle(s) modify the pull force. These conditions would change, of course, in dynamic activities, whereas they are assumed constant in a static effort.

Body segment strength

The output of this complicated chain of controllers, feed-forward signals, controlled elements, and modifying conditions is the " body segment strength" measured at the interface between the body segment involved (i.e., the hand) and the measuring device (or other object against which strength is exerted, such as a hand tool). Body segment strength is what the engineer often needs to know for designing equipment and fitting work tasks to human capabilities.

In measuring "strength," one must ensure that at any moment at least as much resistance is available at the segment–instrument interface as can be exerted by a person. If this were not the case (i.e., if Newton's third law were violated), no reliable strength measurement could be performed.

Muscle Strength is the maximal tension or force that a muscle can develop voluntarily between its origin and insertion.

The best word to refer to this is muscle *tension* (in N/mm^2 or N/cm^2), but the term strength (in N) is commonly used.

Internal Transmission is the manner in which muscle tension is transferred inside the body along links and across joints as torque to the point of application to a resisting object.

This transmission of torque is more complicated under dynamic than static conditions, because of changes in muscle functions with motion and the effects of accelerations and decelerations on masses.

Body Segment Strength is the force or torque that can be applied by a body segment to an object external to the body.

The segment may be the hand, elbow, shoulder, back, foot, or some other body part. (Note that when force is in N, torque is in Ncm or Nm.)

Feedback loops

The model also shows a number of feedback loops through which muscular exertion is monitored for control and modification. The first feedback loop, F_1, is in fact a reflex-like arc that originates at proprioceptors, such as Ruffini organs in the joints (signaling location), Golgi tendon-end organs (indicating changes in muscle tension), and muscle spindles (indicating length). (See Chapter 3.) These proprioceptors (interoceptors) and their signals are not under voluntary control and directly influence the signal generator in the spinal cord for a quick response. The other two feedback loops originate at exteroceptors and are rooted through a comparator that modifies the input signal into the central nervous system. Loop F_2 originates at kinesthetic receptors that are sensitive to events related to touch, pressure, and body position in general. For example, as one pulls on a handle, one's body position is monitored together with the sensations of pressure in the hand and of forces throughout the body, including the sensation of pressure felt in the feet as they press down on the floor. Feedback loop F_3 originates at exteroceptors and signals to the comparator such events as sounds and motions related to the effort. For example, this type of feedback may be the sounds or movements generated in the object on which one pulls, the pointer of an instrument that indicates the strength applied, or exhortations the experimenter or coach shouts to the subject.

Measurement opportunities

The model offers measurement opportunities. Considering the feed-forward section of the model, it becomes apparent that there is (with current technology) no suitable means of measuring the executive programs, their subroutines, or the effects of will or motivation on the signals generated in the central nervous system. (Currently, only very general information can be gleaned from an electroencephalogram, EEG.) Efferent excitation impulses from the motor nerves to the muscles can be recorded through electrodes by an electromyogram, or EMG.

In 1666, Franceso Redi thought that the electric shock generated by a ray (a type of fish) was muscular in origin. This was subsequently demonstrated in 1791 by Luigi Galvani, who depolarized the muscles in frogs' legs. His book De Viribus Electricitatis *(*On the Features of Electricity, *translated into English in 1953) was the first milestone in the development of neurophysiology. Alessandro Volta's erroneous explanation that the electric current could be the result of dissimilar metals in contact with an electrolyte set the ensuing research on a wrong track in the early 1800s until, in 1838, Carlo Matteucci proved that the electricity in fact originated in muscles. During the 19th century, it was popular to stimulate muscles through the skin by applying an external electrical current in the belief that properly applied electricity would perform miraculous cures on a wide variety of ailments (Basmajian and De Luca 1985, pages 1–6).*

EMG interpretation

The interpretation of an EMG, the record of the electrical signals associated with the activation of motor units in muscle in terms of frequency and amplitude, relies on several basic assumptions. One assumption is that the signals stem from the same motor units. A surface electrode can be used for isometric contractions, because in this static case, theoretically, the muscle does not move under the electrode on the skin; in dynamic muscle use, indwelling (wire or needle) electrodes can follow the moving muscle. Another assumption concerns the relationship between EMG amplitude and muscle strength. In the isometric case, calibration often shows a nearly linear increase in EMG intensity from rest to maximal voluntary muscle exertion. By contrast, in dynamic muscle use, the EMG–force relationship is complex and difficult to establish. Another complication comes with the exertion time, when the rate and recruitment coding of motor unit excitation often vary; this is particularly the case with muscle "fatigue," in which the EMG generally shows a shift toward lower frequencies, usually together with an increase in amplitude (Basmajian and DeLuca 1985, Chaffin and Andersson 1991, Soderberg 1992, Perotto 1994, Kumar and Mital 1996). Thus, electromyography is a complex technique for assessing complex muscle strength.

Contraction activities that take place in a muscle can be observed qualitatively, but not quantitatively, in the living human: At this time, no instruments are available to measure directly the tensions within muscle filaments, fibrils, fibers, the muscle itself, or groups of muscles *in situ.*

The conditions of internal transmission, such as mechanical advantages and pull angles, are difficult and often practically impossible to record and control. Yet, a description of the body posture and of positions of various segments during static exertion help us to understand how muscle tension is conveyed within the body to the point of exertion.

Assessment of Human Body Strength

From the foregoing discussion, two obvious conclusions result for strength measurements on living humans:

1. Body strength is what an instrument measures (an unsatisfactory, but realistic, statement).
2. Strength is influenced by motivation and the physical conditions under which it is exerted.

Static = isometric

In mechanics, one distinguishes between statics and dynamics. In physiological terms, the static condition is generated in an isometric muscle contraction in which, presumably, the muscle length remains constant. (The Greek *iso* means "unchanged" or "constant," and *metron* refers to measure or length, here of the muscle.) Because there is no change in muscle length during the isometric effort, there is no motion of the involved body segments; hence, Newton's first law applies, and all forces acting within the system must be in equilibrium.

Because no displacement results from muscular contraction, the physiological isometric case is equivalent to the static condition in physics. This theoretically simple and experimentally well controllable condition has lent itself to rather easy measurement of muscular strength, and most of the information currently available on human strength is limited to static (isometric) muscular effort (Caldwell et al. 1974).

Motion is dynamic

Dynamic muscular efforts are much more difficult to describe and control than static contractions. In dynamic activities, muscle length changes, and therefore, the involved body segments move. Thus, displacement is present, and its time derivatives (velocity, acceleration, and jerk) must be considered. Taking account of these is a much more complex task for the experimenter than that encountered in static testing. A systematic breakdown into independent and dependent experimental variables has been presented for dynamic and static efforts by Kroemer, Marras et al. (1990).

Variables in Experiments

In an experiment, *independent* variables are those that are purposely manipulated in order to assess resulting changes in *dependent* variables.

Isometric

If one sets the displacement (change in muscle length) to zero—the isometric condition—one may measure the force generated and possibly the number of repetitions that can be performed until the force is reduced because of muscular fatigue. (This case is described in Table 1–12.) Of course, with no displacement, its time derivatives—velocity, acceleration, and jerk—are also zero. In the isometric technique, one is also likely to control the mass properties, probably by keeping them constant.

Isovelocity = isokinematic

Besides displacement, one may choose to control velocity—that is, the rate at which muscle length changes—as an independent variable. If velocity is set to a constant value, one speaks of isovelocity or isokinematic (often falsely called isokinetic) muscle-strength measurement. The time derivatives of constant velocity, namely, acceleration and jerk, are zero. Mass properties are usually controlled in isovelocity tests. Displacement, force, and repetition can be chosen as either dependent variable or controlled independent variables. Most likely, force or repetition will be chosen as the dependent variables to assess the result of the testing. Following the scheme laid out in Table 1–12, one also can devise tests in which acceleration or its time derivative, jerk, is kept constant.

Isoforce = isotonic

If one sets the amount of force (or torque) to a constant value, it is most likely that mass properties and displacement (and its time derivatives) are controlled independent variables and repetition is a dependent variable. This isoforce condition, in which muscle tension is kept constant (isotonic), is, for practical reasons, often combined with an isometric condition, such as holding a load motionless (the displacement is zero).

☛☛☛ *The term "isotonic" has often been wrongly applied. Some older textbooks cited lifting or lowering of a constant mass (weight) as typical for isotonics. This is physically false for two reasons. The first is that, according to Newton's law, acceleration and deceleration of a mass require the application of changing (not constant) forces. The second reason lies in overlooking the*

TABLE 1–12. Techniques for Measuring Motor Performance by Selecting Specific Independent and Dependent Variables

Name of Technique	*Isometric (Static)*	*Isovelocity (Dynamic)*	*Isoacceleration (Dynamic)*	*Isojerk (Dynamic)*	*Isoforce (Static or Dynamic)*	*Isoinertial (Static or Dynamic)*	*Free Dynamic*
Variables	*Indep. Dep.*	*Indep. Dep.*	*Indep. Dep.*	*Indep. Dep.*	*Indep. Dep.*	*Indep. Dep.*	*Indep. Dep.*
Displacement, linear/angular	[constant* (zero)]	C or X	C or X	C or X	C or X	C or X	X
Velocity, linear/angular	0	[constant]	C or X	C or X	C or X	C or X	X
Acceleration, linear/angular	0	0	[constant]	C or X	C or X	C or X	X
Jerk, linear/angular	0	0	0	[constant]	C or X	C or X	X
Force, torque	C or X	C or X	C or X	C or X	[constant]	C or X	X
Mass, moment of inertia	C	C	C	C	C	[constant]	C or X
Repetition	C or X	C or X	C or X	C or X	C or X	C or X	C or X

Source: Modified from Kroemer, Marras, McGlothlin, McIntyre, and Nordin 1990.

Legend:

Indep. = independent
Dep. = dependent
C = variable can be controlled
* = set to zero
0 = variable is not present (zero)
X = can be dependent variable

The [boxed] constant variable provides the descriptive name.

changes that occur in the mechanical conditions (pull angles and lever arms) under which the muscle functions during the activity. Finally, it is virtually impossible to generate a constant tension in a muscle when it acts in motion against an external resistance. Hence, practically, a truly isotonic muscle activity occurs only in combination with an isometric one. It is certainly misleading to label all dynamic activities of muscles "isotonic," as is occasionally still done.

Isoinertial

In the isoinertial condition, mass properties are controlled by setting the mass to a constant value. In this case, repeated movement of the mass (as in lifting) may be either a controlled independent or, more likely, a dependent variable. Also, displacement and its derivatives may become dependent outputs. Any force (or torque) that is applied is likely to be a dependent variable, according to Newton's second law (force equals mass times acceleration).

Free dynamic

Table 1–12 also contains the most general case of muscle-performance measurement, labeled "free dynamic." In this case, the independent variables (displacement and its time derivatives), as well as force, are unregulated (ie, left to the free choice of the subject). Only mass and repetition are usually controlled, although they may be used as dependent variables. Displacement and its time derivatives may be dependent variables. Force, torque, or some other performance measure is likely to be chosen as a dependent output.

Experimental control vs. reality

In the scheme developed in Table 1–12, the static isometric test is the simplest test to perform and control. The dynamic tests are more difficult to keep under experimental control—especially the "free dynamic" one, which, naturally, most reflects real-life conditions. All test names indicate the conditions at the muscle, except the term "isoinertial," which refers to the external load.

THE STRENGTH-TEST PROTOCOL

After choosing the type of strength test to be done and the measurement techniques and devices, an experimental protocol must be devised. This includes the selection of subjects and protection of the information obtained from them; the control of the experimental conditions; the use, calibration, and maintenance of the measurement devices; and (usually) the avoidance of training and fatigue effects.

Regarding the selection of subjects, care must be taken to ensure that they are in fact a representative sample of the population about which data are to be gathered. Regarding the management of the experimental conditions, the control of motivational aspects is particularly difficult. It is widely accepted (outside of sports and medical function testing) that the experimenter should not give exhortations and encouragements to the subject. (See Table 1–11.) The so-called Caldwell regimen (Caldwell et al. 1974) pertains to isometric strength testing, but can be adapted for a dynamic test. (See Table 1–12.) Edited excerpts are given in the definition and listing that follow.

Definition: Static body strength is the capacity to produce torque or force by a maximal voluntary isometric muscular exertion. Strength has vector qualities and therefore should be described by both magnitude and direction.

1. Measure static strength according to the following conditions:
   a. Static strength is assessed during a steady exertion sustained for 4 seconds.*
   b. The transient periods of about 1 second each, before and after the steady exertion, are disregarded.
   c. The strength datum is the mean score recorded during the first 3 seconds of the steady exertion.

2. Treat the subject as follows:
   a. The person should be informed about the purpose of the test and the procedures involved.
   b. Instructions should be kept factual and not include emotional appeals.
   c. The subject should be told to "increase to maximal exertion (without jerk) in about 1 second and then maintain this effort during a 4-second count."*
   d. During the test, the subject should be informed about his or her general performance in qualitative, noncomparative, positive terms. Do not give instantaneous feedback during the exertion.
   e. Rewards, goal setting, competition, spectators, fear, noise, etc., can affect the subject's motivation and performance and therefore should be avoided.
3. Provide a minimal rest period of 2 minutes between related efforts—more if symptoms of fatigue are apparent.
4. Describe the conditions existing during strength testing:
   a. Body parts and muscles chiefly used.
   b. Body position (or movement).
   c. Body support or reaction force available.
   d. Coupling of the subject to the measuring device.
   e. Strength measuring and recording device.
5. Describe the subjects:
   a. Population and sample selection, including sample size.
   b. Current health; a medical examination and a questionnaire are recommended.
   c. Gender.
   d. Age.
   e. Anthropometry (at least height and weight).
   f. Training related to the strength testing.
6. Report the experimental results:
   a. Number of data collected.
   b. Minimum and maximum values.
   c. Median and mode.
   d. Mean and standard deviation, for normally distributed data points; for a nonnormal distribution, provide lower and upper percentile values, such as the 1st, 5th, 10th, 25th, 75th, 90th, 95th, or 99th percentiles.

*A shorter period of exertion may be chosen, for example if the strength of fingers is to be measured.

If you can't calibrate, you can't measure.

DESIGNING FOR BODY STRENGTH

The engineer or designer who wants to consider human strength has to make a number of decisions, including the following:

- Is strength use mostly static or dynamic? If it is static, information about isometric strength capabilities can be used. If it is dynamic, other considerations apply in addition, concerning, for example, physical endurance (circulatory, respiratory, metabolic) capabilities of the operator, prevailing environmental conditions, etc. Physiologic and ergonomic texts (e.g. by Astrand and Rodahl 1986, Kroemer 1999b, and Winter 1990) provide such information.
- Is the exertion by hand, by foot, or with the whole body? For each of these situations, specific design information is available. If a choice is still possible, it must be based on physiologic and ergonomic considerations in order to achieve the safest, least strenuous, and most efficient performance. In comparison to hand movements over the same distance, foot motions consume more energy and are less accurate and slower, but they are stronger.
- Is a maximal or a minimal strength exertion the critical design factor?

Maximal user output usually determines the structural strength of the object, in order that the strongest operator not break a handle or a pedal. The design value is set, with a safety margin, above the highest perceivable strength application.

Minimal user output is that exertion expected from the weakest operator which still yields the desired result, so that a door handle or brake pedal can be successfully operated or a heavy object moved.

A range of expected strength exertions is, obviously, that between the specified minimum and maximum. "Average" user strength is usually of no design value.

- Most body-segment strength data are available for static (isometric) exertions. They provide reasonable guidance also for slow motions, although they are probably a bit too high for concentric motions and too low for eccentric motions. As a general rule, strength exerted in motion is less than that measured in static positions located on the path of motion.
- Measured strength data are often treated, statistically, as if they were normally distributed and reported in terms of averages (means) and standard deviations. This dubious procedure is not of great practical concern, however, because usually the data points of special interest are the extremes. Often, the 5th and 95th percentile values are selected. These can be determined easily, if not by calculation, then by estimation.

See Chapter 8 for more details on designing for hand and foot strength.

SUMMARY

Anthropometric and biomechanic variables pertaining to the human body are among the most basic descriptors needed to design "fitting" equipment and work procedures.

Relatively complete information about body size is available only for military populations. However, soldiers are a select sample of the general population: They are relatively young and generally healthy, and only rarely are they persons of unusual body dimensions. Still, until recently, there were not many female soldiers, making estimates for the general civilian population somewhat unreliable.

In North America, a recent survey of the U.S. Army has remedied many of the problems associated with missing anthropometric information. The Army is fairly similar to the general population in its composition, including females. Thus, the information from the Army survey is used in this book to represent and reflect body dimensions of the general population of the United States. A statistical treatment of the measured data allows us to establish correlations among them and, using those correlations, to predict data that were not actually measured.

Statistics applied to anthropometric data provide a wealth of information. For example, the standard deviation divided by the mean is a measure of the variability of the recorded data. A large variability could indicate either a truly diverse sample or problems in data acquisition or treatment. Much variability stems from the fact that most surveys are performed on cross sections of the population, meaning that at one moment in time measurements are taken on persons of different ages. While this reflects interindividual differences, it does not inform us about intraindividual changes occurring with aging or with changes in health.

In recent decades, body dimensions of adults have been increasing in many areas of the earth. In particular, growth in many length dimensions, such as in stature and leg length, has been observed, often accompanied by an increase in body weight. Such information is important for the design of closely fitting equipment, especially clothing.

It is indefensible and inexcusable to design tools, equipment, or workstations for the phantom "average person." Instead, a range of body dimensions and the ways in which certain body dimensions change with respect to each other must be considered.

For example, it is inadequate, though common practice, to establish height–weight indices (such as in desired body weight) when, in fact, the correlation between height and weight is very low. Instead, fairly simple rules apply that consider variations in body dimensions. For the ergonomic design of equipment, a four-step procedure has been established that facilitates proper sizing of designed products.

The human body is often modeled as a rigid skeleton with joints that allow movement. The body members have mass properties and are moved by muscles. The long bones of the segments establish the lever arms at which muscles attach.

Articulations are of different kinds, and some wear out and can be replaced by artificial joints. Their design, construction, implantation, and use has developed into a specific ergonomic subdiscipline and a specialized industry and has helped millions of people to regain mobility they had lost due to accidents, diseases, or wear and tear.

Muscles convert chemically stored energy into mechanically useful force and work. The arrangement of muscles within the human body is rather complex. Usually, agonistic muscle is opposed by antagonistic muscle—pairings of muscle groups that together regulate motion and energy development. This configuration poses a rather interesting challenge for biomechanical modeling. Furthermore, voluntary and involuntary control of muscular exertions, fatigue, posture, motion characteristics, and motivation strongly influence the muscular output.

Until recently, understanding of the musculomechanical conditions was hampered by an inadequate definition of dynamic circumstances. Therefore, isometric (or static) muscle efforts are fairly well researched, while little knowledge exists about dynamic capabilities. Newly developed procedures to test body strength, both statically and dynamically, should provide both theoretical insights and practical information to be used in the ergonomic design of work tasks and equipment.

CHALLENGES

How different are the body dimensions of soldiers from those of civilians?

If differences exist, how important would they be for the design of special tools, equipment, workstations, or work procedures?

Are there "recipes" that describe how to derive "functional" body dimensions from those measured in static, standardized postures?

How do secular changes in body dimensions affect the design of technical products?

Is designing for a "fit from the 5th to the 95th percentile" appropriate?

Are women anthropometrically equivalent to small men?

Which procedures are appropriate to select fair subsamples from a general population for measurement purposes?

How can correlations among body dimensions reduce the needed number of measurements taken on a population sample?

How exactly must body dimensions be known for design purposes?

Can the "staircase effect" be generated by the external electrical stimulation of muscles? If so, might there be a danger of overexertion?

Is the external electrical stimulation of muscles useful for building stronger muscles?

How does cocontraction of agonistic and antagonistic muscles about a joint influence the loading of that joint?

Could wearing of a tight belt around the waist increase intraabdominal pressure? To what, if any, effect?

Which tissues in the trunk determine the posture of the spinal column?

What difference does it make, with respect to forces at the spinal disc, if one gives up the assumption of a straight vertebral column in favor of a column with kyphosis and lordosis?

What are the factors limiting a person's exertion of strength?

What are some practical means to affect (or control) a person's motivation for strength generation?

Is it possible to develop "conversion algorithms" to determine dynamic muscle strength from static strength, and vice versa?

Chapter 2

How the Body Does Its Work

OVERVIEW

In many respects, one may compare the way in which the body generates energy with the functioning of a combustion engine: Fuel (food) is combusted, for which oxygen must be present. The combustion yields energy that moves parts mechanically. The fueling and cooling system (blood vessels) moves supplies (oxygen, carbohydrates, and fat derivatives) to the combustion sites (the muscles and other organs) and removes combustion by-products (lactic acid, carbon dioxide, water, and heat) for dissipation (at the surfaces of the skin and lungs).

☛☛☛ *The Greek physician Alcmaeon (sixth century B.C.) was apparently the first to deliberately and carefully dissect human cadavers. He saw the difference between arteries and veins and noticed that the sense organs were connected to the brain by nerves. Praxagoras (fourth century B.C.) distinguished between arteries and veins, but he thought that arteries carried air, because they were usually found empty in corpses. (The word "artery" is from Greek words meaning "carrying air.") About 280 B.C., Herophilus noticed that the arteries pulsed and thought that they carried blood (Asimov 1989).* ☚☚☚

Body processes are governed by several complex and overlapping *control systems:* the central nervous system, the hormonal system, and the limbic system. Control centers (in the brain and spinal cord—see Chapter 3) rely on feedback from various body parts and provide feedforward signals according to general (autonomic, innate, learned) principles and according to situation-dependent (voluntary, motivational) rules.

3 major systems

The *respiratory system* provides oxygen for energy metabolism and dissipates metabolic by-products. The *circulatory system* carries oxygen from the lungs to cells that consume the oxygen. The circulatory system also brings "fuel," ie, derivatives of carbohydrates and fats, to the cells and removes metabolic by-products from the combustion sites. The *metabolic system* supports the chemical processes in the body, particularly those that yield energy (Kroemer et al. 1997).

THE RESPIRATORY SYSTEM

The respiratory system, which absorbs oxygen and dispels carbon dioxide, water, and heat, interacts closely with the circulatory system, which provides the means of transport of those essentials, as well as of nutrients that nourish the body. Figure 2–1 schematically indicates the

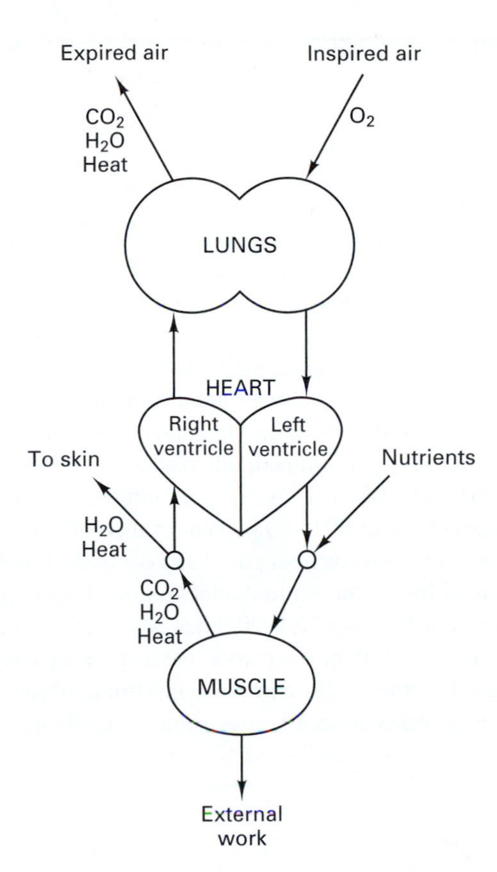

Figure 2–1. The interrelated functions of the respiratory and circulatory systems. (From Kroemer, Kroemer, and Kroemer-Elbert 1990, *Engineering Physiology: Bases of Human Factors/Ergonomics,* 2d ed. With permission of the publisher, Van Nostrand Reinhold. All rights reserved.)

interaction between the two systems. The respiratory system moves air to and from the lungs, where part of the oxygen contained in the inhaled air is absorbed into the bloodstream; it also removes carbon dioxide, water, and heat from the blood into the air to be exhaled. Between 2×10^6 and 6×10^6 alveoli provide an adult with about 70 to 90 m^2 of exchange surfaces in the lungs.

Air exchange is brought about by the pumping action of the thorax. The diaphragm separating the chest cavity from the abdomen descends about 10 cm when the abdominal muscles relax. Muscles connecting the ribs contract and raise the ribs. When the dimensions of the rib cage and its included thoracic cavity increase toward the outside and in the direction of the abdomen, air is sucked into the lungs. When the inspiratory muscles relax, the lung tissue, thoracic wall, and abdomen recoil elastically to their resting positions without involving expiratory muscles. The recoil expelled air from the lungs. When ventilation needs are high, as in heavy physical work, the recoil forces are augmented by activities of expiratory (intercostal) muscles, and contraction of the muscles in the abdominal wall further assists expiration.

The mucus-covered surfaces in the nose, mouth, and throat, adjust the temperature of the inward-flowing air to body temperature, moisten or dry the air, and cleanse it of particles. In a normal climate (see Chapter 5), about 10 percent of the total heat loss of the body, whether at rest or work, occurs in the respiratory tract. The percentage increases to about 25 percent at outside temperatures of about −30°C. In a cold environment, heating and humidifying the inspired air cools the mucosa; during expiration, some of the heat and water is recovered by condensation from the air to be exhaled (hence the "runny nose" in the cold). Altogether, the energy required for breathing is relatively small, amounting to only about 2 percent of the total oxygen uptake of the body at rest and increasing to not more than 10 percent during heavy exercise.

Respiratory Volumes

The volume of air exchanged in the lungs depends on the requirements associated with the work performed. When the respiratory muscles are relaxed, there is still air left in the lungs. A forced maximal expiration reduces this amount of air to the so-called *residual capacity* (or *residual volume*); see Figure 2–2. A maximal inspiration adds the volume called *vital capacity.* Both volumes together are the *total lung capacity.* Only the so-called *tidal volume* is moved, leaving both an inspiratory and an expiratory *reserve volume* within the vital capacity during rest or submaximal work.

Vital capacity and other respiratory volumes are usually measured with the help of a spirometer. The results obtained depend on age, training, gender, body size, and body position of the subject. The total lung volume of highly trained, tall young males is between 7 and 8 liters (L), and their vital capacity is up to 6 L. Women have lung volumes about 10 percent smaller, and untrained persons have volumes of about 60 to 80 percent of their athletic peers.

Breathing frequency

Pulmonary ventilation is the movement of gas in and out of the lungs. It is calculated by multiplying the frequency of breathing by the expired tidal volume. This is called the (respiratory expired) *minute volume.* At rest, one breathes 10 to 20 times every minute. In light exercise, primarily the tidal volume is increased, while with heavier work, the respiratory frequency quickly increases up to about 45 breaths/min, together with the increasing tidal volume. This indicates that the breathing frequency, which can be measured easily, is not a reliable indicator of the heaviness of work performed.

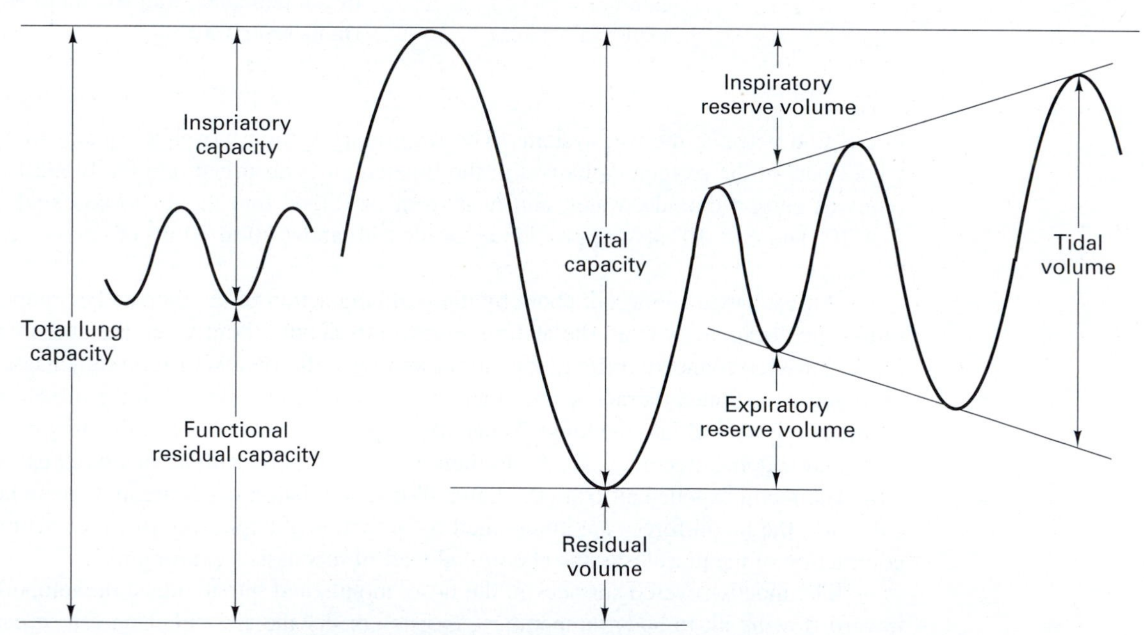

Figure 2–2. Respiratory volumes. (From Kroemer, Kroemer, and Kroemer-Elbert 1990, *Engineering Physiology: Bases of Human Factors/Ergonomics,* 2d ed. With permission of the publisher, Van Nostrand Reinhold. All rights reserved.)

Minute volume

The respiratory system is able to increase the volume of air that it moves and the amount of oxygen that it absorbs by large multiples. The minute volume can be increased from about 5 L/min to 100 L/min or more, an increase in air volume by a factor of 20 or more. Though not exactly linearly related to minute volume, oxygen consumption shows a similar increase.

THE CIRCULATORY SYSTEM

Transport system

The circulatory system carries oxygen from the lungs to the cells, where nutritional materials, also brought by circulation from the digestive tract, are metabolized. Metabolic by-products (CO_2, heat, and water) are dissipated by circulation. The circulatory and respiratory systems are closely interrelated, as shown in Figure 2–1.

Water is the largest component of the body by weight, making up about 60 percent of body weight in men and about 50 percent in women. In slim individuals, the percentage of total water is higher than in obese persons, since adipose tissue contains very little water. The relation between water and lean (fat-free) body mass is rather constant in "normal" adults, about 72 percent. Small changes in hydration are normal, typically on the order of 2 percent, as occurs during the menstrual cycle in women.

Blood

Depending on age, gender, and training, approximately 10 percent of the total fluid volume consists of blood. Four to 4.5 L of blood in women and 5 to 6 L in men are normal. The specific heat of blood is 3.85 joules (J) (0.92 cal) per gram.

In 1658, the Dutch naturalist Jan Swammerdam (1637–1680) used newly improved microscopes and discovered the red blood corpuscle. In the 1840s, the British physician Thomas Addison (1798–1866) studied white blood cells, or leucocytes, from the Greek for "white cell." The third type of objects formed in the blood, called platelets, according to their shape, was studied by the Canadian physician William Osler (1849–1919) who reported on them in 1873. They are called thrombocytes, from the Greek for "clot cells" (Asimov 1989).

Of the total blood volume, about 55 percent is plasma, which is mostly water. The remaining 45 percent consists of formed elements (solids), predominantly red cells (erythrocytes), white cells (leukocytes), and platelets (thrombocytes). The percentage of red-cell volume in the total blood volume is called the hematocrit.

Blood Groups

In 1930, the Austrian physician Karl Landsteiner (1868–1943) showed that human blood could be divided into four types. This knowledge made blood transfusions safe (Asimov 1989).

Blood is classified into four types, according to the content of certain antigens and antibodies: O, A, B, and AB. The importance of this classification lies primarily in the incompatibility reactions between the types in blood transfusions. Another classification categorizes blood according to its rhesus (Rh) factor.

Functions

Blood carries dissolved materials—particularly oxygen and nutritive materials—as well as hormones, enzymes, salts, and vitamins. It removes waste products, including dissolved carbon dioxide and heat.

Oxygen transport

Red blood cells transport oxygen. The oxygen attaches to hemoglobin, an iron-containing protein molecule of the red blood cell. Each molecule of hemoglobin contains four atoms of iron, which combine loosely and reversibly with four molecules of oxygen. Hemoglobin molecules can react simultaneously with oxygen and carbon dioxide. Hemoglobin has a high affinity for carbon monoxide (CO), which takes up space otherwise occupied by oxygen; this property explains the toxicity of CO.

☛☛☛ *The body can lose up to 15 percent of its blood volume without dramatic effects. Yet, at 20-percent loss, blood pressure is reduced and pulse and breathing are affected—gravely so at 30 percent loss, where heart rate is increased as well. At 40-percent loss, death is imminent without a transfusion.* ☚☚☚

Architecture of the Circulatory System

Two subsystems

The circulatory system is nominally divided into two subsystems: the *systemic* and the *pulmonary* circuits, each powered by one half of the heart (which can be considered a double pump). The left side of the heart supplies the systemic section, which branches from the arteries through the arterioles and capillaries to the metabolizing organ (e.g., muscle); from there, the branches combine again from venules to veins that lead to the heart's right side. The pulmonary system starts at the right ventricle, which powers the blood flow through the pulmonary artery, the lungs, and the pulmonary vein to the left side of the heart.

Heart chambers

Each half of the heart has an antechamber (atrium) and a chamber (ventricle)—the pump proper. The atria receive blood from the veins, which is then brought into the ventricles through a valve. In essence, the heart is a hollow muscle that produces the desired blood flow via contraction and with the aid of valves.

☛☛☛ *Based on Galen's (129–199) beliefs, it was thought until the early 1600s that blood was manufactured in the liver and carried to the heart, from which it was pumped outward through arteries and veins alike and consumed in the tissues. The Italian physician Girolamo Fabrici (1537–1619) noticed the valves in the veins that prevent the blood from flowing in the direction that Galen had postulated; yet, Fabrici did not dare to contradict Galen's doctrine. Galen had thought that the heart was a single pump and that there were pores in the thick muscular wall separating the two ventricles. In 1242, the Arabic scholar Ibn an-Nafis wrote that the right and left ventricles were totally separated. He explained how blood pumped by the right ventricle was led by arteries to the lungs, where the blood collected air. The enriched blood then was collected into increasingly larger vessels, which brought it back to the left ventricle, from where blood was pumped to the rest of the body. Unfortunately, an-Nafis' book did not become known outside the Arabic world until 1924. Yet, in 1553, the Spanish physician Miguel Serveto published a book in which he correctly described some features of the circulation. In 1559, the Italian anatomist Realdo Colombo also described major features of the circulatory system. His work was widely used in the medical profession. In 1628, the English physician William Harvey published his book* De Motu Cortis et Sanguinis, *which describes functions of the heart and blood. This book is generally considered the beginning of modern physiology (Asimov 1989).* ☚☚☚

The Heart as a Pump

The ventricle is filled through the valve-controlled opening from the atrium. The heart muscle contracts (called a *systole*), and when the internal pressure is equal to the pressure in the aorta, the aortic valve opens and the blood is ejected from the heart into the systemic system. Continuing contraction of the heart increases the pressure further, since less volume of blood can escape from the aorta than the heart presses into it. Part of the excess volume is kept in the aorta and its large branches, which act as a "windkessel," an elastic pressure vessel. Then, the aortic valve closes with the beginning of the relaxation *(diastole)* of the heart, while the elastic properties of the aortic walls propel the stored blood into the arterial tree, where elastic blood vessels smooth out the waves of blood volume. At rest, about half the volume in the ventricle (the stroke volume) is ejected, while the other half (the residual volume) remains in the heart. During exercise, the heart ejects a larger portion of the volume it contains and increases its contraction frequency. When much blood is required, but cannot be supplied, such as during very strenuous physical work with small muscle groups or during maintained isometric contractions, the heart rate can become very high.

Heart rate

At a heart rate of 75 beats/min, the diastole takes less than 0.5 second and the systole just over 0.3 second; at a heart rate of 150 beats/min, the periods are close to 0.2 second each. Hence, an increase in heart rate occurs mainly by shortening the duration of the diastole.

Specialized cardiac cells (the sinoatrial nodes) serve as "pacemakers," determining the frequency of contractions by propagating stimuli to other cells of the heart muscle. The heart has its own intrinsic control system, which operates at 50 to 70 beats/min in the absence of external influences. Changes in heart action stem from the central nervous system.

The events in the right heart are similar to those in the left, but the pressure in the pulmonary artery is only about one-fifth of those during systole in the left heart.

Since Luigi Galvani's time (1737–1798) it had been known that muscular contractions were associated with small electric potentials. The demonstration that this phenomenon also applied to the heart muscle was made possible by the development of a specific galvanometer by the Dutch physiologist Willem Einthoven (1860–1927) in 1903. The device allowed the recording of the electrocardiogram, often abbreviated as EKG instead of ECG because, in German, cardio *is spelled with a* k. *Einthoven received the Nobel prize for medicine and physiology in 1924 (Asimov 1989).*

EKG or ECG

Myocardial action potentials are recorded in the electrocardiogram (EKG, ECG). The different waves have been given alphabetic identifiers: The P wave is associated with the electrical stimulation of the atrium, while the Q, R, S, and T waves are associated with ventricular events. The EKG is employed chiefly for clinical diagnoses; however, with the appropriate apparatus it can be used for counting and recording the heart rate. Figure 2–3 shows the electrical, pressure, and sound events during a contraction–relaxation cycle of the heart.

Pathways of Blood

Changes in minute volume

Since the available blood volume does not vary, cardiac output can be affected by two factors: the frequency of contraction (heart rate) and the pressure generated by each contraction in the blood. Both determine the so-called (cardiac) minute volume. The cardiac output of an adult at rest is around 5 L/min. During strenuous exercise, this level might be raised by a factor of five, to about 25 L/min, while a well-trained athlete may reach up to 35 L/min.

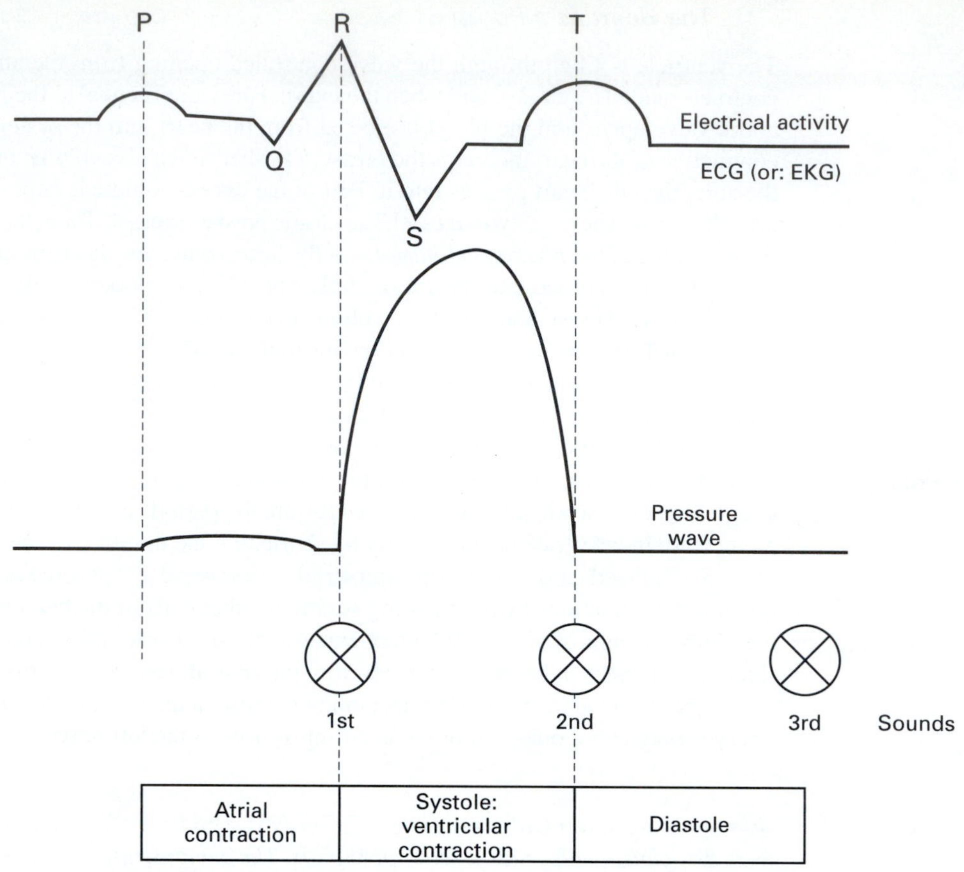

Figure 2–3. Scheme of the electrocardiogram, the pressure fluctuation, and the phonogram of the heart with its three sounds. (From Kroemer, Kroemer, and Kroemer-Elbert 1990, *Engineering Physiology: Bases of Human Factors/Ergonomics,* 2d ed. With permission of the publisher, Van Nostrand Reinhold. All rights reserved.)

Limitation in transport capability

A healthy heart can pump much more blood through the body than is usually needed. Hence, a circulatory limitation is more likely to lie in the transporting capability of the vascular portions of the circulatory system than in the heart itself. As mentioned, the arterial section of the vascular system (before the metabolizing organ) has relatively strong elastic walls that act as a pressure vessel ("windkessel"), thus transmitting pressure waves far into the body, though with much loss of pressure along the way. At the arterioles of the consumer organ, the blood pressure is reduced to approximately one-third its value at the heart's aorta.

Capillary bed

As blood seeps through the consumer organ (e.g., a muscle) via capillaries, the pressure differential from the arterial side to the venous side maintains the transport of blood through the *capillary bed.* Here, the exchanges of oxygen, nutrients, and metabolic by-products between the working tissue and the blood take place. If a lack of oxygen or the accumulation of metabolites requires high blood flow, smooth muscles that encircle the fine blood vessels remain relaxed, allowing the pathways to remain open. The large cross-sectional opening reduces blood flow velocity and blood pressure, allowing nutrients and oxygen to enter the extracellular space of the tissue and permitting the blood to accept metabolic by-products from the tissue.

Reduced blood flow in the capillary bed

Constriction of the capillary bed by tightening the encircling smooth muscle reduces the local blood flow so that other organs in more need of blood may be better supplied. Such compression of the capillary bed can also occur if the striated muscle itself contracts strongly, at more than about 20 percent of its maximal capability. If this contraction is maintained, the muscle hinders or shuts off its own blood supply and cannot continue the contraction. Thus, sustained strong static contraction is self-limiting. (Recall the discussion of muscle endurance in Chapter 1.) A typical example of muscles cutting off their own blood flow is overhead work, wherein muscles must keep the arms elevated. After a fairly short time, one must let them hang down to allow muscle relaxation and renewed blood flow.

The designer of equipment and work tasks should be careful not to require sustained muscular contractions—e.g., for keeping the body in position or in grasping a handle tightly. Instead, the work should permit frequent changes in muscle tension, best achieved by allowing movement.

The venous portion of the systemic system has a large cross section and provides low flow resistance; only about one-tenth of the total pressure loss occurs here. Valves are built into the venous system, allowing blood to flow only toward the right ventricle.

Swollen ankles

Pascal's law states that the static pressure in a column of fluid depends on the height of the column. However, the hydrostatic pressure in, for example, the feet of a standing person is not as large as is expected from physics, because the valves in the veins of the extremities modify the value: In a standing person, the arterial pressure in the feet may be only about 100 mm Hg higher than in the head. Nevertheless, blood, water, and other body fluids in the lower extremities are pooled there, leading to a well-known increase in volume of the lower extremities (swollen ankles), particularly when one stands or sits still.

Regulation of Circulation

Priorities of blood supply

If the concentration of metabolites in a muscle increases, smooth muscles encircling blood vessels will relax, allowing more blood to flow. At the same time, signals from the central nervous system (CNS) can trigger the constriction of other less important vessels supplying blood to organs. This leads to quick redistribution of the blood supply, which favors skeletal muscles over the digestive system (the principle of "muscles over digestion"). However, even in heavy exercise, the systemic blood flow is so controlled that the arterial blood pressure is sufficient for an adequate blood supply to the brain, heart, and other vital organs. To accomplish this, neural vasoconstrictive commands can override local dilatory control. For example, the temperature-regulating center in the hypothalamus can affect vasodilation in the skin if this is needed to maintain a suitable body temperature, even if it means a reduction of blood flow to the working muscles (the principle of "skin over muscles").

Heart blood output

Circulation at the arterial side of a consumer organ is regulated both by local control and by impulses from the central nervous system, the latter having overriding power. The heart increases its output through a higher heartbeat frequency and higher blood pressure. At the venous side of the circulation, the constriction of veins, combined with the pumping action of dynamically working muscles and forced respiratory movements, facilitates the return of blood to the heart. These venous and pulmonary actions make increased cardiac output possible, because the heart cannot pump more blood than it receives.

Heart rate related to oxygen uptake The heart rate generally follows oxygen consumption and hence energy production of the dynamically working muscle in a linear fashion from moderate to rather heavy work. However, the heart rate at a given level of oxygen intake is higher when the work is performed with the arms than with the legs. This reflects the use of different muscles and muscle masses with different lever arms to perform the work. Smaller muscles doing the same external work as larger muscles are more strained and require more oxygen. Also, static (isometric) muscle contraction increases the heart rate, apparently because the body tries to bring blood to the tensed muscles.

Work in a hot environment causes a higher heart rate than work at a moderate temperature, as discussed in Chapter 5. Emotions such as nervousness, apprehension, and fear can affect the heart rate at rest and during light work.

THE METABOLIC SYSTEM

In 1614, the Italian physician Santorio Sanctorius (1561–1636) reported on experiments in which he sat in an elaborate weighing machine while he ate, drank, and eliminated wastes. He found that he lost more weight than the waste alone could account for and attributed this to "insensible perspiration." Sanctorius' experiments were the beginning of the study of metabolism (Asimov 1989).

Homeostasis

The American investigator Walter Cannon introduced the term "homeostasis" to characterize the remarkable internal stability of human body functions. The core temperature, fluid volume, blood pH, and many other functions of the body stay nearly the same even under extremes of our natural environment.

Over time, the human body also maintains a balance between energy input and output. The input is determined by nutrients, from which chemically stored energy is liberated during the metabolic processes within the body. The output is mostly heat and work. Work is measured in terms of physically useful energy, i.e., energy transmitted to outside objects. The amount of such external work performed strains individuals differently, depending on their physique and training. There is close interaction between the metabolic, circulatory, and respiratory systems, as sketched in Figure 2–4.

The Human Energy Machine

Astrand and Rodahl (1977) used an analogy between the human body and an automobile. In the cylinder of the engine, an explosive combustion of a fuel–air mixture transforms chemically stored energy into physical kinetic energy and heat. The kinetic energy moves the pistons of the engine, and gears transfer their motion to the wheels of the car. The engine must be cooled to prevent overheating. Waste products are expelled. This whole process can work only in the presence of oxygen and when there is fuel in the tank. In the "human machine," muscles are analogous to the cylinders and pistons, while bones and joints are equivalent to the gears. Heat and metabolic by-products are generated as the muscles work. Nutrients (mostly carbohydrates and fats) are the fuels that are oxidized to yield energy.

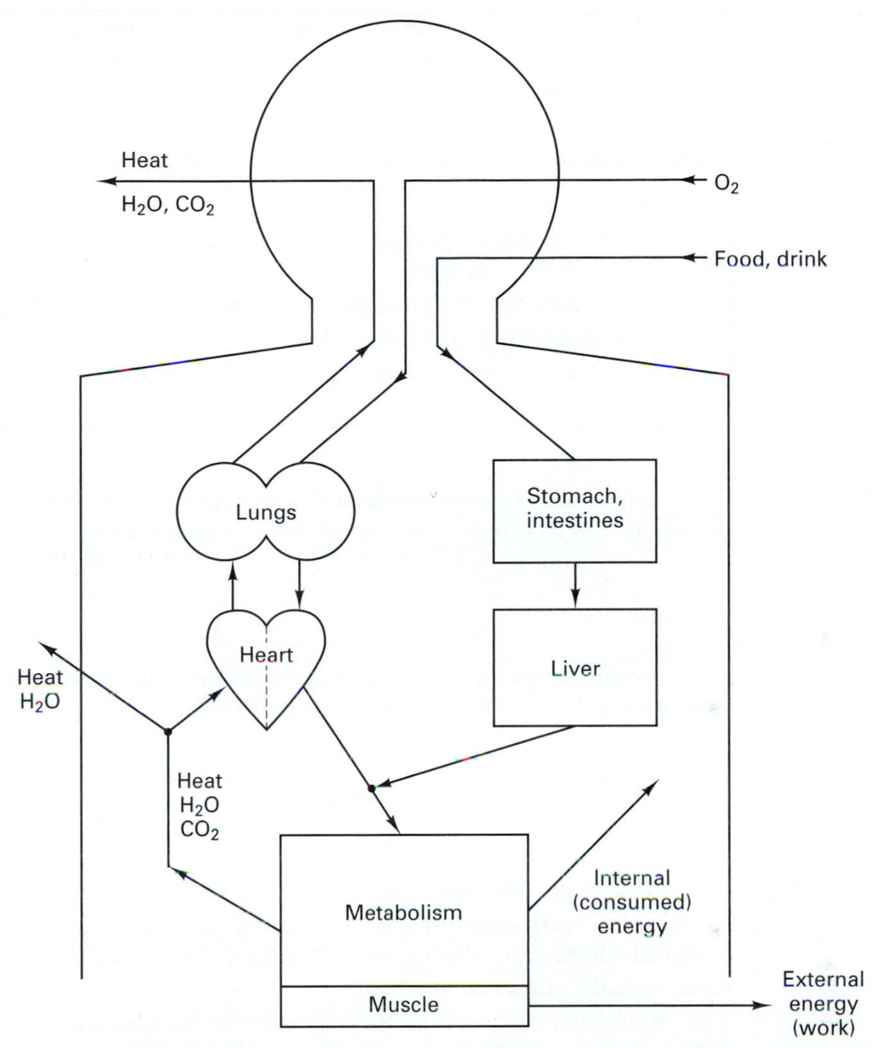

Figure 2–4. Interactions among energy inputs, metabolism, and outputs of the human body. (From Kroemer, Kroemer, and Kroemer-Elbert 1990, *Engineering Physiology: Bases of Human Factors/Ergonomics,* 2d ed. With permission of the publisher, Van Nostrand Reinhold. All rights reserved.)

Human Metabolism and Work

The term "metabolism" refers to all chemical processes in the living body. In a narrower sense, it is used here to describe the (overall) energy-yielding processes.

Input equals output?

The balance between energy input I (via nutrients) and outputs can be expressed by the equation

$$I = M = H + W + S \tag{2–1}$$

where M is the metabolic energy generated, which is divided into the heat H that must be dispelled to the outside, the work W, and the change in energy storage S in the body. (A gain in S is counted as positive, a loss as negative.)

The units of energy (or work) are joules (J) or calories (cal), with 4.2 J = 1 cal. (More precisely, 1 J = 1 Nm = 0.2389 cal = 10^7 ergs = 0.948×10^{-3} BTU = 0.7376 ft lb.)

One uses the kilocalorie (1 kcal = 1 Cal = 1,000 cal) to measure the energy content of foodstuffs.

The units of power are 1 kcal h^{-1} = 1.163 W, with 1 W = 1 J s^{-1}.

Assuming there is no change in energy storage and no heat is gained from or lost to the environment, one can simplify the energy-balance equation to

$$I = H + W \tag{2–2}$$

Human energy efficiency e (work efficiency) is defined as the ratio of work performed to energy input:

$$e \text{ (in \%)} = \frac{100 \text{ W}}{I} = \frac{100 \text{ W}}{M}. \tag{2–3}$$

We are "good heaters" but "bad workers"

In everyday activities, only about 5 percent or less of the energy input is converted into "work"—i.e., energy usefully transmitted to outside objects; under favorable circumstances, highly trained athletes may attain perhaps 25 percent. The remainder, by far the largest portion of the input, is finally converted into heat.

Muscle converts chemical into physical work

Work (in the physical sense) is done by the skeletal muscles, which move body segments against external resistances. For this, the muscle is able to convert (in its mitochondria) chemical energy into physical work or energy. From resting, it can increase its energy generation up to fiftyfold. Such enormous variation in metabolic rate not only requires quickly adapting supplies of nutrients and oxygen to the muscle, but also generates large amounts of waste products (mostly heat, carbon dioxide, and water) that must be removed. Thus, during the performance of physical work, the ability to maintain the internal equilibrium of the body is largely dependent on the circulatory and respiratory functions that serve the muscles involved. Among these functions, the control of body temperature is of particular importance. This function interacts with the external environment, particularly the surrounding temperature and humidity, as discussed in Chapter 5 in more detail.

☛☛☛ *In 1827, the English chemist William Prout (1785–1850) was the first to chemically classify foodstuffs into groups that we call now carbohydrates, fats, and proteins. Of course, this grouping is dietetically important, but not exhaustive, because other substances, such as vitamins, appear in small quantities in our food, but are vital nevertheless. (Asimov 1989).* ☚☚☚

Energy Transformation in the Human Body

In living organisms, such as the human body, energy transformation involves chemical reactions that either liberate energy, most often as heat, or require energy. The first kind of reaction, called catabolism, is *exergonic* (or *exothermic*). The opposite kind of reaction, called anabolism, requires an energy input and is *endergonic* (or *endothermic*). Generally, the breakage of molecular bonds is exergonic, while the formation of bonds is endergonic. Depending on the molecular combinations, bond breakage releases different amounts of energies. Often, reactions do not simply go from the most complex to the most broken-down state, but achieve the process in steps, with intermediate and temporarily incomplete stages.

Food and drink energy

Energy is supplied to the body in food or drink. It is contained in specific chemical compounds that are changed in the course of digestion and then, in the following assimilation, reassembled to different molecular combinations. Their energy content can be released for use by the body; the ergonomist is particularly interested in energy use by muscles to perform physical work.

The Energy Pathways

Ingestion of food and drink

In the mouth, chewing destroys the structure of the food mechanically, and saliva starts its chemical breakdown. Saliva is 99.5 percent water (a solvent) and 0.5 percent salts, enzymes, and other chemicals. The enzyme lysozyme destroys bacteria, thus protecting the mucous membranes from infection and the teeth from decay. Another enzyme, salivary amylase, breaks down starch.

During swallowing, breathing stops and the epiglottis closes for a second or two so that a food "bolus" can avoid the windpipe (trachea) and slide down the gullet (esophagus). Liquids need only about one second, but solid food takes up to eight seconds to glide to the stomach.

Stomach actions

The stomach generates gentle waves, two to four per minute, which mix the bolus with gastric juice. Alcohol is absorbed mostly in the stomach, whose gastric juice starts breaking up proteins chemically, but does little to break down fats and carbohydrates. Carbohydrate-rich foods leave the stomach within two hours, while protein-containing foods are slower; fatty foods stay up to six hours.

☛☛☛ *Digestion was long thought to be a physical action, the result of the grinding of the stomach. In 1752, the French physicist René Antoine Ferchault de Réaumur (1683–1757) showed that digestion was a chemical process (Asimov 1989).* ☚☚☚

Digestion and absorption

The contents of the stomach empty into the top of the small intestine, called the duodenum (Latin for "12 fingers," indicating the length of that section of intestine). Here, the true chemical digestion occurs through the breakup of large complex molecules into smaller ones. These can be transported across cell membranes and absorbed into blood and lymph. It takes the food three to five hours to move through the small intestine, a tube about 3 cm in diameter and 7 m in length. During this time, about 90 percent of all the nutrients are extracted.

In the following large intestine, final processing is completed, and solid waste is eliminated as feces.

Assimilation

The digested and absorbed foodstuffs are assimilated—i.e., reassembled into new molecules that be can used for body growth and repair, or easily degraded to release or store energy.

Altogether, it takes 5 to 12 hours after eating a meal to extract its nutrients and to pass it through the digestive tract.

If you stop smoking, will you gain weight?

Nicotine prolongs the time during which food stays in the stomach before being sent to the intestines—and a full stomach turns off the appetite. Nicotine also dulls the desire for sweets and lessens the activity of the lipoprotein lipase, an enzyme that regulates the storage of fat carried in the blood. In addition, nicotine speeds up the metabolism by increasing the resting metabolic rate (see the discussion later in this chapter) and, possibly, the thermic effects of exercise or work.

When one stops smoking, the nicotine content of the body is greatly reduced. Consequently, sweet food becomes more attractive, the stomach is emptied faster, and fat is stored in the cells more easily. Eight out of 10 persons gain some weight after stopping smoking. Most gain very little, but 20% gather more than 15 kg (Klesges 1995). Not giving in to the desire to eat more—particularly more sweets— and perhaps even introducing nicotine via special medications counteract the potential weight gain.

Foodstuffs

Our food is a mixture of organic compounds (foodstuffs), water, salts, minerals, vitamins, and fibrous material (mostly cellulose). Energy is contained in the primary foodstuffs—*carbohydrates, fats,* and *proteins*—and in alcohol.

> The nutritionally usable energy amounts per gram are, on average, 4.2 kcal (18 kJ) from carbohydrates, 4.5 kcal (19 kJ) from protein, about 7 kcal (30 kJ) from alcohol, and 9.5 kcal (40 kJ) from fat.

Carbohydrate

Carbohydrates range from small to rather large molecules, and most are composed of only three chemical elements: carbon (C), oxygen (O), and hydrogen (H). (The ratio of H to O usually is 2 to 1, as in water—hence the name "carbohydrate," meaning "watered carbon.") Carbohydrates are digested by breaking the bonds between monosaccharides so that the compounds are reduced to simple sugars that can be absorbed through the walls of the intestines into the bloodstream. The blood transports the monosaccharide *glucose* ($C_6H_{12}O_6$) to the liver. From there, glucose is sent either to the central nervous system (the brain and spinal cord) or to the muscles for direct use. Glucose, the primary (but not the largest) source of energy, is used almost exclusively by the central nervous system and provides the quick energy required for muscular actions.

In 1856, the French physiologist Claude Bernard (1813–1878) discovered a form of starch in the liver. Because it was easily broken down into glucose, he called it glycogen (Greek for glucose producer). He showed that glycogen could be built up from glucose to act as an energy store and that it could be broken down to glucose again when energy was needed (Asimov 1989).

The liver may change glucose into a long-chained molecule called *glycogen,* a polysaccharide $(C_6H_{10}O_5)_x$. Glycogen is deposited in the liver and stored as energy near skeletal muscle. When the glycogen storage areas are filled, the liver converts glucose to fat and stores it in adipose tissue. Here, excess energy input becomes felt and seen.

You can "become fat" without eating fat.

Fat

Fat is the carrier of vitamins A, D, E, and K in food. It is also the major energy source for the body. Fat is a triglyceride—a molecule formed by joining a glycerol nucleus to three fatty-acid radicals. Unsaturated fat has double bonds between adjacent carbon atoms; hence, the compound is not saturated with all the hydrogen atoms it could accommodate. Most plant fats are polyunsaturated and therefore liquid, while most animal fats are saturated and thus solid.

When fat is digested, the bonds linking the glycerol with the three fatty acids are broken. Digestion takes place primarily in the small intestine, where the glycerol and fatty-acid molecules can pass through cell membranes. The water-repellent fatty acids are absorbed into the lymph, which finally empties into the bloodstream, which is already carrying the water-soluble glycerol.

Fat storage

Regulated by the liver, fat is usually transported for storage to adipose tissue, from where it will be used when needed. (See later.) Fat also cushions vital organs (such as the heart, liver, spinal cord, brain, and eyeballs) against impact. A layer of fat under the skin insulates the body against heat transfer to and from the environment. As an illustration, note the appearance of swimmers, who usually have more "rounded" bodies than do "skinny" long-distance runners.

Protein

Protein is the third major food component. It consists of chains of amino acids joined together by peptide bonds. Many such bonds exist; hence, proteins come in a variety of types and sizes. In digestion, the protein bonds are broken into amino acids, which are absorbed into the bloodstream. The amino acids are transported to the liver, which disperses some to cells throughout the body, to be rebuilt into new proteins. However, most amino acids become enzymes—organic catalysts that control the chemical reactions between molecules without being consumed themselves. Still others become hemoglobin (the carrier of oxygen in the blood), antibodies, hormones, or collagen. Usually, the body employs protein for these important functions, but it can use its amino acids for energy if other energy carriers are not available.

Stored Energy

Under normal conditions of nutrition or exertion, body fat accounts for most of the stored energy reserves; but glucose and glycogen are the primary and first-used sources of energy at the cell level, particularly in the central nervous system and at muscles.

On average, about 16 percent of body weight is fat in a young man. The percentage increases to 22 percent by middle age and is even more if the man "is fat." Young women average about 22 percent of body weight in the form of fat, a percentage that usually rises to some 35 percent by middle age. Ranges of 12 to 20 percent in men and 20 to 30 percent in women are considered normal. Athletes generally have lower percentages—about 15 percent in men and 20 percent in women. Obesity has been defined as fat content of more than 25 percent in men and greater than 33 percent in women (Weinsier 1995).

☛☛☛ *Assuming 15 percent fat in a 60-kg person as a low value, body fat amounts to 9 kg. Twenty-five percent of a 100 kg person, or 34 percent of a 70-kg person, means approximately 25 kg of body fat. Given that each gram of fat yields about 9.5 kcal,the percentages entail an energy storage in the form of fat of about 85,500 Cal for a skinny lightweight person, and nearly 240,000 Cal for a heavy person.* ☚☚☚

Much less energy storage is provided by glycogen. Most of us have about 400 grams of glycogen stored near the muscles, about 100 grams in the liver, and some in the bloodstream. With an energy value of about 4.2 kcal per gram, we have only some 2,200 Cal as energy available from glycogen.

Energy Release

The enormous amount of energy stored in our bodies as fat must be extracted from it by relatively complex chemical processes that take some time. The energy in glycogen is more easily liberated, while the energy in glucose is most easily and quickly released.

Glucose and glycogen use

The utilization of the energy present in the human body is achieved by catabolism, a destructive metabolic process in which organic molecules are broken down, releasing their internal bond energies. Overall, this is accomplished in the human body by aerobic metabolism, meaning in the presence of oxygen. For example, glucose can be oxidized according to the formula

$$C_6H_{12}O_6 + 6\,O_2 = 6\,CO_2 + 6\,H_2O + \text{Energy}. \tag{2–4}$$

This means that one molecule of glucose combines with six molecules of oxygen, generating six molecules each of carbon dioxide and water, while energy (about 690 kcal/mole) is released.

However, oxidation also occurs by the breakdown of glucose and glycogen molecules into several fragments, which become oxidized by each other. In this case, the process is anaerobic and yields much less energy than the aerobic reactions.

☛☛☛ *In 1913, the British physiologist Archibald Vivian Hill (1886–1977) demonstrated that heat was developed, and oxygen consumed, not during, but after, a muscular contraction, when the muscle was at rest. The German biochemist Otto Meyerhof (1884–1951) independently demonstrated the same fact in chemical terms. This indicated that during muscular contraction, lactic acid is developed from glycogen. When the muscle is at rest again, lactic acid is oxidized, thus paying off the oxygen debt that was incurred during the preceding reaction of "anaerobic glycosis" (Greek for sugar-splitting without air). Hill and Meyerhof were awarded shares of the Nobel prize for medicine and physiology in 1922 (Asimov 1989).* ☚☚☚

Fat use

Fat catabolism takes place in anaerobic steps, except for the last step, which is aerobic. First, fat is split into glycerol and fatty acids. Glycerol is used in a manner similar to glucose, while the fatty acid is re-formed to acetic acid, which enters the Krebs cycle and, finally, becomes oxidized to carbon dioxide and water. The energy yield is 2,340 kcal/mole, more than three times that released from glucose, but the complex fat breakdown takes longer.

Energy Use

Living muscle cells store "quick-release" energy in the form of the molecular compound *adenosine triphosphate (ATP).* Its phosphate bonds can be broken down easily by hydrolysis to adenosine diphosphate (ADP). The anaerobic reaction,

$$ATP + H_2O \rightarrow ADP + \text{energy (output)}, \tag{2–5a}$$

releases energy. While the ATP available in the mitochondria of the muscle can provide energy for a few seconds, ATP must be resynthesized for continuous operation. This is done through creatine phosphate (CP), which transfers a phosphate molecule to the ADP. Energy must be supplied for this endergonic reaction to occur:

$$ADP + CP + \text{energy (input)} \rightarrow ATP + H_2O. \tag{2–5b}$$

The energy needed for the rebuilding of ATP from ADP is liberated in the breakdown of the complex molecules to simpler ones, ultimately CO_2 and H_2O. First, glucose is used, then glycogen, and, finally, fats (and possibly proteins). Thus, the "combustion of foodstuffs" is the ultimate source of the body's energy, keeping the ATP–ADP conversion going.

Muscular Work

The mitochondria are the "cellular power factories" of muscle. (See earlier in this chapter and also Chapter 1.) They provide chemically stored energy in the form of ATP and release it, as just described, so that muscle can contract, thereby converting chemical into mechanical energy.

The first few seconds

At the very beginning of muscular effort, breaking the phosphate bond of ATP releases "quick energy" for muscular contraction. However, the contracting muscle consumes its local supply of ATP in about two seconds.

The first 10 seconds

The next source of immediate energy is CP, which transfers a phosphate molecule to the just-created molecule of ADP, turning the ADP back into ATP. (This cycle of converting ATP into ADP and back to ATP does not require the presence of oxygen.) Since a human has three to five times more ADP than ATP, there is enough energy available for a muscle to perform up to 10 seconds of high activity.

After 10 seconds

After about 10 seconds of ATP–ADP–ATP reactions, energy must be supplied to sustain the re-formation of ATP. Now, the energy absorbed from the foodstuffs comes into play: Glucose (then glycogen, then fat) is broken down, releasing energy for the re-creation of ATP.

Given that only a few seconds have elapsed since the activity started, there simply has not been enough time to use oxygen in the energy-conversion process. Thus, while releasing energy, the breakdown of glucose (generating carbon dioxide and water) is not complete, but other metabolic by-products are generated also, particularly lactic acid. (If this metabolic by-product is not resynthesized within about a minute in the presence of oxygen, the muscles simply cannot continue to work further.)

The anaerobic energy release relies primarily on the breakdown of glucose, although some glycogen is also involved. This is why glucose is called the primary, most easily accessible, and most metabolized energy carrier for the body.

After minutes and longer

If the physical activity has to continue, it must be performed at a level at which oxygen is sufficiently available to maintain the energy-conversion processes. Hence, the energy generated by a quick burst of maximal effort cannot be maintained at a high level for extended periods of time.

In long-lasting work, the energy demanded from the muscles is so low that the oxygen supply at the mitochondria level allows "aerobic" energy-conversion, meaning that sufficient oxygen is available to maintain the energy-conversion processes without the generation of metabolic by-products that would lead to fatigue and force termination of the work.

Aerobic and Anaerobic Metabolism

Without oxygen, a molecule of glucose yields 2 molecules of ATP. With oxygen, the glucose energy yield is 36 molecules of ATP. Even richer in energy is the fat molecule palmitate, which yields about 130 molecules of ATP.

Because the energy yield is so much more efficient under aerobic conditions, in which no metabolic by-products (that cause fatigue and exhaustion) are generated, one can keep up a fairly high energy expenditure, as long as ATP is replaced as fast as it is used up and no metabolite such as lactic acid is developed.

Yet, if a very heavy expenditure of energy is required over long periods of time, such as in heavy physical work or during a marathon run, the interacting metabolic system and the oxygen-supplying circulatory system might become overtaxed. The runner who "hits the wall" has most likely used up the body's glycogen supply and has gone into "oxygen debt."

However, in our everyday activities, the energy output is regulated to conform with the body's abilities to develop energy under a sufficient supply of oxygen. If needed, one simply takes a break. While one is resting, accumulated metabolic by-products are resynthesized, and the metabolic, circulatory, and respiratory systems return to their normal states.

Many of the single intermediate steps in the metabolic reactions are, in fact, anaerobic; but finally, oxygen must be provided. Thus, overall, sustained energy use in the human body is aerobic.

☛☛☛ *We may now refine the earlier used analogy of the "human energy machine". The "human muscle motor" runs on the energy derived from the breakdown of ATP into ADP. The "ATP battery" is recharged by rebuilding ATP from ADP with the help of energy released in the combustion of glucose, glycogen, and fatty and amino acids.* ☚☚☚

Energy Use and Body Weight

Equation (2–1) describes the balance between energy input, energy output, and energy storage. If the input exceeds the output, storage (body fat and hence, weight) is increased; conversely, weight decreases if the input is less than the output. Approximately 7,000 to 8,000 kcal make a difference of 1 kilogram in body weight.

Consistent with the principle of homeostasis, the body tries to maintain a given energy storage. To keep a system in homeostasis, a deviation from the preferred internal state generates a signal that initiates the responses necessary to counter the perturbing influence.

In systems engineering, the *set point* is defined as an independent and adjustable signal against which feedback signals from the controlled system can be compared; the set point "sets" the value of the variable maintained by the system that is controlled. Human body energy appears to regulated by such a set-point mechanism: Body weight is relatively stable, and mechanisms appear to be in place to actively resist changes in this stability (Keesey and Powley 1986). As long as the set point is kept by the body, the body weight remains rather constant. If the body weight declines, adjustments in food intake usually occur via an increased appetite. Also, changes in whole-body metabolism occur that favor regaining the lost weight. Thus, even if one voluntarily reduces the food intake, the body tries to extract enough energy from the reduced food intake to maintain the old body weight. Similar coordinated counteracting adjustments in intake and energy expenditure also occur when one's body weight is elevated from the normally maintained level (Keesey 1995).

Set points can be adjusted voluntarily (e.g., by a change in dietary or exercise habits) or through internally regulated mechanisms (such as fever, central nervous system lesions, hormones, and toxins).

ASSESSMENT OF ENERGY EXPENDITURES AT WORK

The ability to perform physical work is different from person to person and depends on gender, age, body size, health, the environment, and motivation—as sketched in Figure 2–5.

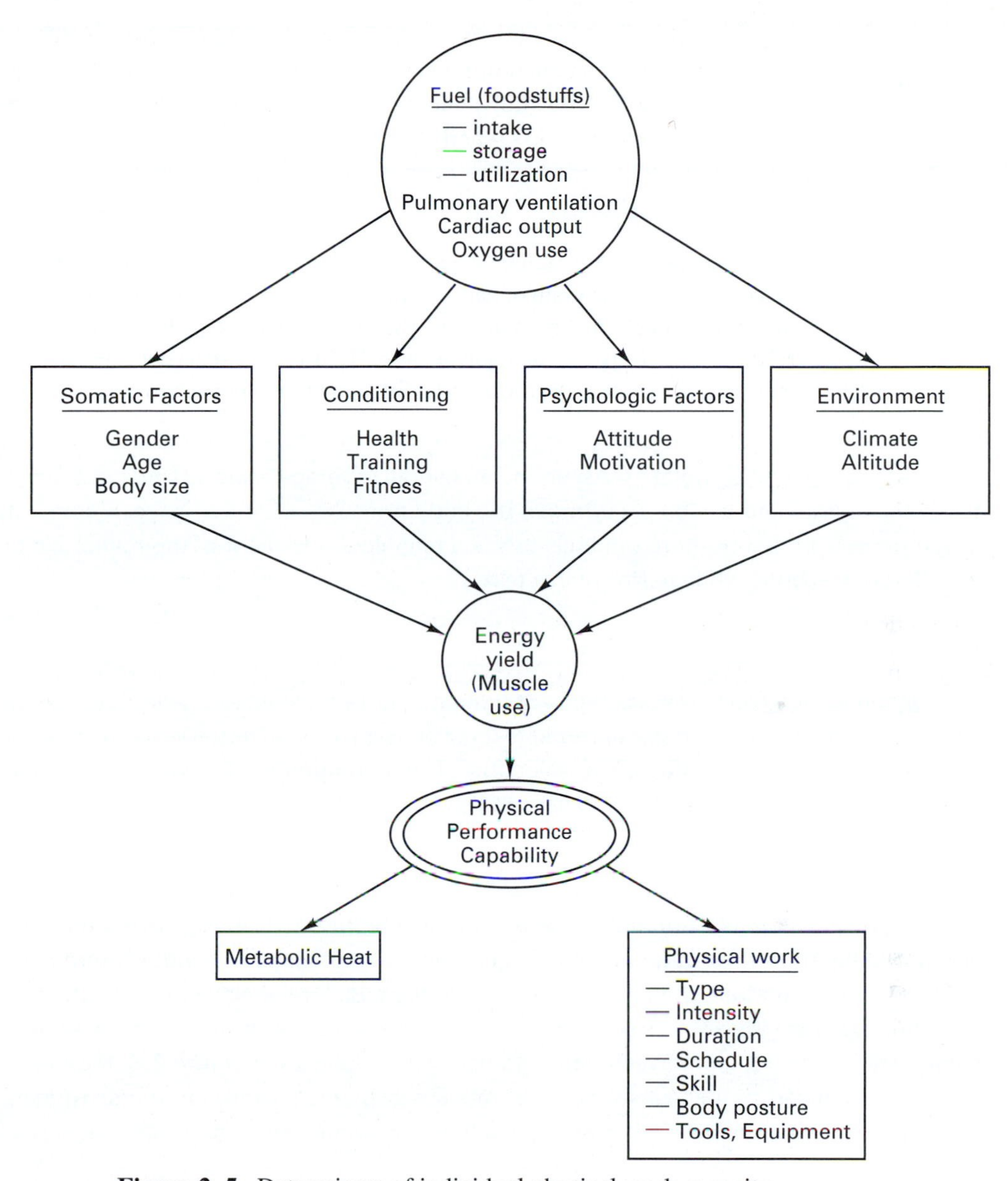

Figure 2–5. Determiners of individual physical work capacity.

Energy requirements of a task To match a person's work capacity with the requirements of a job, one needs to know the individual's energy capacity and how much the job demands from this capacity. To measure an individual's capacity, one makes the person perform a known amount of work (usually on a bicycle ergometer or a treadmill) and measures the subject's reactions. To measure the energy requirements of a given task, one lets a "standardized" person perform the job and again measures the person's reactions to the task. (Since "standard persons" are usually not available, one simply measures the reaction of the workers actually doing the job, assuming that they are "normal.")

<u>O_2 uptake</u> A person's oxygen consumption (and CO_2 release, as discussed earlier) while performing work is a measure of his or her metabolic energy production. The instruments used for this purpose rely on the principle that the difference in O_2 and CO_2 contents between the exhaled and inhaled air indicates the oxygen absorbed and carbon dioxide released in the lungs.

RQ

The respiratory exchange quotient (RQ) compares the carbon dioxide expired with the oxygen consumed. One gram of carbohydrate needs 0.83 L of oxygen to be metabolized and releases the same volume of carbon dioxide. [See Equation (2–4)]. Hence, the RQ is 1 (unit). The energy released is 18 kJ per gram, equivalent to 21.2 kJ (5.05 kcal) per liter of oxygen. Table 2–1 shows these relationships for carbohydrate, fat, and protein conversion.

<u>Caloric value of O_2 uptake</u> Assuming an overall "average" caloric value of 5 kcal/L O_2, one can calculate the energy conversion occurring in the body from the volume of oxygen consumed. Given observation periods of five or more minutes, this is a reliable assessment of the metabolic processes—i.e., of the effort of the body while performing a task.

<u>Standardized tests</u> Assessments of human energy capabilities use various techniques of measuring oxygen consumption in standardized tests with normalized external work, done mostly on bicycle ergometers, treadmills, or steps. The selection of this equipment is based not so much on theoretical considerations as on availability and ease of use.

<u>Heart rate</u> Circulatory and metabolic processes interact closely. Nutrients and oxygen must be brought to the muscle or other metabolizing organ and metabolic by-products removed from it for proper functioning. Therefore, the heart rate (as a primary indicator of circulatory functions) and oxygen consumption (representing the metabolic conversion taking place in the body) have a linear and reliable relationship in the range between light and heavy work, shown in Figure 2–6. If one knows this relationship, one often can simply substitute heart-rate assessments for measurements of metabolic processes, particularly oxygen uptake. This is a very attractive shortcut, since heart-rate measurements can be performed easily.

TABLE 2–1. Oxygen Needed, RQ, and Energy Released In Nutrient Metabolism

	O_2 Consumed (Lg^{-1})	$RQ = \frac{Vol\ CO_2}{Vol\ O_2}$	$kJ\ g^{-1}$	$kJ\ L^{-1}\ O_2$	$Kcal\ L^{-1}\ O_2$
Carbohydrate	0.83	1.00	18	21.2	5.05
Fat	2.02	0.71	40	19.7	4.69
Protein	0.79	0.80	19	18.9	4.49
Average*	NA	NA	NA	21	5

*Assuming the construct of a "normal" adult on a "normal" diet doing "normal" work.

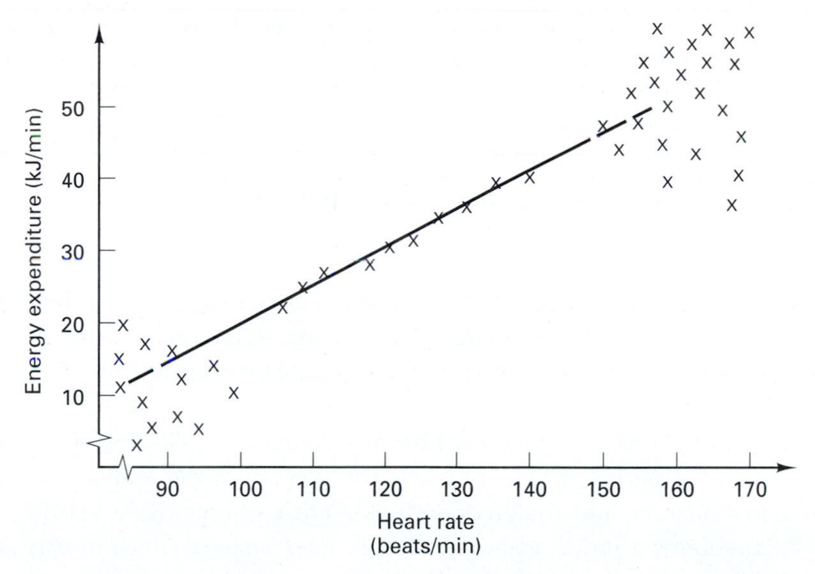

Figure 2–6. Scheme of the relationships between energy expenditure and heart rate. (From Kroemer, Kroemer, and Kroemer-Elbert 1990, *Engineering Physiology: Bases of Human Factors/Ergonomics,* 2d ed. With permission of the publisher, Van Nostrand Reinhold. All rights reserved.)

Heart-rate counting The simplest way to count the heart rate is to feel the pulse in an artery with a fingertip (called palpation), often in the wrist or perhaps in the neck, or to listen to the sound of the beating heart. All the measurer needs to do is count the number of heartbeats over a given period of time (such as 15 seconds) and, from this, calculate an average heart rate per minute. More refined techniques use deformations of tissue due to changes in filling of the imbedded blood vessels as indicators of blood pulses. These "plethysmographic" methods range from measuring the change in the volume of tissues mechanically (for example, in a finger) to using photoelectric techniques that react to changes in the transmissibility of light, depending on the blood filling (such as in the earlobe). Other common techniques rely on electric signals associated with the pumping actions of the heart (EKG) sensed by electrodes placed on the chest.

These measuring techniques are limited in their reliability of assessing metabolic processes primarily by the intra- and interindividual relationships between circulatory and metabolic functions. Statistically speaking, the regression line shown in Figure 2–6 relating heart rate to oxygen uptake (aerobic energy production) is different in slope and intersect from person to person and from task to task. Accordingly, the scatter of the data around the line, indicated by the coefficient of correlation, is also variable. The correlation is low at light loads, under which the heart rate is barely elevated and circulatory functions can be influenced easily by psychological events (excitement, fear, etc.) that may be completely independent of the task proper. With very heavy work, the oxygen–heart-rate relationship may also fall apart—for example, when cardiovascular capacities may be exhausted before metabolic or muscular limits are reached. The presence of a heat load also influences the oxygen–heart-rate relationship.

(For more information on techniques for assessing metabolic processes with commercially available equipment see, for example, Astrand and Rodahl 1986; Bernhard and Joseph 1994; Eastman Kodak 1983,1986; Kinney 1980; Kroemer et al. 1997; Mellerowicz and Smodlaka 1981; and Webb 1985.)

Techniqes of Measuring Oxygen Uptake

Measuring the oxygen consumed over a sufficiently long period of time (such as five minutes) is a practical way to assess the metabolic processes. (A physician or physiologist should supervise this test.) The method is called "indirect calorimetry," since it does not measure expenditures of energy directly.

Classically, indirect calorimetry has been performed by collecting all exhaled air during a certain observation period in airtight (Douglas) bags. The volume of the exhaled air (easily 100 L/min in heavy work) is then measured and analyzed for oxygen and carbon dioxide as needed for the determination of the RQ. This requires a rather elaborate air-collecting system, which mostly limits the procedure to the laboratory.

A major improvement was to divert only a known percentage of the exhaled air into a small collection bag, which means that only a relatively small device has to be carried by the subject. Still, in both cases, the subject must wear a face mask with valves and a nose clip, which can become quite uncomfortable and hinders speaking.

Significant advances have been made through the use of instantaneously reacting sensors that can be placed into the airflow of the exhaled air, allowing a breath-by-breath analysis without air collection and without a constraining face mask. The differences in oxygen measured with different equipment are usually rather small; for example, a comparison between the classical Douglas bag method, the partial (Max Planck) gas meter, and the Oxylog, a small portable instrument, showed variations in the mean of less than 7 percent; the linear regression coefficients were better than 0.90 (Louhevaara et al. 1985). For most field observations, the accuracy of the bagless procedures is quite sufficient.

The use of the heart rate has a major advantage over oxygen consumption as an indicator of metabolic processes, since the heart rate responds faster to work demands and hence indicates more apparently changes in body functions due to changes in work requirements.

Subjective Rating of Perceived Effort

Human are able to perceive the strain generated in the body by a given work task and to make absolute and relative judgments about this perceived effort; that is, one can assess and judge the relationships between the physical stimulus (the work performed) and its perceived sensation. This correlation between the work and its psychologically perceived intensity probably has been used as long as people have sought to express their preference for one type of work over another. Weber, in 1838, and Fechner, in 1860, established formal relationships between the intensity of a physical stimulus and its perceptual sensation. In the 1970s, Borg developed formal techniques to rate the perceived exertion associated with different kinds of efforts using the power function

$$P = e + f(I - g)^n, \tag{2–6}$$

where P is the intensity of the perceived effort, I is the intensity of the physical stimulus (e.g., the work performed), and e, f and n are constants that are specific to the task.

Accordingly, one can assess the perceived effort using a nominal scale ranging from "light" to "hard." Such a verbally anchored scale can be used to "measure" the strain that is subjectively perceived while performing standardized work, thus providing a means of assessing the heaviness of a task. Alternatively, one can use this procedure (similarly to the methods previously described) to assess an individual's capability to perform stressful work.

In 1960, Borg developed a "category scale" for the rating of perceived exertions (RPE). The scale ranges from 6 to 20 (to match heart rates from 60 to 200 beats/min), with every second number anchored by verbal expressions:

THE 1960 BORG RPE SCALE (MODIFIED 1985)

6——no exertion at all
7——extremely light
8
9——very light
10
11——light
12
13——somewhat hard
14
15——hard
16
17——very hard
18
19——extremely hard
20——maximal exertion

In 1980, Borg proposed his "General Scale," which he claimed was a category scale with ratio properties that yields ratios and levels and allows comparisons, but still retains the same correlation (of about 0.88) with the heart rate as the RPE scale, particularly if large muscles are involved in the effort. Normally, the scale (called the CR-10 Scale) has 10 intensity levels, but one might rate the intensity of a sensation such as an ache or a pain higher than 10 (Borg 1990):

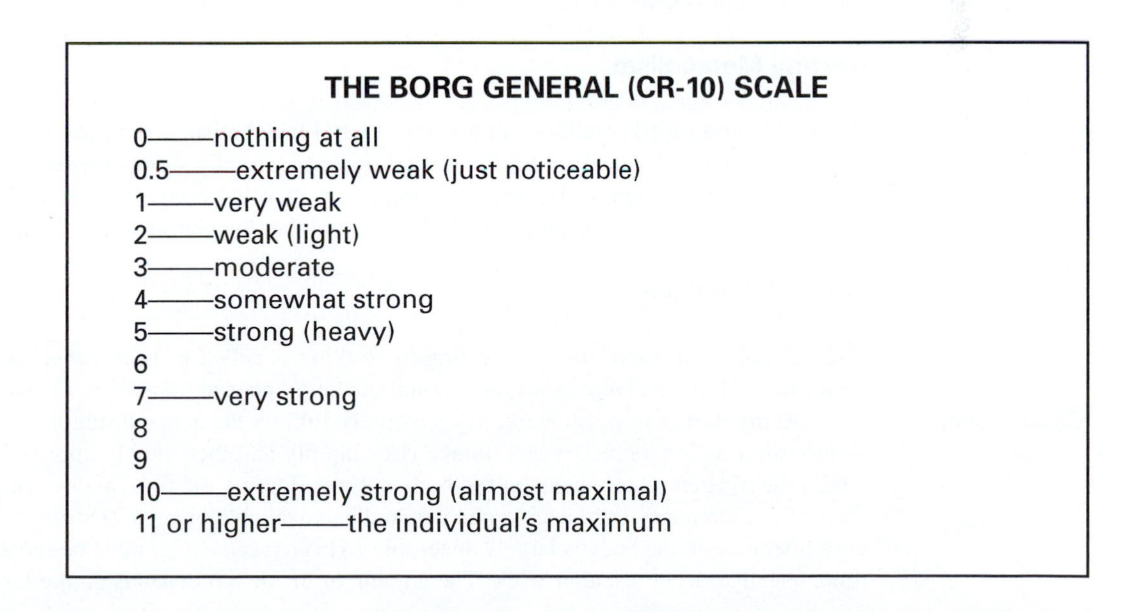

THE BORG GENERAL (CR-10) SCALE

0——nothing at all
0.5——extremely weak (just noticeable)
1——very weak
2——weak (light)
3——moderate
4——somewhat strong
5——strong (heavy)
6
7——very strong
8
9
10——extremely strong (almost maximal)
11 or higher——the individual's maximum

(Note: The terms "weak" and "strong" may be replaced by "light" and "hard" or "heavy," respectively.)

The instructions for the use of the scale are as follows (modified from Borg's publications): While the subject looks at the rating scale, the experimenter says, "I will not ask you to specify the feeling, but do select a number that most accurately corresponds to your perception of [specific symptoms]. If you don't feel anything—for example, if there is no [symptom], you answer zero—nothing at all. If you start feeling something that is just about noticeable, you answer 0.5—extremely weak, just noticeable. If you have an extremely strong feeling of [symptom], you answer 10—extremely strong, almost maximal. This would be the absolute strongest you have ever experienced. The more you feel—the stronger the feeling—the higher is the number you choose. Keep in mind that there are no wrong numbers; be honest, and do not overestimate or underestimate your ratings. Do not think of any other sensation than the one I ask you about. Do you have any questions?"

Let the subject get well acquainted with the rating scale before the test. During the test, let the subject do the ratings toward the end of every work period, i.e., about 30 seconds before stopping or changing the workload. If the test must be stopped before the scheduled end of the work period, let the subject rate the feeling at the moment of stoppage.

ENERGY REQUIREMENTS AT WORK

Basal Metabolism

A minimal amount of energy is needed to keep the body functioning, even if no activities are done at all. This *basal metabolism* is usually measured on a person who has fasted for 12 hours and has restricted her or his protein intake for at least two days, with complete physical rest in a neutral ambient temperature. Under these conditions, the basal metabolic values depend primarily on one's age, gender, height, and weight. Altogether, there is relatively little interindividual variation; hence, a commonly accepted figure for basal metabolism is 1 kcal (or 4.2 kJ) kg^{-1} h^{-1}, or 4.9 kJ min^{-1}, for a person of 70 kg.

Resting Metabolism

The highly controlled conditions under which basal metabolism is measured are rather difficult to accomplish for practical applications. Therefore, one usually simply measures the *resting metabolism* before the working day, with the subject as well at rest as possible. Depending on the given conditions, resting metabolism is 10 to 15 percent higher than basal metabolism.

Work Metabolism

The increase in metabolism from resting to working is called *work metabolism.* This increase above the resting level represents the amount of energy needed to perform the work.

Oxygen debt

At the start of physical work, oxygen uptake follows the demand sluggishly. As Figure 2–7 shows, after a slow onset, oxygen intake rises rapidly and then slowly approaches the level at which the oxygen requirements of the body are met. During the first minutes of physical work, there is a discrepancy between oxygen demand and available oxygen. (During this time, the energy liberation in the body is largely anaerobic.) This *oxygen deficit* must be repaid at some later time, usually during rest after work. The amount of the deficit depends on the kind of work per-

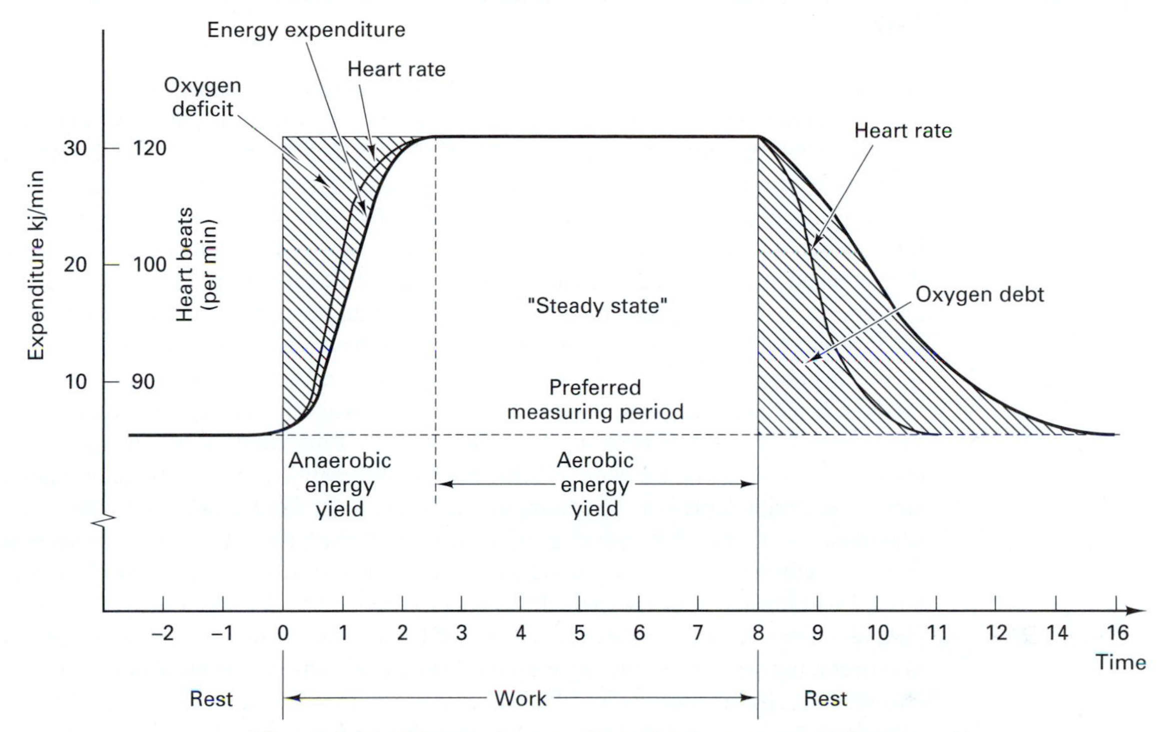

Figure 2–7. Scheme of energy liberation, energy expenditures, and heart rate at steady-state work. (From Kroemer, Kroemer, and Kroemer-Elbert 1990, *Engineering Physiology: Bases of Human Factors/Ergonomics,* 2d ed. With permission of the publisher, Van Nostrand Reinhold. All rights reserved.)

formed and on the person, but the *oxygen debt* repaid is approximately twice as large as the oxygen deficit incurred. Of course, given the close interaction between the circulatory and the metabolic systems, the heart rate reacts similarly, but as it increases faster at the start of work than oxygen uptake does, it also falls back more quickly to the resting level.

Measurement during steady state If the workload stays below about 50 percent of the worker's maximal oxygen uptake, then oxygen consumption, heart rate, and cardiac output can achieve and stay on the required supply level. This condition of stabilized functions at work is called the steady state. Obviously, a well-trained person can attain this equilibrium level between demand and supply even at a relative high workload in a few minutes, while an ill-trained person would be unable to attain a steady state at that high requirement level, but could be in equilibrium at a lower level of demand. Of course, the oxygen uptake, as well as the heart rate, should be measured during the steady-state period.

☛☛☛ *"Fatigue" is operationally defined as a "reduced muscular ability to continue an existing effort."* ☚☚☚

Fatigue

If the energetic work demands exceed about half the person's maximal oxygen uptake, anaerobic energy-yielding metabolic processes play increasing roles. As mentioned earlier in the chapter, this results in accumulations of potassium and lactic acid, which are believed to be the primary reasons for muscle fatigue, forcing the muscle to stop working. The length of time during which a person endures anaerobic work depends on the person's motivation and will to overcome the feeling of fatigue, which usually coincides with the depletion of glycogen deposits in the working muscles, a drop in blood glucose, and an increase in blood lactate. However, the processes involved are not fully understood, and highly motivated subjects may maintain work that requires very high oxygen uptake for many minutes, while other persons feel that they must stop after just a brief effort.

Reduced blood flow causes muscle fatigue

This phenomenon is best researched in regard to maintained static (isometric) muscle contraction. If the effort exceeds about 15 percent of a maximal voluntary contraction (MVC), blood flow through the muscle is reduced—even cut off in a maximal effort—in spite of a reflex increase in systolic blood pressure. Insufficient blood flow brings about an accumulation of potassium ions, and a depletion of sodium, in the extracellular fluid. Combined with an intracellular accumulation of phosphate (from the degradation of ATP), these biochemical events perturb the coupling between nervous excitation and muscle-fiber contraction. This discoupling between CNS control and muscle action signals the onset of fatigue. The depletion of ATP or creating phosphate as energy carriers, or the accumulation of lactate, once believed to be the reasons for fatigue, also occur, but are not the primary reasons. Also, the increase in the number of positive hydrogen ions resulting from anaerobic metabolism causes a drop in intramuscular pH, which then inhibits enzymatic reactions, notably those in the breakdown of ATP (Kahn and Monod 1989).

When severe exercise brings about a continuously growing oxygen deficit and an increase in lactate content in the blood because of anaerobic metabolic processes, a balance between demand and supply cannot be achieved; no steady state exists, and the work requirements exceed capacity levels—as sketched in Figure 2–8. The resulting fatigue can be counteracted by the insertion of rest periods. Given the same ratio of "total resting time" to "total working time," many short rest periods have more "recovery value" than a few long rest periods.

AVOIDING FATIGUE

- Allow short bursts of dynamic work, and avoid long periods of static effort.
- Keep energetic work and muscle demands low.
- Encourage taking many short rest pauses; this is better than taking a few long breaks.

TECHNIQUES FOR ESTIMATING ENERGY REQUIREMENTS

Tables of energy expenditures have been compiled for body postures and for many professional or athletic activities. (See Tables 2–2 and 2–3.)

Instead of using such general tables, one can compose the total energetic cost of given work activities by adding together the energetic costs of the work elements that, combined, make up this activity. If one knows both the time spent in a given element of a certain activity and its metabolic cost per time unit, one can simply calculate the energy requirements of this element by multiplying its unit metabolic cost by its duration.

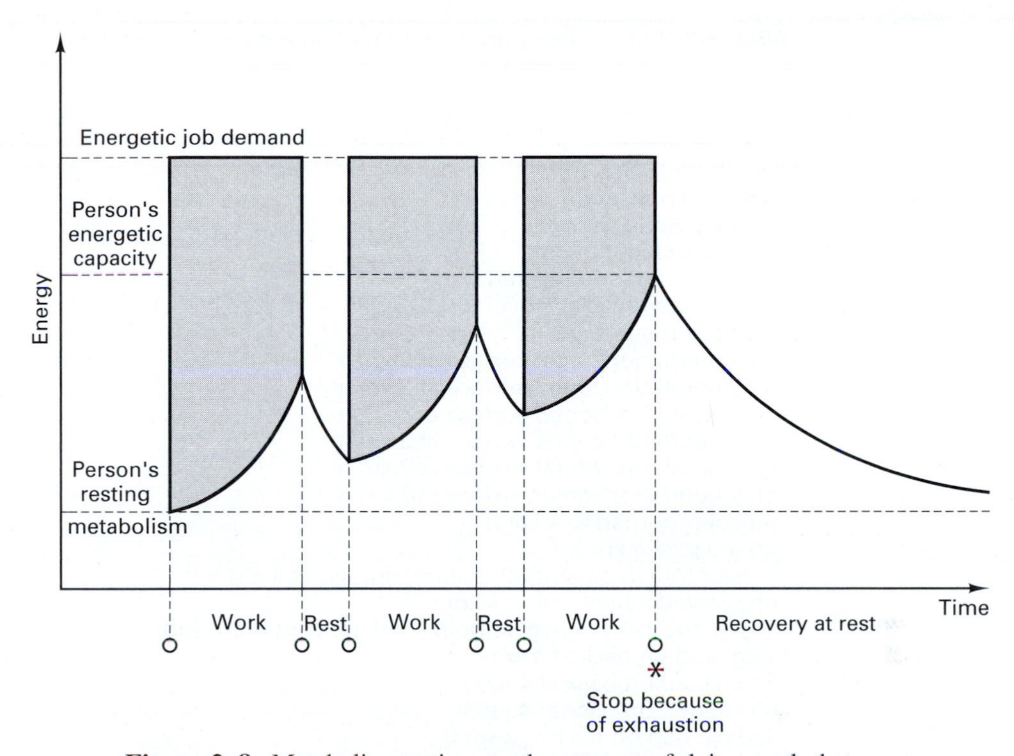

Figure 2–8. Metabolic reactions to the attempt of doing work that exceeds one's capacity even with interspersed rest periods. (From Kroemer, Kroemer, and Kroemer-Elbert 1990, *Engineering Physiology: Bases of Human Factors/Ergonomics,* 2d ed. With permission of the publisher, Van Nostrand Reinhold. All rights reserved.)

☛☛☛ *The following is an example of daily energy expenditure: For a person resting (sleeping) eight hours per day, at an energetic cost of approximately 5.1 kJ/min, the total energy cost is about 2,450 kJ (5.1 kJ* $min^{-1} \times 60$ *min* $h^{-1} \times 8$ *h). If the person then does six hours of light work while sitting, at 7.4 kJ/min, this adds another 2,664 kJ to the energy expenditure. With an additional six hours of light work done standing, at 8.9 kJ/min, and further with four hours of walking at 11.0 kJ/min, the total expenditure during the full 24-hour day would come to about 10,960 kJ (or approximately 2,610 kcal).* ☚☚☚

Knowing the energy requirements of work allows one to judge whether a job is (energetically) easy or hard. Given the largely linear relationship between heart rate and energy uptake, one can often simply use the heart rate to judge work as "light" or "heavy." Of course, such labels reflect judgments that rely very much on the current socioeconomic concept of what is permissible, acceptable, comfortable, easy, or hard. Depending on the circumstances, one finds a diversity of opinions about how "hard" a given job is—but its demands on the body can be objectively measured, such as by heart rate or oxygen uptake.

TABLE 2–2. Energy Consumption (to be Added to Basal Metabolism) at Various Activities

Activity	*Energy Consumed, kJ min^{-1}*
Lying, Sitting, Standing	
Resting while lying	0.2
Resting while sitting	0.4
Sitting with light work	2.5
Standing still and relaxed	2.0
Standing with light work	4.0
Walking without load	
on smooth horizontal surface at 2 km h^{-1}	7.6
on smooth horizontal surface at 3 km h^{-1}	10.8
on smooth horizontal surface at 4 km h^{-1}	14.1
on smooth horizontal surface at 5 km h^{-1}	18.0
on smooth horizontal surface at 6 km h^{-1}	23.9
on smooth horizontal surface at 7 km h^{-1}	31.9
on country road at 4 km h^{-1}	14.2
on grass at 4 km h^{-1}	14.9
in pine forest, on smooth natural ground at 4 km h^{-1}	18 to 20
on plowed heavy soil at 4 km h^{-1}	28.4
Walking and carrying on smooth solid horizontal ground	
1-kg load on back at 4 km h^{-1}	15.1
30-kg load on back at 4 km h^{-1}	23.4
50-kg load on back at 4 km h^{-1}	31.0
100-kg load on back at 3 km h^{-1}	63.0
Walking downhill on smooth solid ground at 5 km h^{-1}	
5° decline	8.1
10° decline	9.9
20° decline	13.1
30° decline	17.1
Walking uphill on smooth solid ground at 2.5 km h^{-1}	
10° incline (gaining height at 7.2 m min^{-1})	
no load	20.6
20 kg on back	25.6
50 kg on back	38.6
16° incline (gaining height at 12 m min^{-1})	
no load	34.9
20 kg on back	44.1
50 kg on back	67.2
25° incline (gaining height at 7.2 m min^{-1})	
no load	55.9
20 kg on back	72.2
50 kg on back	113.8
Climbing stairs 30.5° incline, steps 17.2 cm high, 100 steps per minute (gaining 17.2 m min^{-1}), no load	57.5
Climbing ladder 70° incline, rungs 17 cm apart (gaining 11.2 m min^{-1}), no load	33.6

SOURCE: Adapted from Astrand and Rodahl 1977, Guyton 1979, Rohmert and Rutenfranz 1983, and Stegemann 1984.

Note: While Rohmert and Rutenfranz (1983) claim that intra- and interindividual differences in energy consumption are within ±10 percent for the same activity, a comparison of data presented in various texts shows a much higher percentage of variation, particularly at activity levels requiring little energy.

TABLE 2–3. Total Energy Cost Per Day in Various Jobs and Professions

Occupation	Energy Expenditure, kcal/day Mean	Minimum	Maximum
MEN			
Laboratory technicians	2,840	2,240	3,820
Elderly industrial workers	2,840	2,180	3,710
University students	2,930	2,270	4,410
Construction workers	3,000	2,440	3,730
Steelworkers	3,280	2,600	3,960
Elderly peasants (Swiss)	3,530	2,210	5,000
Farmers	3,550	2,450	4,670
Coal miners	3,660	2,970	4,560
Forestry workers	3,670	2,860	4,600
WOMEN			
Elderly housewives	1,990	1,490	2,410
Middle-aged housewives	2,090	1,760	2,320
Laboratory technicians	2,130	1,340	2,540
University students	2,290	2,090	2,500
Factory workers	2,320	1,970	2,980
Elderly peasants (Swiss)	2,890	2,200	3,860

SOURCE: Adapted from Astrand and Radahl 1977.

Note: The physical job demands may be different today from what they were decades ago, when these data were collected.

DEFINING THE "HEAVINESS" OF WORK:

Light work:	up to 10 kJ (2.5 kcal)	– 90 beats	—	per minute
Medium work:	about 20 kJ (5 kcal)	– 100 beats	—	per minute
Heavy work:	about 30 kJ (7.5 kcal)	– 120 beats	—	per minute
Very heavy work:	about 40 kJ (10 kcal)	– 140 beats	—	per minute
Extremely heavy work:	50 kJ (12.5 kcal)	– 160 beats or more	—	per minute

"Light" work is associated with a rather small energy expenditure (about 10 kJ/min, including the basal rate) and is accompanied by a heart rate of approximately 90 beats/min. In this type of work, the energy needs of the working muscles are covered by the oxygen that is available in the blood and by glycogen at the muscle. Lactic acid does not build up. During "medium" work, with about 20 kJ and 100 beats/min, the oxygen requirement at the working muscles is still covered, and lactic acid that was initially developed is resynthesized to glycogen during the activity. In "heavy" work, with about 30 kJ and 120 beats/min, the oxygen that is required is still supplied if the person is physically capable of doing such work and specifically trained in the job. However, the lactic acid concentration incurred during the initial minutes of the work is not reduced, but remains till the end of the work period, to be brought back to normal levels after cessation of the work.

With light, medium, and even heavy work, metabolic and other physiological functions can attain a steady state throughout the work period (provided that the person is capable and trained). This is not the case with "very heavy work," wherein energy expenditures are in the neighborhood of 40 kJ/min and the heart rate is around 140 beats/min. Here, the original oxygen deficit increases throughout the duration of work, making intermittent rest periods necessary or even forcing the person to stop working. At even higher energy expenditures, such as 50 kJ/min, associated with heart rates of 160 beats/min or more, the concentration of lactic acid in the blood and the oxygen deficit are of such magnitudes that frequent rest periods are needed, and even highly trained and capable persons may be unable to perform the job through a full work shift.

Note that only "dynamic work" can be suitably assessed by energy demands. "Static" efforts, during which muscles are contracted and kept so, hinder or completely occlude their blood supply by compression of the capillary bed. Thus, the heart makes an effort to overcome the resistance by increasing its rate and the blood pressure; but because the blood flow remains insufficient, relatively little energy is supplied to contracted muscles and consumed there. Hence, although such static effort may be exhausting, it is not well assessed by energy measures.

The engineer determines the "effort" required and how work is to be done and also has control over the work environment. To arrange for a suitable match between capabilities and demands, the engineer needs to adjust the work to be performed (and the work environment, discussed in Chapter 5) to the body's energy capabilities.

Ramp, Stair, or Ladder?

The selection of ramps, stairs, or ladders is a good example of assessing "effort" by oxygen consumption, heart rate, and subjective ratings or preferences. Figure 2–9 indicates resulting recommendations, depending on the angle of ascent. For angles of up to 20 degrees, ramps are preferred; between 20 and 50

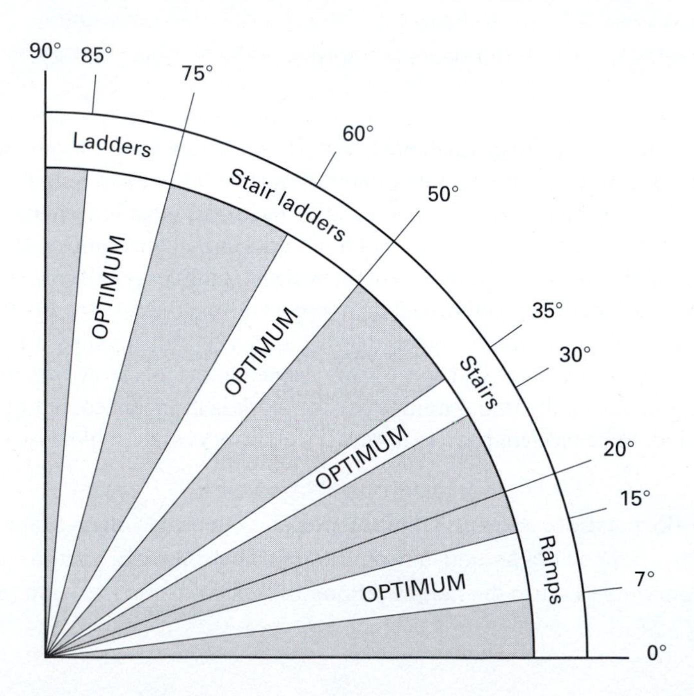

Figure 2–9. Selection of ladders, stairs, or ramps according to the angle of ascent (MIL-STD 1472F).

degrees, stairs; at steeper angles, stair ladders; and for angles above 75 degrees, ladders. The recommended design features are shown in Figure 2–10, but better data for design are desirable (Johnson 1998; McVay and Redfern 1994). On ships, an "alternating tread" ladder is occasionally used instead of a regular ladder (Figure 2–11), because it has been found to be safer and easier to use (Jorna et al. 1989).

☛☛☛ ***What your body does—how and why?***

Sneezing

The sneeze is a very fast expiration of air. It is caused by an irritation in the nasal passage. Its purpose is to clear the nose of cause of irritation.

The sneeze starts when one begins to exhale. The vocal chords come together to plug the windpipe, and a sound evolves. This response is involuntary and difficult to control.

Children sneeze through their noses, which often creates a mess. Adults usually try to send the sneeze on a detour through the mouth.

Yawning

Yawning may be a reaction to either the lack of oxygen, or the accumulation of carbon dioxide, in the lungs or breathing pathways. Yawning occurs particularly when one is inactive, such as when sitting and listening to a boring presentation. The frequency of breathing is diminished, and the "air quality" in the lungs and breathing pathways becomes intolerable. A deep inhalation, facilitated by opening the mouth, flushes all the stale air out, taking carbon dioxide with it and introducing oxygen.

Another theory about yawning stems from the observation that people usually yawn while stretching and that yawning is most frequent in the hour before sleep and after awakening. According to this theory, yawning and stretching were once part of the same reflex; one could even conceptualize yawning as stretching of the face. Indeed, the same drugs that induce yawning also induce stretching.

This does not explain why yawning is "contagious"; perhaps "communal yawning" is a social sign of belonging. The child psychologist Piaget noted that children are susceptible to yawning contagion by age two. People even tend to yawn when reading about yawning or thinking about yawning. Thus, yawning is a stereotyped pattern of action that is itself a releasing stimulus for another stereotyped pattern of action in another individual.

Shivering

As a muscle contracts to generate energy to do work, heat is produced as a side effect. If no work to the outside is done, a muscle contraction generates heat alone. This occurs in shivering as the jaw makes your teeth chatter or other parts of the body are moved by quick muscle contractions. All of this commotion generates heat to keep the body warm in a cold environment.

Muscle cramps

Cramps usually occur in response to a contraction signal sent to a muscle that is stretched. Cramps may arise at in a sports event or while you are asleep or just waking up.

A cramp may involve edema—the accumulation of fluid inside the muscle, but outside its blood vessels. This difference in fluid accumulation generates a pressure, which causes pain. Another explanation of a cramp may be that it is caused by an inadequate supply of oxygen to muscle tissues of persons who suffer from low blood pressure. A third guess is that local controllers or controllers in the central nervous system run amok.

Yet none of these hypotheses explains why the muscle suddenly contracts in a strong and painful manner and maintains that contraction painfully.

Rumbling stomach

Stomach muscles contract and move gas inside the stomach, particularly when it is empty of food. The gases come from air that has been swallowed. The bubbles make audible noises. Putting food into the stomach takes care of the problem.

Snoring

The uvula, a small, soft structure hanging from the palate above the root of the tongue, is at the entrance to the pharynx, which carries air from the nasal cavity to the laryn—the voice box. When the uvula vibrates in the airstream, a snoring sound occurs because the uvula touches other structures. Snoring is most likely to occur when you are lying on your back. Thus, turning over on the side is likely to stop the snoring.

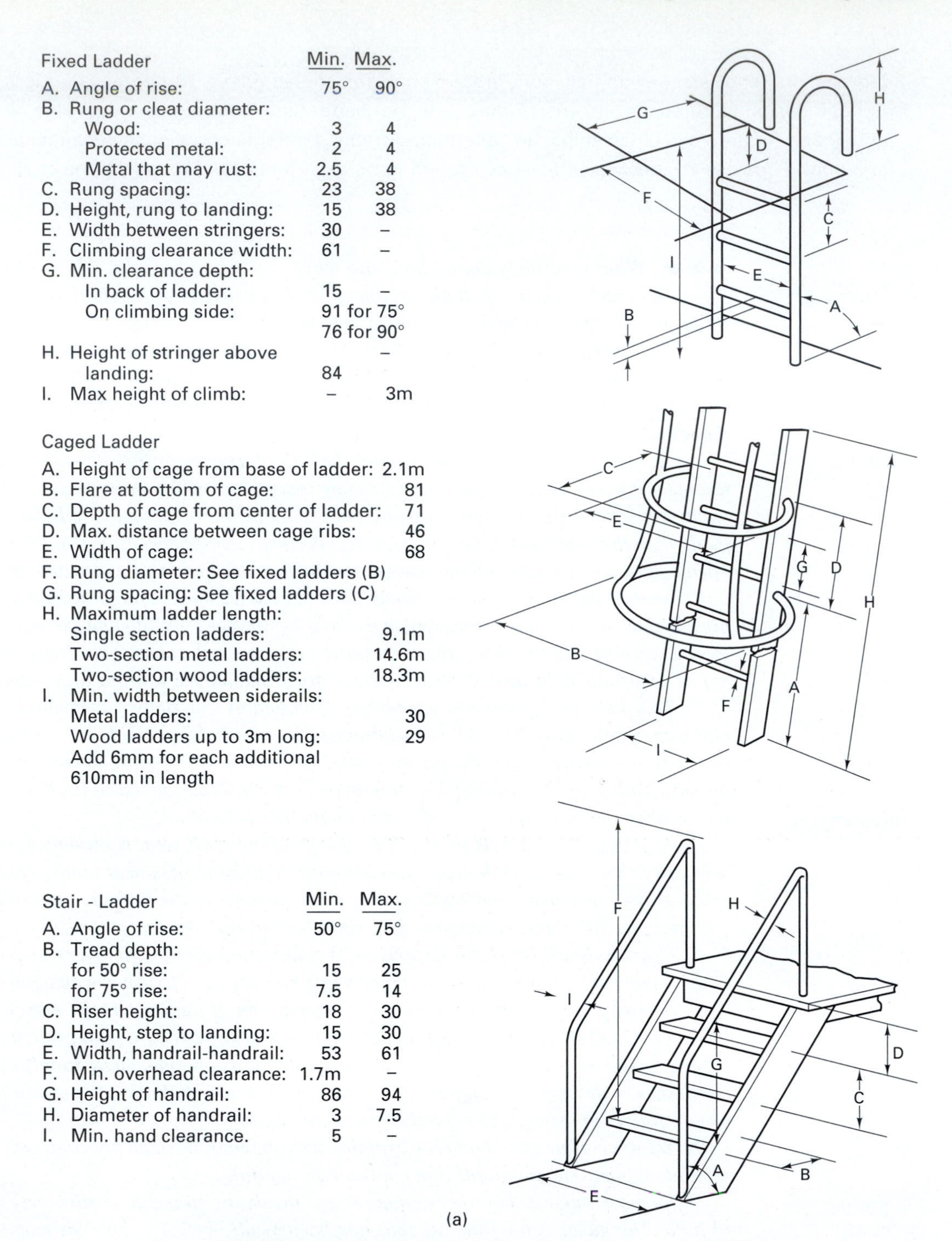

Fixed Ladder	Min.	Max.
A. Angle of rise:	75°	90°
B. Rung or cleat diameter:		
Wood:	3	4
Protected metal:	2	4
Metal that may rust:	2.5	4
C. Rung spacing:	23	38
D. Height, rung to landing:	15	38
E. Width between stringers:	30	–
F. Climbing clearance width:	61	–
G. Min. clearance depth:		
In back of ladder:	15	–
On climbing side:	91 for 75° 76 for 90°	
H. Height of stringer above landing:	84	–
I. Max height of climb:	–	3m

Caged Ladder	
A. Height of cage from base of ladder:	2.1m
B. Flare at bottom of cage:	81
C. Depth of cage from center of ladder:	71
D. Max. distance between cage ribs:	46
E. Width of cage:	68
F. Rung diameter: See fixed ladders (B)	
G. Rung spacing: See fixed ladders (C)	
H. Maximum ladder length:	
Single section ladders:	9.1m
Two-section metal ladders:	14.6m
Two-section wood ladders:	18.3m
I. Min. width between siderails:	
Metal ladders:	30
Wood ladders up to 3m long:	29
Add 6mm for each additional 610mm in length	

Stair - Ladder	Min.	Max.
A. Angle of rise:	50°	75°
B. Tread depth:		
for 50° rise:	15	25
for 75° rise:	7.5	14
C. Riser height:	18	30
D. Height, step to landing:	15	30
E. Width, handrail-handrail:	53	61
F. Min. overhead clearance:	1.7m	–
G. Height of handrail:	86	94
H. Diameter of handrail:	3	7.5
I. Min. hand clearance.	5	–

(a)

Figure 2–10(a). Recommended design values for ladders (in centimeters unless otherwise stated) (MIL-HDBK 759B and STD-1472F).

Regular Stairs	Min.	Max.
A. Angle of rise:	20°	50°
B. Tread depth:	24	30
C. Riser height:	12.5	20
D. Width, (handrail-handrail)		
One-way stairs:	76	–
Two-way stairs:	1.2m	–
E. Min. overhead clearance:	2m	
F. Height of handrail:	84	94
G. Diameter of handrail:	3	7.5
H. Hand clearance:	5	–

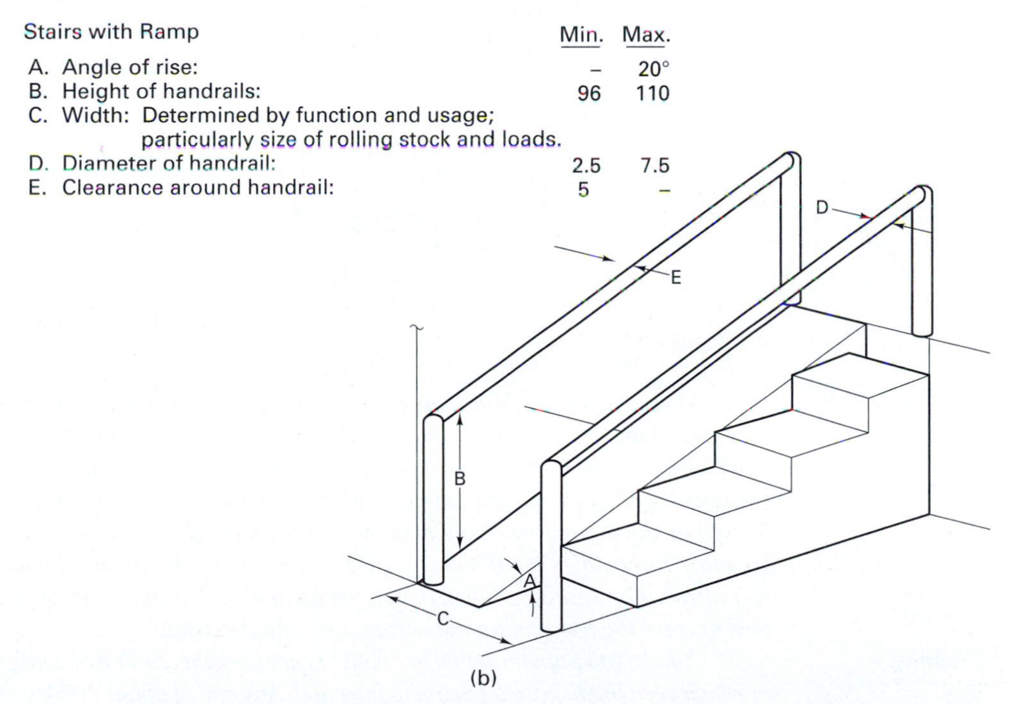

Stairs with Ramp	Min.	Max.
A. Angle of rise:	–	20°
B. Height of handrails:	96	110
C. Width: Determined by function and usage; particularly size of rolling stock and loads.		
D. Diameter of handrail:	2.5	7.5
E. Clearance around handrail:	5	–

Figure 2–10(b). Recommended design values for stairs (in centimeters unless otherwise stated) (MIL-HDBK 759B and MIL-STD-1472F).

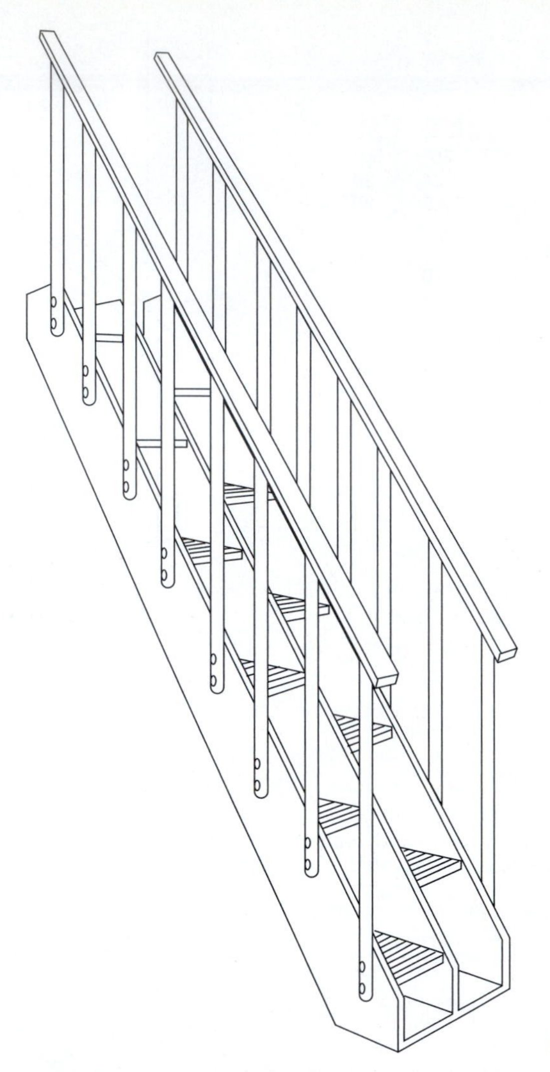

Figure 2–11. Alternating tread stair.

Swelling of ankles

When one sits without moving the legs for a long time, such as in an airplane, swelling (edema) of the lower legs is often experienced, particularly in the feet and in the ankles. The swelling occurs because lymph (a serumlike fluid, mostly plasma) seeps from the capillaries into the interstitial spaces of the connective tissue, making the tissue expand. The resulting pressure caused by the expanding tissue hinders the lymphatic flow, as well as the blood flow through the capillaries and other small blood vessels. To prevent swelling, one should move the lower legs and feet often while sitting, or even better, get up and walk around. After swelling has occurred, it is best to raise the feet and let the accumulated fluid dissipate.

Catching a cold

Contrary to popular belief, a "cold" is not brought about by a chill, wetness, or a draft, none of which even predispose a person to infection. Any one of about 200 known cold viruses can infect the upper respiratory tract, including the nose, sinuses, and throat, at any time of the year. The increased occurrence of colds during the colder weather merely stems from the fact that people spend more time crowded together indoors, making it easy for viruses to spread from person to person, mostly by direct contact such as shaking hands. The "flu" (influenza) develops abruptly from fall through spring; it is also caused by a virus, but is generally more dangerous than a cold.

SUMMARY

The human body converts foodstuffs, in the presence of oxygen, into energy, which then is used to perform work. The respiratory system absorbs oxygen into the bloodstream and extracts from the blood carbon dioxide, water, and heat, which are expelled into the exhaled air. The breathing rate and breathing volume are not suitable for measuring the physical effort of the body.

The circulatory system transports oxygen and energy carriers (mostly glycogen) to various consumer organs (e.g., muscles) and moves metabolic by-products (CO_2, water, and heat) to the lungs and (water and heat) to the skin for dispersion. The heart pumps the blood through the system. The number of heartbeats per minute provides a convenient and reasonably accurate measurement of the circulatory efforts and for the physical workload of the body in general.

The metabolic system breaks the energy carriers of food into molecules that the body can use. Carbohydrate is broken into polysaccharides, glycogen, and glucose. Fat is broken into glycerol and fatty acids. Protein is broken into amino acids. Glucose and glycogen are the most rapidly used energy carriers, while fat serves to store energy and is used mostly in long lasting efforts.

The metabolic process of converting polysaccharides and fats into energy used by the muscles occurs in several steps, some of which are anaerobic; yet, overall, oxygen must be provided. Since it is known how much oxygen must be furnished for the conversion of each energy carrier, the measurement of oxygen consumption is an accurate means of assessing the body's energy needs. Oxygen consumption and heart rate are highly correlated.

Another way to assess the workload of the body is to subjectively rate the perceived effort. This procedure has been well standardized and can be used in combination with, or instead of, the objective measures.

It is one of the tasks of the ergonomist to keep the requirements of a job, in terms of energy demands, within reasonable limits. This can be done if the energy demands of the job are measured and matched with the energy capabilities of the operator.

CHALLENGES

How would special training through exercise, or the development of deficiencies by disease, of any one of the respiratory, circulatory, and metabolic systems affect a person's ability to perform work, either of the physical, strenuous type or of the psychological, mental type?

Which events or functions in the respiratory system are easily measurable by the ergonomist and can be used to assess work-related loading of the body?

What are the specific tasks of the blood in supporting the functions of the working body?

By what means can the heart increase blood flow into the aorta? How is the blood supply to a working organ regulated? What are the functions of an artificial pacemaker for the heart?

Why do well-trained athletes often have a very low resting heart rate?

What specific features limit the use of the heart rate as an indicator of circulatory loading, particularly of metabolic efforts?

The energy-balance equation does not specify either the duration of body processes or the optimal period required to observe them. How does that affect its validity and use?

How would the consideration of energy storage or loss affect the use of the energy-balance equation?

What are the limitations of the energy-balance equation that is purported to be a measure of human energy efficiency? How can one "become fat" if one eats no fat?

What are the important functions of fat deposits in the body?

What are the two major means by which the body stores energy for use as needed?

Which kind of energy storage in the body is used to support a physical effort?

Why does one have anaerobic processes involved if, indeed, the overall process of metabolism is aerobic?

Is it reasonable to assume that "average persons" perform everyday activities?

Various tables of energy consumption for certain activities are found in the literature and also given in this chapter. What limitations and assumptions apply to these tables?

Are there differences in "experimental control" that must be applied if either "objective" or "subjective" measures are taken to assess the workload associated with certain activities?

Chapter 3

How the Mind Works

OVERVIEW

The traditional psychological model of human perception, cognition, and action postulates a sequential system: Input is sensed and then processed, and output follows. While this model has been criticized, no suitable substitute has replaced it yet and, therefore, nearly all currently available information is based on the sequential model.

The nervous system controls body functions. It receives information from body sensors via its afferent peripheral components. The information is then processed, decisions are made, and control signals are generated in the central nervous system. Finally, the signals are transmitted to body organs, particularly muscles, in the efferent peripheral part of the system.

Job and environmental conditions can be very stressful, not only in routine tasks in "normal" environments, but especially in confined and dangerous environments, such as during space exploration. The assessment of the workload is of importance for subsequent design recommendations.

INTRODUCTION

☛☛☛ *"Of course, the brain is a machine and a computer. . . . But our mental processes, which constitute our being and life, are not just abstract and mechanical but personal, as well—and, as such, involve not just classifying and categorizing, but continual judging and feeling also. If this is missing, we become computer-like. . ." (Sacks 1990, p. 20).* ☚☚☚

System

In the system concept of engineering psychology, the human is considered a receptor, processor, and generator of information or energy. Input, processing, and output follow each other in sequence. The output can be used to run a machine, which may be a simple hand tool or a spacecraft. This basic model is depicted in Figure 3–1.

Human-technology system

The actual performance of the combined human–machine system (in the past often called a man-machine system), or human-technology system, is monitored and compared with the desired performance. Hence, one or more feedback loops connect the output side (or one of its elements) with the input side of the system. The difference between output and input is registered in a comparator, and corrective actions are taken to minimize any output–input difference. The human in this system makes comparisons, decisions, and corrections.

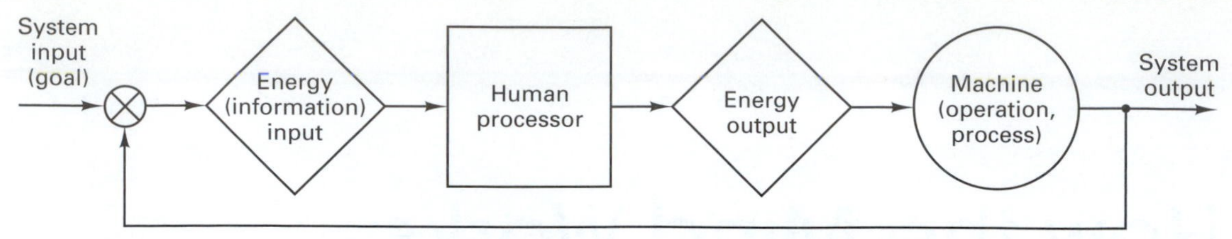

Figure 3–1. The human as energy or information processor.

This "human processor" is the object of research, either to understand basic human functions or to observe human actions and reactions within the system. (Especially in the 1970s and '80s, information theory, including signal channeling and processing, were major research topics. These subjects are beyond the scope of this book, but many related publications in the field of cognitive psychology treat them.) Input and output are the sites of application of ergonomics and human-factors engineering. The design of the machine is the classical engineering task, although with significant help from the ergonomist.

THE "TRADITIONAL" AND THE "ECOLOGICAL" CONCEPTS

Linear sequence of stages

In the traditional concept of engineering psychology, our activities can be described as a linear sequence of stages, from perceiving to encoding to deciding to responding. Research is done separately on each of these stages, on their substages, and on their connections. Such independent, stage-related information is then combined into a linear model to provide information for the engineering psychologist.

Simultaneous stages

"Ecological" psychologists believe that the linear model is invalid; they consider human perception and action to be based on simultaneous, rather than sequential, interactions (Brunswik 1956, Flach 1989, Gibson 1966, Meister 1989, Vincente and Harwood 1990, Meyer and Kieras 1997). Two major concepts in the ecological approach are affordance and its perception.

Affordance is the property of an environment that has certain values to a human. Flach gives the example of a stairway, which affords passage for a person who can walk, but not for a person confined to a wheelchair. Thus, passage is a property of the stairway, but its affordance value is specific to the user. Accordingly, ergonomics or human engineering provides affordances.

The second concept is that information about affordances can be perceived directly and simultaneously by various human senses as part of the intimate coupling of perception and action. Thus, the closed-loop coupling between perception and action of a human in a certain environment is not modeled as a simple linear sequence of the stages perceiving–encoding–deciding–responding; it is not the case that encoding precedes deciding, which in turn precedes responding, but rather, information is distributed throughout the closed-loop system.

If one follows this concept, then indeed, research on stages of behavior that are assumed to follow each other in sequence does not realistically explain human behavior in a technological system. This then leads to "the unfortunate conclusion that not only much of the data produced by traditional academic psychology [are] irrelevant to system design, but [they] may be irrelevant to the science of behavior" (Flach 1989, p. 3). Instead, the study of behavior must be at the level of an "ecological human–environment system." This approach would require fundamentally new models of information, cognition, and performance assessment, different from those associated with traditional psychology. Yet, more than decade after Flach's words, behavioral knowledge is still almost completely based on the traditional sequential-system concept.

ORGANIZATION OF THE NERVOUS SYSTEM

Central and Peripheral Nervous System

Anatomic division

Anatomically, one divides the nervous system into three major subdivisions. The central nervous system (CNS) includes the brain and spinal cord; it has primarily control functions. The peripheral nervous system (PNS) includes the cranial and spinal nerves and transmits signals, but usually does not control anything. The essential function of the PNS is to carry information from receptors throughout the body into the CNS and back out to effectors.

Autonomic system

The *autonomic* nervous system is a specialized system formed from components of both the CNS and the PNS and is responsible for general activation of the body, emergency response, and emotion. The autonomic nervous system is not generally under conscious control. It consists of the sympathetic and the parasympathetic subsystems, which together regulate involuntary functions, such as those of smooth and cardiac muscle, blood vessels, digestion, and glucose release in the liver.

The sympathetic subsystem is generally responsible for arousal mechanisms, the parasympathetic for relaxation mechanisms. When an organism is startled or experiences a "flight or fight" response, the process is mediated by the sympathetic subsystem. Circulatory processes, digestive processes, the heart rate, the respiration rate, blood pressure, and other mechanisms respond to threat via autonomic arousal in a process that has been called a "sympathetic storm." (It has been hypothesized that excess, inappropriate, or maladaptive sympathetic arousal may play a role in maladaptive stress response, chronic anxiety disorders, and even certain medical problems.)

Functional division

Functionally, one divides the nervous system into two major subdivisions: the autonomic system (just discussed), and the somatic (Greek: *soma,* body) nervous system, which controls mental activities, conscious actions, and skeletal muscle.

While at the "Museum" in Alexandria, Egypt, about 280 B.C., Herophilus distinguished sensory and motor nerves. He described the liver and the retina of the eye. Erasistratus (about 250 B.C.) distinguished between the cerebrum and the cerebellum of the brain and thought that the many convolutions of the human brain, more numerous than in any other animals, were related to superior intelligence. In 1810, the German physician Franz Joseph Gall (1758–1828) published the first volume of a treatise on the nervous system. He stated that the gray matter on the surface of the brain and in the interior of the spinal cord was the active and essential part of human thinking and that the white matter was just connecting material. Gall also believed that the shape of the brain had something to do with mental capacity; he went so far as to relate the shape of the brain and the skull to emotional and temperamental qualities, which started the pseudoscience of phrenology. The French anthropologist Pierre-Paul Broca (1824–1880) demonstrated Gall's belief that the brain controlled different parts and functions of the body (Asimov 1989).

Brain and Spinal Cord

Brain

The human brain weighs about 6.6 kg, has a volume of about 1,250 cm^3, and is protected by the bony case of the skull. The brain is suspended within the cerebrospinal fluid, which provides both mechanical protection via shock absorption and certain aspects of nutritional support to the brain.

The brain is usually divided into the forebrain, midbrain, and hindbrain. Of particular interest with regard to the neuromuscular control system is the *forebrain* with the cerebrum, which consists of the two (left and right) cerebral hemispheres, each divided into four lobes: the frontal, temporal, parietal, and occipital lobes. A detailed discussion of lobe-specific brain functions is beyond the scope of this text, but together the lobes control basic attention and consciousness, fine and gross motor control, verbal fluency, language functions, various aspects of intelligence (e.g., comprehension, memory functions, abstract thought, planning, and sequencing), and sensory perception. The multifolded cortex enwraps most of the cerebrum. The cortex controls voluntary movements of the skeletal muscle and interprets sensory inputs. The basal ganglia of the *midbrain* are composed of large pools of neurons, which control semivoluntary complex activities such as walking. Part of the *hindbrain* is the cerebellum, which integrates and distributes impulses from the cerebral association centers to the motor neurons in the spinal cord and thus coordinates muscular activity. Figure 3–2 is a schematical sketch of the major part of the brain.

Spinal cord

The spinal cord is an extension of the brain, and many aspects of human behavior are organized and integrated within it as well. The spinal cord enables communication from the brain outward to the body and receives sensory information that is relayed inward to the brain. It also organizes certain behaviors, largely without utilizing higher functions of the CNS, principally by means of *reflexes.* In a reflexive behavior, sensory information enters a layer (or layers) of the spinal cord and then directly links to a motor nerve, which then exits to the PNS. The role of the CNS in controlling reflexes is generally limited to determination of speed or ease of triggering of a reflex.

Reflex

Reflexes are divided into three categories: *superficial, deep,* and *special.* Superficial reflexes (also called cutaneous reflexes) are elicited by stimulation of the skin. For example, when the sole of the foot is lightly scratched, the toes flex. A deep reflex is demonstrated by the stretching of a tendon after striking the patella with a soft hammer. The stretching of the tendon results in a reflexive muscle contraction, which causes the knee to jerk. Similar deep reflexes are also

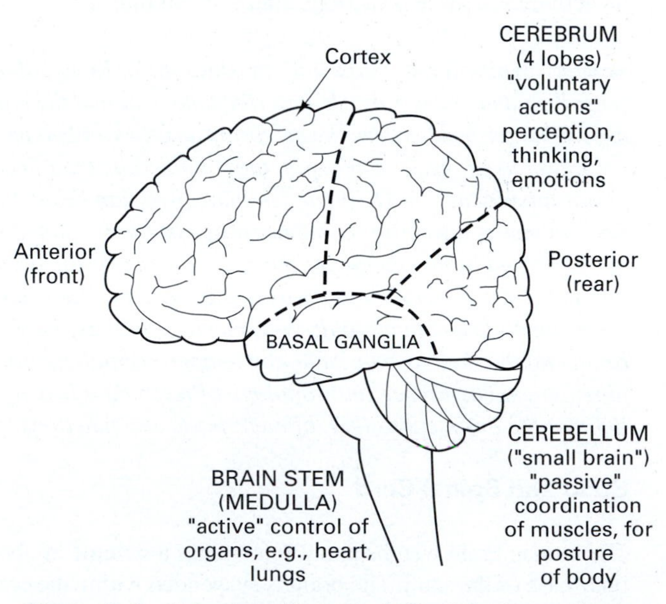

Figure 3–2. Side view (from the left) of the human brain.

elicited in the jaw, biceps, triceps, internal and external hamstrings, pectoralis, adductor, finger flexors, and several other muscle systems. Special reflexes are often used to determine developmentally and clinically significant functions and dysfunctions. For example, the extensor plantar response (Babinski sign) consists of dorsiflexion of the great toe and fanning of the remaining toe when the plantar surface of the foot is stimulated.

The uppermost section of the spinal cord contains the 12 pairs of *cranial* nerves, which serve structures in the head and neck, as well as serving the lungs, heart, pharynx, larynx, and many abdominal organs. The cranial nerves control eye, tongue, facial movements and the secretion of tears and saliva. Their main inputs are from the eyes, the taste buds in the mouth, the nasal olfactory receptors, and touch, pain, heat, and cold receptors in the head. Thirty-one pairs of *spinal* nerves extend from the brain to the appropriate vertebrae and pass out between them to serve defined sectors of the rest of the body. Nerves are mixed sensory and motor pathways, carrying both somatic and autonomic signals between the spinal cord and the muscles, articulations, skin, and visceral organs. Figure 3–3 shows how the spinal nerves emanating from sections of the spinal column (called spinal nerve roots) innervate defined areas of the skin (called dermatomes). Figure 3–4 associates major nerves of the body with their respective areas of cutaneous sensitivity.

Sensors

Somesthetic sensors

The CNS receives information from certain internal receptors, called *interoceptors,* which report on changes within the body—changes in digestion, circulation, excretion, hunger, thirst, sexual arousal, feeling well or sick. *Exteroceptors* respond to sight, sound, touch, temperature, electricity, and chemicals. Since all of these sensations come from various parts of the body (Greek: *soma*), external and internal receptors together are also called *somesthetic* sensors.

Proprioceptors

Other internal receptors called *proprioceptors* include the muscle spindles—nerve filaments wrapped around small muscle fibers—which detect the amount of stretch of the muscle. The Golgi receptors are associated with muscle tendons, detecting their tension, which corresponds to the strength of contraction of muscle. (See Chapter 1.) Ruffini corpuscles are kinesthetic receptors located in the capsules of articulations. They respond to the degree of angulation of the joints (joint position), to change in general, and also to the rate of change.

Vestibular sensors

The sensors in the *vestibulum* also are proprioceptors. They detect and report the position of the head in space and respond to sudden changes in its attitude. This is done by sensors in the semicircular canals, of which there are three, located orthogonally to one another. To relate the position of the body to that in the head, proprioceptors in the neck are triggered by displacements between the trunk and head.

Visceroceptors

Another set of interoceptors, called visceroceptors, report on events within the visceral (internal) structures of the body, such as organs of the abdomen and chest, as well as on events within the head and other deep structures. The usual modalities of visceral sensations are pain, burning sensations, and pressure. Since the same sensations are also provided by external receptors, and because the pathways of visceral and external receptors are closely related, information about the body is often integrated with information about the outside.

External receptors provide information about the interaction between the body and the outside via the senses of sight (vision), sound (audition), taste (gustation), smell (olfaction), temperature, electricity, and touch (taction). (See Chapter 4 for detailed information on the senses.) Several of these sensations are of particular importance for the control of muscular activities: Touch, pressure, and pain, for instance, can be used as feedback mechanism to the body regarding the direction and intensity of muscular activities transmitted to an outside object. (See Chapter 2.) Free nerve endings, Meissner's and Pacinian corpuscles, and other receptors are located

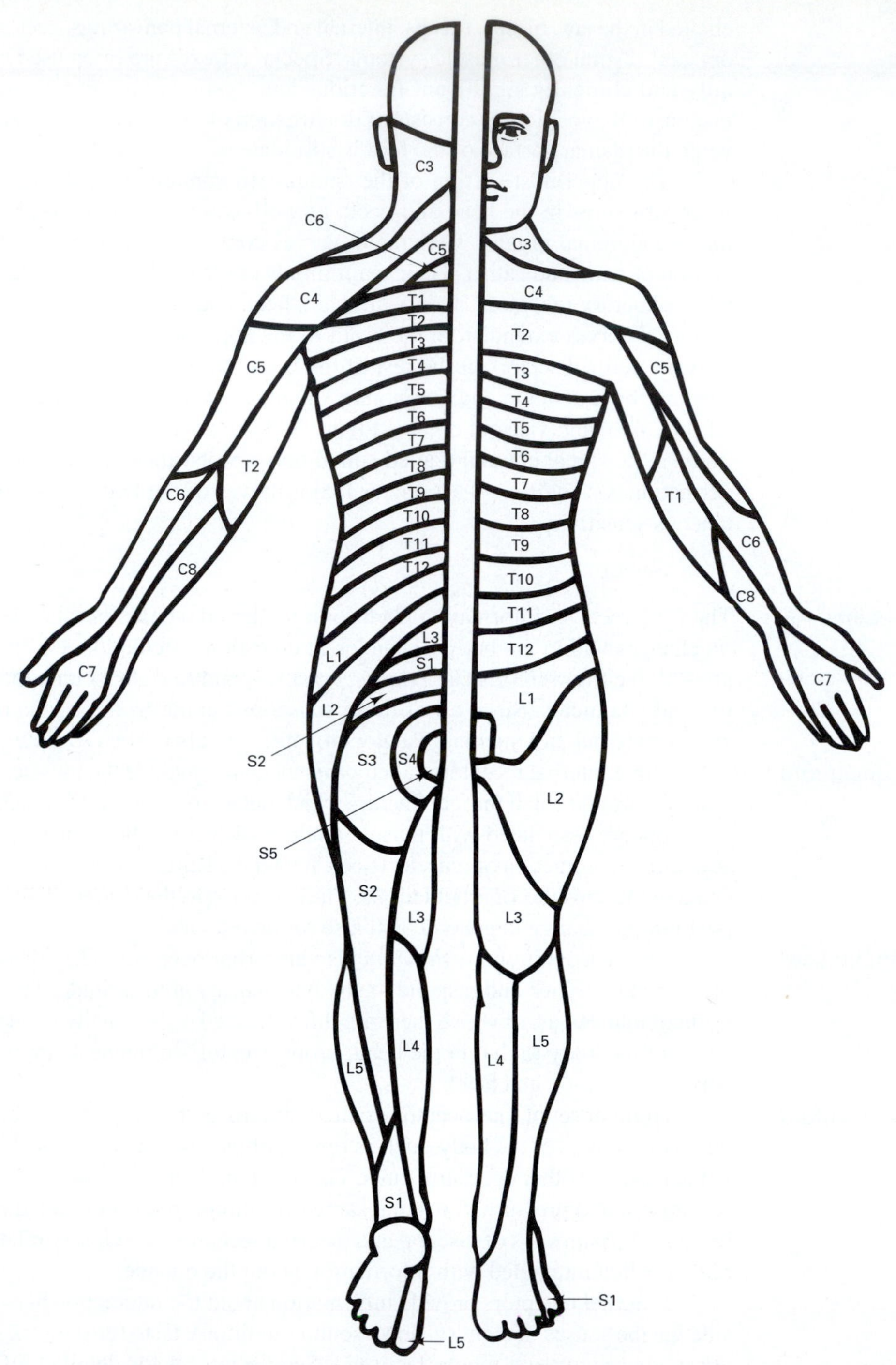

Figure 3–3. Sensory dermatomes with their spinal nerve roots. (Modified from Sinclair 1973.)

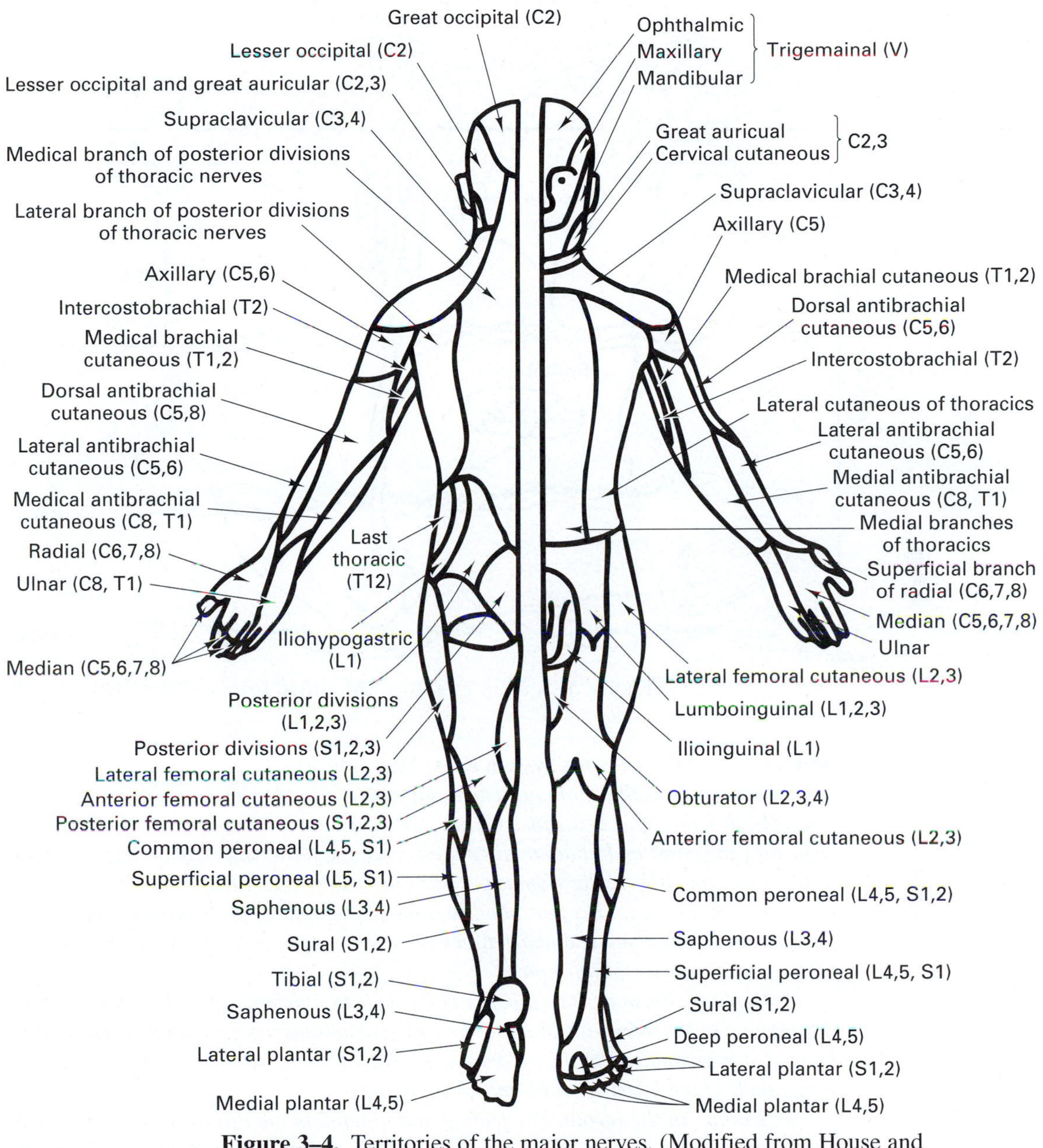

Figure 3–4. Territories of the major nerves. (Modified from House and Pansky 1967.)

throughout the skin of the body, although in different densities. They transmit the sensations of touch, pressure, and pain. Since the nerve pathways from the free endings interconnect extensively, the sensations that are reported are not always specific to a modality; for example, very hot or very cold sensations can be associated with pain, which may also be caused by hard pressure on the skin. Figure 3–5 sketches receptors in the skin.

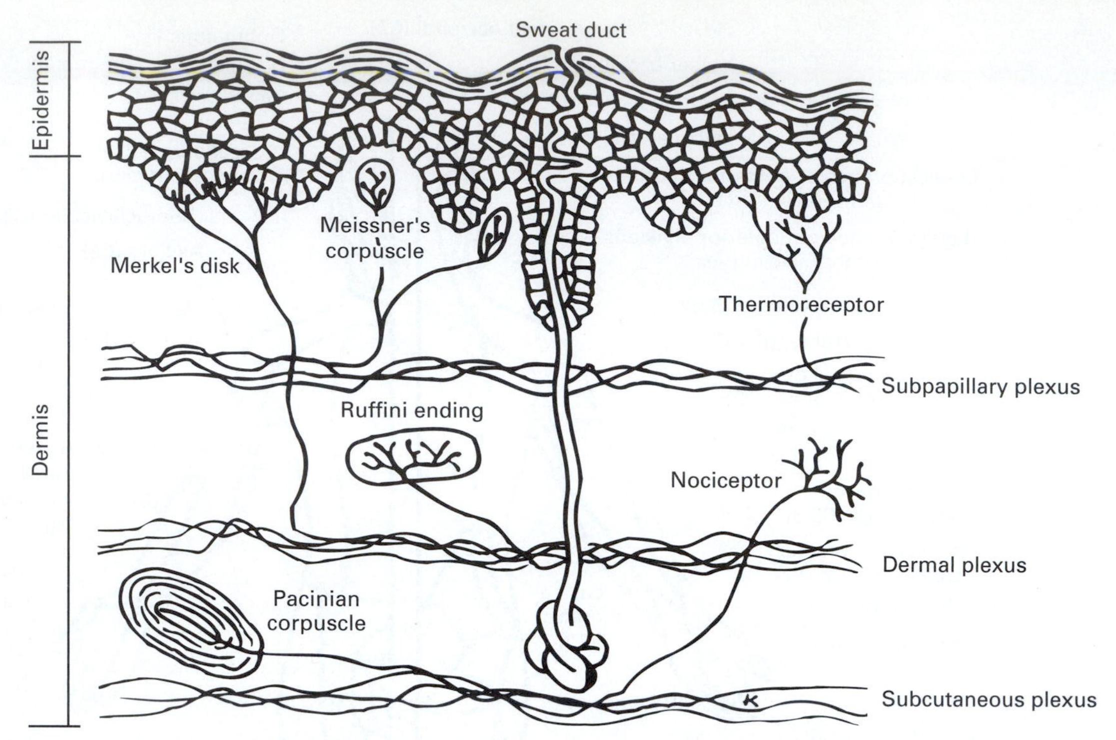

Figure 3–5. Skin receptors. (Modified from Griffin 1990).

What the nervous system does: How and why?

Butterflies in the stomach. *The event underlying "butterflies in the stomach" is not clear. It may be that the heart beats faster when a person is under stress, or perhaps the diaphragm that separates the chest and abdominal cavities might be fluttering. The butterfly feeling may in fact not be an event close to the stomach at all: The ability to localize a sensation in one's own body, especially in the torso, is often poor, and the perception of pain in this area is sometimes "referred pain." For example, a heart attack is sometimes felt in the arm, and shoulder pain may actually come from the diaphragm.*

Blushing. *Blushing is a response of the body to embarrassment by filling the capillaries near the surface of the skin with blood. Why this reaction occurs is unknown. How it occurs is well known: It is stimulated by the autonomic nervous system, which controls involuntary body functions, such as the heart, glands, and blood vessels.*

"Lump" in the throat. *The feeling of a "lump" in the throat is another example of the action of the autonomic nervous system. When a person is threatened, the sympathetic nervous system becomes active and, in this case, makes saliva thick. The thickened saliva is felt as something interfering with swallowing, like a lump in the throat. It can be difficult to distinguish swallowing difficulties due to organic impairment, such as growths or neurological events, from the perception of a lump in the throat due to autonomic arousal in chronic anxiety conditions.*

Hiccups. *Hiccups are breathing gone out of control. In normal breathing, impulses regularly progress along the phrenic nerve leading to the diaphragm that separates the abdomen from the chest cavity. The signals cause the diaphragm to contract and reduce the lung volume for expiration. If there is a sudden burst of nervous impulses, the diaphragm contracts abruptly, generating quick expirations. Hiccups can occur with no apparent stimulus, but tend to happen at* *certain times, such as when a person has just eaten heavily. While there are folk remedies for hiccups, such as a startling noise, hiccups usually go away by themselves.*

Goose bumps. *Goose bumps are the puckering of skin around hair follicles due to the contraction of a muscle at the base of the follicle. The hair stands up and, together with the now enlarged and irregular skin surface, traps air. This reaction is a response to cold when the body tries to maintain a warm and insulating layer of air in order not to lose heat. Goose bumps are involuntary events generated by the sympathetic nervous system. They can appear when a person is stricken by fear. Perhaps they are an archaic reaction to create a "thicker fur," which might protect better against the bite of an animal or against cold. It is not clear, however, why we do not get goose bumps on our faces.*

Head jerk. *When you start to doze off, your head drops. Suddenly it jerks back up. Apparently, a sensor in a joint or in a neck muscle received a signal of excessive stretch and, in a reflex, tightened muscles to reduce the stretch. Why this occurs is not clear.*

Bedtime twitch. *As in the head jerk, muscles relax as you begin to fall asleep. The brain starts off being very active, but apparently, there is a phase in falling asleep when muscle contraction signals suddenly come through. When nerve fibers leading to the leg fire in unison at the onset of sleep, the leg twitches as a whole, resulting in a "hypnic jerk." The purpose of this twitch is unknown. Although some people experience the phenomenon more than others, their experience is unpredictable, unlike "myoclonic jerks," which are spasms that occur at regular intervals during the sleep cycle.*

The "Signal Loop"

Input, process, output

Following the traditional psychological concept, one can model the human as a processor of signals in some detail, as shown in Figure 3–6. Information (energy) is received by a sensor, and a signal is sent along the afferent (sensory) pathways of the PNS to the CNS, where it is compared with information stored in the brain's short- or long-term memory. The signal is processed in the CNS, and an action (or no action at all) is chosen. Appropriate (feedforward) impulses are then generated and transmitted along the efferent (motor) pathways of the PNS to the effectors (the voice, hand, etc.). Of course, many feedback loops exist, although only a few are shown in the model.

Input transducers

Both sides of the processor model can be analyzed further. Figure 3–7 shows how distal stimuli provide information that may be visual, auditory, or tactile. To be sensed, the stimulus must appear in a form to which human sensors can respond; it must have suitable qualities and

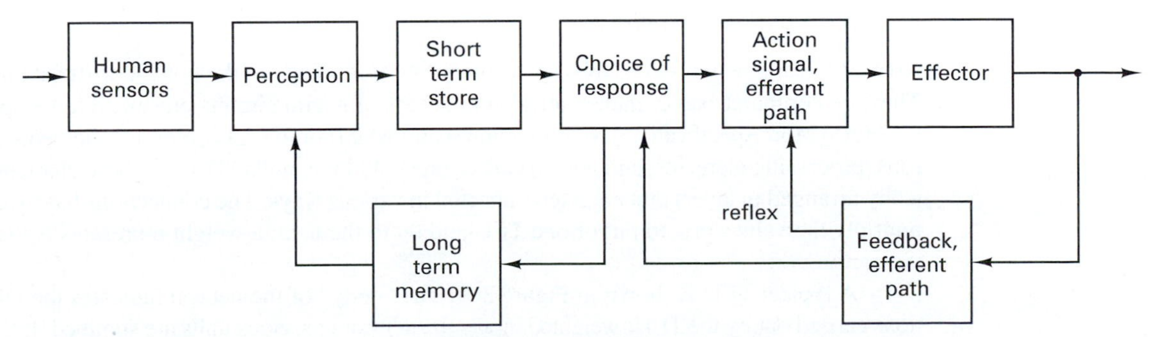

Figure 3–6. The human processor.

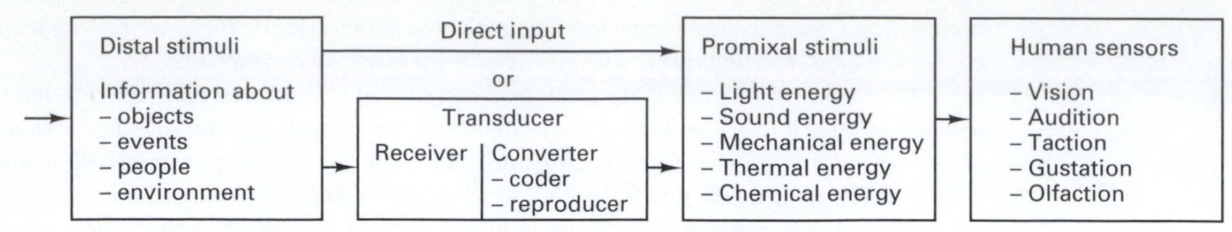

Figure 3–7. Energy input side.

quantities of electromagnetic, mechanical, electrical, or chemical energy. If the distal events do not generate proximal stimuli that can be sensed directly, the distal stimuli must be transformed into energies that can trigger human sensations. To accomplish this, the ergonomist designs transducers. For example, a display of some kind, such as a computer screen, a dial, or a light, can serve as transducer.

Output transducers

On the output side, the actions of the human effector (such as the hand or foot) may directly control the "machine," or one may need another transducer. For example, the movement of a steering wheel by the human hand may be amplified by auxiliary power (power steering). Figure 3–8 portrays the model. Of course, recognizing the need for a transducer and providing information for its suitable design is again a primary task of the ergonomist/human-factors engineer.

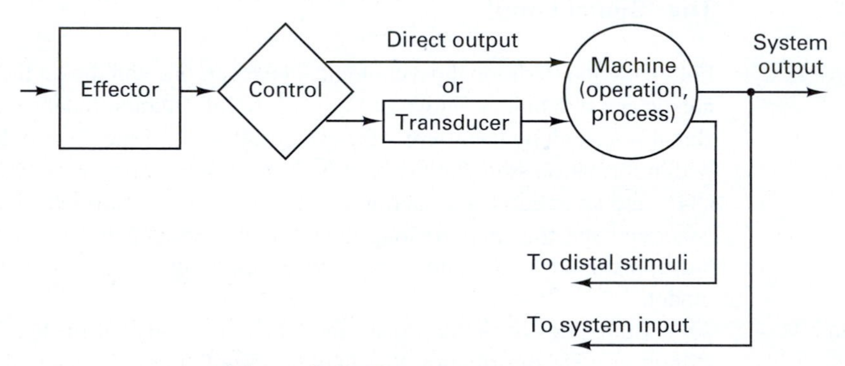

Figure 3–8. Energy output side.

Artificial Neural Networks

Artificial neural networks (ANNs) are a computing technology in the field of artificial intelligence. They are computer-based mathematical simulations of neural circuits presumed to be operating in the brain. More specifically, ANNs are combinations of (nonlinear) computing elements (e.g., neurons, processing elements, and nodes) called threshold logic units (TLUs). These elements are typically arranged in layers that are interconnected in various ways. The connections have weights that multiply the values passed out of one TLU and on to the next; a weight represents the strength of the connection.

A typical TLU is shown in Figure 3–9. The "body" of the neuron indicates the major functions carried out by the TLU: weighted inputs (a_i, ω_i) from previous units are summed. If the weight-

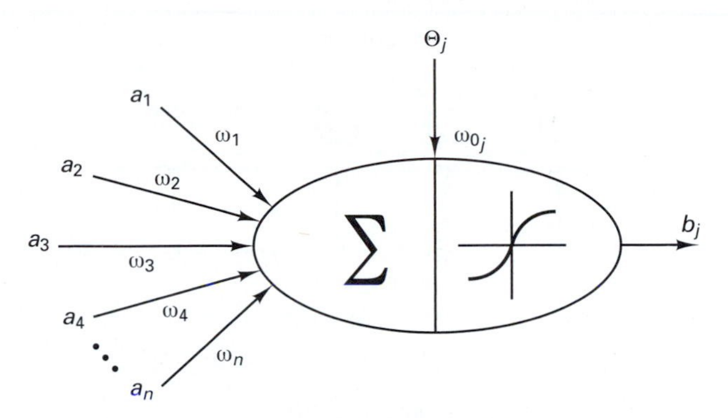

Figure 3–9. Diagram and notation for a typical threshold logic unit of an ANN (Spelt 1991).

ed, summed inputs exceed a threshold value (Θ_j, ω_{oj}), the sum is passed through a ramp function, which creates the output from that neuron, to be passed to the next neuron(s). The activation function can be of various forms, such as binary, linear, or sigmoid. Depending on the activation function, a neuron can be either excitatory (produce a positive output) or inhibitory (produce a negative output).

Typically, a number of TLUs are assembled into an ANN, which serves as a filtering or learning device. The arrangement of processing elements into layers or fields and the topography of the interconnections are two characteristics that define an ANN. Figure 3–10a shows an input vector (**x**) connected to an output vector (**y**) by a single-layer network with bidirectional lateral connections. Such a one-layer configuration could serve, for example, as an ANN with content-addressable memory. Figure 3–10b represents a feedforward network design with an input and an output layer, hence called a two-layer net. Figure 3–10c depicts a fully connected feedforward net with one hidden layer, called a backpropagation (or "backprop") network.

ANNs serve as adaptable memory systems; that is, they learn and then store the acquired knowledge in weight matrices. ANNs can learn to classify either spatial or spatiotemporal patterns. The ability to perform, after training, with only partial input information is one of the strengths of ANNs and is called "fault tolerance." It results from the fact that memory is distributed among the weights in the matrices, in a manner similar to what is presumed to occur in human memory.

Ergonomic Uses of Nervous Signals

Given our present understanding of the human mind, and with the limited technologies that are currently available to pick up and use nervous signals, controlling devices via thought is still a ways off. Yet, some successes based on empirical knowledge have been achieved in rehabilitation engineering. For example, one may electrically stimulate paralyzed muscles for feedforward or use stimulation of skin sensors for feedback control of prostheses.

EEG

Afferent impulses are difficult to identify and separate from efferent signals, mostly because of the anatomical intertwining between the sensory and motor pathways of the peripheral

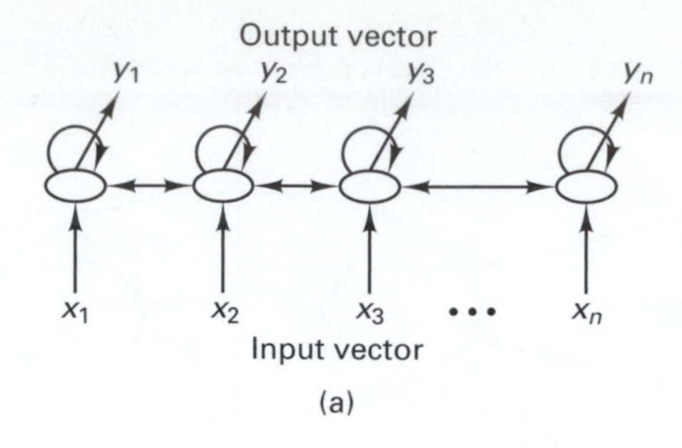

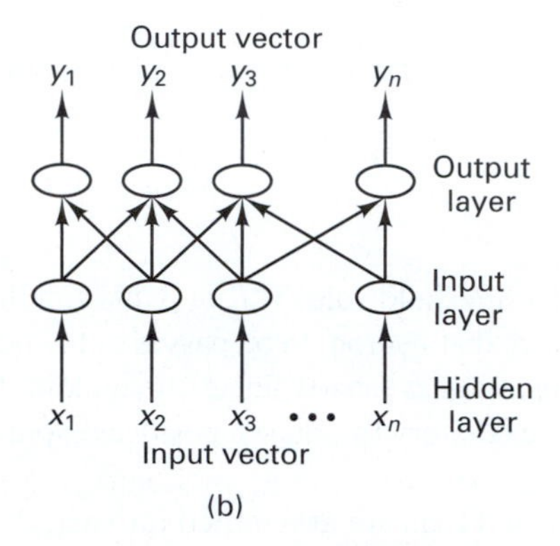

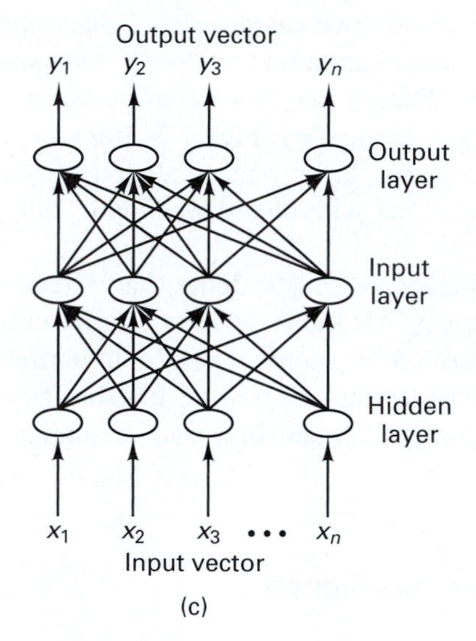

Figure 3–10. Various types of ANN configurations; See text for description. (Reprinted with permission from "Introduction to artificial neural networks for human factors," by Phillip F. Spelt, *Human Factors Society Bulletin,* Vol. 34, No.7, 1991. Copyright 1991 by the Human Factors and Ergonomics Society, Inc. All rights reserved.)

nervous system. Electrical events occurring in the cortex (the *encephalon,* Greek for "wrapping") that covers the central brain are recorded in the electroencephalogram (EEG) by means of surface electrodes. The recorded signals are empirically interpreted and are used particularly to classify sleep stages. (See Chapter 6.) Perhaps further research will lead to better understanding of the meaning of such "brain waves" and of their quantitative values.

EMG

The effects of motoric signals arriving via alpha-motoneurons at muscle fibers are picked up (via indwelling or surface electrodes) as an electromyogram (EMG) at skeletal muscle in general (see Chapter 1), as an electrooculogram at eye muscle (see Chapter 6), and as an electrocardiogram (ECG or EKG) at heart muscle (see Chapter 2). Recording these signals is fairly easily done (Basmajian and DeLuca 1985; Kumar and Mital 1996; Perotto 1994; Soderberg 1992).

Responding to Stimuli

The time that passes from the appearance of a proximal stimulus (e.g., a light) to the beginning of an effector action (e.g., a movement of the foot) is called the *reaction time*. The additional time required to perform an appropriate movement (e.g., stepping on a brake pedal) is called the *motion*, or movement, *time*. Adding the motion time to the reaction time results in the *response time*. (Note that, in everyday use, these terms are often not clearly distinguished.)

Reaction times The experimental analysis of the reaction time goes back to the very roots of experimental psychology: Many of the basic results were obtained in the 1930s, with additional experimental work done in the 1950s and '60s (Wargo 1967). Innumerable experiments have been performed; hence, many different tables of such times have been published in engineering handbooks. Some of these tables apparently have been consolidated from various sources; however, the origin of those data, the experimental conditions (e.g., the intensity of the stimulus) under which they were measured, the accuracy of the measurements, and the subjects who participated are no longer known.

The following table of reaction times is typical of generally used, but fairly dubious, information, often applied without much consideration or confidence:

APPROXIMATE MINIMAL REACTION TIMES

electric shock: 130 ms
touch, sound: 140 ms
sight, temperature: 180 ms
smell: 300 ms
taste: 500 ms
pain: 700 ms

There appears to be little practical time difference in reactions to electrical, tactile, and sound stimuli. The slightly longer reaction times for sight and temperature stimuli may be well within the measuring accuracy or within the variability among persons. However, the time following a smell stimulus appears distinctly longer and the time for taste again considerably longer, while it takes by far the longest to react to the infliction of pain.

Time delays Time passes between the appearance of a signal on the input side at a given section of the nervous system and its re-appearance on the output side. Time delays, or lags, occur at the sensor, in afferent signal transmission, in central processing, in efferent signal transmission, and, finally, in muscle activation. Wargo (1967) concluded from a review of the literature that the fastest possible hand reaction times depended on a series of delays that occur between the start and arrival of a signal at different sections of the nervous system.

ESTIMATED TIME DELAYS

at receptor: 1 to 38 ms
along the afferent path: 2 to 100 ms
in CNS processing: 70 to 100 ms
along the efferent path: 10 to 20 ms
muscle latency and contraction: 30 to 70 ms

Simply adding the shortest times leads to the theoretically shortest possible delay. Of course, in reality, there is little reason to assume a situation in which all the delays are shortest, although that is the condition desired by human-factors engineers. The best chances to reduce delays are in the afferent path length and in the CNS processing time.

SIMPLE AND CHOICE REACTION TIMES

If a person knows that a particular stimulus will occur, is prepared for it, and knows how to react to it, the resulting reaction time (RT) is called the *simple reaction time.* Its duration depends on the modality and intensity of the stimulus.

If one stimulus out of several possible stimuli occurs, or if the person has to choose among several possible reactions, one speaks of the *choice reaction time.* The choice reaction time is a logarithmic function of the number of alternative stimuli and responses; mathematically,

$$\mathrm{RT} = a + b \log_2 N,$$

where a and b are empirical constants and N is the number of choices.

N may be replaced by the probability of any particular alternative, $p = 1/N$, and we have

$$\mathrm{RT} = a + b \log_2 (p^{-1}),$$

which is called the Hick–Hyman equation. (To be exact, $\mathrm{RT} = a + bH$, where H is the transmitted information.)

Simple reaction times Under optimal conditions, simple auditory, visual, and tactile reaction times are about 0.2 second. If conditions deteriorate, so that, for example, there is uncertainty about the appearance of the signal, the reaction slows. For instance, the simple reaction time to hearing tones near the lower auditory threshold (30 to 40 dB) may increase to 0.4 second. Similarly, the visual reaction time is dependent on the intensity, duration, and size of the stimulus. Figure 3–11 indicates these relations, as measured by various researchers; for reasonable sizes of the light source (between 0.5 and 1.7 degrees), the reaction time is shortened by increased luminance and by an increased duration of the flash. The reaction time is minimal with a weak light source (of about 3 cd m^{-2}, which is near the cone threshold), regardless of the duration of the exposure or the size of the stimulus. Reactions to visual stimuli in the periphery of the visual field (such as 45 degrees from the fovea) are about 15 to 30 ms slower than to centrally located stimuli (Boff and Lincoln 1988). (These characteristics of the signal are further explained in Chapter 4.)

Times taken by different body parts upon tactual stimuli vary only slightly, within about 10 percent, for the finger, forearm, and upper arm. When these times are divided into the premotor time (the time from the stimulus to the onset of electromyographic activity in muscles) and the motor time (the time from the onset of EMG activity to the beginning of movement), there are no differences in premotor time

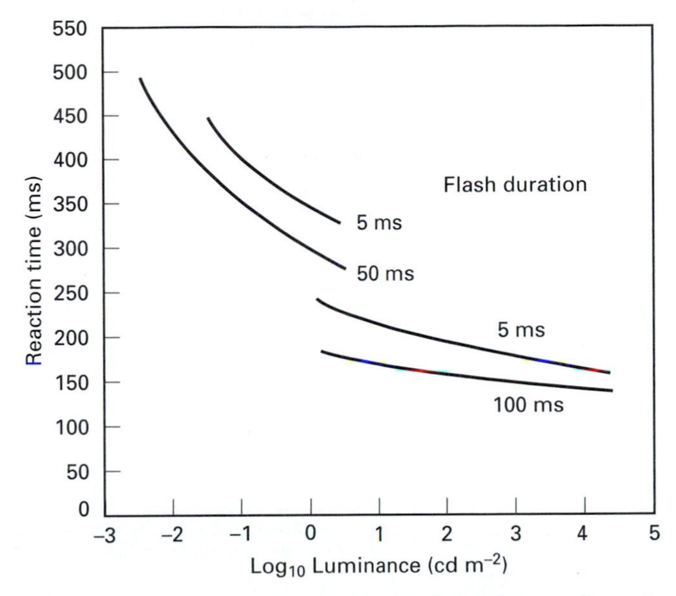

Figure 3–11. Simple visual reaction times, depending on flash duration and flash intensity (illumination). (Modified from "Motor control," by Keele, S. W., in K. R. Boff, L. Kaufman, and J. P. Thomas (eds.), *Handbook of Human Perception and Human Performance,* copyright 1986 by John Wiley & Sons, Inc. Reprinted by permission.

(Anson 1982). This indicates that it takes longer to move a more massive limb than a lighter one—as one would expect from simple mechanical considerations.

The simple reaction time changes little with age from about 15 to 60 years, but is substantially slower at younger ages and slows moderately as one grows old.

Choice reaction time The choice reaction time expands if it is difficult to distinguish between several stimuli that are quite similar, but only one of them should trigger the response—as is expected from the Hick–Hyman formula. Measured reaction times are shown in Table 3–1 and Figure 3–12.

MOTION TIME

Motion time follows reaction time. Movements may be simple, such as lifting a finger in response to a stimulus, or complex, such as swinging a tennis racket. Swinging the racket contains not only more complex movement elements, but also larger body and object masses that must be moved, which takes more time. Motion time also depends on the distance covered and on the precision required. Related data are contained in many systems of time and motion analyses, often used by industrial engineers.

In the early 1950s, Paul Fitts performed well-designed and -controlled studies of motion times, which have become classics. Fitts found that when the precision of the target was fixed, motion time increased with the logarithm of distance. If the distance was fixed, then motion time increased with the logarithm of the reciprocal of the width of the target. Distance and width almost exactly compensated for each other.

TABLE 3–1 Merkel's 1885 Data on Reaction Times for Visually Presented Numerals. Subjects had to press the appropriate one of 10 buttons

Number of Alternatives	*Reaction Time (ms)*
1	187
2	316
3	364
4	434
5	487
6	532
7	570
8	603
9	619
10	622

Source: Adapted from Keele 1986. Reprinted by permission of John Wiley & Sons, Inc.

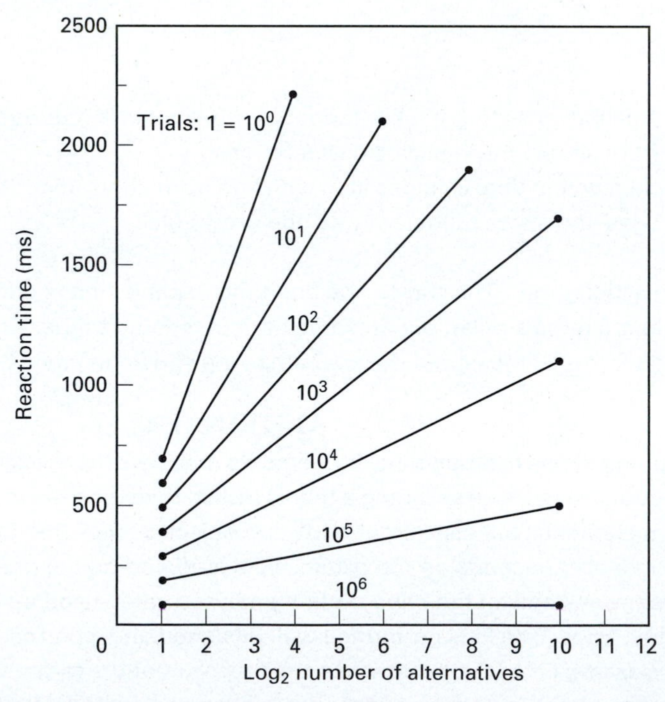

Figure 3–12. Schematic relation between reaction time, practice, and number of alternatives. (Modified from Keele 1986, who used data from Teichner and Kreb 1974. Reprinted by permission of John Wiley & Sons, Inc.)

These relations have been expressed in a motion–time (MT) equation called Fitts' law:

$$MT = a + b \log_2\left(\frac{2D}{W}\right),$$

where D is the distance covered by the movement and W is the width of the target. The expression $\log_2(2D/W)$ is often called the *index of difficulty.* (The factor 2 is used simply to help avoid negative logarithms.) The constants a and b depend on the situation (such as the body parts involved, the masses moved, and the tools or equipment used), on the number of repetitive movements, and on training. (See Wing 1983 or Keele 1986, for more details.) Fitts' law has been found to apply to many movement-related tasks—even to the "capture time" of a moving target.

RESPONSE TIME

The reduction of the response time, the sum of the reaction and motion lags, is a common engineering goal. It can be achieved by optimizing the stimulus and selecting the body member that is best suited to the task.

The best proximal signal is the one that is received quickly (primarily according to modality and intensity) and is different from other signals; in the extreme, one may want to bypass receptors and directly act on nerves. Afferent and efferent transmissions depend on the composition, diameter, and length of nerve fibers and on synaptic connections; practically, the engineer usually selects the shortest distance between sensor location and brain. Yet the best chances for reducing delays are obviously in the processing time needed in the central nervous system (for perceptual tasks such as the detection, identification, and recognition of the signal and for cognitive tasks like deciding and planning). Thus, a clear signal leading to an unambiguous choice of action is the most efficient approach to reducing delays.

Choosing the most suitable body member for the fastest response includes (1) selecting a short afferent distance between sensor and CNS and a short efferent distance to quickly activate muscle and (2) making sure that the minimal segment mass must be moved. Thus, moving an eye is faster than moving a finger, which is faster than moving a leg.

MENTAL WORKLOAD

Resource use The assessment of a workload, whether psychological or physical, commonly relies on the "resource construct," meaning that there is a given (measurable) quantity of capability and attitude available, of which a certain percentage is demanded by the job. If less is required than is available, a reserve exists. (See Figure 3–13.) Accordingly, the workload is often defined as the portion of resource (i.e., of the maximal performance capacity) expended in performing a given task.

Following this concept, one should obviously avoid any condition in which more is demanded from the operator than can be given, because then the performance of the task will not be optimal and the operator is likely to suffer, physically or psychologically, from the overload. However, a task demand that is below the capacity of the operator would leave a residual capacity. Its measurement provides an assess-

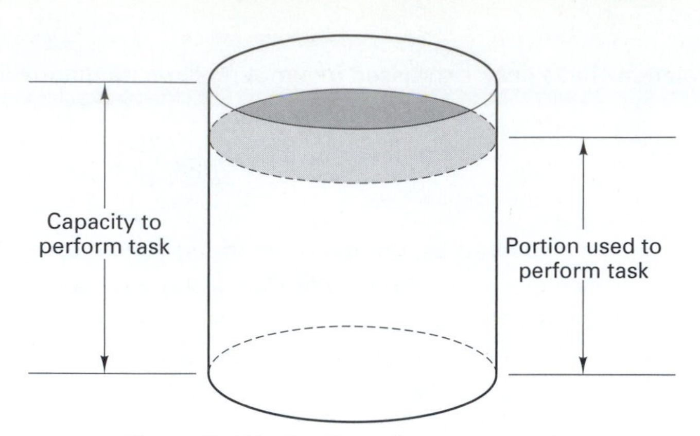

Figure 3–13. Traditional resource model.

ment of the actual workload. Refined models have been proposed—for example, a multiple-resource model in which separate reservoirs are postulated, such as for stages of information processing (afferent, central, and efferent), codes of processing (e.g., verbal or spatial), and input/output modalities (visual, auditory or verbal, and manual). See the overviews by Best (1995), Moray (1988), O'Donnell and Eggemeyer (1986), and Wickens and Carswell (1997) for more details.

MEASURING WORKLOAD

Workload is empirically assessed using four different approaches. Three are objective measures: of primary-task performance, of secondary-task performance, and of physiological events. The other measurement uses subjective assessments. Measures of task performance, as well as of subjective assessment, presume that both "zero" and "full" capacities are known, since those measures assess the portion of capacity loading. Measuring performance on a secondary concurrent task is intended to assess the spare capacity that remains after allocating resources to the primary task. If the subject allocates some of the resources that are truly needed for primary task performance to the secondary task, the secondary task intrudes on the primary task; such an invasive secondary task would modify the workload (Tattersall and Foord 1996).

MEASURES OF WORKLOAD

People are different Of course, individuals differ from each other in their capacities to perform tasks. Thus, the workload imposed by a given task differs from person to person; also, the workload may depend on the temporal state of an individual—for example, on training, fatigue, and motivation.

Use the task performance itself Seven (1989) and Doherty (1991) suggested focusing on the measurement of the primary-task performance by observing how noncritical components of the primary task are performed. The hypothesis is that, as the workload increases, performance changes measurably. Candidates for such unobtrusive measures are the status of a person's speech, the depletion of stock, disorder or clutter at the workplace, and the length of a line of customers. Such embedded measures of workload would not add to the task at hand.

Secondary tasks The following are examples of secondary tasks used in measuring the workload:

- Simple reaction time: Draws on perceptual and response execution resources.
- Choice reaction time: Same as for the simple reaction time, but with greater demands.
- Tracking: Requires central processing and motor resources, depending on the order of control dynamics.
- Monitoring of the occurrence of stimuli: Draws heavily on perceptual resources.
- Short-term memory tasks: Heavy demand on central-processing resources.
- Mathematics: Draws most heavily on central-processing resources.
- Shadowing, i.e., subject repeats verbal or numerical material as presented: Heaviest demands on perceptual resources.
- Time estimations: (a) subject estimates time passed: draws upon perceptual and central-processing resources, (b) subject indicates sequence of regular time intervals by motor activity; makes large demands on motor output resources.

Physiological measures The heart rate, eye movements, pupil diameter, and muscle tension can often be measured without intruding on the primary task. However, these measures may be insensitive to the task requirements or may be difficult to interpret.

Subjective assessments In these tests, we are able to internally integrate the demands of the task, but the subjective assessment of the perceived workload may be unreliable, invalid, or inconsistent with other performance measures. On the one hand, if subjective measures are taken after the task has been completed, they are not real-time evaluations; on the other hand, if performed during the task, they may intrude on the task.

Widely used measures A number of pragmatic measures of the workload have been widely used, even though they have been criticized on both theoretical and technical grounds. They are: The modified Cooper–Harper scale (Wierwille et al. 1985), the overall workload scale (OW, Vidulich and Tsang, 1987), the NASA task load index (TLX, Hart and Staveland 1988), and the subjective workload assessment technique (SWAT, Reid and Nygren 1988, Colle and Reid 1998). The first two are unidimensional rating scales; the last two are subjective techniques using multidimensional scales. While there are good theoretical and statistical reasons for using TLX, SWAT, and OW (Nygren 1991), these assessments are more complex to administer than the Cooper–Harper scale, which is fairly self-explanatory. That scale is shown in Figure 3–14, modified to be applicable to systems other than piloted aircraft.

The ratings obtained from the Cooper–Harper scale are highly correlated with those in the more detailed TLX and SWAT scales, which might be employed after initial ratings have been obtained from the modified Cooper–Harper scale. On the other hand, TLX and OW were found to have superior sensitivity and operator acceptance (Hill et al. 1992). Often, it is advisable to use a combination of workload measurement techniques (Wilson and Eggemeyer 1994, Wierwille and Eggemeyer 1993).

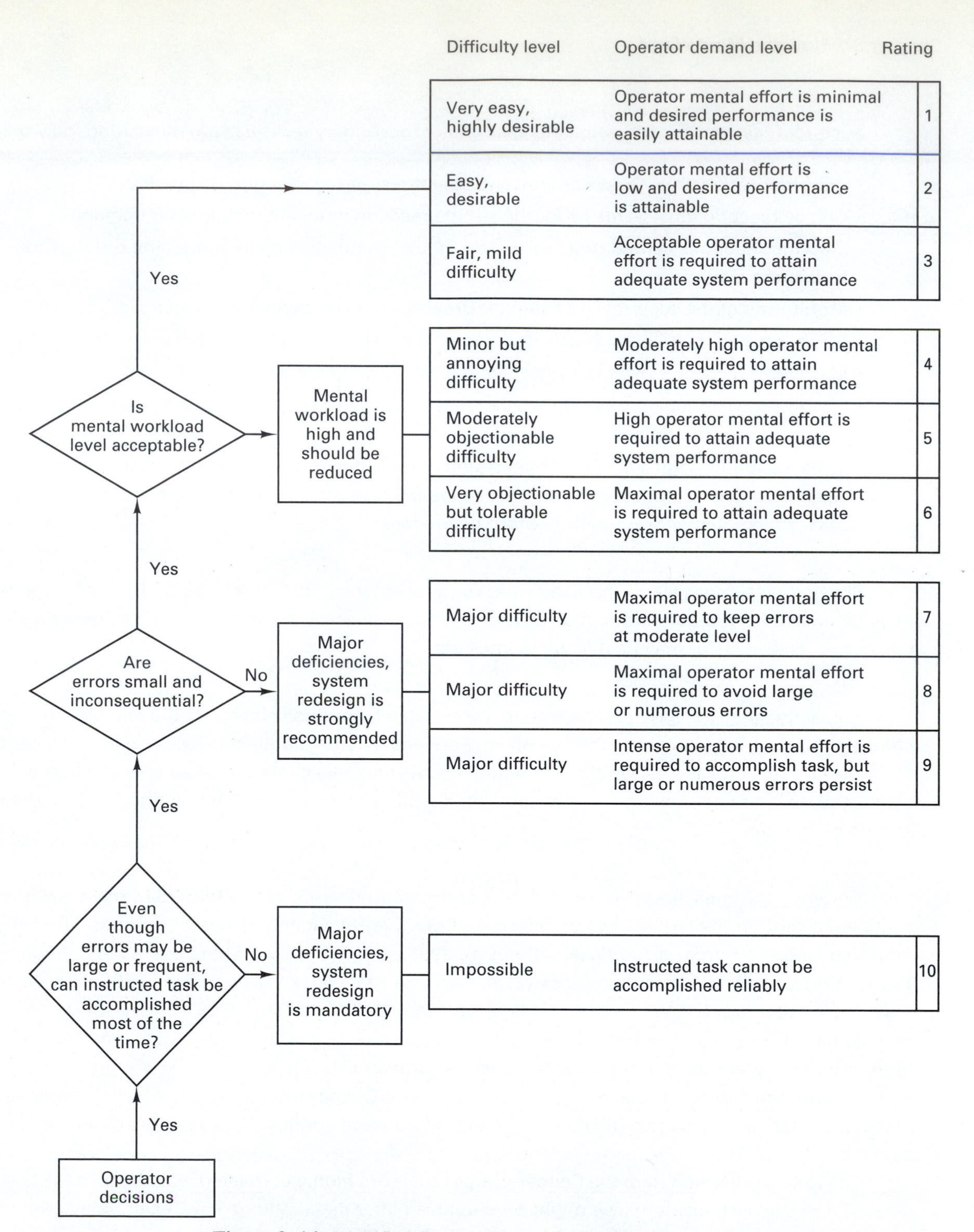

Figure 3–14. Modified Cooper–Harper Scale. (Reprinted with permission from W.W. Wierwille and J.G. Casali (1963). "A validated rating scale for global mental workload measurement application." In: *Proceedings of the Human Factors Society, 27th Annual Meeting*. Copyright 1983 by the Human Factors and Ergonomics Society, Inc. All rights reserved.)

Chicken extract, according to Asian folk wisdom, helps one recover from physical and mental fatigue. Nagai et al. reported in 1996 on tests with twenty male subjects who received either a placebo or a brand of chicken extract on seven days in the morning. The mental performance of the chicken extract users was better, their recovery from mental fatigue was faster, and the concentration of stress-related substances in their blood was reduced. Will chicken soup save the world?

Stress on the Individual and Crews

Stress and anxiety are core concepts of psychopathology: A prevalent general model (called diathesis–stress) assumes that most disorders arise from complex interactions between environmental stressors and (usually biological) predispositions that can make a given individual break down.

Stress can affect health

There is evidence that stress and physical illness are related; for example, the social readjustment rating scale of Holmes and Rahe (1967) lists life events (ranging from the death of a spouse to being fired from a job to taking a vacation and holiday), and subjects are asked to indicate those that they have experienced recently. High scores on that scale have been related to heart attacks, other health problems, and depression. This relationship appears striking, but caution needs to be applied, since some studies investigating the connection utilized only correlational analyses.

The relationship of stress to adverse health may be analogous to that of cigarette smoking to lung cancer: While most people who smoke do not get lung cancer, most cases of lung cancer are related to smoking.

It becomes important, then, to consider multivariate and other measures of the association between stress and disease. Many models have been proposed to better understand these complex associations and examine variations in types of stressors and within subgroups of individuals.

Stress is physical and cognitive

In his classic studies in the 1930s, Selye saw stress primarily as a physical trauma to which the human responded. Today, it is generally accepted that stress is connected not only with physical events, but also with the appraisal of those events by the individual; hence, stress is also a cognitive phenomenon.

Assessment of Stress

There is some debate in the psychological literature over whether objective assessments or subjective reports of the severity of various stressors provide a better measure of the impact of an event on an individual (Mazure 1998). Methods of assessing stress include checklists, semistructured interviews, and structured interviews. In each category, several assessment instruments are widely used. Perhaps the most representative checklist questionnaire is Holmes and Rahe's (1967) Social Readjustment Rating Questionnaire (SRQ). Representative of semistructured interviews are the Life Event Scale (Tennant and Andrews 1976) and the Recent Life Events Interview (Paykel 1997). The Structured Event Probe and Narrative Rating Interview (Dohrenwend et al. 1993) provides a highly organized method of inquiry.

"Mind and body are intimately interconnected, intertwined, and interdependent—of this there is no longer room for doubt` . If, once upon a time, it was possible to believe that physical and mental events belong to essentially different spheres, in contact only in the sense of a spir-

itual mind inhabiting the wondrous mechanisms of the body, that is no longer so. And whilst we may know less in detail than we would like, we do know a great deal about the physiology of emotion and of stress" (Blinkhorn 1988, p. 29). Yet, in psychology, there is no consensus on even a definition of stress. Thus, the "term stress is enshrouded by a thick veil of conceptual confusion and divergence of opinion" (Motowidlo et al. 1986, p. 618).

Job Stress

The concept of stress on the job is both common and elusive. We all have had the experience of being driven to the margin of our physical and psychological capabilities by strenuous physical exertion, a hot climate, the pressure of a schedule, the unreasonable behavior of bosses or colleagues, an oncoming illness, or the feeling that we are expending useless efforts. Some of these stressors are physical, others psychological; some are self-imposed or external; and some are short term or continual.

"Good" stress or distress?

The concept of stress is also elusive because what may be stimulating under one condition may become excessive under other circumstances. The simple "stress produces strain" sequence, which engineers use, may dissolve into the complex relations familiar to psychologists: A stressor may generate a positive "stress" that spurs more activity, or it may result in "dis-stress," which overloads the person and generates ineffectiveness, evasive behavior, anxiety, and even illness.

Demand, Capacity, and Performance

The confusing situation regarding stress and its effects may be clarified by the model shown in Figure 3–15, which reflects three major aspects of ergonomic concern. First, *job demands* depend on the type, quantity, and schedule of tasks; the task environment (in physical or technical terms); and the conditions of the task, including the psychosocial relations existing on the job. These (and possibly other related) work attributes are the job stressors that are imposed on the human.

Second, a *person's capability* to fulfill the demands of the job and, third, the *person's attitude* (influenced by her or his physical and psychological well-being) must be matched with those demands. If the job demands require only a portion of the person's abilities and attention, the person is likely to feel underloaded and underestimated and might eventually become bored, inattentive, and underachieving or, alternatively, might seek more of a challenge. To the contrary, if the job demands exceed the person's capabilities, then the individual is liable to feel overloaded and might seek either to reduce the workload or to increase his or her capabilities. Of course, other sources of stress besides those at work might require a portion of the person's capabilities or attitude, influencing the physical and psychological well-being. In that case, the strain experienced by the individual depends on the sum total of the job and other demands in relation to the person's

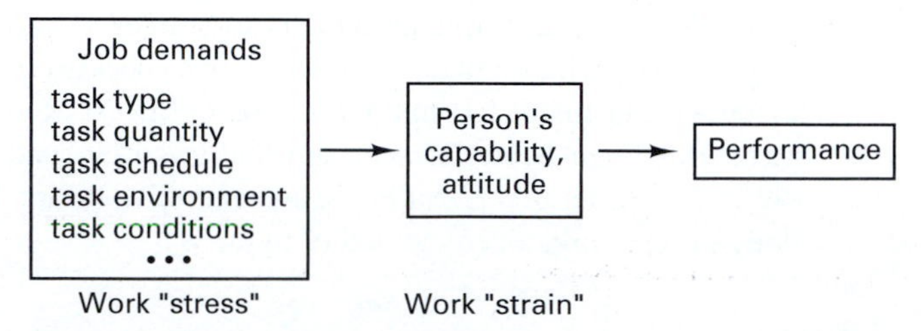

Figure 3–15. Simple model of the relations between job demands ("stress"), human responses ("strain"), and performance.

capabilities and attitude. A buffering factor may be the "hardiness" (or resiliency) of a person—a constellation of behavioral and attitudinal factors believed to be associated with resistance to stress.

All of these attributes, conditions, and reactions affect the *person's performance,* which is, of course, an overriding concern privately, in business, and in our everyday life.

Too much or too little stress? While we usually assume that most "job stresses" are due to overloading, demanding too little of the individual's capacity is not infrequent either. A good match between job demands and a person's capability and attitude is, obviously, a desirable condition. The construct of a U-shaped function shown in Figure 3–16 relates the stress imposed by the work to the resulting strain experienced by the person. According to the "U theory," both too little *and* too much stress produce undue strain—that is, distress.

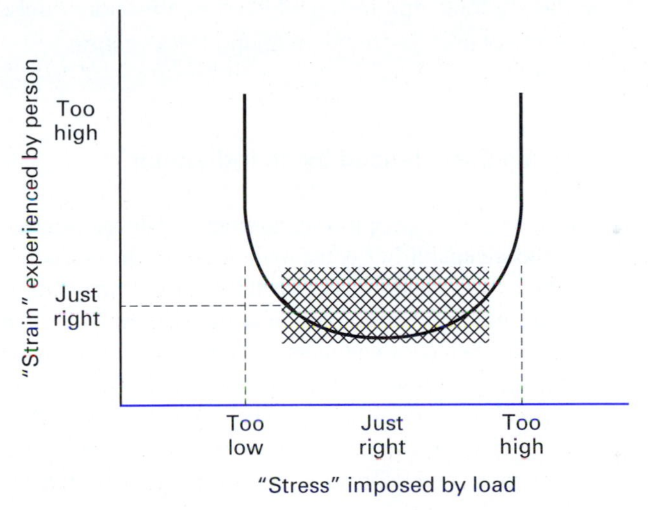

Figure 3–16. Postulated "U-function" relating stress and strain.

Monotony

Monotony is the opposite of variety, either of which can be perceived by an individual as stressful. Monotony is produced by an environment in which either there is no change or else changes occur in a repetitive and highly predictable fashion over which the individual has little control. A varied environment often provokes interest and the human emotion of excitement; in contrast, an unvaried environment produces *boredom,* which also can be considered an emotion. Thus, boredom is an individual's emotional response to an environment that is perceived as monotonous. A bored person often complains of feeling tired or fatigued.

Mental Fatigue

The term "fatigue" is commonly used to indicate a physiological status; however, some psychologists claim that it should be used exclusively to define a subjectively experienced disinclination to continue performing the task at hand (Brown 1994). Of course, *physis* and *psyche* are related, but we must agree on one definition to be able to discuss the topic. Some psychologists now use the term "mental fatigue" to describe a disinclination to continue a task due to psychological factors.

MATCHING DEMANDS TO CAPACITY

According to the U theory, if the job demands are far below the person's abilities, an underload condition exists, and the on-the-job performance is most likely (but not necessarily) diminished. On the other hand, if the work requires more than a person is able and willing to give, an overload condition exists, and both work performance and well-being are likely to suffer. If the job demands (work stress) match the person's capabilities and attitude, the proper amount of strain exists, and the on-the-job performance is satisfying, both objectively and subjectively. In a related concept, Csikszentmihalyi (1990) has proposed that individuals are in a state of "flow" when their skill level is perfectly balanced to the challenge level of a task that has clear goals and provides immediate feedback. This state of flow is thought to occur when people are deeply focused, during which time the individual experiences an almost euphoric state along with high alertness and the sense of losing track of time.

Strain Experienced by an Individual

Stress–strain overloads

Understanding people's capabilities and developing job demands and conditions that are matched to those capabilities is the main focus of the ergonomist. The brief discussion that follows of attributes and conditions of the human being may help to understand proper stress–strain relationships and to avoid in particular occupational overload. Occupational overload, together with a person's behavior and mental and physical attributes, constitutes a primary source of variance in explaining individual and organizational distress. (Many of these attributes fall into the domain of psychopathology, a subject beyond the scope of this text.) For example, overworked nurses exhibit behavioral and emotional symptoms of fear, dread, anxiety, irritation, annoyance, anger, sadness, grief, and depression in response to their occupational conditions (Motowidlo et al. 1986).

Occupational Health Psychology

The extent of strain a person experiences in an occupational setting depends on job experience, age, attitude, self-esteem, and coping ability, among other factors. The growing field of occupational health psychology addresses the convergence of public health and clinical psychology in an industrial or organizational context with an emphasis on preventive management (Quick 1999). Occupational health psychologists focus on both environmental modification and individual behavioral factors that are involved in a work environment. The latter category includes such concepts as "workaholism" (Porter 1996) and cognitive–affective stress propensity (Wofford and Daly 1996). Perhaps one of the most interesting areas of investigation of individual factors has been in the domain of Type A and Type B behavior.

Types A and B

Behavior patterns are highly correlated with the experience of strain. The Type A behavior pattern is exhibited by persons who are engaged in an (often chronic) struggle to obtain (often poorly defined) things from their environment in the shortest possible period of time—if necessary, against the opposing effects of other things or persons. Type A behavior is characterized by aggressiveness, competitiveness, impatience, and urgency in overcoming obstacles. Individuals exhibiting this behavior may also approach obstacles presented by circumstances or other individuals with an attitude of hostility. Type A persons are likely to act in ways that make job events more stressful for themselves, and then they find the resulting strain particularly intense. Many researchers have found these persons to be more susceptible to emotional and cardiovascular disorders than persons showing the opposite type of behavior: Type B individuals are relaxed, unhurried, less aggressive, and not so competitive and usually have fewer disorders (Hackett et al. 1988, Kamarck and Jennings 1991). Recent research suggests that the physiologic or psychologic underpinnings or consequences of *hostility* may be determining factors in the coronary health effects of Type A behavior (Miller et al. 1996, Scheier and Bridges 1995, Helmers et al. 1994), perhaps particularly in men (Guyll and Contrada 1998). The currently favored model is psychophysiologic: It assumes that hostility heightens cardiovascular and neuroendocrine activity and thus initiates and fuels the artherosclerotic process. According to this theory, hostile individuals experience anger more often and more intensely than their nonhostile counterparts, which causes them to experience more frequent and more intense activation of the sympathetic–adrenomedullary system (Williams et al. 1985). Several studies have lent good support to this hypothesis by demonstrating various increases in cardiovascular reactivity or decreases in parasympathetic activity in hostile individuals exposed to different stressors. Other models that explain the link between Type A behavior and coronary heart disease include a psychosocial vulnerability model, a transactional model, and a health behavior model. (For reviews, see King 1997.)

Managing stress?

Stress management is a popular term that has been used by psychologists in a variety of ways. The term "management" implies that stress is felt to be an unavoidable component of living with which individuals can learn to cope. Selye (1978) stated that to talk about being under stress is as pointless as talking about running a temperature. What is really of concern is an excess of stress—a distress. Lazarus and Cohen (1977) identified various categories of stressors, personal ones as well as cataclysmic ones that happen to several people at once (for example, natural disasters and massive corporate layoffs).

Stressors and hassles

Various life events have been used to scale personal stressors. The Social Readjustment Rating Scale, previously mentioned, is a 43-unit checklist of different stressful events, ranging from the death of a spouse to taking a holiday, that have been correlated with symptoms of physical illness and psychological distress. Background stressors, so-called daily hassles, have been ordered into a Hassles Scale and also related to health and well-being (Kanner et al. 1981). Examples of daily hassles in the workplace are inconsiderate coworkers and noisy work environments. On an intuitive level, it is easy to recognize the cumulative impact that such daily annoyances may have on the physical and emotional well-being of an individual, especially if they are perceived as being uncontrollable. Indeed, research bears out the fact that the strain felt by an individual is greater if the stressor is perceived to be outside one's realm of control. Job strain has been operationalized in several research studies by settings involving a high demand and a low control over decisions. A high level of this kind of job strain (or task strain) has been implicated as a risk factor for coronary disease, as well as for hypertension, in a variety of cross-sectional and prospective epidemiological reports (e.g., Kamarck et al. 1998, Karasek et al. 1981, 1988; Schnall et al. 1992). Sustained uncontrollable stress is also thought to be one factor in the development of "learned helplessness," one model for the development of depression. Sustained or chronic stress has been defined in varied ways, but generally assumes a duration of longer than four weeks. Chronic conditions that have been found to increase one's risk of poor health and decreased psychological well-being include poverty, sustained disability, and prolonged marital conflict. In a recent study, subjects enduring severe chronic stressors developed disease upon being inoculated with common cold viruses, while subjects with severe acute stressors did not (Cohen et al. 1998), and these associations were not attributable to personality variables, health practices, the season of the year, the body mass index, demographic variables, or preventive endocrine or immune measures. In addition, individuals are often not adequately aware of the frequency or intensity of background stressors, particularly if they increase gradually over time. An analogy may be found in the animal world: In 1899 (when such studies were not subject to panel review to determine whether they were ethical), Scripture found that frogs would submit to being boiled alive if the water temperature was increased gradually.

Coping

Individual attitudinal variables (such as beliefs and character predispositions) are mitigators of the degree of distress elicited in an individual by a given environmental stressor. Tache and Selye (1986) likened stress reactions to allergy attacks. In an allergic reaction, the overmobilization of the immune system—not the allergen itself—causes individuals to feel ill. People have different coping styles. Folkman and Lazarus (1988) define coping as the cognitive and behavioral efforts to manage specific external or internal demands that are appraised as taxing or exceeding the resources of the person. The resources of an individual may be material, physical, intrapersonal, interpersonal, informational or educational, and cultural. Cognitive components of effective and adaptive coping strategies include rationality, flexibility, and farsightedness (Antonovsky 1979). Effective coping strategies are either problem oriented or emotion focused. Problem-oriented coping is directed at controlling an environmental stressor to reduce its impact and is effective if a stressor is objectively controllable. Examples are time management, environmental control or environmental adaptations, assertive communication, and the setting of limits. Emotion-focused coping strategies include structured relaxation exercises, the use of humor, exercise or hobbies, and guided cognitive reevaluation ("reframing") of a stressor (Lazarus and Folkman 1984). The workplace has become a common setting for stress management programs, which often emphasize developing healthier lifestyles (through smoking cessation, weight management, exercise programs, or treatment for drug and alcohol abuse). Such programs are obviously beneficial to employees who need them, but they also affect the corporate economy if they reduce the number of sick days and the expense of medical insurance coverage.

Emotional intelligence

In a growing body of research on professional leadership, the personality traits of affability, resiliency, and optimism have been noted to be associated with high performance (Cooper and Sawaf 1998). These traits have recently been conceptualized as components of an "emotional (intelligence) quotient" (EQ) in a somewhat controversial theory which postulates that individuals who possess such traits excel at work more so than those with a high intelligence quotient (IQ). Whereas one's IQ is relatively fixed, one's EQ can, to a much larger degree, be built and learned (Goleman 1995). Accordingly, companies can theoretically test an employee's emotional intelligence and teach the employee how to increase it. Indeed, employers may even be able to incorporate EQ theory into more traditional stress management programs. While the validity of the concept of a global emotional intelligence quotient may continue to be debated for some time, it stands to reason that different emotional skills correlate with success in different professions. For example, success in sales requires the ability to gauge a customer's mood and various other interpersonal skills that are quite different from the more self-directed skills required for a person to be a successful athlete or musician. It may be that in the future emotional skills will be taught in workplace training programs, and workers will be assigned tasks that match their emotional skills in addition to their intellectual and experiential skills.

The pursuit of happiness

Typically, 3 in 10 Americans say that they are very happy, most Americans describe themselves as pretty happy, and only 1 in 10 is not happy. Happiness cuts across almost all demographic classifications of gender, age, economic class, ethnic group, and educational level—even wealth. Happy people tend to like themselves, feel personal control over their lives, and are optimistic and extroverted. They are likely to still be happy a decade later (Myers and Diener 1996).

Strain Experienced by Confined Groups

Crews in confined spaces such as spaceships experience a special kind of on-the-job stress. Even during the relatively short missions carried out by U.S. astronauts, severe intraindividual and interindividual problems appeared, ranging from physiological deficiencies (see Chapter 5), to anxiety, to difficulties in interacting with other crew members and disagreements with ground control. Problems of this nature also have arisen under other isolated and confining working conditions, such as in the Antarctic (Harrison et al. 1991).

During one of the U.S. Skylab *missions, Commander Gerald Carr and his fellow astronauts not only "plotted" to hide items from ground control, but also went on a "strike" until some disagreements could be worked out. Similar problems with ground controllers in the former USSR, as well as interpersonal conflicts among cosmonauts, developed aboard the* Salyut *and* Mir *space stations (Holland 1991).*

Changes in environment and physiological functions affect a person's psychological and psychosocial well-being. Among the physiological effects of spaceflight (further discussed in Chapter 5) on astronauts are the following:

- Dimming of vision, peripheral light loss, and, eventually, blackout under increasing plus-Z-direction acceleration.
- Diminished vision, redout, increased accommodation time, and blurring or doubling of vision in minus-Z-direction acceleration.
- Lengthening of visual reaction time by increased *g's*.
- Degradation of estimates of sizes and distances of objects in space because the objects appear sharper and in higher contrast while foreground and background distance cues are absent.
- More time required for the eyes to adapt to the existing wide ranges of illumination and luminance levels when shifting the gaze.
- Reduction of visual performance by vibration, such as during liftoff and landing, when vision is particularly important.
- Shift in perceived colors, reduction in sensitivity to contrast, and in near-field acuity, but improved visual acuity for distant objects (all of these according to anecdotal and contradictory reports).
- Sensitivity to noise appears to increase, whereas auditory functions in general do not change significantly in space.
- Reduced cabin pressure requires crew members to talk louder to be heard, owing to the reduction in transmission of sound.
- The sensitivity of smell is diminished (probably due to the ever-present nasal congestion in microgravity).
- Unpleasant odors (for instance, those associated with medical symptoms, food, or body waste) can become very annoying; yet, pleasant odors are also met with increased responsiveness.

- The sense of taste is degraded, so that food judged to be well-seasoned on earth tastes bland in space. (This effect may be associated with nasal congestion and the upward shift of body fluids.)
- Vestibular effects are evidenced mostly in space sickness and spatial disorientation.
- Larger forces must be applied to generate the same kinesthetic sense stimuli as in a 1 g environment.
- Motor skills are impaired upon entering microgravity, but the deficiency is compensated for after a short period of adaptation.
- Perception of body language and facial expression is changed because crew members appear in different postures and their faces are swollen by the accumulation of fluid in the tissues.
- Anxiety, often present in the initial and final phases of a space mission or during emergency situations, can greatly reduce the effectiveness of both individual crew members—particularly the commander of the spaceship—and the team in general.
- Rigid hierarchical structures (such as those prevalent in the U.S. armed forces) usually become diluted during long missions and develop into a more collegial, democratic sharing of decisions and activities.

While it is known that personal, interpersonal, and psychosocial factors play major roles in the success or failure of long missions, exact information on these factors is still sparse. The former USSR had extensive experience with spaceflights of long duration, but the United States had only flights lasting less than two weeks duration since 1974. The compatibility and cohesion of crew members and the leadership style of the commander of a ship are very important during long periods of confinement, isolation, monotony, and danger, as was experienced in the joint U.S.–Russian *Mir* space expeditions in the mid-1990s.

Three stages in the reacting to isolation have been observed. In the first stage, there is heightened anxiety, apparently dependent on the degree of present danger that a person perceives. Heavy workloads appear to reduce the anxiety level during this time. The second stage is associated with feelings of depression, usually during the segment of time while one settles down to routine duties. These feelings may result from the absence of familiar social roles, such as spouse, father, mother, or club member. The final stage is at the end of the mission and manifests itself in increased expression of affect and anticipatory behavior exhibiting higher levels of anxiety and aggressiveness. In this stage, work performance is commonly degraded, and serious errors of judgment or omission may occur. Table 3–2 summarizes the main issues that affect a crew's behavior and performance. It also lists measures that can be taken to influence behavior and performance positively.

Until the early 1990s, crews of men performed all long space missions. Mixed-gender crews generate new facets of intragroup behavior. One idea for stabilizing relations is to use married couples. Yet, even then, the conditions of confined spaces, working together closely, and social and sexual desires are likely to generate difficult relations among the crew members, with possibly large changes over time. A related problem is that of pregnancy in space: It is at present unknown whether microgravity, radiation, and other space conditions might affect the development of the fetus.

TABLE 3–2. Factors That Influence Behavior and Performance

Psychological, Psychosocial, and Psychophysiological	*Environmental*	*Space System*	*Support Measures*
Limits of performance (perceptual/motor)	Spacecraft habitability	Mission duration and complexity	In-flight psychosocial support
Cognitive abilities	—confinement	Organization for command and control	Recreation
Decision making	—physical isolation	Division of work between human and machine	Exercise
Motivation	—social isolation	Crew performance requirements	Work–rest/avoiding excess workloads
Adaptability	—lack of privacy	Information load	Job rotation
Leadership	—noise	Task load/speed	Job enrichment
Productivity	Weightlessness	Crew composition	Training
Emotions/moods	Artificial life support	Spacecrew autonomy	—preflight environmental adaptation
Attitudes	Work–rest cycles	Physical comfort/quality of life	—social sensitivity
Fatigue (physical/mental)	Shift change	Communications (intracrew and space/earth)	—for team effort
Crew composition	Desynchronization of body rhythms	Competency requirements	—for self-control
Crew compatibility	Hazards	Time compression	In-flight maintenance of proficiency
Psychological stability	Boredom		Earth contacts
Personality variables	Stresses		
Social skills	—single		
Human reliability (error rate)	—multiple		
Space adaptation	—sequential		
Spatial illusions	—simultaneous		
Time compression			

Source: Adapted from Christensen and Talbot, 1986.

ENHANCING PERFORMANCE

There is generally good reason to achieve and perform at one's best. Learning, knowledge, communication, creativity, concentration, and skill, as well as performance under stress are important on the job or in sports; the military is particularly interested in attaining fearlessness, cunning, courage, one-shot effectiveness, reversal of fatigue, and nighttime fighting capabilities.

Training should help attain such performance ideals. Thus, techniques for enhancing human performance have received much attention, particularly in the popular press.

Many entrepreneurs, most of them outside academia, advocate techniques for concentrating on specific targets, accelerating learning, improving motor skills, altering mental states, reducing stress, increasing one's social influence, fostering group cohesion, and performing parapsychological "remote viewing." Examples are biofeedback, a technique for attaining information about internal processes and controlling them; "hemispheric synchronization," or "split-brain learning," based on assumptions about right- and left-brain activities; "neurolinguistic programming," such as procedures for influencing another person; "mind reading;" and even non-tactile "psychokinetic" control of devices.

Some of these techniques are promoted as easy to do, such as by simply watching a videotape or listening to subliminal information during sleep. But the advertised successes—a number of them quite surprising—usually are supported only by personal experiences and testimonial statements, rather than by scientifically acceptable proof. Thus, the National Research Council appointed a committee that investigated some of these techniques and their likelihood of success. The results of that committee's work have been reported in detail by Druckman and Swets (1988), Swets and Bjork (1990), and Druckman and Bjork (1991). The text that follows relies on their statements and findings.

GENERAL FINDINGS

Some theoretical bases quoted by promoters were found to be unproven or questionable. In many cases, advertised results were not substantiated. In other cases (for example, "transcendental meditation"), the topic remains not thoroughly researched and may be difficult to research because underlying assumptions do not conform to currently accepted scientific procedures and understandings.

- Successful learning results from the quality of instruction, the time spent practicing and studying, the motivation of the learner, and the matching of training procedures to the demands of the task. Some of the so-called superlearning programs may be effective if they combine these factors suitably, but there is little or no evidence that they include effective instructional techniques from outside the main realm of accepted research and practice.
- Many measures of the effectiveness of training procedures rely on measurements of performance during training or at the end of the training period, but these assessments are often not indicative of the person's actual performance weeks, months, or years in the future. Little research has been done on how to generate measures of performance at a time when the performance is required.
- Mental practice is, in theory and by some limited experience, effective in enhancing the performance of motor skills. Yet, simply listening to experts or observing the performance of experts on videotapes has not proven to be effective.
- Positive effects of biofeedback on skilled performance remain to be determined. (It should be noted, however, that there is compelling evidence in support of a limited and specific efficacy of biofeedback in the alleviation of certain well-defined somatic disease processes—for example, in the treatment of some forms of tension headaches).
- The literature refutes claims that link differential use of the brain hemispheres to performance.
- Research conducted over a period of 130 years indicates that there is no scientific justification for the existence of parapsychological phenomena.

Specific Findings

Subliminal learning

"Subliminal self-improvement by audiotape" is promoted as contributing to improvement in a person's attitude, confidence, and performance; to a reduction in anxiety; and to help in dieting or stopping smoking. There is some scientific evidence that subliminal learning is possible, such as during sleep, because stimuli can be provided that are not consciously registered (for example, because they are very short), but are perceived nevertheless. Yet, there is apparently no mysterious essence that invades the unconscious mind and generates knowledge or confidence during waking hours.

On the tapes that were investigated, the researchers could not find, either by objective methods or in the subjective actions of subjects, any "embedded" subliminal messages, although the manufacturers claimed that such messages were present. Some users of the tapes maintained that they were affected in the direction of the missing messages, but these listeners probably simply convinced themselves that the messages had the desired effect and hence generated a kind of self-fulfilling expectancy. In such cases, the "subliminal tapes" were successful, self-administered placebos.

Transcendental meditation

Meditation plays a role in several kinds of eastern mysticism, such as Buddhism. Mystical traditions, including Christian and Jewish ones, involve dancing and chanting for long periods of time in order to achieve an altered state of mind. Yet, the effects of self-induced hypnotic states and of focused attention are difficult to assess. In the late 1950s, a yogi from India introduced "transcendental meditation" to the West. Transcendental meditation is said to be a set of techniques for influencing a person's conscientiousness through the regulation of attention. While there is no complete agreement on that concept, most meditation techniques include the need to sit or lie quietly in a particular position, to attend to one's breathing, to adapt a passive attitude, to be at ease, and, sometimes, to repeat a word or phrase (a *mantra* in Hinduism).

Thus, meditation may lead to physiological and physical benefits by distracting a person from stressors, either environmental or self-generated. However, it is difficult to evaluate the effects, because the control group involved in testing procedures must have experience with meditation: Disciples of meditation—particularly followers of yoga—claim that the benefits of meditation are available only to those who practice it. Very few Westerners of scientific background have mastered yoga-related meditation, and there are no scientific publications by Western researchers.

Examining the effects associated with meditation on persons who just "rest," researchers have found their reductions in somatic arousal (such as heart rate, respiration rate, blood pressure, and skin responses) to be smaller than those observed in experienced meditators. But since relaxation and meditation go together, it is difficult to attribute any positive effects to meditation by itself, particularly since following meditation usually includes a change in lifestyle.

Performance under pressure

Sports are a natural laboratory for investigating the ways to perform well under high pressure. A commonsense and appealing concept is that mental health is related to athletic performance, which can most easily be stated in a negative form: Anxious, depressed, hysterical, neurotic, introverted, withdrawn, confused, or fatigued athletes do not perform well, while athletes with more positive attributes do. Some performance gains may be achieved by mental practices such as the imagined rehearsal of activities, by goal setting, by rewards for performance, and by relaxation and biofeedback. These techniques appear most successful when they include multiple approaches, directly administered by a mentor rather than via tape, and are often repeated. Apparently, they are most effective for persons who have problems with precompetitive anxiety or concentration. Yet, the gains from such techniques are apparently small, and they have not been investigated in star athletes.

Preperformance routines are also commonly used to reduce pressure. For example, after a point is played, a tennis player relaxes for a few seconds, then walks back confidently to the serving line, relaxes there for a moment, then concentrates on adjusting the strings of the racquet, and finally, purposely prepares to either serve or receive the next ball. Such patterns of thoughts, actions, and images, consistently engaged in before the performance of a skill, are meant to divert attention from negative and irrelevant information and to establish the appropriate physical and mental state respecting the action that is to be carried out. Performed at a critical time, these cognitive–behavioral techniques appear to focus attention better and result in an improved overall performance.

Apparently, the "ideal mental performance state" includes a clear focus on the requirements of the task, total absorption in the task, high intrinsic motivation, concentration of one's consciousness on the task without experiencing distraction, and a feeling of exercising control by using well-set, rehearsed cognitive and motor routines.

Mens Sana in Corpore Sano?

In the original Latin text, this axiom is followed by the verb *sit,* (so be it), indicating the hope that there be a sane mind in a sane body. Recently, the relationship between aerobic fitness and psychological well-being has received considerable attention, triggered by the recognition that anxiety and depression disrupt the lifestyles of many people. Physicians increasingly prescribe exercise to counteract emotional disorders. Carefully performed exercise can have health benefits (such as weight loss and otherwise improved physiological conditions), which may in turn help to promote a person's overall well-being and adherence to a psychological intervention program involving exercise. Some recent research contradicts earlier studies that failed to find an effect of exercise on mental health. In a review of 17 controlled trials, 13 studies found that exercise alone improved various measures of psychological functioning among cardiac rehabilitation patients. Additional studies in healthy individuals have shown that exercise improves self-confidence and self-esteem, attenuates cardiovascular and neurohumoral responses to stress, and reduces some Type A behaviors. However, the independent effects of exercise on mental health remain difficult to determine, due to methodological considerations and the multiple indices of psychological functioning. Overall, depression appears to be the index that is most consistently improved by exercise (Miller et al. 1997), through both direct and indirect processes.

Improving TeamWork

Enhancing team performance is of interest at work, in sports, or in the military, situations in which one does not perform alone, but rather together with other people who do similar, parallel, or complementary tasks. Teams may be highly coordinated (such as the pilot and copilot in an airplane) or poorly organized (such as academic research teams).

Participants vs. loafers The performance of a team of individuals is not simply the sum of the individual efforts. Rather, a team should provide *greater* resources than an individual does, but teamwork also often involves difficult interpersonal coordination and management problems. It is a common experience (although exceptions exist) that a team's performance falls short of reasonable expectations. The proper selection of members according to their physical, mental, and interpersonal capabilities may enhance the team's performance. If members are not motivated to participate in the team efforts, some will not contribute their share, but will do "social loafing." Yet, not all suboptimal group performance is due to a lowered input on the part of individual members; instead, it may be the result of faulty interpersonal processes for combining individual capabilities.

Brainstorming Long believed to be an effective and idea-stimulating team technique, brainstorming is usually no more successful than individual theorization possibly because of "blocking"—i.e., the inability of a team member to produce ideas while others are talking.

Group decisions Another disappointing result has recently been found for consensus decision making: Group decisions tend to be more "risky" than individual decisions. Popular are the Delphi and Nominal Group techniques. In the Delphi technique, individual judgments or opinions are privately elicited and then summarized, and the results are circulated to all team members for further modification until individual positions stabilize. Thus, anonymity of the individual members' inputs is retained. In the Nominal Group technique, after the initial stages, group members meet face-to-face to exchange information. In spite of the popularity of such techniques, little reliable research has been carried out to determine their effectiveness. However, the evidence on team performance available today is not encouraging: Neither technique appears to improve on the performance of freely interacting groups.

"The review of what we know about group performance is more striking for what is missing than for what is known" (Druckman and Bjork 1991, p. 257).

DETECTING DECEPTION

"White" and other lies

Deception is an important survival skill for animals; it is also common in humans—for example, in giving the impression that one is more knowledgeable than one actually is, in bending the truth in a "white lie," or in intentionally proposing something known to be false. Folk wisdom has it that a liar can be detected by blushing, facial expressions, shifts in one's gaze, and involuntary body movements ("body language"), as well as by changes in loudness of the voice and in other speaking patterns. Yet, a person may not exhibit any of these cues, because an experienced liar can suppress them or because they depend on one's attitude and upbringing. Thus, such cues are, in general, not reliable indicators of either lying or telling the truth. Still, most observers believe that they can detect liars. Studies have shown, however, that professionals such as customs inspectors, police detectives, CIA and FBI agents, and judges do no better than chance in actually detecting deception. Altogether, there is a large discrepancy between scientific evidence and anecdotal subjective assessments (Druckman and Bjork 1991).

Folk wisdom about how liars act (e.g., they don't look you in the eye, they don't smile, they fidget, and they take long to respond) is not a reliable indicator of deception.

Lie detection

Interest in "scientific lie detection" originated in Italy, Germany, Austria, and Switzerland in the late 1880s and persisted until World War II. Psychophysiological measurement techniques were introduced at that time and have been under development ever since (Ben-Shakhar and Furedy 1990, Gale 1988). The polygraph, or lie detector, is based on the concept that lying provokes specific physiological responses and that the associated emotions of the liar can be qualitatively detected and qualitatively interpreted. Most of the physiological events that are recorded by the polygraph are associated with breathing, blood pressure, the heartbeat, and galvanic skin responses, particularly in the palms of the hands.

Use of polygraphs

Polygraphs are in wide use in the United States in three contexts: criminal investigation, security vetting, and the selection of personnel for hiring or promotion. In several states, polygraph tests are admissible in court, while in others they are not. Polygraphs are also widely used in Canada, Japan, Australia, and Israel, and the United Kingdom. In Norway, Sweden, Holland, and Germany (where handwriting analysis is popular), polygraphs are of little interest and, in fact, frowned upon.

Lie-detector procedures rely on an assumed intimate interconnection between mind and body. They purportedly discover facts about the mind through the observation of physical responses to questions (Blinkhorn 1988). The polygraph is simply an instrument that records physiological events in connection with questions asked by the tester. The subject's reactions are then interpreted, and conclusions are drawn regarding the truthfulness of the examinee's statements.

Controversy About the Use of Polygraphy for Lie Detection

Great controversy has been raging regarding several aspects of polygraphic lie detection: the theoretical foundations of the practice, the procedures used, the interpretation of the results, the conclusions drawn, and the future implications for the person who is tested.

Foundations

With respect to the theoretical foundations of lie detection, much discussion has concentrated on whether there is in fact a "guilt reaction and emotion" and if it exists, whether it is reliable and consistent. Apparently, whether one feels such guilt depends on one's character, upbringing, cultural background, and life experience. Social habits play a major role, for example, in concealing the truth about a friend's illness, being defensive regarding a child's behavior, or bad-mouthing a disliked coworker. Some civilizations (as described, for example, by the anthropologist Margaret Mead, 1901–1978) have in fact made lying an art—even built community life on a pattern of lying and cheating. Growing up in a rough neighborhood certainly instills other life criteria than does living in a pampered fashion in a protective environment. Certain psychiatric disorders (e.g., antisocial personality disorder) are indeed characterized by the absence of guilt mechanisms or a conscience. Finally, physiological responses to questions may be able to be controlled; indeed, there are even "manuals on how to beat the polygraph" (Gudjonsson 1988).

Procedures

Polygraphic instrumentation has received relatively little attention in comparison to experimental design and control. The best-controlled situation is in the laboratory, where persons are randomly assigned to either the guilty or not-guilty categories and their reactions to questions tested. Yet, in real life, the status of the examinee is usually not known, which introduces a much more difficult situation. Unfortunately, very few field studies have avoided serious methodological errors (Raskin 1988). The most common test procedure is to determine whether a person is "guilty" or "not guilty," with the label "inconclusive" applied when neither of these conclusions can be drawn. For this, primarily two techniques are used. The first is the so-called control-question procedure, in which irrelevant questions ("Is today Thursday?") are mixed with relevant ones that ask, in essence, "Did you do it"? The second procedure is the "guilty knowledge test," which does not attempt to determine whether the respondent is lying, but rather, seeks to ascertain whether the examinee possesses "guilty" knowledge that would implicate the person (e.g., one question might be "Which knife did you use?") (Lykken 1988). The discussion about the relative merits and problems associated with these tests, and whether or not others should replace them, has been going on without any apparent resolution for decades.

Interpreting the results

The discussion of procedures spills over to the interpretation of the results. Several issues are of great importance. One is the relationship of the measured responses to "guilt," as previously mentioned, and another is the sensitivity of the observed reactions. Still another problem is that of accuracy, which may be measured as either "false-positive errors" (an innocent individual is wrongly classified as guilty) or "false-negative errors" (a guilty person is classified as innocent). Of course, the "inconclusive" finding also may be false, in one direction or the other.

Various measures of accuracy have been made using these criteria, most of them in laboratory settings, but just a few (for obvious reasons) in the real world. The results that were obtained often showed disturbingly low accuracy. In the control-question procedure, guilty subjects were correctly classified in 80 to 84 percent of the studies in both simulations and field tests, but the correct classifications of innocent subjects were only 63 to 72 percent. In guilty-knowledge tests, which were performed only in the laboratory, the average accuracy was about 84 percent for guilty subjects and 94 percent for innocent persons; thus, this method can better protect the innocent.

Reliability

Intra- and intertester repeatability is a further concern. Very few studies have been conducted, almost all laboratory simulations. The investigations found that the false-positive errors were underestimated; it is suspected that false negatives are underestimated in the real world (Ben-Shakhar and Furedy 1990).

There are other problems beyond sensitivity, accuracy, and reliability. For example, some polygraph techniques score certain results higher than others do. Also, varying and nonstandardized procedures are used; some allow the introduction of the tester's personal biases based on impressions gained of the subject that are not derived from the polygraph results.

Perhaps the most important aspect to consider in any discussion of polygraphic assessment is that the conclusions drawn regarding the person's character and the ensuing legal judgments can have far-reaching implications for his or her future. Obviously, then, these conclusions must possess the greatest possible degrees of reliability and validity obtainable.

Aldrich Ames, the "master spy" arrested in the CIA in February 1994, dismissed the polygraphs he regularly was given as a CIA employee as "witch-doctory." He said, with regard to passing the lie detector tests, "Confidence is what does it. Confidence and a friendly relationship with the examiner. . . . [a] rapport, where you smile and you make him think that you like him." Reported by Walter Pinkus in the Washington Post *May 4, 1994.*

SUMMARY

In a "person–machine–environment" system, the human perceives information simultaneously by various senses and plans and executes actions. Thus, the traditional concept of a linear sequence from stimulus to sensory input to perception to processing to effector output is probably overly simplistic and even unrealistic.

Accordingly, the central nervous system, as processor, receives information concerning the outside world (and the inside of the body) from receptors that respond to light, sound, touch, temperature, electricity, chemicals, and acceleration vectors. The information is transmitted in a converging manner along the afferent paths of the peripheral nervous system to the central nervous system, where it is processed and action signals are generated. These signals are sent along the efferent paths of the peripheral nervous system to effectors such as the hand or voice box. The selection of appropriate external stimuli and their transformation so that they can be reliably sensed are among the major tasks of the ergonomic engineer. On the output side, the design task is to select the proper output channels to control a machine.

Job demands and environmental conditions can be highly stressful to the human, both physically and psychologically. Dealing with the stress problem is made difficult by variation in terminology, concepts, and consequences. Different personalities may react differently to the same stressors. Still, in spite of the complexity of the problem, procedures for assessing job demands and for reorganizing and designing the system accordingly are in use. These can help in everyday situations, but are of particular importance in unusual situations, such as in space, where tasks, the environment, and the relations among crew members are novel.

Improving the performance of individuals is the goal of learning and training. To measure the effectiveness of certain practices is often a challenge for two reasons: Performance itself may be difficult to measure, or the specific "treatment variables" of the training may be difficult to define or separate. It is not clear why certain training practices are successful and others are not. Some "unconventional" techniques using, for example, hemispheric training or parapsychological postulates have no value; others, such as presenting subliminal messages, visualization, and observing experts, remain unproven; and some, such as biofeedback, have a highly specific efficacy. Much theoretical methodology still needs to be tested before sound techniques for the successful training of skillful performance can be developed.

CHALLENGES

What difference would it make if we discussed mental functions simply in terms of the brain instead of the mind?

How would our understanding of human mental processing change if we went from the traditional concept to the ecological approach? Would the concept of affordance be thereby modified?

Would a consideration of the various types and sheer number of internal and external sensory receptors make the traditional or ecological concept more plausible?

What transducers are imaginable to make distal stimuli into signals that trigger human sensors?

Consider driving fast in dense traffic. When is a driver more likely to have an accident, when one can either brake or switch to another lane, or when braking is the only possible action?

Why should the presentation of an emergency signal by several modalities (such as light and sound) together be advantageous over presenting the signal to only one type of sensor?

What might be a better design of an automobile braking system than the one currently used?

How could Fitts' law be applied to designing a workstation for manually assembling many small parts?

Why is it difficult to maintain a conversation while looking for somebody's telephone number in a phone book?

What means of measuring the workload of a driver might be employed during rush-hour traffic?

Under what conditions and to what extent should stress management programs be carried out by an employer?

What might be suitable training programs for crews that will go into long, confined missions, such as space exploration?

How can one reduce "end-of-mission" stressors?

State the advantages and problems associated with mixed-gender crews during long, isolated missions, such as space travel to Mars.

What means might be employed to make subliminal learning feasible?

How could one measure the effectiveness of training in terms of how it affects performance?

What carefully planned and executed measures might assure the highest possible team performance?

Chapter 4

Human Senses

OVERVIEW

Seeing, hearing, smelling, tasting, and touching are the "classical" human senses. The human is also sensitive to electricity and to pain and has a posture and motion (vestibular) sense. Of all senses, seeing and hearing are thoroughly understood, and information related to those senses is well suited for ergonomic application. However, information on the sense of touch, including the perception of temperature and sensitivity to electricity and pain, are much less researched, and engineering applications are rather haphazard. The senses of smell and taste are not usually used in engineering.

Proper ergonomic procedures include measures to correct sensory deficiencies, such as in seeing, and to protect sensory functions from damage, such as hearing. Little is commonly done to enhance or protect the other sensory capabilities.

INTRODUCTION

Only five senses?

The human is able to receive signals through several senses, originally classified by Aristotle (384–322 B.C.) into five different categories: seeing (vision), hearing (audition), smelling (olfaction), tasting (gustation), and touching (taction). (Taction might include more than one sense, because one feels several mechanical stimuli, such as contact and pressure, often together with pain, electricity, and temperature.) After more than two millennia, this classification is still commonly used, although it includes neither pain nor the vestibular sense by which we primarily balance the body.

Other classifications are by anatomy, by the sensory organs that perceive different stimuli, and by the stimuli themselves (i.e., the external objects or energies that trigger a sensation).

☛☛☛ *"Consider all the varieties of pain, irritation, abrasion; all the textures of licking, patting, wiping, fondling, kneading; all the prickling, bruising, tingling, brushing, scratching, banging, fumbling, kissing, nudging. Chalking your hands before you climb onto uneven parallel bars. A plunge into an icy farm pond on a summer day when the air temperature and body temperature are the same. The feel of a sweat bee delicately licking moist beads from your ankle. Reaching blindfolded into a bowl of Jell-O™ as part of [an] initiation. Pulling a foot out of the mud. The squish of wet sand between the toes. Pressing on an angel food cake. The near-orgasmic caravan of pleasure, shiver, pain and relief that we call a back scratch" (Ackerman 1990, pp. 80–81).* ☚☚☚

BODY SENSORS

Sensors are transducers

Every human sensor (sensory organ) can be modeled as a *transducer* that consists of two components. The first part is the receptor, which is stimulated by an appropriate proximal stimulus to produce some reaction. The second part is the converter, which codes or reproduces the reaction and generates an electric potential, a signal to be sent along a pathway of nerves.

Stimulation

The receptor must be stimulated by the right kind and quantity of the proximal stimulus (see Chapter 3) so that it can react. How this functions in reality is not fully understood, nor is it known either how the sensor acts as the converter of the signal received by the receptor or how it generates the potential that is sent as a sensory signal to the next neuron and on to the central nervous system.

Convergence of signals

The problem of how signals converge is quite complex, because numerous sensors are distributed over the receptive field—consider, for example, the many tactile sensors in the tip of the finger. The primary receptor cells (transducers) are linked either directly or in branching ways to a first-order afferent neuron, which is the first collecting and switching point that receives the input from all sensors within its receptive field. This neuron may or may not be triggered sufficiently by the incoming signals to send an action potential to the next higher, second-order afferent neuron, which also receives inputs from other first-order neurons. Thus, numerous first-order afferent neurons are linked to second-order neurons, which report to the next level in the hierarchy. The signal flow from several low-order neurons to a few higher neurons in the CNS, possibly repeated several times, follows the "principle of convergence," shown in Figure 4–1.

Stimuli

Some stimuli trigger only one type of sensor, but there are also many nonspecific stimulations. For example, electromagnetic and mechanical energies can trigger sensations of vision, hearing, touch, and pain, as well as sensations of the vestibulum, and thermal stimuli trigger sensations of cold and warmth, as well as pain sensations. The auditory sense responds to pressure changes at frequencies between 20 and 20,000 Hz. Chemical stimuli affect especially the senses of taste and smell.

Adaptation and Inhibition

We are bombarded every minute by a variety of stimuli, which would overwhelm us if we could not filter them out according to their importance. Since each of the afferent neurons may or may not transmit an arriving signal further up the chain toward the CNS, this converging network functions as a filtering or coding system, letting only those signals through that are sufficiently strong, that is of sufficient importance. The sensory system "adapts" by reducing the nervous discharge even while the strength of the proximal stimulus remains the same. A typical example is the adaptation of taction to the feeling of clothes worn. The system may be "inhibited" also by the reduction of nervous discharge, as when a second stimulus appears in the sensory field. A typical example is vigorous rubbing of a stubbed toe: The rubbing triggers new taction sensations which mask the pain that was originally felt. Except for the auditory and vestibular systems, very little is presently known about the excitation or inhibition in the CNS above the first-order neuron.

☛☛☛ *There are several explanations of sensory adaptation. The processes may involve the reduction in the firing rate or intensity of the trigger at the sensor, the reduction in transmission along*

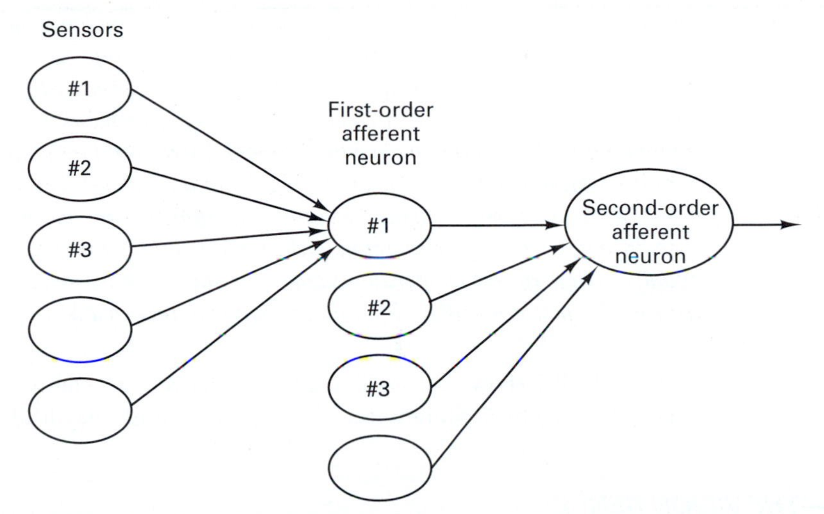

Figure 4–1. Model of converging receptive fields.

the afferent nervous path, or the masking of signals. Different researchers have attributed various meanings to the term "adaptation," and diverse research approaches have been used, none of which have yet yielded coherent results. The field still needs much work. ☚☚☚

Getting used to a background stimulus level so that we distinguish only important signals is probably a mixture of adaptation and inhibition that results from the presence of several stimuli. However, some receptors adapt slowly, keeping their nervous discharges at the same level for a second or more, while others adapt quickly, their discharge rate falling within milliseconds to a low value or zero.

Sensory Thresholds

A stimulus may be of such weak intensity that one cannot sense it: It is below the lower, or absolute, or minimal threshold. The upper or maximal threshold is the limit above which the sensor does not respond any more (such as to sound frequencies above 20 kHz), but at which the stimulus might still annoy or even damage the sensors.

The proportionality of sensation to a change in the stimulus is often of interest. For example, how do we react to a change of 10 dB in sound pressure at low pressure levels as opposed to high ones? The *difference threshold* is the smallest physical difference in the amount of stimulation that produces a just noticeable difference in the intensity of our sensation. This threshold was investigated in detail by Fechner and Weber in the 19th century. Usually, difference thresholds increase with increasing intensity: Weber's law states that the difference threshold is a constant proportion of the intensity of the stimulus. (The law may not hold for extremely small or large intensities of stimuli.)

☛☛☛ *There is an interplay between sensory adaptation and the difference threshold that is strikingly demonstrated in the classical experiment in which a person immerses one hand in warm water and the other in cold water. The initial sensations of warm and cold, respectively, subside slowly as one adapts to the water temperatures. Next, both hands are immersed in lukewarm water.*

Now the warm hand signals the sensation of cold, the cold hand the feeling of warmth. Apparently, the difference in water temperatures makes each hand report the change from the previous state. What would happen if the temperature of the third water container were close to that of one of the others? If the two temperatures were close enough, the hand that moves between them would not signal a noticeable difference, while the other hand would indicate a strong difference in felt temperatures. How would the sensations change if the overall temperature ranges were decreased? If they were increased? Further, what would happen if one got used to higher or lower temperatures through repeated immersion with stepwise increased temperatures? People who work habitually in high or low temperatures can "tolerate" these more easily. Is that a form of adaptation? How do the just-noticeable differences change in this case?

Table 4–1 shows the smallest energies that are detectable and the largest energies that are tolerable or practical. Stimuli near either of these limits may lead to unreliable sensing.

SEEING—THE VISION SENSE

The characteristics of human vision are well researched and described in the literature: Boff et al. (1986) devoted seven chapters to a detailed description of human vision.

Architecture of the Eye

As sketched in Figure 4–2, the eyeball is a roughly spherical organ about 2.5 cm in diameter, surrounded by a layer of fibrous sclera.

When parallel beams of light from a distant target reach the eye, they first encounter the *cornea,* a translucent, bulging, domed section of the sclera at the front of the eyeball, kept moist and nourished by tears. The cornea provides all the refraction needed to focus on an object more than about 6 meters (20 feet) away (if the eye is young and healthy).

Behind the cornea is the *iris,* tissue surrounding a round opening called the *pupil.* The dilator muscle opens, and the sphincter muscle closes, the pupil like the aperture diaphragm of a camera, regulating the amount of light entering the eye.

Having passed through the pupil, light beams enter the *lens.* If a distant object needs to be seen, suspensory ligaments keep the lens thin and flat, so that the light rays are not bent. For close objects (which cannot be focused by the cornea), the ciliary muscle around the lens makes it thicker and rounder, so that the light beams are refracted for suitable focus.

The space behind the lens, the interior of the eyeball, is filled with the *vitreous humor,* a gel-like fluid with refractory properties similar to those of water.

Light focused by the cornea and lens finally reaches the *retina,* a thin tissue that lines about three-quarters of the inner surface of the eyeball, opposite the pupil. The retina is supplied with blood by many arteries and veins and contains about 130 million light sensors.

There are two kinds of light sensors on the retina, named for their shape. The majority, about 120 million, are the *rods,* which respond to even low-intensity light and provide black, gray, and white vision. About 10 million *cones* respond to colored bright light.

Pigments in rods and cones serve to convert light into electrical signals. Each cone contains a pigment that is most sensitive to either blue, green, or red wavelengths. An arriving light beam, if intense enough, triggers chemical reactions in one of the three types of pigmented cones, creating electrical signals that are passed along the *optic nerve* to the brain, which can distinguish among about 150 color hues. Rods contain only one pigment, which, when bleached by light, sets

TABLE 4–1. Characteristics of the Human Senses

	Vision	*Hearing*	*Touch*	*Taste, smell*	*Vestibular*
Stimulus	Light-radiated electro-magnetic energy in the visible spectrum	Sound-vibratory energy, usually airborne	Tissue displacement by physical means	Particles of matter in solution (liquid or aerosol)	Accelerative forces
Spectral range	Wavelengths from ~400 to ~700 nm (violet to red), 10^{-6} mL to 10^4 L	20–20,000 Hz 20 μPa to 200 Pa	Temperature: 3 s exposure of 200 cm^2 of skin, 0 to 400 pulses/s	Taste: salty, sweet, sour, bitter Smell: flowery, fruity, spicy, resinous, burnt, and foul.	Linear and rational accelerations
Spectral resolution	120–160 steps in wavelength (hue), varying from 1–20 nm	~3 Hz for 20–1,000 Hz; 0.3% above 1,000 Hz	~10-percent change in number of pulses/s		
Dynamic range	~90 dB (useful range); for rods = 3×10^{-5}–0.0127 cd/m^2; for cones = 0.127–31,830 cd/m^2	~140 dB	~30 dB (0.01–10 mm displacement)	Taste: ~50 dB ($3*10^{-5}$% to 3% concentration of quinine sulphate) Smell: 100 dB	Absolute threshold is ~0.2 deg s^{-2}
Amplitude resolution ($\Delta I/I$)*	Contrast = 0.015	0.5 dB	~0.15	Taste: ~0.20 Smell: 0.10–50 dB	~0.10 change in acceleration
Acuity	1 min of visual angle	Temporal acuity (clicks) ~0.001 s	Two-point acuity ranges from 0.1 mm (tongue) to 50 mm	?	?
Response rate for successive stimuli	~0.1 s	~0.01 s (tone bursts)	Touches sensed as discrete to 20/s	Taste: ~30 s smell: ~20–60 s	~1-2 s; nystagmus may persist to 2 min after rapid changes in rotation
Best operating range	500–600 nm (green yellow) at 34–69 cd/m^2	34–69 Hz at 40–80 dB		Taste: 0.1% to 10% concentration	~1-g acceleration directed to foot

SOURCE: Adapted from VanCott and Kinkade 1972; Boff and Lincoln 1988.

*I = intensity level; ΔI = smallest detectable change in intensity from I.

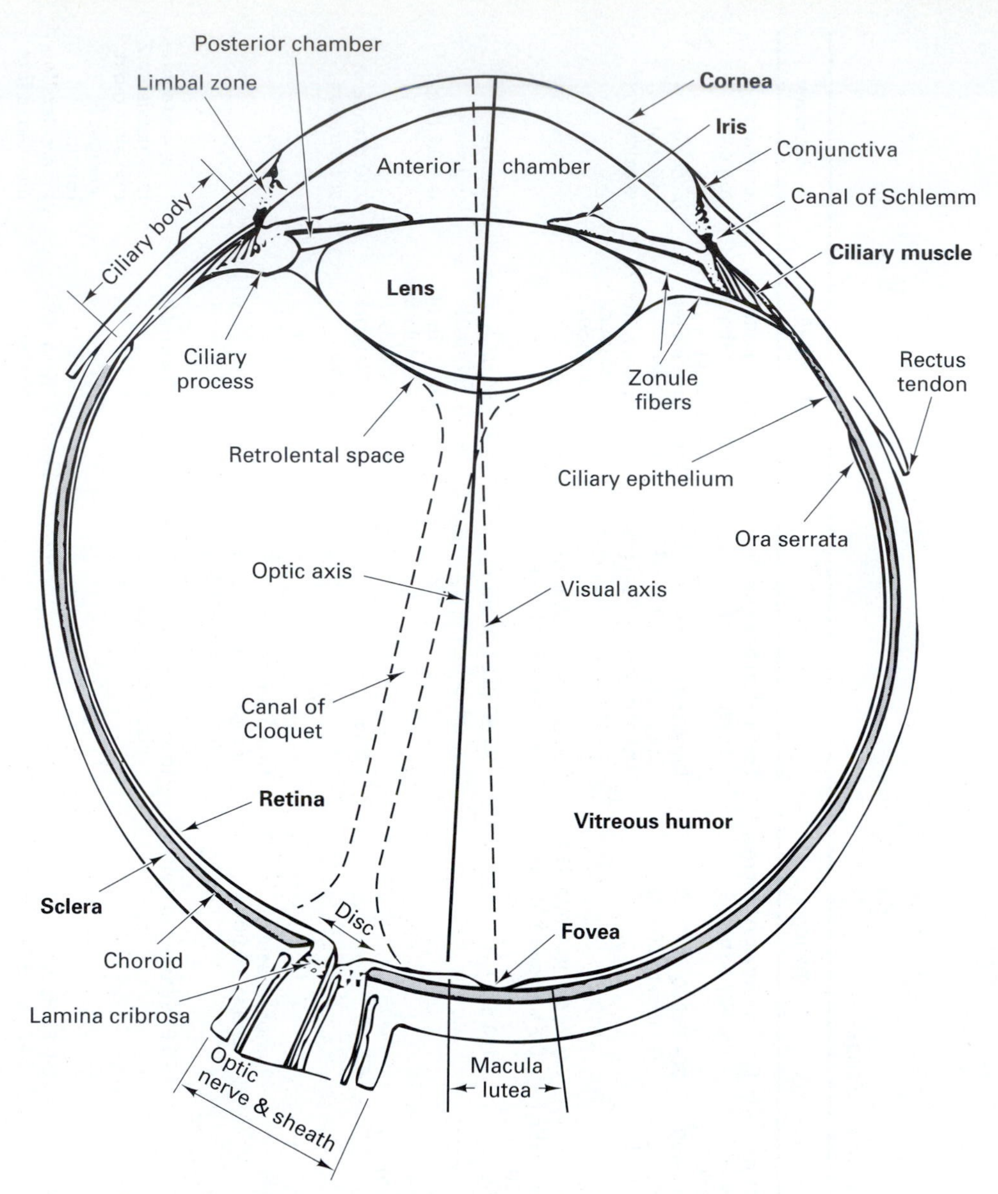

Figure 4–2. Horizontal section through the human eye. (Reprinted with permission from Snyder 1985.)

off electrical impulses that are sent along the optic nerve to the brain for the perception of white, black, and shades of gray.

On the retina, cones are concentrated in the center, directly behind the pupil, where only few rods are found. This area is called the *fovea.* Together with its yellowish surrounding region, known as the *macula,* the fovea is used mostly to read fine print.

The optic nerve exits the eye at its rear, about 15 degrees off center toward the inside. Since there are no light sensors in this area, an image at this *blind spot* cannot be seen. However, since the blind spots of both eyes are medially located, they do not overlap in our field of vision, and therefore, we are unaware of their existence.

☛☛☛ *The Arab physicist Alhazen (965–1039) was the first to discover that vision was made possible by rays of light falling on the eye and was not the result of rays of light sent out by the eyes. Alhazen studied lenses and attributed their magnifying effect to the curvature of their surfaces. His work represents the beginning of the scientific study of optics. At the middle of the 13th century, convex lenses for eyeglasses used by the farsighted—that is, mostly by the elderly—were well known in both China and Europe. In 1451, the German scholar Nicholas of Cusa used concave lenses for the nearsighted (Asimov 1989).* ☚☚☚

Mobility of the Eyes

Each eye is theoretically capable of movements in six degrees of freedom, illustrated in Figure 4–3. Mostly used are rotational movements in *pitch* (around the y-axis), *roll* (around the x-axis), and *yaw* (around the z-axis), but the entire eye also moves linearly—most appreciably, forward and backward (along the x-axis). Thus, the center of rotation of the eye is not fixed, but displacements of the center are fairly small. Therefore, one usually assumes the center of rotation to remain in place, approximately 13.5 mm behind the cornea.

Six striated muscles are attached to the outside of the eye and control its movements. (See Figure 4–4). Since the muscles do not attach at orthogonal directions to each other, they interact. The superior and inferior recti muscles are responsible primarily for pitch, the up-and-down eye rotation. The medial rectus and the lateral rectus muscles provide yaw, the left and right movement. The oblique muscles predominantly provide roll movement.

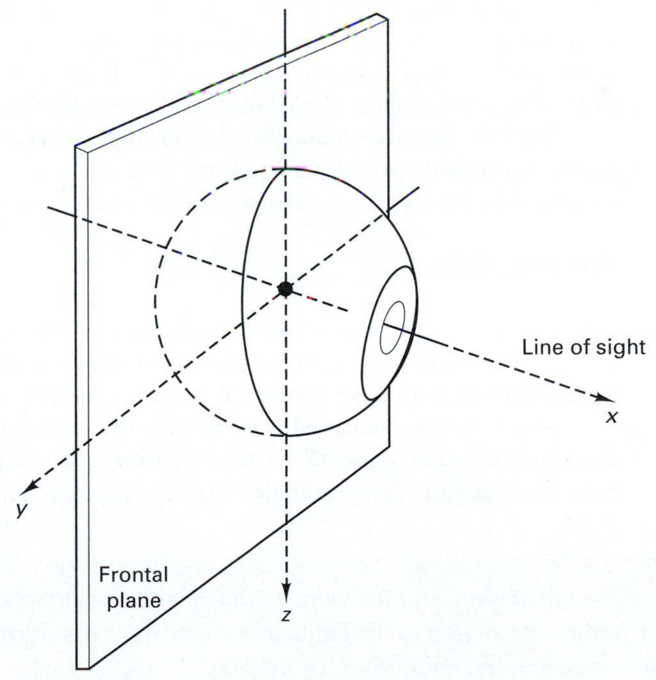

Figure 4–3. Mobility axes of the eye.

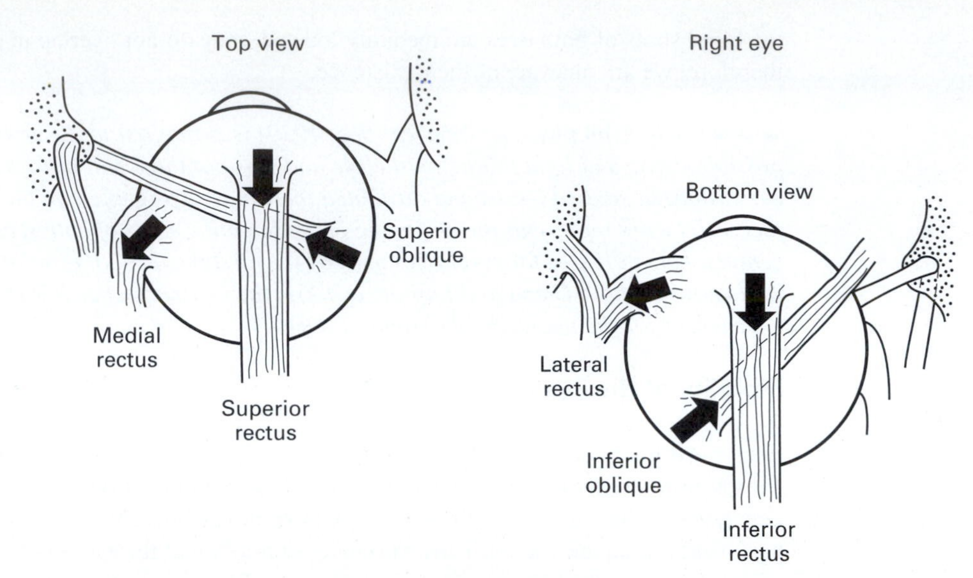

Figure 4–4. Muscles moving the right eyeball.

The six muscles work in concert to produce eye movements. When one looks straight ahead, the eyes are said to be in "primary position." The so-called primary movements of the eye are in the two degrees of freedom associated with up and down gaze motions (pitch) and left and right movements (yaw). Given the complex attachments of the muscles moving the eye, these primary movements are accompanied by some roll. The further the eye pitches and yaws, the more the eye also rolls. The amount of eyeball roll at various pitch and yaw angles away from the primary position is about 1 degree with 10 degrees each of pitch or yaw, 3 degrees with 20 degrees of pitch or yaw, 8 degrees with 30 degrees of pitch or yaw, and 15 degrees with 40 degrees of pitch or yaw. We are not aware of this tilting of the visual image as we move the eyes, because the brain compensates for it.

The eye can track continuously left and right (yaw) a visual target that is moving less than 30 degrees per second or cycling at less than 2 hertz. Above these rates, the eye is no longer able to track continuously, but lags behind and must move in jumps (saccades) to catch up to the visual target.

Line of Sight

If the eye is fixated on a point target, the *line of sight* (LOS) runs from the object, through the lens (pupil), to the receptive area on the retina—most likely, the fovea. Thus, the LOS is clearly established within the eyeball. To describe the LOS direction external to the eye, a suitable reference is needed.

The horizon is often used as reference for the external line-of-sight angle (LOSA) in the medial (*xz*) or a sagittal plane. (See Chapter 1.) Yet, when looking at an object in front of us, we unconsciously adjust two pitch angles: that of the eyes within the head and that of the head itself.

For ergonomic design of work and the workstation, it is important to know each of two angles: LOSA is the eye pitch angle with regard to the head, and P describes the posture of the head as the pitch angle of the head with respect to the neck, trunk, or horizon.

Line-of-sight angle Various approaches have been taken to define the line-of-sight angle LOSA with respect to the head. One is to define it against the so-called Frankfurt plane, which is attached to the skull via the landmarks tragion and the lowest point on the rim of the orbit. (See the glossary.) The orbit is difficult to palpate, whereas it is easy to determine the ear–eye plane because its landmark attachments to the head (the ear canal and the external junction of the eyelids) can be seen and need not be palpated. (For converting angles, the ear–eye plane is, on average, 11 degrees more pitched than the Frankfurt plane (Kroemer 1994). For yaw and roll, perpendicular planes meeting at the eye can be defined. Figure 4–5 illustrates the LOSA and the P angles.

Head angle It used to be difficult to establish the position of the head with respect to the trunk. The links between skull and trunk are ill defined mechanically and consist of at least the cervical vertebrae on top of the thoracic column. These vertebrae have various mobilities in their intervertebral joints. The skull also rotates in three degrees of freedom at the atlas of the first cervical vertebra. A further complication arises from the fact that there is no easily established reference system within the trunk in regard to which one could describe the relative displacements of the neck and head. The use of the ear–eye line and its pitch angle P against the horizon partially circumvents this posture problem.

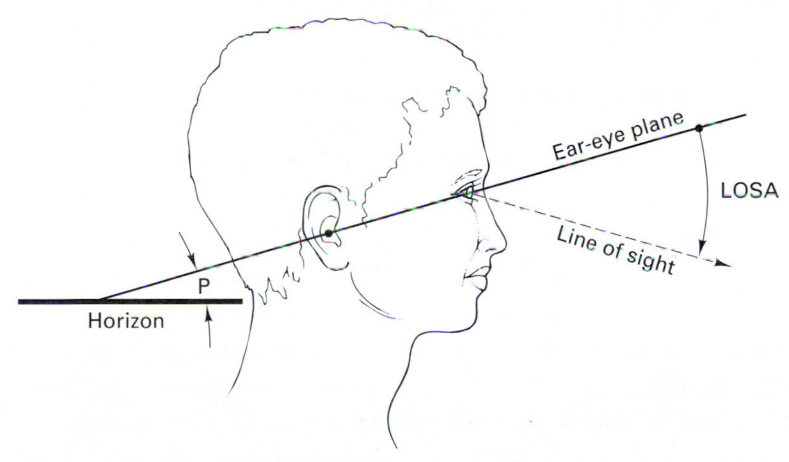

Figure 4–5. The Ear–Eye (EE) Line passes through the right ear hole and the juncture of the lids of the right eye. The EE Line serves as a reference for head posture and for the tilt angle of the line of sight, LOSA.

Visual target size If a visual target is not a point, but can be expressed as the length of a line perpendicular to the line of sight, then the target size is usually expressed as the subtended visual angle—the angle formed at the pupil. The magnitude of this angle depends on the distance D of the object and on its size L. The subtended visual angle α is described in Figure 4–6 and is usually given in degrees of arc (with 1 degree = 60 minutes = 60 × 60 seconds of arc). The equation is

$$\alpha \text{ (in degrees)} = 2\arctan(0.5\ LD^{-1}). \qquad (4\text{–}1)$$

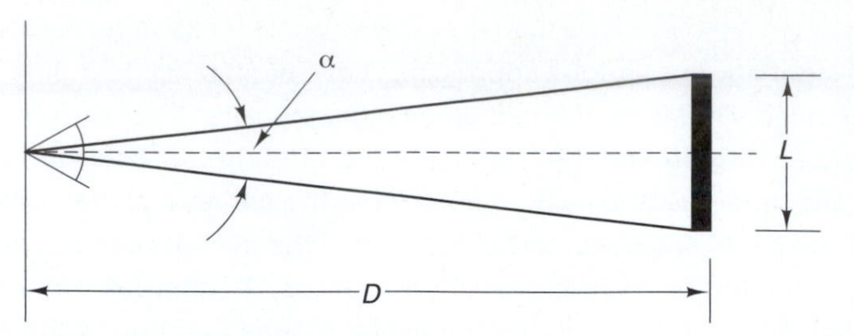

Figure 4–6. The subtended angle.

For visual angles not larger than 10 degrees, this can be approximated by α (in degrees) $= 57.3\ LD^{-1}$ or α (in minutes of arc) $= 60 \times 57.3\ LD^{-1} = 3{,}438\ LD^{-1}$. Note that the equations do not take into account the distance between the pupil and the lens, because that approximately 7-mm distance has no perceptible effect under most conditions.

The human eye can perceive, at a minimum, a visual angle of approximately 1 minute of arc, or 1 second of vernier acuity. Table 4–2 presents the visual angles of familiar objects. For ease of use, technical products should be so designed that the angle subtends at least 15 minutes of arc, increasing to at least 21 min of arc at low light levels.

Diopter

If the target distance D is measured in meters, the reciprocal, $1/D$, is measured in *diopters*. The diopter indicates the optical refraction needed for best focus. Thus, a target at infinity has the diopter value zero, while a target at 1 meter's distance has the diopter value unity (one). Table 4–3 shows values for some typical target distances.

The Visual Field

The "visual field" is the area, measured in degrees, within which the form and color of objects can be seen by both fixated eyes. In its center, the visual field of each eye is occluded by the nose. To the sides, each eye can see a bit over 90 degrees, but only within the inner about 65 degrees can color be perceived.

TABLE 4–2. Visual Angles of Familiar Objects

Object	*Distance*	*Visual Angle (Arc)*
Gage, 5 cm diameter	0.5 m	5.7 degrees
Sun		about 30 minutes
Moon		about 30 minutes
Character on CRT screen	0.5 m	17 minutes
Pica letter at reading distance	0.4 m	13 minutes
U.S. quarter coin at arm's length	0.7 m	2 degrees
U.S. quarter coin	82 m (90 yards)	1 minute
U.S. quarter coin	3 miles	1 second

TABLE 4–3. Target Distances and Associated Focal Points

Target Distance D *(m)*	*Focal Point (Diopter)*
Infinity	0
4	0.25
2	0.25
1	1
0.67	1.5
0.50	2
0.33	3
0.25	4
0.2	5

Farther in the periphery, shades of gray dominate, because of the locations of cones and rods on the retina, as discussed earlier. Upward, the visual field extends through about 55 degrees, where it is occluded by the orbital ridges and eyebrows. However, color can be seen only to about 30 degrees upward. The downward vision is limited by the cheek at about 70 degrees; the area in which color can be seen extends down to about 40 degrees. The "blind spot" on the retina is at approximately 15 degrees to the outside, for each eye.

Rotating the eyeball increases the visual area to the "field of fixation," adding about 70 degrees to the outside of the visual field, but nothing in the upward, downward, or inside directions, because the orbital ridges, cheeks, and nose stay in place. The field of fixation is largest in downward gaze and smallest in upward view.

If, in addition to the eyeballs, the head moves, nearly everything in the environment can be seen, as long as it is not occluded by the body or other structures, when the head and eyes are turned in that direction. Of course, mobility of the head, achieved by neck muscles, is sharply reduced in persons with a "stiff neck," often the elderly. These individuals should locate visual targets close to their naturally chosen line of sight.

Accommodation

Focusing

The action of focusing on targets at various distances is called *accommodation*. (There is some confusion in the meaning and use of optical and visual terms; see Miller (1990) for a thorough discussion.) The normal young eye can accommodate from infinity to very close distances, meaning that a diopter range from 0 to about 10 can be achieved.

Normally, with visual objects at distances of 6 m or more, the relaxed lens refracts the incoming parallel rays (i.e., it focuses them) on the retina. To accommodate closer objects, the lens is made thicker by the pull of the enclosing ciliary muscle: The radius of the frontal lens is reduced from 10 to 6 mm and the radius of the rear surface from 6 to 5 mm. Thus, the "optical refracting power of the eye" is changed by adjusting the curvature of the lens so that, by refraction of the incoming light rays, the image of the target falls focused upon the retina at every distance. (One speaks of the retina being "conjugated to the object" at all accommodations.) If one "looks at nothing," the lens accommodates at a distance of about 1 m, with much variation among individuals.

Amplitude of accommodation

The point that can be focused at the closest distance is called the *near point.* The farthest point that can be focused without conscious accommodation is called the *far point.* The difference between far and near points is called the *amplitude of accommodation.* Most young people can focus at a near point of about 10 cm, but this minimal distance changes to about 20 cm at age 40 and to about 100 cm at age 60, on average.

Convergence

If one aims both eyes at the same point, an angle exists between the two lines of sight connecting each eye with the target. This angle is very small—in fact, negligible—at objects more than 6 m away, but becomes fairly large at short distances. While it takes about 200 ms to fixate on one point at reading distance, the time is reduced to about 160 ms when one focuses at a point 6 m away or farther. If the eyes are not made to converge, an "error of vergence" exists, also called "fixation disparity." This condition commonly occurs when the quality of the binocular stimulus is low—for example, when the observer is greatly fatigued or under the influence of alcohol or barbiturates (Heuer and Owens 1989). A "phoria" exists if the images of one target are not focused on the same spots on the retinas of both eyes, resulting in double images.

Visual Fatigue

People doing close visual work, such as those who often look at the screen of a computer display, frequently complain of eye discomfort, visual fatigue (Chi and Lin 1998), or eyestrain (asthenopia), all of which are vaguely called "subjective visual symptoms or distress resulting from the use of one's eyes" (National Research Council Committee on Vision 1983, p. 153). While the occurrence of such eyestrain (and its intensity) varies much among individuals, it seems often related to the effort of focusing at a distance that is different from the personal minimal "resting distance of accommodation." Apparently, many instances of eye fatigue of which computer users often complain are related to the poor placement of the monitor, source documents, or other visual targets (see Chapters 8 and 9) or to unsuitable lighting conditions at their workplace, as discussed later in this chapter.

For most people whose eyes look ahead, but do not focus on a target, that minimal distance of binocular vergence—also called the "dark vergence" or "dark focus" position, minimal refractive state, or resting distance of accommodation—is about one meter away from the pupils. (This finding is in contrast to the older assumption that the automatic resting position is at "optical infinity"—that is, with parallel visual axes of the eyes.) The actual point of vergence is characteristic of the individual's oculomotor resting adjustment and therefore is also called a "resting tonus posture." The automatic selection of the dark vergence (resting) distance apparently is not only due to the biomechanics of the extraocular muscles, but is also dependent on neural processes that integrate information about head and eye positions. As one lowers the angle of gaze or tilts the head down, the resting point gets closer to the eyes, to about 80 cm for a LOSA of 60 degrees downward. However, the resting distance increases as one elevates the direction of sight, to about 140 cm, on average, at a 15-degree upward direction. Many elderly people find it more difficult to focus on an elevated target than on a lower one, partly because of reduced mobility of the head due to a "stiff neck" (Heuer et al. 1989, 1991; Jaschinski-Kruza 1991; Miller 1990; Owens and Leibowitz 1983; Rabbitt 1991; Ripple 1952; Tyrrel and Leibowitz 1990; von Noorden 1985).

It is natural to look down at close visual targets, such as a written text. Optometrists have always known this and placed the reading section of (bifocal or trifocal) corrective lenses at the bottom of the lens.

Several aspects are of particular interest for the design of work systems that require exact binocular eye fixation.

- First, the binocular resting position and the accommodation distance to which the eyes return at rest are quite different from individual to individual, but are constant for a given person. Thus, one should allow and encourage each person to select the "personal distance" from a target that must be visually fixated, such as a computer screen.
- Second, tilting either the eyes or the head brings about a similar effect on the natural vergence distance. This explains why people find it more comfortable to lean back in a chair while tilting the head forward and down to look at a close object (at about eye height), rather than to sit upright. Doing so reduces the natural vergence distance.
- Third, targets at or near "reading distance" (for example, computer displays) should be distinctly below eye level, particularly if the viewer is elderly.
- Fourth, it is difficult to see a visual target of low optical quality clearly when the observer is fatigued or under the influence of drugs. This finding of practical importance is well known from experience.

Paraphrasing Tyrrell and Leibowitz (1990, p. 342), scientists and grandmothers can now agree that visual fatigue is related to near work, because it is easier to look down on it than to look up to it.

Visual Problems

Worse with age

With increasing age, the accommodation capability of the eye decreases, because the lens becomes stiffer by losing water content. This condition is known as *presbyopia.* The result is difficulty in making light rays converge exactly on the retina. If the convergence is in front of the retina, the condition is called myopia; if the focal point is behind the retina, one speaks of a hyperopic eye.

Better with age?

A nearsighted (myopic) person finds it difficult to focus on far objects, but has little trouble seeing close objects. This condition often improves with age, when the lens remains flattened. In fact, even then, far objects are not exactly in focus, but the rays emanating from them still strike the retina (although not exactly in focus), so that these distant objects can be sufficiently identified.

Worse with age!

In contrast, farsightedness becomes more pronounced with age, meaning that it gets more difficult to focus on near objects. Both problems, myopia and hyperopia, can be fairly easily corrected by either contact lenses or eye glasses. In many people, the pupil shrinks with age. This means that less light strikes the retina, and therefore, many older people need to have increased illumination on visual objects for sufficient visual acuity.

Another problem often encountered with increasing age is *yellowing of the vitreous humor.* The more yellow it gets, the more energy it absorbs from the light passing through; consequently, increased illumination of the visual target is needed to maintain good acuity. Also, light rays are refracted within the vitreous humor, bringing about the perception of a light veil (like mist) in the visual field. If bright lights are in the visual field, the resulting "veiling glare" can strongly reduce one's vision. Obviously, the yellowing problem cannot be corrected with artificial lenses.

Floaters are perceived as small flecks in front of the eye. In reality, they consist of small clumps of gel or cells suspended in the vitreous humor. Floaters are visible to the individual only when they are on the line of sight, casting a shadow on the retina. Frequently, they are not noticed as the eye adjusts to these imperfections. They are more easily perceived when one is looking at a plain background. Occasional floaters are usually harmless, although in rare cases, especially when they occur suddenly, they may be precursors of retinal damage.

Cataracts are pattern of cloudiness inside the normally clear lens. When vision is severely impaired, the cloudy lens may be surgically removed and replaced with an artifical implant.

Glaucoma is the leading cause of blindness in the United States, especially for older people. Glaucoma is a preventable disease of the optic nerve, related to high pressure inside the eye. Regular eye examinations help to detect the beginning of glaucoma and to prevent further damage.

There are other vision defiencies as well, usually not related to aging.

Astigmatism occurs if the cornea is not uniformly curved, so that an object is not sharply focused on the retina, depending on its position within the visual field. Often, the astigmatism is a "spherical aberration," meaning that light rays from an object located at the side are more strongly refracted than those from an object at the center of the field of view, or vice versa. *Chromatic aberration* is fairly common: An eye may be hyperopic for long waves (red) and myopic for short waves (violet or blue). Placing an artificial lens in front of the eye usually solves problems associated with astigmatism.

Night blindness is the condition of a person having less than normal vision in dim light—that is, with low illumination of the visual object.

Color weakness exists if a person can see all colors, but tends to confuse them, particularly in low illumination. Defective color vision is rather common in men, about 8 percent of whom are color-defective, compared with less than 1 percent of women. Some people are *color blind,* meaning that they confuse, for example, red, green, and gray. Only very few people can see no color at all or only one color.

☛☛☛ *Isaac Newton (1642–1727) performed experiments in which he made a beam of light pass through a glass prism. What he saw emerging was a band of colors, with red being the least bent portion of the light, followed by orange, yellow, green, blue, and violet. Making these colorful lights pass through another prism, he showed that they merged to become white light again. In this way, Newton showed that white light was not "pure," but a mixture of different colors (Asimov 1989).* ☚☚☚

Vision Stimuli

Humans are sensitive to light, meaning that they detect changes in visual stimulation, depending on the wavelength, intensity, location, or duration of the stimulus. Usually, discussions of light sensitivity are limited to wavelength and intensity.

The human eye is sensitive and can adapt to increases and decreases in illumination over a wavelength range of about 380 nm to 720 nm—that is, from violet to red. The minimal intensity required to trigger the sense of light perception is 10 photons, or an illuminance* of the eye of about 0.01 lux. At such low intensity, shorter wavelengths (e.g., blue green), are more easily perceived than longer wavelengths; the main perception is of light, not of color.

Viewing Conditions

Cones perceive colors only if an object's illumination* is "bright," above about 0.1 lux. This is called the *photopic* condition. If the illuminance falls between 0.1 and 0.01 lux, both cones and rods respond. This is called the *mesopic* condition and is present at twilight and dawn, when we can see color in the (brighter) sky, but (dimmer) objects appear only in shades of gray. In dim light, below about 0.01 lux, only rods respond. In this *scotopic* condition, only black and white, and gray shades in between, can be perceived.

Figure 4–7 shows, schematically, the sensibility of rods and cones to visible wavelengths, the so-called spectral sensitivity. Rods respond mostly to shorter wavelengths, while cones cover the whole spectrum. However, the schematic is drawn by making the maximum of each curve equal to 100 percent; in reality, the three different pigments of cones have their maximal sensitivities at around 440, 540, and 570 nm, while rods have their single maximal sensitivity at about 510 nm. In terms of luminous intensity (which has been set to 100 percent for the rod and the cone curves in the figure), rods are more sensitive because they respond to lower intensities than cones. This means that, as the ambient light level increases, visual detection shifts from domination by the rod system to predominant use of the cones. In the "dark," the threshold for detecting light of almost any wavelength is determined by the sensitivity of the rods; under "bright" conditions, the cones provide most of the information, independently of the wavelengths of the stimuli. Between these two extremes of illumination, the spectral distribution of light determines (and other parameters codetermine) which system is more sensitive. The transition from using rods to using cones is due largely to a desensitization of the rods by relatively weak light that leaves the cone sensitivity intact. These conditions are shown schematically in Figure 4–8. Throughout the wavelength spectrum, cones need higher intensities of light to function.

☛☛☛ *There are some interesting phenomena associated with vision at night:*

- *In darkness, the color-sensitive cones at the retina are inactive, and what is "seen" is perceived through the rods. Directly behind the lens are only a few rods at the back of the retina. The area constitutes a blind spot, where objects may not be detected at night.*
- *If one stares at a single light source on a dark background, the light seems to move. This is called the* autokinetic phenomenon.
- *If the horizon is void of visual cues, the lens relaxes and focuses at a distance of about 1 to 2 m, making it difficult for a person to notice far objects. This is known as* night myopia.

*See the section titled "Photometry" for terms and units of measurement. Note that the words "bright", "light", "dim", and "dark" indicate subjective judgements, not objective measurements.

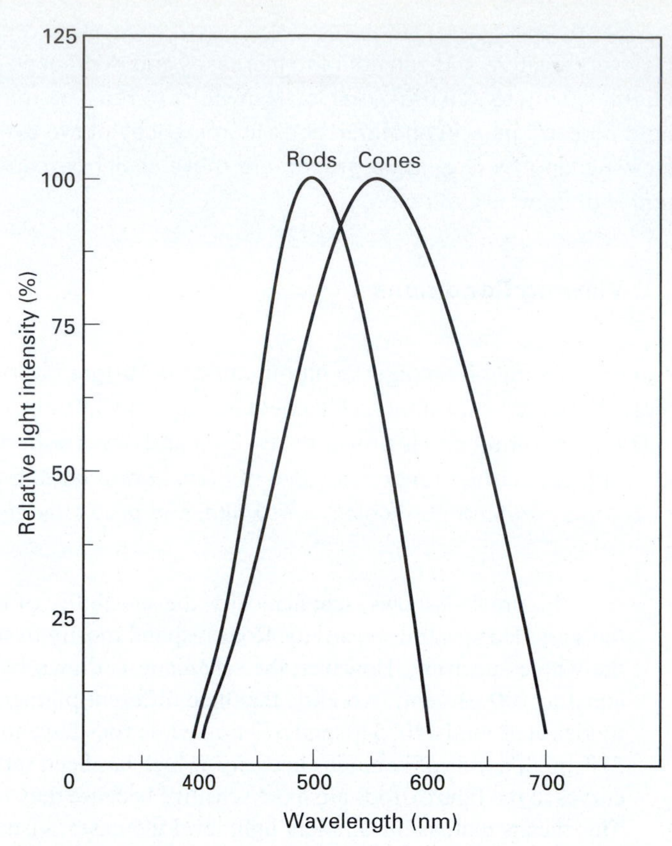

Figure 4–7. Sensitivity of rods and cones to wavelength. The figure shows conditions at illuminations below 0.01 lx (scotopic) and above 0.1 lx (photopic), but not in the intermediate (mesopic) range.

- *Night vision capabilities deteriorate with decreasing oxygen. Thus, at an altitude of 1,300 m (4,000 ft), vision is reduced by about 5 percent; at 2,000 m, the reduction is about 20 percent and up to 40 percent in smokers whose blood has lost some capability to carry oxygen.*

Adaptation

Adaptation to light and dark

The eye can change its sensitivity through a large range of illumination conditions. This property is called visual (or ocular) adaptation, and one commonly distinguishes between adaptation to light and to dark conditions. Adaptation is achieved by several measures: adjustments of the pupil, the spatial summation of stimuli, and photochemical functions, including the stimulation of rods and cones. The actual change in response thresholds of the eye during adaption to the dark or light depends on the luminance and duration of the previous condition to which the eye was adapted, on the wavelength of the illumination, and on the location of the light stimulus on the retina.

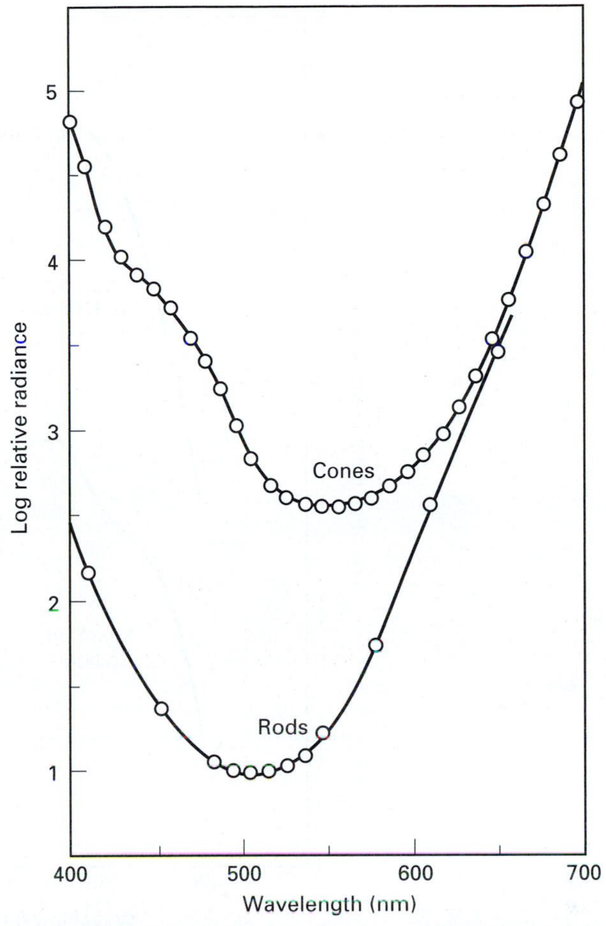

Figure 4–8. Thresholds of cones and rods to light intensity. (Modified with permission from Snyder 1985.)

Full adaptation from light to darkness takes about 30 minutes; during this period, initially the cones are most sensitive, and then the rods. After adaptation, the sensitivity at the fovea (with a preponderance of cones) is only about one-thousandth that at the periphery of the retina (with a preponderance of rods). Therefore, weak lights can be noticed in the periphery of the field of view, but not if one looks directly at them (i.e., when they are refracted onto the fovea).

As Figure 4–9 shows, adaptation to the dark is governed mostly by the change in threshold of the cones. Maximal sensitivity is at about 510 nm, a shift from adaptation to the light, when the highest sensitivity is at about 560 nm. People who suffer from "dark blindness" have nonfunctional rods and can adapt only via cones. Persons who are "color blind" have nonfunctional cones (or only one or two of the three types of cones) and adapt via rods only.

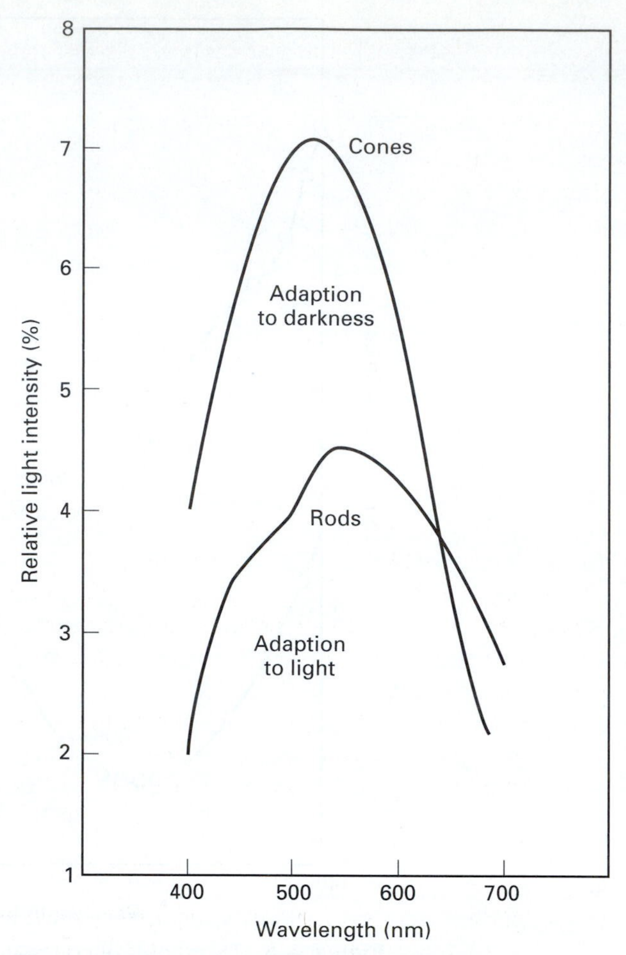

Figure 4–9. Adaptation of cones and rods.

Adaptation from darkness to light is very fast, fully achieved within a few minutes. In contrast, full adjustment to the dark might take up to 30 minutes.

In adaptation to light, wavelengths of about 560 nm (in the yellow region of the visible spectrum) are most easily perceived at the fovea, with a shift to wavelengths of about 500 nm (yellow-green) in adaptation to the dark. Above 650 nm, where there are large differences in response to the different wavelengths at the shorter waves, the spectral sensitivity of the fovea is not very different during adaptation to either the dark or the light: With increasing adaptation to the dark, shortwave stimuli are perceived as having higher luminance. This effect can be an important consideration for the selection of colors in dark rooms. Also, persons who must adapt to the dark after having been in a lighted environment adapt faster to long-wavelength lights (red and yellow) than to short wavelengths (blue).

Human visual response characteristics can be fruitfully considered, for example, in regard to the night illumination of instruments. If the illumination is only by wavelengths longer than about 600 nm (orange), mainly the photopic (cone) system is excited, but the adaptation of the scotopic (rod) system is largely maintained: One can still observe the (mostly black–grey–white) events on the road while driving a car. Therefore, instruments are often illuminated with reddish or yellowish light, with the yellow color probably more suitable for the older eye.

The relative sensitivity of the visual system across the visual spectrum can be measured in various ways, but the variable that is most often used is the intensity needed to detect a stimulus of a given wavelength, called the *threshold intensity.* (Hood and Finkelstein (1986) and Snyder (1985a, b) provide more detailed discussions of this topic.)

Visual Acuity

Visual acuity can be defined in several ways, usually as the ability to detect small details and discriminate small objects with the eye. Visual acuity depends on the shape of the object and on the wavelength, illumination, luminance, contrast, and duration of the light stimulus. Acuity is usually measured at viewing distances of 6 m (20 ft) and a 0.4 m (1.3 ft), since the factors that determine the resolution of an object can differ in the far and near viewing requirements.

Visual acuity may be limited by either optical or neural causes. Optical problems lead to a degraded retinal image, which can often be improved by corrective lenses. Neural limits stem from the coarseness of the retinal mosaic (distribution of receptors) or the sensitivity limits of neural pathways.

Acuity testing To assess acuity, high-contrast patterns are presented to the observer from a fixed distance. The smallest detail detected or identified is taken as the threshold, expressed in minutes of visual arc. Visual acuity is then expressed as the reciprocal of the resolution threshold. It is assumed that a person should normally be able to resolve a detail that subtends about 1 minute of visual arc.

The most common testing procedures use either Landolt rings or Snellen or Sloan letters, all standardized black stimuli against a highly contrasting white background. A measurement of 20/20 on the Snellen chart is regarded as perfect vision. A person is defined as partially sighted if her or his vision (after correction) is worse than 20/70, but still better than 20/200 (meaning that this person can see an object at 20 ft that others can see at 70 or 200 ft, respectively). By definition, a person is legally blind if the vision in the better eye is, after correction, only 20/200 or poorer.

Conditions affecting acuity

Since the results of acuity measures are dependent on the testing conditions, such as the test pattern and its distance from the eyes, the National Academy of Sciences (1980) recommended standards for assessing acuity. The following are among the conditions that affect acuity:

- *Luminance level.* For dark targets on a light background, acuity improves as the background luminance increases (up to about 150 cd/m^2; see later for units of measure).

- *Locus of stimulation on the retina.* The highest resolution is obtained under photopic illumination with the target viewed foveally. At scotopic illumination, acuity is highest when the target is offset approximately 4 degrees from the fixation point.
- *Pupil size.* The highest acuity is observed when the pupil is at an intermediate diameter.
- *Viewing distance.* For a perfect lens, there should be no effect of distance if the visual angle is used to describe the conditions of viewing. However, acuity does change with the viewing distance, because the lens of the eye changes its shape to fixate at different distances. If accommodation errors occur, the image may be blurred. These errors are more pronounced at low luminance levels.
- *Age.* Spatial vision capabilities change considerably with age, generally leading to a decline in acuity starting in the forties.

Visual acuity depends primarily on the ability to see edge differences between black and white stimuli, measured at rather high illuminance levels. Such measurement of static edge acuity is simple, but it is neither the only nor the best measure of visual resolution capabilities. For example, people with perfect Snellen acuity may not do so well in other measures of contrast sensitivity, such as the ability to detect targets from a busy background or to see highway signs at given distances. As one's gaze sweeps, the visual details in the field of view generate an image of ever-changing spatial frequencies and contrasts on the retina. Thus, one can consider the visual world as a constantly changing array of textures composed of varying contrasts and spatial frequencies. Accordingly, contrast sensitivity changes with differing viewing fields, and this "field dependency" is operationally defined by the tests that measure it. Many of these tests use a certain geometric shape—for example, a triangle—that is embedded in a more complex geometric figure.

Contrast

Contrast is defined as the change in luminance of a pattern, or, mathematically, $(L_{max} - L_{min})/(L_{max})$, where L_{max} is the highest luminance and L_{min} the lowest luminance of the alternation pattern. To test perceived thresholds of contrast, it is customary to plot the reciprocal of the contrast threshold function, measured as the modulation $M = (L_{max} - L_{min})/(L_{max} + L_{min})$, and to call this quantity "contrast sensitivity." Complete measures of contrast sensitivity assess visual resolution through ranges of spatial frequencies and contrasts.

Measurement of Light (Photometry)

The measurement of light energies and the perception of those energies by the human observer has been an area of much confusion. The confusion stems from several sources, one being the differences between physically defined and humanly perceived lighting conditions. For example, the subjective descriptors "bright" or "light", and "dim" or "dark" are not solidly related to physical measurements of luminous intensity, illuminance, or luminance, defined shortly. A second source of confusion lies in various competing terminologies, although Boyd (1982) and Pokorny and Smith (1986) have clarified the situation.

Radiometry

Light can be defined as any radiation capable of causing a visual sensation. The natural source of such electromagnetic radiation is the sun. Lamps (also called *luminaires*) are common artificial sources. Measurement of the quantity of radiant energy may be done by determining the rise in temperature of a blackened surface that absorbs radiation.

Table 4–4 lists the four fundamental types of energy measurement:

1. The total radiant energy emitted from a source per unit of time: *radiant flux.*
2. The energy emitted from a point in a given direction: *radiant intensity.*
3. The energy arriving at (incident on) a surface at some distance from a source: *irradiance.*
4. The energy emitted from or reflected by a unit area of a surface in a specified direction: *radiance.*

Lines radiating from a point p define a cone with a solid angle ω with a spherical surface A at a radius r from p. (See Figure 4–10.) The unit of ω is 1 steradian (sr) when $A = r^2$.

TABLE 4–4. Terms, Symbols, and Units of Light Energy

Term	*Symbol*	*Units*
Radiant flux	P_e	watt (W)
Radiant intensity	I_e	$W \times sr^{-1}$
Irradiance	E_e	$W\ m^{-2}$
Radiance	L_e	$W\ m^{-2}\ sr^{-1}$

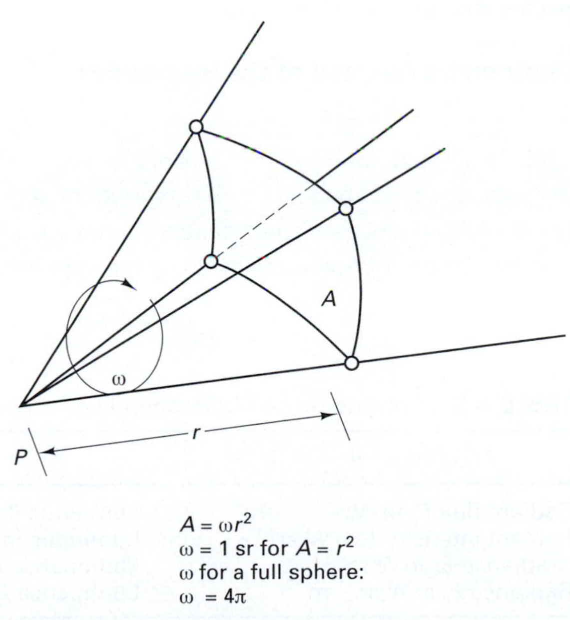

$A = \omega r^2$
$\omega = 1$ sr for $A = r^2$
ω for a full sphere:
$\omega = 4\pi$

Figure 4–10. Definition of the steradian (sr).

The energy emitted from point p is the radiant flux P_e in watt (W). With ω expressed in steradians, the *radiant intensity* per solid angle is

$$I_e = P_e\,\omega^{-1}, \text{ in units of W sr}^{-1}. \tag{4–2}$$

Most artificial light sources do not radiate uniformly in all directions; hence, the radiant intensity may be different in different directions.

The *irradiance* E_e (the radiant flux incident on a unit area of surface) of a sphere is $E_e = P_e\,\omega^{-1} r^{-2}$ in W m^{-2}. Irradiance can also be expressed in terms of the radiant intensity as

$$E_e = I_e\, r^{-2} \text{ in W m}^{-2}. \tag{4–3}$$

The radiant flux emitted by a point source falls on successively greater areas as the distance from the source increases. Since the irradiance E_e changes inversely with the square of the distance r from the source p, the relation between irradiance and distance is called the "inverse-square law of energy flux." Of course, if the surface is flat (instead of being a section of a sphere around p), not all surface elements are at the same distance from the point source p. Thus, areas of the flat surface are irradiated at varying angles, rather than perpendicularly. If a flat surface section is at an angle α from the normal, the irradiance at any point on the section is

$$E_e = I_e\, r^{-2} \cos \alpha. \tag{4–4}$$

The preceding definitions and equations and the inverse-square law are valid only for a true point-source light, such as a star. However, for most calculations, the error made with artificial light sources is negligible. Most light sources are not points, but have finite dimensions; they are called "extended sources." The *radiance* L_e describes the actual density of emitted radiation in a given direction or arriving at a surface, taking into account the projection angle (Pokorny and Smith 1986).

Symbols and units used in radiometry are listed in Table 4–4; their relation to those used in photometry is given in Table 4–5.

Photometry Adapted to the Human Eye

The optical conditions of the human eye, the sensory perception of stimuli, and CNS processing modify the physical conditions described so far. To take account of this fact, in 1924 the Commission Internationale de l'Eclairage (CIE) developed the "standard luminous efficiency function for photometry." In 1951, the CIE adopted a standard luminous efficiency function for dark-adapted (scotopic) vision. Both standards are still valid.

TABLE 4–5. Photometric and Corresponding Radiometric Terms

Radiometry	*Photometry*
Radiant flux P_e in W	Luminous flux F_v in lumen (lm)
Radiant intensity I_e in W sr^{-1}	Luminous intensity I_v in candela (cd) = lm sr^{-1}
Irradiance E_e in W m^{-2}	Illuminance (or illumination) E_v in lux (lx) = lm m^{-2}
Radiance L_e in W sr^{-1} m^{-2}	Luminance L_v in cd m^{-2} = lm sr^{-1} m^{-2}

SOURCE: Adapted from Pokorny and Smith 1986.

Accordingly, photometric energy is radiant energy modified by the luminous efficiency function of the "standard observer." Thus, a set of units parallel to those specifying radiant energy defines photometric energy. (See Table 4–5.) Several other terms are also still in use, a number of them quite unnecessarily, such as "candlepower" instead of "luminous intensity."

With regard to human vision, *luminance,* the light energy reflected (or emitted) from a surface, is the most important variable, unless we look directly into a light source.

Manufacturers of lamps often use the expression "lumens per watt" as a measure of luminaire efficiency, because measurement in watts describes only the electrical power intake of the light source. Typical values are 1,700 lm for a 100W incandescent lightbulb and 3,200 lm for a 40W fluorescent light.

Among the nonmetric units still in use are the following:

Illuminance, the amount of light falling (incident) on a surface, is occasionally measured in footcandles, where 1 footcandle (fc) = 1 lm ft^{-2} and 10.76 fc = 1 lx., Also, 1 phot = 1 lm cm^{-2}; and 10^5 phot = 1 lx. The troland is a unit for the illuminance on the retina from a source with 1 cd m^{-2} luminance, viewed through an artificial pupil of 1 cm^2.

Luminance, the amount of light energy reflected (or emitted) from a surface, is occasionally measured in lamberts (L), or cd $cm^{-2}\,\pi^{-1}$; 3,183 L = 1 cd m^{-2}. Also, we have the footlambert (fL), measured in cd $ft^{-2}\,\pi^{-1}$; 3.426-L = 1 cd m^{-2}. In addition, we have the stilb, or 10^{-5} cd cm^{-2}; 10^5 stilb = 1 cd m^{-2}. And finally, 1 apostilb = 1 cd $m^{-2}\,\pi^{-1}$, and 0.3183 apostilb = 1 cd m^{-2}; and the nit = 1 cd m^{-2} = 10^5 stilb.

Throughout the ages, every nation, and sometimes every region within a nation, developed its own system of measurements. The differences between these were not important as long as trade and communications were slow and infrequent. In France, a commission was appointed in 1790 to develop a system of reasonable measurements. Members of the commission were such great scientists as Laplace, Lagrange, and Lavoisier. The commission generated a system founded on basic natural units, such as the meter (from the Greek, "to measure"), which was equal to 10^{-8} times the distance from the North Pole to the equator. As far as possible, other units were developed that interconnected with the meter. Larger and smaller units were derived by multiplying or dividing by 10. This metric system, so far our most useful and logical system of measurements, was slowly accepted throughout the globe, with the United States one of the last holdouts against it (Asimov 1989).

Color Perception

Complementary colors

Two colors that together appear white are called *complementary.* For example, when yellow and blue lights in proper wavelength proportions are projected together onto a screen, the resultant light appears white.

Color is an experience

Sunlight contains all visible wavelengths, but objects onto which the sun shines absorb some radiation. Thus, the light that an object transmits or reflects has an energy distribution different from the light it received. A human looking at the (transmitting or reflecting) object does not analyze the spectral composition of the light reaching the eyes; in fact, what appears to be of identical color may have different spectral contents. The brain simply classifies incoming signals from different groups of wavelengths and labels them colors by experience. Human color perception, then, is a psychological experience, not a single, specific property of the electromagnetic energy we see as light.

☛☛☛ *"[W]e judge colors by the company they keep. We compare them to one another, and revise according to the time of day, light source, memory." (Ackerman 1990, p. 152) "Not all languages name all colors. Japanese only recently included a word for "blue." . . . Primitive languages first develop words for black and white, then add red, then yellow and green; many lump blue and green together, and some don't bother distinguishing between other colors of the spectrum. . . . [T]he Maori of New Zealand . . . have many words for red—all the reds that surge and pale as fruits and flowers develop, as blood flows and dries" (Ackerman 1990, p. 253).* ☚☚☚

Depending largely on the distribution of cones (and rods) over the retina, not all its areas are equally sensitive to all colors. We see all colors while looking fairly straight ahead, but cannot perceive any colors at the very periphery of our visual field. Green, red, and yellow all can be perceived within an angle of about 50 degrees to the side from straight ahead, while blue can be seen to about 65 degrees sideways and white even at 90 degrees.

Colorimetry

The concept of equivalent-appearing stimuli provides a system for measuring and specifying color (Pokorny and Smith 1986). *Colorimetry* is an experimental technique in which one simultaneously views two spectral fields and tries to adjust the spectral content of one to make both appear identical.

Primary colors: red, green, blue

Color-matching experiments have shown that the human can perceive the same color if one variously mixes three independent adjustable primary colors, viz, red, green, and blue. (This "additive" combination of spectral radiations is not the same as mixing pigments, discussed next.) Lights that contain dissimilar spectral radiations, but that are nevertheless perceived as the same color by the observer, are called *metameric.* Since it is possible to find a metamere for any color by varying only the three primary colors, human color vision is called *trichromatic.*

Mixing pigments

☛☛☛ *When an artist mixes yellow- and blue-pigmented paints to generate green, this is done by "double subtraction." The yellow pigment absorbs all blue, but reflects yellow as well as red and green. The blue pigment absorbs yellow and red, but reflects blue and green. Both pigments mixed together reflect only green, since the blue pigment absorbs the yellow and the yellow pigment the blue. However, if light of a different wavelength were to fall on the pigments, they would each absorb different "colors" and, combined, reflect a different wavelength that might not appear green.* ☚☚☚

Pigments reflect different wavelengths

We see an object as white when it reflects light about equally throughout all wavelengths; but so do gray and black pigments. The differences among gray, black, and white are not of color, but of how much light they reflect. Freshly fallen snow reflects only about 75 percent of the sunlight falling on it, but it appears bright white to us. Black velvet appears very dark black because it reflects only very little of the light that falls on it. Figure 4–11 shows, schematically, how pigments reflect light of different wavelengths (i.e., colors).

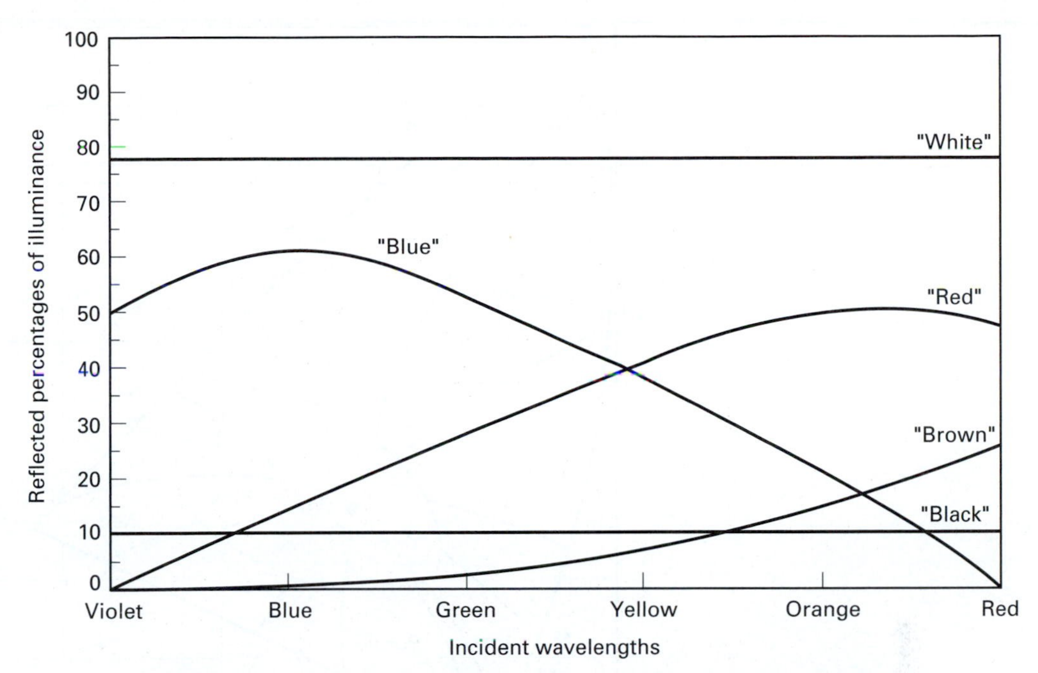

Figure 4–11. Different colors reflected by different pigments (schematic).

Trichromaticity

Trichromaticity is characterized by two facts:

1. Any color can be produced by combining (adding together) a suitable mixture of three arbitrarily selected spectral radiations.
2. Colors (wavelength components) can be specified precisely, making it possible to compare colors.

CIE chromaticity diagram

Thus, the data of a color-matching experiment can be expressed in terms of vectors in a three-dimensional (red, green, and blue) space. The results of the experiment can be shown in a color plane, or chromaticity diagram, which is a two-dimensional representation of what happens when colors are mixed. The diagram is obtained by converting the values of the three stimuli into a form such that the sum of the three equals unity. By convention, X corresponds to the red region of the spectrum, Y to the green region, and Z to the blue region, and $X + Y + Z = 1$.

If the proportions of red and green are plotted along the abscissa X and ordinate Y, respectively, the combined wavelength—the third trichromatic coefficient—falls within a horseshoe-shaped curve called the *spectrum locus.* This curve is shown in Figure 4–12.

The chromaticity diagram, standardized in 1931 by the CIE, plots colors by the amounts of standard red, blue, and green primaries that, when mixed together, yield the given color. The red, green, and blue lights are mathematically specified and called the three primaries.

Because the sum of red, blue, and green required to match any color is expressed as unity, any one standard color component can be determined by subtracting the other two from unity.

The most saturated colors are located on the horseshoe curve. Other perceived colors fall inside this boundary. The white located in the center of the diagram, consisting of one-third red, one-third green, and one-third blue, is designated as "illuminant C" by the CIE; it is about the color of light from a northern summer noon sun on a clear day.

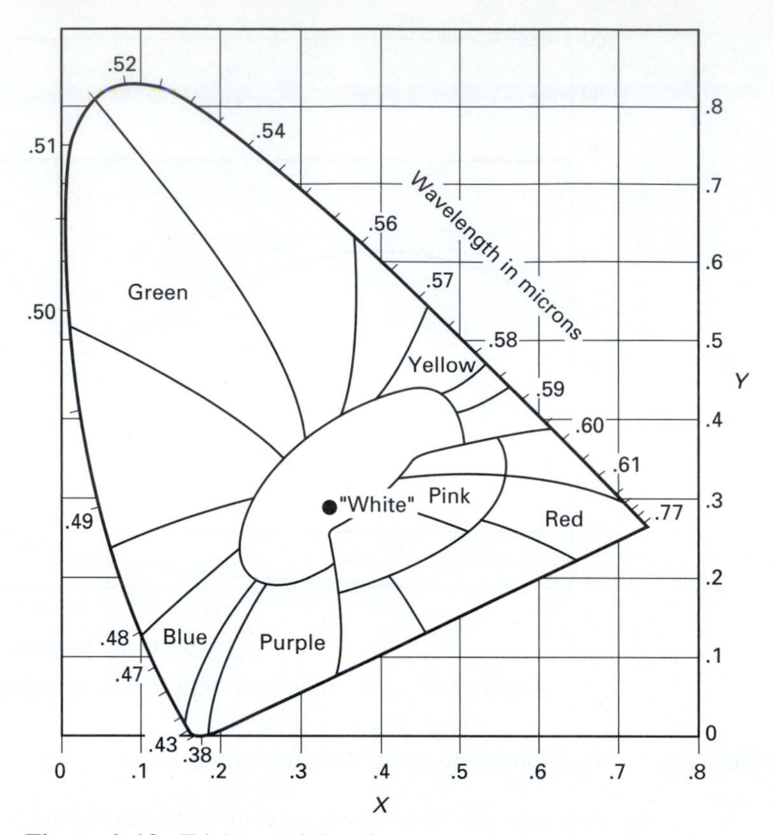

Figure 4–12. Trichromaticity diagram.

The numbers along the curve are wavelengths. A color appears greenish in the upper corner, reddish in the lower right corner, and bluish in the lower left corner. The "purity" of each color is determined by its proximity to the center of the diagram, where it would appear white. Although the CIE diagram is six decades old, it is still in common use.

OTHER THEORIES OF COLOR VISION

The trichromatic theory (Young-Helmholtz Theory), just described, is not the only one attempting to explain human color vision. The main competitors are Hering's Opponent Colors theory and Judd's Zone or Stage theories; but other theories are still being developed and tested, as discussed by Pokorny and Smith (1986). The complex nature of color appearance is not yet fully understood, and the mathematical models available are not entirely satisfactory. A human's judgment of the perceived color of a visual stimulus depends on the person's subjective impressions experienced when viewing the stimulus, and the judgment varies with the viewing conditions and the kind of stimuli.

Color-Ordering Systems

Many systems that order colors according to various variables and criteria are possible; several have been well developed and are often used.

There are three major groups of color systems. The first group is based on principles of additive mixtures of color. Examples are the Maxwell disk and the Ostwald Color System. The second group uses the principles of color subtraction, such as are applied in mixing pigments. This group is widely used in printing. The third group is based on the appearance and perception of colors. Physical standards are selected to present scales of, for example, hue, chroma, and likeness. Examples from this group are the Munsel Color System and the Optical Society of America (OSA) system (Wyszecki, 1986).

Terminology

The following terms are often used, although not in a universal, uniform, or standardized manner (Wyszecki, 1986):

Brightness and dimness are individual perceptions of the intensity of a visual stimulus.

Chroma is the attribute of color perception attained by judging to what degree a chromatic color differs from an achromatic color of the same lightness. The 1942 *Munsell Book of Color* provides many good examples. (See also "saturation," later in this listing.)

Chromatic or achromatic induction is a visual process that occurs when two color stimuli are viewed side by side, when each stimulus alters the color perception of the other. The effect of chromatic or achromatic induction is usually called *simultaneous contrast* or *spatial contrast.*

Hue is an attribute of color perception that uses color names and combinations thereof, such as yellow or yellowish green. The four unique (or unitary) hues are red, green, yellow, and blue, none of which is judged to contain any of the others.

Illuminant color is that color perceived as belonging to an area that emits light as the primary source.

Lightness is the individual perception of how much more or less light a stimulus emits in comparison to a "white" stimulus also contained in the field of view.

Object color is that color perceived as belonging to an object.

Related color is a color seen in direct relation to other colors in the field of view.

Saturation is the attribute of color judgment regarding the degree to which a chromatic color, differs from an achromatic color, regardless of lightness. (See also "chroma," earlier in this listing.)

Surface color is that color perceived as belonging to a surface from which the light is reflected or radiated.

Unrelated color is a color perceived to belong to an area seen in isolation from other colors.

All of the preceding terms rely on perception and judgment. Occasionally, some of them are used inappropriately in place of physically determined concepts. For example, (physical) luminance is not the same as (psychological) brightness. Thus, confusion is rampant.

Esthetics and Psychology of Color

While the physics of color stimuli arriving at the eye can be well described (although often with considerable effort), perception, interpretation, and reaction to colors are highly individual, nonstandardized, and variable. Thus, people find it very difficult to describe colors verbally, given the many possible combinations of individually perceived hue, lightness, and saturation values.

People believe in and describe nonvisual reactions to color stimuli. For example, reds, oranges, and yellows are usually considered "warm" and stimulating. Violets, blues, and greens are often felt to be "cool" and to generate sensations of cleanliness and restfulness. Note, however, that the attraction to certain colors and their combinations, vary culturally and regionally, as travel, say, between Asia and Europe readily shows. Still, pale colors often seem cooler than dark colors, cold colors more distant than warm colors, weak colors more distant than intense colors, soft edges of color patches more distant than hard edges, etc. Although experimental evidence regarding these effects is controversial (Kwallek and Lewis 1990), color schemes are often applied to work and living areas to achieve these stereotypical responses.

ILLUMINATION CONCEPTS IN ENGINEERING AND DESIGN

The characteristics of human vision just discussed provide the bases for engineering procedures to design environments for proper vision. The most important concepts are the following:

1. Proper vision requires sufficient quantity and quality of illumination.
2. Special requirements on visibility, especially the decreased seeing abilities of the elderly, require particular care in the arrangement of proper illumination.
3. Ilumination of an object is inversely proportional to the distance from the light source.
4. Use of colors, if selected properly, can be helpful; but color vision requires sufficient light.
5. What counts most is the luminance of an object, that is, the energy reflected or emitted from it, which meets the eye.
6. Luminance of an object is determined by its incident illuminance, and by its reflectance:

$$\text{luminance} = \text{illuminance} \times \text{reflectance} \times \pi^{-1} \quad (4\text{–}5)$$

   Reflectance is the ratio of reflected light to received light, in percent. (The numerical value for illuminance is in lux and for luminance in cd m^{-2}—see Table 4–5. The factor π^{-1} is omitted when the following nonmetric units are used: luminance in footlamberts (fL), illuminance in footcandles (fc). Figure 4–13 shows luminance levels experienced by humans.
7. The ability to see an object depends largely on the luminance contrast between the object and its background, including shadows. Contrast is usually defined as the difference in luminances of adjacent surfaces, divided by the larger luminance:

$$\text{Contrast (in percent)} = 100\,(L_{max} - L_{min}) / (L_{max}) \quad (4\text{–}6)$$

8. Avoid unwanted or excessive glare. There are two types of glare. Direct glare meets the eye directly from a light source (such as the headlights of an oncoming car). Indirect glare is reflected from a surface into the eyes (such as the headlights of a car in the rearview mirror). Often, kinds of reflected glare are described as either specular (coming from a smooth, polished surface such as a mirror), spread (coming from a brushed, etched, or pebbled surface), diffuse (coming from a matte, nonglossy painted surface), or compound (a mixture of the various types of glare).

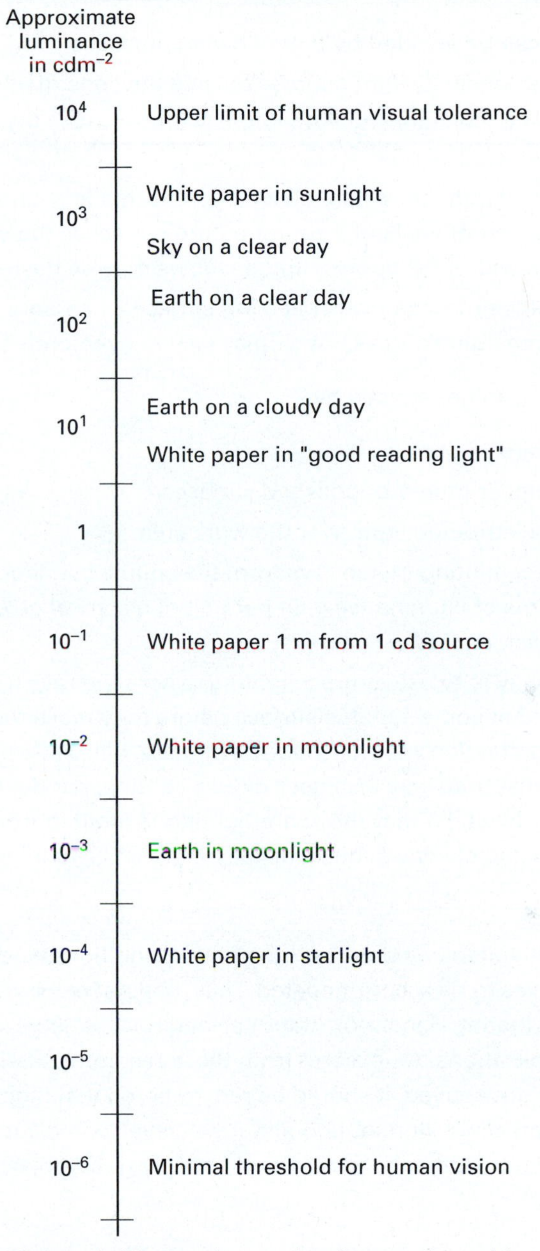

Figure 4–13. Luminance levels experienced by humans. (Adapted from Van Cott and Kinkade, 1972.)

Direct glare can be avoided by the following practices:

- Placing high-intensity light sources outside the cone of 60 degrees around the line of sight.
- Using several low-intensity light sources placed away from the line of sight, instead of one intense source.
- Using indirect lighting, where all light is reflected at a suitable surface (within the luminaire or at the ceiling or walls of a room) before it reaches the work area. This generates an even illuminance without shadows. (But shadows may be desirable to see objects better.)
- Using shields or hoods over reflecting surfaces, or visors over a person's eyes, to keep out the rays from light sources. (A clumsy way to overcome a condition of bad design.)

Indirect glare can be reduced by:

- Diffuse, indirect lighting.
  - Dull, matte, or other nonpolished surfaces.
  - Properly distributed light over the work area.
- Use of direct lighting (when rays from the source fall directly on the work area) is most efficient in terms of illuminance gain per unit of electrical power; but it can produce high glare, poor contrast, and deep shadows.
- The other way is to use indirect lighting, where the rays from the light sources are reflected and diffused at some suitable surface before they reach the work area. This helps to provide an even illumination without shadows or glare, but is less efficient in terms of the use of electrical power. A third way also uses diffuse lighting, but the light source is enclosed by a large translucent bowl, so that the room lighting is emitted from a large surface. This can cause some glare and shadows, but is usually more efficient in using electrical power than indirect lighting is.

What is most suitable depends on the given conditions, such as the task to be done, the objects to be seen, and the eyes to be accommodated. Thus, general recommendations are difficult to compile. Nevertheless, the IES Lighting Handbook (Kaufman and Haynes 1981) contains recommended illuminance values for certain applications. An excerpt from these recommendations is provided in Table 4-6. Before the recommendations are applied, it should be remembered that high reflectances at surfaces such as that of a ceiling or wall can allow illumination at a lower level, while surfaces with low reflectance probably require high illumination. Either condition could supply good object luminance.

Seven Vision Myths

Myth 1: *Straining your eyes can damage your eyesight. Examples of strain are working in dim or glaring light, reading fine print, wearing glasses with the wrong prescription, or staring at a computer screen.*

Truth: *Prolonged use of the eyes under any of the stated conditions can cause eyestrain, since the eye muscles struggle to maintain a clear or unwavering focus. In addition, prolonged staring can dry the front of the eye somewhat, because it reduces blinking, which helps lubricate the cornea. But fatigue and minor dryness, no matter how uncomfortable, cannot permanently harm your vision.*

TABLE 4–6. Recommended Illuminance Values (Examples)

Illuminance categories and illuminance values for generic types of activities in interiors			
Type of Activity	*Illuminance Category*	*Ranges of Illuminances Lux*	*Reference Workplane*
Public spaces with dark surroundings	A	20 to 50	
Simple orientation for short temporary visits	B	50 to 100	General lighting throughout space
Working spaces where visual tasks are occasionally performed	C	100 to 200	
Performance of visual tasks of high contrast or large size	D	200 to 500	Computer workstations
Performance of visual tasks of low contrast or very small size	E	500 to 1,000	Illuminance on task critical
Performance of visual tasks of low contrast and very small size over a prolonged period	F	1,000 to 2,000	
Performance of very prolonged and exacting visual tasks	H	5,000 to 10,000	Illuminance directed on task obtained by a combination of general and local (supplementary) lighting
Performance of very special visual tasks of extremely low contrast and small size	I	10,000 to 20,000	

TABLE 4–6. (continued)

Examples

Area/Activity	*Illuminance Category*	*Area/Activity*	*Illuminance Category*
Auditorium		Barber shop, beauty parlor	E
Assembly	C	Club and lodge room	
Social activity	B	Lounge and reading	D
Bank		Conference room	
Lobby		Conferring	D
General	C	Critical seeing (refer to individual task)	
Writing area	D	Courtroom	
Teller station	E	Seating area	C
		Court activity area	E
		Dance hall, discotheque	B
		Book binding	
Assembly Work		Folding, assembling, pasting	D
Simple	D	Cutting, punching, stitching	E
Moderately difficult	E	Embossing and inspection	F
Difficult	F	Brewery	
Very difficult	G	Brew house	D
Exacting	H	Boiling and key washing	D
Bakery		Filling (bottles, cans, kegs)	D
Mixing room	D	Candy making	
Makeup room	D	Box department	
Oven room	D	Chocolate department	
Decorating and icing		Husking, winnowing, fat extraction crushing and refining, feeding	D
Mechanical	D	Bean cleaning, sorting, dipping,	
Hand	E	packaging, wrapping	
Scales and thermometers	D		D
Wrapping	D	Milling	E

SOURCE: Adapted from Kaufman and Haynes 1981.

Of course, it makes sense to minimize the discomfort, by using the following techniques:

- *"Lighten up." Age tends to cloud the lens of the eye and shrink the pupil, sharply increasing the need for luminance. So if reading strains your eyes, consider installing brighter lights or at least moving the reading lamp closer to the page.*
- *Cut the glare. Position the reading lamp so that light shines from over your shoulder, but make sure it does not reflect into your eyes from the computer monitor. Do not read or do computer work near an unshaded window. And wear sunglasses if you are reading outside.*
- *Stop and blink. When you are working at the computer or reading, pause frequently—say, every 15 minutes—to close your eyes or gaze away from the screen or page, and blink repeatedly. Every hour or so, get up and take a longer break.*
- *Find and maintain the right distance. Keep your eyes at the same distance from the screen as you would from a book. If that is uncomfortable, buy a pair of glasses with a prescription designed for computer work, or use "progressive-addition" bifocals, which have gradually changing power from the top to the bottom of the lens.*

- *Lower the screen. Keep the top of the screen well below eye level. Gazing upward can strain muscles in the eye and neck.*
- *Use a document holder. Put a support for reading matter next to the screen, at the same distance from your eyes.*
- *Clean your screen and your glasses. Dust and grime can blur the images.*
- *If the preceding steps do not reduce the strain from either reading or computer work, have an optometrist or ophthalmologist check whether you need to start wearing glasses or to have your prescription changed.*

Myth 2: *The more you rely on your eyeglasses or contact lenses, the faster your eyesight deteriorates.*

Truth: *This myth is based on the misconception that artificial lenses do the work of the eyes, which then supposedly grow lazy and weak. However, artificial lenses merely compensate for a structural defect of the eye—an improperly shaped eyeball or an excessively stiff lens—that prevents proper focusing, despite the best efforts of the lens muscles. When you wear glasses or contact lenses, the eye muscles no longer need to be tensed or relaxed (depending on whether you are farsighted or nearsighted) any more often than usual. Instead, they work just as hard as the muscles in a normal eye.*

Myth 2: *Eye exercises can help many people see better.*

Truth: *Eye exercises can help some children whose eyes have major binocularity problems, such as crossing, misalignment, or inability to converge. But claims that the exercises can help many children read better are unsubstantiated.*

Some maintain that eye exercises not only help one read better, but also sharpen visual acuity, boost athletic performance, and help correct numerous problems in both children and adults. Such claims are unsupported and implausible.

Myth 4: *The more carrots you eat, the better your eyesight will be.*

Truth: *Carrots are rich in beta-carotene, an orange pigment used by the body to manufacture vitamin A, which is essential for night vision. But a reasonably well-balanced diet supplies enough beta-carotene for the eyes. Extra doses of that nutrient—or of vitamin A itself—do not improve vision.*

However, beta-carotene and other carotenoid nutrients in fruits and vegetables do fight oxidation, a chemical change that can damage cells in the body. Antioxidant nutrients, particularly beta-carotene, may reduce the risk of cataracts and slow the progression of macular degeneration.

Myth 5: *The darker the sunglasses, the better is the protection against harmful ultraviolet light.*

Truth: *In theory, prolonged exposure to the ultraviolet (UV) rays in sunlight increases the risk of cataracts and, possibly, macular degeneration, because UV rays oxidize tissues in the eye. But the darkness or tint of the sunglasses has no effect on how much UV light they absorb. So you need to check the label—or, if possible, have an optician test the lenses—to ensure that they provide adequate protection. At the very least, look for sunglasses labeled "Meets ANSI Z80.3 General-Purpose UV Requirements." These sunglasses block at least 95 percent of the high-energy UVB rays and 60 percent of the lower energy UVA rays. If you spend lots of time in the sun—particularly if you have light-colored eyes—seek stronger protection, indicated by the label "Special-Purpose UV Requirements," which means that the glasses block at least 99 percent of the UVB rays.*

Dark lenses do have one potential advantage over paler ones: They block more visible light, which creates glare and may contribute to macular degeneration. However, nearly all sunglasses block enough visible light for safety and comfort under ordinary conditions. Only people exposed to brilliant sunlight—on ski slopes or tropical beaches, for example—may need extra dark, wraparound sunglasses.

Myth 6: *Once you develop a cataract, it must be removed.*

Truth: *Most cataracts (clouding the lens of the eye) are so minor that they cause little loss of vision. Even when objective tests show a substantial drop in visual acuity, surgery may not be required, because subjective factors matter more: If the loss of acuity does not significantly affect your everyday activities or nightime driving, you do not need surgery, regardless of what any test shows.*

Myth 7: *If you undergo surgery to correct nearsightedness, you will never need glasses again.*

Truth: *Surgery can reduce or eliminate nearsightedness by flattening the cornea. But there is no guarantee that surgery will eliminate the need for glasses. In the traditional operation, called radial keratotomy, the surgeon makes several pie-shaped incisions in the cornea. Even if the operation works perfectly, some patients will eventually need reading glasses, because the lens of the eye stiffens with age. More important, a few percent of all patients cannot see as clearly as they could with glasses before they had the operation. Serious complications, such as infection or rupture of the cornea, are rare, but potentially blinding. Surgery weakens the cornea, so a subsequent blow to the eye theoretically could have devastating results. In addition, surgery can result in problems with glare, and visual acuity may fluctuate for years.*

A newer method, called photorefractive keratectomy, uses a laser beam to shave an automatically preset sliver off the surface of the cornea. The laser reduces the chance of overcorrection or undercorrection, fluctuating vision, and excessive glare, and does not weaken the cornea. But some patients still need glasses for distance vision, at least occasionally, and others will eventually need reading glasses with aging. A person's vision is sometimes hazy for months or longer after the procedure.

(Adapted from *Consumer Reports on Health,* April 1997, pp. 42, 43.)

HEARING—THE AUDITORY SENSE

Acoustics is the science and technology of sound, including its production, transmission, and effects. The acoustical design goal is to establish an environment that

- transmits desired sounds reliably and pleasantly to the hearer;
- is satisfactory to the human regarding noise;
- minimizes sound-related annoyance and stress;
- minimizes disruptions of speech; and
- prevents hearing loss.

Acoustics describes the physical properties of sound, such as frequency or amplitude; psychoacoustics establishes relations between the physics of sound and our individual perception thereof, using descriptors such as pitch, timbre, loudness, noise, and speech comprehension (Gaver 1997).

Sound

Sound is any vibration (passage of zones of compression and rarefaction through the air or any other physical medium) that stimulates an auditory sensation. Unless under water or in space, we are concerned with sound that arrives at the ear by air. A sketch of the ear is given in Figure 4–14.

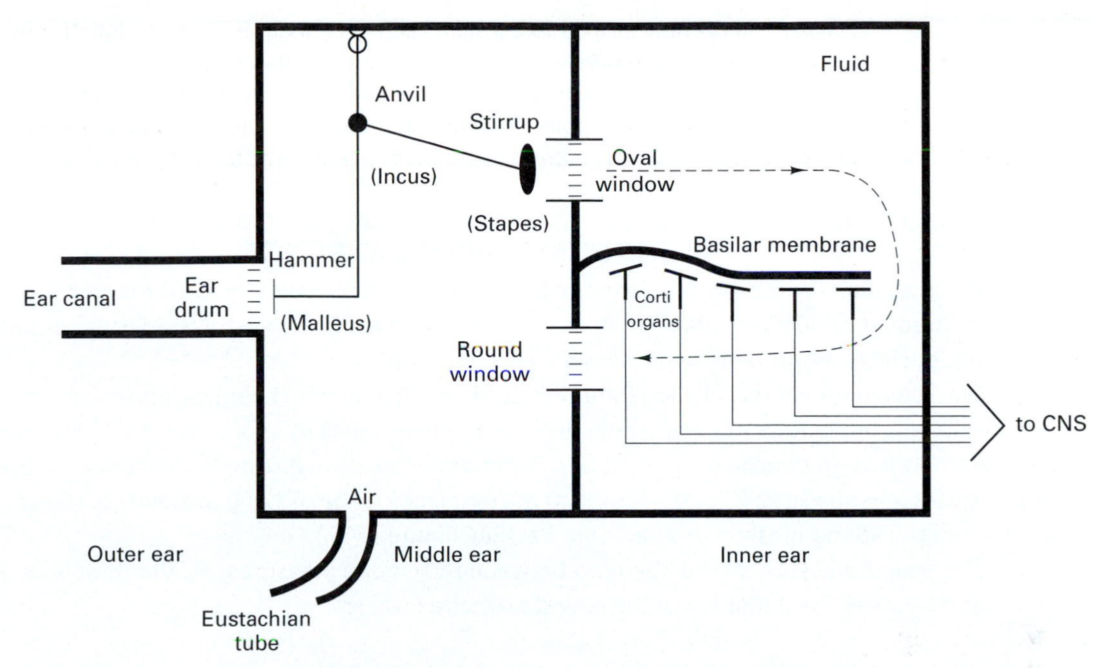

Figure 4–14. Schematic of the human ear. (See also Figure 4–23.)

Ear Anatomy and Hearing

Airborne sound waves arriving from outside the body are collected by the *outer ear* (auricle or pinna) and funneled into the auditory canal (meatus) to the *eardrum* (tympanic membrane), which vibrates according to the frequency and intensity of the arriving sound wave.

Resonance effects of the auricle and meatus have amplified the intensity of sound by 10 to 15 decibels (dB; see later for an explanation of the units) by the time it reaches the eardrum.

In the *middle ear,* the sound that arrived via the eardrum is mechanically transmitted by the *ear bones* (ossicles)—the hammer (malleus), anvil (incus), and stirrup (stapes)—to the *oval window.*

The ear bones mechanically increase the intensity of the sound from the eardrum to the oval window; also, the area of the eardrum is about 25 times larger than the surface of the oval window. Both factors together increase the effective sound pressure by about 22 times.

Air fills the outer and the middle ear, and the *Eustachian tube,* which connects with the pharynx, allows the air pressure in the middle ear to remain equal to the external air pressure.

☛☛☛ *In an airplane, during rapid descent or ascent, a "clogged" Eustachian tube can delay the equalization of pressure between the inner ear and the environs. If you feel ear pressure or pain, or if you cannot hear well, you may try to open the tube by chewing gum or by willful, excessive yawning, but "pumping" your outer ears with your hands will not help your middle ears.* ☚☚☚

The *inner ear* contains the receptors for hearing (and for body position, via the vestibulum, discussed later). The inner ear is filled with a watery fluid (called endolymph or perilymph) that propagates sound waves as fluid shifts from the oval window to the round window through the *cochlea,* an opening shaped like a snail shell with about two-and-one-half turns. The motion of the fluid deflects the *basilar membrane* that runs along the cochlea and stimulates sensory hair cells (cilia) in the *organs of Corti,* located on the basilar membrane. Depending on their structure and location, these organs respond to specific frequencies. Impulses generated in the organs of Corti are transmitted along the *auditory (cochlear) nerve* to the brain for interpretation.

THE HUMAN HEARING RANGE

A tone is a single-frequency oscillation, while a sound contains a mixture of frequencies. The frequencies (frequency distributions) of both are measured in hertz (Hz), their intensities (amplitude and sound pressure levels) in logarithmic units known as decibels (dB).

One reason for the use of a logarithmic scale is that the human being perceives sound pressure amplitudes in a roughly logarithmic manner. Infants can hear tones of about 16Hz to 20 kHz, while old people can rarely hear frequencies above 12 kHz. The minimal pressure threshold of hearing is about 20×10^{-6} $N\ m^{-2}$ (or 20 micropascals; $1\ Pa = 1\ N\ m^{-2}$) in the frequency range of 1,000 to 5,000 Hz. The ear experiences pain when the sound pressure exceeds 140 Pa. (See Figure 4–15.)

The sound pressure level is the ratio between two sound pressures; P_0, the threshold of hearing, is used as reference. The definition of the sound pressure level is

$$SPL = 10 \log(P^2 P_0^{-2}), \quad (4\text{–}7a)$$

or

$$SPL = 20 \log_{10}(P P_0^{-1}) \text{ in dB}, \quad (4\text{–}7b)$$

where P is the root-mean-square (rms) sound pressure for the existing sound. Thus, in decibels, the dynamic range of human hearing from 20×10^{-6} to 200 Pa is

$$20 \log_{10}[200/(20 \times 10^{-6})] = 140 \text{ dB}. \quad (4\text{–}8)$$

In terms of *sound intensity* ("power"), the sound intensity level is similarly defined as

$$SIL = 10 \log(I^2 I_0^{-2}), \quad (4\text{–}9)$$

where I is the rms sound intensity and I_0 is $10^{-12}\ Wm^{-2}$. Ranges of sound intensity levels are shown in Figure 4–16.

ADDING SOUNDS

Because the logarithm is used, doubling the SPL causes an increase of 6 dB; a 1.41-fold increase in sound pressure causes an increase of 3 dB, which means doubling the sound energy. If two sounds (of frequencies and temporal characteristics that are both random) occur at the same time, their combined SPL can be calculated from $SPL_{combined} = 10^{SPL_1/10} + 10^{SPL_2/10} + \cdots + 10^{SPL_n/10}$. Accordingly, one can approximate the combined SPL from the difference d in the intensities as follows: If d is 0 dB, add 3 dB to the

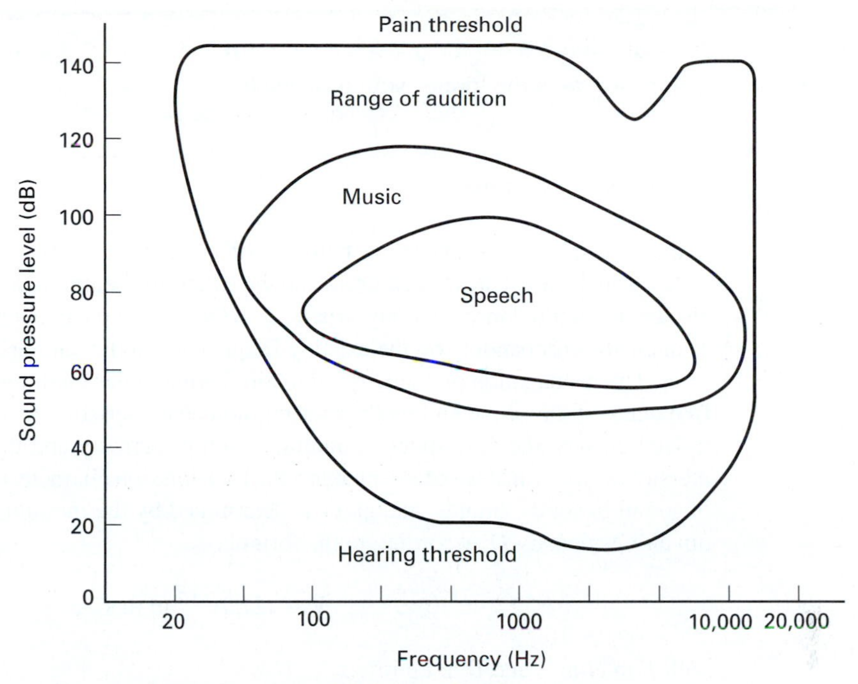

Figure 4–15. Ranges of adult human hearing.

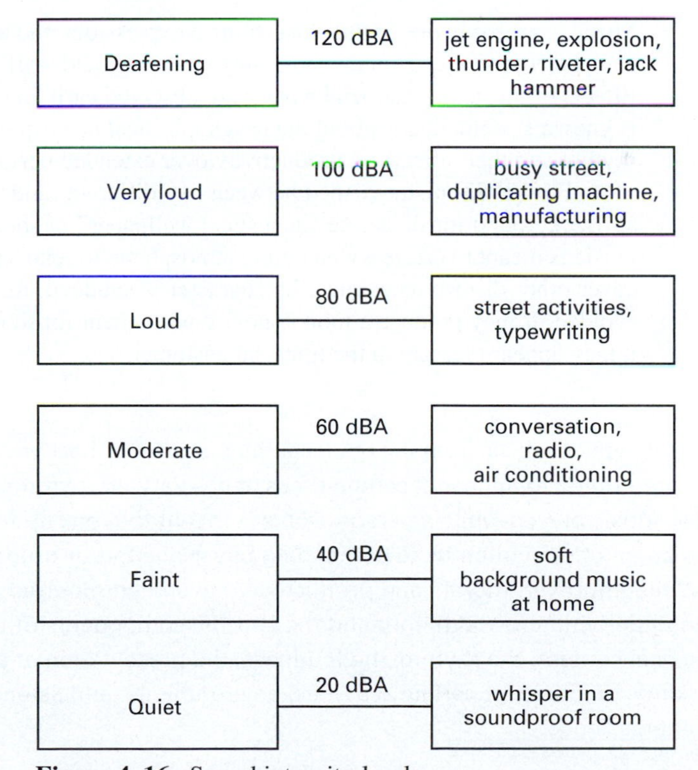

Figure 4–16. Sound intensity levels.

louder sound; if d is 1 dB, add 2.5 dB; for $d = 2$, add 2; for $d = 4$, add 1.5; for $d = 6$, add 1; for $d = 8$, add 0.5; and for $d = 10$ or more, take the louder sound by itself.

Pathways of Sound

In the human body, sound can reach the ear via two different paths. Airborne sound travels through the ear canal and excites the eardrum and the structures behind it, as described earlier. Sound may also be transmitted through bony structures to the head and ear, but this requires 30 to 50 dB higher intensities (depending on the existing frequencies) to be similarly effective.

The propagation of sound through air or some other medium depends on the intensity and frequency of the sound and on the transmission characteristics of the medium. Moving from terrestrial gravity and atmospheric conditions to microgravity and the vacuum of space, or to immersion in water instead of air, changes the transmission characteristics of sound. The velocity V of sound in solids, liquids, and gases is determined by the modulus of elasticity, E, of the medium and its density D, according to the formula

$$V = (E/D)^{-1/2} \text{ in m s}^{-1}, \qquad (4\text{–}10)$$

with E in N m^{-2} and D in kg m^{-3}.

Human Responses to Music

Music is probably one of the oldest human expressions and art forms. Making music has long accompanied activities—for example, singing during field work while marching to rhythmical music. Effects of music on industrial work were observed early in the 20th century, yet even today little is known systematically about the psychophysical consequences of different kinds of music and rhythm and their effects on productivity over extended periods of time.

Fox (1983) distinguished between two different kinds of music: background and industrial. Background music is like "acoustical wallpaper" in shops, hotels, waiting rooms, etc. Such music is meant to create a welcoming atmosphere, to relax customers, to reduce boredom, and to cover other disturbing sounds. Its character is subdued, its tempo intermediate, and vocals are avoided. It may produce a monotonous environment for those continuously exposed to it, while it may appear pleasant to the transient customer.

"Music while you work" is in many respects the opposite of background music. It is not continuous, but programmed to appear at certain times. It has varying rhythms with vocals, and it may contain popular "hits." Music played while a person works is meant to break up monotony—to generate mild excitement and an emotional impetus to demanding physical effort or drudgery in a dreary impersonal environment. While improved morale and productivity have been reported in many circumstances, a clear linking of the underlying arousal theory and the specific components of the music is difficult (Fox 1983). Thus, the musical content, the rhythm, the loudness, the presentation at a certain time, and the selection of particular kinds of music for certain activities, environments, and listener populations are still a matter of art rather than science.

"WHITE" AND "PINK" NOISE

"White" noise is a sound whose spectral density is the same at all frequencies. If one uses an analyzer with filters that are 1 Hz wide, the analysis shows a flat spectrum over the entire bandwidth being analyzed. For the ear, white noise sounds like electronic static.

"Pink" noise sounds more like a waterfall. Pink noise has a continous frequency spectrum and constant power within a bandwidth. In other words, it has equal energy in each octave (or fractional octave) band. The display of an analyzer using octave (or fractional octave) filters would indicate a flat spectrum. However, if pink noise were analyzed using an analyzer equipped with 1-Hz-wide filters (as previously mentioned for white noise), the spectrum would decrease as the frequency increases. The high-frequency filters are wider than the low-frequency filters, and thus, there is less energy per hertz at the higher frequencies.

HUMAN RESPONSE TO NOISE

Definition Noise is psychological and subjective. Single, short tones of low intensity (produced, for instance, by the dripping of water) may be considered noise under certain conditions, just as loud, lasting, complex sounds (e.g., a neighbor's music) may be deemed noise under other circumstances. Noise is defined as any unwanted, objectionable, or unacceptable sound.

Effects Any sound may be annoying and thus be conceived as noise. The threshold for noise annoyance varies, depending on the conditions, including the sensitivity and mental state of the individual. Noise can

1. create negative emotions, feelings of surprise, frustration, anger, and fear;
2. delay the onset of sleep, awaken a person from sleep, or disturb someone's rest;
3. make it difficult to hear desirable sounds;
4. produce temporary or permanent alterations in body chemistry;
5. temporarily or permanently change one's hearing capability;
6. interfere with some human sensory and perceptual capabilities and thereby degrade the performance of a task.

Physiological Effects of Sound

TTS, PTS

Exposure to intense sounds may result in a temporary threshold shift (TTS) from which the hearing eventually returns to normal with time away from the source; or it can cause a permanent threshold shift (PTS), which is an irrecoverable loss of hearing. A PTS may be the result of damage to the cochlear cilia, the organs of Corti at the basilar membrane in the inner ear, or the nerves leading to the CNS. Which of these are damaged depends on the frequency and, of course, intensity of the incident noise. The damage is probably due to an overstimulation of cell metabolism, leading to oxygen depletion and destruction, or to exceeding the elastic capacities of the physical structures. The metabolic activity taking place at the moment of the arrival of noise is important: Heat, heavy work, infectious disease, and other causes of heightened metabolism increase the vulnerability of the sensory organs. Above about 130 dB, the induced turbulence in the ear may also do mechanical damage (Stekelenburg 1982). The timing and severity of the loss are dependent upon the duration of exposure, the physical characteristics of the sound (its intensity and frequency), and the nature of the exposure (whether it is continuous or intermittent). Damage may be immediate, such as by an explosion, or may occur over some time, such as with continuous exposure to noise. The effect of continuous exposure is usually insidious and cumulative. Table 4–7 lists a number of physiological effects of various levels of sound. Following are some other physiological effects of intense sound:

- The muscles of the middle ear contract, mostly affecting the stapes and thereby reducing the transmission of force to the cochlea. However, it takes about 30 ms after the onset of the sound to activate this protective reflex and about 200 ms for complete contraction to occur, which lasts less than a second. Thus, the contraction response may be too late or too short to provide sufficient protection to the inner ear.

TABLE 4–7. Physiological Effects of Noise

Condition of Exposure			
SPL (dB)	*Frequency Spectrum*	*Duration*	*Reported Disturbances*
175	Low frequency	Blast	Rupture of tympanic membrane
167	2,000 Hz	5 min	Human lethality
161	2,000 Hz	45 min	Human lethality
160	3 Hz		Pain in the ears
155	2,000 Hz	Continuous	Rupture of tympanic membrane
150	1–100 Hz	2 min	Reduced visual acuity; vibrations of chest wall; gagging sensations; respiratory rhythm changes.
120–150	OASPL		Mechanical vibrations of body felt; disturbing sensations
120–150	1.6 to 4.4 Hz	Continuous	Vertigo and occasionally disorientation, nausea, and vomiting
135	20-2,000 Hz		Pain in the ears
120	OASPL		Irritability and fatigue
120	300–9,600 Hz	2 s	Discomfort in the ear
110	20 to 31.5 kHz		TTS occurs
106	4,000 Hz	4 min	TTS of 10 dB
100	4,000 Hz	7 min	TTS of 10 dB
100		Sudden onset	Reflex response of tensing, grimacing, covering the ears, and urge to avoid or escape
94	4,000 Hz	15 min	TTS of 10 dB
75	8 to 16 kHz		TTS occurs
65	Broadband	60 days	TTS occurs

SOURCE: Adapted from NASA 1989.
SPL-sound pressure level re 20 μPa.
TTS-temporary threshold shift.
OASPL-overall sound pressure level.

- The concentration of corticosteroids in the blood and brain is increased, and noise also affects the size of the adrenal cortex. In addition, continued exposure is correlated with changes in the liver and kidneys and with the production of gastrointestinal ulcers.
- Electrolytes in the body become imbalanced, and blood glucose levels change. Sex hormone secretion and thyroid activity may also be affected by noise. Changes in cardiac muscle, fluctuations in blood pressure, and vasoconstriction have been reported with 70 dB SPL and above, becoming progressively worse with higher exposure.
- Abnormal heart rhythms have been associated with exposure to noise.

High-intensity sound, whether "pleasant, but loud" or "loud noise," can permanently damage hearing.

EFFECTS OF NOISE ON HUMAN PERFORMANCE

A number of effects of noise on the performance of tasks have been observed (NASA 1989):

1. As noise becomes more intense, we become more aroused and our performance of certain tasks can improve.
2. Beyond a certain level of intensity, however, task performance degrades.
3. Sudden, unexpected noise can produce a startle response that interrupts one's concentration and physical performance of a task.
4. Continuous periodic or aperiodic noise interferes with the performance of complex tasks, such as visual tracking; performance is diminished with increasing noise levels.
5. Psychological effects of noise include anxiety, helplessness, narrowed attention, and other adverse effects that degrade task performance.

Other effects of noise on performance are listed in Table 4–8.

Noise-Induced Hearing Loss

Sounds of sufficient intensity and duration regularly result in temporary or permanent hearing loss. Permanent hearing loss may range from mild to profound, but can largely be prevented or, after it has occurred, alleviated by hearing aids. Yet, if the nervous structure is damaged, the condition is not treatable with current knowledge and technology.

Traumatic exposure

Sound that is of high intensity and brief duration, such as that produced by a cannon or an explosion, can damage any or all of the structures of the ear—in particular, the hair cells in the organs of Corti, which may be torn apart. This type of trauma results in immediate, severe, and permanent hearing loss with a permanent threshold shift (PTS).

Moderate exposure

We are usually exposed to sound levels of less then 100 dB, often over a period of hours, that may initially cause only short-term hearing loss, measured as a temporary threshold shift

TABLE 4–8. Effects of Noise on Human Performance. See Table 4–11 for effects on person-to-person voice communication.

SPL (dB)	Conditions of Exposure Spectrum	Duration	Performance Effects
155		8 h; 100 impulses	TTS 2 min after exposure
120	Broadband		Reduced ability to balance on a thin rail
110	Machinery noise	8 h	Chronic fatigue
105	Aircraft engine noise		Visual acuity, stereoscopic acuity, near-point accommodation all reduced
100	Speech		Overloading of hearing due to loud speech
90	Broadband	Continuous	Decrement in vigilance, altered thought processes, interference with mental work
90	Broadband		Performance degradation in multiple-choice, serial-reaction tasks
85	One-third octave at 16 kHz	Continuous	Fatigue, nausea, headache
75	Background noise in spacecraft	10–30 days	Degraded astronaut performance
70	4,000 Hz		TTS 2 min after exposure

SOURCE: Adapted from NASA 1989.
TTS—temporary threshold shift.

(TTS). During quiet periods, hearing returns to its normal level. Yet, a TTS may include subtle mechanical intracellular changes in the sensory hair cells and swelling of the auditory nerve endings. Other potentially irreversible effects include vascular changes, metabolic exhaustion, and chemical changes within the hair cells.

Repeated exposures to sounds that cause TTSs may gradually bring about a PTS—that is, a noise-induced hearing loss (NIHL). Experiments with animals have shown that with each exposure, cochlear blood flow may be impaired, and some hair cells may be damaged. Often, the damage is confined to a special area on the cilia bed on the cochlea, related to the frequency of the sound. With continued exposure to noise, more hair cells are damaged, which the body cannot replace; also, nerve fibers to that region in the ear degenerate, a process that is accompanied by corresponding impairment within the central nervous system.

NIHL frequencies

Impairment of hearing ability at special frequencies indicates exposure to noise at those frequencies. In western countries, NIHL usually occurs initially in the range of 3,000 to 6,000 Hz—particularly at about 4,000 Hz—then through higher frequencies, culminating at around 8,000 Hz. Yet, reduced hearing near 8000 Hz is also characteristic of aging, which often makes it difficult to distinguish between environmental and age-related causes. With continued exposure to noise, NIHL increases in magnitude and extends to lower and higher frequencies. NIHL increases most rapidly in the first years of exposure; after many years, it levels off in the high frequencies, but continues to worsen in the low frequencies (National Institutes of Health 1990).

Audiometry

Hearing ability (especially loss) is assessed by measuring the auditory thresholds (sensitivity) at various frequencies. Such pure-tone audiometry is often combined with measures of an individual's understanding of speech.

Understanding speech

NIHL is associated mostly with problems in differentiating speech sounds in their high frequency ranges. Especially affected is one's ability to make out the relatively high-frequency, low-intensity consonants. With NIHL occurring so often in the higher frequency ranges, important informational content of speech is often unclear, unusable, or inaudible. Also, other sounds, such as background noise, competing voices, or reverberation, may interfere with the listener's ability to receive information and to communicate.

In Western countries, noise-induced hearing loss usually starts in the range of 3,000 to 6,000 Hz—particularly around 4,000 Hz. Then it extends into higher frequencies, culminating at about 8,000 Hz. Yet, reduced hearing at (and above) 8,000 Hz is also brought about by aging. This may make it difficult to distinguish between environment- and age-related causes.

Sounds That Can Damage Hearing

Some sounds are so physically weak that they are not heard. Other sounds are audible, but have no temporary or permanent aftereffects. Some sounds are strong enough to produce a temporary hearing loss. Sounds that are sufficiently strong or long lasting and that involve certain frequencies can damage one's hearing.

The exact distinctions among these sounds cannot be stated simply, because not all persons respond to sound in the same manner. Yet, in general, about 85 dBA of sound level is potentially hazardous. (See below for the meaning of the letter "A" in "dBA.") The particular hazard depends on the actual frequency spectrum and duration of the sound. Most environmental sounds include a wide band of frequencies above and below the range from 20 Hz to 20 kHz that humans can hear.

The level, frequency, and duration of exposure to a sound are critical for determining whether the sound can damage one's hearing.

It appears that sound levels below 75 dBA do not produce permanent hearing loss, even at about 4,000 Hz, a frequency to which people are particularly sensitive. At higher intensities, however, the amount of hearing loss is directly related to the sound level (for comparable durations). In the United States, current OSHA regulations allow 16 hours of exposure to 85 dBA, 8 hours to 90 dBA, 4 hours to 95 dBA, etc. In Europe, also 8 hours at 90 dBA are allowed, but 4 hours at 93 dBA or 16 hours at 87 dBA—the energy trade-off is 3 dBA for 4 hours. If the sound level is about 140 dB, damage does not follow the simple energy concept; apparently, impulse noise above that level generates an acoustic trauma from which the ear cannot recover.

Simple subjective experiences can indicate whether one is being exposed to a hazardous sound—for example, a sound that is appreciably louder than conversational level, a sound that makes it difficult to communicate, ringing in the ear (tinnitus) after having been exposed to a noisy environment, or experiencing muffled sounds after leaving a noisy area.

Individual Susceptibility to NIHL

In young children, there is little difference in hearing thresholds between girls and boys. Yet, between ages 10 and 20, males begin to show reduced high-frequency auditory sensitivity. Women continue to have better hearing than men into advanced age. (These differences may be due to a greater exposure of males to noise, not to any inherent susceptibility to hearing loss.) There is a broad range of individual differences and sensitivities to a given noise exposure. The biological reasons for this are unknown, but, for example, a TTS or PTS in response to a given noise may differ as much as 50 dB among individuals. It is suspected that the anatomy and mechanical characteristics of the individual ears play a role, as may the use of ototoxic drugs or previous exposure to noise (National Institutes of Health 1990).

PREVENTING NIHL

Exposure to excessive sound, in both level and duration, contributes to the risk of NIHL. Common sources of noise are guns, power tools, chain saws, airplanes, farm vehicles, firecrackers, automobile and motorcycle races, music (whether heard live or through loudspeakers or headphones), and many occupational environments. NIHL can occur whether one likes the sound or not.

Three strategies There are three major approaches to countering the damaging effects of noise:

1. Avoid generation The first, fundamental, and most successful strategy is to reduce or avoid the generation of sound by properly designing machine parts such as gears or bearings, reducing rotational velocities, changing the flow of air, or replacing a noisy apparatus with a quieter one. "Active countermeasures" are being developed in which sounds with the same frequency and amplitude, but in the direction opposite (180 degrees off phase) that of the noise source, physically erase the source noise. Currently, this works best at frequencies below 1 kHz.
2. Impede transmission The second strategy is to impede the transmission of sound from the source to the listener. In occupational environments, one might try to put mufflers on the exhaust side of a machine, encapsulate the noise source, put sound-absorbing surfaces in the path of the sound, or physically increase the distance between source and ear.
3. Leave the area The third strategy is to remove humans from noisy places altogether, at least for parts of the work shift.

HEARING PROTECTION DEVICES

The last resource for protecting the hearing (in fact, a subclassification of the second strategy) is to wear a hearing protection device (HPD) that reduces the harmful or annoying subjective effects of sounds. HPDs are either worn externally (e.g., sound-isolating helmets or muffs, generally called "caps") or inserted into the ear canal ("plugs"). These hearing protectors are variously effective, depending partly on the intensity and the frequency spectrum of the sound arriving at the ear and partly on the "fit" of the protector to the wearer's ear. An inappropriate initial fit, loosening of the device during activity, and, of course, failure to wear the equipment reduces its effectiveness.

Unfortunately, commercially available HPDs cannot differentiate and selectively pass speech versus noise energy. Therefore, the devices do not directly improve the signal-to-noise relation. However, they can occasionally improve intelligibility in intense noise by lowering the total energy of both speech and noise that is incident on the ear, thereby reducing distortion due to overload in the cochlea. An HPD has little or no degrading effect on intelligibility in noise above about 80 dBA, although an HPD can cause considerable misunderstanding at lower levels (at which protection usually is not needed anyway). Some of the negative effects of the device may be due to the tendency to lower one's own voice because the bone-conductive voice feedback inside the head is amplified by the presence of the protector, mostly at low frequencies. Therefore, one's own voice is perceived as louder in relation to the noise than is actually the case, often resulting in a compensatory lowering of the voice by 2 to 4 dB. Thus, one should make a conscientious effort to speak louder when one wears an HPD (Berger and Casali 1992).

Nonverbal signals, such as warning sounds or sounds of machinery, are also affected by the wearing of an HPD. Signals above 2,000 Hz are most likely to be missed, due to the high-frequency properties of conventional HPDs, particularly if the person wearing one has impaired hearing at these or higher frequencies. This indicates that warning signals should be specifically designed to penetrate the device—for example, by using low frequencies (below 500 Hz). Such frequencies diffract easily around barriers, which is a positive side effect. Yet, for a person with normal hearing, wearing an HPD does not usually compromise the detection of signals. Altogether, the wearing of an HPD is highly advisable if the ambient noise arriving at the human ear cannot be lowered otherwise.

Audible noise with a constant sound level of 85 dB(A) or greater is hazardous. (The notation "dB(A)" is explained in the next section.) If humans must be subjected to such noise, hearing protection devices are necessary. People should not be exposed to continuous noise levels exceeding 115 dB(A) rms in overall sound pressure level under any circumstances. (Note that this OSHA requirement pertains only to the United States; other countries have different requirements.)

INFRASOUND AND ULTRASOUND

Below and above the regular hearing limits, sound levels, though inaudible, may still have vibrational effects on the human body. To protect the human, the following intervention strategies can be adopted:

- Infrasound pressure levels shall be less than 120 dB in the frequency range of 1 to 16 Hz, for 24-hour exposure. See table 4–9 (NASA 1989). To achieve this goal, well-fitted ear plugs provide attenuation at frequencies below 20 Hz similar to that in the 125-Hz band; in contrast, earmuffs are not effective. (Berger and Casali 1992).
- Hearing conservation measures shall be initiated when the ultrasonic criteria listed in Table 4–9 are exceeded (NASA). Earmuffs and plugs provide protection with an attenuation of at least 30 dB at frequencies between 10 and 30 kHz (Berger and Casali 1992).

TABLE 4–9. Airborne High-Frequency and Ultrasonic Limits

One-third Octave Band Center Frequency, kHz	*One-third Octave Band Level, dB*
10	80
12.5	80
16	80
20	105
25	110
31.5	115
40	115

SOURCE: NASA 1989.

Pychophysics of Hearing

While physical measurements can explain acoustical events, people interpret and react to them in very subjective ways—for example, finding certain sounds either attractive or noisy. The sensation of a tone or complex sound depends not only on its intensity and frequency, but also on how we feel about it.

Loudness

The subjective experience of the combined frequency and intensity of a sound is called *loudness*. Compared with the intensity at 1,000 Hz, at lower frequencies the sound pressure level must be increased to generate the feeling of "equal loudness." For example, the intensity of a 50-Hz tone must be nearly 75 dB to sound as loud as a 1,000-Hz tone with about 50 dB. However, at frequencies in the range of approximately 2,000 to 6,000 Hz, the intensity can be lowered and still sound as loud as at 1,000 Hz. Yet, above about 8,000 Hz, the intensity must be increased again above the level at 1,000 Hz to sound equally loud. "Equal-loudness contours" (called "phon" curves) were originally developed by Fletcher and Munson in 1933 and revised by Robinson and Dadson in 1957. (See Figure 4–17.) The actual shapes of the curves are slightly different if the stimulus sounds are presented either by loudspeaker or by earphone.

"A" Filter The differences in human sensitivity to tones of different frequencies are imitated by filters that are applied to sound-measuring equipment. These filters, today often in the form of software, "correct" the physical readings to what the human perceives. Different filters are identified by different let-

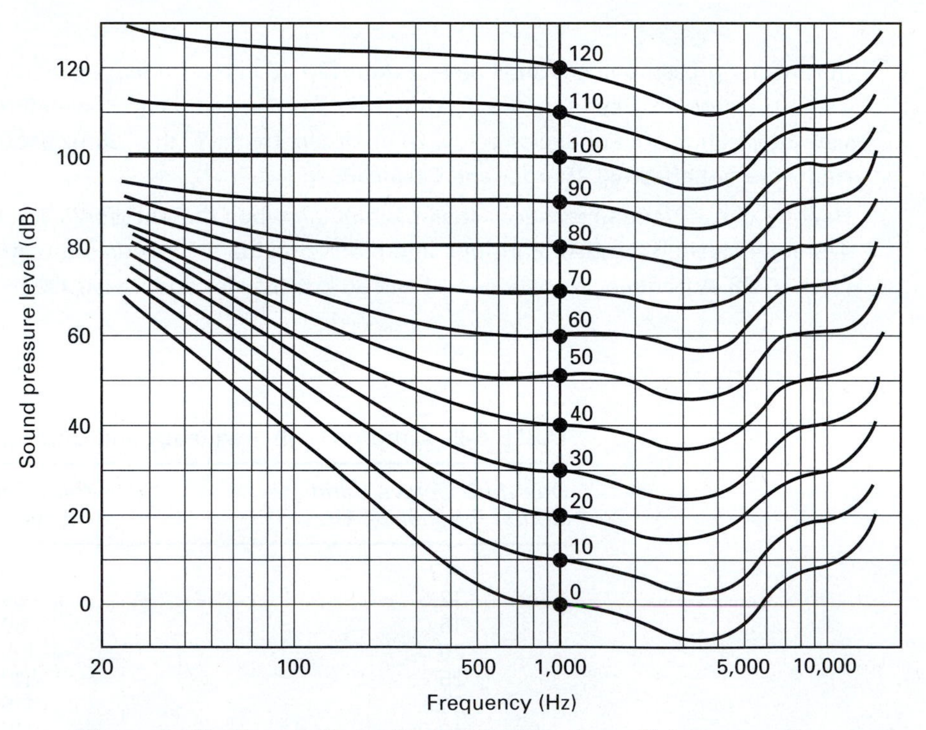

Figure 4–17. Curves of equally perceived loudness ("phon curves") are in congruence with the sound pressure level at 1,000 Hz.

ters of the alphabet. The "A" filter is most often used, because it corresponds best to the human hearing response at 40 dB. A-corrected SPL values are identified by the notation dBA or dB(A).

Pitch and timbre

Perceptions of equal loudness indicate that there are nonlinear relationships between *pitch* (the perception of frequency) and loudness (the perception of intensity). *Timbre* is even more complex, because it depends on changes in frequency and intensity over time.

To be heard, a low-frequency sound must be louder than a high-frequency sound.

Voice Communications

Intelligibility

The ability to understand the meanings of words, phrases, sentences, and entire speeches is called *intelligibility,* which is, obviously, a psychological process that depends on acoustical conditions. For the satisfactory communication of most voice messages over noise, at least 75-percent intelligibility is required. Direct face-to-face communication provides visual cues that enhance the intelligibility of speech, even in the presence of background noise. Indirect voice communications lack the visual cues. The distance from speaker to listener, background noise level, and voice level are important considerations. The ambient air pressure and gaseous composition of the air affect the efficiency and frequency of the human voice, and consequently, of speech communication.

Speech-to-noise "ratio" The intensity of a speech signal relative to the level of ambient noise is a fundamental determinant of the intelligibility of speech. The commonly used *speech-to-noise ratio* S/N is really not a fraction, but a signed difference; for a speech of 80 dBA in noise of 70 dBA, the S/N is simply +10 dB. With an S/N of +10 dB or higher, people with normal hearing should understand at least 80 percent of spoken words in typical broadband noise. As the S/N falls, intelligibility drops to about 70 percent at 5 dB, to 50 percent at 0 dB, and to 25 percent at −5 dB. People with noise-induced hearing loss may experience even larger reductions in intelligibility, while persons used to talking in noise do better.

Predicting Speech Intelligibility

Speech communication is highly dependent on the frequencies and sound energies of the interfering noise. Therefore, for predicting intelligibility, one should use a narrow band instead of a broadband dosimeter to measure the intensity of both the signal and noise sounds. Several techniques predict speech intelligibility on the basis of narrowband measurements.

Articulation Index The articulation index (AI) is an often-used metric for assessing intelligibility. It requires that noise levels in the centers of 15 one-third-octave bands, which range from 200 to 5,000 Hz, be measured in the work environment and compared with speech peaks in the same bands. The noise level in each band is subtracted from the speech level. The differences are weighted, with the highest

TABLE 4–10. Sample Calculations of the Articulation Index AI

Band Centers	*Observed Speech Peaks Minus Noise, dB*	*Weight Factor*	*Result*
200	30	0.0004	0.0120
250	26	0.0010	0.0260
315	27	0.0010	0.0270
400	28	0.0014	0.0392
500	26	0.0014	0.0364
630	22	0.0020	0.0440
800	16	0.0020	0.0320
1,000	8	0.0024	0.0192
1,250	3	0.0030	0.0090
1,600	0	0.0037	0.0000
2,000	0	0.0038	0.0000
2,500	12	0.0034	0.0408
3,150	22	0.0034	0.0758
4,000	26	0.0024	0.0624
5,000	25	0.0020	0.0500
			AI = 0.4738

SOURCE: Sanders and McCormick 1987.

weight given to the most critical voice frequencies. The weighted differences are then summed to produce a single AI value. This technique is described in Table 4–10. Excellent to very good intelligibility can be expected with 1 < AI < 0.7, good intelligibility with 0.7 < AI < 0.5, acceptable with 0.5 < AI < 0.3, and marginal or unacceptable with AI < 0.3 (NASA 1989). Figure 4–18 facilitates comparisons between AI and other measures of intelligibility.

Speech Interference Level The AI is a relatively complete, but fairly complex, means of predicting intelligibility. A more convenient alternative is the speech interference level (SIL), which requires less complex instrumentation and fewer data points. The SIL is simply the average of the decibel values of the existing noise levels of the octave bands 600 to 1,200, 1,200 to 2,400, and 2,400 to 4,800 Hz.

Preferred Speech Interference Level The SIL is now often used in a slightly modified form, called the preferred speech interference level (PSIL). The PSIL is computed as the arithmetic average of the three noise measurements taken in the octave bands centered at 500, 1,000, and 2,000 Hz. The higher the PSIL, the poorer the communication; for a typical distance between speaker and listener of 1 meter, a PSIL of 80 or above would indicate difficulties in communication. The PSIL should *not* be used if the noise has powerful low- or high-frequency components.

Preferred Noise Criteria Similar to, or even simpler than, the SIL or PSIL are preferred noise criteria (PNC) curves, which also use an octave-band analysis of noise. The highest PNC curve penetrated by

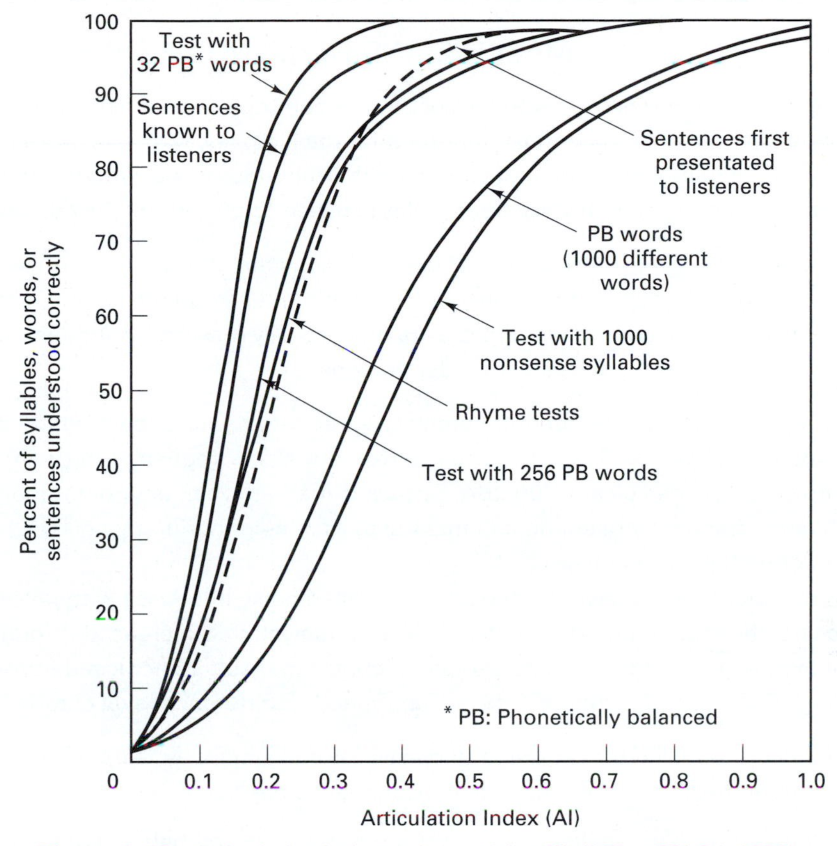

Figure 4–18. Articulation index and speech intelligibility (NASA 1989). These relations are approximate and depend upon the type of spoken material and the skill of talkers and listeners.

the noise spectrum is the value indicative of a given situation. Rooms in power plants, for example, have recommended PNCs of 50 to 60, while offices should be at 30 to 40.

SHOUTING IN NOISE

People have a tendency to raise their voices to speak over noise and to return to normal when the noise subsides. (This tendency is called the *Lombard reflex,* although it is probably not a reflex, but a conditioned response.) In a quiet environment, males normally produce about 58 dBA, in a loud voice 76 dBA, and, when shouting, 89 dBA. Women normally have a voice intensity that is 2 or 3 dBA less at lower efforts and 5 to 7 dBA less at higher efforts. Thus, people can fairly easily increase the S/N by raising their voices fairly easily at low noise levels, but the ability to compensate lessens as the noise increases. Above about 70 dBA, raising one's voice becomes inefficient, and it is insufficient at 85 dBA or higher. Furthermore, this forced effort of a shouting voice often decreases intelligibility, because articulation becomes distorted at the extremes of voice output. The S/N is a rough estimate for predicting the effectiveness of communication, but the loss of intelligibility by "masking" through noise (see next) depends not only on the intensity, but also on the frequency, of both the signal and the noise. In general, at small S/N differences, low-frequency noise causes more speech degradation than high-frequency noise does.

MASKING AND FILTERING OF SPEECH

Speech frequencies The frequencies in voice communications range from about 200 to 8,000 Hz, with the range of about 1 to 3 kHz most important for intelligibility. Men use more low-frequency energy than women do. Intelligibility is little affected by either filtering or masking frequencies below 600 or above 3,000 Hz, but interfering with voice frequencies between 1,000 and 3,000 Hz drastically reduces intelligibility.

Consonants In speech (as in written text), consonants are more critical for understanding words than are vowels. Unfortunately, consonants have higher frequencies and, concurrently, generally less speech energy than vowels and therefore are more readily masked by ambient noise; hence, they are more difficult to understand, especially for older persons.

Masking Noises of a predominantly low-frequency nature, particularly those with single-frequency components, have masking effects that spread upward in frequency, intruding upon speech bandwidths if the noise is between 60 and 100 dBA. Speech is masked when environmental sounds inhibit its perception. A given frequency of sound can mask signals at neighboring (especially higher) frequencies, possibly rendering them inaudible.

The efficiency of crews is often impaired when noise interferes with voice communication. If masking occurs, the time required to accomplish communication is increased through slower, more deliberate verbal exchanges. Not only is this annoying, but it can result in increased human error due to misunderstandings. Table 4–11 and Figure 4–19 indicate speech-interference-level criteria for voice communications.

Filtering For the transmission of speech—for example, by telephone—frequency or amplitude filtering is often used, and the following considerations apply:

1. *Frequency clipping* affects vowels if the clipping occurs below 1,000 Hz, but has little effect if it is above 2,000 Hz. *Peak clipping* is usually not critical if it is below 600 Hz or above 4,000 Hz. *Center clipping* is highly detrimental, particularly in the range from 1,000 to 3,000 Hz.

TABLE 4–11. Speech Interference Level (Noise) Criteria for Voice Communications

Speech Interference Level (SIL), dB	*Person-to-Person Communication*
30–40	Communication in normal voice satisfactory.
40–50	Communication satisfactory in normal voice at 1 to 2 m; need to raise voice at 2 to 4 m; telephone use satisfactory to slightly difficult.
50–60	Communication satisfactory in normal voice at 30 to 60 cm; need to raise voice at 1 to 2 m; telephone use slightly difficult.
60–70	Communication with raised voice satisfactory at 30 to 60 cm; slightly difficult at 1 to 2 m. Telephone use difficult. Earplugs or earmuffs can be worn with no adverse effects on communications.
70–80	Communication slightly difficult with raised voice at 30 to 60 cm; slightly difficult with shouting at 1 to 2 m. Telephone use very difficult. Earplugs or earmuffs can be worn with no adverse effects on communications.
80–85	Communication slightly difficult with shouting at 30 to 60 cm. Telephone use unsatisfactory. Earplugs or earmuffs can be worn with no adverse effects on communications.

SOURCE: Adapted from NASA 1989.

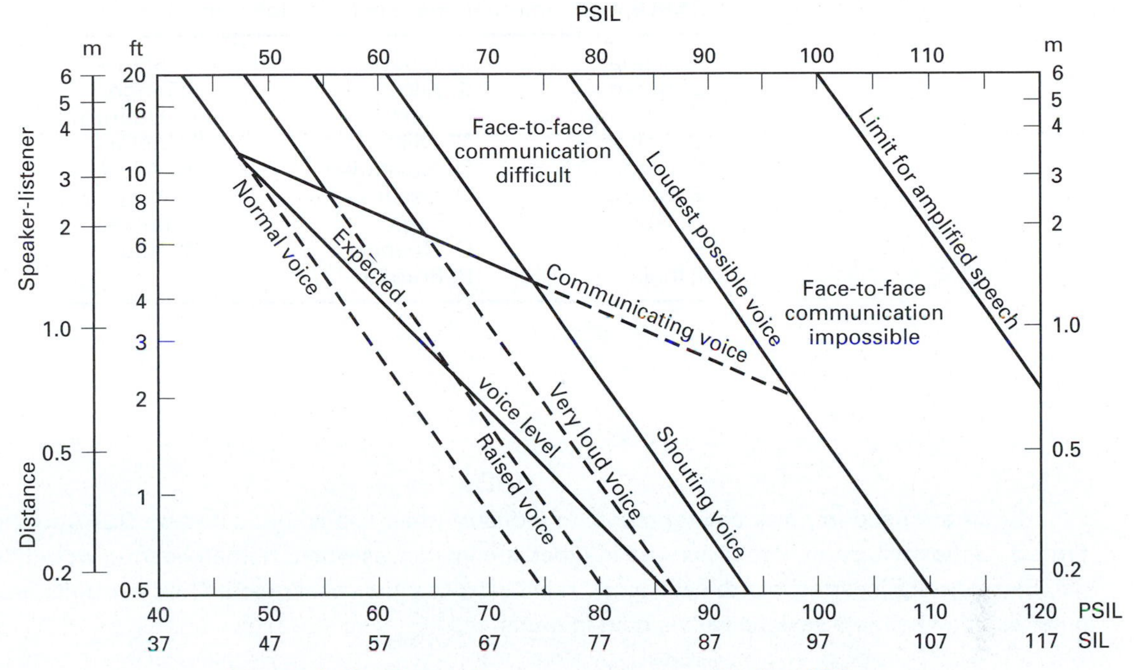

Figure 4–19. Face-to-face communications (NASA 1989).

2. In *amplitude clipping,* cutting the peaks affects primarily vowels and reduces the quality of transmission in general, but is not usually a great problem. Surprisingly, peak clipping and then reamplifying improves the perception of consonants, which carry most of the message. (One can read most written messages even if all vowels are missing.) *Center clipping,* in contrast, garbles the message because it affects primarily consonants.

COMPONENTS OF SPEECH COMMUNICATION

Speech communication has five major components: the message itself, the speaker, the means of transmission, the environment, and the listener.

The message itself becomes clearest if its context is expected, its wording is clear and to the point, and the ensuing actions are familiar to the listener.

The speaker should speak slowly and use common and simple vocabulary with only a limited number of terms. Redundancy can be helpful (for example, "Boeing 747 jet"). Phonetically discriminable words should be used. The International Spelling Alphabet is shown in Table 4–12.

Speech should be transmitted by a "high-fidelity system" that has little distortion in frequency, amplitude, or time.

The hearer's ability to understand the message is, of course, affected by noise, which can be assessed by the speech interference level, discussed earlier. Wearing an HPD (a plug, muff, or helmet) produces some filtering and reduces the overall sound level. Yet, it may be advisable to use special ear protection that is penetrable by specific frequencies, or one may have to adjust the signal so that it will not be masked by noise. (See above and also Chapter 11.)

TABLE 4–12. International Spelling Alphabet

A: Alpha	J: Juliet	S: Sierra
B: Bravo	K: Kilo	T: Tango
C: Charlie	L: Lima	U: Uniform
D: Delta	M: Mike	V: Victor
E: Echo	N: November	W: Whiskey
F: Foxtrot	O: Oscar	X: Xray
G: Golf	P: Papa	Y: Yankee
H: Hotel	Q: Quebec	Z: Zulu
I: India	R: Romeo	

HPDs

HPDs are used in noisy environments in industry (their use is mandated by OSHA in the United States), in the military, at auto races, and in other circumstances where human hearing would be in danger. Earplugs are inserted into the ear canal, canal caps seal the canal near its rim, earmuffs encircle the outer ear, and helmets seal the ear via built-in muffs.

Passive HPDs In the past, HPDs were "passive," their attenuation achieved by making sounds pass through material that absorbs, dissipates, or otherwise impedes energy flow. These conventional HPDs can be highly efficient when properly selected and correctly worn; they have little or no degrading effect on the wearer's understanding of speech and other sounds in ambient noise above about 80 dBA, but they do cause misunderstandings during quieter periods.

Conventional HPDs do not selectively pass speech versus noise at given frequencies; thus, they do not improve the *S/N* ratio. In fact, most passive devices are designed to attenuate high-frequency sound more than low-frequency sound, thereby reducing the power of consonants and distorting speech. Persons who already have suffered a hearing loss at higher frequencies experience a further elevation of their hearing threshold when they use the conventional HPD (Casali and Berger 1996).

Active HPDs New HPD designs incorporate electronics to improve communication and the reception of signals by the wearer. These "active" devices can

- provide diminished attenuation in low-level noise and increased protection during loud periods;
- reduce noise by destructive interference at selected frequency bands;
- let pass or boost desired critical bands, especially those needed for speech; and
- transmit desired signals, such as those for speech, warnings, or music, via built-in loudspeakers.

Combining the desired features of active and passive devices is expected to lead to effective HPDs for special applications and, probably, general use.

Improving Hearing

Many people, especially as they become old, experience reduced hearing ability, predominantly in the higher frequency ranges. Devices to aid impaired hearing can be a great improvement toward understanding speech, especially when they utilize electronics to amplify certain bandwidths of sound and filter out ambient noise.

Yet, many people with impaired hearing are greatly frustrated when they try to use these hearing aids, because the device was not properly selected in the first place, was not replaced by a new and better one, or is difficult to use.

Hearing and motor skills often worsen over the years. Barnes and Wells (1994) recommended various ergonomic design applications to overcome the great dissatisfaction of nearly three-quarters of American hearing-aid users because of the following problems:

- *They had to remove the hearing aid to use the telephone.*
- *They could not maintain the device at the proper volume.*
- *They were not alerted to the battery running low.*
- *Changing the battery was troublesome.*
- *They experienced discomfort while wearing the hearing aid.*
- *The device whistled at certain volumes.*

REVERBERATION

Reflection of sound from surfaces is called *reverberation.* Reverberation time in a room is defined as the time it takes the SPL to decrease by 60 dB after the source of sound is shut off. A certain amount of reverberation is desirable, because it makes speech sound alive and natural. But too much reverberation is undesirable if reflections arrive at the same time a new word is uttered and hence interfere with its perception. A room with little or no reverberation is called "dead." In such a room, there is little interference between words, and intelligibility is near 100 percent, but because the sound of the word decays before it can propagate through the room, verbal communication may bc difficult. If the delay between reflected and original sounds is long, separate sounds (the original and its echo) are heard. Long room reverberation times can produce a bouncing or booming sound. As the reverberation time increases, intelligibility decreases in a nearly linear fashion. If the reverberation time increases beyond 6 seconds, intelligibility is cut in half. A highly reverberant room is called "live" or "hard."

COMMUNICATION AT ALTITUDE OR UNDER WATER

Communication that takes place at a high altitude or under water requires specific technical means. At high altitudes, where the ambient pressure is low, the human voice and earphones, as well as loudspeakers, become less efficient generators of sound, and microphones become less sensitive at certain frequencies. This combination of conditions requires amplification to be incorporated into either the source, the transmitter, or the receiver of signals.

Under water, hearing is limited by the reverberation of sound and noises made by movement and by an increased minimal threshold of hearing, which may be raised about 40 dB by the impedance mismatch between water and air at the ear. The difficulties can be overcome by amplifying the transmitted sound and by using a directional receiver to discriminate against sounds coming from other directions. One can talk under water, but the noise that accompanies the bubbles that are emitted masks speech, so that the listener hears mainly vowel sounds. If a face mask with a built-in microphone is not available, the diver should wait for all bubbles to die away before saying the next word and use simple vocabulary—mostly vowels.

Acoustic Phenomena

Directional Hearing. *Humans are able to tell where a sound is coming from by using the difference in arrival times (phase difference) or intensities (as a result of the inverse-square law of energy flux) to determine the direction of the sound. Yet, the ability to use stereophonic cues varies among individuals and is greatly reduced when earmuffs are worn.*

Distance hearing. *The ability to determine the distance of a source of sound is related to the fact that sound energy diminishes with the square of the distance traveled, but the human perception of energy depends also on the frequency of the sound, as just discussed. Thus, a source of sound appears more distant when it is low in intensity and frequency and appears closer when it is high in intensity and frequency.*

Difference and summation tones. *Two tones that are sufficiently separated in frequency (so that they excite separate areas on the basilar membrane) are perceived as two distant tones. When the tones are very loud, one may hear two supplementary tones. The more distinct tone is at the frequency difference between the two tones, while the quieter tone is the summation of the original frequencies. For example, two original tones, at 400 and 600 Hz, generate a difference tone at 200 Hz and a summation tone at 1,000 Hz.*

Common-difference tone. *When several tones are separated by a common frequency interval (of 100 Hz or more), one hears an additional frequency based on the common difference. This affect explains how one may be able to hear a deep bass tone from a sound system that is physically incapable of emitting such a tone.*

Aural harmonics. *One may perceive a pure sound as a complex one, because the ear can generate harmonics within itself. These "subjective overtones" are more pronounced with low- than with high-frequency tones, especially if these are about 50 dB above the threshold for human hearing.*

Intertone beat. *If two tones differ only slightly in their frequencies, the ear hears only one frequency, called the* intertone, *that is halfway between the frequencies of the original tones. The two tones are in phase at one moment and out of phase at the next, causing the intensity to wax and wane; thus, one hears a beat. Beating occurs at a frequency equal to the difference in frequencies of the two tones. If the beat frequency is below six per second (that is, if the frequencies of the original tones are close together), the beat is very distinct, appearing as a variation in loudness; only the intertone is then heard. When the beat is above eight per second (i.e., with an enlarged frequency interval between the two original tones), the intertone appears to be pulsating or throbbing, and one may hear the original tones as well. When the beats occur more often than about 20 times per second (with an even larger frequency separation between the original tones), the intertone becomes faint, and the two original tones are predominant.*

Doppler effect. *As the distance between the source of sound and the ear decreases, one hears an increasingly higher frequency; as the distance increases, the sound appears lower. The larger the relative velocity, the more pronounced is the shift in frequency. The Doppler effect can be used to measure the velocity at which source and receiver move against each other.*

Concurrent tones. *When two tones of the same frequency are played in phase, they are heard as a single tone, its loudness being the sum of the two tones. Two identical tones exactly opposite in phase cancel each other completely and cannot be heard. This physical phenomenon (called* destructive interference *or* phase cancellation*) can be used to suppress the propagation of acoustical or mechanical vibrations and, with current technology, is particularly effective at frequencies below 1,000 Hz.*

SMELLING—THE OLFACTORY SENSE

☛☛☛ *Smells may be dear to us, but we have no names for them. Instead, we compare them with other smells, such as of flowers; or we describe how they make us feel, such as intoxicating, sickening, pleasurable, delightful, or revolting. Some smells are fabulous when they are diluted, but truly repulsive when they are not. Odors can play major roles in eating, drinking, health, therapy, stress, religion, personal care and relations (Ackerman 1990, Ballard 1995).* ☚☚☚

Odor Sensors

In the upper rear part of each human nostril, several million smell receptors (bipolar olfactory neurons) are located in a patch of 4 to 6 cm^2 (called the olfactory epithelium). Each neuron has a dendrite that ends in an olfactory knob, from which hairlike cilia protude into the nasal mucus. The nasal airways are bent, so, normally, little airflow passes along the olfactory cilia, but the flow can be increased by sniffing. In still unknown ways, certain molecules trigger the sensors, which then send signals directly to the olfactory bulbs of the brain.

Also distributed throughout the mucus of the nasal cavity (and parts of the oral cavity) are free endings of the fifth (trigeminal) cranial nerve, which are connected to a different region of the brain. The trigeminal receptors provide the so-called common chemical sense and are triggered by, among other odorants, substances that generate irritating, tickling, and burning sensations, which then initiate protective reflexes, such as sneezing, or interrupt one's breathing (Cometto-Muniz and Cain 1994; National Research Council 1979). In high enough concentrations, most, if not all, odorants stimulate both the olfactory and common chemical sensors.

Odorants

Not all odorants are external: The body generates its own smells. Body odor arises from the apocrine glands, which are small in infants, but develop during puberty. Most of these glands are at the armpits, face, chest, genitals, and anus.

☛☛☛ *Odorants are chemical substances and can be analyzed by chemical methods. Odors are sensations and must be assessed by measuring human responses to them. If the physical and chemical determinants of odor were fully understood, it would be possible to predict the sensory properties of odorous materials from their chemical analysis—in practical terms, one could construct an "odor meter" analogous to a decibel meter for sound. Such an understanding, however, is not yet at hand, nor is any such device available. Nonetheless, various instrumental and sensory methods of measurement have been developed and have been applied to sources of odor and to the ambient atmosphere. Still, many of the available techniques are costly and time consuming, and not all have been validated.* ☚☚☚

Many atmospheric contaminants are odorless or very nearly so; carbon monoxide is a notorious example. By contrast, many other substances are readily detectable even in minute concentrations. For example, an organic sulfur compound at a concentration of one molecule per billion molecules of air is likely to be smelled.

Most existing odorants are mixtures of basic components, generating complex odors to which different people, in different environments and at different lengths of exposure, react quite differently. Accordingly, various theories have been proposed to explain the sensitivity of our sense of smell, among them Amoore's (1970) stereochemical model, which assumes that the dimensions and shapes of certain odor molecules must fit certain receptor sites to generate the sensation of an odor.

While we use the olfactory sense daily, little is known systematically about it, partly because a smell is not easily quantified. The usual test procedure is to present the substance that has an odor in varying concentrations; the threshold concentration is found when it provokes a sensation in 50 percent of the cases. The sensation (the "smell") is described in relation to other smells; it may be considered pleasant or unpleasant. The assessment of "odor annoyance" has been attempted in a manner similar to that of noise classification (Hangartner 1987).

Describing Qualities of Odors

Several systems to describe and distinguish the qualities of odors have been used in the past. The classic Linacus–Zwaardemarker nine-category system has been largely displaced by either of two other procedures:

1. The Crocker–Henderson approach ranks every smell in four categories along a nine-point scale. The categories are fragrant, acid, burnt, and caprylic ("soapy").
2. Henning's "smell prism" (see Figure 4–20) employs six primary odors (flowery or fragrant, spicy, fruity, resinous, burnt, and foul or putrid or rotten), arranged in the form of a prism. Stimuli that typically evoke olfactory responses in these categories are violets, cloves, oranges, balsam, tar, and rotten eggs, respectively.

Effects of Odors

The effects of odors can be physiological and independent of the actual perception, or perceived, psychologic, and even psychogenic. *Physiological effects* may stimulate the central nervous system, especially the hypothalamus and the pituitary gland, eliciting changes in body temperature,

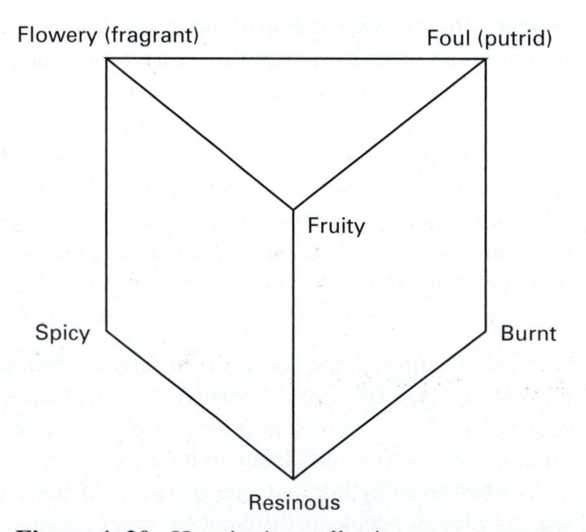

Figure 4–20. Henning's smell prism.

appetite, arousal, and other physiologic reactions, by modifying olfaction and by triggering (potentially harmful) reflexes. Unfavorable responses include nausea, vomiting, and headache; shallow breathing and coughing; problems with sleep, stomach, and the appetite; irritation of the eyes, nose, and throat; destruction of the sense of well-being and the enjoyment of food, home, and the external environment; and feelings of disturbance, annoyance, or depression. Exposure to some odorous substances may also lead to a decrease in heart rate, constriction of the blood vessels of the skin and muscles, the release of epinephrine, and even alterations in the size and condition of cells in the olfactory bulbs of the brain.

Psychologic and psychogenic effects concern especially one's attitude and mood, such as anger or benevolence toward others, cooperation, creativity, self-perception, and the ability to perform. These effects may be part of the "sick building syndrome" (Ballard 1995).

However, no quantitative relationship between the intensity or duration of the exposure to odor and the magnitude of the symptoms has yet been established.

Since ancient times, it has been supposed that pleasant aromas preserve health and that unpleasant odors are injurious. These suspicions formed the basis for the use of aromatic Eau de Cologne and of pomanders stuffed with balsams and for the attribution of diseases to atmospheric "miasmas." Thus, the word "malaria" is derived from the Italian expression for "bad air," mala aria (National Research Council 1979, p. 3). Despite the vast influence of odor on daily life, surprisingly little evidence has been collected to quantify its effects on people. There is no scientific odor classification system, nor is there a clear understanding of how odor is sensed or processed. . . . Although the odor environment has little importance as an agent of disease or dysfunction, it does have an effect on how people perform in tasks, remember, feel, make social judgments, and act towards other people. . . . Researchers are unsure of how odor can cause psychogenic illness or sick building syndrome, but bad odors have been known to trigger these problems. (Ballard 1995, pp. 191,192).

The fact that only minute quantities of some stimulating agents are required to bring about a sensation of smell helps the designer who wishes to use smell as a means of engineering. For example, the addition of methyl mercaptan in the amount of only 25×10^{-6} gram to each liter of natural gas suffices to render the gas detectable by smell. Similarly, the addition of 3 to 10 ppm of pyridine to argon, another inert gas, endows the argon with an odor (Cain et al. 1987). Another application is spraying sweetener in the ambient air: If the wearer of a respiratory mask or hood smells the sweetener, the seal on the mask is proven inadequate.

Only 20 percent of the perfume industry's income comes from making perfumes to wear; the other 80 percent comes from perfuming the objects around us. Thus, used-car dealers have a "new-car" spray, real-estate dealers sometimes spray the aroma of cake-baking around the kitchen of the house before showing it to a client, and shopping mall managers add the smell of food to their air-conditioning systems to put shoppers in the mood to visit their restaurants (Ackerman 1990).

Using the sense of smell for engineering purposes has some attraction; for example, smell can penetrate vast and complex areas, such as underground mines, where a "stench" can signal an emergency evacuation. But the olfactory sense easily adapts and is quite different from person to person. (After spending months in Bombay, India, a city in which strong odors abound, one of the authors could no longer smell the "wild" animals in a circus.) Also, smells change with concentration and time, and many odors can easily be masked by other odors. Interestingly, most children and men cannot smell exaltolide, while many women find it very strong; but sensation varies with the menstrual cycle. Smoking and aging affect one's smelling capabilities. A blockage of the nasal passages—for example, from a cold—can temporarily eliminate the ability to smell.The magnitude of the human sensory response to odor (i.e., the perceived intensity of the odor) decreases as the concentration of the odorant gets smaller. (This effect is used to control indoor odors by ventilation or outdoor odors by the use of tall stacks.) However, the relationship between the intensity and concentration of an odorant is not a direct proportionality: When odorous air is diluted with odor-free air, the perceived odor decreases less sharply than the concentration. For example, a tenfold reduction in the concentration of amyl butyrate in air is needed to reduce its perceived odor intensity by half. Nor do all odorants respond by the same ratios: Some, like amyl butyrate, show sluggish changes in odor intensity with changes in concentration, while others change more sharply (National Research Council 1979).

How we delight our senses varies greatly from culture to culture. . . . Masai women, who use excrement [cattle dung] as a hair dressing, would find American women's wishing to scent their breath with peppermint equally bizarre (Ackerman 1990, pp. xvi, 24).

TASTING—THE GUSTATION SENSE

Like the sense of smell, the sense of taste is poorly understood. A human can readily distinguish primary qualities (i.e., categories) of taste, but cannot distinguish quantitative differences so easily. This is partly due to the fact that the sense of taste interacts strongly with sensations of smell, temperature, and texture, all of which are present in the mouth. Pepper, for example, tastes as it does because it stimulates several types of receptors. Food becomes almost tasteless if a cold causes a temporary loss of the sense of smell.

"How strange that we acquire taste as we grow. Babies don't like olives, mustard, hot pepper, beer, fruits that make one pucker, or coffee. . . . No two of us taste the same plum. Heredity allows some people to eat asparagus and pee fragrantly afterward (as Proust describes in "Remembrance of Things Past"), or eat artichokes and then taste any drink, even water, as sweet. Some people are more sensitive to bitter tastes than others and find saccharin appalling, while others guzzle diet sodas. Salt cravers have saltier saliva. Their mouths are accustomed to a higher sodium level, and foods must be saltier before they register as salty" (Ackerman 1990, p. 141).

TASTE SENSORS

Nearly ten thousand taste buds, the receptors for taste qualities, are located mostly on the human tongue, but are also found at the palate, pharynx, and tonsils. The taste buds (really a budlike collection of cells), which are continuously replaced every two weeks, are arranged in clusters. It appears that some taste buds react only to one stimulus of our four taste qualities, while others respond to several or all of the stimuli. The tip of the tongue is particularly sensitive to sweet, the sides to sour, the back to bitter, and all to salty stimuli. The number of taste buds seems to diminish with aging after the middle forties, when the remaining buds may also atrophy. Taste sensitivity differs from person to person and decreases with age. Some people cannot taste certain substances; for example, 3 of 10 persons cannot perceive phenylthiocarbamide.

To taste a substance, it must be soluble in the saliva. Taste sensitivity depends on several interactive variables, including the nature of the stimulus, its concentration, its location of activity, and its time of application. Furthermore, the sensitivity depends on the previous state of adaptation to the taste in question, on the chemical condition of the saliva, on temperature, and on other variables.

Taste Stimuli and Qualities

Our sensations of taste qualities are evoked by stimuli: sodium chloride (salty), acid hydrogen ions (sour), nitrogenalkaloids (bitter), and inorganic carbon (sweet). Our perception of all of them is strongly affected by temperature. Of the four qualities, bitter and sweet are even less understood than salty and sour. One attempt to describe the relations between taste qualities is shown in Figure 4–21, where the solid lines represent interactions and the dotted lines indicate no relations.

☛☛☛ *The following excerpt is from an article on wine: "'Number one was a youngster: fresh, light, passionfruity, straightforward, not oakey, fruit-sweet, fine and understated, well-balanced and clean—a modern style with low-key oak and showing cool-fermentation high-tech winemaking, but not very complex, and needing age. Number two was much more complex, showing some maturity: very smoky, lightly leesy, toasted nut aromas; strongly constituted, tightly structured, powerful, spicy with plenty of wood-derived complexity, very long in flavour and excellent drinking now, but in no danger of falling apart. Number three was, for me, even bigger and richer, fatter and more complex than number two. Strong aromas of butter, grilled nuts, toasted bread, smokiness, very full-bodied, ripe, and alcholic, quite fat and reminding everybody of a good 1978 white Burgundy.' Let's just take this slowly. By my count, and excluding what might be repetitions, there are 28 different qualifiers in that quotation, implying 28 scales of assessment. This writer is asking me to believe that it is possible to discriminate 28 taste scales, let alone assess points along each one? And to do this reliably?"* (Applied Ergonomics *1988, 19(4), 324).* ☚☚☚

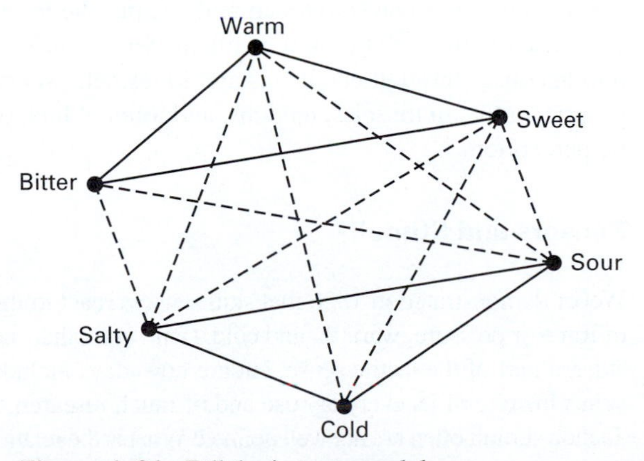

Figure 4–21. Békécy's taste model.

TOUCHING—THE CUTANEOUS SENSES

Sensors and Stimuli

The sensory capabilities located in the skin are called *cutaneous* (from *cutis,* Latin for skin) or *somesthetic* (from *soma,* Greek for body). They are commonly divided into four groups:

1. The *mechanoreceptors,* which sense taction (i.e., contact or touch), tickling, pressure, and related commonly understood, but theoretically ill-defined, stimuli.
2. The *thermoreceptors,* which sense warmth or cold, relative to each other and to the body's neutral temperature.
3. The *electroreceptors,* which respond to electrical stimulation of the skin. (It is disputed whether such specific sensors exist.)
4. The *nocireceptors* (from the Latin *nocere,* to damage), which sense pain. We feel what is commonly, but imprecisely, called sharp or piercing pain, usually associated with events on the surface, and dull or numbing pain, usually felt deeper in the body. (The point is controversial, because some researchers state that there are no specific pain sensors, but that pain is transmitted from other sense organs.)

The study and engineering use of the cutaneous senses is hampered by many uncertainties. First, stimuli are often not well defined, particularly in older research. Second, sensors are located in different densities over the body. Third, the functioning of sensors is not understood exactly: Many sensors react to two or more distinct stimulations simultaneously and produce similar outputs, and it is often not clear whether or which specific sensors respond to a given stimulus. Fourth, the pathways of signal conduction to the central nervous system are complex (as discussed earlier) and may be joined by other afferent paths from different regions of the body. Finally, arriving signals are interpreted in unknown ways at the CNS. Given all these uncertainties, it is obvious that many of our engineering applications are based mostly on everyday experience and rely on very limited experimental data. Much more research needs to be performed to provide the ergonomist with complete, reliable, and relevant information on the cutaneous senses for the design of systems.

Sensing Taction

The taction sense reacts to touch at the skin. The term *tactile* is often used if the stimulus is received solely through the skin, while the term *haptic* is applied when information is obtained simultaneously through cutaneous and kinesthetic senses—that is, through the skin and through proprioceptors in muscles, tendons, and joints. Much of our everyday perception is actually haptic perception.

Sensors and Stimuli

Weber demonstrated in 1826 that skin sensors react to the location of the stimulus and to the stimuli of force or pressure, warmth, and cold. (Pain and other more diffuse feelings were originally not considered part of the tactile sense, but are nowadays included insofar as they respond to stimuli on the skin.) In spite of its everyday use and of much research, the sense of taction is not fully understood. Taction stimuli often are not well defined: What is the relation between pressure, force, and touch? Which sensors respond to each stimulus in what way? Do the sensors respond singly or in groups? If in groups, in what patterns do they respond? Does one sensor respond to only one stimulus or to several stimuli?

Questions exist regarding the association of specific nervous functions with the stimulation of specific sensors and how the two interact to provide a signal to the brain. Often, one postulates that specialized nerve endings are present, one for each sensation, and that nerves are connected to specific centers in the brain. However, pattern theory assumes that the particular experience of a sensation depends on the coaction of several separate and elementary nervous events, but not all need to be present in each pattern. An overview of the unsettled state of theories and knowledge is provided by Sherrick and Cholewiak (1986).

Architecture of the Taction System

All tactile sensors are triggered to discharge signals by a stimulus of the appropriate type and intensity. The signal that is generated varies in frequency and amplitude and travels toward the CNS. Several types of sensors have been identified. The most common is a *free nerve ending,* a proliferation of a nerve that distally dwindles in size and then disappears. Thousands of such tiny fibers extend through the layers of skin. They respond particularly to mechanical displacements and are very sensitive near hair follicles, where *basket endings* surround hair bulbs and respond to displacement of the hair shaft.

In glabrous (smooth and hairless) skin, encapsulated receptors are also common. Among these are *Meissner* and similar *Merkle corpuscles,* which are particularly numerous in the ridges of the fingertips. These receptors transmit transient electrochemical surges to the nerves in response to ambient pressures. A single Meissner corpuscle may connect with up to nine separate nerves, which may also branch to other corpuscles. This is an example of simultaneous convergence and diversion of the neural pathways. (How such an arrangement can reliably code neural signals is not yet understood.)

The *Pacinian corpuscle* is an encapsulated nerve ending of a single, dedicated nerve fiber. These highly responsive tactile receptors are located in profusion in the palmar sides of the hand and fingers and in distal joints. They are also prevalent near blood vessels, at lymph nodes, and in joint capsules. *Krause end bulbs* are particularly sensitive to cold, but probably respond to other stimuli as well.

Nerves from the various receptors are colligated in peripheral cutaneous nerve bundles, which proceed with their neighbors to the dorsal roots of the spinal cord. Fibers originating at the same receptor may pass on to separate dorsal roots (i.e., dermatomal segments). It is not yet understood how such a complex nervous signal pattern is interpreted by the brain.

Nerve fibers have been differentiated according to their conduction velocities. Conduction speeds in the human peripheral nerves range from 1 to 120 m/s. In general, conduction velocity is greater with a higher fiber threshold, a thicker fiber, and the presence of myelin sheathing.

Tactile Sensor Stimulation

Classic experiments concerned a subject's ability to perceive the presence of a stimulus on the skin, to locate one stimulus or two simultaneous stimuli, and to distinguish between one stimulus or two stimuli applied at the same time. While earlier research procedures were highly individual, modern research uses mostly three classes of stimulation:

1. step functions, in which a displacement is produced quickly and held for a period of a second or so;
2. impulse functions, in which a transient of some given waveform is produced for a few milliseconds; and

3. periodic functions, which displace the skin at constant or variable frequencies for several milliseconds.

These forms of mechanical energy can be imparted to the skin by different transducers. Most research has used skin displacement, but other experiments rely on units of force or on measures of energy transmitted. (Again, much work remains to be done.) A compilation of the "classic" tactile sensitivity of body parts was done by Boff and Lincoln (1988). Fransson-Hall and Kilbom (1993) carefully measured the tactile sensibility of the inner surface of the hand and found the thumb region most sensitive, the palm of the hand average, and the finger region least responsive.

Sensing Temperature

Sensors and Stimuli

While there is no question that temperature can be sensed, there is no agreement on what its sensors are in the body. (Among them are the previously mentioned Krause end bulbs.) Research is hampered by the conventional experimental procedure of applying pointed metallic cylinders of different temperatures ("thermodes") to the skin, a technique that unfortunately generates both touch and temperature sensations. Cooling or warming by air convection or radiation may be better ways to stimulate the receptors involved. An additional complication arises from the fact that temperature sensations are relative and adaptive.

Objects at skin temperature are judged as neutral or indifferent, a value of temperature called "physiological zero." A temperature below this level is called cool, a temperature above, warm. Slowly warming or cooling the skin near physiological zero may not elicit a change in sensation. This range in which no change in sensation occurs even though the temperature varies is called the *zone of neutrality*. For the forearm, the neutrality zone ranges approximately from 31° to 36° C. Neutrality zones are different for different body parts.

Cold and Warm Sensations

Apparently, some nerve sensors respond specifically to cold and falling temperatures, while others react to heat and increasing temperatures. The two scales may overlap, which can lead to paradoxical or contradictory information. For example, spots on the skin that consistently register cold when stimulated at less than physiological zero may also report cold when they are stimulated by a warm thermode of about 45°C. (An opposite paradoxical sensation of warmth has also been observed.) Also, the sensation of warmth can be aroused, in some instances, by applying a pattern of alternating warm and cold stimulation. This occasionally generates the sensation of heat even if a cold probe is applied.

Changes toward warm temperatures from physiological neutral are more easily sensed than changes toward cold, at a ratio of about 1.6 to 1. The longer the stimulus is applied and the larger the area of the skin to which it is transmitted, the smaller is the temperature change that can be discerned. Warm sensations adapt within a short period of time, except at rather high temperature levels. Adaptation to cold is slower and does not seem to occur completely, perhaps because both vasodilation and sweating are responses to a rise in skin temperature, while vasoconstriction is the only countermeasure to cold. Rapid cooling often causes an "overshoot" phenomenon; that is, for a short time, one feels colder than one physically is.

Chapter 5 discusses, in detail, interactions of the body with the environments.

Sensing Cutaneous Pain

Touch, pressure, electricity, warmth, and cold can arouse unpleasant, burning, itching, or painful sensations. At present, it is not clear whether or not modality-specific pain sensors exist *per se.* It is also questionable whether there are distinct "pain centers" in the central nervous sytem; in fact, some researchers even question whether pain is a modality separate from other sensory experiences.

☛☛☛ *"The full array of devices and bodily loci employed in the study of pain would bring a smile to the lips of the Marquis De Sade and a shudder of anticipation to the Graf von Sacher-Masoch" (Sherrick and Cholewiak 1986, pp. 12–39).* ☚☚☚

Many research results are difficult to interpret because of the various levels and categories that fall under the label "pain." Pain can range from barely felt to unbearable. The threshold for pain is a highly variable quantity, probably because pain is so difficult to separate from other sensory and emotional components. Besides cutaneous pain, discussed here, there is visceral, tooth, head, or nerve trauma pain. One can even adapt to pain, at least under certain circumstances and to certain stimuli. Some people have experienced so-called second pain, which is a new and different pain wave following a primary pain after about two seconds. "Referred pain" indicates the displacement of the location of the pain, usually from its visceral origin to a more cutaneous location; an example is cardiac anginal pain, which may be felt in the left arm.

Pain: All in the Brain?

More than three centuries ago, René Descartes taught that pain is a purely physical phenomenon: Tissue injury stimulates specific nerves that transmit an impulse to the brain, causing the mind to perceive pain. This was the medical model until 1965, when the psychologist Ronald Melzak and the physiologist Patrick Wall proposed their "gate-control" theory of pain. According to this theory, sensory signals must go through a gating mechanism in the dorsal horn of the spinal cord, which would either let them pass or stop them. Melzak and Wall's most startling suggestion was that individual emotions controlled the gate. Indeed, studies have shown different pain thresholds and tolerances in individuals and groups of people: Women appear to be more sensitive than men to pain (except during the last few weeks of pregnancy), extroverts have greater pain tolerance than introverts have, and training can diminish one's sensitivity to pain.

Today, it is evident that the brain is actively involved in the experience of pain. Gate-control theory accepts Descartes' view that what you feel as pain is a signal from tissue injury transmitted by nerves (now named A-delta and C fibers) to the brain, yet it adds the notion that the brain controls the gateway for such a signal. In 1994, the neurosurgeon Frederick Lenz noticed that areas of the brain governing ordinary sensations could become abnormally sensitized, generating extreme pain in response to perfectly harmless sensations. According to current theory, the brain generates the experience of pain, which, together with other sensations, is conceived as a set of "neuromodules" akin to individual computer programs on a hard drive. The neuromodule is not a discrete anatomical entity, but a network linking components from many regions of the brain. It gathers inputs from sensory nerves, your memory or your mood—and if the signals reach a certain threshold, they trigger the neuromodule. When you feel pain, it is your brain running a neuromodule that generates your personal pain experience (Gawande 1998).

Sensing Electrical Stimulation

While one feels electrical currents in the skin, whether one does so through the presence of specific sensors for electricity is an open question. Currently, there is no known receptor that is specialized to sense electrical energy or that is particularly receptive to it. In fact, electricity apparently can arouse almost any sensory channel of the peripheral nervous system. Figure 4–22 schematically shows the sensitivity to two-point electrical stimulation at various body sites.

Research has shown that the threshold for electrical stimulation depends heavily on the configuration and location of the electrodes that are used, on the waveform of the electric stimulus

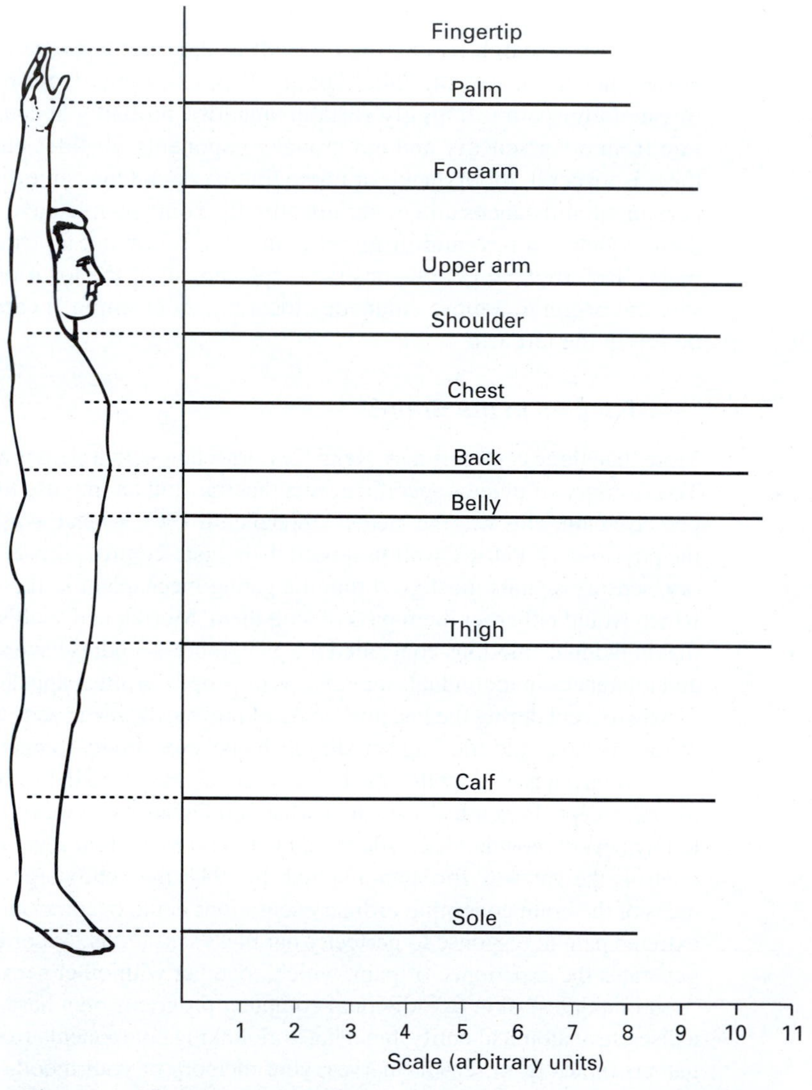

Figure 4–22. Relative sensitivity of body sites to two-point electrical stimulation. (Modified from Boff and Lincoln 1988.)

the rate of stimulus repetition and on the individual subject. Generally, the threshold is about 0.5 to 2 mA with a pulse of 1-ms duration. During shorter pulse durations, a temporal summation of single pulses occurs (Boff and Lincoln 1988).

BALANCING THE BODY—THE VESTIBULAR SENSE

Sensors and Stimuli

Located next to the cochlea in the inner ear, on each side of the head, are three semicircular canals with two sacklike otolith organs: the utricle and the saccule, shown schematically in Figure 4–23. These nonauditory organs are called the *vestibulum*.

The arches of the three vestibular canals are at about right angles to each other, with one canal horizontal and two vertical when the head is erect. Each canal functions as a complete and independent fluid (endolymph) circuit, in spite of the fact that all of the canals share a common cavity in the utricle. Thus, the three canals are sensitive to different rotations of the head. Each canal, near its junction with the utricle, has a widening (an *ampulla*) that contains a protruding ridge (the *crista ampullaris*) which carries cilia, sensory hair cells, that respond to displacements of the endolymph. Cilia also are located in both the utricle and saccule.

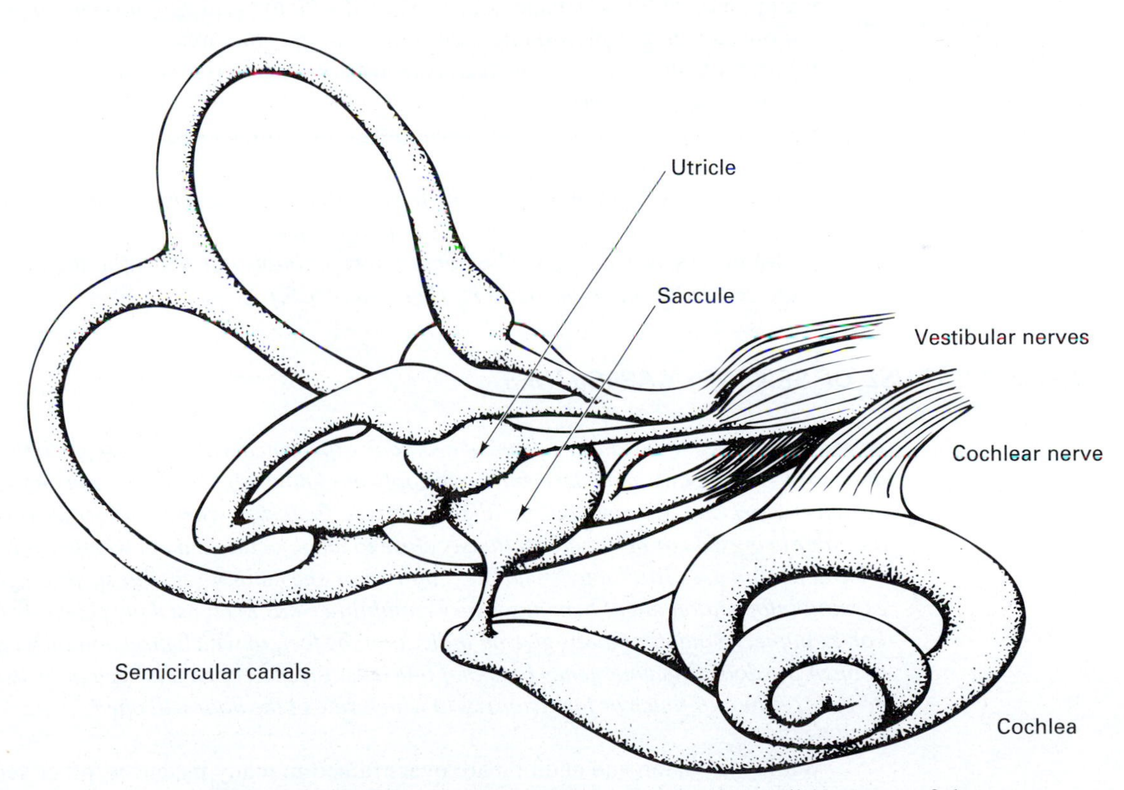

Figure 4–23. The three semicircular canals, the otolith organs of the vestibulum, and the cochlea, with their nervous connections to the eighth cranial nerve. (Adapted from a 1934 drawing by Max Broedel.)

Response to Accelerations

The vestibular system responds to the magnitude and direction of accelerations, including the acceleration due to gravity. The sack-shaped utricle and saccule are sensitive to gravity and other linear accelerations of the head. The response latency to *linear acceleration* is fairly long, about 3 s to 0.1 G, diminishing asymptotically to approximately 0.4 s at accelerations of 1 G or higher. The response time to *rotational acceleration* is about 3 s at accelerations slower than $1° s^{-2}$, falling to roughly 0.3 s for accelerations exceeding $5° s^{-2}$. However, the system adapts to constant acceleration, and small changes in acceleration may not be perceived. Very little is known about acceleration thresholds in different planes of rotation (Howard 1986).

A number of peculiar features are associated with the pea-sized vestibular system. Bringing the head into various postures requires that the brain compare signals not only with a new spatial reference system, but also with new reference inputs from the sensors, because the endolymph now loads the cilia in different ways. Sideways rotation (roll or yaw) induces the two vestibular systems in the head to generate different base signals, which must then be consolidated in the brain.

☛☛☛ *The complex signals from the vestibulum interact with other sensory inputs arriving simultaneously at the cerebellum and cerebral cortex. Consequently, several "vestibular illusions" can occur (Boff and Lincoln 1988), including the following:*

- Illusionary tilt. *Interpretation of linear acceleration as body tilt.*
- Unperceived tilt. *When the body is aligned with the gravitation vector, a person in a airplane that is performing a bank does not feel this roll.*
- Inversion illusion. *A person in zero gravity, or lying in a prone position may feel as if being upside down.*
- Elevator illusion. *A change in gravitational force produces an apparent rise or lowering of seen objects.*
- Coriolis cross-coupling effect. *The feeling of falling sideways when the head is tilted forward while the body rotates about a vertical axis.*
- Motion or space sickness. *This condition is probably due to conflicting inputs from the vestibular and other sensors. (See Chapter 5.)* ☚☚☚

ENGINEERING USE OF SENSORY CAPABILITIES

☛☛☛ *In spite of thousands of years of everyday experience with the senses, fairly little has been done to systematically and conscientiously apply our knowledge of human sensory capabilities to design engineering. True enough, we all feel with our fingertips whether a surface is smooth or not, a welder brings his or her hand cautiously close to an object to find out whether it is still hot, and a blind person uses Braille to "read" text that he or she cannot see. But surprisingly little of the existing information about human sensory capabilities has been used purposefully by engineers. For example, round doorknobs give no indication, by feel, in which direction they must be turned to open the door, and emergency bars that one must press to open a hinged door usually provide no cues (either for touch or for vision) as to which side of the door will open.* ☚☚☚

While our vision and audition are overstrained in many instances, other senses, such as taction, olfaction, and gustation, are underused. The cutaneous senses of the hands are commonly utilized, but other body segments that have the same or similar sensitivities are hardly ever employed.

The CNS receives signals from the various human senses simultaneously, allowing us to get a general picture of the events taking place inside and outside the body. For example, the sensation of exerting a force on an object in order to move it (such as is felt in lifting) is presented to the CNS by muscle spindles, which report on muscle stretch (length and change in length), by Golgi tendon organs, which report on muscle tension (the development of the force), by Ruffini joint organs, which report on the location of the limbs and angles of the joints, and by the cutaneous senses, because the bending of any joint stretches some regions of skin around the joint and relaxes others. Furthermore, the sense of vision provides information about the movement of object and body segments, and sounds associated with the movement supply additional information.

REDUNDANT INFORMATION

The body and various equipment and procedures that we use often provide redundant information to us, in several sensory modalities at the same time. For example, the pilot feels acceleration through body sensors, an instrument displays the attitude of the plane, and a recorded voice tells the crew to pull up if the aircraft noses down.

"KNOBS AND DIALS"

The most widespread use of human tactual capabilities has been in the coding of controls by shape. In the decade after World War II, much research on controls and displays was performed. Jokingly, that time has been called the "knobs and dials era" by human-factors engineers. Nearly all the shape-coding knowledge that we use today was derived at that time. Knobs on controls were formed like wheels or like airplane wings. Shape and size coding was used to indicate what would happen after activating these controls, with the information conveyed both by vision (if one looks at the control handle, which requires sufficient illumination) and by feel (as one touches the control, which might be too late). For specific design recommendations on controls and displays, see Chapter 10.

Incomplete research findings are one explanation for the fact that so little systematic engineering use has been made of human sensory skin capacities. To provide the needed information, research must be based on solid theories regarding the various receptors, their stimulation, and their responses; regarding the screening and propagation of the signals to the central nervous system; and regarding the interpretation of the signals there. Theories of somesthetic sensitivity require testable models of the absorption and propagation of various forms of energy in the path from the surface of the skin to the receptors. In 1986, Sherrick and Cholewiak found the condition of research regarding both theoretical underpinnings and methodological procedures still to be in a "primitive state."

ENHANCING HUMAN PERCEPTION

Just as hair in the human skin amplifies mechanical surface distortion so that the associated sensor may be more easily activated, a number of engineering means exist to make perception more intense. For taction, a thin layer of cloth between the object's surface and the fingertips enhances the perception of unevenness (probably by filtering out tactile noise). For taste and smell, the concentration of an active substance may be enlarged to ensure and hasten its perception. Emergency signals can simultaneously provide sound, light, color, and smell cues to enhance the speed and accuracy with which the accompanying information is received, recognized, and processed.

CHANGING SENSORY MODALITIES

It is difficult to submit sensory information that is habitually conveyed by one sense through another sense, such as transmitting visual information through taction, as is often done for blind persons. The coding of the signal from one sensory system to the other is difficult, particularly if one has never experienced the first kind of sensory input. Kantowitz and Sorkin (1983) found that it took nine days of extensive practice for a blind person to triple his or her reading performance from an initial rate of about 10 words per minute, using an optical-to-tactile converter. However, two sighted persons became highly proficient on the same converter after about 20 hours of experience, at which time they were able to read 70 to 100 words per minute. It might be better, at least in certain circumstances, not to use a natural code that preserves the actual spatial and temporal relationships between the original character (signal) and the one displayed in the other sensory mode; perhaps an artificial code should be employed that allows better and more exact perception. Thus, in the earlier example of the of the blind person, instead of using Braille to convey the shape of letters, it would be advantageous to have the text spoken.

USING THE TACTION SENSE

Given the lack of reliable experimental information, current guidelines for design applications of the sense of taction still rely much on extrapolations of previous knowledge, commonsense experiences, and guesses. On this basis, the following engineering recommendations are made, though with much caution:

Touch information, transmitted through mechano-receptors, can be differentiated by the human regarding

- the magnitude of mechanical deformation,
- the temporal rate of change, and
- the size and location of the area of the skin that is stimulated (i.e., the number of receptors stimulated).

Vibration at the fingertips shows the following minimal thresholds:

- 200 Hz with a displacement of about 2×10^{-4} mm and 800 Hz with a 10^{-3} mm deformation.
- Below 10 Hz and above 1,000 Hz, only general pressure, not vibration, is sensed.
- The highest sensitivity for vibrations appears to be at about 250 Hz, but "flying by the seat of the pants" may be rather dangerous, since sensitivity at the buttocks is very low.

One may prefer certain body areas for tactile input sites. For static two-needle point stimulation, two-point resolution starts at about 2 mm separation at the tip of the finger, at about 4 mm separation at the lips, and at about 40 mm separation at the forearm, but requires about 70 mm separation on the back. Sensitivity to any taction stimulation is highest in the facial area and at the fingertips, and fair at the forearm and lower leg. Some body areas, such as the eyes, though even more sensitive, are out of bounds.

Tactile sensitivity is highly dependent on

- the strength of the stimulus,
- the rate of change of the stimulus, and
- the temperature.

For example, if a stimulus of low intensity appears slowly on cold skin, it may not be noticed.

Measured by step-function inputs, the minimal threshold for force varies from 5×10^{-5} N on the face to 35×10^{-5} N on the big toe. Such stimuli, if they last but 1 ms each, must be separated by at least 5.5 ms to be perceived as two stimuli at the fingertip. If the break between stimuli is too short, they fuse into the sensation of a single stimulus; this phenomenon is called temporal fusion.

Under normal conditions of vigilance, auditory signals are better detected than weak mechanical vibratory signals and electrocutaneous signals. In response to electrocutaneous signals, a long response latency, many misses, and false alarms occur. However, in complex environments requiring a heightened level of vigilance, electric signals will provide redundancy. Experiments discussed by Sherrick and Cholewiak (1986) indicate that the cutaneous system has utility as either a sole or an additional channel for information input.

USING THE TEMPERATURE SENSE

For several reasons, our sense of temperature is difficult to use for communicative purposes: It has a relatively slow response time, a poor ability to identify location, the capability to adapt to a stimulus over time and it may integrate several stimuli that are distributed over a certain area of skin. Furthermore, interactions exist between mechanical and temperature sensations (for example, a colder weight feels heavier on the skin than a warm weight) and thermal sensations can be stimulated chemically (for instance, by applying menthol, alcohol, or pepper to the skin), but not mechanically.

The strength of thermal sensation depends on the location and size of the sensing body surface. The temperature sensation is made stronger by increasing

- the absolute temperature of the stimulus, and its difference from physiological zero,
- the rate of change of temperature, and
- the area of the exposed surface (e.g., immersion of the whole body in a bath, compared to only partial immersion).

Assuming a "neutral" skin temperature of about 33°C, the following "rules of thumb" apply for naked human skin:

- A skin temperature of 10°C appears "painfully cold", 18°C feels "cold", and 30°C still feels "cool." The highest sensitivity to changes in coolness exists between 18° and 30°C.
- Heat sensors respond well throughout the range of about 20° to 47°C.
- Thermal adaptation—that is, "physiological zero"—can be attained in the range of approximately 18° to 42°C, meaning that changes are not felt when the temperature difference is less than 2°C.

- The ability to distinguish between different temperatures is best just below the range of "physiological zero" for cold sensations and above for warm sensations. Distinctly cold temperatures (near or below freezing) and hot temperatures (above 50°C) also provoke sensations of pain. At about 45°C, both cold and heat fibers are stimulated, which may result in the paradoxical sensation of cold when the stimulus is hot. Another interaction may occur between the sensation of pressure and temperature: A force applied under cold conditions appears to be greater than when it is applied under hot conditions.
- Compared to warmth, cold is sensed more quickly, particularly in the face, chest, and abdominal areas. The body's ability to feel warmth is less distinct, but it is best in hairy parts of the skin, around the kneecaps, and at the fingers and elbows.

USING THE SMELL SENSE

Olfactory information is seldom used by engineers, because few research results are available, because people react quite differently to olfactory stimuli, because smells can be easily masked, and because olfactory stimuli are difficult to arrange. As mentioned earlier, among the few industrial applications is adding smelling methylmercaptan to natural gas and pyridin to argon to allow people to smell leaking gas.

USING THE TASTE SENSE

Information on gustation is not used in engineering applications at present, but is of much importance to the food and beverage industry.

USING THE ELECTRICAL SENSE

Electricity is only seldom used as an information carrier, although it has great potential for transmitting signals to the human.

Attaching electrodes is convenient. The energies that are transmitted are low, requiring only about 30 microwatts at the electrode–skin junction, up to a tolerable limit of about 300 milliwatts. Coding can be via placement, intensity, duration, and pulsing. Electrical stimulation can provide a clear, attention-demanding signal that is resistant to masking. Its major drawback, which it shares with mechanical stimulation, is the problem of pain: Aching of deep tissues, stinging or burning can appear if electrodes or energies are improperly applied, and there may be fear of electrical shock.

USING THE PAIN SENSE

Pain does not lend itself to engineering applications, primarily because one is ethically bound not to cause pain, but also because the sensation of pain follows the damage already done too slowly to prevent more damage.

☛☛☛ *"Fighting the enemy, boredom, Romans staged all-night dinner parties and vied with one another in the creation of unusual and ingenious dishes. At one dinner a host served progressively smaller members of the food chain stuffed inside each other: inside a calf, there was pig, inside the pig a lamb, inside the lamb a chicken, inside the chicken a rabbit, inside the rabbit a dormouse, and so on. Another host served a variety of dishes that looked different but were all made from the same ingredient. . . . Slaves brought garlands of flowers to drape over the diners, and rubbed their bodies with perfumed unguents to relax them. The floor might be knee-deep in rose petals. Course after course would appear, some with peppery sauces to spark the taste buds, others in velvety sauces to soothe them. Slaves blew exotic scents through pipes into the room, and sprinkled the diners with heavy, musky animal perfumes like civet or ambergris. Sometimes the food itself squirted saffron or rose water or some other delicacy into the diner's face." (Ackerman 1990, p 144).*

SUMMARY

The eye plays a major role in collecting information. Thus, the provision of proper visual signals, through the selection of proper illumination and contrast, and the avoidance of unbecoming circumstances, such as glare, are important ergonomic tasks to provide sensory inputs. The eye is seldom used as a means for output, "eye tracking" being one of the exceptions.

The sense of hearing provides inputs only. Recommendations for the design of communication and sound-signal systems are based on complex, but well-researched, relations between the frequency and intensity of sounds. Protection of the ear from damaging noises is also well understood, and related technical means are documented and ready to be applied.

Surprisingly and disappointingly, the human sense of touch is not well researched, and much of the existing information is dubious or difficult to apply. Nevertheless, the sense of touch (for pressure, vibration, electricity, and temperature) is often employed in human-operated systems. Temperature sensing is rather well understood, although not very reliable as input. The vestibular sense provides information about head and body posture, but is largely dependent on the existence of a well-defined gravity vector (which is missing in space). Furthermore, the human processor is easily confused by conflicting information from the vestibular and other senses—leading, for example, to nausea. The senses of smell and taste are not currently used for many engineering applications, but are widely used in industry.

CHALLENGES

What might be reasonable approaches to classifying the various human senses?

How can one avoid interactions between different senses during experiments? Or would it be reasonable to present signals that simultaneously cover two or more modalities?

What might be a practical and easily used, but still scientifically exact, reference plane or system for identifying the direction of the line of sight?

Why do many older people need higher illumination to see objects clearly?

Does one person see a given color—say, green—exactly the same as another person? If not, what can be done, in a practical sense, to compare the two perceptions?

Does the definition of the border between dark and light objects have any effect on the perception of contrast?

Do different physical explanations underlie the mixing of three independent primary colors and the artist's mixing of pigments in paint to generate a given color?

How might one determine whether colors have specific effects on a person's mood and on the performance of a task?

How can one assess the effects of different kinds of music on mood and productivity?

How might the generation of specific smells (such as in perfumes) be made a systematic science, instead of remaining an individual art?

What is needed to turn "making things taste good" into a scientific or technological systematic process instead of remaining an art?

Does the replacement of metal objects at different temperatures (thermodes) by the flow of air at different temperatures solve the problem of the simultaneous stimulation of temperature and tactile sensors?

Which sensory modalities should be combined, under certain conditions and for certain signals, to increase the likelihood of perception, to increase the speed of recognition, and to enhance CNS processing?

Chapter 5

How the Body Interacts with the Environment

OVERVIEW

Most of us work in a "normal" environment—that is, in a moderate climate and on solid ground not much above sea level. However, experiencing changes in climate from summer to winter, working at a high altitude, being subjected to vibrations and impacts, diving underneath the sea, or flying an airplane can challenge the body in many ways. Various technical and behavioral means are available to avoid or minimize negative effects on human comfort, health, and performance.

In industry, dust, toxic fumes, and chemicals can harm the worker. Recognizing and avoiding such conditions is the domain of *industrial hygiene*. Besides a number of textbooks in the discipline (e.g., DiNardi 1997; Plog 2001, in press), in the United States, AIHA and ASHRAE (see the "Addresses" section in the Introduction) have published recommendations and guidelines.

THERMOREGULATION OF THE HUMAN BODY

The human body generates energy and, at the same time, exchanges (gains or loses) energy with the environment. Since a rather constant core temperature must be maintained, the body must dissipate heat in a hot climate, while excessive heat loss must be prevented in a cold environment.

There is some controversy about the best modeling of energy production in the body and the exchange of energy with the environment, given an individual's conditions of work and clothing. Specific topics of concern are whether one may assume a "core" that must be kept at nearly constant temperature (as we shall assume in this chapter; see also Kroemer, et al. 1997; Youle 1990) or whether it is reasonable to postulate an average skin temperature (as in ASHRAE 1985).

Maintain core temperature

The human body has a complex control system for maintaining the deep-body core temperature very close to 37°C (about 99° F), as measured in the intestines, the rectum, the ear, or (most often) as estimated by the temperature in the mouth. While the temperature of the body fluctuates slightly throughout the day, due to diurnal changes in body functions (see Chapter 6), the main task of the temperature-control system is to regulate the energy exchange between (metabolic) heat generated within the body and external energy; the body may absorb heat in hot surroundings or lose heat in a cool environment.

Keeping the core temperature close to 37°C is the primary task of the human thermoregulatory system, in cold or hot environments. Changes in core temperature of plus or minus 2° from 37°C affect body functions and task performance severely, while deviations of plus or minus 6°C are usually lethal. At the skin, the human temperature-regulation system must keep temperatures well above freezing and below the 40°C or so in its outer layers, but there are major differences from region to region. For example, the toes may be at 25°C, the legs and upper arms at 31°C, and the forehead at 34°C. Combined, these temperatures make us comfortable (Youle 1990).

The Energy Balance

Equation (2–1) in Chapter 2 describes the energy exchange between inputs to the body and outputs as

$$I = M = H + W + S, \tag{5–1}$$

where I is the energy input via nutrition that is transformed into metabolic energy M, which is the sum of the heat H that must be dispelled to the outside, the external work W done, and the energy storage S in the body.

Heat balance

The energy to be exchanged with the thermal environment is given by

$$H = I - W - S. \tag{5–2}$$

The system is in balance with the environment if all heat energy H is dissipated to the environment while the stored energy remains essentially unchanged. The amount of it is much larger than that of W—see Chapter 2, equation 2–3.

Energy Exchanges with the Environment

Energy is exchanged with the environment through radiation R, convection C, conduction K, and evaporation E.

Radiation

Heat exchange by radiation R is a flow of electromagnetic energy between two opposing surfaces—for example, between a windowpane and a person's skin. The amount of heat radiated depends primarily on the temperature difference between the two surfaces, but not on the temperature of the air between them. Heat always radiates from the warmer to the colder surface; hence, the human body can either lose or gain heat through radiation.

The amount of radiating energy Q_R lost from or gained by the body through radiation depends essentially on the size S of the participating body surface and on the difference Δ between the fourth power of the temperatures T (in degrees Kelvin) of the surfaces:

$$Q_R = f(S, \Delta T^4). \tag{5–3}$$

Equation (5–3) is approximated by

$$Q_R \approx S h_R \, \Delta t, \tag{5–4}$$

where h_R is the heat-transfer coefficient and t is the temperature in degrees Celsius. (For more details, see Kroemer et al. 1997 and Youle 1990.)

Heat exchange through convection C and conduction K both follow the same thermodynamic rules. The heat transferred is again proportional to the area of human skin participating in the process and to the temperature difference between skin and the adjacent layer of the external medium. Hence, in general terms, heat exchange by convection or conduction is given by

$$Q_{C,K} = f(S, \Delta t), \quad (5\text{–}5)$$

which is approximated by

$$Q_{C,K} \approx Sk\,\Delta t, \quad (5\text{–}6)$$

where k is the coefficient of conduction or convection.

Convection

Heat exchange through convection C takes place when human skin is in contact with air or some other gas and with water or another fluid. Heat energy is transferred *from* the skin to a layer of colder gas or fluid next to the skin's surface, or *to* the skin if the surrounding medium is warmer. *Convective heat exchange* is facilitated if the medium moves quickly along the surface of the skin, thus maintaining a temperature differential. There is always some natural movement of air or fluid, as long as a temperature gradient exists; this is called *free convection.* More movement can be produced by forced action, such as by a fan or while swimming in water rather than floating motionless; this is called *induced convection.*

Conduction

Conductive heat exchange K exists when the skin is in touch with a solid body. As long as there is a difference in temperature, heat flows, especially if the conductance of the object is high; less energy flows if the skin touches an insulator having a low k value.

☛☛☛ *Cork and wood "feel warm" because their heat-conduction coefficients are below that of human tissue. Metal of the same temperature accepts body heat easily and conducts it away; therefore, it feels colder than cork or wood even when they are all at the same temperature.* ☚☚☚

Evaporation

Heat exchange by evaporation E is in only one direction: The human being loses heat by evaporation. Water never condenses on living skin; if it did, heat would be added to the body. Evaporation requires energy of about 580 cal per cm^3 of evaporated water, which reduces the heat content of the body by that amount. Some water is evaporated in the respiratory passages, but most evaporation occurs on the skin (as sweat).

The heat lost by evaporation, Q_E, from the human body depends on the participating wet body surface S and on the humidity h of the air; that is,

$$Q_E = f(S, h), \quad (5\text{–}7)$$

with higher humidity making evaporative heat loss more difficult than dryer air. Movement of the layer of air nearest the skin increases heat loss through evaporation (similar to the effect of such movement in convection), since it replaces humid air by drier air. Some evaporative heat loss occurs even in a cold environment, because in the lungs water evaporates into the air to be exhaled (the volume of which increases with enlarged ventilation at heavier work) and sweat continuously diffuses onto the skin.

Temperature difference

Figure 5–1 schematically shows the body's heat loss via radiation, convection (and, similarly, conduction), and evaporation in a warm environment. These heat exchanges all depend, directly or indirectly, on the difference in temperature between the participating body surface and the environment.

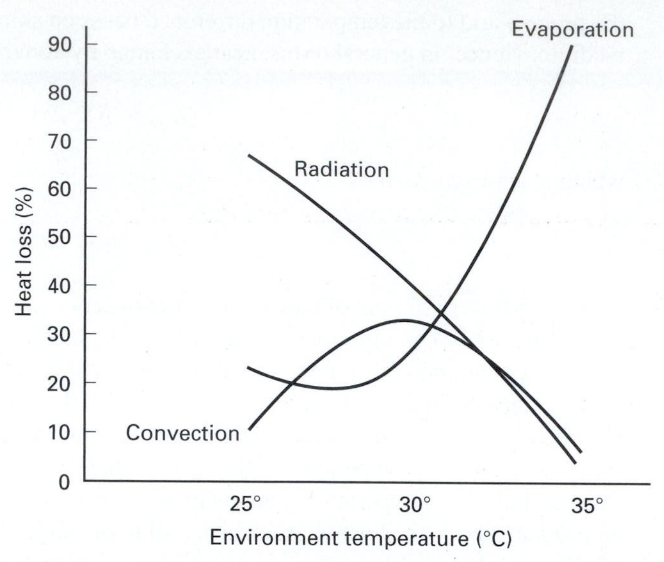

Figure 5–1. Contributions of the different kinds of heat transfer in cool and warm environments.

Heat balance

Heat balance exists when the heat H developed in the body [see Equation (5–2)] and heat exchange with the environment by radiation R, convection C, conduction K, and evaporation E are in equilibrium (ie, they all add up to zero). This relationship can be expressed as

$$H + R + C + K + E = 0. \qquad (5\text{–}8)$$

Note that the quantities R, C, K, and E are negative if the body loses energy to the environment and positive if the body gains energy from the environment; E is always negative.

Temperature Regulation and Sensation

Body heat flow

Heat is produced in the body's "metabolically active" tissues—primarily the skeletal muscles, but also internal organs, fat, bone, and connective and nerve tissue. Heat energy is circulated throughout the body by the blood. The actions of constriction, dilation, and shunting of blood vessels (see Chapter 2) modulate the blood flow. Heat exchange with the environment takes place at the body's respiratory surfaces and, of course, through the skin, as just discussed.

In a cold environment, body heat must be conserved, which is done primarily by reducing blood flow to the skin and by increasing insulation via clothing. In a hot environment, body heat must be dissipated and gain from the environment prevented. This is done primarily by increases in blood flow to the skin and by the production and evaporation of sweat.

Various body temperatures

Temperatures in the human body are not uniform throughout; there are large differences between "core" and "shell" temperatures. Under normal conditions the average temperature gradient between skin and deep body is about 4°C at rest, but in the cold the difference in temperature may be 20°C or more. Thus, the temperature-regulation system, located in the hypothalamus, has to maintain various temperatures at different body parts under various conditions.

Body heat and work

If the body is about to overheat, it must generate less internal heat. Therefore, muscular activities will be reduced, possibly to the extent that no work is being performed anymore. In the opposite case, when more heat must be generated, the work or exercise level will be augmented by increased muscular activities. (Given its low efficiency, muscular work generates much heat; see Chapter 2.)

Muscles can generate more heat or less heat, but cannot cool the body. In contrast, the production of sweat influences the amount of energy lost, but cannot bring about a heat gain. Vascular activities affect the heat distribution through the body and control heat loss or gain, but they do not generate energy. Muscular, vascular, and sweat-production functions combine to regulate body heat as the human interacts with the external climate.

The German physicist Daniel Gabriel Fahrenheit (1686–1736) worked first with alcohol-filled thermometers and then, from 1714 on, with mercury thermometers. He noted the height of the column of mercury in a mixture of ice, water, and ammonium chloride, the lowest temperature he could achieve, and called that zero. A mixture of ice and water he set at 32°, and the temperature of boiling water at 212°. The Fahrenheit mercury thermometer and scale were the first that could measure temperature with an accuracy sufficient for scientists. In 1742, the Swedish astronomer Anders Celsius (1701–1744) suggested that the freezing temperature of water be set to zero and the boiling temperature to 100°. This centrigrade scale was called the Celsius scale by international agreement in 1948 and is officially used in all countries except the United States (Asimov 1989).

Temperature sensors

Various temperature sensors are located in the core and the shell of the body (Chapter 4). In the skin of the forearm, for example, the sensors can completely adapt in the range of approximately 30° to 35°C, as "physiological zero." Temperatures below that range are perceived as cool or cold, above, as warm or hot. There is some overlap in sensations of cool and warm in the intermediate range. Below 15° and above 45°C, the temperature sensors are less discriminating, but also less adaptive. A "paradoxical" effect is that, around 45°C, sensors again signal "cold," while in fact the temperature is rather hot. (Boff and Lincoln 1988).

Achieving Thermal Homeostasis

Two temperature gradients

The human thermoregulatory system must achieve two suitable temperature gradients: from the core to the skin and then from the skin to the surroundings. The gradient from the core to the skin is internally the most important, because overheating or undercooling of the key tissues in the brain and the chest must be avoided, even at the cost of overheating or undercooling the shell.

First, blood flow

Thermal equilibrium, called *homeostasis,* is achieved primarily by regulation of the blood flow from deep tissues and muscles to skin and lungs. By far, most heat is exchanged with the environment at the skin; but in the lungs, from 10 to 25 percent of the total body heat is dissipated.

Second, muscle activities

Secondary actions to establish thermal homeostasis take place at the muscles, which generate heat by *voluntary* or *involuntary* efforts: work and exercise, or shivering. If the goal is to gain heat, the regulatory system initiates skeletal muscle contractions; but if a heat gain must be avoided, it reduces or abolishes muscular activities.

Third, clothing and shelter

Changes of clothing and shelter are our tertiary, conscientious actions to establish thermal homeostasis. Together with blood-flow regulation and muscle activities, they achieve the appropriate temperature gradient between the skin and the environment. Changes of clothing and shelter affect radiation, convection, conduction, and evaporation. Light and heavy clothes have different permeability and differing abilities to establish stationary insulating layers. Clothes affect conductance (i.e., energy transmitted per surface unit, time, and temperature gradient), and their color determines how much external radiation energy is absorbed or reflected.

ASSESSING THE THERMAL ENVIRONMENT

The thermal environment is determined by four physical factors: the air (or water) temperature, the humidity, the air (or water) movement, and the temperatures of surfaces that exchange energy by radiation. The combination of these four factors determines the physical conditions and our perception of the climate.

Temperature Air temperature is measured with thermometers, thermistors, or thermocouples. Whichever of these devices is used, one must ensure that the ambient temperature is not affected by the other three factors—particularly humidity, but also air movement and surface temperatures. To measure the so-called dry temperature of ambient air, one keeps the sensor dry and shields it with a surrounding bulb that reflects radiated energy. Hence, air temperature is often measured with a dry-bulb thermometer.

Humidity Air humidity may be measured with a psychrometer, a hygrometer, or some other electronic device. All of them usually rely on the fact that the cooling effect of evaporation is proportional to the humidity of the air, with higher vapor pressure making evaporative cooling less efficient. Therefore, we can measure humidity using two thermometers—one dry, the other wet. The highest absolute content of water vapor in the air is reached when any further increase would lead to the development of water droplets. The amount of possible vapor depends on the pressure and temperature of the air, with lower pressure and higher temperature allowing more water vapor to be retained than lower temperatures. One usually speaks of "relative humidity," which indicates the actual vapor content in relation to the maximum possible content (the "absolute humidity") at the given air temperature and pressure.

Air movement Air movement is measured with various types of anemometers that employ mechanical or electrical principles. We can also measure air movement with two thermometers—one dry and one wet (similar to the way we assess humidity)—relying on the fact that the wet thermometer shows more increased evaporative cooling with higher air movement than the dry thermometer shows.

Radiation Radiant heat exchange depends primarily on the difference in temperature between the surfaces of the person and the surroundings, on the emission properties of the radiating surface, and on the absorption characteristics of the receiving surface. One easy way to assess the amount of energy acquired through radiation is to place a thermometer inside a black globe that absorbs practically all arriving radiated energy.

Thermal comfort A person's feeling of comfort is not determined solely by the physics of thermal balance according to Equation (5–8): In a warm environment, skin wetness plays a major role, and in a cold environment, skin temperature does. Two scales, the Bedford and the ASHRAE scales, are widely used to assess individual thermal comfort. As shown in Table 5–1, they yield similar results (Youle 1990). In the past, various techniques were used to assess the combined effects of some or all four environmental factors and to express these in one model, chart, or index. They resulted in several empirical thermal indices, which are based on data compiled from the statements of subjects who were exposed to various climates. Most establish a *reference* or *"effective"* climate that "feels the same" as various combinations of the several component climates.

TABLE 5–1. Scales for Assessing Subjective Thermal Comfort

Bedford		*ASHRAE*	
Much too warm	7	+3	Hot
Too warm	6	+2	Warm
Comfortably warm	5	+1	Slightly warm
Comfortable	4	0	Neutral
Comfortably cool	3	−1	Slightly cool
Too cool	2	−2	Cool
Much too cool	1	−3	Cold

Source: Adapted from Youle 1990.

ET A well-known example is the effective temperature ET, which reflects combinations of dry temperature, humidity, and air movement with various levels of activities and clothing. (See Figure 5–2.) Such a climate index can be provided by instruments specially arranged to respond to the components of climate as a human does. For example, a dry thermometer placed inside a black globe responds both to air temperature and radiation; a wet thermometer responds to both air velocity and air humidity.

WBGT The wet-bulb globe temperature (WBGT) index is generated by an instrument with three sensors whose readings are automatically weighted and then combined. The combined effects of all climate parameters are weighted as follows:

Outdoors,

$$WBGT = 0.7WB + 0.2GT + 0.1DB. \quad (5\text{–}9)$$

Indoors,

$$WBGT = 0.7WB + 0.3GT. \quad (5\text{–}10)$$

where WB is the wet-bulb temperature of a sensor in a wet wick exposed to a natural air current, GT is the globe temperature at the center of a black sphere 15 cm in diameter, and DB is the dry-bulb temperature measured while the thermometer is shielded from radiation.

The WBGT is commonly applied to assess the effects of warm or hot climates. Depending on the activity level (expressed in watts), the WBGT temperatures given in Table 5–2 are considered safe for most healthy people, although there is some concern about the adequacy of the WBGT for combinations of high humidity and little air movement (Parsons 1995, Ramsey 1995).

REACTIONS OF THE BODY TO HOT ENVIRONMENTS

In hot environments, the body produces heat and must dissipate it. To achieve this, the skin temperature should be above that of the immediate environment in order to facilitate energy loss through convection, conduction, and radiation.

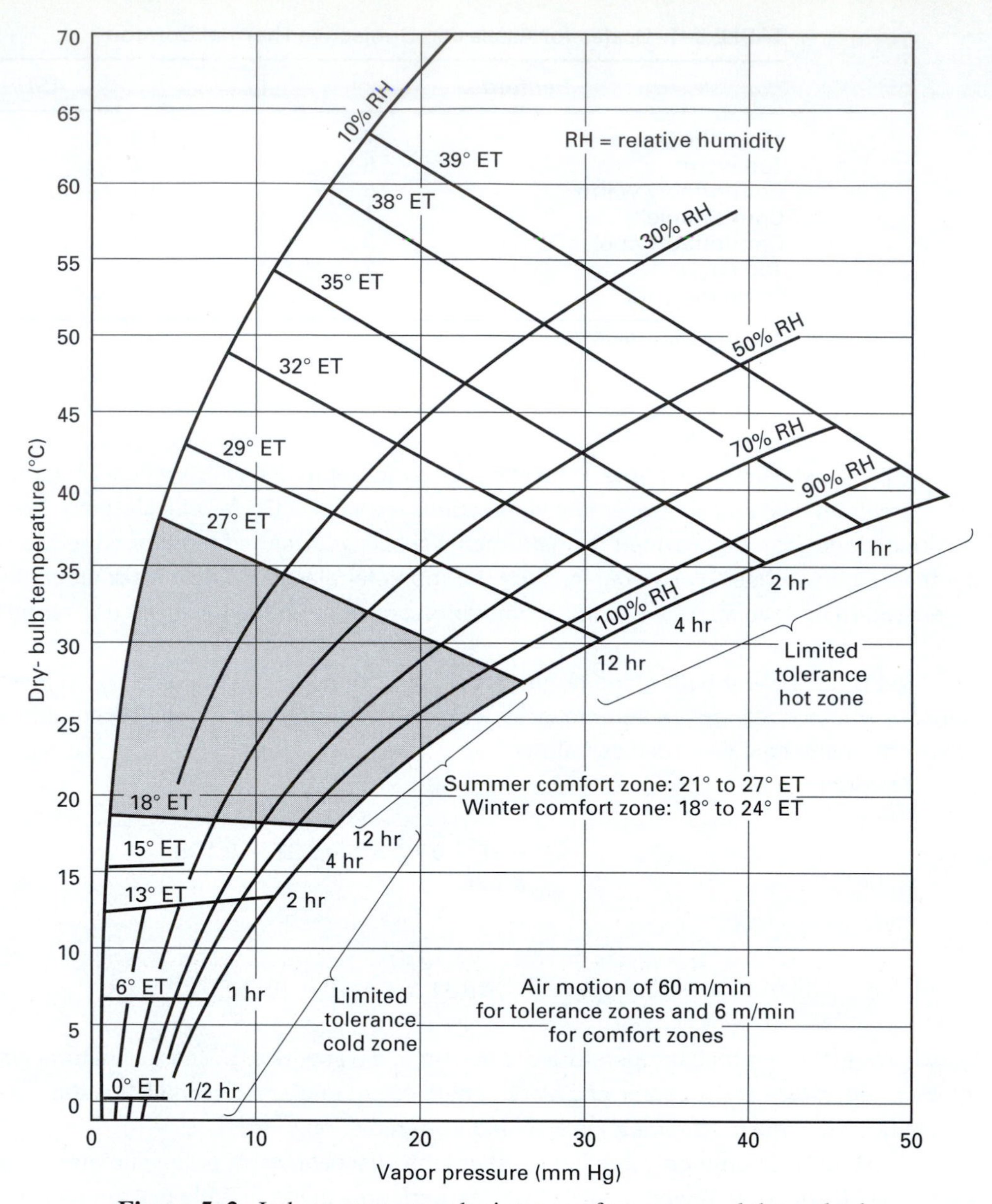

Figure 5–2. Indoor summer and winter comfort zones and thermal tolerance for appropriately dressed sitting persons doing light work (MIL-HDBK 759).

Sweat evaporation

If heat transfer is not sufficient, sweat glands are activated, and the evaporation of the sweat that is produced cools the skin. The recruitment of sweat glands from different areas of the body varies among individuals. The overall amount of sweat developed and evaporated depends very much on clothing, the environment, work requirements, and the individual's acclimatization.

TABLE 5–2. "Safe" WGBT Values

*Metabolic rate M (W)**	*"Safe" WBGT (°C)* Person is acclimatized to heat	*Person is not acclimatized to heat*
$M \leq 117$	33	32
$117 < M \leq 234$	30	29
$234 < M \leq 360$	28	26
$360 < M \leq 468$	No air movement: 25 With air movement: 26	No air movement: 22 With air movement: 23
$M > 468$	No air movement: 23 With air movement: 25	No air movement: 18 With air movement: 20

Source: Abbreviated from ISO 7243, 1982.
*Assuming a skin surface area of 1.8 m^2.

If heat transfer by blood distribution and sweat evaporation is still insufficient, muscular activities must be reduced to lower the amount of energy generated. In fact, this is the final and necessary action of the body; otherwise the core temperature would exceed a tolerable limit.

☛☛☛ *If the body has to choose between unacceptable overheating and continuing to perform physical work, the choice will be in favor of maintaining the core temperature, which means that work activities must be diminished or stopped.* ☚☚☚

SIGNS OF HEAT STRAIN

Excessive sweating There are several signs of heat strain on the body. The first is the *sweat rate.* In strenuous exercises and hot climates, several liters of sweat may be produced in an hour. Sweat begins to drip off the skin when the generation of sweat reaches about one-third of the maximal evaporative capacity. Of course, sweat running down the skin contributes very little to heat transfer.

Heart rate and core temperature Heat strain increases the circulatory activities. Cardiac output must be enlarged, which is brought about mostly by a higher *heart rate.*

Increased core temperature Another sign of heat strain is a *rise in core temperature,* which must be counteracted before the temperature exceeds the sustainable limit.

Drink water! The water balance within the body provides another sign of heat strain. Severe *dehydration,* as indicated by the loss of 1 or more percent of body weight, can critically affect the ability of the body to control its functions. Therefore, the fluid level must be maintained, best by frequent drinking of small amounts of water. Normally, it is not necessary to add salt to drinking water, since, in western diets, the salt in the food is more than sufficient to resupply salt lost through sweating.

Among the first reactions to heavy exercise in excessive heat are sensations of discomfort and perhaps skin eruptions, so-called prickly heat, which is associated with sweating. As a result of sweating, heat cramps may develop, which are muscle spasms related to a local lack of salt. Cramps may occur after one has quickly drunk large amounts of fluid, which dilutes the body fluids.

Heat exhaustion is a combined function of dehydration and overloading of the circulatory system. Associated effects are fatigue, headache, nausea, and dizziness, often accompanied by giddy behavior.

Heat syncope is a loss of consciousness (commonly called fainting) due to a failure of the circulatory system. Heatstroke is an overloading of both the circulatory and sweating systems and is associated with hot, dry skin, an increased core temperature, and mental confusion. Table 5–3 lists some symptoms, causes, and treatments for heat-related disorders.

TABLE 5–3. Heat Disorders

Disorder	*Symptoms*	*Causes*	*Treatments*
Transient heat fatigue	Decrease in productivity, alertness, coordination, and vigilance.	Not acclimatized to hot environment.	Gradual adjustment to hot environment.
Heat rash ("prickly heat")	Rash in area of heavy perspiration; discomfort or temporary disability.	Perspiration not removed from skin; sweat glands inflamed.	Periodic rests in a cool area; showering, bathing; drying skin.
Fainting	Blackout, collapse.	Shortage of oxygen in the brain.	Lay down.
Heat cramps	Painful spasms in used skeletal muscles.	Loss of salt; large quantities of water consumed quickly.	Adequate salt with meals; salted liquids (unless advised differently by a physician).
Heat exhaustion	Extreme weakness or fatigue; giddiness; nausea; headache; pale or flushed complexion; body temperature normal or slightly higher; moist skin; in extreme cases, vomiting or loss of consciousness.	Loss of water or salt; loss of blood plasma; strain on the circulatory system.	Rest in cool area; salted liquids (unless advised differently by a physician).
Heat stroke	Skin is hot, dry, and often red or spotted; core temperature is 40°C (105°F) or higher and rising; mental confusion; delirium; convulsions; unconsciousness. Death or permanent brain damage may result unless treated immediately.	Thermoregulatory system breaks down under stress, and sweating stops. The body's ability to remove excess heat is almost eliminated.	Remove to cool area; soak clothing with cold water; fan body; call physician or ambulance immediately.

Source: Adapted from Spain, Ewing, and Clay 1985.

Skin burns

When human skin touches a hot surface, a burn may follow. The actual critical contact temperature depends on the duration of the contact and on the material of the object or fluid that is contacted. The shorter the contact, the higher is the temperature that can be tolerated, as listed in Table 5–4.

TABLE 5–4. Maximal Surface Temperature That Can be Tolerated by Human Skin Without Risk of Burn

Material	*Maximal Surface Temperature (°C) for Contact Times of* 1 s	*4 s*	*1 min*	*10 min*	*8 h*
Metals					
Uncoated, smooth surface	65	60			
Uncoated, rough surface	70	65	50		
Coated with varnish, 50 μm thick	75	65			
Concrete, ceramics	80	70		all	all
Glazed ceramics (tiles)	80	75	55	48	43
Glass, porcelain	85	75			
Plastics					
Polyamid with glass fibers	85	75			
Duroplast with fibers	95	85	60		
Teflon, Plexiglas	NA	85			
Wood	115*	95	60		
Water	65	60	50		

SOURCE: Siekmann 1990.
*Up to 25°C or higher for very dry and very light woods.

REACTIONS OF THE BODY TO COLD ENVIRONMENTS

The human body has few natural defenses against a cold environment. Most of our counterresponses are behavioral, such as putting on suitably heavy clothing, covering the face, seeking shelter, or using external sources of warmth.

Conserving body heat

In a cold climate, the body must conserve heat while producing it. To conserve heat, the temperature of the skin is naturally lowered, reducing the temperature difference between the skin and the outside. Keeping the circulating blood closer to the core, away from the skin, accomplishes heat conservation; for example, the blood flow in the fingers may be reduced to 1 percent of that in a moderate climate. This results in cold fingers and toes, with possible damage to the tissue if the temperatures get close to freezing.

The development of "goose bumps" helps to retain a relatively warm layer of stationary air close to the skin. The stationary envelope acts like an insulator, reducing energy loss at the skin.

Shivering

The other major reaction of the body to a cold environment is to increase metabolic heat generation. Shivering is the involuntary mechanism that accomplishes such an increase (called thermogenesis).

☛☛☛ *Shivering usually begins in the neck, apparently because warmth is important in supplying blood to the brain. The onset of shivering is normally preceded by an increase in overall muscle tone in response to body cooling. With increased firing rates of the motor units, but no actual limb movements generated, a feeling of stiffness is generally experienced. Then shivering suddenly begins, caused by muscle units firing at different frequencies of repetition (rate coding) and out of phase with each other (recruitment coding). Since no energy due to mechanical work is transferred to the outside, the total activity is transformed into heat production, allowing an increase in the metabolic rate of up to four times the resting rate. If the body does not become warm, and if motor-unit innervations become synchronized so that large-muscle groups are contracted, shivering may become rather violent. While such shivering can generate heat that is five or more times that generated at the resting metabolic rate, it can be maintained only for a short period.*

Another mechanism that may produce heat is called *non-shivering thermogenesis:* Body organs, particularly in the liver and the viscera, increase their metabolism. The existence of this response in humans is debated.

Muscle contractions

Of course, muscular activities can be performed voluntarily, such as by increasing the dynamic muscular work performed or by moving body segments, contracting muscles, flexing the fingers, etc. Such dynamic muscular work may easily increase the metabolic rate to 10 or more times the resting rate.

Vasoconstriction and -dilation

Activation of cutaneous vasoconstriction is apparently under the control of the sympathetic nervous system, in addition to local reflex reactions to direct cold stimuli. An interesting phenomenon associated with cutaneous vasoconstriction is the so-called hunting reflex (a cold-induced vasodilation): After initial vasoconstriction has taken place, a sudden opening of blood vessels allows warm blood to return to the skin, rewarming that section of the body (e.g., the hands). Then, constriction returns, and the sequence may be repeated several times. If vasoconstriction and metabolic rate regulation cannot prevent serious energy loss through the body surfaces, the body will suffer some effects of cold strain.

SIGNS OF COLD STRAIN

To reduce the temperature difference between the body and the outside, the body lowers the skin temperature in a cold environment. Thus, the fingers, toes, face, and ears are usually first exposed to cold damage, while the body core is protected as long as possible.

Reduced dexterity As the skin temperature is lowered to about 20° to 15°C, manual dexterity begins to diminish. Tactile sensitivity is severely diminished as the skin temperature falls below 8°C.

Frostbite If the temperature approaches freezing, ice crystals develop in the cells and destroy them, a condition known as frostbite.

Nervous block At local tissue temperatures of 10° to 8°C, the peripheral motor nerve velocity is decreased to near zero, generating a nervous "block." Hence, severe cooling of the skin and, subsequently, the body, increasingly diminishes one's ability to perform activities, even if they could be lifesaving. First, the person may be unable even to light a match, and then he or she may become apathetic, desiring only to sleep, and finally plunge into hypothermia.

Serious problems with lowered core temperature Severe reductions in skin temperatures are usually accompanied by a fall in core temperature. Lowering the core temperature has very serious consequences, because vigilance drops when the core temperature is below 36°C. At core temperatures below

35°C, one may not be able to perform even simple activities. When the core temperature drops even lower, the mind becomes confused, with loss of consciousness occurring around 32°C. At core temperatures of about 26°C, the heart may fail. At very low core temperatures, such as 20°C, vital signs disappear, but the oxygen supply to the brain may still be sufficient to allow revival from hypothermia.

Immersion in cold water

Immersion in cold water can quickly bring about hypothermia by convection. While one can endure up to two hours in water at 15°C, one is helpless in water of 5°C after only 20 to 30 min. Wearing clothing that provides insulation can increase the survival time in cold water; also, obese people with much insulating adipose tissue have an advantage over thin persons. Floating motionless results in less metabolic energy being generated and spent than when swimming vigorously, which also removes the insulating envelope of warmed water around the body.

How Cold Does it Feel?

Perceived coldness

A person's decision to stay in the cold or to seek shelter depends on the subjective assessment of how cold the body surface or core actually is. If one's body is becoming very cold, it is dangerous not to perceive and react to the body's signals, or else the body temperature may fall so low that further cooling is below the threshold of perception.

The perception of the body's decreasing temperature depends upon signals received from surface thermal receptors and from sensors in the core. As skin temperatures decrease below 35°C, the intensity of the cold sensation increases, becoming strongest near 20°C, but at lower temperatures, perception of the intensity decreases. It is often difficult to separate feelings of cold from pain and discomfort.

Under what conditions a person is exposed to the cold may greatly influence how cold the person perceives him- or herself to be. It can make quite a difference whether one is exposed to cold air (whether or not it is moving) or to cold water, whether one is wearing protective clothing, and what one is actually doing.

When the temperature plunges, each downward step can generate an "overshoot" sensation of cold receptors that react very quickly not only to the difference in temperature, but also to the rate at which the temperature changes. Yet, if the temperature stabilizes, the cold sensations become smaller as one adapts to the condition. Exposure to very cold water accentuates the overshoot phenomenon observed in cold air. The reason may be that the thermal conductivity of water is about a thousand times greater than that of cold air at the same temperature. Thus, cold water causes a convective heat loss that may be 25 times that of cold air (Hoffman and Pozos 1989).

☛☛☛ *In experiments, subjects wearing a flotation suit were immersed in cold water at 10°C. Their temperatures at the groin, back, and rectum were continuously recorded, and the subjects rated how cold they perceived those areas to be. The results of the experiment showed that the subjects were unable to reliably assess how cold they actually were. Neither their core nor their surface temperatures correlated with their cold sensations (Hoffman and Pozos 1989).* ☚☚☚

Poor judgment of coldness

The results of many experiments and experiences indicate that the subjective sensation of cold is a poor, unreliable, and possibly dangerous indicator of the core and surface temperatures of the body. Measuring the ambient temperature, humidity, air movement, and exposure time and reacting to these physical measures is probably a better strategy than relying on subjective sensations.

Acclimatization

Continuous or repeated exposure to hot and, less so, cold conditions brings about a gradual adjustment of body functions, resulting in a better tolerance of the climatic stress.

Acclimation to heat

Acclimatization to heat is demonstrated by increased sweat production, a lowered skin and core temperature, and a reduced heart rate, compared with a person's first reactions to heat exposure. The process (called acclimation) is very pronounced within about a week, and full acclimatization is achieved within about two weeks. Interrupting heat exposure for just a few days reduces the effects of acclimatization, which, upon a person's return to a moderate climate, is entirely lost after about two weeks.

How to acclimate to heat

A healthy person can adjust to a dry or humid heat. Acclimatization to heat does not depend on the type of work performed or whether the work is heavy and of short duration, or moderate and continuous. A healthy and well-trained person acclimates more easily than someone in poor physical condition, but training cannot replace acclimatization. However, if physical work must be performed in a hot climate, then such work should also be included in the acclimation phase. Since the body can adapt to heat, but not to dehydration, liberal drinking of water is helpful during acclimation and then throughout heat exposure, in order to replace fluid lost through the evaporation of sweat.

Acclimation to cold?

Acclimatization to cold is much less pronounced; in fact, scientists doubt that true physiological adjustment to moderate cold takes place when appropriate clothing is worn. There are so-called local acclimatizations, particularly in the flow of blood in the hands and face. However, the adjustment to the cold lies predominantly in choosing proper clothing and work, with the result that in "normally cold" temperatures, the body has little or no need to change its rate of heat production or, relatedly, food intake.

Women and men do not exhibit great differences in their ability to adapt to either hot or cold climates, though women may be at a slightly higher risk for heat exhaustion and collapse and for injuries to their extremities from the cold. However, these slight statistical tendencies can be easily counteracted by ergonomic means and may not be obvious at all when one observes only a few persons of either gender.

WORKING STRENUOUSLY IN HEAT AND COLD

Hot and cold climates (as well as air pollution and high altitudes, discussed later) affect human abilities to perform short or long, moderate or heavy, work. The following discussion, useful for engineers and managers, is a synopsis of the known effects of heat and cold on the human body (Kroemer 1991).

Effects of Heat

It is normal to have hot skin in the heat

When exposed to whole-body heating, the human body must maintain its core temperature near 37°C. It does so by raising its skin temperature, increasing blood flow to the skin, accelerating the heart rate, and enlarging cardiac output. This change in routing reduces the amount of blood that can be supplied to muscles and internal organs. Yet, if muscles must work, their raised metabolism increases the demand on the cardiovascular system.

Cardiovascular Effects. The pumping capacity of the heart is between about 25 (in "average" adults) and 40 (in star athletes) liters per minute. The blood vessels in skin and internal organs can accept up to 10 liters, and all muscles together up to 70 liters, per minute. Since the available cardiac output is half or less of these 80 liters, the ability of the heart to pump blood is the limiting factor for muscular work in a hot climate.

Effects on Muscles. An increase in muscle temperature above normal does not affect the maximal isometric contraction capability of muscle tissue, but the power output of muscles is reduced at higher (and lower) temperatures. Overheating of muscles accelerates the metabolic rate, which can make a muscle ineffective if it must work over some period of time. Lowering the muscle temperature before exercising can counteract the loss of power and endurance stemming from excessive muscle temperature. Lowering the temperature reduces the cardiovascular strain and lactic acid concentration in the blood and depletes muscle glycogen at a slower rate.

Dehydration. When working in a hot environment, the body loses water by sweating (i.e., it gets dehydrated). Even acute water loss, incurred in a short time (say, a few hours or less), called *hypohydration,* does not reduce isometric muscle strength (or reaction times) if the water loss is less than 5 percent of the body weight. However, fast and large water losses (such as those induced by diuretics) generate the risk of heat exhaustion, which comes about primarily through a depletion of fluid volumes in the body. Dehydration reduces the body's capacity to perform aerobic or endurance-type work.

To counteract water loss, one must drink fluid. Plain water is best. If strenuous activities last longer than one or two hours, diluted sugar additives may help to postpone the development of fatigue by reducing the body's utilization of muscle glycogen and improving fluid-electrolyte absorption in the small intestine. Regular, liberally salted food (as is customary in the United States) is normally sufficient to counteract salt loss. In fact, salt tablets have been shown to generate stomach upset, nausea, or vomiting in up to 20 percent of all athletes who took them.

Acclimatization. As already discussed, most heat acclimatization takes place during the initial week of exposure. First, the cardiovascular system adjusts, enlarging the blood plasma volume and decreasing the heart rate from their levels at first exposure to heat. The body core temperature returns to normal after 5 to 8 days in the heat. The chloride concentration in the sweat takes up to 10 days to adapt, as does the production of sweat volume and its controlled evaporation on the skin. Thus, within two weeks of staying in a hot climate, the body acclimates completely.

Effects on Mental Performance. It is difficult to evaluate the effects of heat (or cold) on a person's mental or intellectual performance because of large subjective variations and a lack of practical, objective testing methods. However, as a rule, mental performance deteriorates with rising room temperatures, starting at about 25°C for those not acclimatized to the heat. The threshold increases to 30°C or even 35°C if the individual has acclimatized. Brain functions are particularly vulnerable to heat; keeping the head cool improves one's tolerance to elevated deep-body temperatures. A high level of motivation may also counteract some of the detrimental effects of heat. Thus, in laboratory tests, mental performance is usually not significantly affected by heat as high as 40°C WBGT.

WORKING IN THE HEAT: SUMMARY

The short-term maximal exertion of muscle strength is not affected by heat or water loss. But the ability to perform high-intensity endurance-type physical work is severely reduced during acclimatization to heat, which takes normally up to two weeks. Even after one is acclimatized, the demands on the cardiovascular system for heat dissipation and for blood supply to the muscles continue to compete. The body prefers heat dissipation, with a proportional reduction in a person's capability to perform. Dehydration further reduces the ability of the body to work; hypohydration poses acute health risks. One's mental performance is usually not affected by heat as high as 40°C WBGT.

Effects of Cold

It is normal to have cold skin in the cold

As in a hot climate, the body must maintain its core temperature near 37°C in a cold environment. When exposed to cold, the human body first responds by peripheral vasoconstriction, which lowers the skin temperature in order to decrease heat loss through the skin. Such reduction in blood flow occurs in all exposed areas of the body with the exception of the head, where up to 25 percent of the total heat loss can take place. If the flow of blood away from the periphery is insufficient to prevent heat loss, shivering sets in. Shivering is a regular muscular contraction mechanism, but one that generates no external work, since all energy is converted to heat. The muscular activities of shivering and physical work require an increased oxygen uptake, which is associated with increased cardiac output.

Cardiovascular Effects. Enlarging the stroke volume of the heart brings about most of the increase in cardiac output associated with physical work while the heart rate remains at low levels. Yet, keeping the heart rate low as a reaction to exposure to the cold opposes the response associated with physical exercise—that is, to increase the heart rate in order to help increase cardiac output.

Effects on Body Temperature. The two opposing cardiac responses to cold and exercise affect the body temperature. With light work in the cold, the core temperature tends to fall after about one hour of activity. Cold sensations in the skin regularly initiate reactions leading to a lowered skin temperature, yet areas over active muscles can remain warmer due to the heat generated by muscle metabolism. Thus, in the cold, relatively much heat is lost through convection (and evaporation). Which of the opposing physiological cold responses predominates depends on various special conditions (e.g,. the ambient temperature, the type of body activity, and the insulation by clothing).

While one can feel the coldness of air in the upper respiratory tract, warming the air in the upper respiratory passages is sufficient to preclude cold injuries to lung tissues under normal conditions, even if one feels discomfort and constriction of the airways when inspiring very cold air through the mouth. Yet, the air temperature is seldom too cold for exercise and physical work.

Effects on Energy Cost. For submaximal work in the cold, oxygen consumption is increased compared with the amounts consumed when working at normal temperatures. Some of this increased oxygen cost at low work levels may relate to shivering, some to the extra effort required to "work against" heavy clothing worn to insulate the body against heat loss. At medium exercise intensities, oxygen cost in the cold is about the same as at normal temperatures.

Fairly little experimental work has been performed regarding maximal exercise levels. The limited information that is available indicates that a cold climate does not affect the ability to perform maximal exercise, as long as the exposure does not exceed about five hours. In this case, the physiological stimuli provoked by exercise appear to override those of cold. However, if the core temperature gets lower, one's maximal work capacity is reduced, apparently mostly by suppressing the heart rate and thus reducing the transport of oxygen to the working muscles in the bloodstream.

Little is known about the effects of exposure to cold on endurance. However, a decrease in muscle temperature diminishes one's muscle contraction capability, inducing an early onset of fatigue.

Dehydration. Dehydration occurs surprisingly easily in the cold, partly because sweating is increased in response to the higher energy demands of working in the cold, and partly because the sensation of thirst is suppressed. Also, urine production is increased in the cold, which can trigger water loss through more frequent urination. While dryness of cold air may cause respiratory irritation and discomfort, severe dehydration through the lungs does not occur, since exhaled air is cooled on its way out to nearly the temperature of the inhaled air, returning water vapor by condensation onto the surface of the airways. (This explains the common experience we have of a "runny nose" in the cold.)

Acclimatization. As mentioned earlier, the human body acclimates to a prolonged stay in the cold much less effectively than it adjusts to heat; mostly, decreases in shivering have been observed. It is uncertain whether fitness training facilitates adaptation to the cold.

Most of the counterresponses to exposure to the cold are taken to improve one's insulation. For example, one can wear proper clothing and stay within sheltered areas. Thus, the body usually carries a fairly normal microclimate while in the cold and hence has little or no need to acclimate.

Effects on Mental Performance and Dexterity. As already mentioned, if the core temperature of the body drops below about 35°C, vigilance is reduced, nervous coordination suffers, and apathy sets in. Loss of consciousness occurs in the low thirties. Manual dexterity is reduced if finger skin temperatures fall below 20°C. Tactile sensitivity is diminished at about 8°C, and near 5°C skin receptors for pressure and touch cease to function and the skin feels numb. While muscle spindles are initially more active as the muscle temperature drops, at about 27°C their activity is reduced to 50 percent, and it is completely abolished at about 15°C. These effects explain the difficulty of performing finely controlled movements in the cold.

WORKING IN THE COLD: SUMMARY

Strong isometric muscle exertions are impaired only if the muscles are cold. The ability to do light work is reduced in the cold. Endurance activities are impaired only if core or muscle temperatures are lowered and if dehydration occurs. Clothing worn for insulation may hinder work. Dexterity and mental performance suffer in extreme cold.

DESIGNING THE THERMAL ENVIRONMENT

There are many ways to generate a thermal environment that suits the physiological functions of either acclimatized or non-acclimatized persons. The primary approach is to adjust the physical conditions of the climate (namely, temperatures, humidity, and air movement) so as to influence heating or cooling of the body via radiation, convection, conduction, and evaporation. These interactions, listed in Table 5–5, must be carefully considered in designing and controlling the environment.

TABLE 5–5. Designing the Thermal Environment to Increase (+) or Decrease (−) Body Heat Content

					Temperatures Compared to Skin Temperature					
	Air Humidity		*Air Movement*		*Air, Water*		*Solids*		*Opposing Surface*	
Heat Transfer	*Dry*	*Moist*	*Fast*	*Calm*	*Hotter*	*Colder*	*Hotter*	*Colder*	*Hotter*	*Colder*
Radiative	No direct effect		No direct effect		NA	NA	NA	NA	+	−
Convective	No direct effect		−	(−)	+	−	NA	NA	NA	NA
Conductive	NA	NA	NA	NA	NA	NA	+	−	NA	NA
Evaporative	−	(−)	−	(−)	−	(−)	NA	NA	NA	NA

The negative sign in parentheses indicates a relatively small heat loss.
NA means not applicable.

Windchill

Heat loss by convection is increased if air moves swiftly along exposed surfaces. Therefore, with increased air velocity, body cooling becomes more pronounced. During World War II, experiments were performed on the effects of air movement on the cooling of water at different temperatures. These physical effects were also assessed psychophysically in terms of the windchill sensation at exposed human skin. This resulted in a table of "wind-chill (equivalent) temperatures" widely used until the year 2001 when the U. S. National Weather Service developed a new chart that reflected the cooling of exposed human skin more realistically. The new wind chill chart, reprinted as Table 5-6, indicates how air movement can increase energy loss from naked human skin. Such wind acts as if the actual air temperatures were reduced to the "wind-chill temperatures" listed in Table 5-6. For example, a wind of 30 km/h at an actual air temperature of −5 degrees C generates an equivalent calm-air temperature of −13 degrees C. With stronger wind and and colder air, the cooling of exposed skin and hence the danger of frostbite increases. Obviously, simply covering the skin, for example with gloves or a face mask, increases insulation against heat loss by wind chill.

☛☛☛ *What is of importance to the individual is not the climate in general—the so-called macroclimate—but the climatic conditions with which one interacts directly. Every person prefers an individual microclimate that feels "comfortable" under given conditions of adaptation, clothing, and work.* ☚☚☚

A suitable microclimate is variable

The suitable microclimate is not only highly individual, but also variable. It depends, for example, on age and gender: Older persons tend to be less active, to have weaker muscles, to have a reduced caloric intake, and to start sweating at higher skin temperatures. The suitable microclimate also depends on the body's surface-to-volume ratio, which in children is much higher than in adults, and on the fat-to-lean body-mass ratio, which is generally larger in women than in men.

Thermal comfort depends largely on the type and intensity of work performed. Physical work in the cold may lead to increased heat production and hence to decreased sensitivity to the cold environment. While hard physical work in a hot climate can become intolerable if an energy balance cannot be achieved.

Clothing

Of course, clothing also affects the microclimate. Air bubbles contained in clothing material or between layers of clothing provide insulation, against both hot and cold environments. Permeability to fluid (sweat) and air plays a role in heat and cold as well (Bensel and Santee 1997). The colors of clothes are important in a heat-radiating environment, such as sunshine, with darker colors absorbing radiated heat and light colors reflecting incident energy.

The insulating value of clothing is measured in clo units, with 1 clo $= 0.155°C\ m^2\ W^{-1}$. This is approximately the value of the "normal" clothing worn in the United States by a sitting subject at rest in a room at about 21°C and 50-percent relative humidity.

The particular clothing worn determines the surface area of exposed skin. More exposed surface areas allow better dissipation of heat in a hot environment, but can lead to excessive cooling in the cold. Fingers and toes need special protection in the cold because they have large surfaces with small volumes and are away from the warm, larger body masses. The head and neck have warm surfaces that release much heat, which is desirable in a hot environment, but not in the cold.

Acclimatization

Thermal comfort is also affected by acclimatization [i.e., the status of the body (and mind) of having adjusted to changed environmental conditions]. A climate that felt rather uncomfortable and reduced one's ability to perform physical work during the first day of exposure may be quite agreeable after two weeks. Seasonal changes in climate, unusual work, different clothing, and attitude have major effects on what we are willing to accept or even consider comfortable. In the summer most people find warm, windy, and rather humid conditions comfortable, while during the winter we feel that cool and dry weather is normal.

Various combinations of temperature, humidity, and air movement can subjectively appear similar. The WBGT discussed earlier is most often used to assess the effects of warm or hot climates on human beings; various similar approaches have been proposed for a cold climate, but are not universally accepted yet. [See Youle (1990) for a critical overview and for details.]

Standards, recommendations

Information specific to an artificial environment, such as that in offices, is contained in the latest edition of ANSI-ASHRAE Standard 55. For outdoor activities, recommendations in military and ISO standards are applicable.

WORKING IN HOT OR COLD CLIMATES: SUMMARY

Skin temperatures in the range of 32° to 36°C, associated with core temperatures between 36.7° and 37.1°C, are agreeable. Preferred ranges of relative humidity are between 30 and 70 percent. Deviations from these zones are uncomfortable, can make work difficult, or may even become intolerable.

With appropriate clothing and light work, comfortable temperature ranges for the environment are about 21° to 27°C ET in a warm climate or during the summer, but just 18° to 24°C ET in a cool climate or during the winter. Proper clothing and ergonomic management of work–rest ratios allow physical work to be performed in hotter and colder climates.

TABLE 5–6. Wind Chill Temperature Equivalents (metric)

Wind *km/h*	Actual Air Temperature in degrees C												
Calm	10	5	0	−5	−10	−15	−20	−25	−30	−35	−40	−45	−5
Measured at 10 m elevation	**Equivalent Wind Chill Temperature in degrees C with Calm Air**												
5	9	4	−2	−7	−13	−19	−24	−30	−36	−41	−47	−53	−58
10	9	3	−3	−10	−15	−21	−27	−33	−39	−45	−51	−57	−63
15	8	2	−4	−11	−17	−23	−29	−35	−41	−48	−51	−60	−66
20	7	1	−5	−12	−18	−24	−31	−37	−43	−49	−56	−62	−68
25	7	1	−6	−12	−19	−25	−32	−38	−45	−51	−57	−64	−70
30	7	0	−7	−13	−20	−26	−33	−39	−46	−52	−59	−65	−72
35	6	0	−7	−14	−20	−27	−33	−40	−47	−53	−60	−66	−73
40	6	−1	−7	−14	−21	−27	−34	−41	−48	−54	−61	−68	−74
45	6	−1	−8	−15	−21	−28	−35	−42	−48	−55	−62	−69	−75
50	6	−1	−8	−15	−22	−29	−35	−42	−49	−56	−63	−70	−76
55	5	−2	−9	−15	−22	−29	−36	−43	−50	−57	−63	−70	−77
60	5	−2	−9	−16	−23	−30	−37	−43	−50	−57	−64	−71	−78

The Shaded area lists the "Wind Chill Temperatures" which indicate the cooling effects of combined wind velo
and air temperture on naked human skin. For example, a wind of 5 km/h at 0°C cools exposed skin at the sa
rate as a temperture of −2°C with no wind; at a measured air temperature of −20 degrees, a 60-km/h wind g
erates an equivalent calm-air temperature of −37°C. Frostbite can occur to exposed warm human skin belo
wind chill temperature of −25°C. Frostbite is possible in 10 minutes to warm skin that is suddenly bared to w
chill below −35°C, and frostbite is possible in less than 2 minutes at and below −60°C. These times are sho
if the skin is already cool at the start of the exposure. (Adapted by Kroemer from information received 2001/2
from the U. S. NOAA National Weather Service, courtesy Mark Tew.)

Indoors, air temperatures at floor level and at head level should differ by less than about 6°C. Differences in temperatures between body surfaces and sidewalls should not exceed approximately 10°C. The velocity should not exceed 0.5 m/s and, preferably, should remain below 0.1 m/s.

For outdoor activities, proper clothing can generate a suitable microclimate for the body, and with an appropriate work regimen, physical work may be performed in the cold; heat often limits the intensity and duration of work because it is difficult to dissipate body heat to the hot environment.

WORKING IN POLLUTED AIR

Natural events such as forest fires, dust storms, and volcanic eruptions can fill the air with contaminants, mostly smoke, soot, and dust. Air pollution is often a human-made problem, well known from the smog that frequently blankets Los Angeles, Mexico City, London, Beijing, and other cities. Primary pollutants in the air are carbon monoxide, oxides of sulfur and nitrogen, and particulates. All directly affect the respiratory system of the body and hence, indirectly, circulation and metabolism. (See Chapter 2.)

CO. Carbon monoxide (CO) is the pollutant that lowers physical work performance most strongly. Hemoglobin in the human blood has an affinity 230 times greater for CO than for oxygen. Consequently, CO easily attaches to hemoglobin and takes the place of oxygen, thus reducing the ability of blood to provide cells with oxygen. Furthermore, CO attached to hemoglobin causes the remaining binding sites on the hemoglobin molecule to develop a high affinity for oxygen, thus making it more difficult to release oxygen to the cells that need it.

SO_2 and NO_2. Sulfur dioxide (SO_2) increases the flow resistance in the upper respiratory tract. This effect is bothersome for asthmatics, but does not appear to decrease the submaximal exercise capability of healthy individuals. The effect on maximal exercise capabilities has not been studied. Nitrogen dioxide (NO_2) is potentially harmful to humans, but does not seem to affect submaximal exercise capabilities, although it can be an irritant in the upper respiratory tract. Inhaling particulates from the soot of cigarette smoke or of dust can also irritate the respiratory tract. The effects of such particulates on maximal exercise capabilities have not been studied.

Ozone. Secondary pollutants evolve from the interactions of primary pollutants with each other and with water, salt, and ultraviolet light. Secondary pollutants include ozone, as well as peroxyacetyl nitride and other aerosols. Ozone is formed by the interaction of oxygen, nitrogen dioxide, hydrocarbons, and ultraviolet light; thus, ozone formation is closely tied to sunlight. Ozone is a potent irritant of airways, but no clear physiological impairment of submaximal or maximal performance capability has yet been demonstrated, although it is suspected.

Exhausts and Aerosols. Automobile exhausts are the primary source of atmospheric peroxyacetyl nitride. While blurred vision and eye irritation are known symptoms of exposure, the effect of peroxyacetyl nitride on submaximal or maximal work efforts has not been studied sufficiently. Aerosols, formed by the interactions of various acids and salts, can cause discomfort, but have not been linked to decrements in work-performance capabilities.

WORKING IN POLLUTED AIR: SUMMARY

Only carbon monoxide shows a clear detrimental effect on maximal aerobic performance capabilities. Other compounds can cause irritation, but currently there is no evidence that they decrease work capabilities. However, the lack of definitive studies is a serious problem.

WORKING STRENUOUSLY AT HIGH ALTITUDES

It is known that cognitive functions in persons climbing high mountains (4,200 m or higher) are impaired: In comparison to their performance at sea level, these individuals show diminished success in standardized perceptual, cognitive, and sensorimotor tasks, as well as reduced learning and retention of new perceptual and cognitive skills (Kramer et al. 1993).

The ability to perform strenuous work depends significantly on the supply of oxygen to the working muscles. (See Chapter 2.) While the oxygen content in the ambient air remains constant at about 21 percent to an altitude of at least 100 km above the earth's surface, the barometric pressure falls considerably with increasing height. Multiplying the percentage of oxygen by the barometric pressure yields the partial pressure of oxygen. At sea level, where the barometric pressure is 760 torr, the partial pressure of oxygen in the ambient air is 159 torr. At 3,000 m height (nearly 10,000 feet), the barometric pressure is about 252 torr; hence, the partial pressure of oxygen is about 110 torr, a reduction of nearly 30 percent from the figure at sea level.

Effects on Oxygen Transfer. Like any other gas, oxygen moves from higher to lower concentrations. According to this general rule, a reduction in the partial pressure of oxygen in inspired air at high altitudes must decrease the ability of the body to supply its cells with oxygen. This is of critical importance for the mitochondria in muscle tissue, where most of the energy for physical work is generated.

A series of processes determines the ability of the body to bring oxygen to the mitochondria. Breathing (ventilation) moves air in and out of the lungs. In the lungs, oxygen is transferred from the air across lung tissue into the bloodstream, where it combines with hemoglobin. Oxygen-carrying hemoglobin is then transported in the bloodstream to the muscle cells, where the oxygen diffuses out of capillaries into the cell and, finally, to the mitochondrion.

Effects on Breathing. The first process, ventilation, is in fact easy to augment (for both physical and physiological reasons) and automatically increases at altitudes above 3,000 meters. Physical work also increases ventilation. Thus, there are no effects on the ability of human beings to breathe that would limit their work capacity at higher altitudes.

Effects on Blood Oxygenation. The second process is the diffusion of oxygen from the lungs to the blood. The reduction in the partial oxygen pressure in the lungs at high altitudes generates a smaller difference between the air pressure in the lungs and the blood pressure, thereby slowing diffusion. Since the velocity of blood flow is increased during physical work (owing to increased cardiac output), the time available for oxygen diffusion in the lungs to each passing hemoglobin cell is reduced. Hence, blood oxidation falls with increasing altitude.

The next step in the process is to make oxygen-rich blood available in the arterial (systemic) branch of circulation. The oxygen content of arterial blood depends on the hemoglobin concentration in the blood and on the ability of the hemoglobin to attract oxygen. Within the first few hours of a person's exposure to high altitudes, the ability to carry oxygen remains the same as at sea level, but the actual oxygen content in the blood is reduced, owing to reduced diffusion in the lungs, as just discussed. But after a few hours of exposure at altitudes above 3,000 m, the distribution of fluids in the body shifts. The blood plasma volume in the circulatory bloodstream decreases, because up to 30 percent of the volume moves into cells. But with acclimatization, the blood volume may increase slightly, and the production of red cells is stimulated, resulting in an increase in hemoglobin concentration in the blood. Thus, in spite of the shift in volume, and because of the increase in hemoglobin concentration in the flowing blood, the ability of arterial blood to carry oxygen remains at approximately sea-level values.

Effects on Oxygen Supply to Muscles. The next step in the oxygen-transport process is the provision of oxidized blood to the working muscles—specifically, the mitochondria. During the first few days of exposure to high altitudes their oxygen supply is diminished according to the reduced oxygen content in the blood (which is the result of reduced diffusion in the lungs). As the exposure to high altitudes continues and acclimation is achieved, the oxygen content in arterial blood returns to sea-level values.

The final process in the oxygen-transport chain is the oxidation of tissue. At high altitudes, hemoglobin releases oxygen more easily to the tissues than at sea level. With acclimatization to such altitudes, the capillaries in muscle tissues become enlarged, and hence, the diffusion of oxygen from the bloodstream to cells becomes easier.

Cardiovascular Effects. A person's ability to perform strenuous aerobic work depends on the heart's ability to move blood through the body, so that oxygen can be provided to metabolizing muscles and so that metabolic by-products (e.g., lactic acid, carbon dioxide, heat, and water) can be removed from them. The cardiac output, or minute volume, is essentially the product of stroke volume and heart rate. Cardiac output is not much affected at lower altitudes, but shows marked changes in heights above about 1,500 m. The blood volume that is pumped actually increases at rest and during submaximal efforts, but after about two days' exposure to high altitudes, the volume becomes progressively reduced. Following about two weeks of staying at a high altitude, cardiac output is lowered—at all levels of effort—to a minute volume below that at sea level. It then remains at that low volume for the duration of the stay at the high altitude. This reduction is primarily due to a diminished stroke volume, which, in turn, mostly follows from the reduced blood-plasma volume.

Physiological Adjustments to Altitude. The initial response of the body to high altitudes is to increase ventilation, both the number of breaths taken per minute and the depth of each breath. The increased ventilation enlarges the pressure of oxygen within the lungs and facilitates the release of carbon dioxide to the air.

Another adjustment is the redistribution of fluid in the body, as just discussed. The reduction in blood-plasma volume occurs within hours of arriving at altitudes of more than 3,000 meters. With long high-altitude stays, some of the blood is redistributed, but it does not return to its sea-level distribution.

During the first few hours of exposure to a high altitude, reduced oxygen content in the arterial blood is caused by the difficulty of diffusing sufficient oxygen from the air in the lungs into the blood. Adaptation to altitude begins after a few hours of exposure, mostly via a relative and an absolute increase in hemoglobin in the circulating blood, as well as by a slight increase in the blood volume. These increases restore the capability of the blood to bring oxygen to the working cells. However, the cardiac output capability remains suppressed (see earlier) throughout the stay at higher altitudes. Thus, the capability of performing physically highly demanding activities remains reduced at altitudes above 1,500 m, even after adaptation.

"Altitude Sickness." The rapid change in fluid distribution at high altitudes is associated with several well-known medical problems. Acute mountain sickness (AMS) commonly occurs at heights above 3,000 meters. Symptoms include headaches, lassitude, nausea, insomnia, irritability, and depression. The appearance of AMS is directly related to the rate of ascent and to the final altitude. Symptoms become apparent after several hours of exposure and reach their peak within one or two days, after which they recede over the next several days. AMS can be reduced or eliminated by gradual ascent and by medication. While AMS is often debilitating, it is self-limiting. But in some persons who gain an altitude rapidly, an excessive accumulation of fluid in spaces between cells or in the cells themselves can occur in the brain or lungs. Both edema conditions may be life threatening, but can be counteracted by an immediate return to lower elevations and by medical aid.

WORKING AT HIGH ALTITUDES: SUMMARY

- Brief high-intensity activities ("explosive efforts") do not suffer with increasing altitude, because they are anaerobic and hence do not depend on oxygen transport. Likewise, the short exertion of muscle strength is not reduced during acute exposure to high altitudes.
- The ability to perform (submaximal or maximal) aerobic work remains at its sea-level value up to an altitude of about 1,500 m.
- Submaximal work capacity is not affected up to about 3,000 m, but any given task requires a larger percentage of the available (reduced) maximal capacity than at sea level. Hence, the ability to endure such submaximal efforts is also reduced at higher altitudes. The longer the effort, the greater is the decrement.
- Above 1,500 meters, maximal work capacity decreases at a rate of approximately 10 percent per 1,000 m (with much variability among individuals). This reduction persists for one's entire stay at a given altitude.
- Cognitive performance is reduced at high altitudes, such as above 4,200 m.

THE EFFECTS OF VIBRATION ON THE HUMAN BODY

Vibration is defined as oscillatory motion about a fixed point. A vibration is called periodic when the oscillation repeats itself.

There are two types of vibration on which much research has been done. In the first type of vibration, the body continues to vibrate at the same frequency over a considerable period of time. The simplest way of describing this motion is by a sinusoidal equation. The other type of vibration is that of one-time shocks and impacts, called nonperiodic vibrations.

Body parts vibrate differently

The human body reacts to the different kinds of vibration in various ways. If the body were rigid, all its parts would undergo the same motion, but only if the driving movement were translational (linear). When the body is rotated, even if it is rigid, not all its parts have the same motion. Of course, the human body is not rigid, and different body parts vibrate differently even if they are under the influence of the same linear vibration.

Of particular interest are the responses of the spinal column, the head, and the hands. Measurements at the vertebrae are difficult to perform. Previously, X rays were used, but they have now been largely abandoned because of radiation danger. Also, chemically inert nails have been driven through the skin of the back so that the motion of vertebrae could be observed on the protruding shafts; understandably, it is difficult to find volunteers for this procedure. It is easier to measure motion of the head, because one can firmly hold between the teeth a dental mold from which a rigid bar extends outward between the lips; this allows skull motions to be observed. Responses to vibration at the hand also are easily observed: The transmitted vibrations should be measured at the interface between the tool and the hand of the operator. This is often impossible to do, because available instruments affect the handling of the tool. Therefore, one has to attach measurement devices, such as accelerometers, to either the object or the hand or forearm (i.e., away from the body–object interface). Doing so, however, is likely to distort the actual conditions. (An example of this problem is the assessment of impacts between finger and key when one is operating a keyboard.) In spite of the difficulty, extensive information is available from experimental measurements and epidemiological investigations. (See, for example, Griffin 1990, 1997; Guignard 1985; Putz-Anderson 1988; and Wasserman and Wilder 1999.)

Jackhammer disease

In 1862, the French physician Maurice Raynaud described a vibration-induced disease of jackhammer operators: Their fingers may become pale and cold because the vibrations cause sphincter muscles around blood vessels to contract, cutting off the blood supply to the hand. Thus, the condition was called "white" or "dead" finger or hand; it now carries the name "Raynaud's disease."(See Chapter 8.)

Back and stomach problems

Another major area of interest for both research and application is the effects of impacts and vibrations on the spinal column, prodded in particular by complaints of truck and bus drivers, and of operators of earthmoving equipment, about back and stomach problems.

Head and hand vibrations

Another topic of interest is that of head vibrations, with respect to both the ability to see visual targets and the ability to avoid motion sickness. Even some overuse disorders related to keyboarding (see Chapter 8) might be induced by repeated vibrations due to and impacts from operating the keys.

Measuring Vibration

Direction

The coordinate system used to describe the *direction* of mechanical vibration of the human is, unfortunately, different from systems used in other applications, such as astronautics. In vibration research, the x-axis is indeed in the forward direction in reference to the vibrating body, but the y-axis goes to the left and the z-axis upward. (See Figure 5–3.) In aerospace work, different conventions are used for directions. (See Figure 5–14.)

The *magnitude* of vibration is described by displacement and its derivatives, velocity and acceleration, over time. (See Figure 5–4.)

Displacement

The *displacement* of a mass can be described as the maximal amplitude of the mass above or below the stationary location; this is the *peak amplitude.*

Acceleration

Another way of describing the magnitude of oscillation is by velocity. However, while this is an appropriate way to describe vibration, it is seldom used, mostly because the instrumentation for measuring acceleration is, at present, more convenient.

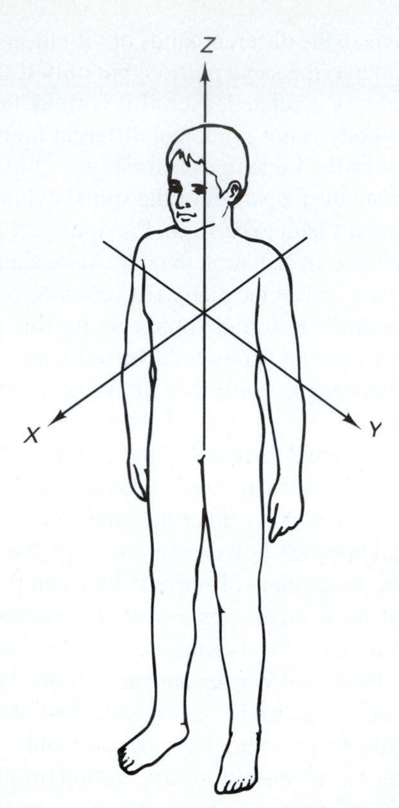

Figure 5–3. Convention on directions of vibrations and impacts. Compare with Figure 5–14.

Acceleration units

A common measure of *acceleration* is the "rms" value—the square *r*oot of the *m*ean of the *s*quared values. The use of this measure avoids the problem associated with the mean (or average) acceleration, whose numerical value over time is zero for a periodic vibration. Acceleration can be stated in "*G*" units, with 1 G = 9.80665 m s^{-2}, the average acceleration on earth due to gravity. Occasionally, the logarithmic decibel (dB) scale is used (see Tables 5–7 and 5–8), although the range between the threshold of perception and the threshold of pain is only about 1,000 : 1 for vibration, much smaller than for sound (Griffin 1990).

TABLE 5–7. Reference Quantities Defined in ISO 1683 (1983) (lg = $\log_{10}$)

Description	*Definition (dB)*	*Reference Quantity*
Sound pressure level in air	20 lg (p/p_0)	2×10^{-5} Pa
Other-than-air sound pressure level	20 lg (p/p_0)	10^{-6} Pa
Vibration acceleration level	20 lg (a/a_0)	10^{-6} m s^{-2}
Vibration velocity level	20 lg (v/v_0)	10^{-9} m s^{-1}
Vibration force level	20 lg (F/F_0)	10^{-6} N
Power level	10 lg (P/P_0)	10^{-12} W
Intensity level	10 lg (I/I_0)	10^{-2} W m^{-2}
Energy density level	10 lg (w/w_0)	10^{-12} J m^{-3}
Energy level	10 lg (E/E_0)	10^{-12} J

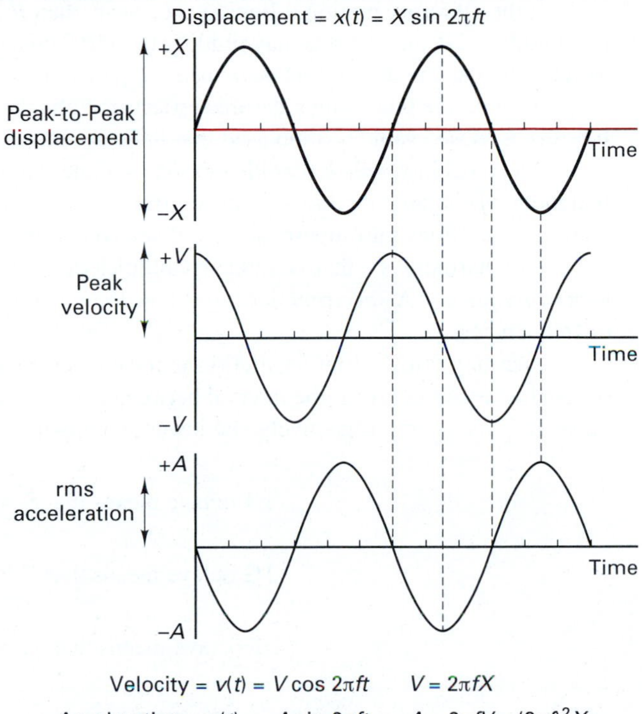

Figure 5–4. Sinusoidal vibration described by displacement, velocity, and acceleration over time.

TABLE 5–8. Conversions Between Decibels and Other Units of Acceleration and Velocity (Reference Levels Defined in Table 5–7)

Decibel (dB)	*Acceleration ($m\ s^{-2}$)*	*Velocity ($m\ s^{-1}$)*
−20	10^{-7}	10^{-10}
0	10^{-6}	10^{-9}
20	10^{-5}	10^{-8}
40	10^{-4}	10^{-7}
60	10^{-3}	10^{-6}
80	10^{-2}	10^{-5}
100	10^{-1}	10^{-4}
120	1	10^{-3}
140	10	10^{-2}
160	10^{2}	10^{-1}
180	10^{3}	1
200	10^{4}	10

Amplitudes

If the vibratory motion follows a sine wave, then P is its peak magnitude, the peak-to-peak magnitude is $2P$, and the rms magnitude is $0.707P$. Thus, in discussing magnitude, it is necessary to carefully use the descriptors *peak, peak-to-peak,* or *rms.*

Effects of time

None of the preceding measures reflects the effects associated with the duration of motion. For this, a "dose" value is often used that indicates the effects of the time of exposure.

Mixtures of frequencies

An object in *simple harmonic* motion oscillates sinusoidally at a single frequency, but most real motions contain vibrations of several frequencies. Fourier analysis can decompose even such complex vibrations into superpositions of sinusoidal motions. In some cases, the frequencies involved are harmonics—that is, integer multiples of the lowest frequency, also called the *fundamental frequency.* A *spectrum* describes how the magnitude of the vibration varies over a range of frequencies.

Octave bands

The magnitude of an oscillation is most often measured in either octave or third-of-an-octave bands. An octave is the interval between two frequencies when one frequency is twice the other. If f_1 and f_2 are, respectively, the lower and upper frequencies of the bands, then,

$$1/1 \text{ octave means that } f_2 = 2f_1,$$

$$1/2 \text{ octave means that } f_2 = 2^{1/2}f_1,$$

$$1/3 \text{ octave means that } f_2 = 2^{1/3}f_1,$$

and

$$1/6 \text{ octave means that } f_2 = 2^{1/6}f_1.$$

For example, when centered on 1 Hz, the octave band is from 0.707 to 1.414 Hz, while the third-octave band is from 0.891 to 1.122 Hz. In these cases, the bandwidth increases in proportion to the frequency. Another method is to determine the frequency content using a constant bandwidth (such as 0.1 or 1.0 Hz) at all frequencies.

Frequency

In the discussion that follows, we assume that the waveform of the vibratory motion is sinusoidal. The reciprocal of the *period* T is the *frequency* f, the number of cycles of motion per second, expressed in hertz. One often uses the angular *frequency* ω, expressed in radians per second; since a complete cycle (360 degrees) corresponds to 2π radians, $\omega = 2\pi f$, in rad s^{-2}.

Instantaneous values

At the maximum displacement, the velocity is zero and the acceleration is at a minimum; when the displacement is zero, the velocity is maximal and the acceleration is zero, as shown in Figure 5–4.

At any time t, the *instantaneous displacement* x is described by

$$x(t) = X \sin(2\pi ft + \phi),$$

where X is the peak displacement and ϕ is the phase angle (time delay).

The *instantaneous velocity* v of the motion is the first time derivative of the displacement; thus,

$$v(t) = 2\pi f X \cos 2\pi ft = V \cos 2\pi ft,$$

where $V = 2\pi f X$ is the peak velocity.

The *instantaneous acceleration a* of the motion is the time derivative of velocity:

$$a(t) = -(2\pi f)^2 X \sin 2\pi ft = -A \sin 2\pi ft;$$

analogously to the case of the instantaneous velocity, $A = (2\pi f)^2 X = 2\pi f V$ is the *peak acceleration.*

The *jerk* is the time derivative of acceleration; the *peak jerk* equals $(2\pi f)^3 X$, or $(2\pi f)^2 V$, or $2\pi f A$.

Table 5–9 applies to sinusoidal vibration and shows how to convert the measures of displacement, velocity, acceleration, and jerk; the table also gives the conversions among peak, peak-to-peak, and rms values.

Time window

All of the methods of measuring vibration that we have discussed incorporate the duration of the vibration only implicitly. For example, there are many possible rms values for the same vibration, depending on the period over which the rms value is determined. Thus, the "time window" used must be defined.

Effects of Vibration

Good and bad "vibes"

Vibration can produce a wide variety of effects, depending on its intensity and direction of vibration and on the body parts to which it is transmitted. Commonly, one groups the observed or suspected effects of vibration into interference with comfort, with activities, and with health.

While vibration is most often unwelcome, there are also good vibrations, such as those associated with the pleasant feeling of shaking hands, sitting on a rocking chair or a swing, or laughing. Vibration can be a source of excitement—for instance, on the fairground, on sailboats or skis, or in motor cross racing. Treatment using vibration has been advocated for improving joint mobility of athletes or of patients suffering from arthritis (Griffin 1990).

TABLE 5–9. Conversions of Vibration Parameters (For Sinusodial Motions)

	Displacement X	*Velocity* V	*Acceleration* A
Displacement X	X	$X = V/(2\pi f)$	$X = A/(2\pi f)^2$
Velocity V	$V = X2\pi f$	V	$V = A/(2\pi f)$
Acceleration A	$A = X(2\pi f)^2$	$A = V2\pi f$	A

	PEAK	*PEAK to PEAK*	*rms*
PEAK	Peak	Peak = 0.500 × (Peak to Peak)	Peak = 1.414 rms
PEAK to PEAK	Peak to Peak = twice Peak	Peak to Peak	Peak to Peak = 2.828 rms
rms	rms = 0.707 Peak	rms = 0.3535 × (Peak to Peak)	rms

The cause–effect relationships between vibrations and human responses are complex and often difficult to research. Even in the well-controlled laboratory environment, equipment used to generate vibrations often cannot produce pure sinusoidal motions; or the equipment may not be capable of generating, at the same time, high accelerations and large displacements together with considerable forces. The reactions of the human body are quite different from person to person and may depend on the muscle tension and posture of the subject, as well as on any restraining devices that are employed. In the real world, vibrational effects often are not the sole stressors of humans; for example, it may be quite difficult to determine specifically how vibrations or impacts experienced throughout the day by a truck driver affect that person's performance and well-being.

Whole-body Vibration

Whole-body vibration occurs when one stands, lies, or sits on a vibrating surface. Vibration is then transmitted in some way throughout the entire body; yet, in certain situations—for example, when one is sitting on a vibrating seat—the feet may not experience much motion. Thus, the distinction between "whole-body" and "local" vibration is not always clear. Furthermore, seated persons are often simultaneously exposed to local vibrations (e.g., the head from a headrest, the back from a backrest, the hands from a steering wheel, and the feet from the floor).

BIOMECHANICAL RESPONSES TO VERTICAL VIBRATION

Under vertical oscillation at frequencies below 2 Hz, most parts of the human body move together. The associated sensation is that of alternately being pushed up and then floating down. The eyes are able to follow objects that either move with the body or are stationary. Yet, free movements of the hand may be disturbed, which can cause problems in activities that require exact positioning of the hands. If the vibration has a frequency below 0.5 Hz, it may cause motion sickness.

Vertical oscillation at frequencies above 2 Hz brings about amplification of the vibration within the body. Yet, the frequencies with greatest amplification (i.e., the resonance frequencies) are different for different parts of the body, for different individuals, and for different body postures. At frequencies between 4 and 5 Hz, resonances occur, for example, in the head and hands; discomfort is strongly felt. At frequencies above 5 Hz, the force required to generate a given vertical acceleration falls rapidly with increasing frequency; thus, the vibration reaching the head, as well as its associated discomfort, decrease. But at frequencies between 10 and 20 Hz, the voice may warble, and vision may be affected, particularly at frequencies between 15 and 60 Hz because of resonances of the eyeballs within the head. Table 5–10 lists other human-body resonance frequencies.

When one is standing, keeping the knees straight or bent can greatly influence the effects of frequencies above 2 Hz. As regards sitting, the design of the seat has fairly little influence at frequencies below 2 Hz, but "soft" seats, like those used in many automobiles, can greatly amplify vertical vibrations, doubling the experienced frequency, for example. Thus, the design of seats that will be affected by vibration geatly influences the vibrational effects that will be experienced by the seated person.

BIOMECHANICAL RESPONSES TO HORIZONTAL VIBRATION

Sideways, or fore-and-aft, vibration of the seated body below 1 Hz sways the body, even if muscle action resists the motion. Between 1 and 3 Hz, it is difficult to stabilize the upper parts of the body, a situation that is associated with great discomfort. With increasing frequency, horizontal vibration is less readily

TABLE 5–10. Examples of Resonances of the Body and its Parts in Response to Vibrations in *Z*-Direction*

Body Part	*Resonances (Hz)*	*Symptoms*
Whole body	4 to 5, 10 to 14	General discomfort
Upper body	6 to 10	
Head	5 to 20	
Eyeballs	1 to 100, mostly above 8 20 to 70 strongly	Difficulty seeing
Skull, jaw	100 to 200	
Larynx	5 to 20	Change in pitch of voice
Shoulders	2 to 10	
Lower arms	16 to 30	
Hands	4 to 5	
Trunk	3 to 7	
Chest	5 to 7	Chest pain
Heart	4 to 6	
Chest wall	60	
Stomach	3 to 6	
Abdomen	4 to 8	Abdominal pain
Bladder	10 to 18	Urge to urinate
Cardiovascular and respiratory systems	2 to 20	Reactions similar to those in response to moderate work
Brain	below 0.5	
Motion sickness	1 to 2	Sleepiness

*Note that displacement and acceleration interact with the frequency of vibration to generate specific effects on sensation, performance, and health. Excitation in one direction (e.g., *z*) may produce body responses in other directions (such as *x* or *y*). Body posture, muscle tension, or a restraint system may strongly affect responses.

transmitted to the upper body, so that at frequencies above 10 Hz the vibration is felt mostly at the surface of the seat. A backrest can greatly influence the effect of horizontal vibration, helping to stabilize the upper body at low frequencies, but strongly transmitting the vibration to the upper body at high frequencies, primarily in the anterior–posterior direction, of course.

Subjective Assessment of Vibration Effects

Stevens' power law can be used to establish subjective ratings of the sensations associated with vibrations. The law relates the perceived sensation P to the magnitude of the physical stimulus I by the formula $P = KI^n$, where K is a constant. The values for n have been found to be mostly in the range between 0.9 and 1.2 (Griffin 1990). Semantic scales have been developed by several authors, notably for the assessment of vibrations felt while riding in a vehicle (SAE 1973). These and similar approaches have been used to establish "comfort contours," similar to those developed to assess noise.

Overall, persons of different body size, age, or gender report the same sensations associated with vibration, although larger subjects tend to be a bit less sensitive to frequencies below 6 Hz. ISO Standard 2631 (1985) offers guidance for the evaluation of whole-body vibration with regard to comfort boundaries. Figure 5–5 is an excerpt therefrom. Although the standard needs improvement (Griffin 1990), it indicates the time-dependent comfort boundaries for combinations of frequencies and accelerations in whole-body vibrations. For example, vertical vibrations at a frequency of 4 Hz are said to be comfortable for a period of eight hours if the rms acceleration is about 0.1 m s^{-2}, but the time is reduced to about one hour if at about 0.4 m s^{-2}.

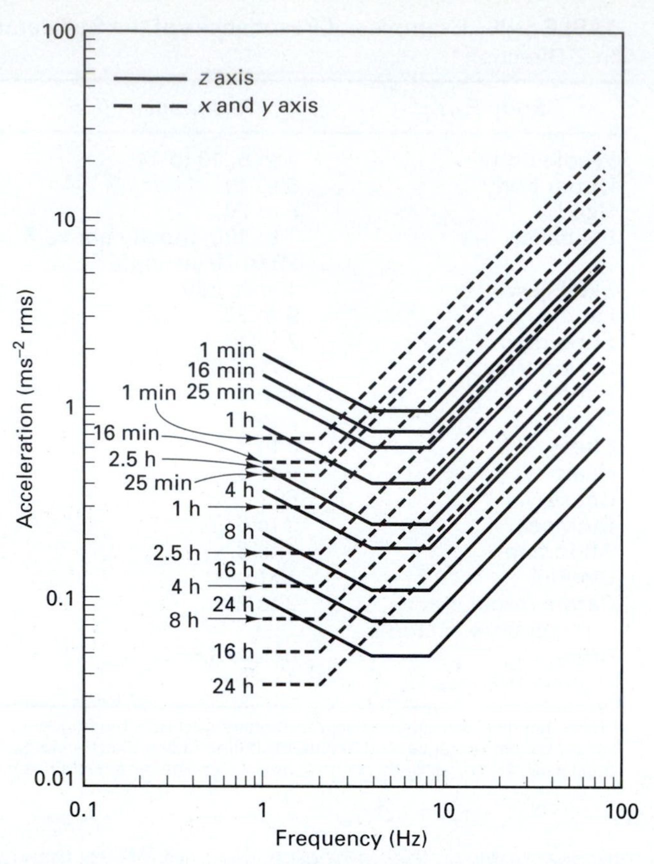

Figure 5–5. Comfort and discomfort boundaries for vertical and horizontal whole-body vibrations (ISO 2631, 1985).

Combined Vibration and Noise

In many experiments, the vibratory equipment generates noise. Subjects' responses indicate a similarity between the sensations of noise and vibration, as expressed by Stevens' power law, which may allow formulating conditions of equivalence. Thus, under conditions in which both mechanical and acoustical vibrations are present, the reduction in one stimulus may not only improve the overall perception of the vibratory environment, but might in fact reduce the perception of the other, unchanged stimulus (Griffin 1990).

Effects of Vibration on Performance

The human may be modeled as a system that receives information through sensors, makes decisions about the information, and then performs activities based on these decisions. (See Chapter 3.) Little research has been done on the effects of mechanical vibrations on a person's mental activities, state of arousal, ability to make decisions, or attitudes.

Biomechanical effects of body vibration concern the eyes, as principal input ports, and the hands and mouth, the chief means of output. One's performance can be similarly affected when the body oscillates while the outside remains stationary versus when the visual or manual target vibrates while the body is still.

An operation error related to vibration (called "breakthrough" or "feedthrough") is dependent on combined, often complex, transfer functions of the biodynamic system, of the control, and of the machinery that is controlled (Griffin 1990). Output may be discrete, as, for instance when one is pressing a button, or continuous, such as when one is driving an automobile. Automobile driving is also an example of a "pursuit-tracking task," which is in contrast to "compensatory tracking," where only the difference between the existing and the desired location of the vehicle is controlled.

TRACKING

The negative effects of vibration on performance are similar for pursuit and compensatory tracking. Vibrations between 6 and 10 Hz induced in the upper body and between 2 and 10 Hz in the shoulder region cause a reduction in performance. The hands have resonances at 4 to 5 Hz, depending on the direction of the vibration, the seating conditions, the posture, etc. There is also a strong effect of acceleration, with larger accelerations affecting control operations more than smaller accelerations. The effects of (induced or resulting) vibration that occurs in several axes simultaneously are complex and difficult to model.

MANUAL TASKS

Fortunately, the characteristics of many controls and controlled machines attenuate vibration errors considerably at frequencies above 1 Hz, but highly sensitive systems or tasks that require small or precise movement may be strongly affected by vibratory environments. Examples are the activation of push buttons and handwriting.

VISUAL TASKS

If a visual target oscillates slowly, the eyes pursue the movement and maintain a stable image on the retina. (See Chapter 4.) This is called a *reflex response.* The human eyes are able to perform pursuit reflexes for display oscillations of up to about 1 Hz. At higher frequencies, the saccadic movements are too slow and the image becomes blurred. When the observer is vibrating, the head and eyes experience both translational and rotational movements. The complex motoric compensatory activities become increasingly insufficient as the frequencies exceed 8 Hz; vision problems occur when the apparent displacement of the visual object gets larger than one minute of arc (Griffin 1990; Oborne 1983). In general, the effects of translational vibration decrease with a large viewing distance, while the effects of rotational oscillation are independent of distance.

VOICE

The airflow through the larynx, as well as breathing irregularities and related changes in general body tension, may change the pitch of the voice, particularly with regard to vertical oscillations at frequencies between 5 and 20 Hz.

Vibrations Causing Injuries and Disorders

Whole-body vibration with a peak magnitude below 0.01 m s^{-2} is hardly felt, while accelerations of 10 m s^{-2} rms or higher may be assumed hazardous. The effects of intermediate accelerations depend on the actual frequency, direction, and duration of the vibration. Unfortunately, the available information is still insufficient to establish causal relationships; only a small number of physiological responses to whole-body vibration are well documented (Griffin 1990). Exposure to vibration often results in short-lived changes in various physiological parameters, such as the heart rate: Vertical vibration in the range of 2 to 20 Hz can produce a cardiovascular response similar to that experienced during moderate exercise (Guignard 1985). At the onset of exposure to vibration, increased muscle tension and initial hyperventilation have been observed (Dupuis and Zerlett 1986). The musculoskeletal system is often strongly affected by the motions and energies that it must resist or counteract. Reflex responses can be inhibited (Martin et al. 1986).

While motions between skin and underlying structures make it difficult to interpret electromyographic (EMG) recordings taken with surface electrodes, EMGs have been used widely to record the activity of back muscles during whole-body vibrations. (See Chapters 1 and 11.) The observed muscular contractions do not necessarily protect the body and may in fact enhance the strain beyond that of a passive system (Seidel 1988), for example, because of untimely contraction due to phase lags.

Among the responses of the sensory system, vision is easily disturbed by motions of the body, as discussed earlier. Since vibration and noise often occur together, detrimental effects on hearing have been reported, in some cases supposedly even in the absence of noise. Tactile perception as well is affected in many circumstances (Griffin 1990).

In spite of numerous uncertainties, standards have been published that purport to describe acceptable exposures to vibration. Figure 5–6 gives vertical acceleration conditions.

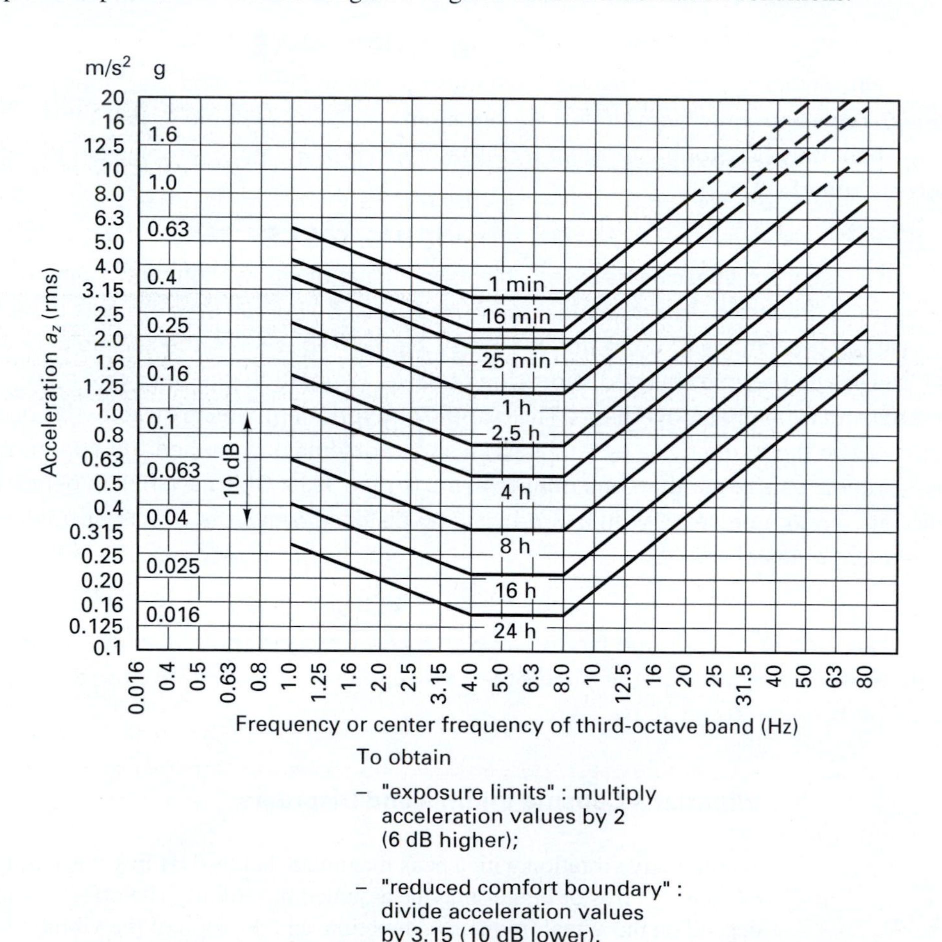

Figure 5–6. Vertical-acceleration limits in terms of frequency and exposure time (ISO 2631, 1985).

"The respectable scientist may wish to avoid the danger of compromise upon the uneven, unproven, and often undefined, ground on which the democratic production of national, international and other standards takes place. Imperfect guidance can be beneficial, but attempts to 'whitewash' the imperfections in a standard are more likely to create a prestigious white elephant than to preserve the standard in pristine condition. Environmental standards can be immensely valuable, but they should be evolved today in the recognition that they must assist the evolution of, and eventually make room for, improved guidance tomorrow. Consequently, users of any standard would be wise to assess the areas of expertise, the interests and the knowledge available to those who formulated the standard" (Griffin 1990, p. 633).

While chronic degenerative effects on bone or cartilage have been reported after prolonged exposure to vibration, vibration may be beneficial for the skeletal system of persons who experience only weak stimuli for tissue maintenance or growth (Wolff's law), such as paraplegics and astronauts. On the other hand, prolonged vibration may lead to degeneration of the spinal column—for instance, in tractor and truck drivers—or to "vibration white-finger disease" in workers using vibrating tools such as jackhammers (Dupuis and Zerlett 1986; Oborne 1983; Wasserman et al. 1986).

Motion Sickness. The vestibular system is of major importance with respect to the appearance of vibration-induced "motion sickness," often accompanied by nausea or vomiting. One may attempt to describe the condition functionally as *Motion sickness is caused by motions that are erroneously perceived,* but this description identifies neither the factors that cause or aggravate motion sickness—the direction, frequency, amplitude, and duration of motion and the absence or presence of visual information—nor their interactions. A prevailing theory is that the problem arises from a conflict among the sets of information received from two or more sensory systems, particularly involving vestibular and visual sensations. The involvement of the vestibular system is critical, as is indicated by the strong tendency toward motion sickness caused by head movements alone during body oscillation. Motion sickness is particularly prevalent on ships, in aircraft, and in automobiles, but it also exists in flight simulators. Many astronauts suffer from motion sickness, which may or may not subside within hours or days after they are launched into space.

Neither the sensory conflict theory nor other proposed explanations can predict the degree of motion sickness expected from different parameters. A complete theory is still needed that encompasses both causes and effects and their relations. At present, the following advice can be given (*CSERIAC Gateway,* Vol. 4, No.1, 1993): Beware of

- wide fields of view, especially involving time lags in visual perception;
- head movements during acceleration and oscillation;
- head movements while wearing distorting optical devices, including magnification lenses.

There is no experimental evidence that the relation between exposure to motion and meals eaten has any effect on the occurrence of motion sickness, except that there may be truth in the saying that motion sickness thrives on an empty stomach. Consumption of fluids may be advisable, even if little is retained. Mental activity may be beneficial for minimizing sickness, but head movements should not accompany such activity. As Griffin (1990, pp. 327–328) put it, "Fighting at sea and singing have both been said to suppress symptoms, although only the latter can be recommended here!"

Effects of Impacts and Sustained Gs

Many kinds of impacts

While one intuitively understands the transitions from impact to shock to bump to vibration, their actual delineations are arbitrary. Different approaches to describing these events and their effects on the human body have been proposed, but none have been generally accepted. They have been described in terms of triangular or trapezoidal acceleration over time, but the actual profiles of the events are often different from the idealized profiles and remain difficult to measure.

Impacts are often defined as events with a sudden onset, a duration of less than one second, and high acceleration. Human tolerance to impact depends, among other factors, on the direction of impact that is experienced, the magnitude of deceleration, and the total time of deceleration. Linear impacts occurring at right angles to the spinal axis are better tolerated than those parallel to the spine. Among life-threatening skeletal fractures, damage to the vertebrae is most common, while at high impact, head injury is most frequent and severe. Of course, a wide variety of conditions affect a person's survival. For example, a 2-meter, head-first fall of a child onto a flat, solid surface can result in skull fracture; and the same can happen to an adult who falls backwards on the head from an upright position when there is not enough time for the body to assume a protective posture. Human skull tolerance limits appear to be about 150 to 200 G for three ms^{-2} of average acceleration and 200 to 250 G for peak accelerations (R. G. Snyder, personal communication, February 18, 1991). Figure 5–7 lists some conditions under which humans have survived impacts; of course, more often, people do not survive them.

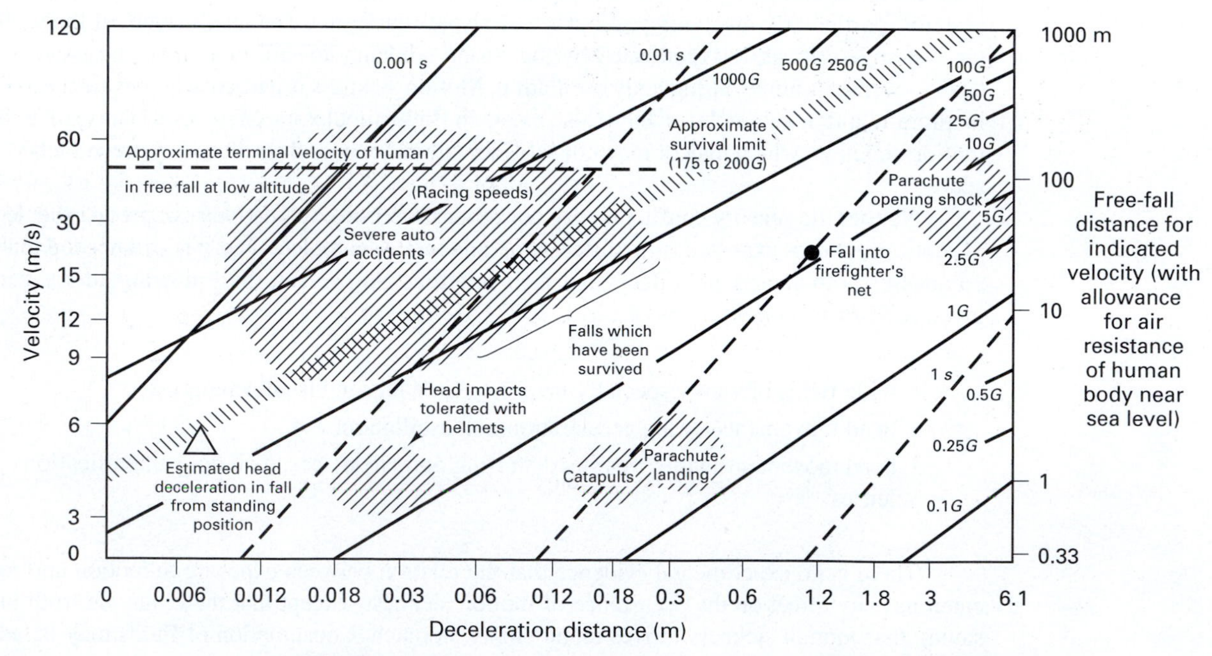

Figure 5–7. Approximate conditions under which humans have survived impacts (modified from Snyder 1975).

Direction affects tolerance to impact Human tolerance to multiple-G accelerations sustained for a few seconds depends on the direction of the resulting force relative to the body. Experience and experimentation have shown that such sustained G forces are best tolerated in the plus or minus *x*-direction—that is, when the body is supported either on its belly (prone) or on its back (supine), perpendicular to the direction of impact. Fairly little is known about sideways actions (i.e., in the plus or minus *y*-direction).

Sustained acceleration in the *z*-axis is difficult to tolerate, particularly when it acts downward, because even a few Gs hinder the blood supply to the brain and, therefore, may lead to vision disturbances, "grayout," or blackout. This kind of acceleration has caused many problems and accidents for pilots of aircraft. Early astronauts were put into reclined body-contoured seats during their blastoff from earth so that they could tolerate acceleration in the *x*-direction. For the same reason, some military aircraft have a pilot seat that automatically reclines during high-G flight maneuvers.

Models of the Dynamic Response of the Human Body

Understanding and modeling the response of the human body to impact or vibration is a difficult enterprise. First, there are various intensities and types of impacts and vibrations, in various directions. Second, there is great inter- and intraindividual variability in responses, also depending on posture, body support, and restraint systems. Third, the response of the body cannot be explained simply in terms of resonances, because, while body elements, if taken in isolation, do show specific natural frequencies, those frequencies are highly damped. Also, the interactions between differently vibrating body segments may generate a complex network of causal chains and temporal sequences.

Transfer functions

In most studies, the dynamic response of the body is assumed to be *linear*—that is, proportional to the excitation (which is either sinusoidal or the actual or simulated motion of a vehicle). Dynamic responses of the body or its parts are usually described by *transfer functions* determined at certain frequencies.

Transfer impedance and transmissibility

There are two groups of transfer functions. One group, called *transfer impedance functions,* relates two measures obtained at different points of the body. If one is interested only in the magnitude, for example, of head motion to seat motion, one speaks of *transmissibility.* "Comfort" curves are assumed to be the inverse of transmissibility. Figures 5–8 and 5–9 present examples of observed transmissibilities.

Mechanical impedance

The other group of transfer functions refers to the ratio of two different measures obtained at the same point of the body, *called mechanical impedance.* This group describes relations between the force that drives a system at a particular frequency and the resulting movements, viz., displacement, velocity, and acceleration.

Damping and stiffness

A force F applied to an object produces an acceleration a that is proportional to, and in phase with, F. The constant of proportionality is called the *mass* and is given by $m = F/a$. (This is, of course, a restatement of Newton's second law.) If the force F is applied to a (massless, ideal) damper, it produces a velocity v that is also proportional to, and in phase with, the force. This constant of proportionality is called the *damping* and is given by $c = F/v$. The application of a force F to a (massless, ideal) spring produces a displacement x. The constant of proportionality is called the *stiffness* of the spring and is given by $k = F/x$.

The human body is not rigid

In a rigid body, force and acceleration are always in phase, and thus, at any frequency, the ratio of their rms magnitudes indicates the mass of the object. At most frequencies, however, the human body does not behave as if it were rigid, and force and acceleration are out of phase in a manner that depends on the stiffness and the damping at each frequency. Of course, one can still calculate the ratio of force to acceleration, but it no longer equals the static mass of the object. Therefore, the term "apparent" or "effective" mass is used. Similarly, the properties of dampers

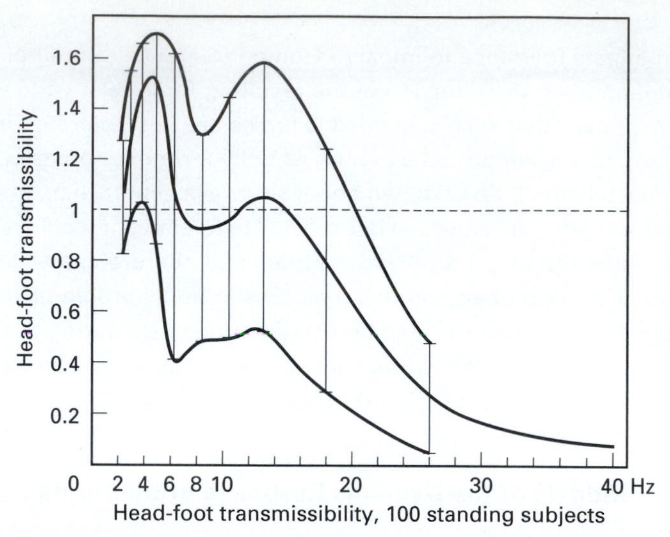

Figure 5–8. Typical head–foot transmissibility: 100 standing subjects, mean and 67-percent boundaries. Note the resonances at about 4 and 14 Hz. Source: Modified from "Vibration at Work" by Oborne, D. J., in D. J. Oborne and M. M. Gruneberg (eds.)., *The Physical Environment at Work.* Copyright 1983 by John Wiley & Sons, Ltd. Reprinted by permission of John Wiley & Sons, Ltd.

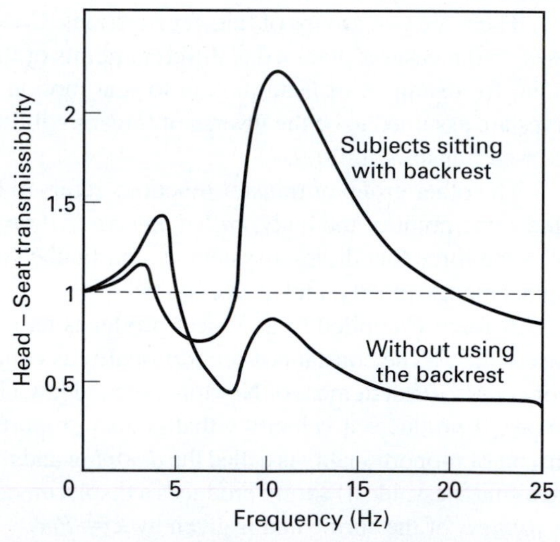

Figure 5–9. Example of head–seat transmissibility of a sitting subject, either leaning against a backrest or without using the backrest (modified from Griffin 1990).

and springs have different effects on the movements of masses at changing frequencies; thus, the term "impedance" is used instead of "damping," and "dynamic stiffness" or "dynamic modulus" replaces "stiffness."

The inverse ratios also have distinct names: Acceleration divided by force is called *accelerance* (or *inertiance*), velocity divided by force is called *mobility,* and displacement divided by force is known as *dynamic compliance* (Griffin 1990).

Phase lags

In a simple system, there are fixed relations among acceleration, velocity, and displacement. Changing from velocity to acceleration adds a 90-degree phase lag, and the values change by $2\pi f$. Table 5–11 displays these relationships.

TABLE 5–11. Dynamic Responses of Pure Masses, Dampers, and Springs

Element	*Modulus*		*Phase*
Mass m	Apparent mass	$= m$	a and F are in phase
	Mechanical impedance	$= i\omega m$	v lags F by 90°
	Dynamic stiffness	$= -\omega^2 m$	d and F are 180° out of phase
Damper c	Apparent mass	$= c/i\omega$	a leads F by 90°
	Mechanical impedance	$= c$	v and F are in phase
	Dynamic stiffness	$= i\omega c$	d lags F by 90°
Spring k	Apparent mass	$= -k/\omega^2$	a and F are 180° out of phase
	Mechanical impedance	$= k/i\omega$	v leads F by 90°
	Dynamic stiffness	$= k$	d and F are in phase

Source: Griffin 1990.

Sets of subsystems One often models a complex system as a set of simple subsystems with discrete components (of mass, damping, and stiffness) that, together, should have the same mechanical impedance as the complex body. A typical model of the sitting body involves two or three masses, as sketched in Figure 5–10: m_1 is the mass of the body that moves relative to the platform supporting the seated body, and m_2 is the partial mass of the trunk and the legs that does not move relative to the platform. If the feet and lower legs do not move in phase with the seat, the model must be extended by m_3. Properly determining these masses and their stiffness and damping parameters is essential to making the model realistic.

Standing person For a person standing on a vibrator, the model that attempts to describe the reactions of the different body parts is more complex. Figure 5–11 shows it according to ISO Standard 7962 (1987), which is meant to be applicable up to a frequency of 31.5 Hz.

Such models as those just discussed are only estimates of the actual vibration responses. There are large interindividual variations, and variations also exist in the same subject due to posture changes. Often, horizontal and vertical impulses are present at the same time. Body parts may move in several planes even if their motion is stimulated only in one direction; for example, the head performs pitch motions in the medial plane even if the vibration applied to the body is strictly vertical. The main head movements induced by vibrations in the x, y, and z directions and transmitted through a rigid seat without a backrest are listed in Table 5–12. Resonances in response to vibrations in the z direction are compiled in Table 5–10.

With his special candor and humor, Griffin stressed the difficulty of correctly modeling the vibrational responses of the human body. He cautioned that the database is too meager to allow,

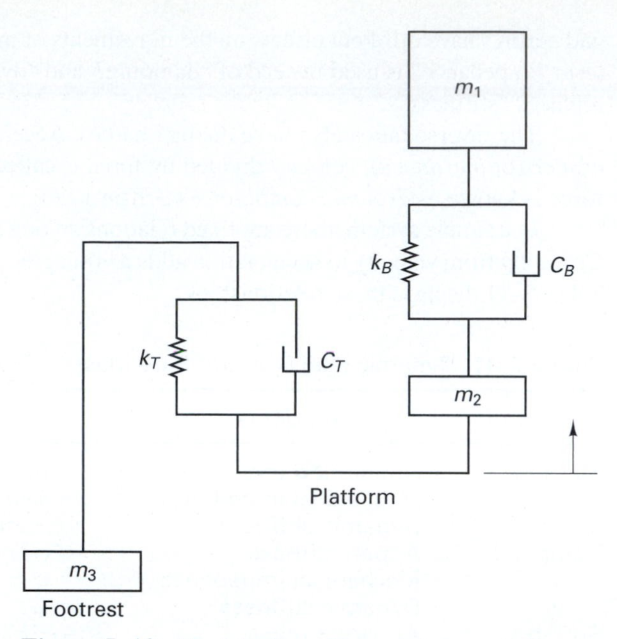

Figure 5–10. Model of a subject sitting on a vibrator (modified from Griffin 1990).

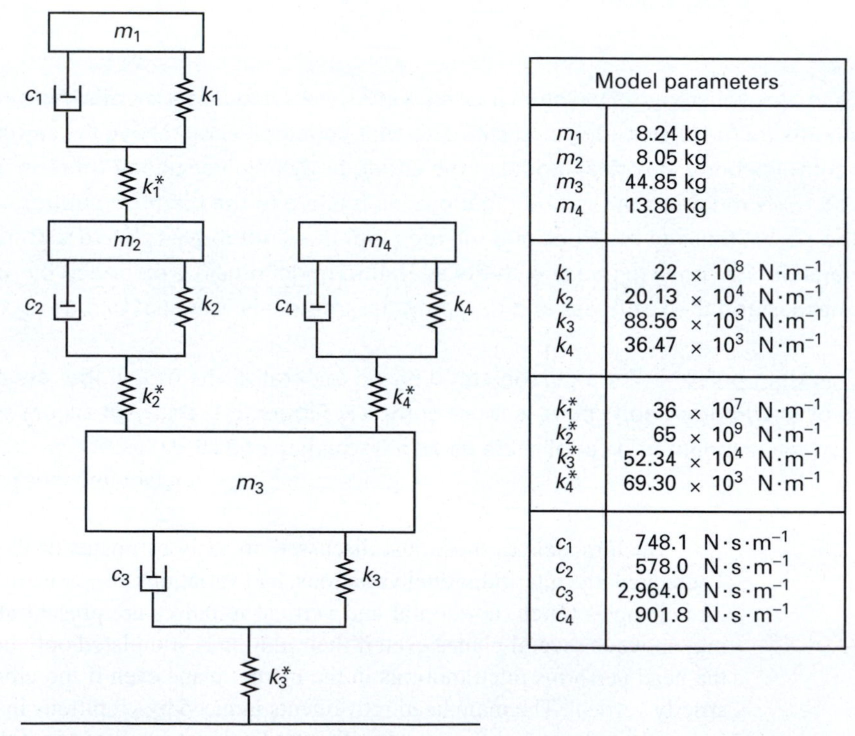

Model parameters	
m_1	8.24 kg
m_2	8.05 kg
m_3	44.85 kg
m_4	13.86 kg
k_1	22×10^8 N·m^{-1}
k_2	20.13×10^4 N·m^{-1}
k_3	88.56×10^3 N·m^{-1}
k_4	36.47×10^3 N·m^{-1}
k_1^*	36×10^7 N·m^{-1}
k_2^*	65×10^9 N·m^{-1}
k_3^*	52.34×10^4 N·m^{-1}
k_4^*	69.30×10^3 N·m^{-1}
c_1	748.1 N·s·m^{-1}
c_2	578.0 N·s·m^{-1}
c_3	2,964.0 N·s·m^{-1}
c_4	901.8 N·s·m^{-1}

Figure 5–11. Four-degree-of-freedom model for calculating the vertical transmissibility of the human being, either sitting or standing (ISO 7962, 1987).

TABLE 5–12. Major Head Movements Caused by Certain Vibration Frequencies Transmitted Through a Rigid Seat without a Backrest

	Linear Displacement in			*Rotational Displacement in*		
Direction of Exciting vibration	z	x	y	*Yaw (About* z*)*	*Roll (About* x*)*	*Pitch (About* y*)*
z	5–10 Hz*	5–12 Hz*				5–12Hz 5 Hz
x	2–12 Hz*	about 4 Hz				1 Hz and above
y	about 2 Hz	below 4 Hz	below 5 Hz		below 6 Hz	

Source: Data from Griffin 1990.

*Transmissibility is reduced if the back is not in contact with backrest, particularly at the higher excitation frequencies. Transmissibility is also much affected by body posture and muscle tension.

with certainty, the establishment of forces on, and movements of, the body, as well as insufficient to predict the effects of vibration on comfort, health, or performance. Thus, currently available standards must be applied carefully because "Many models have achieved complexity without representing the known behavior of the body" (Griffin 1990, p. 181).

SUMMARY OF VIBRATION

Much empirical and anecdotal experience exists regarding the reactions of the human (with respect to comfort, performance, or health) to vibrations and impacts. In experimental research, vibrations are usually modeled as single or combined sinusoidal displacements of the body, while impacts are modeled in terms of short-term, high-onset single events.

Responses of the body are often expressed in terms of resonances at certain frequencies. Yet, the dynamic responses of the body are complex mechanical functions of mass properties, dampers, and springs, together with other physiological and psychological reactions. Actual responses to vibration show large interindividual variations and depend on body posture, restraint systems, and the actual directions and kinds of impulses that exist in the real world—for example, in a vehicle driven over rough terrain. Thus, currently available standards and design recommendations must be used with great caution. Nevertheless, the existing information provides guidelines with respect to the vibrations and impacts that have major effects on human comfort, health, and performance, as well as providing information on technical means of avoiding those effects as much as possible.

ASTRONAUTS AND "WEIGHTLESSNESS"

Being able to fly through space is an ancient human desire. While flying by airplane had become feasible around 1900, well into the 1950s humans could only dream of seeing Mother Earth from far away, of circling the moon, or of landing on another celestial body. With advances in rocket technology (see Table 5–13), visits to space became a reality in the 1960s.

In the early years of preparation for space travel, researchers tried to extrapolate from experiences gained in other long, isolated, perhaps dangerous situations, such as journeys in submarines, polar exploration, confinement in mental hospital wards, prison stays, missions of bomber

TABLE 5–13. The Development of Spaceflight

1926	Robert Goddard launches his first fluid-powered rocket.
1931	Auguste Piccard and Paul Kipfer ascend by balloon to 16 km altitude.
1934	Wernher von Braun and coworkers start developing rockets.
1942	A V2 rocket attains 90 km altitude.
1944	German V2 missiles bombard allied cities.
1947	Chuck Yeager flies faster than sound in the X-1 rocket plane.
1957	*Sputnik 1* carries a sphere into orbit around the earth. *Sputnik 2* follows with the dog Laika aboard.
1958	NASA shoots its first satellite into space.
1959	*Lunik 3* spacecraft transmits pictures of the back side of the moon.
1961	Yuri Gagarin is the first human being to circle the earth.
1963	Valentina Tereshkova is the first woman in earth orbit.
1968	*Apollo 8* spacecraft circles the moon.
1969	Neil Armstrong and Edwin Aldrin land on the moon in the *Apollo 11* spacecraft.
1970	The crew of *Apollo 13* survives an explosion on board the spacecraft
1971	Upon return from the space station *Salyut 1,* three cosmonauts perish in their capsule.
1975	*Venera 9* spacecraft lands softly on Venus and transmits pictures of its surface.
1976	*Viking* spacecraft land on Mars and transmit pictures of its surface.
1977	*Voyager* spacecraft flies toward Jupiter, Saturn, Uranus, and Neptune.
1981	First reusable Space Shuttle started.
1986	*Challenger* shuttle explodes shortly after start; seven astronauts are killed.
1986	*Mir* space station is launched and continually used by international crews.
1998	The first modules of the International Space Station ISS are launched into earth orbit.

crews, experiences of prisoners of war, participation on professional athletic teams, and shipwrecks and other disasters (Holland 1991). From the 1960s on, actual experiences were gained, especially by Soviet space crews with several international guests. Cosmonauts stayed aboard the *Salyut* and then the *Mir* space stations for many months, some a year or longer. American astronauts did not perform long-term space missions for about two decades after their last *Skylab* activity in early 1974, but they joined their Russian colleagues on *Mir* in the 1990s. So the International Space Station (ISS), whose first sections were launched into earth orbit in 1998, could be designed taking into account extensive experiences of humans actually living and working in space (though still near the earth).

Live and perform

During the 1960s and 1970s, the decision was made to send humans into space, instead of just shipping remote or self-controlled machines. The two main reasons were that humans respond better than machines to unexpected situations and that it behooved us to satisfy the sheer curiosity and love of adventure which are so fundamental to our nature.

☛☛☛ *In order of priority, a human in space must survive, perform tasks within the spacecraft, and carry out extravehicular activities for construction and maintenance of the space station and for space exploration. The major problems to be overcome are psychosocial, physiological, and medical in nature. Their magnitudes increase with the duration of the space mission.* ☚☚☚

The technical challenges are to construct a shell and a method of propulsion that is safe and suitable for space travel. Inside the space station, we must keep the people functioning and healthy by providing a breathable atmosphere, a suitable macro-climate, and appropriate management of food, drink, and waste. (See Table 5–14.) The specific requirements are to protect the residents of the station from radiation and, possibly, to generate artificial gravity, both discussed in more detail shortly (Albery and Woolford 1997). Altogether, this is a typical example of a human-centered systems-engineering task (Blanchard and Fabrycky 1996) to be executed in a new and dangerous environment.

Habitability

The space available to each individual must be of sufficient volume, provide privacy and room for storing personal items, allow personalization of decor, and give shelter from uncomfortable temperatures and airflow and from intrusive noises, lights, and odors. In addition to the psychosocial aspects of living under "space stress" (discussed in Chapter 3), special problems can arise if crews of male and female astronauts are put together for spaceflights of long duration (Frazer 1991) and if medical problems appear.

A space crew builds its own microsociety, which is particularly necessary on long missions—for example, to Mars. Crew members are separated from their loved ones and from Mother Earth and are confined to a small, crowded space. They experience sensory deprivation, especially in the vestibulum, and their senses may change, especially as regards their sensitivity; sounds and smells appear changed and more intrusive, for example. The circadian rhythm may be

TABLE 5–14. Technology Needs for Long-Duration Spaceflights

Function	*Technology Needs*
Atmospheric pressure and composition control	• Order-of-magnitude improvement in the reliability of the hardware to be utilized in long-duration missions for total and partial pressure control and monitoring, fire detection and suppression, etc.
Temperature and humidity control	• Improvement in current thermal control technology, including nontoxic heat-transfer fluids, metal hydride heat pumps, and rotating bubble membrane radiator.
Atmospheric revitalization (CO_2 control, removal, and reduction, O_2 and H_2 makeup, trace gas monitoring and control)	• Minimization of the weight, volume and power demands in closing the CO_2 loop by electrical, chemical, absorption/desorption, or molecular sieve processes. • Development of long-duration quality monitoring, including sensors and control technology for trace contamination, toxic compounds, and pathogens.
Food supply (storage, processing and preparation, growth chambers)	• Development of closed-loop bioregenerative food production systems. • Development of automated systems for harvesting and processing edible biomass or space crops.
Water management (wastewater collection and processing, water quality monitoring, storage and distribution of recovered water)	• Development of closed-loop portable water recycling systems • Investigation of alternative ways to minimize the weight, volume, and power demands of waste-processing equipment (eg, vapor compression and distillation, vapor-phase catalytic ammonia reduction, supercritical water oxidation). • Development of techniques for long-duration quality monitoring.
Waste management	• Development of waste-management systems (physical and chemical, or bioregenerative) to collect and process urine, collect and store fecal matter, and recycle waste.
Portable life-support systems	• Development of automated control concepts, EVA heat-storage concepts, and atmospheric control concepts to minimize weight, pressure, volume, and power demands.
Health maintenance	• Development of concepts of occupational medicine to include personal hygiene, exercise, diagnostics and therapeutics, and surgery and medical aid capabilities for utilization in reduced-gravity environments.

Source: Adapted from Jenkins 1991.

altered and sleep and relaxation patterns changed. Crews of mixed gender and different cultural backgrounds must get along, which may be difficult, as illustrated by the experiences of international crews during long missions aboard by those aboard the Russian *Mir* space station.

One list set up by NASA established the following important factors for mixed space crews (CSERIAC Gateway, *Vol .7, No.1, 1996):*

- *Language*
- *Nonverbal communication styles*
- *Task- and relationship-oriented behavior*
- *Patience and tolerance*
- *Decision-making processes*
- *Assertiveness*
- *Interpersonal interest*
- *Respect for other cultures*
- *Personal hygiene and cleanliness*
- *Gender roles and stereotypes*
- *Conflict management and resolution*
- *Trust in people*
- *Scheduling and time management*
- *Sense of humor*

Radiation

Radiation can be life threatening, is unpredictable, and increases as the astronaut moves further away from earth. There are three primary natural sources of radiation: the earth's magnetosphere, solar flares and wind, and cosmic radiation. Radiation from the magnetosphere is strongest between about 2,400 and 19,000 km above the ground. Standard spacecraft shielding can protect space crews from exposure, particularly since they usually pass quickly through the radiation zone. More shielding is necessary to protect humans against solar winds, which are composed of high-energy particles.

rem

The most serious threats are from solar flares and cosmic radiation. Solar flares run in about 11-year cycles, corresponding to sunspot activity. During each cycle, between 20 and 30 flares expose the astronaut to up to 100 rem of radiation, while up to 5 flares may generate up to 1,000 rem, and 2 flares may be even more powerful than 5,000 rem. (NASA has set a radiation exposure limit of 400 rem that is not to be exceeded during an astronaut's career. A standard X-ray exposes a patient to about 0.01 rem; Denver, Colorado, residents are exposed to approximately 0.2 rem per year. Exposure to 100 rem causes acute radiation sickness, and 300 rem can be lethal.) Another major problem is that solar flares are difficult to predict: Particles that precede a flare do so by only about one hour, giving little time for space crews to seek shelter.

Solar flare protection

Spaceships, as well as space stations such as those on the moon, must be provided with radiation shielding to protect against the most intense bursts of solar radiation, which usually last for less than a day. Solar storms must be monitored at all times, and astronauts on the moon must stay within one hour of travel time to a radiation shelter.

Cosmic radiation

It is very difficult to protect astronauts against cosmic radiation, which has very high energy particles. Cosmic radiation not only damages or destroys cells, but can also damage genetic material, with consequences for reproduction. Very dense shielding is required. This may be fairly easily accomplished on a moon station, because one can use natural topographic features, such as valleys, which provide some shelter, or one can burrow the station into the moon rock or heap lunar material on the roof of a shelter.

Onboard radiation

While in the spaceship, the crew must be protected not only from solar and cosmic high-energy particles, but also from any radiation that is generated onboard by a nuclear power generator. (A nuclear rocket engine is currently the most promising technique for generating propulsion for extended space exploration.) This need for protection adds mass to the spaceship; using current technology, the ship's mass would be at least 500 tons.

Pollution and Contamination

Onboard pollution is a serious problem. Besides radiation, just discussed, it may consist of chemical releases, dust, noise, microbes, and particulate debris. Toxic chemicals or disease-causing organisms can endanger space crews. Human beings can be breeding grounds for bacteria and viruses. Recycled air, water, and wastes can carry contamination.

Protective Spacesuits

During extravehicular activities, the crew must be protected against reduced atmospheric pressure, such as on Mars, and against the vacuum conditions in space. All the space suits used by Russians and Americans provide such protection, as well as a breathable internal atmosphere and cooling or heating. In earth's orbit, temperatures can range from approximately −130° (in darkness) to 120°C (in sunlight). If the suit fails or is punctured, for example, by a micrometerorid, the results can be catastrophic.

Impact with Objects in Space

Human-made debris is in orbit about the earth, and meteoroids and asteroids may be anywhere in space. Impact with an object of sufficient size can have disastrous consequences for spacecraft and crew, either through the loss of pressure and air or by direct physical injury. The risk near earth is relatively small, since larger objects can be tracked from earth, and thus, the crew may be able to evade damage or, if not, perhaps help could be sent. In interplanetary space, there are relatively few objects, and even if the spaceship is large, the likelihood of a collision is still small, but if it occurs, no help or rescue can be expected from earth.

Microgravity

Many technical aspects of the construction of space vehicles are fairly well mastered. Yet, protecting astronauts from the slow effects of the absence of gravity is still a largely unsolved technical challenge.

Weightlessness

Physically speaking, some weak "gravity pull" exists everywhere in space, either because adjacent celestial masses exert forces of attraction or because the spacecraft is being accelerated, linearly or angularly. Since there is no true "zero gravity" (except in the center of a spacecraft in a stable orbit), one calls any acceleration level below 10^{-4} G "microgravity," popularly known as weightlessness. (The unit value of G is the average acceleration of gravity on earth, 9.80665 m s^{-2}.)

Space sickness

Microgravity has detrimental physiological effects. However, on trips into space, many crew members become accustomed to working and living there within just a few hours and are able to perform well in about three days. Complete adjustment, so that one feels "at ease," takes about three weeks—and then, upon returning to earth, about as long to readjust again to gravity. (Personal communication by astronauts Carr and Lousma, January 11, 1990.) However, many astronauts suffer from nausea initially, and a few are affected throughout a space mission. This can hinder the execution of tasks, possibly even severely.

For long-duration spaceflight—that is, lasting 60 days or longer—physiological problems associated with microgravity become serious and need to be counteracted by technical and behavioral means (Jenkins 1991). Most problems can be grouped under the categories of radiation, the musculoskeletal system, fluids and blood, and nervous control.

Musculoskeletal System

Muscle atrophy

A major problem associated with spaceflight of long duration is the deconditioning of the human musculoskeletal system. Because of their diminished use in low gravity, muscles lose some of their volume. Left untreated, muscle atrophy creates a condition similar to that experienced by paraplegics, whose muscles deteriorate from lack of use. Such loss of muscle mass, together with cardiovascular deconditioning, can cause significant problems. For example, after a 211-day mission, Soviet cosmonauts could not maintain a standing position on earth for more than a few minutes.

Demineralization of bones

To counter the effects of disuse in low gravity, the muscles must be exercised extensively. Exercising muscles also helps against another skeletal problem, the demineralization of bones in space. The loss of calcium, through urine and fecal excretions, increases the likelihood of a bone fracture. Bones that have been found to be most susceptible are those in the legs, those in the hips, and the lower lumbar vertebrae. These bones support much of the human body weight in an environment with gravity. In long-duration spaceflights, cosmonauts have lost up to 11 percent of the mass of certain bones, a demineralization process that raises other concerns. For example, increased calcium flow through the kidneys may add to the risk of kidney stones. Altogether, the cumulative effect on the bones of people who have been in space for long periods may be similar to osteoporosis in the elderly.

Exercise!

Loss of both muscle mass and bone minerals can be counteracted by a daily, extensive regimen of exercise, using, for example, resistance machines and motion-restricting suits. The effectiveness of this measure was demonstrated when cosmonauts who had spent a full year in orbit were able to walk several yards without aid upon landing on earth.

Blood and Fluid Distribution

Fluid redistribution

Another major problem that occurs in low gravity is the redistribution of body fluids. The body's circulatory system operates well in a 1G environment. For example, valves in the veins prevent blood from pooling in the feet and legs and instead force it toward the right ventricle of the heart. The antipooling system still operates in space, although there is little gravity to work against. This results in a reduction of fluid volume in the legs and an accumulation in the head, neck, and chest.

Fluid imbalance

The redistribution of fluid in a low-gravity environment has important ramifications. Internal sensors indicate that there is too much blood in the upper body, for which the body compensates by reducing the blood flow. This aids the upper body, but worsens the problem in the lower body, where the fluid level is already low: Up to 20-percent reductions in leg fluid volume after more than 200 days of spaceflight have been encountered. As a side effect, the astronaut's thirst is decreased. Drinking insufficient fluid then reduces the overall fluid level in the body, which, in turn, adds to the demineralization of bones and increases the retention of sodium.

Because the brain perceives an excess of blood in the body, the production of new red blood cells drops. The result may be anemia in the astronauts, but probably only after about 100 days in microgravity.

Reduced heart size

Changes in fluid distribution within the body also have a negative effect on the heart. The increased volume of fluid in the upper body raises the blood pressure in the head and neck veins. Sensing this heightened pressure, the heart actually reduces its output by lowering its contracting rate and shortening the length of the diastole. In cosmonauts who stayed in space for months, actual decreases in heart size were noted.

Dizziness and fainting when back in gravity

After a time spent weightless, the body establishes a new circulatory pattern, which seems to have no ill effects on the astronaut while he or she is in space. However, upon returning to earth, astronauts experience frequent episodes of dizziness and even fainting. This is probably due to the reduced heart rate and the decreased total blood volume. Back in gravity, the antipooling circulatory mechanisms of the body are again utilized, and the available blood is redistributed, which involves its moving blood from the upper to the lower body. The brain is then insufficiently supplied by blood, resulting in dizziness and fainting. The problem is usually cured within a week or so on earth, since more blood and red blood cells are produced, thus supplying enough blood throughout the body, but the recovery may be much slower in a low-gravity environment, such as that on the moon or Mars. During this period of recovery, astronauts are quite sensitive to rapid changes in posture, for example, when they stand up rapidly.

One method of countering the redistribution of body fluids is the use of a reduced-pressure suit around the legs, which has been successful in aiding blood flow there.

Changes in blood composition

Besides the changes in the cardiovascular system, the blood composition also is altered. The plasma volume, the mass of erythrocytes, and the concentration of phosphorus decrease, the cholesterol level increases. The red blood cell content also decreases, reducing the amount of oxygen in the astronaut's blood.

Reduced ATP level

A long stay in microgravity decreases the level of ATP and the intensity of glycolysis (Chapter 2). This is of little concern in microgravity, where less energy is needed to move the body than on earth. However, the diminution in the ability to generate energy through metabolic processes (for which ATP, glycolysis, and available oxygen are important) can reduce an astronaut's overall level of work capacity that is needed upon reentry into an environment with gravity. Such a decrease in capacity could be quite critical after a long trip to Mars, which has only one-third of the earth's gravity.

Nervous Control

Disturbed vestibular sense

During the first few days of spaceflight, disturbances in the vestibular system (Chapter 4) have been experienced, possibly mostly related to the otolith receptors. In low gravity, the disturbed vestibular sense, particularly if combined with conflicting stimuli from the visual and proprioceptive systems, can cause disorientation, vertigo, dizziness, and postural and movement illusions. These symptoms, in turn, bring about a deficiency in sensorimotor coordination during the first few days, often accompanied by nausea. About every second space crew member suffers profound discomfort and motion sickness over some time, which varies by individual from hours to days, until the body adapts and reorientates itself (Nicogossian et al. 1989).

Disturbed sensorimotor coordination, interpretation errors

Similar symptoms occur when astronauts reenter an environment with gravity to which the body (and mind) must adjust. Doing so usually reduces the astronaut's work capacity, causes errors (such as movement illusions), in interpreting the visual environment, and brings about internal discomfort. (See Chapter 3.) Such symptoms can be expected for the first few days after landing on extraterrestrial bodies.

Circadian rhythm

Depending on the work–rest cycle and external time cues, an astronaut's circadian rhythm (see Chapter 6) may be altered. So far, astronauts near the earth have been kept on a 24-hour cycle, with 8 to 10 hours of work and 8 hours of sleep. This has avoided problems with conflicting internal rhythms; it might even be advisable to maintain this 24-hour cycle in spaceships traveling to Mars.

Sleeping and eating

Microgravity affects both sleeping and eating. Sleep disturbances are common in the low gravity of spaceflight, but other conditions may play a role as well. Helped by improved food technology, eating and drinking practices have been developed that are not substantially different from those on earth.

Vibrations and sounds

A problem that must have more to do with psychological than physiological events is an increased sensitivity to vibrational and acoustic stimuli. After about 30 days in low gravity, the tolerance level to sounds and vibrations has been found to be greatly reduced, which may affect the sleep patterns and communications of astronauts and which can bring about general annoyance.

Affected immune system

Another problem in space is the suppression of the body's immune system. The reasons for this are not well understood, but cosmonauts returning to earth after more than 200 days in space displayed severe allergic reactions that they had not had before leaving earth. The problem needs to be studied, because currently no countermeasures are known. The problem might be quite severe if, for example, astronauts were sent to Mars. Even if there was nothing on Mars that could harm humans, upon their return to earth after two or more years, the astronauts could be susceptible to many diseases here.

Affected endocrine system

The endocrine system also suffers ill effects from long exposure to microgravity. The primary effects are changes in plasma composition and decreases in hormone levels. Together with other changes, these effects are likely to stem from a lack of both physiological and psychological stress during orbital flights. Whether the changes are even more pronounced on long-duration spaceflights and what the consequences thereof would be are not known at this time.

Training

Long-term spaceflight without countermeasures produces major negative changes in the cardiovascular, respiratory, musculoskeletal, and neuromuscular systems of the space crew. Physical exercises are key countermeasures to many of these unwanted adaptations.

Since endurance and strength seem to be most important for performance of space tasks, relatively high aerobic fitness and muscular strength, especially of the upper body musculature, are expected to be a criterion for selection and training of astronauts, particularly of those who are involved in extravehicular activities. Yet, astronauts who have built up unusually large (hypertrophied) capacities, such as marathoners and weight lifters, will suffer significant losses of these.

Performing extravehicular activities at regular intervals will probably be sufficient to maintain the endurance and strength required to perform such work effectively. Yet, if insufficient EVAs occur, a minimum of one maximal aerobic exercise every 7 to 10 days during spaceflight may be all that is necessary for maintenance of normal cardiovascular functions and replacement of body fluids.

Means for such exercise may be in electro-myo-stimulation, though conventional exercises appear to be generally appropriate. These are likely to include workouts on some exercise equipment. Soviet cosmonauts wore an elasticized suit, called a pigeon suit, for 12 to 16 hours a day. This suit resisted motions of its wearer, who therefore had to exert extra energy to perform actions; thus the wearer was forced to exercise continually.

Effects of Microgravity on Performance

Movements

The absence of gravity is, in general, a bonus for locomotion in space. Once one is accustomed to microgravity, motion is accomplished with minimal effort, and acrobaticlike maneuvers are done with ease. However, to exert force, the body must be braced so that counterforces can be developed. For example, persons doing EVA assembly work must be tethered or otherwise anchored to the machinery on which they are working, so that the forces or moments that arise in reaction to those that they are applying do not hurl the workers into space.

Motor performance

A problem usually encountered only in the early stages of spaceflight is a decrease in motor performance. Crew members find it difficult to accurately estimate the amount of physical work required to perform certain tasks, and also, it takes them longer than on earth to perform those activities, owing to the lack of gravity-created references and resistances. After a few days in low gravity, the movements and activities become more precise, and the perceived level of difficulty decreases.

Upon return to earth after the Skylab 3 *flight in 1973, the astronaut Jerry Carr said, "I didn't faint, but I felt pretty clumsy. My head felt like a big watermelon and I had to work hard to support it. I'd been a butterfly for 84 days and suddenly weighed something again" (*Final Frontier *2(3), 1989, p. 29).*

Stamina and stability

Upon returning to earth gravity after a prolonged stay in space, astronauts' vertical stability was found to have declined, together with the reduction in muscle strength already discussed. These problems disappeared in about a week upon return to full gravity. It is unknown, however, how long it would take to recover in a 1/3G environment, as on Mars.

It is somewhat amusing to note that astronauts who have returned to earth tend to drop things for the first few days. They seem to be carrying on the habit of simply letting an item go in low gravity, where it floats.

Body posture and size

In space, there is also a change in body posture and in some body dimensions. If relaxed, the body assumes a semicrouching position, with the knee and hip angles at about 130°; flattening of the curvature of the lumbar and thoracic spine section also occurs, and the pelvic angle changes (resulting in an extension of the body of up to 10 cm). The head and cervical column become bent forward, and the upper arms "float up" against the trunk to about 45°. This posture, reported by NASA in 1978 and shown in Figure 5–12, is quite similar to one found in relaxed underwater postures, reported by Lehmann in 1962 and shown in Figure 5–13. The posture seems to result mostly from a new balance of muscular and other tissue forces acting about the various body joints. While the relaxed space posture by itself has no ill effects, it needs to be considered in the design of workstations. For example, it is difficult to work at waist level, since the astronaut must continually force the arms down to that level. Also, bending forward requires an effort by the abdominal muscles, and it is difficult to sit or stand erect. Obviously, the design of space clothing must take into account the "neutral" space posture.

Artificial Gravity

Rotational acceleration

Practically all the problems associated with the musculo-skeletal and body-fluid systems could be alleviated or avoided in spaceships if artificial gravity similar to that on earth could be provided. Two techniques come to mind. The first is to rotate the space structure, possibly by using the metabolic energy exerted by the crew's exercising, thereby generating a force that acts centrifugally from the center of rotation. This force is the product of the mass of the astronaut, the radial distance from the center of rotation, and the square of the angular velocity. Thus, the greater the velocity and the farther away from the center of rotation, the greater is the centrifugal force experienced by a person. A further complication arises with rotational acceleration: According to physical laws, a person moving within the rotating environment also experiences Coriolis forces, which act perpendicularly to both the vector of rotation and the vector of the person's motion. This confuses the vestibular system and is likely to cause severe motion sickness, which, unfortunately, would probably last for the duration of the mission.

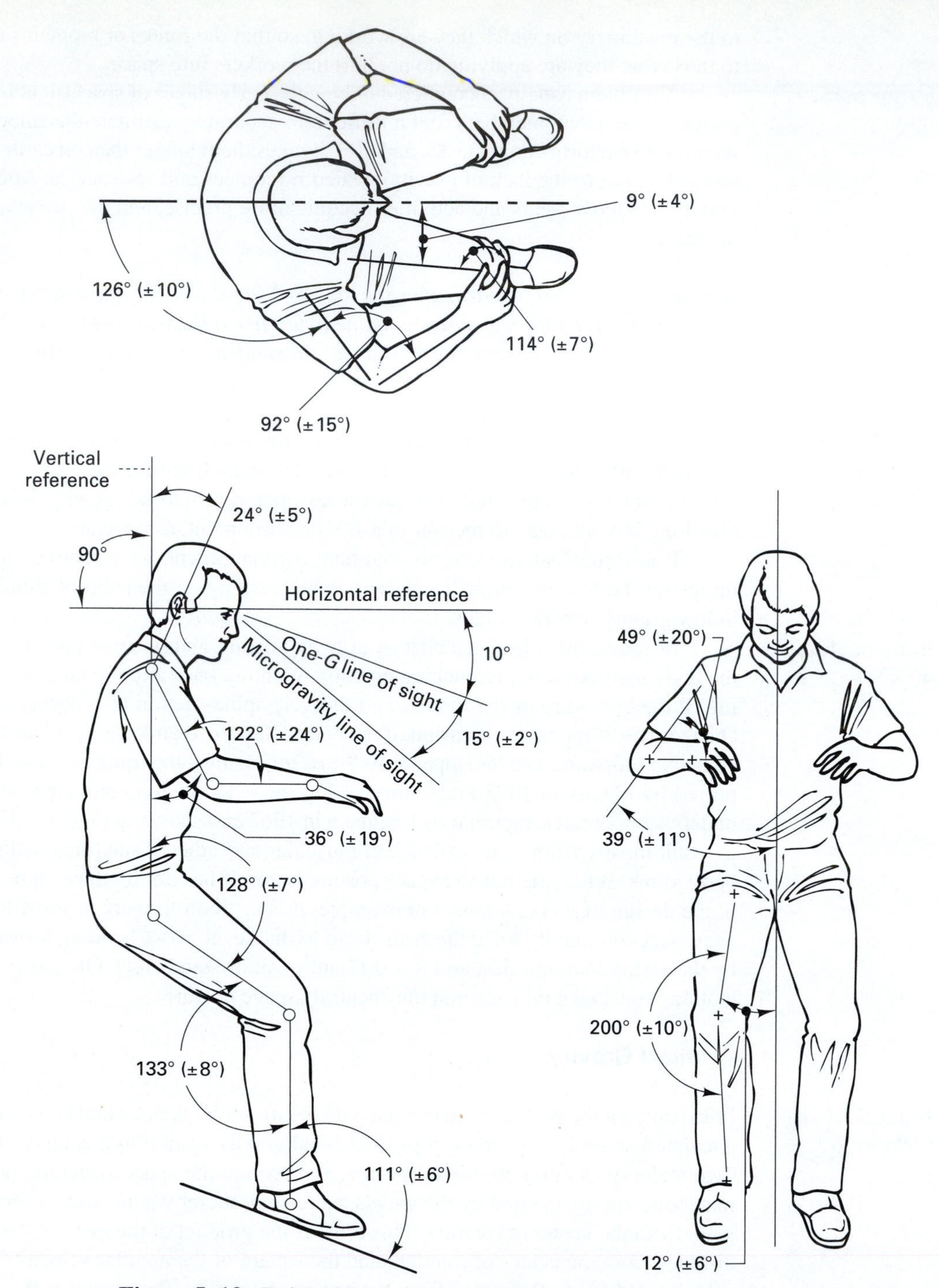

Figure 5–12. Relaxed posture assumed in space (NASA 1989).

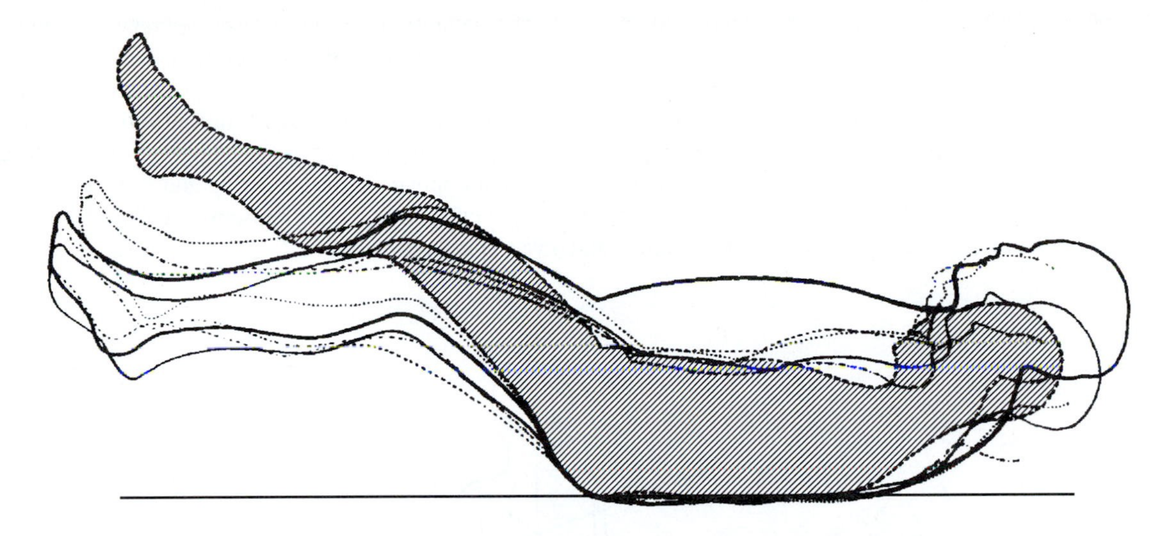

Figure 5–13. Relaxed postures assumed under water (with permission from Lehmann 1962).

☛☛☛ *In 1835, the French physicist Gaspard-Gustave de Coriolis (1792–1843) examined motion on a spinning surface. An object on the Earth's equator travels tangentially to it at about 1,600 km per hour. If the object also moves in a northern (or southern) direction away from the equator, it retains that speed, but the ground underneath moves more slowly—in fact at zero speed on the poles. The object thus gains on the ground and therefore, in effect, curves off eastward. Similarly, an object moving south (or north) toward the equator, in effect, curves westward. This curving motion is called the* Coriolis effect *and explains the rotation of water and air currents above the earth surface, which is in opposite direction in the southern and northern hemispheres. The Coriolis effect is noticeable, for example, in the directions of rotation of high- and low-pressure weather systems and must be considered in the calculation of the path of artillery fire or the launching of a satellite (Asimov 1989).* ☚☚☚

Linear acceleration

Another, better solution to the problem of artificial gravity is to generate a linear-acceleration environment. For example, on a trip to Mars, the spaceship could be linearly accelerated in the desired direction for half the distance. Then, it would be turned around and, through the use of the same accelerating engines, be decelerated until it arrived at Mars with zero speed. The condition of constant linear acceleration would avoid all problems associated with a rotating environment. Unfortunately, present technology provides neither engines nor fuel to generate the required linear acceleration over the necessary long period. Technological progress may eventually solve this problem.

ACCELERATIONS IN AEROSPACE

The pilot of a high-performance aircraft or a space crew departing from earth or arriving at earth or some other celestial body is usually subjected to linear or rotational acceleration. The acceleration may be sustained for some time, or it can be of the impact variety—that is, with a quick onset and cessation.

Another direction convention

In aerospace work, the direction of acceleration is usually described relative to the direction in which the eye (or another body organ) is displaced by the acceleration. This system has a positive acceleration in the x direction from back to chest (eyeballs in), a positive acceleration in the y-direction from left to right (eyeballs left), and a positive acceleration in the z-direction from head to toe (eyeballs up). The system is depicted in Figure 5–14 (cf. Figure 5–3) and described in Table 5–15. There is a special convention of opposite signs for acting and resultant accelerations in the z-axis. A similarly descriptive system exists for angular motions: A roll (cartwheel) is about x, a pitch (somersault) about y, and a yaw (pirouette) about z.

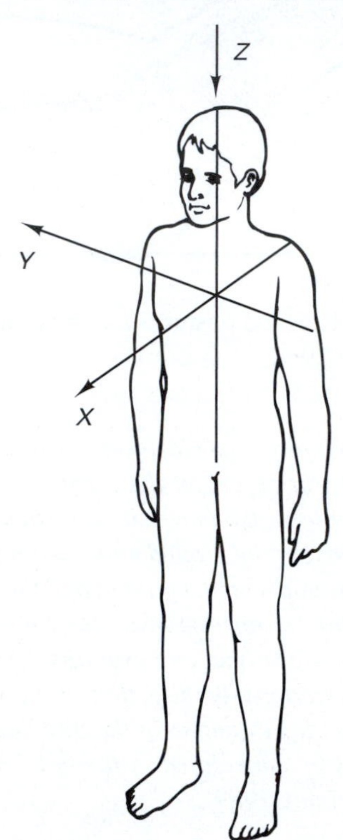

Figure 5–14. Aerospace convention of acceleration directions (NASA, 1989). Compare with Figure 5–3.

TABLE 5–15. Aerospace Taxonomy

	Direction of Acting Acceleration		*Inertial Resultant of the Acceleration of the Body*	
Linear Motion	*Action*	*Description of Acceleration*	*Reaction*	*Vernacular Description*
Forward	$+a_x$	Forward	$+G_x$	Eyeballs in
Backward	$-a_x$	Backward	$-G_x$	Eyeballs out
To right	$+a_y$	Right lateral	$+G_y$	Eyeballs left
To left	$-a_y$	Left lateral	$-G_y$	Eyeballs right
Upward	$-a_z$	Headward	$+G_z$	Eyeballs down
Downward	$+a_z$	Footward	$-G_z$	Eyeballs up

Source: Adapted from NASA 1989.

Impacts

Fairly high impact accelerations may be encountered as a result of thrusters firing, seats ejecting, capsules firing, and escape devices deploying. Flight instability, air turbulence, and crash landings also may result in such high-impact accelerations, some of whose G values are as follows:

- Violent maneuvers—up to 6G, omnidirectional;
- Parachute opening shock—approximately $10G_z$;
- Ejection firings—up to $17G_z$;
- Crash landings—from 10 to more than 100G, omnidirectional.

G levels

For all practical purposes, on earth there is a constant 1G level, pointing toward the center of earth. Within a spacecraft built with current technology, up to $+2G_x$ may be experienced during the separation of a stage, but during launch, entry, and abort operations, up to $+6G_x$ can exist. In transorbital flight, sustained accelerations are very low, such as 10^{-6}G, omnidirectional. During orbital maneuvers, angular accelerations up to plus or minus 1.5° s^{-2} are possible in all directions.

Effects of *G*

Depending on the magnitude and direction of a given acceleration, the effects may be imperceptible and normal, they may generate discomfort and impairment, or they may be dangerous or even fatal. During $+g$ acceleration, one experiences increased weight. (According to Newton's second law, $f = mg$.) Drooping of skin and body tissue occur at $+2G_z$, inability to raise body parts and dimming of vision at $+3G_z$, and blackout and loss of consciousness at 5 to $6G_z$. Downward acceleration effects include facial congestion and reddening of vision at $-3G_z$. Few persons can tolerate $-5G_z$ for more than 5 s.

g-LOC

Flying a fast airplane, such as a fighter plane, on a path other than horizontal and straight can cause $+g$-induced loss of consciousness, called g-LOC for short. This has happened to nearly one out of three fighter pilots (Gawron 1997). The cause is a reduced oxygen supply to the brain, which is in turn caused by the inability of the heart to generate a sufficiently high pressure in the bloodstream to the head. Blacking out occurs when the needed blood pressure of at least 22 mm Hg at the brain cannot be maintained because the flying maneuver generates a large opposing g_z-force on the blood and all other body components. Some oxygen is stored in the brain, so the pilot can go to very high g_z-levels lasting no more than 5 s without suffering from *g*-LOC. An early sign of impending *g*-LOC is tunnel vision. (But no visual warning symptoms appear with a steep *g* onset). Then unconsciousness sets in, usually lasting about 15 s, followed by amnesia for another 10 s. Of course, during *g*-LOC, the pilot is not able to perform any tasks. After regaining consciousness, physiological impairments follow and performance is severely degraded. The onset of *g*-LOC, its duration, and how soon one recovers from it depend on the magnitude and onset rate of g_z, the time spent at the g_z offset rate, and the number of earlier exposures.

Countering *g*-LOC

Of course, the principal way to counteract *g*-LOC is to avoid severe flight maneuvers, but that may not be possible, especially for fighter pilots. Another solution would be to fly the plane automatically while the pilot is incapacitated, but this endangers the person. Or one can select individual pilots who have a higher than common $+g_z$ tolerance; one can train them to expect, recognize, and act upon early symptoms. Anti-*g* straining maneuvers (there are several AGSMs; see Winnery and Murray 1990) include a pilot-induced increase in arterial blood pressure as a primary tactic; such an increase can also be achieved by positive pressure breathing and by using an anti-*g* suit (Goodman et al. 1995; Nunnely et al. 1995). A design solution is to change the direction of the $+g_z$ vector relative to the pilot by placing the person in a different body position, at least during the severe flight path; this is often done by reclining the back of the seat and elevating the pedals.

THE HUMAN CAN OVERCOME BAD DESIGN—BUT SHOULDN'T HAVE TO!

In the mid-1970s, some high-performance U.S. fighter aircraft were equipped with reclining seats that would support the pilot's body in a half-supine position during high-acceleration maneuvers, such as tight turns. Yet, experience showed that declining the body 35° or 40° backwards is often not effective and in fact can slightly decrease instead of improve the tolerance to high G forces with sharp onsets. While this experience is in contrast to theoretical considerations, it had already been reported as the result of experiments performed during World War II in Germany and Canada. Since a completely supine position (which would effectively counteract "grayout" and blackout) could not be assumed in these aircraft, the temporary recommendation was to sit upright, and even slightly crouched forward, on the seat originally designed for reclined sitting, in preparation for and during high-G maneuvers (Wood et al. 1990).

Further information about linear and rotational accelerations is given in the preceding section on vibration and is contained in the newest edition of NASA Standard 3000, which includes the guidelines for designing the aerospace equipment shown in Table 5–16.

TABLE 5–16. Impact Design Limits for Accelerations Lasting Up to 1 S

Direction of Impact Acceleration	*Impact Limit*	*Rate of Impact*
$\pm G_x$	20 G	10,000 G/sec
$\pm G_y$	20 G	1,000 G/sec
$\pm G_z$	15 G	500 G/sec
45° off axis (any axis)	20 G	1,000 G/sec

Source: NASA 1989.

SUMMARY OF AEROSPACE HUMAN ENGINEERING

The technology to construct and launch space vehicles exists, but protection of the crew from solar and cosmic radiation is still an unsolved problem. There are also slow, lingering, and dangerous effects of the lack of gravity in space: the body's musculoskeletal, vascular, cardiac, and sensory systems suffer in substance and performance. This, in turn, is detrimental to health and the ability to perform tasks both in space and on the ground, either upon returning to earth or upon landing on another celestial body. The generation of artificial gravity in the space vehicle would alleviate these problems.

WORKING AND DIVING UNDER WATER

Functioning under water presents excitement and challenges to humans; physically, physiologically, and emotionally. The ergonomic issues fall into two main groups: challenges to the senses and effects of water pressure.

Sensory Inputs and Perception

Sight

Water absorbs energy predominantly in the red-orange end of the visible spectrum, so objects under water appear mostly in shades of blue green. Also, light is dispersed by suspended particulates. The combination of reduced radiant energy and color contrast leads to decreased visual acuity.

Human eye tissue tolerates high external water pressure poorly, so lenses have been developed that can be worn under water, but whose frame often severely limits peripheral vision. This eye gear creates an air–water interface, through which light rays pass and change their trajectory. The refraction creates distortions in the perception of the size and distance of underwater objects.

Sound

Compared with transmission in air, the speed of sound and the distance over which it is conveyed are markedly increased in water, owing to its higher density than that of air. One's ability to perceive the direction of sound depends on one ear (the more distant from the origin) receiving the sound waves fractions of a second later than the other (closer) ear. The human auditory system cannot, however, easily make this distinction under water because of the increased speed of the waves. The combination of relatively amplified and seemingly directionless sound can be quite disconcerting to the visually hampered novice diver, who may fear injury from the propeller of a boat that is actually far away.

Taste and Smell

Except for the taste of aging rubber from the mouthpiece of the breathing apparatus and, perhaps, the "smell of fear," the senses of taste and smell are effectively functionless in the underwater world.

Touch

The relative cold of the underwater environment tends to blunt the senses of fine touch and, to a lesser extent, pain. Protective or warming gloves can reduce the loss significantly. The combination of cold and gloves decreases fine motor coordination.

Spatial Orientation

We orient ourselves in space using internal cues (e.g., intracranial and intravascular pressures), the pull of gravity, and external cues such as the sky and the normal positions of objects in our environment. In deep water, much pressure is applied to the whole body of a diver, the effects of gravity are minimized, the sky may not be visible, and surrounding objects may not be present or noticeable. Yet, the diver quickly learns to use one cue that is present underwater, but not on land: bubbles of exhaled air, which always travel up.

Effects of Water Pressure

Most of us can stay underwater by holding our breath for less than a minute, but people who dive regularly for sponges and pearls are able to stay underneath for a couple of minutes longer. This time limit is set by the need to breathe. A snorkel allows us to get air if we remain very close to the surface, but it does not work at any greater water depth, because of the increased water pressure.

The body under water

To understand the basic physics and physiology of the human being under water, we can model the body (for the moment disregarding solid-bone structures) as compressible tissues permeated with cavities that are filled with watery fluids or, as in the middle ear and lungs, with air. Water is not compressible, but the gas is.

At sea level, the human body is under "one atmosphere" of air pressure (1 atm = 10.13 N/cm^2 = 101.3 kPa = 14.7 psi). With each 10 m (33 ft) of depth in saltwater (about 34 ft in freshwater), the body is subjected to an additional atmosphere of external pressure.

This compresses its few gas-filled spaces, such as the lungs. Breathing from an air supply becomes impossible with depth, unless the air is also under pressure.

Gas expansion with ascent

After having stayed at depth, and with the body's air volumes compressed accordingly, upon ascent, the gas expands. Excess gas volume in the human gastrointestinal tract can be expelled by belching or flatulence, but air in the lungs must be constantly exhaled on ascent, or an embolism of air can rupture the lungs and cause death.

Increased pressure can have other serious effects on human physiology as well. Oxygen and nitrogen are dissolved in body fluid in proportion to the atmospheric pressure and to the composition of the air. At extreme pressures in the ocean depths, air can actually be toxic to tissues, by virtue of the high oxygen levels dissolved therein. Deep-sea divers therefore must breathe a modified gas mixture to compensate for the potential toxicity.

Nitrogen makes up about 78% of air. Increased ambient pressures cause more nitrogen to be dissolved in human tissues. High brain concentrations of nitrogen lead to impaired cognition and a clouded sensorium, a condition known as "nitrogen narcosis." Because nitrogen does not dissolve easily in body fluids, this condition takes time to develop and is unlikely to occur with the depth–time combinations achieved by recreational divers.

Because of its low affinity for dissolution in body fluids, nitrogen rapidly exits tissues as the pressure decreases with one's ascent from depth. But if the amount of dissolved nitrogen or the speed of ascent exceeds the ability of the body to exhale the nitrogen, bubbles of the gas form in organs, causing tissue damage; in joints, causing pain and destruction of tissue; and in the blood, causing brain ischemia.

Caisson

Staying and working under water, to build quays, bridges, and tunnels, for example, was facilitated in the early 1800s by the use of an air pump to fill a diver's helmet or a multiperson chamber called a caisson. Laborers enter a caisson through an air lock to work in compressed air that allows breathing and prevents flooding. But when improperly decompressed to one atmosphere, they often developed joint pain and more serious problems, including numbness and paralysis, and even death in some cases.

☛☛☛ *The name "bends" was applied to decompression sickness in the early 1870s: Caisson workers building the St. Louis Bridge walked with a stoop, which fashionable women then employed as the "Grecian Bend." Of the approximately 600 caisson employees in St. Louis, 119, including the workers' physician, were seriously affected, and 14 died. At about the same time, the Brooklyn Bridge was constructed; among the caisson laborers there, 20 men died, and many had lasting health problems. Even the chief engineer, Mr. Roebling, became paralyzed and had to supervise the building of the bridge from his bed. Today, about 900 cases of decompression illness are reported annually among recreational divers. (Moon et al. 1995).* ☚☚☚

Decompression illness

Not only caisson workers and divers are at risk, but so are pilots and astronauts—all who experience quick pressure drops in their environment. The general term *decompression illness* applies to two ailments loosely related to the development of gas bubbles: arterial gas embolism and decompression sickness (Bennett and Elliott 1993; Edmonds et al. 1994).

Arterial gas embolism

When a diver holds the breath during a rapid ascent, when the water pressure is quickly decreasing, the obstructed airway prevents the escape of gas that is expanding within the lungs and bronchial airways. The expanding air can rupture organs and enter the blood. Air bubbles are then carried with the blood flow into the arterial circulation, often to the brain. Possible results are a sudden loss of consciousness, convulsions, and paralysis.

Decompression sickness

For deep dives, an inert gas (usually nitrogen or helium) is put into divers' air tanks together with the oxygen they need for breathing. At a depth at which the pressure is increased, the gas dissolves from the lungs into the blood, from where it diffuses into tissues, especially in the spinal cord and the brain, which are well supplied with blood. Given enough time, the gas taken up by the tissues during the dive is washed out by the blood, taken to the lungs, and exhaled. However, when a diver surfaces too quickly, the pressure of the dissolved gas in the tissues exceeds ambient pressure, and bubbles can form. The various combinations of intra- and extravascular gas bubbles can lead to limb pain, coughing, shortness of breath (called "chokes"), numbness, and paralysis.

Ergonomic solutions

Knowledge of the underlying physics and physiologic mechanisms allows human-engineered equipment to supply suitable breathing air and proper diving procedures, especially for descent and ascent. A self-contained underwater breathing apparatus (SCUBA) filled with compressed air is widely used by recreational divers to depths less than 40 m. With this gear, they can stay under water for up to 10 minutes without special control of their descent or ascent. To go lower, much more complicated equipment and strictly controlled procedures are necessary: With today's technology, a dive 90 m deep requires a number of separate tanks with their own breathing rigs (called regulators) for air, oxygen, helium, and nitrogen, as well as a carefully observed regimen for the diver's descent, stay under water, and ascent. A dozen or so decompression stops, each lasting several minutes, can make for a long time under water: With current equipment, diving down to 70 m takes at least 2½ hours, of which only half an hour is spent at the desired depth.

The design of proper diving equipment itself is a challenging ergonomic task. Of course, the apparatus must function reliably under great water pressure, but it must also be safely and quickly usable (1) under conditions of limited visibility and reduced mobility due to underwater protective clothing, (2) in coldness, (3) in a confined space when one works inside a shipwreck or when one is caving, and (4) when a diver fears something in the water or is otherwise in danger. The procedures employed in training novice divers are critical and must be carefully designed and executed.

SUMMARY

The human body interacts with its environment. Among the features of the earth environment, the climate is characterized by temperature, humidity, and air movement. Exchanges of heat energy between the environment and the body follow the physical processes of radiation, convection, conduction, and evaporation. Their effectiveness depends on several factors, including the clothing worn, the energy content of the work conducted, and the exposure time. The temperature difference between exposed skin and the environment is very important, but humidity and airflow also play major roles. Within reasonable limits, the human can function in both hot and cold environments, given suitable job demands and clothing.

Air pollution is a detriment not only to health, but also to the ability to perform physically demanding work, particularly if carbon monoxide is present. Further studies may show the deleterious effects of other pollutants.

A person's arrival at a higher altitude after having lived near sea level influences his or her ability to perform physical work in various ways. Up to about 1,500 meters, the ability to perform aerobic work is not much affected; but at higher altitudes, both the short-term maximal capacity and the long-term submaximal ability are diminished, and they remain reduced for the entire stay at a high altitude.

The effects of vibration on the human depend very much on the amplitude, frequency, direction, and point of application of the vibration. Vibration in one direction—for instance, from foot to head—can bring about responses in other directions, such as head nodding. Whole-body vibration can have consequences quite different from the vibration of only body parts, such as the hands. Body posture and body restraint systems can greatly affect the results of vibrations. Some information is available about suitable ergonomic measures, depending on circumstances. However, more research and better modeling are needed.

Within the last few decades, flying airplanes at very high speed has become feasible. Effects of sharp accelerations upon the supply of oxygen to the brain can lead to unconsciousness and can be controlled by ergonomic means.

Space is an environment that is new to humans. So far, its exposure effects have been experienced by just a handful of persons, most over periods of days and some over months. The general finding is that microgravity can diminish the functioning of the body in many ways, particularly regarding the musculo-skeletal and circulatory systems. Countermeasures to avoid health and performance problems, especially upon re-entry into gravity, are being developed. The design of equipment, tasks, and work environments suitable for long-term space missions is still a great challenge.

Deep-sea diving and working under deep water can result in great danger to humans, but the several kinds of decompression illness can be avoided by proper use of equipment and an ascent regimen.

CHALLENGES

Consider the effects, in theory and practice, of the concept of either "constant core temperature" or "average skin temperature."

Why is it difficult to control heat transfer through the head?

The energy-balance equation given in this chapter does not consider the time domain. What are the consequences of that omission?

While it is true that the temperature of air between a radiating surface and the body does not affect the energy transferred by radiation, it should have an affect on energy transfer by convection. How?

Why is there no energy transfer between the human body and the environment through condensation?

Why is it more important to avoid overheating and undercooling of the body's core than of the shell?

Is it conceivable that procedures other than the current subjective ones will be used to establish indices of "effective climate"?

Which engineering means exist to control the environmental climate in (a) offices, (b) workshops, such as a machine shop or a foundry, and (c) outdoors?

Are sensations of feeling hot or of feeling cold reliable indicators of climate strain?

Is exercising a practical means of acclimatizing oneself to working in heat, in cold, or at a high altitude?

Through which body parts is vibration most likely transmitted?

Is it reasonable to assume that vibration is transmitted to the body either only horizontally or only vertically, as is done in most research?

How would the descriptors of vibration change if the acting vibration were not sinusoidal?

What professionals other than truck drivers are particularly exposed to vibrations or impacts?

What might be the effects of several sources of vibration arriving simultaneously at different body parts?

Could one imagine that, under certain circumstances, vibrations on the job might be helpful in the performance of that job?

What explanations other than "conflicting CNS information" might be applied to the problem of motion sickness?

What factors are likely to affect the likelihood of withstanding impacts?

Which body postures and physical behaviors might be suitable to combat the effects of sustained strong accelerations?

Why is it difficult to make crash dummies anthropomorphic?

What are some reasons for or against sending humans into space?

Discuss the aspects of "systems engineering" associated with space engineering.

Consider details that make a confined living (or working) space "habitable."

What means are there for astronauts to exercise in space?

Chapter 6

Body Rhythms, Work Schedules, and Effects of Alcohol

OVERVIEW

The human body changes its physiological functions throughout the day and night. During waking hours the body is prepared for physical work, while during the night sleep is normal. Human attitudes and behaviors also change rhythmically during the day. Work should be arranged to least disturb physiological, psychological, and behavioral rhythms in order to avoid negative health and social effects and to maintain one's performance. Alcohol usually reduces work performance.

INTRODUCTION

The human body follows a set of daily fluctuations, called *circadian rhythms* (from the Latin *circa,* "about," and *dies,* "the day," or *diurnal* rhythms, from the Latin *diurnus,* "of the day"). These rhythms, regular physiological events that are observable, for example, in body temperature, heart rate, blood pressure, and hormone excretion, are systems of temporal programs within the human organism. Each is thought to be controlled within the body by a self-sustained pacemaker or internal clock, which runs on a daily cycle. Several rhythms, such as core temperature, blood pressure, and sleepiness, are coupled to each other, entrained and synchronized by time markers, often called *zeitgeber* (from the German *Zeit,* "time," and *geber,* "giver"). Among the zeitgebers are daylight and darkness, true clocks, and temporally established activities such as work tasks, office hours, mealtimes, etc. Human social behavior (the inclination to do certain activities, as well as to rest and sleep) both follows and reinforces biological rhythms.

The female menstrual cycle is a well-documented chronobiological rhythm. Seasonal mood changes have been noted for hundreds of years. Other cycles, such as a hypothetical dependency phase on the moon, are mythical.

☛☛☛ *So-called biorhythms were a fad a few decades ago; they were said to be regular waves of physiological and psychological events, starting at birth, but running in different phases and wavelengths. Whenever "positive" sections of any of these rhythmic phenomena coincided, a person was believed to be able to perform exceptionally well. In contrast, if "negative" phases concurred, the person was supposedly doing badly. Research showed that these rhythms were myths or artifacts.*

FEMALE MENSTRUAL CYCLE

The female menstrual cycle is regulated through synchronization of the activities of the hypothalamus, pituitary, and ovary. The 28-day period is usually divided into five phases:

1. preovulatory or follicular,
2. ovulatory,
3. postovulatory or luteal,
4. premenstrual, and
5. menstrual.

The main hormonal changes occur in the release of estrogen and progesterone around the 21st day of the cycle; estrogen shows a second peak at ovulation. Hormonal release is low during the premenstrual phase.

Hormones and behavior

It has been commonly assumed that hormonal changes during the menstrual cycle have profound effects on a woman's psychological and physiological states. However, after reviewing the existing scientific work, Patkai (1985) concluded that whatever correlation there was between certain behavioral or physiological events and the menstrual phase was weak and did not indicate causality.

Performance during the menstrual cycle

Most of the existing research on the subject relies on self-reported changes in mood and physical complaints in the course of the menstrual cycle. Fairly little information is available on changes in arousal and on objective measures of performance. While there is neurophysiological evidence that estrogen and progesterone affect brain function, these two hormones have antagonistic effects on the central nervous system, with estrogen stimulatory and progesterone inhibitory. Varying hormone production during the menstrual cycle may affect the capacity to perform certain tasks, but the extent to which hormones actually determine a person's performance depends on how a decrease in total capacity may be offset by increased effort. Patkai cites a study in which secretaries showed the highest typing speed before the onset of menstruation and during the first three of the cycle. The suggestion that they were making a stronger effort on those days was rejected by the secretaries, since they considered themselves to be working at full capacity all the time.

The occurrence of negative moods and physical complaints in the majority of women before and during menstruation is fairly well established, but the precise nature of this so-called premenstrual syndrome is not yet determined. While there is evidence that menstruation can bring about behavioral changes, they can be mediated by social and psychological factors. Medication can often counteract discomfort and a negative mood.

CIRCADIAN RHYTHMS

Cycle length of 24 hours

Human health is maintained by a suitable balance among physiological variables in spite of external disturbances. This state of balanced control is called *homeostasis.* However, a close look at this supposedly "steady state" of the body reveals that many physiological functions are in fact not constant, but show rhythmic variations. Rhythms with a cycle length of 24 hours are called *cir-*

cadian (or *diurnal*) rhythms, those which oscillate faster than once every 24 hours are called *ultradian,* and those which repeat less frequently are known as *infradian.*

Day high, night low?

Among the circadian rhythms, the best-known physiological variables are body temperature, heart rate, blood pressure, and the excretion of potassium. (See Figure 6–1.) Most of these variables show a high value during the day and lower values during the night, but hormones in the blood tend to be more concentrated at night, particularly in the early morning hours. The amount by which the variables change over a day and the times at which extremes occur are quite different among individuals and even within the same person.

Self-governed diurnal rhythms

One way to observe the diurnal rhythms and assess their effects on performance is simply to follow the activities of a person. During the daytime one is normally expected to be awake, active, and eating, while at night the human sleeps and fasts. Physiological events do not exactly follow that general pattern. For example, body core temperature falls even after the person has been sleeping for several hours and is usually lowest between 3 and 5 A.M. The temperature then begins to rise, more quickly upon awakening. It continues to increase, with some variations, until late

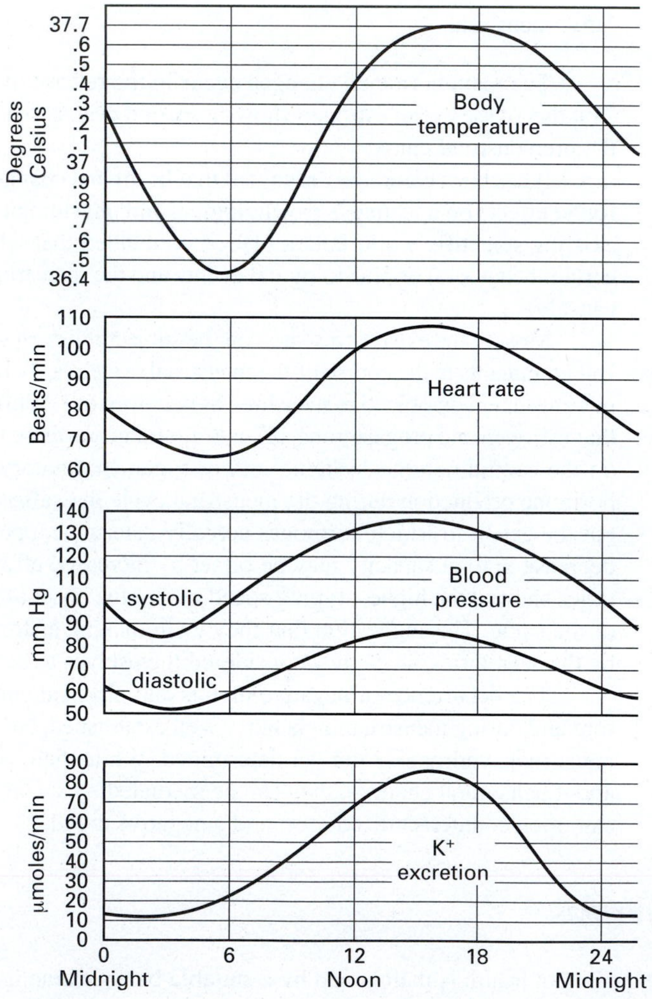

Figure 6–1. Typical variations in body functions over the day (adapted from Colligan and Tepas 1986).

in the afternoon. Thus, body temperature is not a passive response to our regular daily behavior, such as getting up, eating meals, performing work, and doing other social activities, but is self-governed.

Circadian pacemaker in the SCN

In the human being, the circadian pacemaker is located in the suprachiasmatic nucleus (SCN) of the hypothalamus. Under constant living conditions, the underlying physiological rhythms of the body are solid, self-regulated, and remain in existence even if one's daily activities change.

Masking

Yet, variations in observed rhythmic events (due to exogenous influences) may mask internal regular fluctuations. For example, skin temperature (particularly at the extremities) increases with the onset of sleep, regardless of what time it occurs, and turning the lights on increases the activity level of caged birds at any time. Thus, a person's observed skin temperature or activity level does not necessarily indicate the individual's internal rhythm, but may in fact mask it.

External activities and internal events coincide

Still, under regular circumstances, signs of external activity and internal events do tend to coincide. During the night, the low values of physiological functions—for example, core temperature and heart rate—are due primarily to the diurnal rhythm of the body; however, they are further aided by nighttime inactivity and fasting. During the day, peak activity usually coincides with high values of the internal functions. Thus, normally, the observed diurnal rhythm is the result of internal (endogenous) and external (exogenous) events that concur. If this balance of concurrent events is disturbed, consequences in health or performance may appear.

The number of deaths due to cardiac arrest is highest during the morning hours, as is the number of human births.

Running free

When a person is completely isolated from external factors (zeitgebers), including regular activities, the internal body rhythms "run free." This means that the circadian rhythms are free from external time cues and are only internally controlled. Experiments have consistently shown that circadian rhythms persist when running free, but their periods are slightly different from the regular 24-hour duration. Most rhythms run freely at about 25 hours, and some take longer.

Desynchronization and entrainment

If rhythms run at different phases, they are desynchronized. However, if a person is put again under regular (such as 24-hour) time markers and activities, the internal rhythms resume their coordinated steady cycles. The strongest entraining factor appears to be the light–dark sequence (Aschoff 1981): Very bright lights and sunshine are effective in advancing or retarding the diurnal rhythm (Czeisler et al. 1989; 1990a, b). Activity is also a strong enforcer (Turek 1989).

Providing new time markers and enforcing related new activity times can shift the body's internal rhythm. This happens when one travels to and remains in a new time zone. It takes most people from three days to a full week to adjust their circadian rhythms by five or six hours when they cross the Atlantic by airplane. The time to entrain a new diurnal rhythm depends on the individual, the magnitude of the time shift, and the intensity of the new time markers. Intensive lights at the appropriate time are strong entrainers.

The manipulation of zeitgebers allows the laboratory simulation of jet lag or shift work. Entraining or synchronizing the internal rhythms so that they follow periodic cues has been demonstrated at cycle durations between 23 and 27 hours. With shorter or longer periods of cues, the circadian rhythms are free running, though often not completely independent of the cues.

It appears to be a bit easier to set one's internal clock "forward," such as occurs in the spring, when daylight saving time is introduced, or during a west-to-east flight, such as from North America to Europe. However, a study (by Duchon et al. 1989) of shift work did not indicate a significant beneficial effect of forward compared to backward rotation.

INDIVIDUAL DIFFERENCES

Some people have consistently shorter (or longer) free-running periods than others. Females have, on the average, a free-running period that is about 30 minutes shorter than that of males.

Doves and owls?

There are at least three scales, based on questionnaires, that purport to assess one's diurnal type of "morningness" or "eveningness." However, evidence suggests that these properties may depend on situations or habits, that they may vary with culture, age, or the work schedule, and that the scales may have different reliability and validity (Greenwood 1991).

Aging

Circadian rhythms change with a person's age; rhythm amplitudes especially diminish with increasing age. This phenomenon is particularly obvious in the case of body temperature. In the elderly, there is also a pronounced shift toward morning activity. The oscillatory controls seem to lose some of their power with aging, which means that the person is more susceptible to rhythm disturbances.

Daily Performance Rhythms

Given the systematic changes in physiological functions during the day, one expects corresponding changes in mood and performance. Of course, attitudes and work habits are also, affected—and often strongly—by the daily organization of getting up, working, eating, relaxing, and going to bed. Experimentally, one can separate the effects of internal circadian rhythms and of external daily organization. But for practical purposes, one wants to look at the combined results of the internal and external factors as they affect performance.

Early in this century it was thought that the morning hours would be best for mental activities, with the afternoon more suitable for motoric work. Research has not supported this thesis, either for everybody or for all activities.

Postlunch dip "Fatigue" arising from work already performed was believed to reduce one's performance over the course of the day. The performance of simple mental work, such as recording numbers, showed a pronounced reduction not only between 2 and 5 A.M., but again early in the afternoon. This afternoon decline was labeled the "postlunch dip." However, the reduction in performance is not regularly paralleled by similar changes in physiological functions; for example, body temperature remains fairly constant during the afternoon. Hence, it was postulated that performance was affected by the interruption of activities by the noon meal and that subsequent digestion of the meal would lower the personal's psychological arousal level, bringing about an increased lassitude, together with increased blood glucose and changed blood distribution in the body. Accordingly, the postlunch dip appears to be caused by the exogenous masking effect of the food intake and activity break, rather than by endogenous circadian events.

In activities that require a medium to heavy amount of physical work, no such dip has been found (except when the food and beverage ingestion was very heavy and when true physiological fatigue had built up during the prelunch activities).

Many different activities can be performed during the day. They may follow a circadian rhythm or may differ because of external requirements. They may be physical or mental in nature. Personal interest, fatigue or boredom, and rewards or urgency usually have stronger effects on performance than diurnal rhythms have during daytime hours. For example, information processing in the brain (including immediate or short-term memory demands), mental arithmetic activities, and visual searches are strongly affected by personality, by the length of the activity, and by motivation.

Thus, it appears that one is not justified in making normative statements about variations in diurnal performance during regular working hours.

SLEEP

Two millennia ago, Aristotle thought that during wakefulness some substance ("warm vapors") in the brain built up that needed to be dissipated during sleep. In the19th century, there were two opposing schools of thought: that sleep was caused by congestion of the brain by blood and, contrariwise, that blood was drawn away from the brain.

Theories

Behavioral theories were common in the 19th century. According to some of those theories, sleep was the result of an absence of external stimulation, or sleep was not a passive response, but an activity to avoid fatigue from occurring. Early in the 20th century, it was thought that various sleep-inducing substances accumulated in the brain, an idea taken up again in the 1960s. During the 1930s and 1940s, various neural inhibition theories were discussed which included the assumption that the brain possessed sleep-inducing centers, in the form of a reticular structure, for its arousal.

More theories

Restorative theories about the function of sleep focus on various types of recovery from the wear and tear of wakefulness. Alternative theories claim that sleep is not restorative, but simply a form of instinct or diminution of behavior to occupy the unproductive hours of darkness; the relative immobility of the body during sleep can be a means of conserving energy. It appears that these three aspects of sleep function—restoration, occupying time, and energy conservation—all explain certain characteristics of sleep, but none of them does so completely or sufficiently.

Two central clocks?

Horne (1988), who discussed these points in detail, found it convenient to model the regulation of alertness, wakefulness, sleepiness, sleep, and many other physiological functions as being under the control of two "central clocks" of the body. One clock controls sleep and wakefulness, the other physiological functions, such as body temperature. Under normal conditions the internal clocks are linked together, so that body temperature and other physiological activities increase during wakefulness and decline during sleep. However, this congruence of the two rhythms may be disturbed, for instance, by night-shift work, in which one must be active during nighttime and sleep during the day. As such patterns continue, the physiological clocks adjust to the external requirements of the new sleep–wakefulness regimen. This means that the formerly well-established physiological rhythm flattens out and, within a period of about two weeks, reestablishes itself according to the new schedule.

Sleep Phases

Sleep is not homogeneous, but has several stages that repeat, more or less regularly, during the night. These are commonly observed and labeled according to brain and muscle activities. The brain and muscles show large changes from wakefulness to sleep, which can be observed fairly easily by electrical means (polysomnography).

EEG

Electrodes attached to the surface of the scalp pick up *electrical activities of the brain*—specifically, the cortex, which is also called the encephalon, because it wraps around the inner brain. Thus, the measuring technique is named electro-encephalography, or EEG. The signals can be described in terms of amplitude and frequency. Certain constituents of EEG waves that appear to be regularly associated with sleep characteristics have been labeled vertices, spindles, and complexes. The EEG amplitude is measured in microvolts, and it rises as consciousness falls from alert wakefulness through drowsiness, to deep sleep. EEG frequency is measured in hertz; the frequencies observed in human EEG range from 0.5 to 25 Hz. Sleep researchers call frequencies above 15 Hz fast waves, those under 3.5 Hz slow waves. Frequency falls as sleep deepens; slow-wave sleep (SWS) is of particular interest to sleep researchers.

Certain frequency bands have been given Greek letters. The main divisions are as follows:

1. Beta, above 15 Hz. These fast waves of low amplitude (under 10 microvolts) occur when the cerebrum is alert or even anxious.
2. Alpha, between 8 and 11 Hz. These frequencies appear during relaxed wakefulness, when little information is input to the eyes—particularly when the eyes are closed.
3. Theta, between 3.5 and 7.5 Hz. These frequencies are associated with drowsiness and light sleep.
4. Delta, under 3.5 Hz. These are slow waves of large amplitude, often over 100 microvolts, and occur more often as sleep becomes deeper.

EMG, REM

As discussed in Chapter 1, muscle activities can be recorded via surface electromyography, or EMG. In observing sleep, the electrical activity of the muscles that move the eyes (monitored via electrooculography, or EOG) and those in the chin and neck regions are of particular interest. After a person falls asleep, the sleep stages become progressively "deeper," indicated by more synchronous and less frequent brain activity. In deep sleep, the heart rate and respiration are slow, and muscles retain their tonus. Yet, in its deepest sleep phase, the brain becomes nearly as active as in wakefulness, the heart rate and breathing vary, dreams are frequent, and the eyes move rapidly under closed lids. Accordingly, this phase is called "rapid-eye-movement" (REM) sleep. The REM phase becomes longer and the light-sleep phases get shorter as the sleep-stage sequence repeats itself, about every 90 to 100 minutes throughout the night.

The importance given to EMG and EEG events by sleep researchers has changed over the last few decades. Currently, EOG outputs of the eye muscles are most often used as the main measure identifying REM and non-REM sleep stages. Non-REM sleep is further subdivided into four stages according to their associated EEG characteristics. (See Table 6–1.)

Value of sleep phases

Only the human brain assumes a physiological state during sleep that is unique to sleep and cannot be attained during wakefulness. While, for example, muscles can rest during relaxed wakefulness, the cerebrum remains in a condition of "quiet readiness," prepared to act on sensory input, without any diminution in responsiveness. Only during sleep do cerebral functions show marked increases in their thresholds of responsiveness to sensory input. In the deep-sleep stages associated with slow-wave non-REM sleep, the cerebrum apparently is functionally disconnected from

TABLE 6–1. Sleep Stages

Condition	*Muscle EMG*	*Brain EEG*	*Sleep stage*	*Average percent of total sleep time*
Awake	Active	Active, alpha and beta	0	—
Drowsy, transitional "light sleep"	Eyelids open and close, eyes roll	Theta, loss of alpha, vertex sharp waves	1, non-REM	5
"True" sleep		Theta, few delta, sleep spindles K-complexes	2, non-REM	45
Transitional "true" sleep		More delta SWS (< 3.5 Hz)	3, non-REM	7
Deep "true" sleep		Predominant delta SWS (< 3.5 Hz)	4, non-REM	13
Sleeping	Rapid eye movements, other muscles relaxed	Alert, much dreaming, alpha and beta	REM	30

SOURCE: Adapted from Horne 1988.

subcortical mechanisms. The brain needs sleep to restitute, a process that cannot take place sufficiently during waking relaxation (Horne 1985, 1988).

Five to six hours needed

It seems that, on average, the first five to six hours of regular sleep (which happen to contain most of the slow-wave non-REM sleep and at least half of the REM sleep) are required for a person to keep performing at normal psychological levels. Further sleep may be called facultative or optional, because it serves mostly to "occupy unproductive hours of darkness," with dreams being the "cinema of the mind" (Horne 1988, pp. 54, 313).

Napping after lunch Many people have taken up the habit of lying down after lunch for 10 or 15 minutes ("Nur ein Viertelstündchen" as the Germans say), often falling asleep briefly. In laboratory tests, this kind of common nap has been shown to have little effect on the performance of subsequent work, yet many people say they simply need that nap after lunch.

Sleep Loss and Tiredness

If a person does not get the usual amount of sleep, he or she gets tired, and the obvious remedy is to get more sleep. Figure 6–2 shows the effects of sleep loss on body temperature—the temperature is elevated, but keeps its phase.

Why do we need sleep?

As mentioned earlier, it is not entirely clear why humans (or animals, for that matter) need sleep. The traditional opinion is that sleep has recuperative benefits, allowing some sort of restitution or repair of brain tissue following the "wear and tear" of wakefulness. However, what is meant by these terms is usually neither clearly expressed nor fully understood. Certainly, sleep is accompanied by rest and thus by energy conservation; but a human being can attain similar relaxation while awake, when not forced to be active. Many experiments have failed to show any restorative physiological effects of sleep; in fact, even moderate sleep deprivation has little physiological effect, as discussed by Horne (1985, 1988). For example, sleep deprivation does not

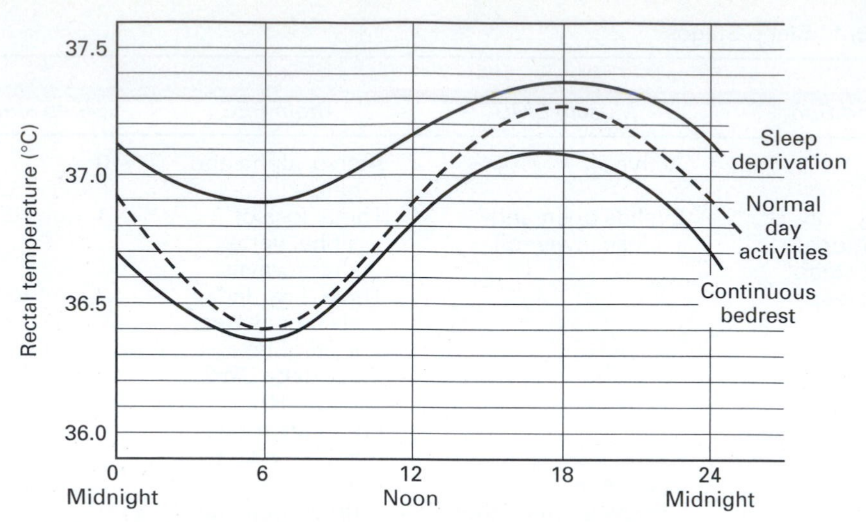

Figure 6–2. Changes in body temperature associated with normal activities, bed rest, and sleep deprivation (schematic from Colligan and Tepas 1986).

impair muscle restitution or the ability to perform physical work. Apparent reductions in the capability to do physical exercise related to sleep deprivation (such as those reported by Froeberg 1985) may be due mostly to reduced psychological motivation, rather than a decrease in physiological capabilities.

The effects of moderate sleep deprivation on body functions are not clear, but may be less consequential than often believed. In contrast, the restorative benefits of sleep to the brain are fairly well researched. Two or more nights of sleep deprivation diminish both a person's actual psychological performance and the motivation to perform (but apparently, not the inherent cognitive capacity to perform). Irritability, suspiciousness, and slurred speech are the most common side effects.

The brain needs sleep

However, while changes due to sleep deprivation indicate some impairment of the central nervous system (apparently, the brain needs to sleep), Horne states that the deficits are not as extensive as one might expect if a person needed eight hours of sleep per day for brain restitution. After up to two days of sleep deprivation, even though a person "feels tired," his or her mental performance is still rather normal on stimulating and motivating tasks; boring tasks, however, show a reduction in performance.

All task performance is reduced after more than two nights of sleep deprivation.

It is not surprising to learn that nighttime performance levels during long-term sleep deprivation are much lower than levels that are normal during the day, when the body and brain are usually awake. Harrowing tales are told by U.S. hospital interns and residents, many of whom are forced to routinely work 120-hour weeks, including 36 hours at a stretch. Some admit that mistakes are frighteningly common. A California resident fell asleep while sewing up a woman's uterus—and then toppled onto the patient. In another California case, a sleepy resident forgot to order a diabetic patient's nightly insulin shot and instead prescribed another medication. The man went into a coma. Compassion can also be a casualty: One young doctor admitted to abruptly cutting off the questions of a man who had just been told he had AIDS: "All I could think of was going home to bed." (Time, *December 17, 1990, p. 80).*

Normal Sleep Requirements

While there are, as usual, variations among individuals, certain age groups in the Western world show rather regular sleeping habits. Newborns sleep 16 to 18 hours a day, mostly in sets of a few hours' duration. Young adults sleep, on the average, 7.5 hours, with a standard deviation of about 1 hour. Some adults are well rested after 6 or 7 hours of sleep, or even less, while others habitually sleep 8 hours or more. The amount of slow-wave sleep in both short and long sleepers is about the same, but the amounts of REM and non-REM sleep periods differ considerably.

Many people who sleep for just a few hours per day are able to keep up their performance levels even if their total sleep time is shorter than normal. A common low limit seems to be around five hours of sleep per day, with even shorter periods still being somewhat beneficial.

PROLONGED HOURS OF WORK AND SLEEP DEPRIVATION

Long work, little sleep

There are conditions in which persons must work continuously for long periods, such as 24 hours or more. Not only are they then working without interruption, but also, they are deprived of sleep. Hence, any deleterious effects they experience are partly a function of the long working hours and partly a function of sleepiness.

What kinds of effects they are depends on the types of tasks performed, on the motivation of the worker, and even on timing, because wakefulness and sleepiness appear in cycles during the day. As a general rule, the performance of a task is influenced by three factors:

- the internal diurnal rhythm of the body,
- the external daily organization of work activities, and
- the individual's motivation and interest in the work.

Each factor can govern, influence, or mask the effects of the others on task performance.

The performance of different types of work is affected differently by long periods of work with concurrent sleep loss. Froeberg (1985) found the following relationships:

- A task that must be performed uninterruptedly for half an hour or longer will be more affected by sleep loss than a shorter work task will.
- If such a task must be replicated, the performance is likely to become worse with each successive repetition.
- The performance of monotonous tasks is highly diminished by sleep deprivation, whereas the performance of a task that is new to the operator is less affected.
- On the other hand, doing a complex task is affected more than doing a simple one.
- Work that is paced by the work itself deteriorates more with sleepiness than does an operator-paced task.
- A person's accuracy in performing a job may still be quite good even after losing sleep, but it takes longer to perform the job.
- A task which is interesting and appealing, even if it does include complex decision making, can be performed rather well even over long periods of time. But if the task is disliked and unappealing, decision making is prolonged.

In general, loss of sleep (particularly if it is associated with long periods of work, such as a full day) diminishes a person's performance of a task, prolonging the individual's reaction time, causing the person to fail to respond or produce a false response, slowing cognition, and disturbing short- and long-term memory capibilities. Most of the deterioration occurs when the circadian rhythms are set for a night's sleep. Performance is further diminished following two or three nights of sleep deprivation. After missing four nights, very few people are able to stay awake and to perform, even if their motivation is very high.

Thus, the performance of all mental tasks (except brief, interesting ones) diminishes with long hours of work associated with sleep loss. The longer the work period and the more monotonous, repetitive, uninteresting, and disliked the task, the more the performance degrades.

The coffee or tea pickup

Caffeine is a stimulant (of the chemical family called methylated xanthines) that is quickly absorbed into the bloodstream. For about half an hour after drinking a strong cup of coffee, most people feel more awake and better able to pay attention; the heart rate is increased by 2 to 10 beats per minute over a period of 5 to 15 minutes. Drinking 5 to 10 cups of strong coffee is likely to have an "overdosing" effect, generating a condition called caffeinism, whose symptoms include lightheadedness, tremor, headache, palpitation, and difficulty in falling asleep.

Caffeine is contained in coffee, tea, cocoa, and most chocolate products. It is added to many soft drinks and is a component of some medications for headache, cold, and allergy. Cocoa and chocolate products often contain theobromine, which has effects similar to those of caffeine on behavior and body functions. Theophylline, with analogous effects, is contained in tea. The amounts of caffeine and related substances are as follows:

In coffee, 60 to 150 mg per 250 cm^3 (cup), with instant coffee usually in the low range.

In tea, 8 to 50 mg per 250 cm^3 (cup), with instant tea usually in the middle range.

In cocoa, about 15 mg per 250 cm^3 (cup).

In soft drinks, between 40 and 70 mg per 355 cm^3 (12 fl. oz.).

In chocolate, about 12 mg caffeine and 120 mg theobromine per 50 g (2 oz, on average), while most baked chocolate goods have about half these amounts.

Sleep Deprivation and Recovery

During long working times after sleep deprivation of at least one night, short periods of reduced arousal or even of light sleep, known as *lapses* or *gaps*, occur. With increasing time at work, so-called *microsleeps* happen increasingly often: The person falls asleep for a few seconds, but these short periods (even if frequent) do not have much recuperative value, because one still feels sleepy and performance still degrades.

Gaps, microsleeps

Short sleeps called *naps* lasting one to two hours improve one's subsequent performance during long working spells after a sleepless night. (However, if a person is awakened from napping during a deep-sleep phase, *sleep inertia* with low performance can appear and may last up to 30 minutes.) The temporal placement of a nap may have differing effects: Naps of at least two hours taken in the late evening or at night, when the diurnal rhythm is low, have beneficial effects on subsequent performance lasting several hours, while a daytime nap may be of little value.

Napping

If long periods of mental work are unavoidable, one may try to intersperse the work with physical activities or exercises. Also, "white noise" (see Chapter 4) may improve performance slightly. Hot snacks are often welcome, as are hot and cold (usually caffeinated) beverages. Occasional "stirring" music may help. Bright illumination of 2,000 lux or more helps to suppress the production of the hormone melatonine, which causes drowsiness. Drugs, particularly amphetamines, can restore one's performance to nearly normal level, even when given after three nights without sleep* (Froeberg 1985).

While the authors of this book report scientific findings, they do not recommend the use of drugs to overcome the effects of unbecoming work schedules.

A good night's sleep

Recovery from sleep deprivation is quite fast. A full night of undisturbed sleep lasting several hours longer than usual restores one's efficiency almost fully.

SHIFT WORK

Definition

One speaks of shift work if two or more persons, or teams of persons, work in sequence at the same workplace. Often, each worker's shift is repeated in the same pattern over a number of days. For the individual, shift work means attending the same workplace either regularly at the same time (continuous shift work) or at varying times (discontinuous, including rotating, shift work).

Shift work is not new

In ancient Rome, by decree, deliveries were to be made at night to relieve street congestion. Bakers habitually have worked through the late night hours. Soldiers and firefighters, too, always have been accustomed to night shifts. With the advent of industrialization, long working days became common, with teams of workers relaying each other to maintain blast furnaces, rolling mills, glassworks, and at other workplaces where continuous operation was desired. The 24-hour period was covered with either two 12-hour works shifts or three 8-hour work shifts.

Workdays

Since the industrial revolution, when 12-hour shifts on six or seven days of the week were common, drastic changes in work schedules have occurred. In the first part of the 20th century, the then-common six-day workweeks with 10-hour shifts were shortened. Today, work schedules generally are 8 hours per day, five days a week, a practice introduced into many countries during the 1960s. Within less than a century, the number of days worked per week was reduced as was the number of hours worked per day, and two weekend days became free of work. Yet, working "overtime" has become a fairly regular feature in many jobs.

SHIFT SYSTEMS

Working in shifts is different from "normal" day work in either (or both) of two ways:

1. Work is performed regularly during times other than morning and afternoon.
2. At a given workplace, more than one shift is worked during the 24-hour day. A shift often lasts 8 hours but may be shorter or longer.

Though estimates vary, probably about one of four workers in developed countries is on some kind of shift schedule.

Features of shift systems

The main features of the many shift systems are shown in Figure 6–3. For convenience, they can be classified into several basic patterns, but any given shift system may contain aspects of several patterns. Kogi (1985) asked four particularly important questions regarding the features of shift systems:

1. Does a shift extend into hours that would normally be spent asleep?
2. Is the shift worked throughout the entire seven-day week, or does it include days of rest, such as a free weekend?
3. Into how many shifts are the daily work hours divided? Are there two, three, or more shifts per day?
4. Do the shift crews rotate, or do they work the same shifts permanently?

Other important features are as follows:

- The starting and ending time of a shift.
- The number of workdays in each week.
- The hours of work in each week.
- The number of shift teams.
- The number of free days per week or per rotation cycle.
- The number of consecutive days on the same shift, which may be a fixed or variable number.
- The schedule by which an individual either works or has a free day or free days.

All of these various aspects may affect the welfare of the shift worker, the performance of her or his work, and the work schedule of the organization.

Scheduling

The most common schedules contain either a permanent shift assignment or a weekly rotation schedule. Several examples of such schedules are shown in Table 6–2. In most systems

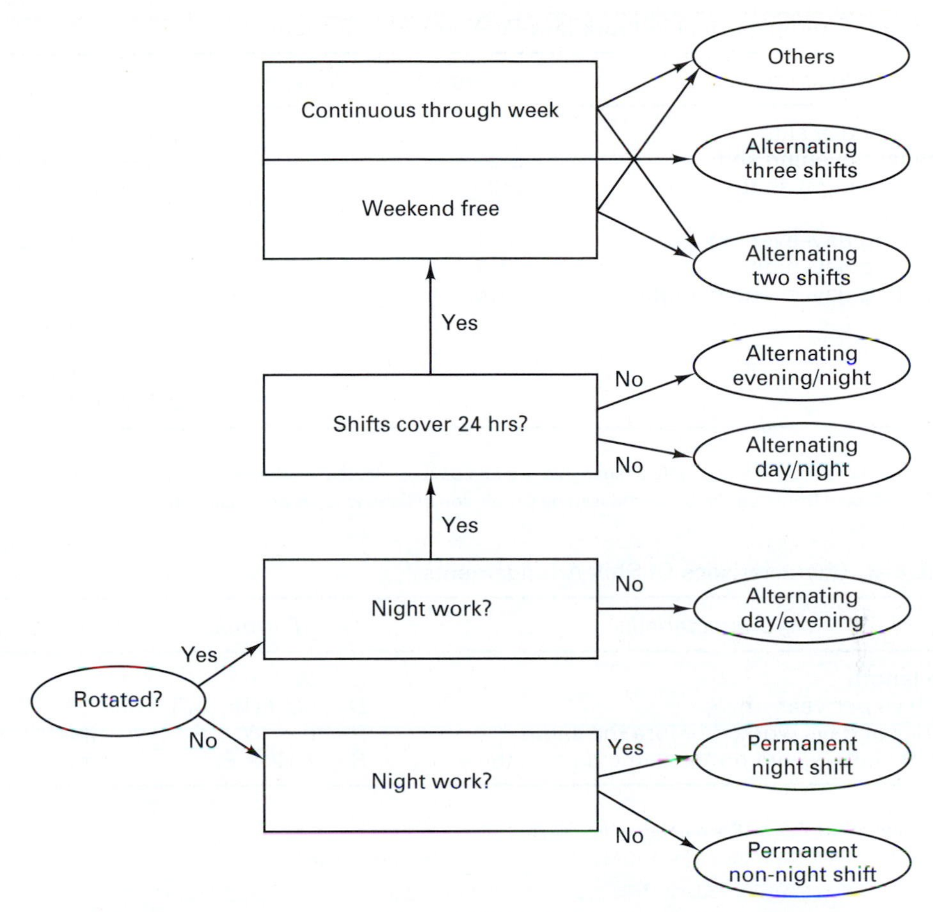

Figure 6–3. Flowchart of key features of shift systems. Note that other shift attributes are possible. *Source:* Adapted from K. Kogi, "Introduction to the Problems of Shiftwork" (1985) in S. Folkard and T. H. Monk (eds.), *Hours of Work.* Reproduced by permission of John Wiley and Sons, Limited.

used today, the same shift is worked for five days and is followed by two free weekend days. This regimen does not, however, evenly cover all the 168 hours of the week; additional crews are needed to work on weekends or other "odd" periods. If one uses three shifts a day with four teams, the shift system (for one team) is 1–1–2–2–3–3–0–0 with a 6:2 workday–free-day ratio and a cycle length of eight days; this is known as the *metropolitan rotation.* The *continental rotation,* which also has three shifts per day and four crews, has the sequence 1–1–2–2–3–3–3, 0–0–1–1–2–2–2, 3–3–0–0–1–1–1, 2–2–3–3–0–0–0; its workday–free-day ratio is 21:7, with a cycle length of exactly four weeks.

Many shift arrangements

The ratio of workdays versus free days in a complete cycle is an important characteristic of any shift system. Table 6–3 presents several features that describe different shift schedules. Obviously, a great number of arrangements are possible. Many are presented and discussed in the literature: see, for example, the publications by Colligan and Tepas 1986; Colquhoun 1985; Eastman Kodak Company 1986; Folkard and Monk 1985; Johnson et al. 1981; Kogi 1991; Knauth 1996; and Smith et al. 1998.

TABLE 6–2. Examples Of Five-Workdays-Per-Week Shift Systems

System	*Workdays: free days*	*Shift sequence*
Permanent day shift	5:2	1–1–1–1–1–0–0, 1–1–1–1–1–0–0, . . .
Permanent evening shift	5:2	2–2–2–2–2–0–0, 2–2–2–2–2–0–0, . . .
Permanent night shift	5:2	3–3–3–3–3–0–0, 3–3–3–3–3–0–0, . . .
Rotations		
Alternating day–evening	10:4	1–1–1–1–1–0–0, 2–2–2–2–2–0–0, . . .
Alternating day–night	10:4	1–1–1–1–1–0–0, 3–3–3–3–3–0–0, . . .
Alternating day–evening–night	15:6	1–1–1–1–1–0–0, 2–2–2–2–2–0–0, . . . 3–3–3–3–3–0–0 (forward rotation) . . . or 1–1–1–1–1–0–0, 3–3–3–3–3–0–0, . . . 2–2–2–2–2–0–0 (backward rotation) . . .

Legend

1 represents day shift, 2 evening shift, 3 night shift, 0 a free day, i.e., without a scheduled shift.

SOURCE: Adapted from Kogi 1985. Reproduced by permission of John Wiley & Sons, Limited.

TABLE 6–3. Characteristics Of Shift Arrangements

Characteristic	*Formula*	*Condition*
Cycle length	$C = W + F$	
Free days per year	$D = 365F(W + F)^{-1}$	
Number of days worked before the same set of shifts recurs on the same days of the week	$R = C = W + F$	if $(W + F)$ is a multiple of 7
	$R = 7(W + F)$	if $(W + F)$ is not a multiple of 7

Legend

W = number of work days, F = number of free days

SOURCE: Kogi 1985. Reproduced by permission of John Wiley & Sons, Limited.

COMPRESSED WORKWEEK AND EXTENDED WORKDAY

One speaks of a *compressed workweek* when the required hours of work per week (such as 40) are performed in only four or even three days, allowing the worker to have three or four free days each week. Apparently, this is an attractive idea to many persons: It reduces the number of trips to and from work, and there are fewer occasions on which one has to set up and close down shop at work. However, there are also concerns about increased fatigue due to long workdays and about reduced performance and safety because compressing the weekly working hours into a few days means extending the duration of each work shift.

Long shifts vs. long work

The type of work to be performed is a major determinant of whether long workdays can or should be done. Extended working days have been used on jobs where people remain on "stand-by" much of the shift, as firefighters do. Also, long shifts have been employed when only few or minimal physical efforts are required that, though diverse and interesting, fall into routines. Examples of such occupations are nursing, clerical work, administration, technical maintenance, computer supply operations, and the supervision of automated processes. While some manufacturing, assembly, machine operation, and other physically intensive jobs are sometimes performed during long shifts, whatever little information about them that exists has been gathered mostly from subjective statements of employees, some limited batteries of psychological tests, and the scrutiny of performance and safety records in industry.

Varied effects of long workdays

Data gathered on the effects of long workdays are contradictory, spotty, and apparently dependent on the given conditions. In some cases, production and performance are high shortly after the introduction of a compressed workweek, but fall off after prolonged periods on such a schedule (Ong and Kogi 1990). However, other observations have not shown this trend. The people involved often indicate significantly enhanced satisfaction, probably due more to easily and better arranged leisure time rather than to any perceived improvements carrying out the work itself (Dunham et al. 1987; Lateck and Foster 1985).

Working very long shifts, such as 12 hours, is likely to cause drowsiness, with concomitant reductions in cognitive abilities, motor skills, and, consequently, performance during the course of each shift and as the workweek progresses. A fatigued worker may take "careless shortcuts to [the] completion of a job," and work practices may be less safe in tasks "that are tedious because of high cognitive or information-processing demands, or [in tasks] with extensive repetition" (Rosa and Colligan 1988, p. 315). Table 6–4 lists some advantages and disadvantages of compressed workweeks and extended workdays.

TABLE 6–4. Advantages and Disadvantages of Compressed Workweeks/Extended Workdays

Advantages
Generally appealing
Increases possibilities for two or more consecutive days away from the job
Reduces commuting problems and costs
Affords more time per day for scheduling meetings or training sessions
Incurs fewer start-up and warm-up expenses
Increases production rates
Improves in the quantity or quality of services to the public
Offers better opportunities to hire skilled workers in tight labor markets
Disadvantages
Decreases job performance due to long work hours or to moonlighting on free days
Requires overtime pay
Tends to fatigue workers
Increases tardiness and early departure from work
Increases absenteeism
Increases employee turnover
Increases on-the-job and off-the-job accidents
Decreases production rates
Poses scheduling problems if the operations of the organization are longer than the workweek
Presents difficulty in scheduling child care and family life during the workweek
Runs counter to traditional objectives of labor unions
Increases energy and maintenance costs
Exposes workers to physical and chemical hazards for a longer period of time (acceptable doses are usually for an eight-hour exposure)

Adapted from Kogi (1991) and Tepas (1985).

FLEXTIME

A flexible arrangement of work hours during the day allows the employee to distribute the prescribed number of working hours per shift (for example, 8 hours) over a longer block of time, such as 10 or 12 hours. Often, a "core" time (of, say, 6 hours) when all workers must be present must be covered on every workday. One can "slide" or "float" the working time across the core time so that the start of work is anytime before the core and the end of work anytime after the core. Table 6–5 lists some advantages and disadvantages of flextime.

TABLE 6–5. Advantages and Disadvantages of Flextime

Advantages
Appeals to many workers
Does not result in any loss in base pay
Increases day-to-day free time
Reduces commuting problems and costs
Enables employers to adjust size of workforce to short-term fluctuations in demand
Is less fatiguing for workers
Reduces job dissatisfaction and increases job satisfaction
Increases democracy in the workforce
Recognizes and utilizes employees individual differences
Reduces tardiness
Reduces absenteeism
Reduces employee turnover
Increases production rates
Provides better opportunities to hire skilled workers in tight labor markets
Disadvantages
Necessitates irregular work hours produced by short-term changes in demand
Poses difficulty covering some jobs at all required times
Poses difficulty in scheduling meetings or training sessions
Leads to poorer communication within the organization
Leads to poorer communication with other organizations
Increases energy and maintenance costs
Increases buffer stock for assembly-line operations
Requires more sophisticated planning, organization, and control
Reduces quantity or quality of services to the public
Requires special recording of time
Requires additional supervisory personnel
Extends health and food service hours

SOURCE: Adapted from Kogi (1991) and Tepas (1985).

Many flexible work arrangements

During the last few years, quite a few work arrangements more flexible than the regular eight hours a day, five days a week have become popular. Among these arrangements are averaging hours over days or weeks, staggered work hours, seasonally adapted hours of work or lengths of shifts, part-time work (often combined with job sharing), and telecommuting, whereby a person works at home on a computer connected to the office. All of these arrangements decouple the individual from strict job requirements in time or location.

Flexibility requires that the employer change from traditional organizational patterns regarding supervision, communication, job requirements, work equipment, and scheduling; on the other side, the employee must increase her or his self-reliance, independence, responsibility, and skills. These major changes require a careful assessment of the nature of the job, as well as extensive cooperation between management and employees.

WHICH SHIFT SYSTEMS ARE MOST SUITABLE?

Humans are used to being active during the day and sleeping at night. This inherent feature of our work regimen is governed by the internal clocks of diurnal rhythms. Night work, then, is "unnatural," since it requires one to work when one should sleep. For any individual, this may just be difficult to do, or it may offset the person's circadian rhythm and affect his or her well-being and performance. Also, the person's social and domestic interactions may be impeded, since our current Western societal structure has the norm of "work during the day, play in the evening, and sleep at night." Deviating from the norm does not necessarily mean that night work is harmful, but it may be stressful, depending on the circumstances and the person.

Organizational criteria by which to judge the suitability of shift systems include the number of shifts per day, the length of every shift, the times of the day during which no work is done, the coverage of the week by shifts, and whether there is shift work on holidays. These "independent variables" (to use the terminology of experimental design) have been discussed earlier.

An important "dependent variable" is the *health of the shift worker.* Do certain shift regimes affect a person's physiological or psychological well-being? For example, will the fact that a worker doesn't sleep at night, during a night shift, imperil the worker's health? And what is the effect of shift work in terms of personal interactions with family, friends, and society in general?

Another "dependent variable" is the performance of workers on shift schedules. Is the same output to be expected, regardless of the time of work during the 24-hour day? Are specific activities better performed during certain shifts? Do changes in schedule affect the shift worker's output? Finally, are there more accidents during certain shifts than others?

Monk and Wagner (1989) examined mining accidents that occurred during a full decade of regular seven-day shifts. While one should have expected a steadily decreasing number of accidents as the body adapted to the schedule, accidents were highest on Sunday nights, the fourth night shift. Apparently, the workers did not get enough sleep during the day on Sunday, which would indicate that social obligations overrode the body's need for rest.

Several aspects of shift work—the worker's health and well-being, performance and accidents, and psychological and social adjustments—interact with each other, but not always in the same direction. There are no general, simple rules for choosing among the many possible shift systems: The selection of a suitable work regimen depends on many conditions.

Health and Well-being

Just one shift?

Because the human circadian system is so stable, theoretical findings, common sense, and personal experiences suggest that the normal synchrony of behavior in terms of alternating periods of rest and activity should be maintained as much as possible. Thus, work schedules should be arranged in accordance with the internal system, or if this is impossible (for example, if night work is necessary), they should disturb the internal cycles as little as possible. One logical conclusion is that work activities that conflict with the internal rhythm of the body should be kept as short as feasible, so that one can return to the normal cycle as quickly as possible. For example, one should schedule a single night shift followed by normal workdays, instead of requiring a worker to do a series of night shifts, as shown in Table 6–2. Such a series upsets the internal clock, while a single night shift, then a day of rest, may not severely disturb the entrained cyclicity.

Many shifts in series?

Another solution, also theoretically sound and supported by experience, is to entrain new diurnal rhythms. It takes regular and strong time markers to overpower the normal signals, especially light and darkness. For shift work, this can mean that the same setup (such as working the night shift) should be maintained for long periods (several weeks or even months) and not be interrupted by different arrangements (not even by weekends). It appears that some people are more willing and able than others to conform to such regular nonday shift regimes (for example, on offshore oil wells).

Sleep

Health complaints of shift workers are often voiced, and effects are suspected, but difficult to prove. Night-shift workers have, on the average, about half an hour shorter sleep time than persons who are permanently on a day shift. However, Carvalhais et al. (1988) found that persons who permanently work the *evening* shift actually sleep about half an hour *longer* than persons on the day shift. Many persons on night shift complain about the reduced quality of sleep that they get during the day, with noise often mentioned as particularly disturbing.

Health problems

Some researchers found that shift workers have (with statistical significance) more health complaints (particularly digestive and other gastrointestinal disorders) than persons working day shifts while other researchers have failed to find statistical significance between complaints (Alfredsson et al. 1991; Costa 1996; Folkard and Monk 1985; Monk et al. 1996). No differences have been found in the mortality of night-shift workers compared to workers on other shifts. However, it appears fairly clear that persons who suffer from health disturbances are more negatively affected by night shifts than by day or evening-shift arrangements. It also appears that older workers, perhaps due to deteriorating health and insufficient restful sleep (both phenomena that seem to increase with age), may be more negatively affected by shift work than younger workers.

Performance

The reduced quantity and quality of sleep experienced by night workers led researchers to believe that many of those workers suffer from a chronic state of partial sleep deprivation. Deleterious effects of sleep deprivation on a person's behavior have been well demonstrated, as discussed earlier. For some tasks, the interaction of circadian discrepancies between the demands of work and those of the body, resulting in sleep deprivation, may be significantly detrimental to one's night work performance, including affecting safety (Monk 1989). (Accident statistics are usually confounded by many variables in addition to the shift factor, such as the work task, the age and skills of the worker, the shift schedule, etc.)

During the first or first few night shifts, performance is likely to be impaired between midnight and the morning hours, with the lowest performance around 4:00 A.M. Such impairment, which may be absent or minimal in the performance of cognitive tasks, varies in level, but is similar to that induced by legal blood alcohol limits (Monk 1989). However, as the worker continues on night shifts, her or his internal clock realigns itself with the new activity rhythms, and a daily routine of social interactions, sleeping, and going to work is established.

Tolerance for shift work differs from person to person and varies over time. Three out of 10 shift workers have been reported to leave shift work within the first three years, due to health problems (Bohle and Tilley 1989). The tolerance of those remaining on shift work depends on various personal factors (e.g., age, personality, troubles, and diseases, as well as the ability to be flexible in one's sleeping habits, in order to overcome drowsiness), on social–environmental conditions (e.g., family composition, housing conditions, and social status), and, of course, on the work itself (the workload, shift schedule), and the worker's income and other compensation. These factors interact, and their importance differs widely from person to person and changes over one's work life.

Evening-shift workers suffer particularly in their social and domestic relations, while night-shift workers are more affected by the conflict between the requirement to work while being physiologically in a resting stage, and to sleep during the day. However, physiological and health effects are not abundant in shift workers who have been on their assignments for years, possibly because persons who cannot tolerate those conditions abandon shift work soon after trying it.

Social Interaction

Needs for social interactions are individually and culturally different. For example, parents of small children want to be home with the family and are unlikely to accept unusual work assignments that keep them away, while older persons who do not need to interact with their children so intensely may be more inclined to work non-normal hours.

A major problem associated with shift work is the difficulty of maintaining normal social interactions when the work schedule forces one to sleep at times during which social relations usually occur. Family relations, as well as interactions with friends and participation in social activities, are all affected. Common daily activities, such as shopping or watching television, may not be easily carried out. Care must be taken, however, not to automatically transfer one's own living conditions and social expectations to different countries and cultures: Certain events or conditions may or may not be present, may be regularly scheduled at different times, or may be of different value to individuals and the society at large. The Latin American siesta time, for example, is not commonly observed in North America, and some kinds of stores that are open continuously in the United States close for an hour or more at midday, remain open only until late afternoon, and stay closed on weekends in Europe. Family ties are much more important in some cultures than in others and may vary among individuals. Television plays a large role in the daily life of some groups of people, but not others. Thus, statements regarding the effects of shift work on a person's social interactions may apply in one case, but may not be pertinent in others. However, it is generally true that shift work and its consequences to the individual worker often interfere with social relations. Whether this has a demonstrable effect on well-being and performance then depends on the individual.

HOW TO SELECT A SUITABLE WORK SYSTEM

Daytime work is best. The foregoing discussions should have made it clear that, if at all possible, working hours should be from morning to afternoon. However, in many cases, this normal arrangement is replaced by shift work, covering either the late afternoon and evening hours or the night.

Performance appears to invariably fall during overnight work, a phenomenon probably related to the circadian rhythm. This drop is of particular concern with regard to long periods of duty and when the overnight work period follows poor sleep. Reports of reduced performance during the night are numerous, as summarized by Rogers et al. (1989) and Monk et al. (1996).

Permanent shift assignment? The argument for permanent assignment to either an evening or a night shift is well founded on the grounds that such a lasting arrangement allows the internal rhythms to become entrained to the new rest–work pattern. However, that reasoning is not as convincing as it might appear: Most shift arrangements are not truly consistent or permanent, because the weekend interrupts the cycle. Furthermore, strong zeitgebers (such as light and dark during the 24-hour cycle) remain intact even for the person on regular evening or night shifts, thus hindering a complete entrainment of the internal functions.

Only one shift? This approach leads to the opposite conclusion, also well founded in theory. That is, it is better to work only occasionally outside the morning–afternoon period and to work only one such evening or night shift. In this case, most people are able to perform their work without suffering many detrimental effects for this one unusual period, while they remain entrained on the usual 24-hour cycle. Of course, some individuals are able and willing to adjust fairly easily to different work patterns. Thus, unusual work patterns may be more acceptable to those who volunteer for the assignment.

Shift length One general recommendation is that physically demanding work should not be expected over periods longer than 8 hours unless frequent rest pauses are possible; but even an 8-hour shift may be too long for very strenuous work. The same applies to work that is mentally very demanding, requiring complex cognitive processes or much attention. For other, everyday work, durations of 9, 10, or even 12 hours per day are quite acceptable.

Flextime, compressed time Flextime arrangements often are welcomed by employees, possibly in combination with compressed workweeks, particularly if they allow extended free weekends.

Other criteria As a part of the decision to select one of the many possible shift plans, criteria such as the following allow justifiable, systematic judgments:

- The daily working time should not last longer than eight hours.
- The number of consecutive night shifts should be as few as possible; best is one single night shift, followed by a free day, interspersed with the other work shifts.
- Each night shift should be followed by at least 24 hours of free time.
- Each shift plan should contain at least two consecutive work-free days, with the weekend preferred.
- The number of free days per year should be at least as many as for the continual day worker.

Using these criteria, one can design shift plans that comply with organizational, physiological, and psychological requirements. Any plan that is selected will inevitably have drawbacks, but these can be identified and counteracted by providing special health care services (Knauth 1996).

Making evening or night shifts easier High illumination levels, such as 2,000 lux or more, should be maintained at the workplace during evening and night shifts. These levels help to suppress production of the hormone melatonin, which causes drowsiness. Furthermore, environmental stimuli, such as occasional stirring music, hot snacks, and caffeinated hot and cold beverages, should be employed to keep the worker alert and awake. The work should be interesting and demanding, since boring and routine tasks are difficult to perform efficiently and safely during the night hours.

Coping strategies The shift worker must develop coping strategies for setting the biological clock, obtaining restful sleep, and maintaining satisfying social and domestic interactions. Unless the shift worker is on a very rapidly rotating schedule, the aim is to reset the biological clock appropriately to the shift-work regimen. For example,

- Sleep should be taken directly after a night shift, not in the afternoon.
- Sleep time should be regular and kept free from interruptions.
- Shift workers should seek to gain their family's and friends' understanding of their rest needs.
- Certain times of the day should be set aside specifically and regularly to be spent with family and friends.

". . . the times, they are a'changin. . ."

The widespread use of the computer and the handy cell phone has hastened a fundamental change in white-collar working habits and conditions: Many people do most of their work away from the company office and outside regular working hours. So-called telecommuting, in its various forms, has become popular (not only among parents of small children) because it allows the workload to be spread over the days as the worker finds it convenient to do. It may in fact be that many overwork themselves in this fashion. Numerous blue-collar workers, such as truck drivers and repair and troubleshooting crews, also have long work shifts with often irregular hours.

Kahn and Rowe (1998) describe our current employment structure as age graded and rigidly compartmentalized; that is, it assigns many people to 20 or more years of education, followed by 40 years of intensive work, followed by 20 years of relatively unproductive retirement. To keep pace with the societal trends of continuing education, two-career families, and a productive old age, these researchers propose to divide the workday into four-hour periods instead of the common eight hours. Such a four-hour module might aid high school and college students seeking work experience, employees with young children, and senior citizens who are not yet ready to retire, but who seek a reduced workload.

While some studies have indeed found that flextime, telecommuting, and part-time work reduce absenteeism and increase job satisfaction, one of the obstacles to implementing alternative arrangements may be a resistant corporate culture. Both supervisors' and peers' attitudes in larger corporate environments may disfavor flexible scheduling options and reduce employees' willingness to accept such options when they are available. Some managerial staff and employees themselves may believe that taking advantage of flexible work options means not beeing serious about their careers or that using such options will harm their careers.

The New York Times *metropolitan editor Joyce Purnick told the Barnard College graduating class,*

"If I had left the Times *to have children, and then came back to work a four-day week the way some women reporters on my staff now do, or I had taken long vacations and leaves to be with my family, or left the office at 6 o'clock instead of 8 or 9—I wouldn't be the metro editor" (quoted from the* APA Monitor, *Vol. 29, No. 7, of July 8, 1998)*

ARRANGEMENTS FOR AIRPLANE CREWS

Airplane flight crews that must cross time zones during their long-distance flights and catch some sleep at their destination before returning have a number of problems. The first is that the quality and length of sleep at the stopover location is frequently often much less satisfactory than at home. The resulting tiredness is often masked or counteracted by the use of caffeine, tobacco, and alcohol. The second problem is the extended time of duty, which includes preflight preparations, the flight period itself, and the wrap-up after arrival at the stopover. Detrimental effects are substantially stronger after an eastward flight than a westward one; also, crew members over the age of 50 are more affected than their younger colleagues (Graeber 1988).

Recommendations for the shift arrangement for flight crews are fairly well established. In general, but particularly when flying eastward, flight crews should adhere to well-planned timing that should duplicate, as far as possible, the sleep–wakefulness periods at home, meaning that crew members should try to go to bed at their regular home times and get up at their regular times. In this manner, they will maintain their regular diurnal rhythm. Of course, their next flight duty should be during their regular time of wakefulness.

BODY RHYTHMS AND SHIFT WORK: SUMMARY

Human body functions and social behavior follow internal rhythms. Aside from the female menstrual cycle, the best-known rhythms are a set of daily fluctuations called circadian or diurnal rhythms, which appear in such functions as body temperature, heart rate, blood pressure, and hormonal excretions. Under regular living conditions, these temporal programs are clearly established and persistent.

The well-synchronized rhythms and the associated behavior of sleeping (during the night) and being active (during the day) can be desynchronized and put out of order if the time markers during the 24-hour day are changed and if activities are required at unusual times. The resulting sleep loss and tiredness influence human performance in various ways. One's performance of mental tasks, attention and alertness usually are diminished, but most physical activities may still be carried out without any degradation in performance. Some researchers are concerned that disturbing the internal rhythm by certain types of shift work, might be detrimental to a person's health; indeed, gastrointestinal problems are proven to be more frequent with workers on a night shift than with persons on day work. A sociopsychological effect of shift work is that it prevents those who are on it from participating in some family and other social activities.

The following recommendations, drawn from a review of physiological, psychological, and social behaviors associated with performance, present acceptable regimes of working hours and shift work:

- Job activities should follow entrained body rhythms.
- It is preferable to work during the daylight hours.
- Evening shifts are preferred to night shifts.
- If shifts are necessary, two opposing rules apply: (1) Either work only one evening or night shift per cycle, then return to day work, and keep weekends free, or (2) stay permanently on the same shift.

- A duration of 8 hours of daily work per shift is usually adequate, but shorter times for highly (mentally or physically) demanding jobs may be advantageous, and longer times (such as 9, 10, or even 12 hours) may be acceptable for some types of routine work.
- Compressed workweeks—for example, four 10-hour days per week—often are acceptable for routine jobs.
- Together with changing work tools and habits, new and much less rigid time schedules may become common, requiring different management attitudes.

EFFECTS OF ALCOHOL ON PERFORMANCE

Alcohol affects the body

Alcoholic beverages contain many chemical substances; approximately 200 congeners have been identified in wine. It is still not clear which of these (in addition to ethanol) are responsible for the undesired physiological and psychological effects or for the feelings of temporal euphoria and freedom from inhibition associated with drinking alcohol. While there is some evidence that alcoholic beverages, if taken in limited doses, may be beneficial for the cardiovascular system, taken excessively, alcohol harms human beings.

Alcohol in the human bloodstream has neurological effects:

1. First, it impairs the functioning of the cerebral cortex, which houses intelligence.
2. Then the limbic system is affected, which, among other functions, controls mood.
3. Finally, alcohol impairs the brain stem, where the "fight-or-flight" response is generated and where such functions as heart rate, blood pressure, and respiration are controlled.

Thus, alcohol impairs the central nervous system—the greater the blood alcohol level, the more the impairment in

- judgment;
- language;
- insight;
- memory;
- the ability to understand and make plans; and
- motor control and body posture.

In addition, the reception and perception of sensory inputs and appropriate responses are diminished. Large emotional swings may occur, typically ranging from laughter and giddiness ("the life of the party") to sadness and tearfulness ("crying in one's beer").

Long-term excessive users of alcohol (called alcoholics) are likely to show pathological effects. Toxic changes occur in organs such as the brain and in muscles. In the intestinal system, metabolic processes suffer, since alcohol interferes with absorption, digestion, and with the metabolism and utilization of nutrients and vitamins.

Two out of three Americans consume alcohol; fourteen percent of men and 4% of women are estimated to be heavy drinkers.

BLOOD ALCOHOL CONTENT

BAC

The effects of alcohol are usually stated in relation to the blood alcohol content (BAC, in percent). It is most accurate to measure alcohol content in a blood sample, but BAC is often approximated from a sampling of exhaled air (e.g.,with a Breathalizer™). To predict BAC in fasting persons, Price et al. (1986) derived equations that can be simplified as follows (Price, personal communication 31/03/2001):

For males,

$$\text{BAC}(\%) = 0.03d + 0.17d^2 - 0.10d^3$$

and for females,

$$\text{BAC}(\%) = 0.17d - 0.08d^2 + 0.04d^3$$

where d is the dosage of pure ethanol in mL per kilogram of body weight.

ABSORPTION

Alcohol absorption

Consumed orally, alcohol is absorbed into the bloodstream by simple diffusion through cell boundaries in the linings of the stomach and the gastrointestinal tract, with large intra- and interindividual differences in the resulting BAC.

Once absorbed, the alcohol is distributed by the blood. Since ethanol freely mixes with water, it quickly reaches all of the body's cells. (About half the blood volume is water, and the cells are bathed in it.) Hence, the alcohol content of organs with a good blood supply (such as the brain, lungs, liver, and kidneys) quickly becomes the same as that of the blood. The highest alcohol content in the blood occurs about half an hour after ingestion. Full absorption may take up to six hours with a heavy meal.

☛☛☛ *Distilled alcohol (hard liquor) is absorbed faster than beer. The more alcohol that is drunk (over a given time), the* slower *its absorption, because alcohol diminishes the stomach's capability to empty itself, slowing the arrival of alcohol at the lining of the intestinal tract. (This effect may explain the higher tolerance—ie, slower absorption rate—of drinkers than of abstainers.)*

Absorption is also slowed by a "full stomach," also due to delayed gastric emptying. Absorption is

- *slowest after a meal high in carbohydrates;*
- *moderate after a meal high in fats;*
- *fastest after a meal high in proteins.* ☚☚☚

Who gets more drunk?

Alcohol is freely diffused in the body. Thus, people who have less body water, such as women, obese persons, and the elderly, generally have a higher BAC than those with more body water, like men, slim persons, and younger people, who have drunk the same dosage of alcohol. Women taking oral contraceptives metabolize ethanol more slowly than other women or men, meaning that they could stay intoxicated longer. Also, the highest BAC occurs directly before the menstrual flow date.

The strength of the effects of the alcohol depends on the time of day, ie, on the circadian rhythm, discussed earlier in this chapter. For example, the effects of alcohol are stronger in the early afternoon than in the evening.

ELIMINATION OF ALCOHOL

Coffee, fresh air, and work don't help

Alcohol is eliminated from the body at a uniform rate until its concentration is very low. The rate per hour is about 0.015 percent for men and 0.019 percent for women. Alcohol is oxidized, and the by-products—carbon dioxide and water—excreted by breathing; the water that is produced increases the need to urinate. The elimination of alcohol is little affected by physical work and is not sped up by drinking coffee, since oxidation is not affected by caffeine.

Hangover

Too much alcoholic good cheer can result in a hangover. Headache, nausea, and stomach irritation are caused by undigested by-products of alcohol, particularly acetaldehyde and lactic acid, that build up in the bloodstream as the liver falls behind in digesting alcohol. Since the speed of alcohol conversion cannot be changed, either by drinking coffee or by breathing fresh air, the simple cure is to wait long enough to give the liver the time it needs to eliminate the alcohol from the body. Symptoms of a hangover may be relieved by over-the-counter medications, including antacids, aspirin, and other pain relievers; yet these remedies can cause further stomach irritation.

EFFECTS OF ALCOHOL ON THE NERVOUS SYSTEM

PNS

Alcohol has several effects on the *peripheral nervous system*:

- Nerve excitation is increased by a low BAC level, but inhibited by a high one.
- The transmission of neural impulses at the synaptic junction may be reduced by the depressant effects of alcohol.

CNS

Alcohol affects the *central nervous system,* including the brain. The usual effects are as follows:

- At BAC levels below 0.05 percent, inhibitions are reduced and judgment is impaired. (The reduction of inhibitions is responsible for the illusion of stimulation by alcohol.)
- At 0.1 percent BAC, sensory and motor functions are depressed.
- At 0.2 percent BAC, control of emotion is lost.
- At 0.3 percent BAC, cognition is affected and stupor occurs.
- At 0.4 to 0.5 percent BAC, the drinker becomes comatose.
- At 0.6 percent BAC, breathing and heartbeat are depressed, and death can occur.

Alcohol tolerance

Becoming "tolerant" to alcohol does not increase the lethal threshold: The LD-50 (a lethal dose in 50% of those who drink) remains 500 milligrams per deciliter.

EFFECTS OF ALCOHOL ON THE SENSES

Vision

Human *vision* is variably affected by alcohol. Visual acuity is relatively insensitive to the BAC, as is adaptation to light and dark. But alcohol increases the sensitivity to dim lights and decreases the ability to discriminate among various bright lights. Resistance to glare is also reduced, and color sensitivity is affected: Light red, green, and yellow are less easily discerned from white, but blue and violet hues are more easily discriminated.

Alcohol causes the eyes to converge at long viewing distances and diverge at short ones. Depth perception is impaired at a rather high BAC, such as 0.1 percent. The ability to judge distances is reduced. Visual accommodation is impaired and eye-movement latency increased. The critical fusion frequency (the highest rate at which light is perceived as flashing) is decreased by large dosages of alcohol (BAC levels of about 0.1 percent), but not by smaller doses. Peripheral vision is somewhat reduced by alcohol, but only under a heavy general information load. Together, these findings lead to the conclusion that alcohol impairs the ability to see rapidly changing events.

Hearing

Like vision, *hearing* is variably affected by alcohol. Auditory acuity seems to be rather insensitive to alcohol, but the ability to glean information from auditory stimuli is impaired.

Touch

Both *smell* and *taste* sensitivity are diminished with as little as 0.01 percent BAC.

The sensitivity of *touch* is reduced, particularly with respect to two-point discrimination. Sensitivity to *pain* is diminished by alcohol.

Effects of Alcohol on Motor Control

Wobbliness

Motor control is greatly reduced by alcohol. For example, standing without swaying, touching the index fingers together, and other measures of hand steadiness and gait control show much diminution, even at only 0.1 percent BAC.

Reaching

The *simple reaction time* is increased by alcohol, but only at BAC levels of 0.07 percent or more; a BAC of 0.08 to 0.1 percent increases reaction time by about 10 percent. The *choice reaction time* is even more affected by alcohol, and the incidence of wrong choices is increased as well. Response to an auditory signal seems to be more prolonged than to a visual signal.

Effects of Alcohol on Cognition

Regarding *verbal performance,* alcohol increases superficial, egocentric, and inappropriate associations. Alcohol also reduces fluency and mastery of words. *Arithmetical calculations* are impaired, and *time* seems to pass more quickly to a person who has drunk much alcohol.

The retrieval of information from *memory* does not seem to be affected, but alcohol leads to memory deficiencies regarding events that took place when the person became intoxicated, especially if much information had to be stored.

Judgment

Simple auditory or visual *vigilance* tasks are not affected by alcohol, but complex ones are impaired. The impairment of *judgment* under alcohol is well documented, regarding judging one's own performance as well as somebody else's. Alcohol increases the willingness to *take risks.*

Alcohol Abuse and Dependence

Contrary to the popular image of the down-and-out unemployed drug or alcohol abuser, more than 70% of people who use illicit drugs or abuse alcohol are employed, according to data from the 1996 National Household Survey on Drug Abuse (NHSDA). It is also estimated from this study that 11% of the American workforce uses illicit drugs or consumes alcohol excessively. These workers increase employer health care costs and decrease overall corporate productivity. Programs such as mandatory drug testing and employee assistance programs (EAPs) have been found to reduce medical claims and absenteeism related to alcohol and substance abuse.

Alcohol is a contributing factor to deaths in 30 to 50% of motor vehicle accidents, 40% of falls, 26% of fires, 49 to 70% of homicides, and 25 to 37% of suicides. Medical complications of excessive alcohol consumption include dementia, anemia, pancreatitis, cirrhosis, gastritis, insomnia, impotence, peripheral neuropathy, myopathies, encephalopathy, and Korsakoff's syndrome.

Problematic drinking

The acronym "CAGE" represents a quick screening method (and is an easy mnemonic) that is useful in clinical and employment settings for identifying problematic drinking patterns. To apply the technique, just ask a person, "Have you ever

C thought you should **cu**t back on your consumption of alcohol?
A felt **a**nnoyed by those who criticize your drinking?
G felt **g**uilty or bad about your drinking?
E had a morning "**e**ye-opener" to relieve a hangover?

Answering yes to two or three of these questions invokes high suspicion of alcohol abuse, while answering yes to all four indicates a serious condition.

Alcohol Abuse

Abuse is a pattern of use that does not entail withdrawal effects or a buildup of tolerance, but may develop into dependence. Typically, one observes the following behavior in an individual who abuses alcohol or another substance:

- failure to fulfill important roles;
- repeated use of alcohol when it is physically dangerous to do so (e.g., drunk driving or operating machinery when drunk);
- use of alcohol despite resultant recurrent legal problems;
- use of alcohol despite resultant social or interpersonal problems.

Substance abuse

Alcohol dependence is called alcoholism if at least three of the following criteria apply over a 12-month period (the same is true for dependence on any substance)

- Tolerance to alcohol is markedly increased: More is needed to achieve the same effect, or the same amount has markedly less of an effect.
- The characteristic withdrawal effect of alcohol is experienced, or alcohol is used to avoid or relieve withdrawal symptoms.
- The amount or duration of use is often greater than intended.
- The user repeatedly tries, without success, to control or reduce her or his consumption
- The user spends much time drinking alcohol and recovering from its effects.
- The user reduces the time she or he spends at work or pursuing social or leisure activities, or abandons these activities altogether, because of alcohol consumption.
- The user continues to imbibe, despite objective ongoing physical or psychological problems associated with the use of alcohol.

[For more information, see the *Diagnostic and Statistical Manual of Mental Disorders* (*DSM-IV*) published by the American Psychological Association.]

EFFECTS OF ALCOHOL ON PERFORMANCE OF INDUSTRIAL TASKS

Price (1988) has documented trends indicating performance deficits at various alcohol dosages. The evidence indicates that psychomotor performance is least impaired and perceptual–sensory performance most impaired. Cognitive performance is moderately impaired.

Alcohol reduces the ability to perform industrial work tasks. Errors are increased and output is decreased with increasing BAC. In assembly tasks, productivity was reduced up to 50 percent at 0.09 BAC. Deleterious effects on the quantity and quality of work performed were also demonstrated in operation of machine-tools such as punch or drill presses, and in welding. While different operators react differently to different alcohol dosages over different application times and with different work tasks, the falloff in performance with increasing BAC is a clear trend (Hahn and Price 1994).

EFFECTS OF ALCOHOL ON AUTOMOBILE DRIVING

The driver who is under the influence of alcohol appears to have a shrunken visual field; in particular, information is collected at shorter viewing distances and less frequently. Response latency and response errors are increased. Exact steering, such as that required to stay in lane or to park, is impaired, as is the ability to judge one's driving speed. Willingness to take risks is increased.

☛☛☛ *Here are some signs that a driver is impaired by alcohol:*

- *approaches a red traffic light fast and then makes a sudden stop.*
- *changes lanes often ("weaves").*
- *straddles the centerline.*
- *changes speed often.*
- *drives very fast.*
- *drives in darkness without headlights.*
- *does not dim the vehicle's bright lights for oncoming traffic.*
- *drives very measuredly and slowly.* ☚☚☚

EFFECTS OF ALCOHOL ON PILOTS

In 1985, the Federal Aviation Agency adopted a rule that no person with a BAC of 0.04 percent or higher may act as a crew member of a civilian aircraft.

☛☛☛ *In an experiment using a flight simulator, both younger (mean age 25 years) and older (mean age 42 years) pilots showed reduced flying performance, including communication, for at least two hours after having reached 0.10 percent BAC. Their overall performance remained impaired for up to eight hours (Morrow and Jerome 1990).*

Twelve male pilots, all with relatively few flying hours (50 to 315) and without instrument rating, performed simulated flight activities either under placebo conditions or with alcohol dosages that brought about a BAC of about 0.04 percent. In many aspects of the task, these pilots' performance was reduced with alcohol, but the main flying tasks were relatively unaffected. Pilots who consumed alcohol were often inattentive to important secondary tasks and violated safety procedures. The researchers (Ross and Mundt 1988) concluded that, even under low alcohol levels, pilots' performance would reduce the margin of safety in routine flying conditions and that, in circumstances of increased demands on the pilot, the probability of an accident would be increased significantly.

EFFECTS OF ALCOHOL: SUMMARY

In most people, even at relatively small blood alcohol levels, motor performance is diminished. Cognitive performance is even more reduced, while sensory perception, decision making, and response time are most severely impaired. Unfortunately, the affected person usually is not aware of these impairments, because alcohol also reduces the ability to make judgments about one's own performance.

CHAPTER SUMMARY

The human body functions in patterns that, in essence, reflect resting at night and being active during the day. The circadian rhythms of the body should not be interrupted by work requirements; yet, on occasion, it is necessary to work over long periods or during the night. In that case detriments in certain kinds of performance are likely, and health consequences may ensure.

For shift work, it is advisable to keep the body on the same natural rhythm and intersperse only one evening or night shift. Another solution is to adjust the internal rhythms to working regularly at the times.

Alcohol, even in small doses, impairs nervous system functions, the senses, motor control, and cognitive processes. Thus, the performance of work tasks is hampered in proportion to the alcohol content in the blood. Various legal requirements apply, such as for driving automobiles or piloting aircraft.

CHALLENGES

Discuss the interactions between activities of the body according to internal rhythms and the effects of time markers.

What may explain the large individual differences in daily rhythms, sleep, and activities?

Should one try to design different work schedules for "morning" or "evening" persons?

What would you do if you were forced to take a noon break or forced not to take such a break?

Evaluate the theory that it is only the brain, and not the body, that needs sleep.

Two extreme theories of dreaming are that dreams are expressions of mental states (Freud, Jung, and others) and that dreams are merely a "cinema of the mind" (Horne). Evaluate each of these theories.

Consider the interactions between having to work extremely long periods and missing sleep, as they affect performance.

How would your work be affected if you were forced to get along with, say, five hours of sleep per night?

Given certain types of jobs involving mental and physical work and combinations thereof, which means might be appropriate to help one's performance during long work periods? Accordingly, which kinds of activities should, and which should not, not, be required during a night shift?

Suppose you divide a long periods, such as a year, into divisions other than seven-day weeks, 24-hour days, and weeks with weekends. Might people consent to work on a basis different from the "five days on, two days off" arrangement now common?

How might these arrangements affect one's social interactions.

Is the absorption of alcohol the only factor that explains the higher tolerance of habitual drinkers, as opposed to that of abstainers?

Does drinking beverages with caffeine have a beneficial effect on a person's behavior, even if caffeine does not influence the oxidation of alcohol?

How important is it to be aware of one's impaired judgment after imbibing alcohol.

What can be done to counter the effects of alcohol on one's work performance and behavior?

part II

DESIGN APPLICATIONS

Chapter 7

Ergonomic Models, Methods, and Measurements

OVERVIEW

Models

We use *models* to understand our own physiology and psychology, as well as our roles while cooperating with other people, doing tasks, and working equipment. Models characterize us individually and as representatives of humankind—for example, in our biomechanical features. Models describe how we drive our cars along a grid of roads and how we do, or should do, our jobs. The designer of human-operated machinery (such as spacecraft, airplanes, and automobiles) utilizes computerized models of the human being (as pilot, driver, or passenger) to design proper shells and interfaces so that we are safe, feel comfortable, and are competent in using the equipment. Other, less complex models are useful for the design and evaluation of everyday workplaces.

Methods

The model by means of which we try to understand our functioning provides the underpinnings for the *methods* that we use to assess our performance as individuals or as parts of social or human-technology systems. In the description of body strength and endurance, for example, we use biomechanical techniques to evaluate the effects of body posture, physiological procedures to ascertain muscular effort, and psychological methods to judge the effects of motivation and stress. In the same vein, models also determine the methods and procedures through which we design human-technology systems, complex or simple, for usability, efficacy, and safety.

Measurements

Models and methods lead to *measurements*—the specialized ways in which we assess specific parameters of our functioning and performance. For instance, the strength exerted by our hand on a tool can be measured in terms of acceleration, force, or torque, in their magnitudes and directions, all varied over time; by electromyographic signals, oxygen consumption or heart rate; or by the effects of exhortation, by rating our perceived exertion, or by comparison with other stressful situations.

INTRODUCTION

We all have ideas, images, concepts, constructs, and patterns that help us to understand our roles or other people's roles in the private or work environment, while performing tasks and operating equipment. Ergonomists and human-factors engineers prefer well-organized and objectively describable patterns, often in mathematical form. These patterns, called *models* (or *paradigms*), describe or imitate, in a systematic manner, the appearance and the behavior of the human, often in some stressful situation.

Humans vs. machines

Regarding the human being's role in the modern work world or in space exploration, we often distinguish between what people can and should do and what machines do better. A general distinction is that people can think and feel and are intuitive and vulnerable, whereas machines are strong and logical, lack a personality, and may be discarded when they have served their purpose. A more detailed listing of the respective roles of people and machines is presented in Table 7–1.

MODELS

Definition of a model

In ergonomics and human-factors engineering, the term *model* is often defined as follows: *A model is a mathematical or physical system that obeys specific rules and conditions and whose behavior is used to understand a real (physical, biological, human–technical, etc.) system to which it is analogous in certain respects.*

Two aspects of this definition deserve special attention:

- The model "obeys specific rules and conditions." This means that the model is itself restricted. For example, the model may be a simple design template that is only two dimensional, displaying the static outline of one specific single-percentile size.
- The model is "analogous in certain respects to the real system." This means that the model is limited in its validity (or fidelity), with its boundaries often so tight that they barely overlap the actual conditions. For instance, a two-dimensional, static, average design template does not represent the bodies of all office employees.

Thus, when using the model, one needs to keep in mind its internal limitations and its limits of applicability. In some cases, the model is relatively simple—for example, showing the outline of the human body. Yet, even basic anthropometric models must represent the fact that humans come in many sizes in different proportions, they move and don't maintain frozen postures, and that they have different strengths, capabilities, skills, training, posture, and motivation. (See Chapter 1.) It is disturbing to still see, even in the year 2000, anthropometric computer-aided design (CAD) tools being advertised and used that do not reflect human variability.

> We should pause here to ponder that drawing distinctions, making classifications and developing models usually imposes artificial divisions of our own choosing upon a universe that is, in many ways, all in one piece. We do so because it helps us in our attempted understanding of the universe. It breaks down a set of objects and phenomena too complex to be grasped in their entireties into smaller bits that can be dealt with one by one. There is nothing objectively 'true' about such classifications, however, and the only proper criterion of their value is their usefulness. Adapted from Asimov, I. (1963). *The Human Body: Its Structure and Operation* (p. 13). New York, NY: Signet.

TABLE 7–1 People or Machines?

Capability	*Machines*	*People*
Speed	Much superior to humans.	Lag about 1 second.
Power	Consistent and as large as designed.	1.5 kW for about 10 seconds, 0.4 kW for a few minutes, 0.1 kW for continuous work over a day.
Manipulative abilities	Specific.	Great versatility.
Consistency	Ideal for routine, repetitive, precise/work.	Not reliable.
Complex activities	Multichannel, as designed.	Single (or few) channel(s).
Memory	Best for literal reproduction and short-term storage.	Large store, long term, multiple access. Best for principles and strategies.
Reasoning	Good deductive.	Good inductive.
Computation	Fast and accurate, but poor at error correction.	Slow and subject to error, but good at error correction.
Input sensitivity	Can be outside human senses; depends on design.	Wide range and variety of stimuli perceived by one unit (e.g., the eye deals with relative location, movement, and color).
	Insensitive to extraneous stimuli.	Affected by heat, cold, noise, and vibration.
	Poor for pattern detection.	Good at pattern detection. Can detect signals in high noise levels.
Overload reliability	Sudden breakdown.	Can function selectively, may degrade.
Intelligence	None(?)	Can deal with expected and unpredicted events. Can anticipate. Can learn.
Decision making	Dependent on program and sufficient inputs.	Can decide even on the basis of incomplete and unreliable information.
Flexibility, improvision	None.	Have.
Creativity, emotion	None.	Have.
Expendable	Yes.	No.

SOURCE: Modified from W. E. Woodson and D. W. Conover (1964), *Human Engineering Guide for Equipment Designers.* Berkeley, CA: University of California Press, pp. 1–23.

TYPES OF MODELS

Every model represents a (proven) theory or a (tentative) hypothesis that incorporates the current state of knowledge and that can be verified (or proven false) by consulting available data or by conducting new experiments.

Models and submodels

The first stage in the formulation of a model is the identification of relevant subsystems or of independent and dependent variables. The next step, the modeling stage, is the formulation of the relations among the subsystems or variables. The final stage is that of validation or disconfirmation.

Submodels

Commonly, one constructs a submodel of the human operator and another submodel of the equipment or processes with which he or she is working; then the two component models are linked together to show the interface between them and how they interact. Thus, a general model of the behavior of the human–equipment–process system is generated. Occasionally, one even models the user of the operator–machine system: the office manager, air traffic controller, military officer. In this case, there are three submodels: the operator, the "machine," and their supervisor.

Open and closed

An *open* model is affected by circumstances outside it, such as climatic conditions, vibrations, impacts, or changes in workload. A *closed* model excludes these external effects, functioning within its own "cocoon."

Loops

An open-loop model does not consider the effects of the activities of the model on itself. An example is a person firing a gun in the dark: After the bullet has left, the shooter does not know whether it hit the target or not, and hence firing a second shot is not affected by the first because there is no feedback about the outcome. A *closed-loop* system, in contrast, utilizes feedback about previous actions to modify subsequent activities.

Normative and descriptive

A *normative model* assumes that there is some appearance or behavior that is normal, perfect and ideal, in a standardized and unvarying way; often, this is a singular appearance or behavior that the model represents. Thus, a normative model is frequently deterministically constructed, presupposing that the effects of variables within the system or acting upon the system from the outside can be clearly predicted and hence modeled. The opposite is a *descriptive model,* which reflects actual changes in behavior due to variations (often assumed to be stochastic) in internal or external variables.

Simulation

Descriptive models are often used for *simulation,* in which the model's variables are run throughout their given ranges in order to exercise the model.

Mathematical

While in the past most models were physical (such as templates or manikins), today they are often mathematical and computerized. A *mathematical model* has the advantage of being precise, formal, and often general: The variables can be manipulated easily and parameters in the equations assumed freely. But the rigid mathematical structure can be a disadvantage, especially when the form and nature of the equations need to be presumed without testing them and the equations cannot be changed without changing the model itself. Thus, some mathematical or statistical models fit reality poorly, often being either too general or too specific. Furthermore, if the boundary conditions are not explicitly and carefully stated, computerized mathematical models in particular can be extrapolated inappropriately.

Good and bad models

The value of a model is judged against a set of criteria:

- *Validity* is the agreement of outputs of the model with the performance of the actual system that the model represents. (This criterion is also called fidelity or realism.)
- *Utility* is the model's ability to accomplish the objectives for which it was developed.
- *Reliability* is the repeatability of the model, in the sense that the same or similar results are obtained when the model is exercised repeatedly. Reliability may also be considered the ability to apply the model to similar, but not identical, systems.
- *Comprehensiveness* is the applicability of the model to various kinds of systems.
- *Ease of use* (usability) is, obviously, a very important criterion. If highly trained and skillful capabilities are required from the user, a model is not likely to be used often or by many people. On the other hand, oversimplification of a model to achieve ease of use is not desirable either, for the model may then fail to describe complex actual conditions.

☛☛☛ *A computer-aided design and manufacturing (CADAM) model was described in 1991 (CSERIAC Gateway 2(3), 11–12) in which the body contours of human submodels, called ADAM and EVE, are based on those of hypothetical 95th-percentile human beings. The modeled contours are then multiplied by 0.93 to allegedly represent "average persons" or multiplied by 0.8725 to purportedly depict 5th percentile phantoms. Thus, ease of use was achieved by sacrificing validity because linear scaling does not realistically describe the human body.* ☚☚☚

ERGONOMIC MODELS

Starting in the mid-1990s, the G-13 committee of SAE undertook a major effort to model the human, geometrically and behaviorally, as she or he interacts with equipment and the environment. McDaniel (1998) characterized this "ergonomic model" as

- a computer-based simulation of individual human beings,
- work-related characteristics and performance,
- required for the operation, production, and maintenance of equipment, and
- having the purpose of influencing equipment design.

Such modeling requires collaboration between researchers, developers, and users, because

- programmers may not understand the science of ergonomics,
- ergonomists may not understand the possibilities and limitations of software, and
- users are often neither programmers nor ergonomists, but want
  - —valid tools that are easy to learn and use and
  - —a simple solution for a complex task.

The need for ergonomic models has been apparent since the mid-1980s. (See the discussion of modeling in the first edition of this book.) Then, as today, two solutions are conceivable: the *integrated model* (called then *supermodel*), which has a complete set of analytical functions, shared by all of its components, and the *shell model,* with a wide modular architecture allowing various modelers and users to insert their own submodels. McDaniel (1998) listed the advantages and drawbacks of these two types of models:

Integrated Models

Pros: High fidelity, compatible integrated functions, a shared structure that makes for efficient processing, and total control of input and processing, preventing misuse of the model.

Cons: Difficult to modify, limited applicability (primarily because needed detailed information is available only to specific groups of persons and for particular tasks).

Shell Models:

Pros: Users can insert their own models and data, excellent flexibility, capability for third parties to develop their own models for insertion.

Cons: Dissimilar definitions of variables, incompatible functions, amalgamation of incompatible models and data, flexibility that allows misuse of model.

Inadequate Models

☛☛☛ *Some models are simply false: The proverbial "average person," for example, does not exist, except in the minds of some journalists and politicians, but nevertheless, it was used in early biomechanical modeling.* ☚☚☚

Models may be false, inaccurate, misleading, or inadequate because of misunderstood human (and system) behavior or because of misuse of modeling procedures (or both). Some models seem to have been developed by persons who know a lot about how to program a computer, but too little about the human and how he or she functions with and within the system. Their models of humans are likely to be inaccurate, unrealistic, and overly simplified, but probably "work well" in terms of mathematics and computerization.

"A person with a lot of common sense and no technology may not arrive at the optimum solution; however, a person with a lot of technology and no common sense is really dangerous." *Johnson, S. L. (1998). "Selecting Computer-Based Ergo Tools" (p. 42).* IIE Solutions, July 1998, *40–45.*

Animation

For example, some models incorporate human motions based on *animation:* the creation of patterns of movement observed under certain conditions and then applied, often inadequately, to other conditions. A typical case is taking a sequence of static positions observed in strength testing and "morphing" them into an apparently dynamic motion pattern allegedly depicting the actual lifting of an object. Smooth as many of these animations appear, they are often simplified or exaggerated and are therefore likely to mislead the user of the model.

Linear behavior

A related fallacy is the assumption, born from the desire to keep the model simple, of linear behavior, meaning that if one variable (say, the workload) increases, the associated dependent variable (say, the speed of human activities) will change linearly as well. But many human behavior traits simply do not respond proportionally to changes in the task. If, therefore, a system is based on linear algorithms, then the system behavior is unlikely to be truly descriptive (or predictive) of human behavior—and increasingly so the more extreme (nonlinear) the conditions are.

Current models of the human have come a long way from simplistic assumptions, such as a representation by static two-dimensional contour templates. Yet, realistic human behavior, such as passively reacting to external forces or actively performing tasks (or both combined), is still only incompletely understood and modeled.

Misuse of Modeling

Simplistic assumptions

In 1986, a biomechanical model of the human body was developed to explain the stresses exerted on the spinal column while one was performing lifting tasks. The model was more advanced than its predecessors, because it included more details and it attempted to explain dynamic activities and their effects on the body, whereas previous models were static in nature. However, many ad-hoc assumptions were made in the development of the model, including the following:

- The dimensions were those of the mythical 50th-percentile male.
- The body was assumed to move at a constant velocity (after initial acceleration and before final deceleration).
- Body segments were formed as cylinders.
- The curvature of the lumbar spine was held constant under all conditions.

- The locations of joints were taken from an erect standing posture.
- A constant lever arm of the posterior back muscles about the spinal column was assumed, as was a constant lever arm of the abdominal muscles at 10 percent of stature.
- All muscle forces involved in lifting were summed to a minimal total effort. (That is, no coactivation of muscles was assumed.)

Obviously, these assumptions are overly simplistic—in fact, unrealistic—and severely limit the model's validity.

Overextended application

Yet, the temptation is strong to overlook or disregard some of the basic assumptions of a model and the limitations they impose, in order to expand the application of the model to wider boundaries. Thus, in 1988, the foregoing model was described (not by the original author) in a shortened text with the titillating title "A Knowledge-Based Model of Human . . . Capability." In this publication, the application of the model to a variety of actual working conditions was proposed, some of which were clearly outside the stated boundaries of the original model.

Incorrect inputs

Another misuse is the feeding of incorrect data to the model. Unable to evaluate the correctness of the input data, the model spits out results anyhow. A related problem is hidden under the euphemism "fitting input data to the model." This may simply mean that the data need to fit certain formatting requirements on the input—in which case there is no problem. However, if fitting data really means modifying the data, their meaning, or their "behavior," then such fitting is really falsification of the data.

False use of output

Finally, one may misinterpret and misapply the outputs, of a model—for example, by transferring static strength calculations to dynamic activities.

Three main misuses of models:

1. The model itself is inappropriate: "Whatever in, garbage out."
2. Inputs to the model are false: "Garbage in, garbage out."
3. Outputs are misapplied: "Garbage use."

User, beware!

To avoid these problems, the user of the model must be able to judge the appropriateness of the model to the situation. The user also must be able to assess the validity of the input data, and the user must refrain from applying outputs of the model to conditions outside of its model constraints.

Validation is one way to check whether the model represents reality. Validation of the model means, in the simplest case, feeding "true" data into the model and comparing the output of the model with the behavior of the "true" system (i.e., using the model to make predictions). If the model does not describe reality, then either its internal algorithms or its basic structure is insufficient.

Neglecting to assess the validity of a model is like buying an airplane or a car without trying it out.

METHODS

The model that we have of the human being and her or his performance as part of a technical or social system determines the *methods* by which we describe human performance and also determines how we design equipment for ease of use and performance. If the model is static, its algorithms are static, and consequently, the measures that we use as inputs to the model and that we receive as outputs are static as well. Yet, reality is dynamic: Humans move and change, so do their systems, and so does system performance.

The automotive industry, for example, relied into the 1990s on two-dimensional design templates—static models of the occupants' bodies—to lay out the interiors of cabs. The templates were placed with their heels at the "package design origin," the location of the gas pedal. Different sizes of the driver were approximated with rigid body templates described in SAE standards. One template displayed the so-called eye ellipse, which itself had been determined by placing visitors to automobile exhibitions into a stylized seat and marking the location of their eyes with reference to the seat. However, observations of actual car driving in the 1980s showed that the drivers' eyes were often outside the standardized eye ellipses. Thus, the model and method had become outdated.

Depending on the model, on the given goal and conditions, and on the affordable time and effort, various methods exist to determine possible links between treatments (changes in independent variables, such as positioning the driver in a vehicle) and outcomes (changes in dependent variables, such as simulating the visibility of instruments). The proper use of these methods is an everyday challenge for the human factors engineer/designer and is the bread and butter of the ergonomic researcher.

Cross section

The easiest task is to measure a given condition—for example, the interior space of automobiles or the body sizes of the user population. This effort is called a *cross-sectional* survey, because all specimens (or at least a representative sample of them) are measured at the same time, without regard to age. The single-point-in-time measure does not allow any direct conclusion about ongoing effects, such as aging.

Numerous methods exist for determining relations between treatments and their effects. The main divisions (which partly overlap) are between observations and experiments.

Observational research

Observational studies (e.g., recording the incidence of repetitive trauma in keyboarders), also called descriptive studies, are common in medical research, where they are often called epidemiological studies. These types of investigations can identify powerful associations (such as that between noise levels and hearing loss), but are retrospective and often require a long duration and large numbers of observations. The two main types of epidemiological studies are case-control and cohort studies. With *case control,* the researcher compares factors (such as exposure time) found in one group of persons with an affliction (say, repetitive trauma among keyboarders) with factors in a comparable group without that affliction (keyboarders without repetitive trauma). In a *cohort* study, one simply follows large groups of people over long periods of time with the intention of identifying factors (possible causes and preventives) associated with certain outcomes, such as loss of hearing or cumulative trauma.

Experimental research

In *planned experiments with treatments* (also called clinical trials, especially in medicine), one assigns people to groups with and without experimental treatments and observes the effects of the treatments in terms of outcomes in the dependent variables. This type of test is the current "gold standard" of experimental research if the subjects are volunteers and are informed and safeguarded; if the recruitment of participants is unbiased and assignment to treatment groups random; if neither the subjects nor the researchers know who is treated and who is not; and so forth.

Lab experiments

Laboratory experiments tightly control and often even eliminate extraneous variables that could confound or even falsify results. Also the laboratory environment permits a careful manipulation of the specific independent variables.

Field studies

The sterile conditions in the laboratory commonly are far different from the real world of ergonomics, so *field studies* are the realistic approach to assess the effects that changes in an independent variable (say, noise) have on a dependent variable (say, understanding speech)—although extraneous conditions may interfere.

Turning a Question into a Testable Hypothesis

Design of experiments

A lack of rigor in many ergonomic studies has been reported by Heacock et al. (1997), who compiled a checklist that is useful both for planning experiments and for assessing past studies. Investigations of, and experiments with, human beings and their performance must be carefully planned, executed, evaluated, and reported in order to test a theory or, more often, a hypothesis. The usual approach is to state a hypothesis (e.g., "Of two repeated measurements of hand strength, the second exertion is the strongest") and then determine whether it is true or false based on the experimental results.

Null hypothesis

Testing is commonly done in terms of the "null hypothesis: There is no difference between the outcome of the tests" (in the preceeding example, the first and second hand exertions yield the same results). Whether the null hypothesis is rejected or not is determined by statistical evaluation of the experimental data. Weimer (1995) discussed, in a down-to-earth manner, how to develop and carry out a research project; this is also the topic of many more theoretical treatises on methodological and statistical aspects in human-factors research. (See Williges' 1995 listing of relevant literature.)

Thorough guides to the design of experiments and the analysis of their results have been compiled by Williges (1995) and Han et al. (1997), with the latter dealing with complex studies with multiple variables. A major aspect of their discussion is how to control individual differences among the subjects who participate in experiments.

Assessing Performance

Assessing the performance of an existing human-operated system (including the prototype) is often a complicated and experimentally expansive task, especially if the system is complex and interacts with many other variables (i.e., if it is an open system); on the other end of the spectrum, the overall usability of new system can often be tested with just a few users (Lewis 1994; Scerbo 1995; Virzi 1992).

MEASUREMENTS

☛☛☛ *In measuring, we to aim for the scientifically desirable while employing the art of the feasible.* ☚☚☚

To measure something requires that one know the relationships between the *dependent variables* (e.g., strength exertion, task performance, or heart rate) that respond to the selected *independent variables* (e.g., pilot workload or flight path) that are to be experimentally manipulated. Hence, the experimenter must have an ergonomic model of cause–effect relations and, accordingly, select a measuring method and measuring instruments in order to obtain suitable outcome parameters.

Objective vs. subjective?

A measurement may be in objective or subjective units. Performance measures, force, torque, work or energy, power, distance or displacement (and its time derivatives, speed and acceleration), heart rate, oxygen consumption, incidents, accidents, errors, and the like *objective* parameters are preferred by most researchers because of their apparent independence from human subjectivity.

Subjective assessments are self-reports (judgments) of states within the subject or observer and appear prone to many sources of errors. Subjective assessments are abhorred by advocates of objective measurements. However, as Muckler and Seven stated in 1992, every measurement in science is dependent on human beings, either by the selection of the measure or by data collection, analysis, and interpretation. In fact, several scientific disciplines, including psychology and sociology, both significant parts of ergonomics, rely mostly on human observation. Psychologists in particular have developed procedures of training and execution that make subjective assessments reliable, precise, and valid. (cf., for example, such tools as the interview, the questionnaire, ratings, rankings, paired comparisons, observation, and verbal protocol analysis.) Muckler and Seven (1992) argued that objectivity and subjectivity are not useful ways of distinguishing among measures of human performance. The distinction becomes even more blurred as we recognize that many outcome quantities that we measure "objectively" (e.g., keying performance, in words per minute, and hand strength, in newtons) are codetermined by inherent subjective processes (e.g., motivation) that are difficult to control and quantify.

Of course, there may be as many measurement classes and units therein as there are models, methods, and independent variables, past, present, and future. This is especially true with regard to measuring the performance of humans and human–equipment systems. Yet, to assess relatively simple outcomes, certain groups of measurement have become well established, and using them has the advantage of having values available for comparison in the literature. A review of the first six chapters of this book reveals a fairly limited number of measures that serve ergonomists; most rely on physics in terms of length, time, and temperature, as well as on basic physiologic and psychologic assessment tools, with some biochemical indicators used as well (Vivoli et al. 1993). If any one measurement has the most widespread use, it must be the *heart rate,* which appears to be a truly transdisciplinary tool of psychophysical significance (Bedny and Zeglin 1997; Jorna 1993; Roscoe 1993). The selection of measurements and instrumentation is thoroughly discussed by Radwin (1997), Radwin et al. (1996) and Rodgers (1997).

EXAMPLES OF MODELS, METHODS, AND MEASUREMENTS

This book contains a large number of models and related methods for design and assessment.

In the first chapter, for example, the human body is modeled, according to Borelli's 17th-century concept, as a skeleton of links that are articulated in the joints and that is powered by muscles playing the role of engines that move the links about the joints. The functioning of muscles to produce a strength output is modeled in form of a flow system with feedforward and feedback. Common measures referred to in Chapter 1 are distance, angle, circumference, diameter, force, and torque (moment).

In the second chapter, the body's energy production is compared to the processes in a combustion engine, and models are presented of the interactions among the circulatory, respiratory, and metabolic subsystems. Common measures are oxygen uptake, heart rate, blood pressure, and endurance time.

In the third chapter, the control of the human body is modeled in terms of neural networks. The nervous system is constructed of the central and peripheral subsystems. Inputs of information to the body occur in the various sensors. A signal that is triggered is transmitted along the efferent pathways to the brain, where the information content is processed and decisions are made regarding actions. The actions are initiated by signals sent along the efferent pathways to output effectors, such as the mouth or hands. Within this complex model, components such as the eye or the ear can be modeled. Typical measurements are in bits, bytes, volts, amperes, and time. In the assessment of performance, discussed at the end of Chapter 3, judgments of various kinds include the forced decision, true–false determinations, the recognition of speech patterns; also mentioned are force exertion, time measurements, body movements, heart rate, breathing patterns, sweat rate, and skin conductivity.

In Chapter 4, paradigms for the human sensory facilities are presented, including the scheme of converging perceptive fields. Among the common measures are duration, angle, distance, acceleration, force, energy, power, frequency, wavelength, amplitude, flux, illuminance, temperature.

In Chapter 5, the interaction of the body with the environment is specified in terms of energy, temperature, humidity, motion, and endurance; distance, velocity, acceleration, and pressure; the effects of chemicals; and a variety of physiological measures mentioned earlier.

Chapter 6 describes models of human body rhythms, with measurement units taken from biochemistry, physiology and physics as well as medicine and psychology; performance in the widest sense is assessed using common engineering and managerial tools.

Certainly, the more complete and realistic the model of the human operating a technical system is, the better the designer can "fit" both equipment and the task to the operator. Given the complexities of the human body and mind, one should expect that written information and physical models (such as two-dimensional templates and three-dimensional dummies) will soon be replaced by computer-based models, which can incorporate complex information about the human and allow fast and multiple use of that information. Computer-based modeling will facilitate the ergonomic design of equipment (tools, a workstation, or a vehicle) for ease of use and for safe and efficient operation.

SUMMARY

The more systematic information about the human body and mind and their functioning within systems becomes known, the better one is able to express that knowledge in formal models. Simple body-sized analogues are mostly descriptive, while more complex models are based on well-established theories and make allowances for the effects of the environment and for adaptability and learning through feedback. Models are useful if they are valid (realistic) and reliable. It is also essential that they do not require excessive specific knowledge and experience from the user.

To achieve simplicity in a model, too often the criteria of validity and reliability are neglected, such as by using false ideas of proportions of the human body or unrealistic animation for body movements. While these problems can be fairly easily recognized and corrected, more serious ones incorporate complex, though unproven, hypotheses, especially in behavioral models. Fortunately, these hypotheses can be spotted by their simplistic details, such as assumed linear relations between variables.

Realistic modeling has permitted many improvements in the design of simple and complex human–machine systems. For example, machine parts that need maintenance in equipment can be designed to be accessible, restraint systems for people in automobiles have become very effective, and workplaces can be designed to avoid bending and twisting body movements of the worker.

The intent in all modeling is to develop a model that guides research. Toward this end, information is gathered in a systematic manner that will focus the research approach. Significant steps have been made in that direction, but many more remain. Progress in advanced modeling requires systematic knowledge that, in may instances, still does not exist. For example, little is known about human cognitive processes, decision making, and instinctive actions during an emergency.

CHALLENGES

Examine the exceptions and boundary conditions made in physical models versus behavioral models of the human.

What are the consequences of either limiting or enlarging the numbers of "specific rules and conditions" that a model obeys?

How can one test a "concept model"?

What are the practical advantages of using a closed versus an open model?

Consider the models employed in Chapters 4 and 5. Are they normative or descriptive? Also, test whether these models are valid, reliable, and comprehensive.

What is more important to the modeler, mastery of the model's computer aspects or knowledge of the functioning of the human being?

How does modeling give guidance to future research?

Chapter 8

Designing to Fit the Moving Body

OVERVIEW

Most of us work while walking or standing or while sitting on a chair or stool, but we often kneel and reach, bend, or twist our bodies. Some work is done with the body lying supine or prone—for instance, in low-seam mining and repair work. In many non-Western countries, sitting and kneeling on the ground are common during work.

The sitting posture is particularly useful when a relatively small workspace must be covered with the hands and finely controlled activities performed. For this, the workstation and work object must be suitably designed. Fewer restrictions apply for the ergonomic design of the workspace in which we are on our feet, walking and standing.

The workspace of the hands depends on the posture of the trunk and on the work requirements. Hence, various suitable workspace envelopes can be described. Yet, vision requirements at work often codetermine the suitable workspace.

Controls are operated usually either with the hands or the feet. Foot operation is stronger, but slower and, if frequent, should be required from seated operators only. Hand operation of controls is faster and more versatile, but weaker.

Tools and equipment should be designed to fit properly into the hand. This requires not only suitable sizing of a handle, but also its arrangement so that the wrist or arm is not be brought into a straining position.

Repeated and forceful operations, especially when accompanied by improper postures, may lead to overuse disorders, often associated with the repetitive use of hand tools, particularly if they vibrate. Keyboarding is another common source of repetitive injuries.

Ergonomic recommendations for the proper design of workstations make work easy and efficient.

MOVING, NOT STAYING STILL

Design for Motions

We want to move our body at work and not keep a static posture. Therefore, workstations are ergonomically designed for body movement, not for body positions that are maintained over time. One convenient way of doing so is to select the extreme body positions that are expected to occur and design for motion between them.

AN EXCERPT FROM ISO INTERNATIONAL STANDARD 6385

1. The work area shall be adapted to the operator:
- Height of the work surface shall be adapted to body dimensions and work performed.
- Seating arrangements shall be adjusted to the individual.
- Sufficient space shall be provided for body movements.
- Controls shall be within reach.
- Grips and handles shall fit the hand.

2. The work shall be adapted to the operator:
- Unnecessary strain shall be avoided.
- Strength requirements shall be within desirable limits.
- Body movements should follow natural rhythms.
- Posture, strength and movement should be harmonized.

3. Particular attention shall be paid to:
- Alternating in and between sitting and standing postures.
- Sitting is preferable to standing (if one must be chosen).
- Keeping the chain of force vectors through the body short and simple.
- Allowing suitable body posture, providing appropriate support.
- Providing auxiliary energy if strength demands are excessive.
- Avoiding immobility [and] preferring motions.

Our body is built to move about, not to remain still. It is uncomfortable and tiresome having to maintain any body position without change over extended periods. We experience this discomfort while driving a motor vehicle where the location of the trunk on the seat, of the head in order to see, and of the hands and feet on controls constrains us to a nearly immobile posture.

Lie, sit, stand

One traditionally distinguishes three categories of body postures: lying, sitting, and standing. Yet, there are many other positions, not just transient ones between the three major positions, but postures that are independently important—for example, kneeling on one or both knees, squatting, or stooping, all of which are often employed during work in confined spaces, such as loading cargo into an aircraft, in agriculture, and in many daily activities. Reaching, bending, and twisting of body members are usually short-term efforts.

Work is seldom done lying down; but even that occurs—for example, in repair jobs or in low-seam underground mining. Prone or supine positions have been employed purposely in high-performance aircraft to better tolerate the acceleration forces experienced in aerial maneuvers. (See Chapter 5.) For example, during World War II, experiments were performed to use a pilot who was lying down, which reduces the profile of the aircraft, and in some current fighter airplanes and tanks, the pilot or driver is in a semireclining posture.

0–90–180 postures

Sitting and standing are usually presumed to be associated with a more or less erect posture of the trunk and fairly straight legs while standing; sitting at work was thought to be properly done when the lower legs were, in essence, vertical, the thighs horizontal, and the trunk upright. This convenient model of all major body joints at zero or 0, 90, or 180° is suitable for standardization of body measurements, but the "0–90–180 posture" is not commonly employed, not subjectively preferred, and not especially healthy. (See the discussion of computer workstation design in Chapter 9.)

SUITABLE BODY MOTIONS AND POSITIONS AT WORK

By the rules of *experimental design,* body movements and postures and their changes can be considered independent variables. If all other conditions and variables (such as the work task and the environment) are controlled, which often requires a laboratory setting, the effects of posture changes on postulated dependent variables can be observed, measured, and evaluated. Various dependent variables have been selected in different disciplines, such as the following:

- In physiology, oxygen consumption, heart rate, blood pressure, the electromyogram, and fluid collected in the lower extremities.
- In medicine, acute or chronic disorders, including cumulative trauma injuries.
- In anatomy and biomechanics, X rays, CAT scans, changes in stature, disc and intraabdominal pressure, and model calculations from models.
- In engineering, observations and recordings of posture; forces and pressures on the seat, backrest, or floor; amplitudes of body displacements; and "productivity."
- In psychophysics, structured or unstructured interviews and subjective ratings by either the experimental subject or the experimenter.

These techniques have become standard procedures, but they are not always appropriately used and interpreted.

EMG use

For example, electromyography has been employed extensively to assess the strain generated in the trunk muscles of seated persons. Most EMG techniques inherently assume static (isometric) tension of the observed muscles, with the maximal voluntary contraction used as a reference amplitude (Basmajian and DeLuca 1985; Soderberg 1992; Kumar and Mital 1996). Most sitting is not static, however, and the EMG amplitudes are small. The importance of small changes in electromyographic events is questionable; Wiker et al. (1989) warned against the use of EMG information alone for the design and evaluation of working postures if the EMG signals suggest that less than 15 percent of maximal voluntary muscle strength is required.

Spinal compression

Results of measurements and calculations from models of the pressure within the spinal discs of sitting and standing persons have been put in doubt by Jaeger (1987). Adams and Hutton (1985) pointed out that the contributions of the facet joints (apophyseal joints) of the spinal column to load bearing have been largely neglected. Aspden (1988) calculated much less pressure within a bent spine than in a straight column.

IAP

Boudrifa and Davies (1984) investigated the relationships between intraabdominal pressure (IAP) and back support, and McGill and Norman (1987), as well as Marras and Reilly (1988), researched the relationship between IAP and spinal compression when lifting: McGill (1999a,b) concluded that abdominal pressure may not relieve the spinal column, as had been assumed.

Fidgeting

Observations of body motions (voluntary and unconscious) while sitting are difficult to interpret. Much movement may occur because of discomfort or because a chair facilitates changes; similarly sitting still may be enforced by a confining chair design or may indicate that a comfortable posture is being maintained. In the same vein, seat pressure distribution is not well related to perceived seat comfort in cars (Gyi and Porter 1999).

Subjective assessments supreme?

Table 8–1 lists a number of observational and recording techniques for assessing body motions and postures. K. H. E. Kroemer judged their status and usefulness: Some techniques are well established and easy to use, while others are not. In nearly all the techniques, however, the threshold values that would separate suitable from unsuitable conditions are unknown or variable. Thus, the interpretation of the results obtained by many of the techniques that are listed is difficult, to say the least: Most currently useful techniques are based on subjective assessments by a person or persons. Their judgments presumably encompass all phenomena addressed in physiological, biomechanical, and engineering measurements and appear to be most easily scaled and interpreted.

Discomfort and pain questionnaires

Many studies use subjective judgments (by the subject or the experimenter) that assess the suitability of an existing situation. While the procedures vary widely among researchers, most rely on the initial work by Shackel et al. (1969) and by Corlett and Bishop (1976). "Discomfort" or "pain" questionnaires have been developed, for example, by Occhipinti et al. (1985) in Italy, Kuorinka et al. (1987) in Scandinavia, and Chaffin and Andersson (1991) in the United States. By administering a questionnaire, one may generate a heightened awareness of problems among employees that should be in everybody's interest, although management sometimes sees the issue with apprehension. The manner of administering the survey may affect the outcome (Andersson et al. 1987), but the reliability of the information obtained can be quite high if the questionnaire is well instructed and administered (Booth-Jones et al, 1998).

Nordic Questionnaire

An often-used, well-standardized inquiry tool is the Nordic Questionnaire, developed by Kuorinka et al. (1987) and modified by Dickinson et al. (1992). This well-structured survey requires forced binary or multiple-choice answers. It consists of two parts, one asking for general information, the other specifically focusing on the low back, neck, and shoulder regions. The general section uses a sketch of the human body, seen from behind, divided into nine regions. The interviewee indicates whether she or he has any musculoskeletal symptoms in these areas. Figure 8–1 is a sketch of the body area in question. If needed, further body sketches can be used, showing the body in side or frontal views or giving further details (van der Grinten and Smitt 1992).

A specific section of the Nordic Questionnaire concentrates on body areas in which musculoskeletal symptoms are most common, such as the neck and low back. The questions probe deeply with respect to the nature of complaints, their duration, and their prevalence. Chaffin and Andersson (1991) modified the test for use in the United States; further changes may be advantageous to check on particular conditions. Still, using the Nordic Questionnaire provides internationally standardized information.

In terms of physical effort, such as that measured by oxygen consumption or heart rate, *lying* is the least strenuous posture. Yet, lying is not well suited to performing physical work with the arms and hands, because, for most activities, they must be elevated, which is strainful by itself.

Walking and *standing* are much more energy consuming than lying, but they generate large work spheres for the arms and hands, and if one walks around, much space can be covered. Furthermore, walking or standing facilitates a dynamic use of the whole body and its limbs and thus is suitable for the development of large energies and impact forces, such as those attained in splitting wood with an axe.

TABLE 8–1. Methods for Assessing Posture

Observation/techniques	*Measurement procedures*	*Assessment criteria*	*Threshold values*	*Relevant?*
Oxygen consumption	E	E	V	probably
Heart rate	E	E	V	yes
Blood pressure	E	E	V	yes
Blood flow	E–D	E–D	V	yes
Innervation	D–V	D–V	U	yes
Leg and foot volume	E	V	U	perhaps
Temperature, skin or internal	E–D	U	?	yes
Muscle tension	E–D	V	U	yes
Electromyography	D	D	V	yes
Joint diseases	E–D	D	V	yes
Musculoskeletal disorders	D	D	V	yes
Cumulative trauma disorders	D	D	V	yes
Spinal phenomena:				
—disc pressure	E–V	D	V	yes
—disc disorders	D–V	D	V	yes
—disc shrinkage	D	D	U	yes
—facet disorders	D–V	V	U	yes
—alignment of vertebrae	D–V	V	V	yes
—spine curvature	D–V	V	U	yes
—mechanical stresses, including model calculations	D–V	V–U	V	yes
Intraabdominal pressure	E–D	V	U	perhaps
Surface (skin) pressure at				
—buttocks	E–D	V	U	yes
—back	E–D	V	U	yes
—thighs	E–D	V	U	yes
Upper extremity posture	E–D	V	U	yes
Posture				
—of head, neck, trunk, and legs	E–D	V	?	yes
—changes in posture	E	?	?	yes
—change in stature	E	E	V	perhaps
Sensations (ratings) of				
—ailments	E–D	V	D	yes
—pain	E–D	V	D	yes
—discomfort	E–D	V	D	yes
—comfort, pleasure	E–D	D	D	yes

Source: Modified from Kroemer 1991a.

Legend

E: Well established

D: Being developed

V: Variable

U: Unknown

?: Questionable

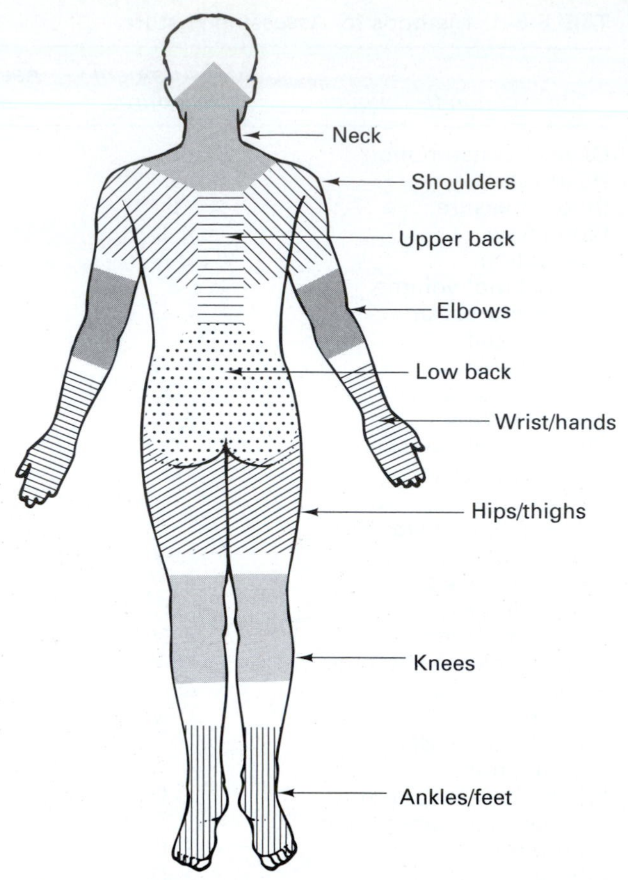

Figure 8–1. Body sketches used in the Nordic Questionnaire (Kuorinka et al. 1987).

Sitting is, in most aspects, between walking or standing and lying. Since a person's body weight is partially supported by a seat, energy consumption and circulatory strain are higher when the person is sitting than when lying, but lower than while standing. The arms and hands can be used freely, but their workspace is limited if one remains seated. The energy that can be developed is less than that attained when standing, but, given the better stability of the trunk supported on the seat, and possibly by using armrests, it is easier to use the fingers to perform activities that must be finely controlled. Operating pedals or controls with the feet also is easy in the sitting posture, because the feet are hardly needed to stabilize the posture and support the body weight and, thus, are fairly mobile.

The two most important working postures are walking or standing and sitting. Commonly, in either condition, the most easily sustained posture of the trunk and neck is one in which the spinal column is straight in the frontal view, but follows the natural S-curve in the side view [i.e., with lordoses (forward bends) in the cervical and lumbar regions and a kyphosis (backward bend) in the thoracic area]. Yet, maintaining that trunk posture over long periods becomes very uncomfortable, mostly because of the muscle tension that must be maintained to keep the body in such a position. Also, the inability to move the legs and feet when staying still is quite disadvantageous, because the feet and lower legs swell as a result of the accumulation of body fluid— a problem to which many women are particularly prone. Thus, either standing still or sitting still is "unphysiologic"; instead, the posture should be changed often, by incorporating interludes of walking by the standing operator and, at least occasionally, by the seated person, as well as by moving head, trunk, arms, and legs.

Sitting and back pain

Because so many persons suffer from discomfort, pain, and disorders in the spinal column (particularly in the low back and neck areas), the posture and movements of the spinal column have been of great concern to physiologists and orthopedists. Explanations for all these troubles have been sought in the human body's "not being built for long periods of sitting or standing," not being fit because of lack of exercise, or having undergone degeneration, particularly of the intervertebrate discs; the last two factors could be counteracted by physical activities and special exercises. To change the posture of sitting persons, various devices have been proposed, such as pulsating cushions of the seat and backrest, or frequent readjustments of the chair, particularly of the angles of the seat pan and backrest. Many suggestions have been made regarding the shape and angle of the seat pan and the angle of the backrest. It even has been proposed not to provide support for the back at all, so that the trunk muscles would be employed to stabilize the body: yet, this suggestion contradicts the evidence that the use of a suitable backrest takes some of the load away from the spinal column. The muscles that stabilize the spinal column run in essence between the pelvic and shoulder areas. Since they can only contract, not push, their intensive use increases the compression force on the spinal column, as sketched in Figure 8–2. To summarize these considerations,

1. changes in posture help to avoid continued compression of the spinal column as well as muscular fatigue, and
2. a seat should be designed so that the sitting posture can be changed frequently. If sitting for a long time is required, then a tall backrest that reclines helps to support the back and head and allows the person to take a relaxing break.

Chapter 9 provides more information on sitting and seats.

Comfort

The concept of comfort, as related to sitting, was elusive as long as it was defined, simply and conveniently, but falsely, as the absence of discomfort. Helander and Zhang (1997) showed that, in reality, these two aspects are not extremes of a single scale of values. Instead, there are two scales, one for comfort and the other for discomfort, that are not even parallel, but partly overlap. Apparently, discomfort is related mostly to biomechanically poor seat design features and to circulatory problems and fatigue that increase with long periods of sitting; in contrast, comfort is associated with the feeling of well-being, with support, plushness, softness, and even esthetics. Figure 8–3 sketches these features; for more detail, see the discussion in Chapter 9.

RECORDING AND EVALUATING POSTURES AT WORK

Postulated postures

There are two approaches to recording a person's posture. One is to postulate given postures and observe how often they actually occur or by how much actual postures deviate from them. Often, certain postures are preferred for some reason and then labeled as good or healthy. For example, in the late 1800s, the "upright" or "erect" back was thought to be healthy and therefore was promoted; a hundred years later, the so-called neutral posture became an idol, often with little explanation or scientific foundation (more on this in Chapter 9).

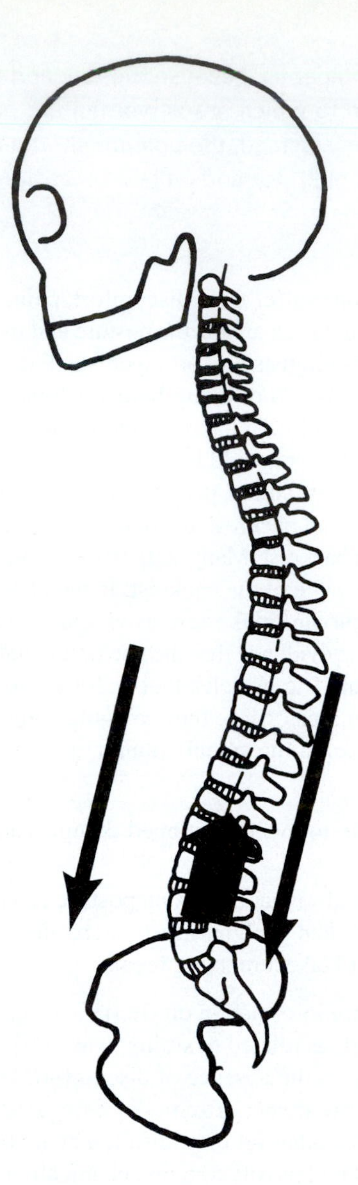

Figure 8–2. Activation of longitudinal trunk muscles generates compression of the spine.

Actual postures

The other technique is to observe and record, in detail, the actual positions of body members. This procedure is facilitated if one concentrates on particularly important body parts and records their positions, using standardized terminology either in descriptive terms (Occhipinti et al. 1991) or in angles of deviation measured against a reference (Priel 1974; Corlett et al. 1979; Gil and Tunes 1989). Some techniques provide a set of predrawn body-segment positions from which the observer selects those that best represent the actual conditions. The observations may be recorded at the workplace or taken from a photograph, movie, or videotape at a later time. Continuous recordings allow the description of motions (instead of one-shot postures); some recording techniques utilize markers attached to the subject to describe body movements exactly (Genaidy et al. 1994).

Many procedures

Various methods and techniques have been developed to describe body postures and make judgments about their suitability. Priel (1974) developed a "posturegram" that records the levels at which the limbs and joints are located and the direction and magnitude of their movements. This

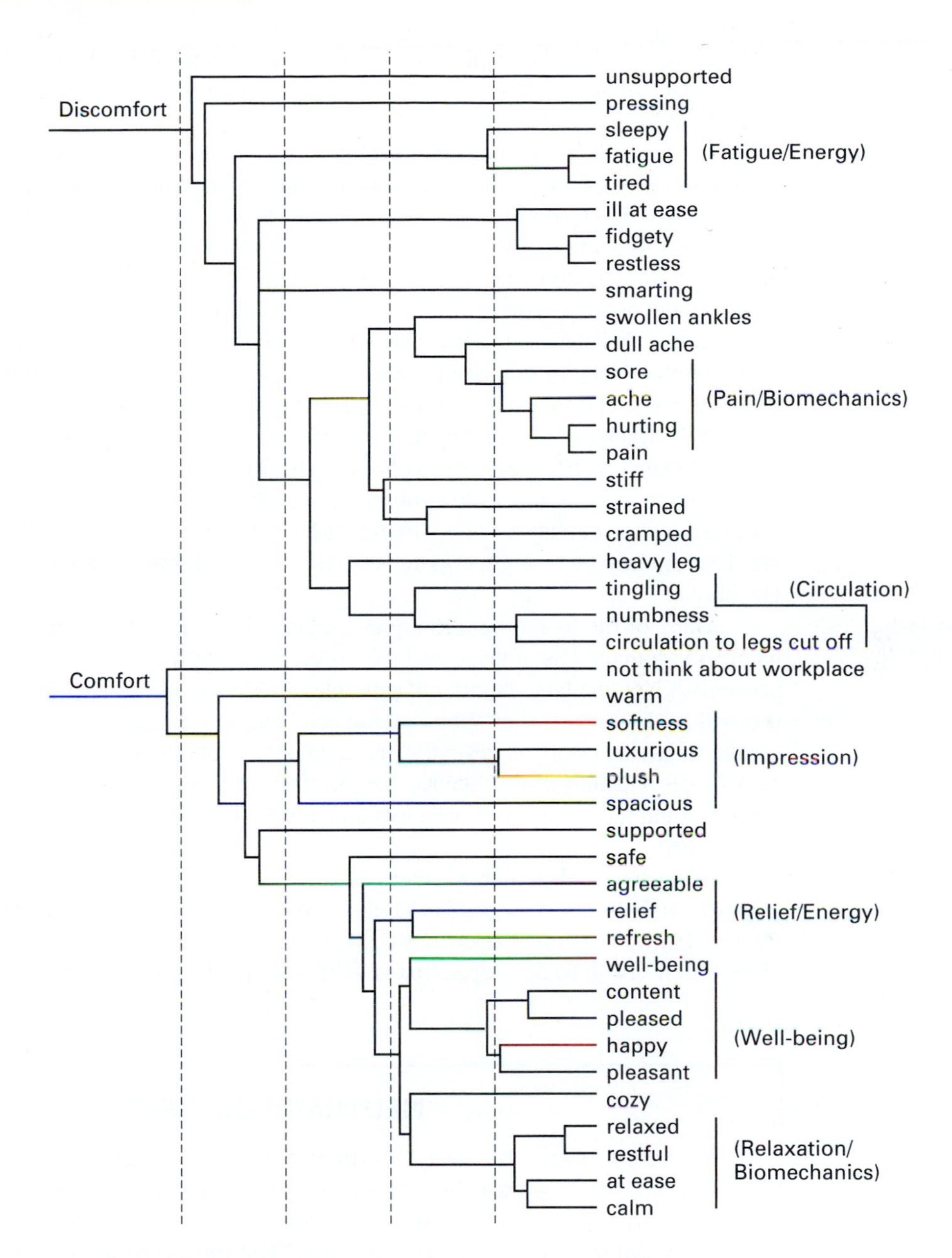

Figure 8–3. Attributes of comfort and discomfort while sitting (courtesy of M. Helander).

system relies on "basic" postures of the body, deviations from which are recorded. The Ovaco Working Posture Analysis System (OWAS) is similar in that it postulates basic body-segment positions which one compares with actually observed working postures. The system also includes an evaluation of the suitability of the working postures (Karhu et al. 1981). OWAS was computerized by Mattila et al. (1993) and checked for reliability by de Bruijn et al. (1998). Based on OWAS, the Rapid Upper Limb Assessment (RULA) procedure was developed to investigate workplaces with a greater frequency of work-related upper limb disorders (McAtamney and Corlett 1993). In the TRAM system (Berns and Milner 1980) and in the PATH system (Buchholz et al.

1996), postures at the workplace are recorded at regular intervals on prepared forms, as with OWAS. TRAM also includes estimates of forces and of the suitability of various body positions. The ARBAN system (Holzmann 1981) stresses the ergonomic analysis of work conditions, uses videotaping to code the posture and workload, and computerizes the results of the analysis and their evaluation. The computer routine and evaluation are based on heuristic rules. "Posture targeting" (refined by Corlett, Madeley, and Manenica 1979) uses prepared sketches of the human body, with targets (like in shooting competitions) associated with each major body joint. On the targets, deviations from a standard posture are recorded so that they show the direction and magnitude of displacement of body segments. These posture codes can be combined with assessments of the postural loading (Wilson and Corlett 1990). Keyserling (1986a,b; 1990) used a computer-aided procedure with a menu of standard postures for the trunk and shoulders. Gross deviations from the standards are read from videotapes and recorded with respect to their frequency of occurrence. Malone (1991) developed a taxonomy that also used diagrammatic depictions to record posture, while Paul and Douwes (1993) employed photographs. The PEO technique uses a portable (e.g., hand-held) computer for in situ (on-line) observation at the workplace; the program calculates the number and direction of body postures (Fransson-Hall et al. 1995). Several of the preceding techniques are described by their originators in Karwowski and Marras's 1999 Occupational Ergonomics Handbook.

Awkward postures?

Many of the foregoing techniques evaluate the suitability of observed postures. Unfortunately, the foundations of these evaluations are generally ill defined and ill supported—not surprisingly, given the large number of possible criteria and circumstances on which an evaluation is based, as discussed earlier. While certain body movements and awkward postures (such as twisting the trunk) are clearly undesirable, the suitability of others depends very much on the given circumstances (Genaidy et al. 1995; Keyserling 1998). Thus, incorporating an evaluation of observed postures (and motions) into procedures that record postures is a rather difficult and, so far, unresolved task.

Improved techniques needed

According to their authors, many of these techniques have been employed with some success, although to varying degrees of fidelity, repeatability, and amount of time consumed (Burt and Punnett 1999; Fisher and Tarburtt 1988; Malone 1991; Ziobro 1991). Yet, a truly satisfactory technique still needs to be developed that is valid, repeatable, reliable and also easy to use.

"NEUTRAL" POSTURE?

In recent years, the term "neutral posture" has become popular, usually suggesting a healthy or desirable body position. (See Chapter 9.) Unfortunately, it is often not clear what "neutral" means: Is it the middle of the total range of motion of a joint? That would make some sense for the wrist, indicating that the hand is straight (i.e., in line with the forearm). But there is no obvious significance to the "middle" joint position in the elbow or knee, shoulder or hip, or spinal column. Does the term "neutral" suggest that all tissue tensions about a joint are balanced, so that the position is stable? Does the term imply a minimal sum of tissue tensions (torques) around a body joint, or about several joints, or all body joints? Does "neutral" imply minimal joint discomfort, or a relaxed posture or a posture instinctively assumed for a task, to generate high body strength or to avoid fatigue?

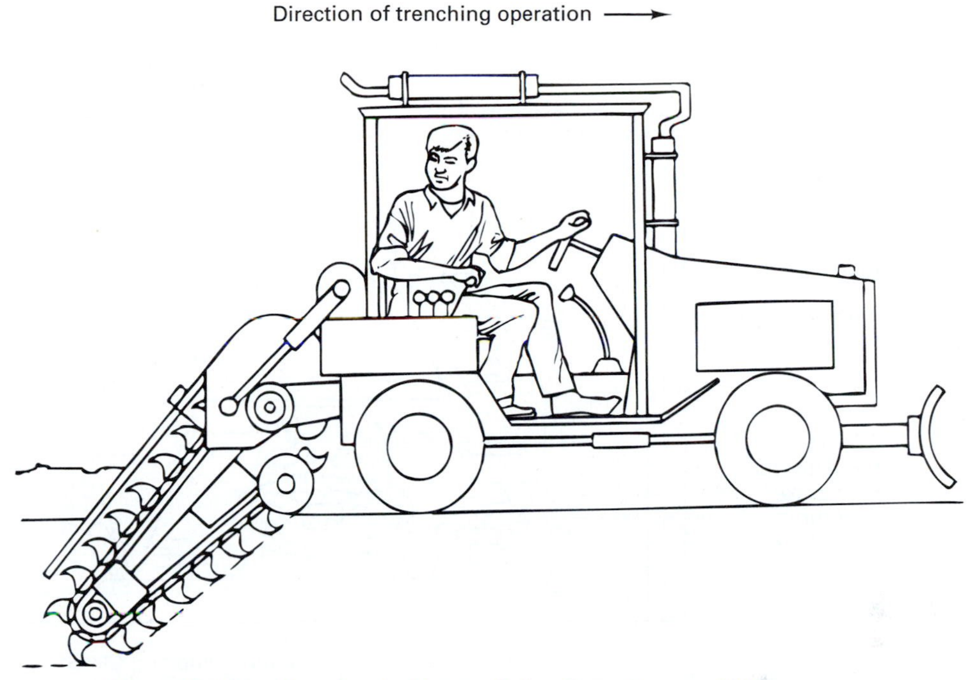

Figure 8–4(a). Trencher (with permission from Brennan 1987).

Brennan (1987) described the ergonomic design challenges associated with the posture of the driver of a trenching vehicle. The operator sits looking forward, the direction in which the machine moves, but the trenching tool is attached to the rear of the machine. [See Figure 8–4(a).] To observe the trenching operation, the driver must rotate her or his trunk and neck nearly 180°. [See Figure 8–4(b).] While all the controls that move the vehicle are commonly located in front of the operator, the controls for operating the trenching attachment are located to the side. [See Figure 8–4(c).] This arrangement is ergonomically faulty because, among other things, it enforces a much-twisted posture on the operator during the trenching operation and it is likely to result in mistakes in operating the machine. Unfortunately, similar arrangements are often found in underground mining equipment, earthmoving machinery, and motorized lift trucks: Contorted body postures are imposed on the operator; controls are improperly located; the operator's vision is blocked; and noise, jolts, and impacts from the ground are transmitted to the operator.

DESIGNING FOR THE STANDING OPERATOR

Move, don't stand still

Standing is used as a working posture if sitting is not suitable, either because very large forces must be exerted with the hands or because the operator has to cover a fairly large work area. In the latter case, the person walks about, which is much preferable to standing still; standing in place should be imposed only for a limited period. Forcing a person to stand simply because the work object is customarily put high above the floor or outside sitting reach is not a sufficient justification. For example, in automobile assembly, car bodies have been turned or tilted and parts redesigned so that the worker did not have to "stand and bend" or "stand and reach" in order to do

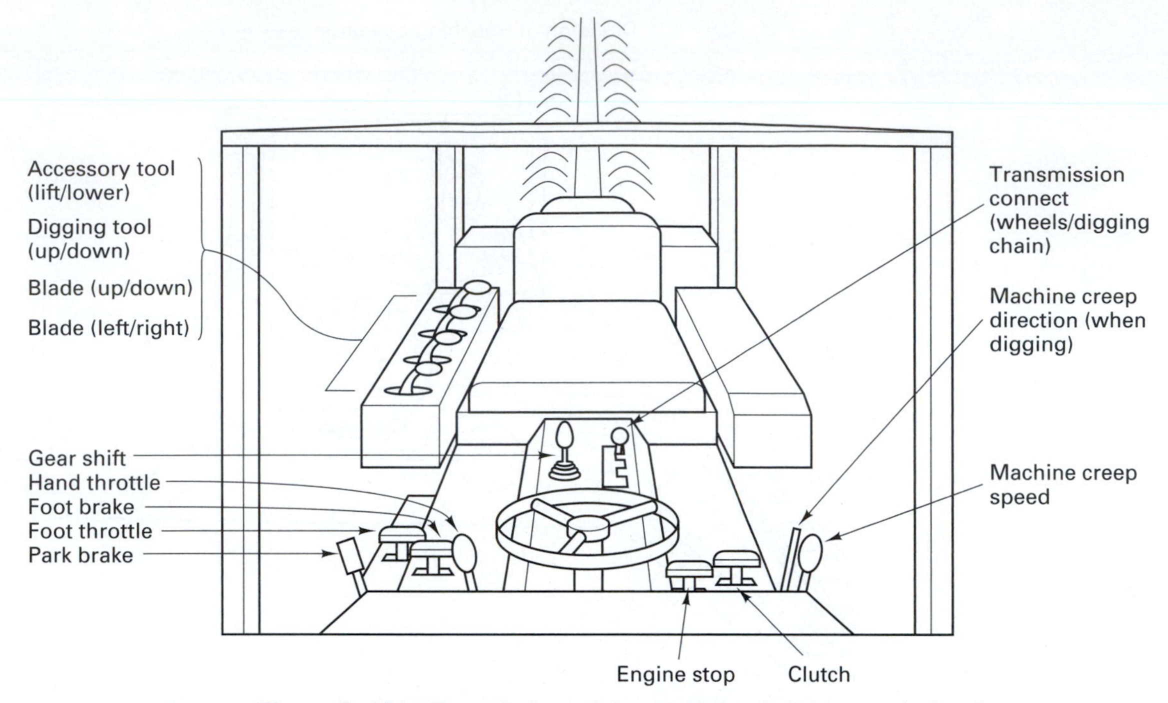

Figure 8–4(b). Frontal view of the trencher cab (with permission from Brennan 1987).

the work. Other examples of workstations designed for standing operators are shown in Figure 8–5, those workstations were designed in accordance with the need to exert large forces over large spaces, to make expansive, strong exertions under visual control, or to work with large objects.

Work near elbow height

The height of the workstation depends largely on the activities to be performed with the hands and the size of the object. Thus, the main reference point is the elbow height of the worker. Related anthropometric information, presented in Chapter 1, must be adjusted to account for a slumped, instead of upright, body. As a general rule, the strongest hand forces and most useful mobility are between elbow and hip heights. Thus, the height of the support surface (for example, a bench or table) is determined by the working height of the hands and the size of the object with which a person will work.

Shoes and walking surface

Sufficient room for the feet of the operator must be provided, including toe and knee space to move up close to the working surface. Of course, the floor should be flat and free of obstacles. Elevated platforms should be avoided if possible, because one may stumble over an edge. Elastic floor mats and soft shoe soles can reduce foot, leg, and back discomfort (Hansen et al. 1998; Krumwiede et al. 1998; Stuart-Buttle et al. 1993). Appropriate friction between soles and the walkway surface helps to avoid slips and falls (Chiou et al. 1996; Groenquist and Hirvonen 1994; Jones et al. 1995; Lin and Cohen 1995; Leclercq et al. 1995; Redfern and Bidanad 1994).

No twists or bends

Basically, body movements associated with work while walking or standing are desirable in physiological respects, but they should not involve excessive bends and reaches and especially should not include twisting motions of the trunk. People should never be forced to stand (or even worse, to stand still) at a workstation just because the equipment was originally ill designed or badly placed, as is, unfortunately, too often the case with punch or drill presses used in continuous work. Other machine tools, especially lathes, are constructed so that the operator must stand and lean

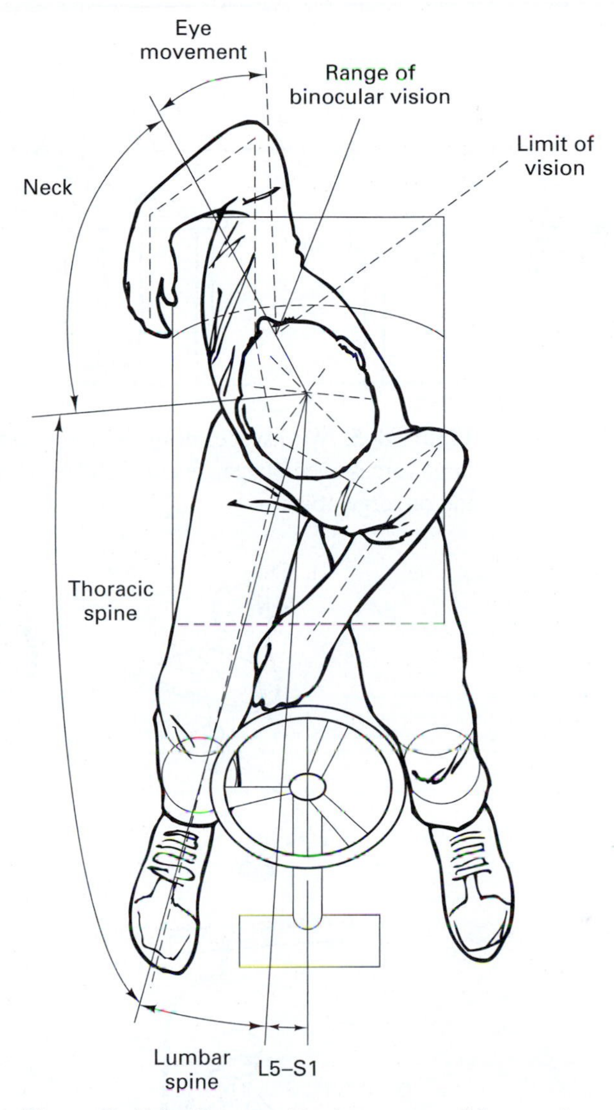

Figure 8–4(c). Contorted body posture of the trencher operator looking at the trenching equipment (modified from Brennan 1987).

forward to observe the cutting action and, at the same time, extend the arms to reach controls on the machine.

Semisitting

So-called stand seats may allow the operator to assume a somewhat supported posture somewhere between sitting and standing. Examples of these contraptions are shown in Figures 8–6 and 8–7. Occasionally, high stools can be employed to allow (rather uncomfortable) sitting at workstations at which the operator otherwise would stand. Such semiseats usually do not have full backrests and do not support the body fully; therefore, and for reasons of stability, much weight remains on the feet. Thus, semisitting, although better than standing in place, is by no means satisfactory.

Figure 8–5. Workspaces designed for a standing operator: Required are large forces over a large area, forceful exertions with the hand, and working on large objects.

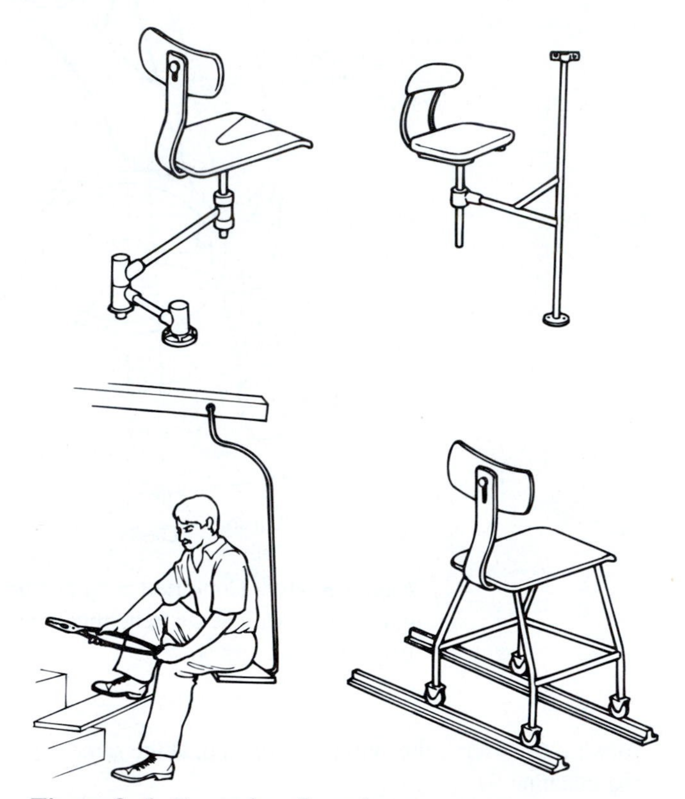

Figure 8–6. Examples of stand-seats.

DESIGNING FOR THE SITTING OPERATOR

Sitting is a much less strenuous posture than standing, mostly because it requires fewer muscles to be contracted to stabilize the body, which, in turn, is largely due to the support that the body enjoys at its midsection through the seat pan and seat back. Sitting allows better-controlled hand movements, but coverage is of a smaller area, and the hands exert less force. Suitably seated, a person can operate controls with their feet and can apply much force with them. In designing a workstation for a seated operator, one must consider in particular the free space required by the legs and feet. If this space is severely limited, very uncomfortable and fatiguing body postures result, as shown in Figure 8–8. Some persons who sit at work complain of low back pain and foot swelling, usually because the same posture has been maintained for a long time.

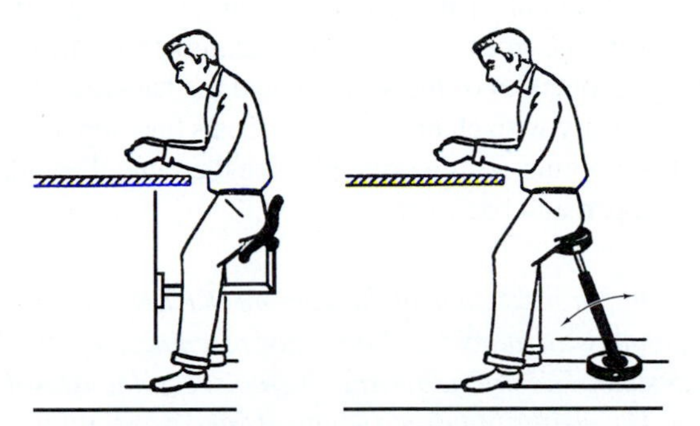

Figure 8–7. Stools used to temporarily allow (rather uncomfortable) sitting at workstations designed for a standing operator.

Figure 8–8. Leg space for the seated operator must be provided.

The preferred working area is in front of the body, at about elbow height with the upper arm hanging (as for a standing operator). Exact and fast manipulations are most easily done in this area. Many activities of seated operators require close visual observation, which codetermines the proper height of the area to be used for manipulation, depending on the operator's preferred visual distance and direction of gaze. (See Chapter 4.)

In Western civilization, it has become customary to provide chairs that are at about the popliteal height of the sitting person. (See the anthropometric tables in Chapter 1.) Thus, seat heights range from about 35 to 50 centimeters. The comfort of a seat is determined largely by the design properties of the seat pan and the backrest. This is discussed in detail in Chapter 9 which deals mostly with chairs in offices. Seats for shop use are usually sturdier and simpler, but the general ergonomic design principles are the same: Provide suitable height, width, breadth, and shape for the pan and backrest.

☛☛☛ *In India, one of the authors (KHEK) was invited to see a modernized small manufacturing plant with many hand-operated machines, mostly drill presses and punch presses, elevated on pedestals. The operators sat on stools. Yet, the overall impression was that of a "staged" situation. The visitor inquired steadfastly and was finally told that all of these machines originally had been placed directly on the floor, without pedestals and the operators sat, squatted, and kneeled on the floor as well, as they were accustomed to doing. Then, management decided to "improve" the working conditions according to Western images and put the machines on pedestals and the operators on stools. When left alone, the operators would assume their regular traditional postures on these stools, with their feet at seat height. For visitors, however, the operators were exhorted to put their feet down.* ☚☚☚

"SITTING" IN NON-WESTERN COUNTRIES

The tourist visiting southeast Asian countries notices readily that people's sitting behavior is different from that in North America or Europe. Sitting cross-legged on the ground, in a manner somewhat similar to the "lotus position," is quite widespread. The feet and legs are crossed in front of the body, and the body weight is transmitted mostly through the buttocks, while the legs and feet serve to stabilize the posture—but some people even sit on their feet. Another common position is the "squat" posture, in which the soles of the feet are on the ground, the knees bent severely, and the thighs close to the trunk, with the person nearly sitting on the heels. (See Figure 8–9.) Often, one leg is extended away from the body, while the other is kept close. Also, quite frequently the kneeling position is used, occasionally with the feet rotated outward. This posture is often assumed if work requires that the upper body be bent forward. These "low" postures reduce the heights of the eyes and elbows with respect to the ground. For workplace and equipment design, this means that the working-height dimensions of sitting Westerners often do not suit Asian operators.

How Chairs Became Used in China

Chinese cultural relics from the Shang through the Han dynasties (1600 B.C. to 220 A.D.) show that people sat and slept on mats in small rooms with low ceilings. Kneeling or sitting with crossed legs were common postures, considered proper according to contemporaneous etiquette and ritual. (See Figure 8–10.) Sitting with extended legs with the feet thrust forward as well as squatting was considered impolite or immoral. (It is interesting to note that "pointing" with the feet toward a person or showing the sole of the foot is still considered rude in Thailand.)

The opening of the Silk Road allowed envoys of the Chinese Han dynasty to visit western Asia. There, they saw stools and other chairlike furniture, which they then introduced to China. But it took until the 3rd century A.D. before folding stools appeared in the Chinese imperial court. Yet, by the 4th century, traditional rituals and formalities had been changed. Houses had higher ceilings and rooms were more spacious. New items of furniture, including stools that were about the same height as those in the West, gradually came into use, as stone carvings and paintings indicate. Persons until then depicted as sitting cross-legged, with robes covering their feet, now were shown seated on hourglass-shaped stools, on four-legged stools, or on beds raised on legs. (See Figure 8–11.) During the 7th to the 10th centuries, the traditional life on mats gradually disappeared, and the folding stool became popular.

Around the year 1200, complete sets of raised furniture, including stools, chairs, tables, screens, dressing tables, and racks, existed in China. Drawings of the Ming dynasty (1368–1644) show a variety of styles, often elevated on legs, with forms of classical simplicity. However, in noble households, it was still regarded improper for women to sit on chairs (Xing 1988).

Figure 8–9. Working postures often encountered in Asia.

Figure 8–10. For thousands of years, many Chinese lived on mats.

Figure 8–11. By the Tang dynasty (618–907 A.D.), stools and chairs had appeared in China and were gaining in popularity.

DESIGNING FOR WORKING POSITIONS OTHER THAN SITTING OR STANDING

Semisitting is one example of a working posture that is neither sitting nor standing. In many cases other postures must be assumed at work, although often only briefly, such as when reaching to a barely accessible object, stooping in a low-ceilinged compartment, or straining to perform repair work inside a narrow opening. Little can be done in terms of systematically designing body supports related to such unusual and awkward postures, except not to design equipment that requires them.

A visitor to a high-tech manufacturing facility was impressed by the manager's explanations of how highly automated the production was and how few people were needed to run it. "Why," the visitor asked, "are there so many cars parked outside?" The answer was that they belonged to the repair people who did the old-fashioned bloody-knuckles dirty repair work on the automated manufacturing machinery.

Awkward postures

Some jobs, though, habitually require bent, stooped, and twisted working positions—for example, loading and unloading luggage of aircraft passengers, both behind the check-in counter and in aircraft cargo holds. Repair, maintenance, and cleaning jobs often demand awkward body postures. Low-seam mining is notorious for requiring bent, stooped, kneeling, and even crawling and lying working postures from the miners (Gallagher 1999). In the building and construction trades, bent and twisted body postures are frequent as Buchholz et al. (1996), Helander (1981) and Schneider and Susi (1994) described. Other examples are in agriculture (Meyers et al. 1997), in manufacturing and assembly work (Duquette et al. 1997; Helander and Nagamachi 1992; Kragt 1992), and at grocery checkout counters (Johansson et al. 1998; Estill and Kroemer 1998). Notoriously bad is entering and exiting of rear seats in automobiles (Giacomin and Quattrocolo 1997).

Avoid awkward postures

These and other examples indicate a need for a systematic ergonomic design approach. First, it must be established whether or not such postures are indeed necessary. If not, better solutions for the work can be found that no longer include them. If the postures in question cannot be avoided, special body supports must be designed; for example, military standards and specifications describe the body supports of tank crews and of pilots in fighter aircraft. Helander (1981) has compiled recommendations for the construction industry. A semireclining chair for overhead tasks has been proposed by Lee et al. (1991), and space requirements for operator compartments in low-seam mining equipment have been described by Conway and Unger (1991) (See Figure 8–12.)

WORK IN RESTRICTED SPACES

In some cases, work must be performed in restricted spaces, such as cargo holds of airplanes or underground mines. The primary restriction usually lies in the lowered ceiling of the workspace. Work becomes more difficult and stressful as the ceiling height forces workers to bend their necks and backs, or requires squatting or even lying. Thus, if work must be done in low-ceilinged spaces, equipment and mechanical aids should be developed that alleviate the task. For example, in aircraft baggage handling, it is advantageous to first collect the luggage in containers and then put these in place within the cargo hold, rather than handling individual pieces.

Body Position	Height H (cm)		Depth D (cm)	
	Minimum	Preferred	Minimum	Preferred
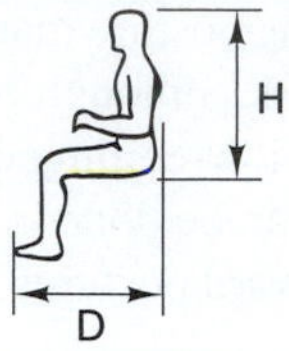 H D	103	110	94	100
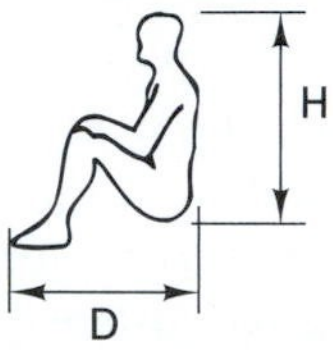 H D	98	110	94	100
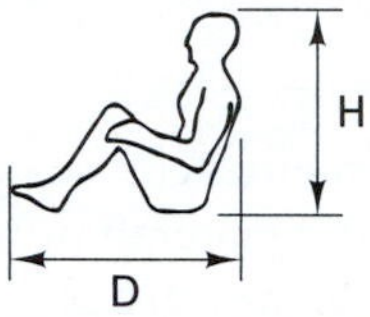 H D	98	110	88	90
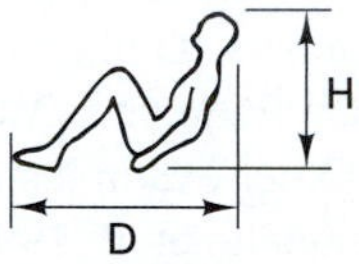 H D	64	65	125	140
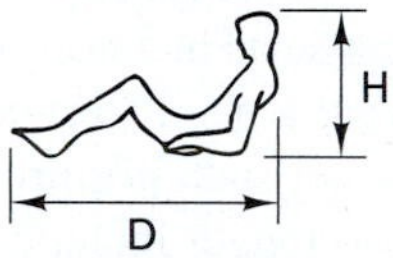 H D	60	62	150	160
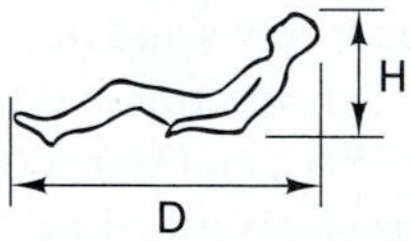 H D	50	54	170	180
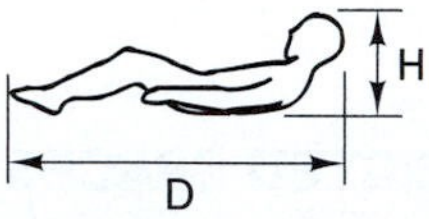 H D	48	52	190	195
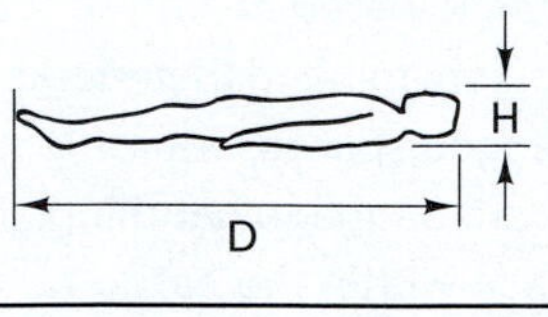 H D	38	46	200	210

Figure 8–12. Spaces required to accommodate U.S. coal miners (adapted from Conway and Unger 1991).

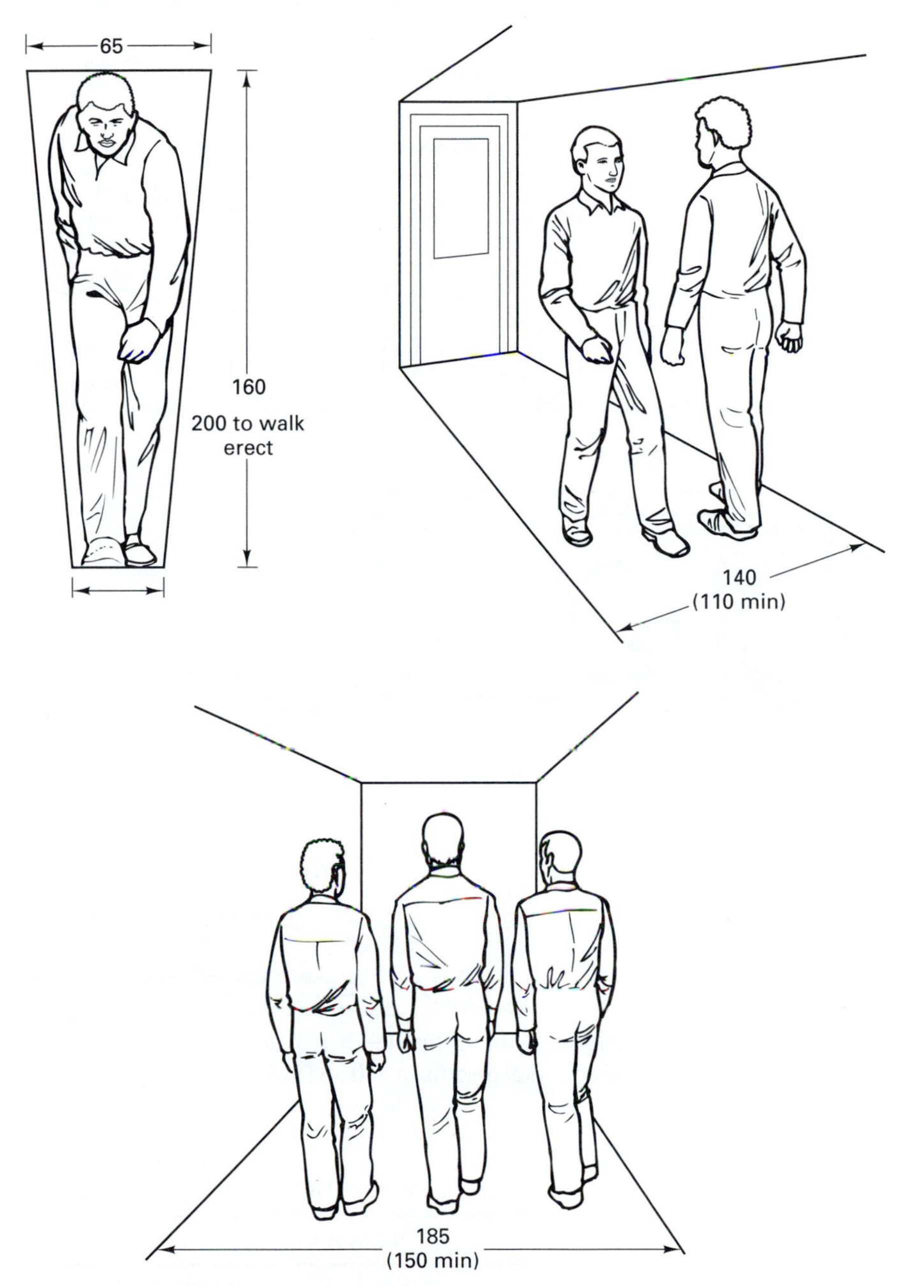

Figure 8–13. Minimal dimensions (in cm) for passageways and hallways (adapted from VanCott and Kinkade 1972).

Other restricted spaces are passageways, walkways, hallways, and corridors. Minimal dimensions for these are given in Figure 8–13. Dimensions for tight places, where one may have to squat, kneel, or lie on the back or belly, are given in Figure 8–14 and Table 8–2. Dimensions for escape hatches, shown in Figure 8–15,

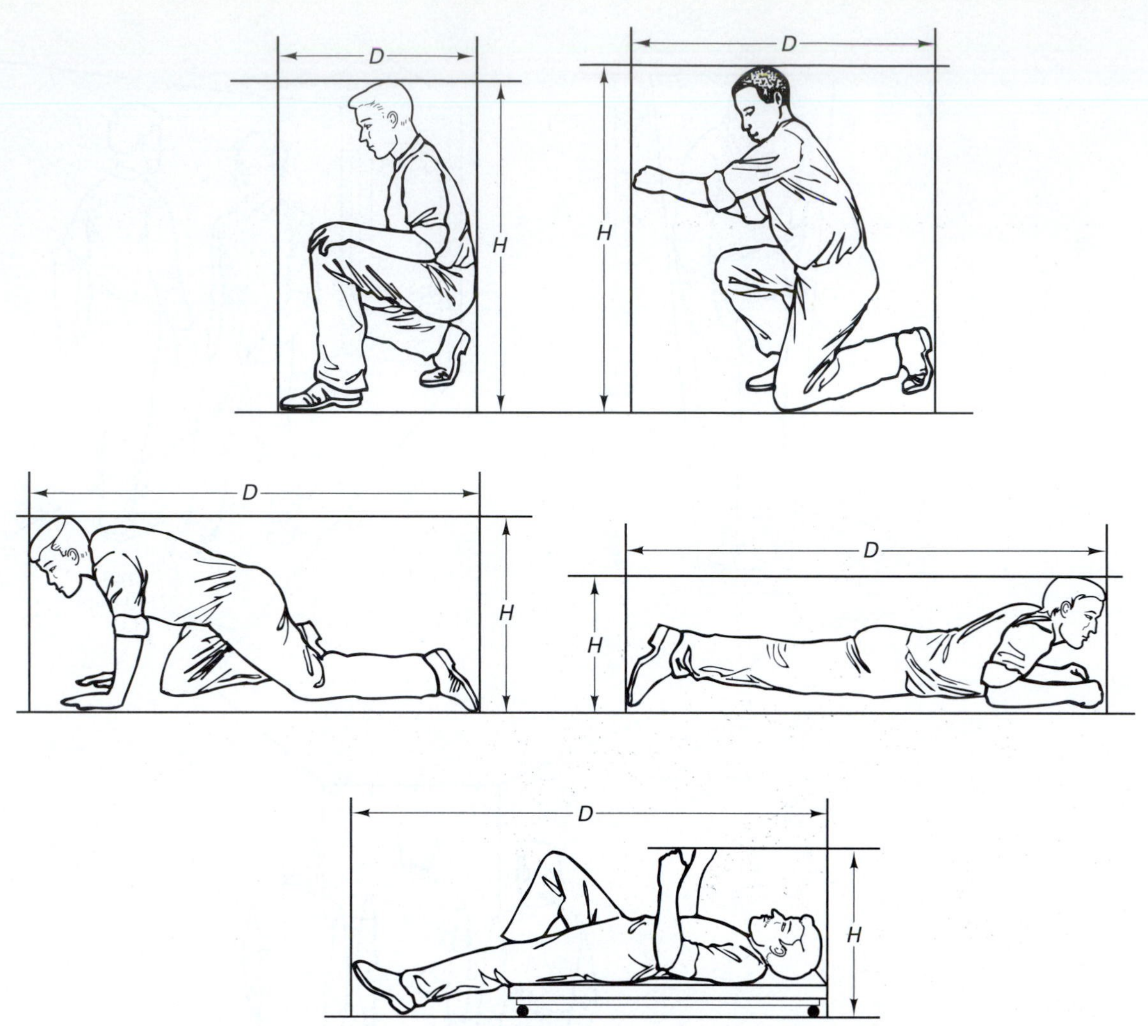

Figure 8–14. Minimum height and depth dimensions for "tight" workspaces (adapted from MIL-STD 759).

TABLE 8–2. Dimensions (in cm) for "Tight" Workspaces

	Height H			*Depth D*		
	Minimal	*Preferred*	*Arctic clothing*	*Minimal*	*Preferred*	*Arctic clothing*
Stooped or squatting	66	—	130	61	90	—
Kneeling	140	—	150	106	122	127
Crawling	79	91	97	150	—	176
Prone	43	51	61	285	—	—
Supine	51	61	66	186	191	198

Source: Adapted from MIL-STD 759.

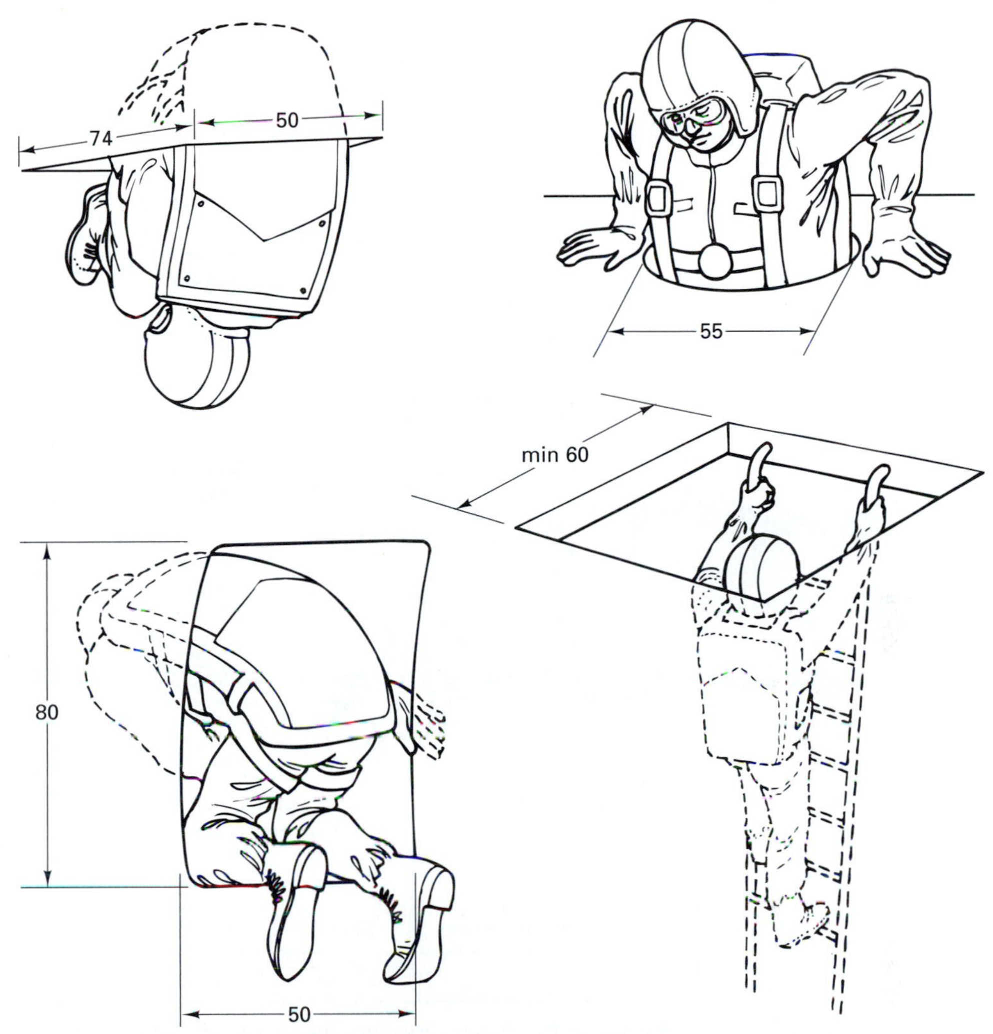

Figure 8–15. Minimal openings for escape hatches (adapted from Van Cott and Kinkade 1963).

need to accommodate even the largest persons wearing their work clothes and possibly using equipment. These openings can be made somewhat smaller for maintenance workers who need to get through access openings in enclosures of machinery; recommended dimensions are shown in Figure 8–16. The size for openings through which one hand must pass, holding and operating a tool, depends on the given circumstance, yet some recommended dimensions are shown in Figure 8–17. These dimensions need to be modified if the operator also must see the object through the opening and if special tools must be used and movements performed with one hand. In some cases, both hands and arms must fit through the opening, which then needs to be about shoulderwide. For further information, see the standards issued by ISO, NASA, the U.S. military, and various design handbooks [e.g., by Eastman-Kodak (1986) or Woodson et al. (1991)].

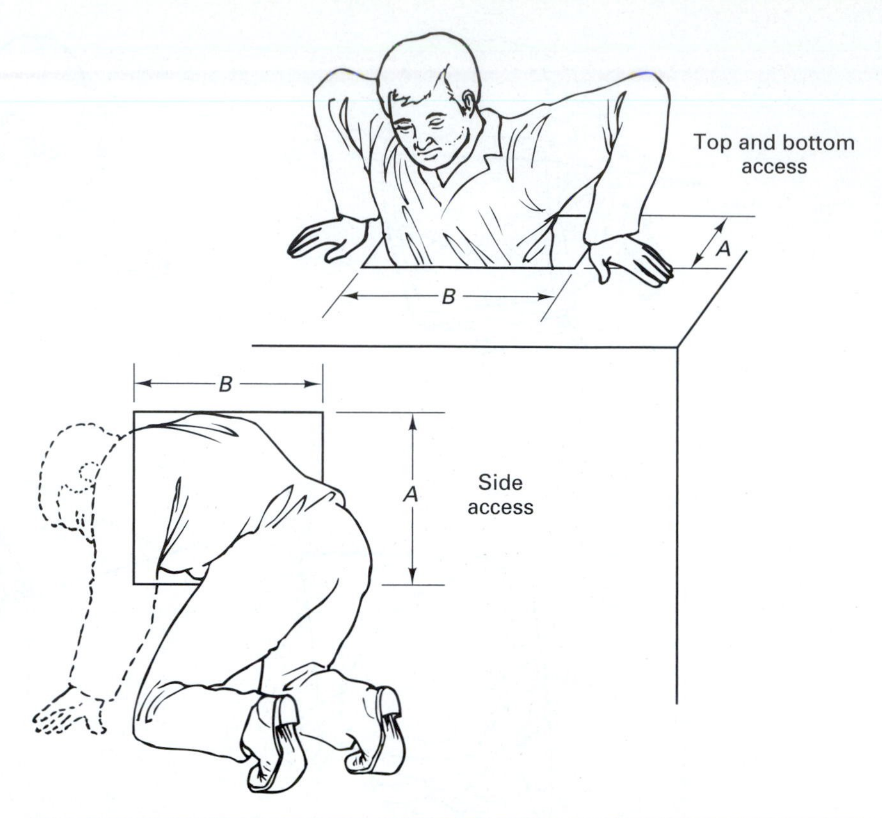

Dimensions	A, depth		B, width	
Clothing	Light	Bulky	Light	Bulky
Top and bottom access	33 cm	41 cm	58 cm	69 cm
Side access	66 cm	74 cm	76 cm	86 cm

Figure 8–16. Access openings for enclosures (adapted from MIL-HDBK 759).

DESIGNING FOR FOOT OPERATION

In comparison to hand movements over the same distance, foot motions consume more energy, are less accurate, and are slower; but they are more powerful, as one would expect from biomechanical considerations.

If a person stands at work, fairly little force and fairly infrequent operations of foot controls should be required, because, during these exertions, the operator has to stand on the other leg alone. For a seated operator, however, the operation of foot controls is much easier, because the body is largely supported by the seat. Thus, the feet can move more freely and, given suitable conditions, can exert large forces and energies.

	Approximate dimensions (cm) A	B	Task
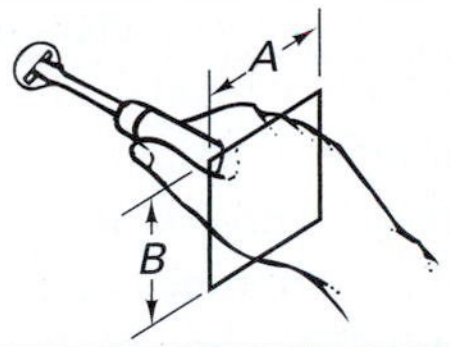	11	12	Using common screwdriver, with freedom to turn hand through 180°
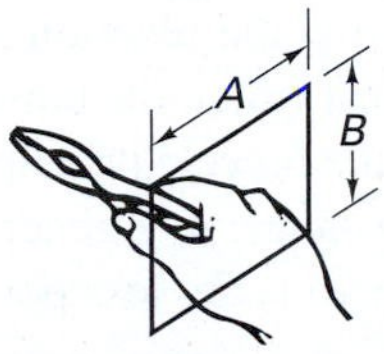	13	12	Using pliers and similar tools
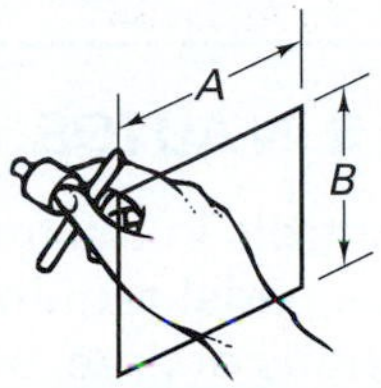	14	16	Using "T" handle wrench, with freedom to turn hand through 180°
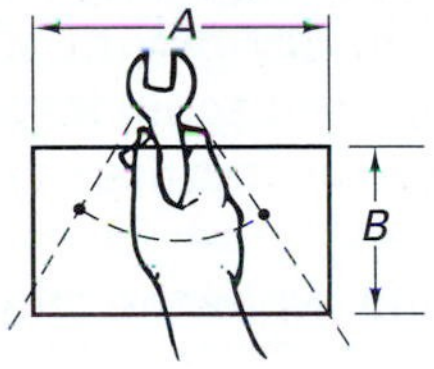	27	20	Using open-end wrench, with freedom to turn wrench through 60°
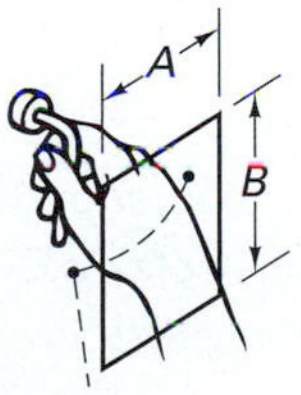	12	16	Using Allen-type wrench with freedom to turn wrench through 60°
	9	9	Using test probe, etc.

Figure 8–17. Minimal opening sizes (in cm) to allow one hand holding a tool to pass (adapted from MIL-HDBK 759).

Bicycling

A good example of exerting energy with the feet is pedaling a bicycle: All energy is transmitted from the leg muscles through the feet to the pedals. The pedals should be located directly underneath the body, so that the body weight above provides the reactive force to the force transmitted to the pedals. The crank radius should be about 15 centimeters for short-legged persons and up to 20 centimeters for persons with long legs. Suitable pedal rotation is usually between 0.5 and 1 Hz, but depends on such factors as the gear ratio of the bicycle (which is often a variable), the ground surface, and the purpose of bicycling—for leisure, exercise, or competition. In some cases, it is desirable to lower the center of mass of the combined person–bicycle system. Then the cranks may be moved forward and upward, to nearly the height of the seat. Placing the pedals forward makes the body weight less effective in generating the reaction force to the pedals' effort; hence some sort of backrest should be provided, against which the buttocks and back press while the feet push forward on the pedals. Instead of using the bicycle principle to propel the body, one can employ that approach to generate energy—for instance, in a generator of electricity. (In this case, "pedaling" may be done with the hands, but the arms are less powerful than the legs.)

TERRIBLE PEDALS IN AUTOS

The traditional arrangement of foot controls in the automobile is, by all human-factors rules, atrocious: The gas pedal requires that the foot be kept in the same position over long periods of time, and the brake pedal must be reached by a complex and time-consuming motion of the foot from the gas toward the body, to the left (or right, depending on which side of the car the driver's seat is on) and then again forward. It makes no sense that, although pushing forward on the gas accelerates the vehicle, pushing forward on the brake decelerates the car. The arrangement also encourages the use of only one foot alone, while the other foot is left idle.

Small forces, such as those used for the operation of switches, can be generated in nearly all directions with the feet, with the downward or down-and-forward direction preferred. The largest forces can be generated with extended or nearly extended legs, limited in the downward direction by body inertia and in the more forward direction by both inertia and the provision of buttock and back support surfaces. These principles are typically applied in automobiles. For example, a clutch or brake pedal can normally be operated easily while the knee is at about a right angle. But if the power-assist system fails, very large forces must be exerted with the feet. In that case, one must thrust one's back against a strong backrest and extend the legs to generate the needed pedal force.

Figures 8–18 through 8–23 provide information about the forces that can be generated with the legs via the feet, depending on the body support and the hip and knee angles. The largest forward thrust can be exerted with the nearly extended legs, which leaves very little room to move the foot control further away from the hip. Actual force data are compiled, for example, in NASA and U.S. military standards and by Eastman-Kodak (1986) and Woodson et al. (1991). However, caution is necessary when applying these data, because they were measured on different populations under varying conditions.

As with the hands, there is a preferred workspace for the feet. It is shown in Figure 8–24.

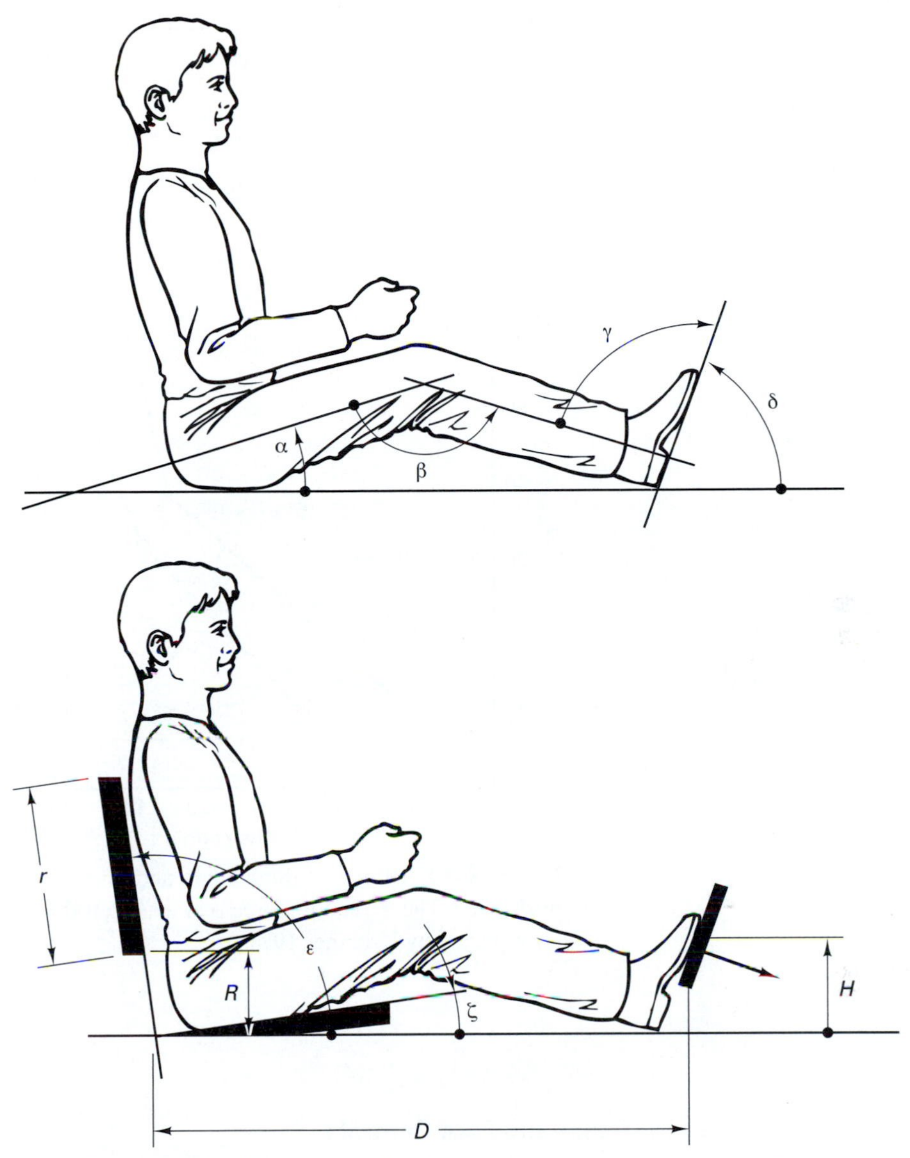

Figure 8–18. Conditions affecting the force that can be exerted on a pedal: body angles (top) and workspace dimensions (bottom).

In automobiles, power-assisted brakes and steering systems generate a difficult design problem. As long as auxiliary power is available, brakes can be operated easily, in almost any conceivable leg posture. Yet, if the auxiliary system fails, forces are suddenly required from the operator that are 3 to 10 times as high for regular braking (or steering). Accordingly, the driver must not only recognize that much more effort is required, but also must develop this effort quickly and often in a body posture—for example, with a strongly bent knee—that is not favorable for such a large exertion.

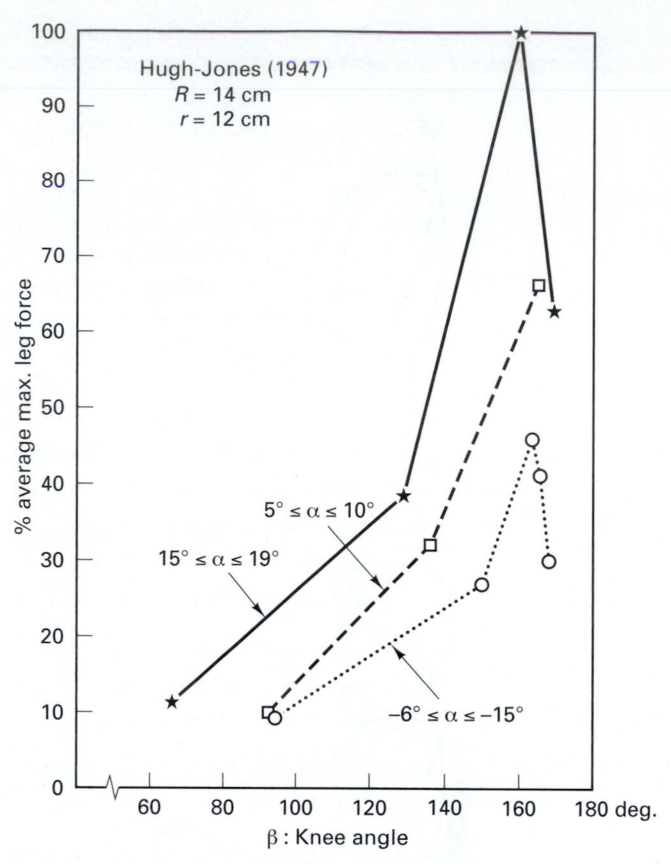

Figure 8–19. Effects of thigh angle and knee angle on pedal push force. The maximum force is at least 2,100 N (three studies reported by Kroemer 1971).

Rules for Designing Foot Controls

1. Require repeated operation only from a seated operator.
2. Design for pushing roughly in the direction of the lower leg.
3. Have the person exert small forces by tilting the foot about the ankle.
4. Have the person apply large forces by pushing with the whole leg, preferably with a solid back support from the seat.
5. Do not require fine control, continuous operation, or quick movements. (Note, however, that all of these are foolishly expected for speed control in today's automobiles.)

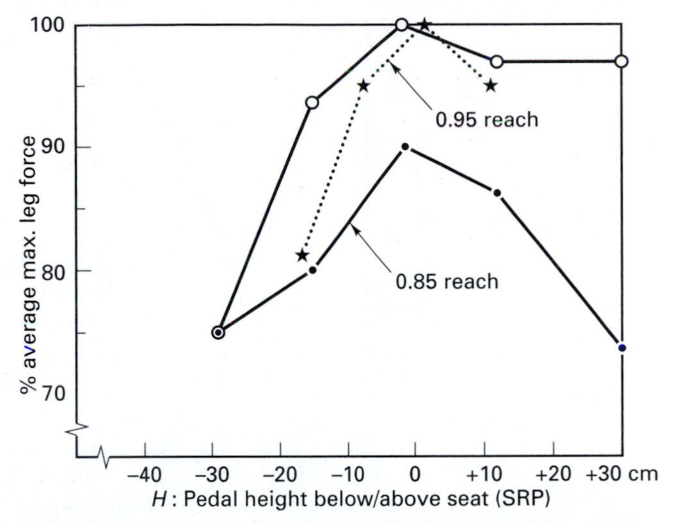

Figure 8–20. Effects of pedal height H on pedal push force. The maximal force is about 2,600 N (two studies reported by Kroemer 1972).

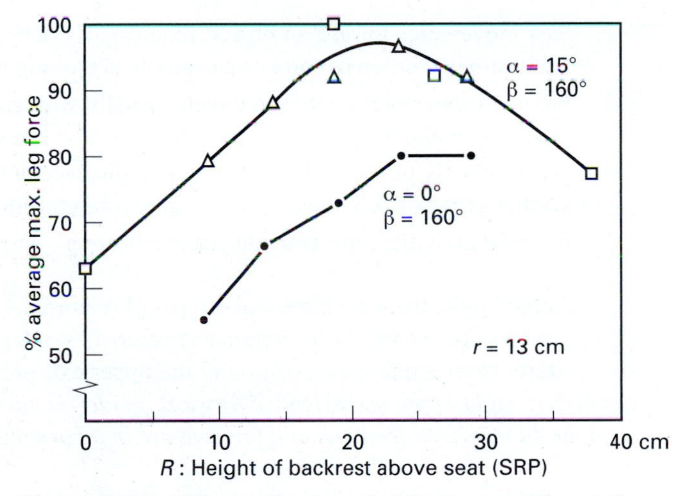

Figure 8–21. Effects of backrest height R on pedal push force. The maximal force is about 1,700 N (Kroemer 1972).

DESIGNING FOR HAND USE

The human hand is able to perform a large variety of activities, ranging from those that require fine control to others that demand large forces. (But the feet and legs are capable of more forceful exertions than the hand; see previous section.)

One may divide hand tasks in the following manner:

1. Fine manipulation of objects, with little displacement and force. Examples are writing by hand, assembling small parts, and adjusting controls.

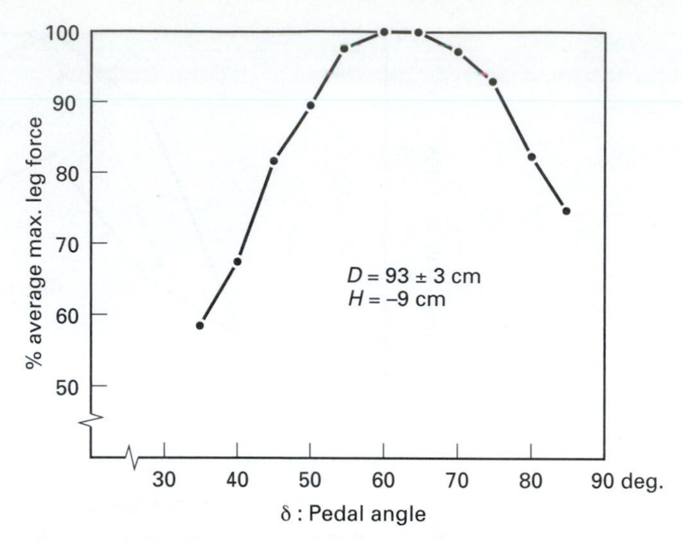

Figure 8–22. Effects of pedal (ankle) angle on foot force generated by ankle rotation. The maximal force is about 600 N (Kroemer 1971).

2. Fast movements toward an object, requiring moderate accuracy to reach the target, but a fairly small exertion of force. An example is moving to a switch and then operating it.
3. Frequent movements between targets, usually with some accuracy, but little force. An example is an assembly task wherein parts must be taken from bins and assembled.
4. Forceful activities with little or moderate displacement (such as are required for many assembly or repair activities—for example, when turning a hand tool against resistance).
5. Forceful activities with large displacements (e.g., when hammering).

Accordingly, there are three major types of requirements: for accuracy, for strength, and for displacement. For each of these, certain characteristics of hand–arm movements can be described if one starts from a reference position at the upper extremity: *The upper arm hangs down; the elbow is at right angle and extended forward, rendering the forearm horizontal; and the wrist is straight. In this case, the hand and forearm are in a horizontal plane at approximately umbilical height.*

Accurate and fast movements

Fitts' law provides guidance for accurate and fast movements (see Chapter 3): The smaller the distance traveled and the larger the target, the more accurate is a fast movement. Thus, finger movements are the fastest and most accurate, followed by movements of the forearm only. Among forearm movements (when the upper arm remains still), a horizontal forearm sweep (seemingly rotating about the elbow, but in fact turning about the shoulder joint in which the upper arm twists) is faster, more accurate, and less tiring than when the forearm flexes or extends in the elbow. The least accurate and most time-consuming and fatiguing movements are those in which the upper arm is pivoted out of its vertical reference location. This establishes the preferred manipulation space mentioned in the first chapter. Its location is sketched in Figure 8–25.

Forceful exertions

Exerting force with the hands is a complex matter. Of the digits, the thumb is the strongest and the little finger the weakest. The gripping and grasping strengths of the whole hand are larger then exerted with any digit alone, but depend on the coupling used between the hand and the

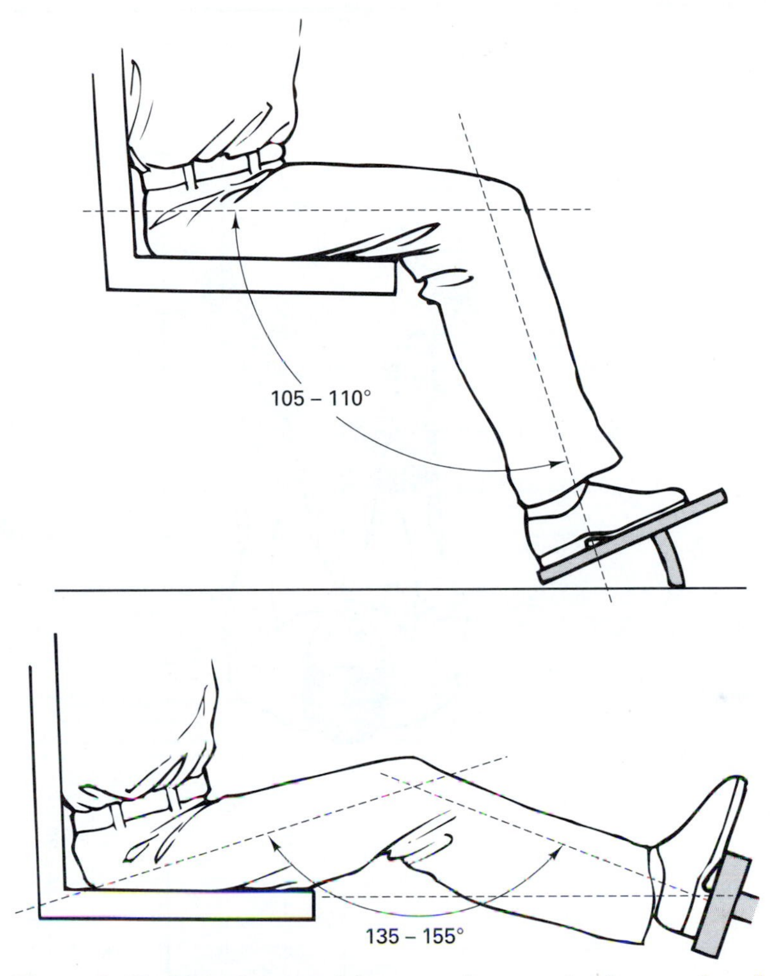

Figure 8–23. Light downward forces can be exerted at knee angles of about 105 to 110°, while strong forward forces require knee extension at 135 to 155° (adapted from VanCott and Kinkade 1972).

handle that is gripped or grasped. (See Figure 8–26.) The forearm can develop fairly large twisting torques. Large force and torque vectors are exertable with the elbow bent at about a right angle, but the strongest pulling or pushing forces toward or away from the shoulder, respectively, can be exerted with the extended arm, provided that the trunk can be braced against a solid structure. Torque about the elbow depends on the elbow angle, as depicted in Figure 8–27. Thus, forces exerted with the arm and shoulder muscles are largely determined by body posture and body support, as shown in Figure 8–28. Likewise, finger forces depend on the angle of the finger at the joint that is crooked, as listed in Tables 8–3 and 8–4.

Table 8–5 provides detailed information about manual strength capabilities measured in male students and male machinists. Female students developed between 50 and 60 percent the digit strengths of their male peers, but 80 to 90 percent in tip or pad "pinches". Yet, the data presented in the table, as well as those found elsewhere, must be applied with much caution, because they

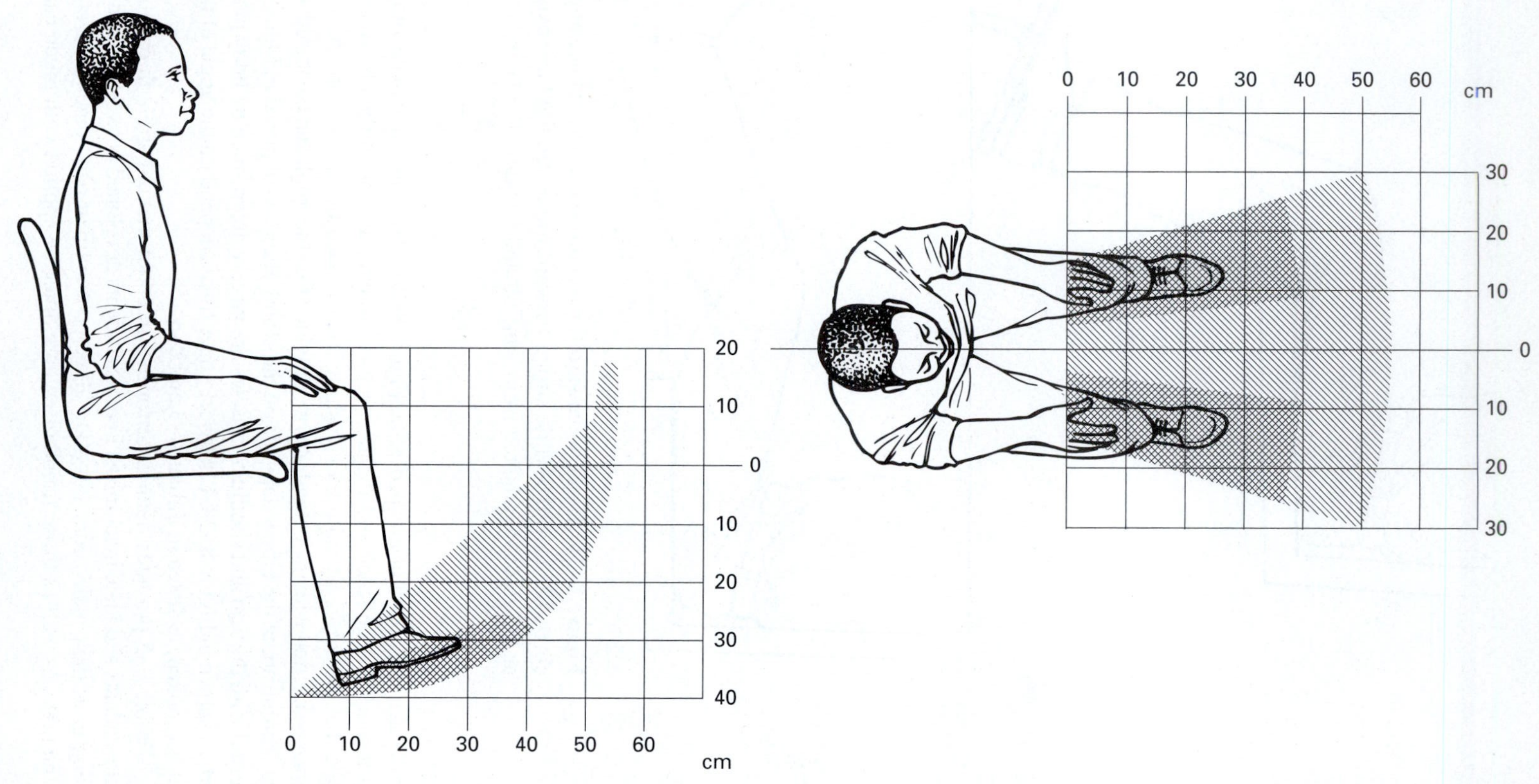

Figure 8–24. Preferred (crosshatched) and regular workspaces for the feet, assuming a seated operator.

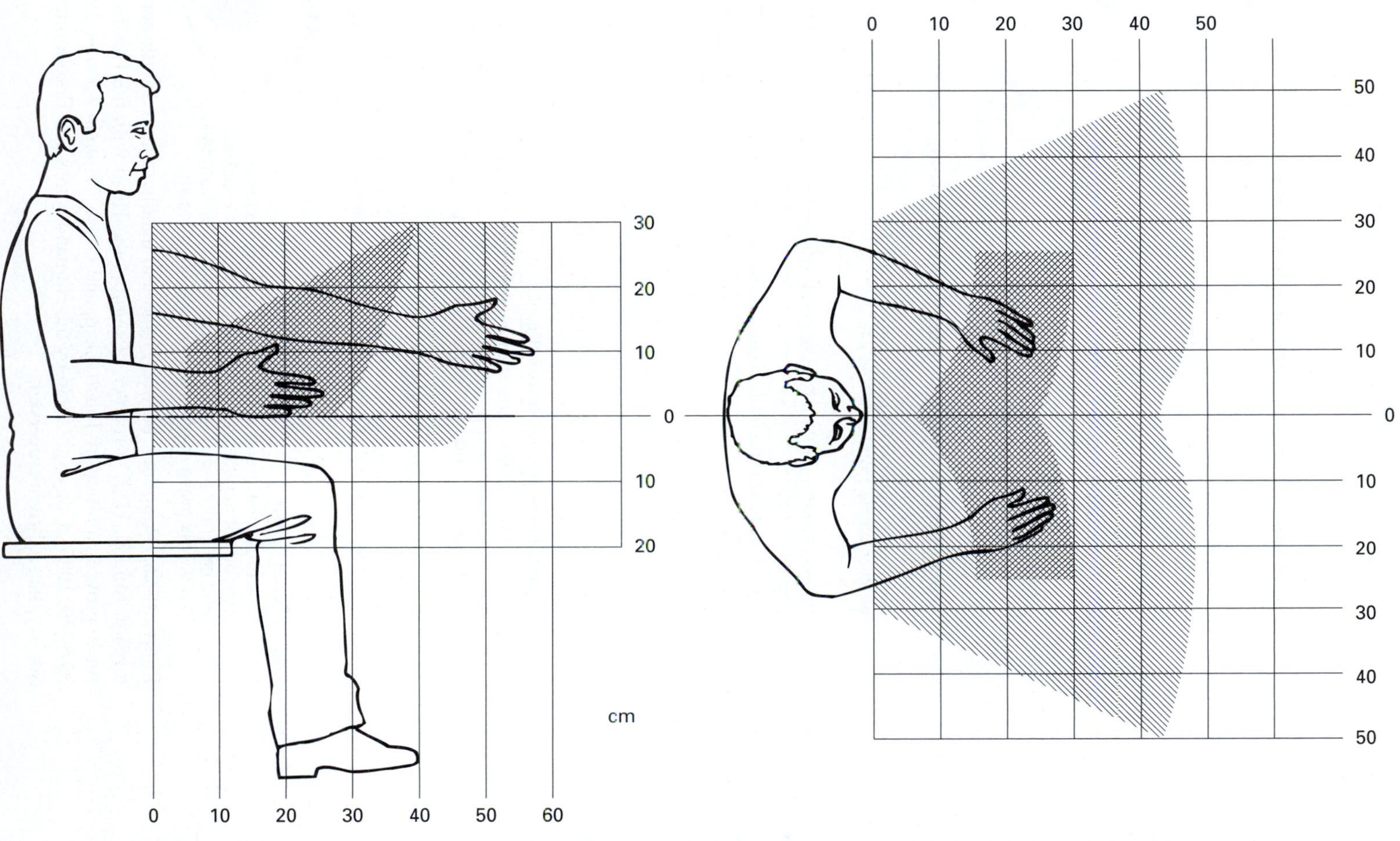

Figure 8–25. Preferred (crosshatched) and regular manipulation spaces within the overall reach envelope of the hands.

Coupling #1. Digit Touch:
One digit touches an object.

Coupling #2. Palm Touch:
Some part of the palm (or hand) touches the object.

Coupling #3. Finger Palmar Grip (Hook Grip):
One finger or several fingers hook(s) onto a ridge, or handle. This type of finger action is used where thumb counterforce is not needed.

Coupling #4. Thumb–Fingertip Grip (Tip Pinch):
The thumb tip opposes one fingertip.

Coupling #5. Thumb–Finger Palmar Grip (Pad Pinch, or Plier Grip):
Thumb pad opposes the palmar pad of one finger (or the pads of several fingers) near the tips. This grip evolves easily from coupling #4.

Coupling #6. Thumb–Forefinger Side Grip (Lateral Grip, or Side Pinch):
Thumb opposes the (radial) side of the forefinger.

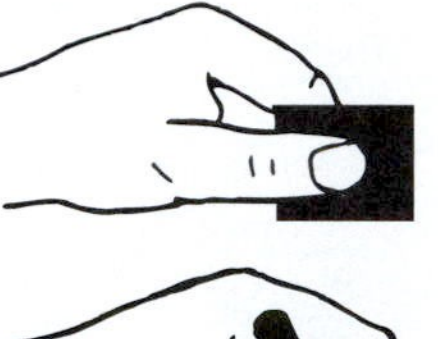

Coupling #7. Thumb–Two–Finger Grip (Writing Grip):
Thumb and two fingers (often forefinger and middle finger) oppose each other at or near the tips.

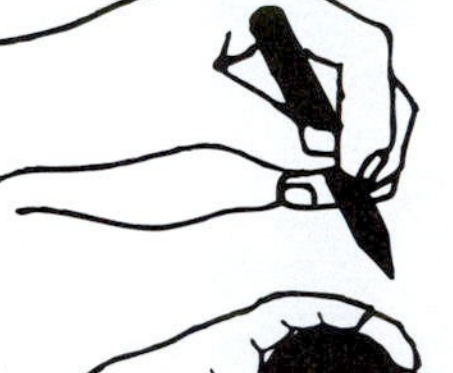

Coupling #8. Thumb–Fingertips Enclosure (Disk Grip):
Thumb pad and the pads of three or four fingers oppose each other near the tips (object grasped does not touch the palm). This grip evolves easily from coupling #7.

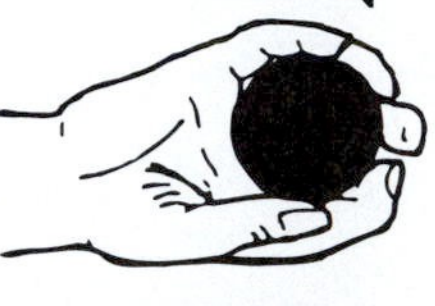

Coupling #9. Finger–Palm Enclosure (Collet Enclosure):
Most, or all, of the inner surface of the hand is in contact with the object while enclosing it. This enclosure evolves easily from coupling #8.

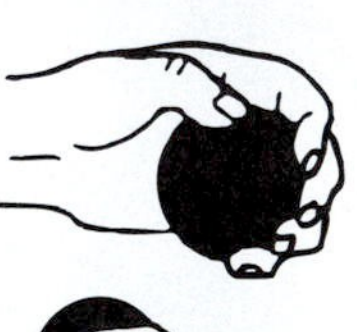

Coupling #10. Power Grasp:
The total inner hand surfaces is grasping the (often cylindrical) handle which runs parallel to the knuckles and generally protrudes on one or both sides from the hand. This grasp evolves easily from coupling #9.

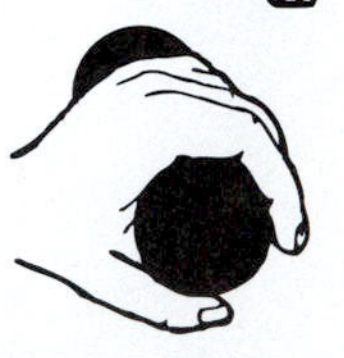

Figure 8–26. Couplings between hand and handle (adapted with permission from K. H. E. Kroemer, "Coupling the Hand with the Handle: An Improved Notation of Touch, Grip, and Grasp," *Human Factors 28,* 337-339, 1986). Copyright 1986 by the Human Factors and Ergonomics Society, Inc. All rights reserved.

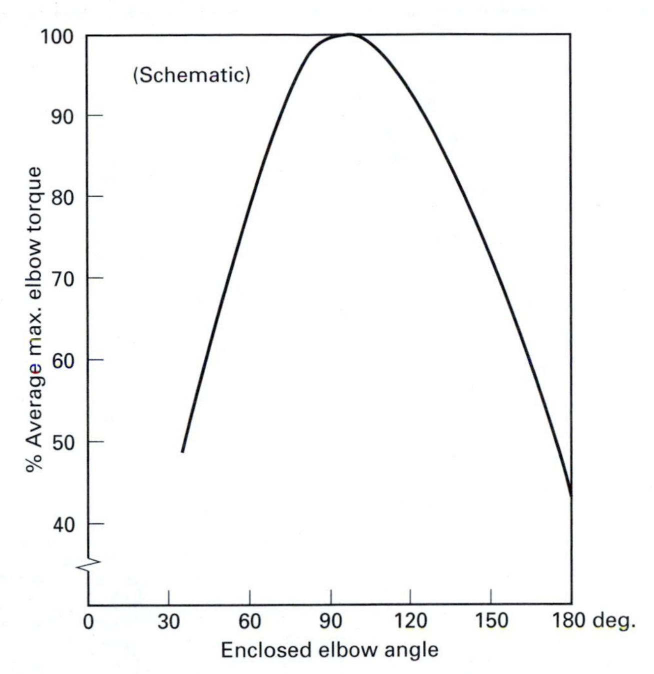

Figure 8–27. Effects of elbow angle on elbow torque.

are likely to have been obtained from different groups of subjects, with different techniques and under different physical and psychological conditions than in the case at hand (Astin 1999). Furthermore, the users to which the data are to be applied may be fatigued or, conversely, may be particularly trained or motivated, with possibly major effects on their strength. (See Chapter 1.)

DESIGNING HAND TOOLS

Hand tools need to fit the contours of the hand, need to be held securely with straight wrist and suitable arm postures, and must utilize strength and energy capabilities without overloading the body. Hence, the design of hand tools is a complex ergonomic task.

The use of hand tools is as old as humankind. The art developed from simple beginnings—using a stone, bone, or piece of wood that fitted the hand and performed a certain function—to the purposeful design of modern implements (such as the screwdriver, cutting pliers, or power saws) and of controls in airplanes and power stations. (See Chapter 10.) A vast literature is available on tool design, recently summarized by Bullinger et al. (1997) and by Freivalds (1999).

Fit the tool

The tool must fit the dimensions of the hand and utilize the strength and motion capabilities of the hand–arm–shoulder system. Some dimensions of the human hand were given in Chapter 1; further information can be found in the publications by Gordon et al. (1990), NASA/Webb (1978), Wagner (1988), and, particularly, Greiner (1991).

Many methods of gripping

Some hand tools, such as surgical instruments, screwdrivers used by optometrists, and writing instruments, require a fairly small force, but precise handling. Commonly, the manner of holding these tools has been called "precision gripping." Other instruments must be held strongly between large surfaces of the fingers, thumb, and palm. Such holding of the hand tool allows large

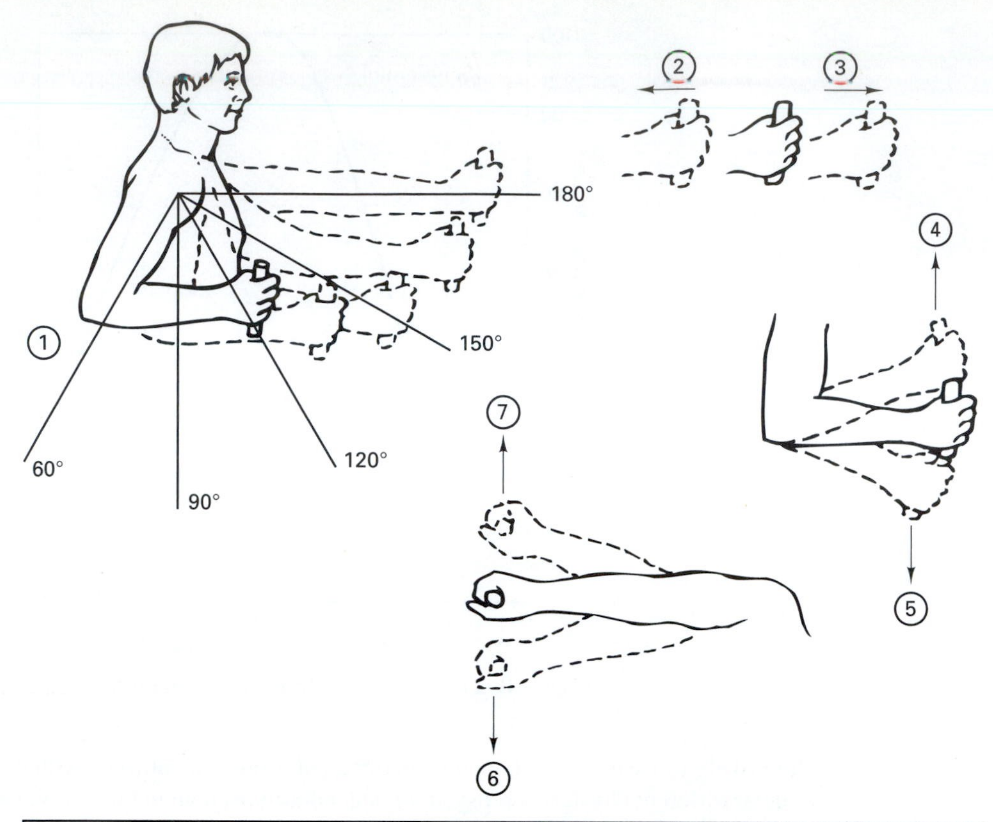

Fifth-percentile arm strength (N) exerted by sitting men													
(1)	(2)		(3)		(4)		(5)		(6)		(7)		
Degree of elbow flexation (deg)	Pull		Push		Up		Down		In		Out		
	Left	Right	L	R	L	R	L	R	L	R	L	R	
180	222	231	187	222	40	62	58	76	58	89	36	62	
150	187	249	133	187	67	80	80	89	67	89	36	67	
120	151	187	116	160	76	107	93	116	89	98	45	67	
90	142	165	98	160	76	89	93	116	71	80	45	71	
60	116	107	96	151	67	89	80	89	76	89	53	71	

Figure 8–28. Fifth-percentile arm strengths exerted by sitting men (adapted from MIL-HDBK 759).

forces and torques to be exerted and hence has commonly been called "power grasping." Yet, there are many transitions between these two extremes, from merely touching an object with a finger (such as pushing a button) to pulling on a hooklike handle and from holding small objects between the fingertips to transmitting large energy from the hand to the handle. One attempt to systematically arrange the various couplings of hand with handle is shown in Figure 8–26.

Shaping the grip

For the touch-type couplings (numbers 1 through 6 in Figure 8–26), relatively little attention must be paid to fitting the surface of the handle to the touching surface of the hand. Yet, one may want to put a slight cavity into the surface of a pushbutton so that the fingertip does not slide off, to hollow out the handle of a scalpel slightly so that the fingertips can hold on securely, and

TABLE 8–3. Average Forces (and Standard Deviations), in Newtons, Exerted by Nine Subjects in Fore, Aft, and Down Directions with the Fingertips, Depending on the Angle of the Proximal Interphalangeal PIP Joint

	PIP joint at 30°			*PIP joint at 60°*		
Direction	*Fore*	*Aft*	*Down*	*Fore*	*Aft*	*Down*
DIGIT						
2 Index	5.4	5.5	27.4	5.2	6.8	24.4
	(2.0)	(2.2)	(13.0)	(2.4)	(2.8)	(13.6)
2 *n*	4.8	6.1	21.7	5.6	5.3	25.1
	(2.2)	(2.2)	(11.7)	(2.9)	(2.1)	(13.7)
3 Middle	4.8	5.4	24.0	4.2	6.5	21.3
	(2.5)	(2.4)	(12.6)	(1.9)	(2.2)	(10.9)
4 Ring	4.3	5.2	19.1	3.7	5.2	19.5
	(2.4)	(2.0)	(10.4)	(1.7)	(1.9)	(10.9)
5 Little	4.8	4.1	15.1	3.5	3.5	15.5
	(1.9)	(1.6)	(8.0)	(1.6)	(2.2)	(8.5)

Source: Kroemer, unpublished data.
n: Nonpreferred hand.

TABLE 8–4. Mean Digit Poke Forces (and Standard Deviations), in Newtons, Exerted by 30 Subjects in Direction of the Straight Digits (See Also Table 8–5.)

Digit	*Ten mechanics*	*Ten male students*	*Ten female students*
1, Thumb	83.8 (25.2)	46.7 (29.2)	32.4 (15.4)
	A	C	D
2, Index Finger	60.4 (25.8)	45.0 (30.00)	25.4 (9.6)
	B	C	DE
3, Middle Finger	55.9 (31.9)	41.3 (21.6)	21.5 (6.5)
	B	C	E

Source: From Williams 1988.
Entries with different letters are significantly different from each other ($p \leq .05$).

to roughen the surface of a dentist's tool and not make it round in cross section, which would allow it to turn in the hand. Thus, design details that facilitate holding onto the tool, moving it accurately, and generating force or torque play important roles even for small hand tools.

Strong couplings

Considerations of "secure tool handling" become even more important for the more powerful enclosure couplings (numbers 8, 9, and 10 in Figure 8–26). These types of interlocking hand and handle are typically used when large energies must be transmitted between the hand and the tool. The design purpose is to hold the handle securely (without fatiguing muscles unnecessarily, and avoiding pressure points) while exerting a linear force or rotating torque at the working end of the tool.

Transmitted energy

It is important to distinguish between the energy transmitted to the work object and the energy transmitted between the hand and the handle. In many cases, the energy transmitted to the external object is not the same in type, amount, or time as that generated between hand and handle. Consider, for example, the impulse energy transmitted by the head of a mallet, compared to the way energy is transmitted between hand and handle, or the torque applied to a screw with the tip of a screwdriver, compared to the combination of thrust and torque generated by the hand. Thus, the ergonomist must consider both the interface between tool and object and the interface between tool and hand.

TABLE 8–5. Forces of Digits, and Grip and Grasp Forces Exerted by 21 Male Students* and 12 Male Machinists. Means (and Standard Deviations) in Newtons

Couplings (see Figure 8–26)	*Digit 1 (thumb)*	*Digit 2 (index)*	*Digit 3 (middle)*	*Digit 4 (ring)*	*Digit 5 (little)*	
Push with digit tip in direction of the extended digit ("poke")	91 (39)* 138 (41)	52 (16)* 84 (35)	51 (20)* 86 (28)	35 (12)* 66 (22)	30 (10)* 52 (14)	See also Table 8–4
Digit touch (coupling #1) perpendicular to extended digit	84 (33)* 131 (42)	43 (14)* 70 (17)	36 (13)* 76 (20)	30 (13)* 57 (17)	25 (10)* 55 (16)	—
Same, but all fingers press on one bar	—	digits 2, 3, 4, 5 combined: 162 (33)				
Tip force (as in typing; angle between distal and proximal phalanges about 135°)	—	30 (12)* 65 (12)	29 (11)* 69 (22)	23 (9)* 50 (11)	19 (7)* 46 (14)	—
Palm touch (coupling #2) perpendicular to palm (arm, hand, and digits extended and horizontal)	—	—	—	—	—	233 (65)
Hook force exerted with digit tip pad (coupling #3, "scratch")	61 (21) 118 (24)	49 (17) 89 (29)	48 (19) 104 (26)	38 (13) 77 (21)	34 (10) 66 (17)	all digits combined: 108 (39)* 252 (63)
Thumb–fingertip grip (coupling #4, "tip pinch")	—	1 on 2 50 (14)* 59 (15)	1 on 3 53 (14)* 63 (16)	1 on 4 38 (7)* 44 (12)	1 on 5 28 (7)* 30 (6)	—
Thumb–finger palmar grip (coupling #5, "pad pinch")	1 on 2 and 3 85 (16)* 95 (19)	1 on 2 63 (12)* 34 (7)	1 on 3 61 (16)* 70 (15)	1 on 4 41 (12)* 54 (15)	1 on 5 31 (9)* 34 (7)	—
Thumb–forefinger side grip (coupling #6, "side pinch")	—	1 on 2 98 (13)* 112 (16)	—	—	—	—
Power grasp (coupling #10, "grip strength")	—	—	—	—	—	318 (61)* 366 (53)

Source: Higginbotham and Kroemer, unpublished data.

Manually driven tools may be classified as follows, using, in part, Fraser's 1980 listing:

1. Percussive (eg, an ax or a hammer); human task: swing and hold a handle.
2. Scraping (saw, file, chisel, or plane); human task: push or pull and hold a handle.
3. Rotating or boring (borer, drill, screwdriver, wrench, or awl); human task: push or pull, or turn and hold, a handle.
4. Squeezing (pliers or tongs); human task: press and hold a handle.
5. Cutting (scissors or shears); human task: pull and hold a handle.
6. Cutting (knife); human task: pull or push and hold a handle.

Note that in each case the operator must hold the tool.

Power-driven tools may use an electric power source (saw, drill, screwdriver, sander, grinder) compressed air (saw, drill, wrench), smoothed internal combustion (chain saw), or explosive power (bolter, cutter, riveter).

Impacts from power tools

When using manual tools, the operator generates all the energy and therefore is always in full control of the energy exerted (with the exception of percussive tools, such as hammers or axes). By contrast, with power-driven tools, the operator usually just holds or moves the tool, whose energy is mostly generated and applied to the outside by the auxiliary power source. Yet, if that tool suddenly encounters resistance, the reaction force may directly affect the operator, often via a jerk or impact that can lead to injury. This is frequently the case with powered chain saws, screwdrivers, wrenches, and augers. Another major problem with many powered hand tools, such as jackhammers, riveters, and power wrenches and sanders, is that vibrations may be transmitted to the operator. Vibrations and repeated impacts transmitted to the operator often lead to various kinds of overuse disorders, discussed later, particularly if the vibrations or impacts are associated with improper postures.

"Handy rule"

Proper posture of the hand–arm system while using hand tools is very important. As a rule, the wrist should not be bent, but must be kept straight to avoid overexertion of such tissues as tendons and tendon sheaths and compression of nerves and blood vessels.

Oblique thrust line

Normally, the grasp centerline, or thrust line, of a straight handle is at about 70 degrees to the forearm axis. (See Figure 8–29.) For example, the use of common straight-nosed pliers often requires a strong bend in the wrist, and neither the direction of thrust nor the axis of rotation corresponds with that of the hand and arm. This condition frequently results in wrist bending which reduces the force that can be applied (Zellers and Hallbeck 1995). "Bending the tool and not the wrist" improves the situation, as shown in Figure 8–30.

Formfitting

Another technique for avoiding unsuitable postures and unnecessary muscle exertions in keeping the tool in the hand is to fit the form of the handle to the human hand. Instead of using a straight, uniform surface (see Figure 8–31), the surface may be formed to fit the hand parts that touch—for example, the thenar pad and the rest of the palm, as shown in Figures 8–30, 8–32, and 8–33. Bulges and restrictions along the handle generate a "form-fit," which, in addition to friction, prevents the handle from sliding out of the hand. Strong notchings or serrations, however, or other extreme form-fits can make the handle very uncomfortable if it is held differently than anticipated by the tool designer.

Friction between hand and handle

Formfitting and surface treatment of the handle can be very important if dirt, dust, oil, or sweat reduces the coefficient of friction between the handle and the hand. In such a case, special shapes and surface treatments can either keep these media away or alleviate their effects (Bobjer et al. 1993).

The thrust force T that can be developed on a handle is a function of the coefficient of friction, μ, between the handle and the hand (or the glove) and the grasping force G (which is perpendicular to T):

$$T = \mu G. \qquad (8\text{–}1)$$

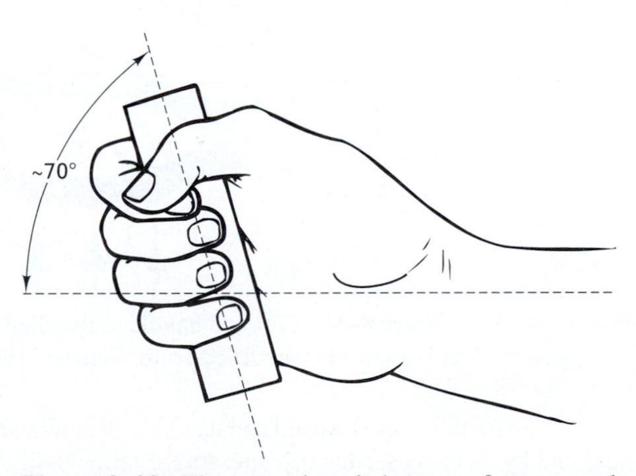

Figure 8–29. The natural angle between forearm and grasp center is about 70°.

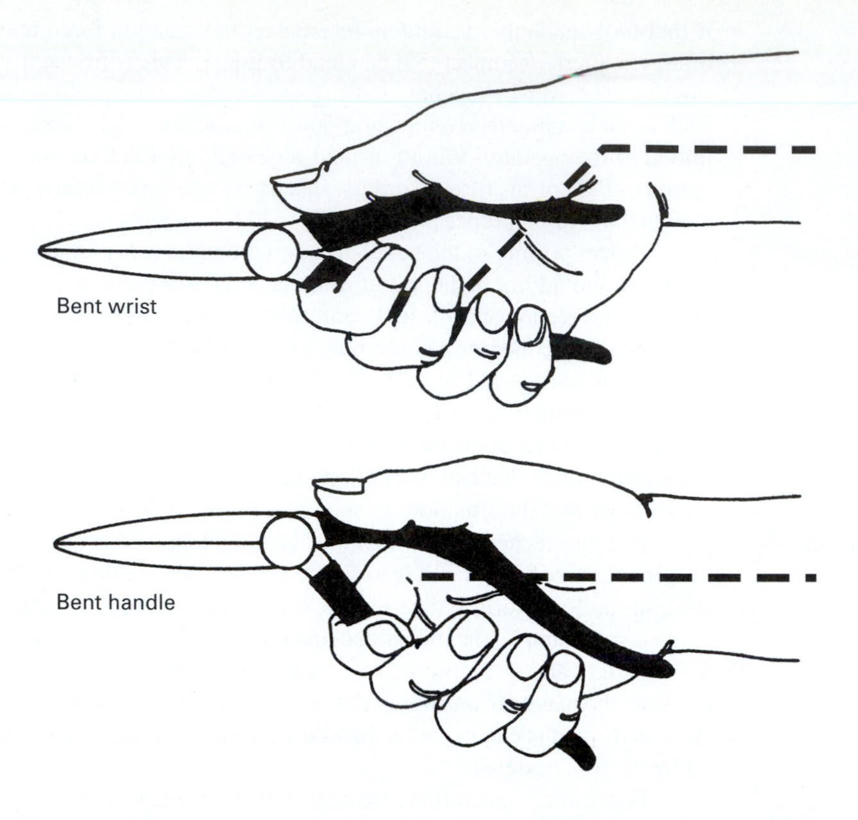

Figure 8–30. Common use of straight-nosed pliers is often accompanied by a strong bend in the wrist (modified from Tichauer 1973).

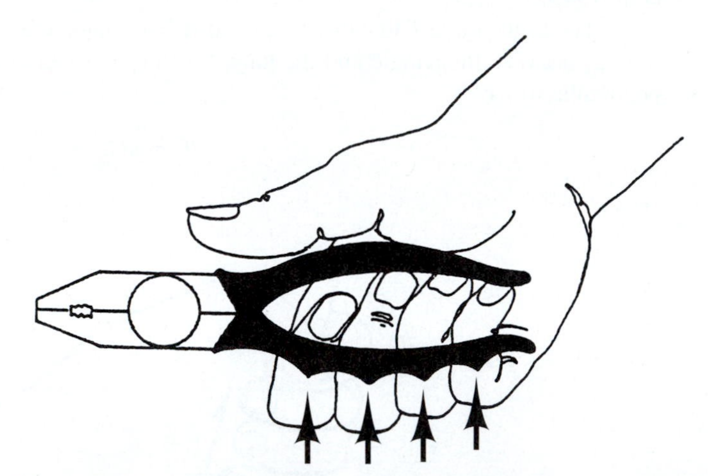

Figure 8–31. Fitting a handle to the shape of the hand can be helpful or painful (modified from Tichauer 1973).

The friction between hand and handle is determined largely by the surface texture of the handle and by its cross-sectional shape. As the terms indicate, "smooth" surfaces provide little friction, while it is difficult to slide on "rough" surfaces. Grooves, ridges, and serrations hinder sliding perpendicularly to them, the more so with more protruding and sharper edges. Of course, care

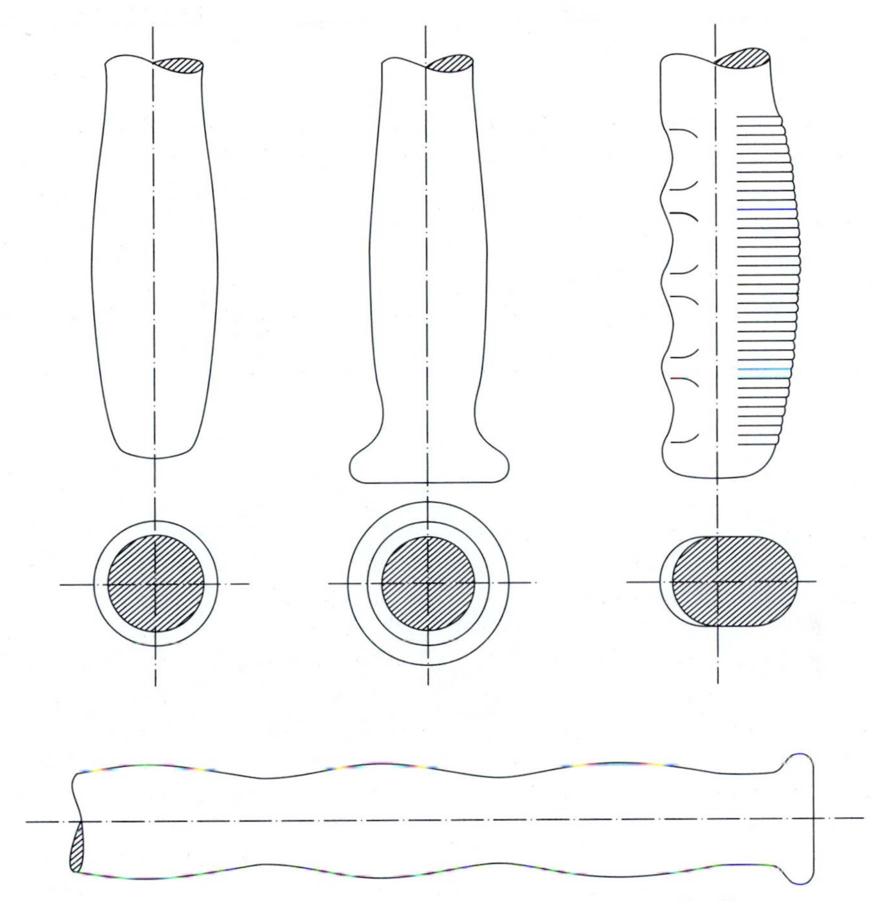

Figure 8–32. Suitable bulges and constrictions along a handle allow many hand positions without severe tissue compression. A flange at the end prevents the hand from sliding off the handle.

must be taken not to exert too much pressure that might damage hand tissues or even cut into the skin, a problem that can be alleviated by wearing gloves (which by themselves might increase the coefficient of friction). Thus, in the shaping of a handle, a proper balance must be found between easy sliding because of a low coefficient of friction and mechanical interlocking due to a unity coefficient of friction.

Thrust force

The greater the grasping force and the friction, the greater is the thrust force. The grasp force depends on the strength of a person's hand, which in turn depends on the sizes of the handle and hand. The coefficient of friction depends on the texture and shape of the handle, on the skin conditions of the hand (eg, soft vs. callous), on the surface of a glove or mitten if that is worn, and on intermediate materials: Sweat or grease increases slipperiness, while dust, sand, and other abrasive materials impede gliding.

Wearing gloves

Often, gloves or mittens are worn, either because of hot or cold temperatures that must be kept away from the skin, or because mechanical injuries must be avoided, or to alleviate the transmission of vibrations and shocks from the tool to the hand. Usually, wearing suitable gloves increases the friction between hand and handle. However, gloves must fit the hand well; too often, though, employees are supplied working gloves that are too large. The mismatch can make it difficult to grasp a handle securely, and excessive glove material can get caught, for instance, in a grinding wheel and lead to injury.

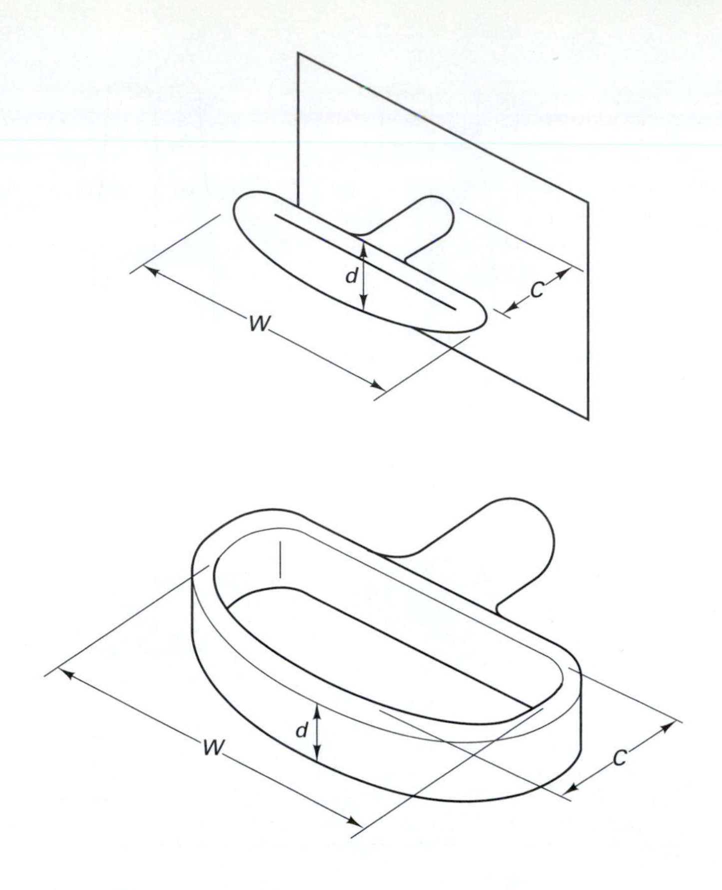

	W Width (cm)	d Thickness (cm)	C Clearance (cm)	D Displacement (cm)
Minimum	10	1,6	3,8	2,5
Maximum	13*	3,8	4,4**	Preferred 5

* Gloved operation of D handle
** Gloved operation of T handle

Figure 8–33. The T-handle is suitable for pushing and pulling, the D-handle mostly for pulling (MIL-HDBK 759). Either can be used with the left or right hand.

Space suit gloves

The design of gloves or mittens, and the materials they are made of, affect the amount of force that can be exerted. For example, gloves on space suits have been shown to strongly reduce not only the mobility and tactile sensitivity of the hand, but also endurance in exerting strength and the amount of force available to the outside. This is so because the hand inside the glove must expend much energy to bend and move the stiff glove itself before energy can be exerted to the object in space.

Hand strength and endurance

In terms of hand strength, the biomechanical details already discussed in Chapter 1 apply—in particular, whether intrinsic or extrinsic muscles of the hand are used, how well these muscles are developed, and the skill or experience of the user. It is also important to consider the duration of each grasp: Endurance is enhanced and fatigue reduced if the muscular effort is short and if it requires only a small percentage of the strength actually available.

The relation between handle size and hand size is important in two ways. If the handle is too small, not much force can be exerted, and large local tissue pressures might be generated (such as when one is writing with a very thin pencil). By contrast, if the handle is too large for the hand, the hand muscles must work at disadvantageous lever arms (such as when one is trying to squeeze the caulking gun common in the United States). Numerous studies on grip strength have been conducted. [See Reith's (1982) review.] In most cases, more-or-less cylindrical handles have been used in tests. With these handles, diameters between 3 and 4 centimeters have been found to allow the largest grasp or compression force, with up to 6-cm diameters being suitable for persons with large hands. Yet, assessing the grasp force is not as simple as one might believe, since the contribution of each digit or of sections of the palm should be measured separately (because each contributes its own "force times coefficient of friction"). A few studies shed light on the contributions of hand sections (eg, Bishu et al. 1992; Bjoering et al. 1998; Lowe and Freivalds 1998; Yun and Freivalds 1995), mostly, though, an overall, averaged grip strength is measured (Imrhan 1998).

Lefties

In North America and Europe, nine out of 10 people, men or women, are right handed, and some tools are designed to fit only the right hand. Many tools can be used with either the left or right hand, but about 1 in 10 persons prefers to use the left hand and has better skills and more strength available there. Thus, it is advisable to provide these individuals with hand tools specifically designed for use with the left hand—unless the tool suits both hands equally.

DESIGN RULES FOR HAND TOOLS

1. Push or pull in the direction of the forearm, with the handle directly in front of it: keep the wrist straight.
2. Provide good coupling between hand and handle by shape and friction.
3. Avoid pressure spots and "pinch points."
4. Round edges and pad surfaces.
5. Avoid tools that transmit vibrations to the hand.
6. Do not operate tools frequently and forcefully by hand; a robot or other machine is better suited for such activities.

DESIGNING FOR HUMAN STRENGTH

Data on forces and torques often are reliable only in terms of orders of magnitude; exact numbers should be viewed with great caution, because they were obtained from measurements on various groups of subjects under widely varying circumstances. (See, for example, Tables 8–3, 8–4, and 8–5 and Figure 8–34.) In many cases, instead of relying on published data, it is advisable to take strength measurements on a sample of the intended user population under actual conditions of use of the tool to verify that a new design is practicable.

Force depends on body posture

The body force or torque that is exerted depends on a person's strength, training, experience, skills, and body posture. For example, note how the finger forces listed in Table 8–3 depend decidedly on finger posture. Hand forces depend on wrist position (Zellers and Hallbeck 1995; Imrhan 1998) and on arm posture, as shown in Figure 8–28. Exertions with the arm, leg, and shoulder or back depend considerably on the posture of the body and on the support provided to the body

	Force-plate[1] height	Distance[2]	Force, N Mean	Force, N SD
	Percent of shoulder height		With both hands	
	50	80	664	177
	50	100	772	216
	50	120	780	165
	70	80	716	162
	70	100	731	233
	70	120	820	138
	90	80	625	147
	90	100	678	195
	90	120	863	141
	Percent of shoulder height		With one shoulder	
	60	70	761	172
	60	80	854	177
	60	90	792	141
	70	60	580	110
	70	70	698	124
	70	80	729	140
	80	60	521	130
	80	70	620	129
	80	80	636	133
	Percent of shoulder height		With both hands	
	70	70	623	147
	70	80	688	154
	70	90	586	132
	80	70	545	127
	80	80	543	123
	80	90	533	81
	90	70	433	95
	90	80	448	93
	90	90	485	80

Figure 8–34. Maximal static horizontal push forces of males (adapted from NASA 1989).

(ie, on the reaction force, in the sense of Newton's third law) in terms of friction or bracing against solid structures. Figure 8–34 and Table 8–6 illustrate this relationship; both were derived from the same set of empirical data, but were extrapolated to show the effects of friction at the feet, body posture, the location of the point at which force is exerted, and the use of body parts on horizontal push and pull forces applied by male soldiers.

Exoskeleton to augment human capability

An exoskeleton is a structure that is manufactured from strong material and powered to move in synchrony with the body. When it is distant from the human body, but mimics human motions, it is called a teleoperator. When it surrounds the body and moves with it as a unit, it can provide superhuman strength (Crowell 1995).

	Force-plate[(1)] height	Distance[(2)]	Force, N	
			Mean	SD
Force plate		Percent of thumb-tip reach*	With both hands	
	100 percent of shoulder height	50	581	143
		60	667	160
		70	981	271
		80	1285	398
		90	980	302
		100	646	254
			With the preferred hand	
		50	262	67
		60	298	71
		70	360	98
		80	520	142
		90	494	169
		100	427	173
		Percent of span**	With either hand	
	100 percent of shoulder height	50	367	136
		60	346	125
		70	519	164
		80	707	190
		90	325	132

[(1)]Height of the center of the force plate – 20 cm high by 25 cm wide – upon which force is applied.
[(2)]Horizontal distance between the vertical surface of the force plate and the opposing vertical surface (wall or footrest, respectively) against which the subjects brace themselves.

*Thumb-tip reach – distance from backrest to tip of subject's thumb as arm and hand are extended forward.
**Span – the maximal distance between a person's fingertips when arms and hands are extended to each side.

Figure 8–34. (cont.)

TABLE 8–6. Horizontal Push and Pull Forces Capable of Being Exerted Intermittently or for Short Periods by Male Soldiers

*Horizontal force**	*Applied with***	*Condition (μ: coefficient of friction at floor)*
100 N, push or pull	both hands, one shoulder, or the back	with low traction, $0.2 < \mu < 0.3$
200 N, push or pull	both hands, one shoulder, or the back	with medium traction, $\mu \sim 0.6$
250 N, push	one hand	if braced against a vertical wall 51–150 cm from, and parallel to, the push panel
300 N, push or pull	both hands, one shoulder, or the back	with high traction, $\mu > 0.9$
500 N, push or pull	both hands, one shoulder, or the back	if braced against a vertical wall 51–180 cm from, and or if anchoring the feet on a perfectly nonslip ground (like a footrest)
750 N, push	the back	if braced against a vertical wall 60–110 cm from, and parallel to, the push panel or if anchoring the feet on a perfectly nonslip ground (like a footrest)

Source: Adapted from MIL-STD 1472.

*These are minimal forces that may be nearly doubled for two, and less than tripled for three, operators pushing simultaneously. For the fourth and each additional operator, not more than 75 percent of their push capability should be added.

**See Figure 8–34 for examples.

RULES FOR DESIGNING FOR OPERATOR STRENGTH

The engineer or designer who wants to take operator strength into account has to make a series of decisions, including answering the following questions:

1. Is the strength exertion *static or dynamic?* If it is static, information about isometric capabilities can be used. If it is dynamic, additional considerations apply, concerning, for example, physical (circulatory, respiratory, or metabolic) endurance capabilities of the operator or prevailing environmental conditions. Chapters 1, 2, and 5 provide such information.

Most body-segment strength data are available for static (isometric) exertions and provide reasonable guidance for slow motions as well, although they are probably too high for concentric motions and a bit too low for eccentric motions.

Of the little information available regarding dynamic strength exertions, much is limited to isokinematic (constant-velocity) cases. As a general rule, strength exerted in smooth motion is less than that measured in the static positions located on the path of motion.

2. Is the *exertion by hand, by foot,* or with other body segments? In each case, specific design information is available from the text in this chapter. If the choice is free, it must achieve the safest, least strenuous, and most efficient performance.
3. Is a *maximal or a minimal exertion* the critical design factor? Maximal user strength determines the structural strength of the object—for example, so that a handle or a pedal is not broken by the strongest operator. The design value is to be set above the highest perceivable application of strength, with a margin of safety, of course.

Minimal user strength is that strength exerted by the weakest operator which still yields the desired result—for instance, so that a door handle or brake pedal can be successfully operated or a heavy object be moved.

The range of expected strength exertions is, obviously, that between the minimum and the maximum. The infamous "average-user" strength is usually of no design value. (See Chapter 1.)

Measured strength data are often treated, statistically, as if they were normally distributed and reported in terms of averages (means) and standard deviations. This treatment allows the use of common statistical techniques to determine data points of special interest to the designer, as discussed in some detail in Chapter 1. Yet in reality, body/segment strength data are often in a skewed rather than a bell-shaped distribution. This is of no great concern, however, because usually the data points of special interest are the extremes—maximal or minimal, as just discussed. If these values cannot be calculated, they can be estimated.

Maximal exertions are near (and, indeed, frequently, above) the 100th percentile.

Minimal strength is at a given percentile value at the low end of the distribution: often one selects the fifth percentile.

DESIGNING FOR VISION

In many work tasks, the eyes must focus on the work object or the tool, or they must at least provide general guidance for manipulation. This is often a problem in repair work or in some assembly tasks either when only a small opening is available for manipulation and vision or when other objects may interfere with one's vision.

Microscopes

A particularly difficult ergonomic problem has to do with the use of a microscope. Most traditional microscopes are designed so that the eye must be kept close to the ocular. This design forces one to maintain the same posture, often over extended periods. In addition, the microscope may be designed or placed so that the operator must bend the head and neck to locate the eye properly in relation to the eyepiece. Large forward bends of the head and cervical column, exceeding 25 degrees from the vertical, are particularly stressful, because neck extensor muscles must be tensed to prevent the unbalanced head from pitching forward even more. The neck bend also affects the posture of the trunk; thus, both neck and trunk muscles must be kept in tension over long periods. Consequently, complaints of microscope operators about pains and aches in the neck and back are frequent. Furthermore, some microscopes have hand-operated controls located high in front of the shoulders. This arrangement requires that the hands be lifted to that position and kept there, which in turn requires that the muscles controlling the arm and hand posture be kept tensed. Microscopes that allow a variation in the eye position with respect to the eyepiece, locating the latter so that the operator need not bend forward, and hand controls that are properly located can alleviate many of these problems.

Line of sight

In general, visual targets that require close viewing should be placed in front of the operator in or near the medial plane, at "reading distance" (40 to 80 cm) away from the eyes. The angle of the line of sight (from pupil to target) is preferably between 20 and 60 degrees below the ear–eye plane—(that is, 0 to 40 degrees below the horizon if the head is held "straight up." (See Chapters 4 and 9.)

If the vision requirements are less stringent—for example, if the person must look at the target only occasionally—the visual targets may be placed on a partial sphere surrounding the operator. (See Figure 8–35.)

DESIGNING TO AVOID OVERUSE DISORDERS IN THE SHOP AND OFFICE

In Australia during the early 1980s, an epidemic of so-called repetition strain injuries (RSIs) occurred among keyboard operators. Figure 8–36 shows the frequency of injuries reported in one large company. Similar events associated with keyboarding, though not on such large scale, have since

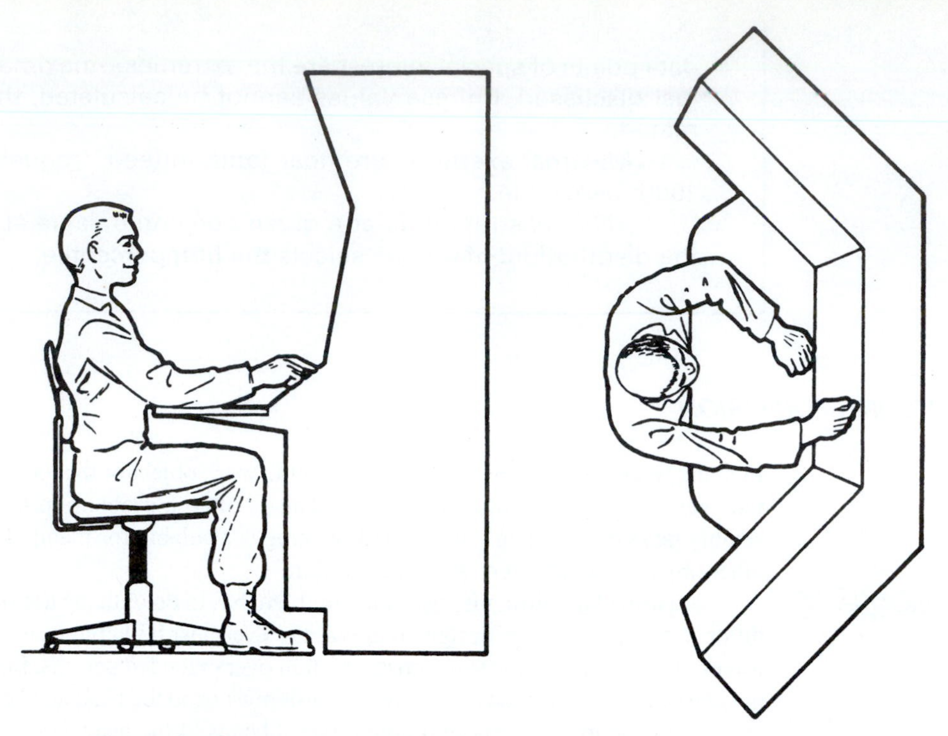

Figure 8–35. Console space suitable for placement of displays and controls (MIL-STD 1472).

occurred in Japan and the United States. A review of the literature yields many reports on injuries related to repetitive activities or occupations since 1713 (Kroemer 1998). Among such injuries are "writers' cramp," "washerwomen's sprain," "carpenters' arm," "bricklayers' hand," "shovelers' hip," "telegraphists' cramp," "typists' tenosynovitis," "tennis elbow" and "golfers' elbow." Overuse disorder is common among musicians (Fry 1986; Lockwood 1989). Understanding the underlying injury mechanisms enables ergonomic measures to prevent overuse disorders from occurring.

Excessive repetition

An overuse disorder is the result of excessive use of a body element—often a joint, muscle, or tendon. In contrast to a single-event injury, called acute or traumatic, the overuse disorder stems from oft-repeated actions that are not injurious when they occurr once or seldom, but whose cumulative effects over time finally result in an injury. These effects are usually related to body motion or posture, energy or force exerted, and duration or repetitiveness.

Many names for ODs

Different terms have been used to describe injuries stemming from repetitive motion, such as occupational overuse disorder (or injury or syndrome), regional or work-related musculoskeletal disorder (MSD), repetitive motion or stress or strain injury (RSI), osteoarthrosis rheumatic disease, and cumulative trauma disorder (CTD). In this text, we shall use the term "overuse disorder" (OD).

☛☛☛ *In his famous 1713 book* De Morbis Artificum *(*Diseases of Workers, *translated by Wright 1993), Bernardino Ramazzini reported on diseases associated with distinct occupations and trades. He described ODs that appeared in workers who did "violent and irregular motions" and assumed "unnatural postures of the body" (Wright 1993, p. 43). He also reported that ODs oc-*

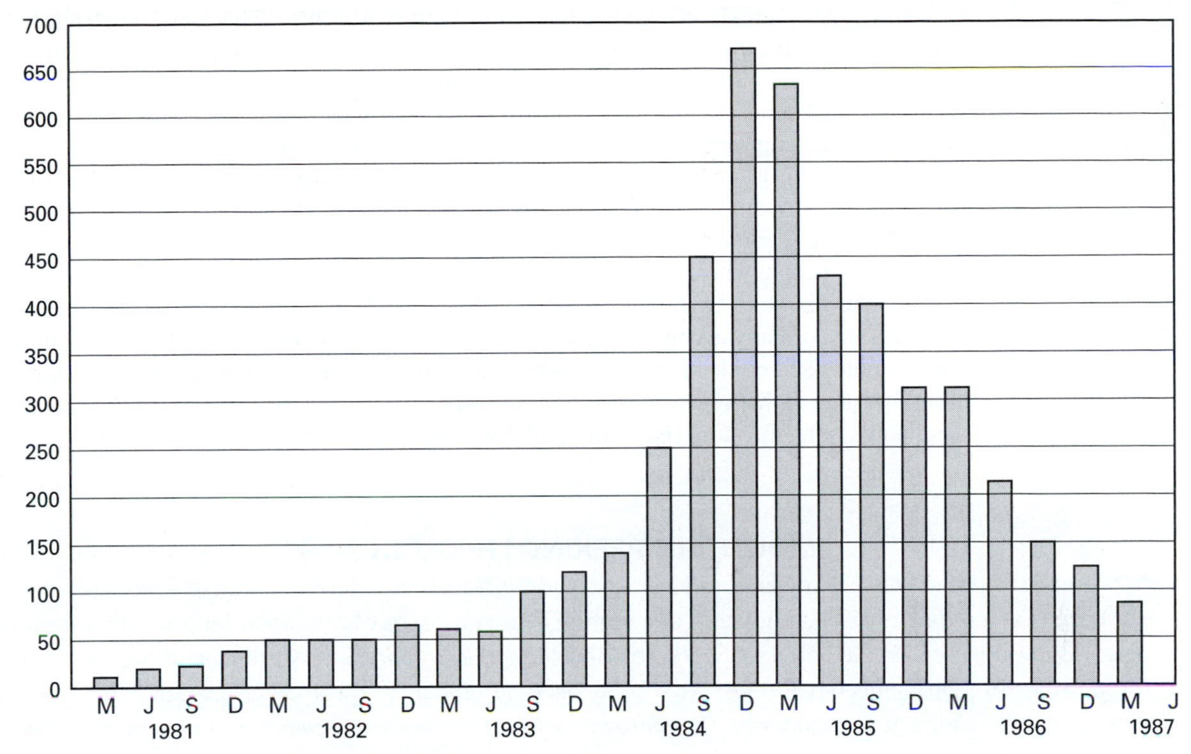

Figure 8–36. Overuse disorders at Telecom Australia. (Adapted from B. Hocking, "Epidemiological Aspects of Repetition Strain injury in Telecom Australia," *Med. J. Aust.* 1987 *147,* 218–222. Copyright 1987, *Medical Journal of Australia,* reproduced with permission.)

curred among secretaries and office clerks; he said that their diseases were caused by "incessant movement of the hand and always in the same direction," by a constant sitting posture, and by a prevailing "strain on the mind" (Wright 1993, p. 254). Since Ramazzini's treatise (which is one of the bases of today's occupational medicine and industrial hygiene), OD cases have been reported to occur with specific kinds of agricultural, industrial, and office work, as well as in musicians and athletes.

Causes of ODs

ODs have long been diagnosed in the medical profession. For example, in the 19th-century, Raynauld's phenomenon (also called "dead finger" or jackhammer disease) was known to be caused by an insufficient blood supply associated with repetitive impacts. In 1893, Gray described inflammations of the extensor tendons of the thumb in their sheaths after excessive exercise. And in 1934, Hammer stated that human tendons cannot tolerate more than 1,500 to 2,000 exertions per hour. Inflammation of tendons and tendon sheaths was often reported in typists during the 1930s through the 1950s. Comprehensive reports on carpal tunnel syndrome appeared in the 1950s and 1960s.

Avoiding ODs

Around 1900, occupation-related diseases associated with repetitive activities were well reported and known in the medical literature. In the 1940s, engineers and physicians knew and described how ODs were related to the design and operation of work equipment and work schedules. From the 1960s on, specific physiological and biomechanical strains of human tissue—especially tendons and their sheaths—were linked to repetitive activities. This new knowledge led to recommendations for the design, arrangement, and use of tools and equipment, especially when vibrating, to alleviate or avoid ODs, particularly in the hand–arm region.

Why do ODs occur?

While diagnoses and medical treatments of ODs were well established, their specific relations to work equipment and occupational activities have been hotly contested, for example, in massive and largely unsuccessful lawsuits against U.S. keyboard manufacturers in the 1990s. One point of view is that in body "usage within reason," ODs should not occur in healthy persons and that, therefore, a connection between work and pathology is inadvertently introduced by a physician. Some people who claim that they have an OD are suspected to malinger, to have compensation neurosis, or to be victims of mass hysteria. The fear of contracting an OD could lead one to lowering normally acceptable discomfort threshold, a possibility that may have contributed to the so-called RSI epidemic of the 1980s in Australia. Yet, the prevalent position taken in the current literature is that repetitive activities at work, daily living, and recreation are causative, precipitating, or aggravating (Bernard 1997; National Research Council 1998, 1999).

"MOUNTAIN PEEKING THROUGH FOG"

The appearance of health complaints related to cumulative trauma may be compared to a mountain. Its wide base is analogous to an accumulation of common everyday cases of tiredness, fatigue, uneasiness, and discomfort during or after a long day's work. The next higher layer represents instances of occasional movement or postural problems beyond just weariness, often accompanied by small aches and pains that, however, disappear after a good night's rest. The narrower levels above are like cases of soreness, pain, and related persistent symptoms; they are present throughout most of the day and do not go away completely during the night or over a weekend. Above this level, smaller again, is a layer which represents symptoms that make it difficult to continue related activities and that may lead one to seek advice from friends and co-workers as to how to alleviate the problems. The very pronounced symptoms and health complaints symbolized by the next higher level prompt discussions with nurses or physicians, who may recommend managerial and engineering changes at work. On top of this layer appear disorders, injuries, and diseases that need medical attention and often cause short-term disability. Near the peak of the mountain are the injuries and diseases that require acute medical treatment such as surgery. The very tip of the mountain represents a (fortunately small) number of disabilities that medical interventions cannot alleviate.

The top of this "mountain of problems" with its broad base and small point is visible above the "fog of psychosocial perception" that usually shrouds its base and lower sections. What becomes visible depends on the individual's and society's sensitivity to discomfort and pain, willingness to disclose problems to supervisors and health-care givers, and awareness level with regard to recognizing problems. If the fog reaches high, only the peak of the mountain, with the severe cases, is visible. The lower the fog, the more problems become evident. In clear conditions, even the basic and most widespread layers of the "mountain of cumulative trauma" are in sight.

Figure 8–37 shows this analogy. The less serious concerns about health or work appear at the low levels. With increasing height, the risk to the person's health becomes more pronounced, and the performance of the task is affected. Changes in engineering or managerial aspects of work, including work–rest arrangements, often can still alleviate the conditions. If persistent symptoms or acute aches and pains appear, medical advice is usually sought. At the high levels, pathological conditions exist that require medical intervention. If these are not completely successful, disability results.

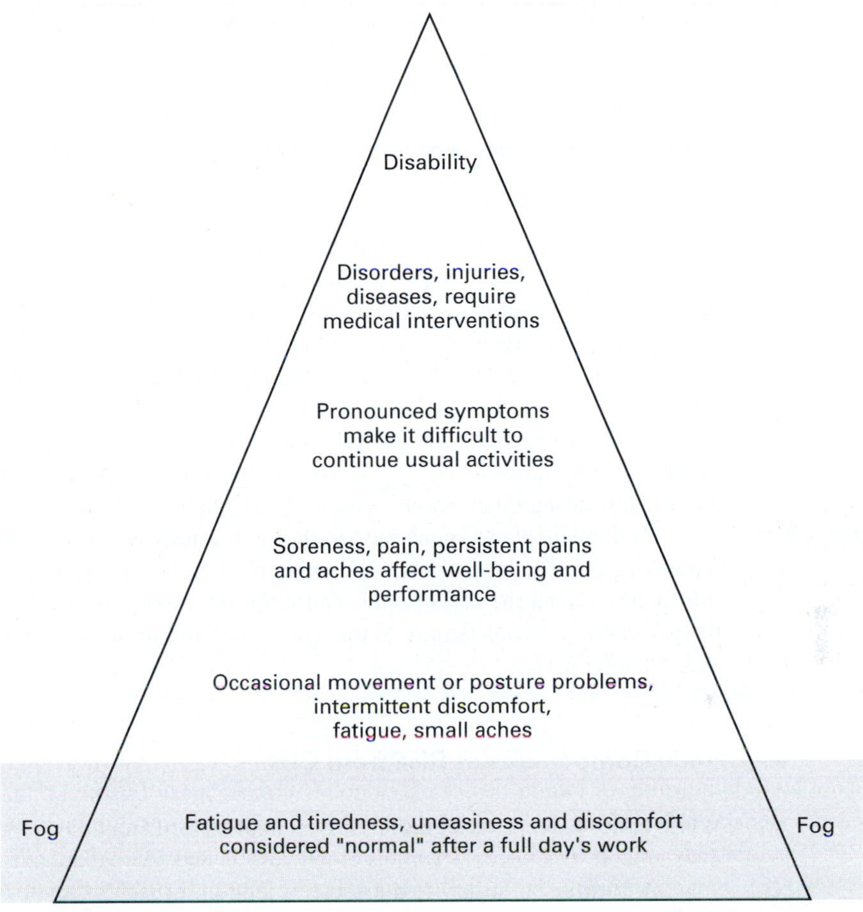

Figure 8–37. Analogy of the "mountain of cumulative trauma partly obscured by fog," the level of which indicates perception and awareness of symptoms.

BIOMECHANICAL STRAINS OF THE BODY

The human body can tolerate low-level mechanical strains (eg, of motion, posture, force, or vibration) from single or repeated mechanical stresses. The risk of injury to tissue rises with increasing physical demands of one's work and lifestyle (called extrinsic factors) and with reductions in a person's physical capacity (intrinsic factors, such as age, anatomy, and pain response). All soft tissues—muscles, tendons, synovial sheaths, ligaments, fascia, cartilage, intervertebral discs, and nerves—fail under excessive loading. But even at strain levels clearly below the failure range, injuries to soft tissue include a complex cascade of responses with (muscle) fatigue, inflammation, and structural degeneration that usually can heal, but only when the strain is taken off. Given the biomechanics of the human body, internal strains are generally several times higher than the external loads that generate them (National Research Council 1999; Wasserman 1998).

Overuse disorders occur particularly in muscles, their tendons, and their sheaths and are often associated with impeded blood flow, swelling, and a buildup of pressure inside tissue spaces, such as inside the carpal tunnel. Any of these conditions may irritate or damage nerves. ODs are frequent in the hand–wrist–forearm area, in the shoulder and neck, and in the low back. Repetitive loadings may even damage bone, such as the vertebrae of the spinal column.

Borelli's model

Biomechanically, one can describe the human body as consisting of a bony skeleton whose segment *links* connect in *joints* and are powered by *muscles* that bridge the joints. (The Borelli model is discussed in Chapter 1.) The muscle actions are *controlled* by the nervous and hormonal systems and *supplied* through a network of blood vessels, which also serves to remove metabolic by-products such as CO_2, lactic acid, heat, and water (See Chapter 2.).

Structure

Bones provide the stable internal framework for the body. Bones can be shattered or broken through sudden impacts, and they can be damaged through continual stresses, such as in vibration. Bones are connected to each other and to other elements of the body through the connective tissues of cartilage, ligaments, and tendon–muscle units. Tendons are often encapsulated by a fibrous tissue, the sheath, which allows gliding motion of the tendon, facilitated by the synovia, a viscous fluid that reduces friction with the inner lining of the sheath. (See Chapter 1.)

Supply

Blood provides oxygen and nutrients through arteries and their ever-more-branching network to the working muscles and organs. Blood also removes metabolic waste products from muscles and organs through the venous system. (See Chapter 1.)

Control

Action signals are generated in the central nervous system and sent through *nerves* of the feedforward (efferent) pathways of the peripheral nervous system to the muscles. Sensors collect information about the actions and send it through nerves of the feedback (afferent) pathways of the peripheral nervous system to the spinal cord and brain for appropriate reactions and actions. (See Chapter 3.)

Body Components at Risk from ODs

While bones (except vertebrae) usually do not suffer from ODs, joints, muscles, and tendons and their related structures, as well as blood vessels and nerves, are at risk.

Strain

A *strain* is an injury to a muscle or tendon. Muscles can be excessively stretched, which is often associated with aching and swelling. A more serious injury occurs when a group of fibers is torn apart. If the blood or nerve supply to the muscle is interrupted for an extended length of time, the muscle atrophies. Tendons contain collagen fibers, which neither stretch nor contract; if overly strained, they can be torn. Scar tissue may then form, which is easily reinjured and may create chronic tension. Also, tendon surfaces can become rough, impeding their motion along other tissues. The gliding movement of a tendon in its sheath, caused by muscle contraction and relaxation, may be quite large—for example, 5 cm in the hand when a finger is moved from fully extended to completely flexed. Synovial fluid in the tendon sheath, acting as a lubricant to allow easy gliding, may be diminished, which causes friction between the tendon and its sheath. The first signs are feelings of tenderness, warmth, and pain, which may indicate inflammation.

Inflammation

Inflammation of a tendon or its sheath, often of both, is a protective response of the body, its purpose being to limit bacterial invasion. The feeling of warmth and the swelling stem from the influx of blood. The resulting compression of tissue produces pain. Movement of the tendon within its swollen surroundings is impeded. Repeated forced movement may cause additional fiber tissue to become inflamed, which, in turn, can establish a permanent (chronic) condition.

A *bursa* is a fluid-filled sac lined with a synovial membrane, a slippery cushion that prevents rubbing between bone and muscle or tendon. An often-used tendon, particularly if it has become roughened, may irritate its adjacent bursa, setting up an inflammatory reaction (similar to the inflammation in tendon sheaths) that inhibits the free movement of the tendon and hence reduces joint mobility.

Sprain

A *sprain* occurs when a joint is displaced beyond its regular range and fibers of a ligament become strongly stretched, torn apart, or pulled from the bone. Sprains can result from a single trauma, but also by repetitive actions. Injured ligaments may take weeks or even months to heal, because their blood supply is poor. A ligament sprain can bring about lasting joint instability and increase the risk of further injury.

Pressure

Nerve compression may stem from pressure by bones, ligaments, tendons, and muscles within the body or from hard surfaces and sharp edges of workplaces, tools, and equipment. Increased pressure within the body can occur if the position of a body segment reduces the size of the opening through which a nerve runs. An additional source of compression is the swelling of other structures within this opening—often, irritated tendons and their sheaths. Carpal tunnel syndrome (see later) is a typical case of nerve compression.

Impairment of nerves

Impairment of a motor nerve reduces the ability of the nerve to transmit signals to innervated muscle motor units. Thus, motor-nerve impairment impedes the controlled activity of muscles and hence reduces the ability of those muscles to generate force or torque to tools, equipment, and work objects. *Sensory-nerve impairment* reduces the information that can be brought back from sensors in the body to the central nervous system. Sensory feedback is important for many activities, because it contains information about force and pressure applied, positions assumed, and motion experienced. Sensory-nerve impairment usually brings about sensations of numbness, tingling, or even pain. The ability to distinguish hot from cold may be reduced as well. *Impairment of an autonomic nerve* reduces the ability to control such functions as sweat production in the skin. A common sign of autonomic-nerve impairment is dryness and shininess of skin areas controlled by the nerve in question.

Reduced flow of blood

Blood-vessel, or vascular, compression, often of an artery, reduces blood flow through the supplied area. The result is a reduced supply of oxygen and nutrients to such tissues as muscles, tendons, and ligaments, as well as diminished ability of the body to remove metabolic by-products, such as lactic acid from tissues. Vascular compression produces ischemia, which limits the duration of muscular actions and impairs the recovery of a "fatigued" muscle after activity. Vascular compression is often found in the neck, shoulder, arm and hand.

Vasospasm

Vibrations of body members, particularly of the hand and fingers, may result in *vasospasms* that reduce the diameter of arteries. Of course, blood flow to the body areas supplied by the vessels is then impeded, and the area becomes blanched. In the hand, this condition is known as the White-finger (Raynaud's) phenomenon. Exposure to cold may aggravate the problem, because it can also trigger vasospasms, particularly in the fingers. Associated symptoms include intermittent or continued numbness and tingling, with the skin turning pale and cold, and eventually loss of sensation and control. In the fingers, the condition is often caused by vibration transmitted from tools such as pneumatic hammers, chain saws, power grinders, and power polishers. The frequent operation of keys on keyboards might be a source of vibration strain to the hand–wrist area.

Carpal Tunnel Syndrome

In 1959, Tanzer described several cases of carpal tunnel syndrome. Two of his patients had recently started to milk cows on a dairy farm, three worked in a shop in which objects were handled on a conveyor belt, two had done gardening with considerable hand weeding, and one had been using a spray gun with a finger-operated trigger. Two patients had been working in large kitchens, stirring and ladled soup twice daily for about 600 students.

Among the best known ODs is carpal tunnel syndrome (CTS), first described 125 years ago. In 1964, Kroemer observed the occurrence of ODs in typists and surmised that the condition might be related to the force with which they struck the keys on their keyboards, the displacement of the keys, the frequency with which the keys were struck, and the posture of the arms and hands. In 1966 and 1972, Phalen described the typical gradual onset of numbness in the thumb and the first two-and-a-half fingers of the hand, all supplied by the median nerve. In 1975, Birkbeck and Beer described the results of a survey they took of the work- and hobby-related activities of 658 patients who suffered from CTS. Four out of five patients were employed in work requiring light, highly repetitive movements of the wrists and fingers. In 1980, Huenting et al. found frequent impairments in the hands and arms of operators of accounting machines. In 1983, the American Industrial Hygiene Association acknowledged the prevalence and importance of CTS by publishing Armstrong's ergonomic guide for avoiding the condition, which he called an occupational illness (Kroemer 1998).

Thus, in the 1970s and early 1980s, CTS was well recognized as an often-occurring, disabling condition of the hand that can be caused, precipitated, or aggravated by light activities in the office or heavy work in the shop or on the construction site. Naturally, leisure activities may be involved as well. Critical postures include a flexed or hyperextended wrist—especially in combination with forceful exertions—in highly repetitive activities and vibrations. Of course, case studies and epidemiological surveys have inherent attributes (such as the absence of a control group and the possibility of confounding variables) that make it difficult to establish a causal connection between activities and disorders, but observations of the two strongly suggest a cause-effect relationship (Bernard 1997).

Carpal tunnel

In 1963, Robbins described the anatomical conditions that explain the tunnel syndrome. On the palmar side of the wrist, near the base of the thumb, the carpal bones form a concave floor that is bridged by three ligaments (the radial carpal, intercarpal, and carpometacarpal ligaments), which in turn are covered by the transverse carpal ligament, firmly fused to the carpal bones. Thus, a canal is formed with the carpal bones as the floor and sides, and ligaments covering this canal like a roof. (See Figure 8–38.) The structure is called the carpal tunnel. Its cross section is roughly oval. Flexor tendons of the digits pass through the carpal tunnel, as do the median nerve and the radial artery. This crowded space is diminished in size if the wrist is flexed, extended, or laterally pivoted. Swelling of the tendons or of their sheaths reduces the space available for tendons, nerves and blood vessels.

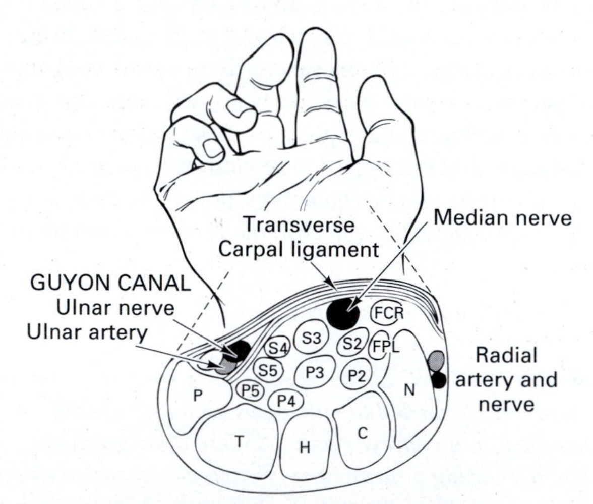

Figure 8–38. Schematic view of the carpal tunnel with the tendons of the superficial *(S)* and profound *(P)* finger flexor muscles, flexors of the thumb *(FCR, FPL),* nerves and arteries, carpal bones *(P, T, H, C, N)* and ligaments (adapted from Kroemer 1989b).

CTS

The median nerve, which passes through the carpal tunnel, innervates the thumb, much of the palm, and the index and middle fingers as well as the radial side of the ring finger. Outside pressure on the tunnel ligament, as well as tendons or sheaths becoming inflamed and swelling, reduces the tunnel space, as does a deviation of the wrist from straight. The resulting pressure on the median nerve and blood vessels leads to carpal tunnel syndrome, medically defined as a reduced conduction velocity of electric impulses along the affected section of the medial nerve.

OCCUPATIONAL ACTIVITIES AND RELATED DISORDERS

Table 8–7 lists those conditions that are most often associated with overuse disorders. Of course, the list is neither complete nor exclusive: New occupational activities occur all the time, such as use of a computer mouse beginning in the late 1900s, and several activities may be part of the same job—for example, use of the mouse and keyboard.

The major activity-related factors in repetitive strain injuries are rapid, often-repeated movements, forceful exertions, static muscle loading (frequently to maintain posture), vibrations, and cooling of the body. The detrimental effects of these elements are, or may be, aggravated by inappropriate organizational and social factors, discussed later (Bernard 1997; Kurorinka and Forcier 1995; National Research Council 1998, 1999; Nordin et al. 1997).

Repetitiveness

Repetitiveness is a matter of definition ("more than once per time unit"), and what is low or high depends on the specific activities or body part involved. For industrial work, Silverstein (1985) proposed that high repetitiveness be defined as a cycle of less than 30 seconds' duration, or as more than 50 percent of the cycle time spent performing the same fundamental motion.

Forcefulness

Forcefulness is also a matter of definition, and what is low or high depends on the specific activities or body part involved. Again, Silverstein (1985) suggested that high force exerted with the hand (e.g., more than 45 N), may be a cause of OD. Yet the force applied to a computer key is only about one newton which, however, still translates to a multiple thereof in terms of the tension force in the tendon of the (Rempel et al. 1997, 1999).

Tension related to posture

Static muscle tension, often generated to maintain body posture, is stressful when high enough. If muscles must remain contracted at more than about 15 percent of their maximal capability, circulation is impaired. This can result in tissue ischemia and delayed dissipation of metabolites. (See Chapter 1.) Body position can also affect the space for passage of blood, nerves, and tendons. For example, "dropped" or "elevated" wrists, shown in Figure 8–39, reduce the available cross section of the carpal tunnel and hence generate a condition that may cause CTS, particularly in persons with small wrists (Morgan 1991). In fact, any strong deviation of the wrist from the neutral position, as well as the pinch grip and any inward or outward twisting of the forearm, especially with a bent wrist, is stressful.

"Seven sins."

Seven conditions in particular need to be avoided:

1. Activities with many repetitions.
2. Any activity that requires a prolonged or repetitive exertion of more than about one-third of the operator's static muscular strength available for that activity.
3. Putting body segments in extreme positions. (See Figures 8–39 and 8–40.)
4. Making a person maintain the same body posture for a long time. (See Figure 8–39.)
5. Pressure from tools or work equipment on tissues (skin, muscles, or tendons), nerves, or blood vessels. (See Figures 8–40 and 8–41.)
6. A tool vibrating the body or a part of the body.
7. The exposure of working body segments to cold, including airflow from pneumatic tools.

TABLE 8–7(A). Common Overuse Disorders*

Disorder	*Description*	*Typical job activities*
Carpal tunnel syndrome (writer's cramp, neuritis, median neuritis) (N)	The result of compression of the median nerve in the carpal tunnel of the wrist. This tunnel is an opening under the carpal ligament on the palmar side of the carpal bones. Through this tunnel pass the median nerve, the digit flexor tendons, and blood vessels. Swelling of the tendon sheaths reduces the size of the opening of the tunnel, pinching the median nerve, and blood vessels. The tunnel opening is also reduced if the wrist is flexed or extended or ulnarly or radially pivoted. Tingling, numbness, or pain in all digits but the little finger.	Buffing, grinding, polishing, sanding, assembly work, typing, keying, operating a cash register, playing musical instruments, surgery, packing, housekeeping, cooking, butchering, hand washing, scrubbing, hammering.
Cubital tunnel syndrome (N)	Compression of the ulnar nerve below the notch of the elbow. Tingling, numbness, or pain radiating into ring or little finger.	Resting forearm near elbow on a hard surface or sharp edge; also when reaching over an obstruction.
DeQuervain's syndrome (or disease) (T)	A special case of tendosynovitis that occurs in the abductor and extensor tendons of the thumb, where they share a common sheath. The condition often results from combined forceful gripping and hand twisting, as in screw driving.	Buffing, grinding, polishing, sanding, pushing, pressing, sawing, cutting, surgery, butchering, using pliers, operating a turning control such as that on a motorcycle, inserting screws in holes, forceful hand-wringing.
Epicondylitis ("tennis elbow") (T)	Tendons attaching to the epicondyle (the lateral protrusion at the distal end of the humerus bone) become irritated. This condition is often the result of the impact of jerky throwing motions, repeated supination and pronation of the forearm, and forceful wrist extension movements. The condition is well known among tennis players, pitchers, bowlers, and people hammering. A similar irritation of the tendon attachments on the inside of the elbow is called medical epicondylitis, also known as "golfer's elbow."	Turning screws, small-parts assembly, hammering, cutting meat, playing musical instruments, playing tennis, pitching, bowling, golfing.
Ganglion (T)	A tendon sheath swelling that is filled with synovial fluid, or a cystic tumor at the tendon sheath or a joint membrane. The affected area swells and causes a bump under the skin, often on the dorsal or radial side of the wrist. (Since, in the past, the condition was occasionally smashed by striking the swelling with a Bible or heavy book, it was also called a "Bible bump.")	Buffing, grinding, polishing, sanding, pushing, pressing, sawing, cutting, surgery, butchering, using pliers, operating a turning control such as that on a motorcycle, inserting screws in holes, forceful hand-wringing.
Neck tension syndrome (M)	An irritation of the levator scapulae and trapezius group of muscles of the neck, commonly occurring after repeated or sustained overhead work.	Belt conveyor assembly, typing, keying, small parts assembly, packing, carrying a load in the hand or on the shoulder.

*Type of disorder: N, nerve; T, tendon; M, muscle, V, vessel.

TABLE 8–7(B). Common Overuse Disorders* (continued)

Disorder	*Description*	*Typical job activities*
Pronator (teres) syndrome (N)	Result of compression of the median nerve in the distal third of the forearm, often where it passes through the two heads of the pronator teres muscle in the forearm; common with strenuous flexion of elbow and wrist.	Soldering, buffing, grinding, polishing, sanding.
Shoulder tendonitis (rotator cuff syndrome or tendonitis, supraspinatus tendinitis, subacromial bursitis, subdeltoid bursitis, partial tear of the rotator cuff) (T)	This is a shoulder disorder, located at the rotator cuff, which consists of four tendons that fuse over the shoulder joint where they pronate and supinate the arm and help to abduct it. The rotator cuff tendons must pass through a small bony passage between the humerus and the acromion, with a bursa as cushion. Irritation and swelling of the tendon or the bursa are often caused by continuous muscle and tendon effort to keep the arm elevated.	Punch press operations, overhead work, assembly, packing, storing, construction work, postal "letter carrying," reaching, lifting, carrying load on shoulder.
Tendonitis (tendinitis) (T)	An inflammation of a tendon. Often associated with repeated tension, motion, bending, being in contact with a hard surface, or vibration. The tendon becomes thickened, bumpy, and irregular in its surface. Tendon fibers may be frayed or torn apart. In tendons without sheaths, such as within the elbow and shoulder, the injured area may calcify.	Punch press operation, assembly work, wiring, packaging, core making, using pliers.
Tendosynovitis (tenosynovitis, tendovaginitis) (T)	This is a disorder of tendons that are inside synovial sheaths. The sheath swells. Consequently, movement of the tendon within the sheath is impeded and painful. The tendon surfaces can become irritated, rough, and bumpy. If the inflamed sheath presses progressively onto the tendon, the condition is called stenosing tendosynovitis. DeQuervain's Syndrome is a special case occurring in the thumb, while the trigger finger condition occurs in flexors of the fingers.	Buffing, grinding, polishing, sanding, punch press operation, sawing, cutting, surgery, butchering, using pliers, operating a turning control such as that on a motorcycle, inserting screws in holes, forceful hand-wringing, keyboarding.
Thoracic outlet syndrome (neurovascular compression syndrome, cervicobrachial disorder, brachial plexus neuritis, costocalvicular syndrome, hyperabduction syndrome) (V, N)	A disorder resulting from compression of nerves and blood vessels between the clavicle and the first and second ribs, at the brachial plexus. If this neurovascular bundle is compressed by the pectoralis minor muscle, blood flow to and from the arm is reduced. This ischemic condition makes the arm numb and limits muscular activities.	Buffing, grinding, polishing, sanding, overhead work, keying, operating a cash register, wiring, playing musical instruments, surgery, truck driving, stacking, material handling, postal "letter carrying," carrying heavy loads with extended arms.

*Type of disorder: N, nerve; T, tendon; M, muscle, V, vessel.

TABLE 8–7(C). Common Overuse Disorders* (continued)

Disorder	*Description*	*Typical job activities*
Trigger finger (or thumb) (T)	A special case of tendosynovitis wherein the tendon becomes nearly locked, so that its forced movement is not smooth, but occurs in a snapping or jerking manner. This is a special case of stenosing tendosynovitis crepitans, a condition usually found with digit flexors.	Operating a finger trigger, using hand tools that have sharp edges pressing into the tissue or whose handles are too far apart for the user's hand, so that the end segments of the fingers are flexed while the middle segments are straight.
Ulnar artery aneurysm (V, N)	Weakening of a section of the wall of the ulnar artery as it passes through the Guyon tunnel in the wrist; often caused by pounding or pushing with the heel of the hand. The resulting "bubble" presses on the ulnar nerve in the Guyon tunnel.	Assembly work.
Ulnar nerve entrapment (Guyon tunnel syndrome) (N)	Results from the entrapment of the ulnar nerve as it passes through the Guyon tunnel in the wrist. The condition can occur from prolonged flexion and extension of the wrist and repeated pressure on the hypothenar eminence of the palm.	Playing musical instruments, carpentering, brick-laying, use of pliers, soldering, hammering.
White finger ("dead finger," Raynaud's syndrome, vibration syndrome) (V)	Stems from insufficient blood supply and brings about noticeable blanching. Finger turns cold, gets numb, and tingles, and sensation and control of finger movement may be lost. The condition is due to closure of the digit's arteries caused by vasospasms triggered by vibrations. A common cause is continued forceful gripping of vibrating tools, particularly in a cold environment.	Chainsawing, jackhammering, use of vibrating tool, sanding, scraping paint, using a vibrating tool too small for the hand, often in a cold environment.

*Type of disorder: N, nerve; T, tendon; M, muscle, V, vessel.

FORERUNNERS OF OVERUSE DISORDERS

- Rapid and often-repeated actions
- Exertion of finger, hand, or arm forces
- Pounding and jerking
- Contorted body joints
- Polished-by-use sections of the workplace or clothing; custom-made padding
- Blurred outlines of the body owing to vibration
- The feeling of cold and the hissing sound of fast-flowing air

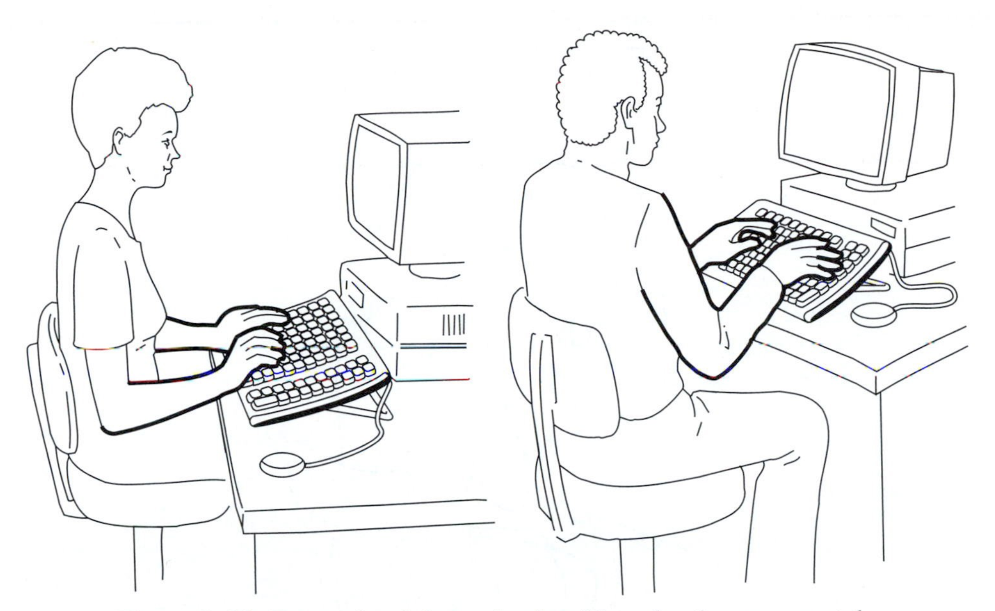

Figure 8–39. Dropped and elevated wrists. Note also the pressure at the edge of the table. Courtesy of Herman Miller, Inc.

STAGES OF OVERUSE DISORDERS AND THEIR TREATMENT

ODs appear in various, variable, and often confusing ways. The onset of signs and symptoms can be gradual or sudden. Three stages are commonly distinguished:

> Stage 1 is characterized by aches and "tiredness" during the working hours. Usually, the condition resolves itself overnight and over days off. There is no reduction in work performance. Stage 1 may persist for weeks or months.
>
> Stage 2 is characterized by tenderness, swelling, numbness, weakness, and pain that start early in the work shift and do not resolve themselves overnight. Sleep may be disturbed, and the capacity to perform the repetitive work is often reduced. Stage 2 usually persists over months.
>
> Stage 3 is characterized by symptoms that persist when the person is at rest. Pain occurs even with nonrepetitive movements, and disturbs sleep. The patient is unable to perform even light duties and experiences difficulties in daily tasks. Stage 3 may last for months or years.

Stage 1 is often reversible through work modification and rest breaks. Exercise as a precautionary measure (eg, "work hardening") or for rehabilitation must be applied with great caution and is often of questionable value (Silverstein et al. 1988; Williams et al. 1989; Lee et al. 1992). In Stage 2 or Stage 3, the most important factor is abstinence from performing the causative activities, often combined with rest. This treatment may mean major changes in one's working habits and lifestyle. Further care may involve physiotherapy, drugs, and other medical treatments, including surgery. Medically, it is important to identify an OD case early, at a stage that allows effective treatment. Ergonomically, it is even more important to recognize potentially injurious activities and conditions early and to alleviate them through work (re)organization and work (re)design before harm is done to a person.

Figure 8–40. Contorted working posture at a punch press.

NONBIOMECHANICAL FACTORS IN ODS

There is no doubt that individual, organizational, and social factors can affect the physiological pathways leading from tissue loading to overload, impairment, and, possibly, disability (Moon and Sauter 1996). Yet, how and whether they do so are difficult to assess. Such individual factors as age and medical status relate strongly to the biological mechanisms. So do body mass and gender, but the mechanism is less clear. Such factors as genetics and general conditioning are even less well established. These relationships, however, when found statistically significant in studies, rarely show a high predictive value (National Research Council 1999).

Psychosocial factors include social support and workplace stress, as well as the content, variety, demands, control, satisfaction, and enjoyment of the job. Social support appears to have a mediating effect on stress. Among the organizational variables, poor job content (eg, in terms of identifying and integrating) tasks and high demand have been found to be related to higher rates of musculoskeletal disorders; to a lesser extent, this also seems to be true for job control. Psychological stress has some causal plausibilty but it is weaker than for individual and biomechanical variables. Altogether, social and organizational variables, even when found statistically significant for the generation of musculoskeletal disorders, are usually not important factors thereof (National Research Council 1999).

Figure 8–41. Edge of desk presses into forearm. Courtesy of Herman Miller, Inc.

ERGONOMIC INTERVENTIONS

A variety of management actions can reduce or eliminate the risk of ODs. Of major significance are engineering designs or redesigns; work methods, including tools and personal protective equipment; and administrative controls. Engineering and work methods designs are the most reasonable and successful of these actions because they address the origin of the problem and can eliminate any hazard. Administrative controls and training, if suitably done, may have beneficial effects by reducing the exposure and hazard (National Research Council 1999).

ERGONOMIC MEANS OF COUNTERING ODS

Of course, it is best to avoid conditions that may lead to an OD. Work objects, equipment, and tools used should be suitably designed; instruction on, and training in, proper postures and work habits are important; and managerial interventions such as work diversification (the opposite of job simplification and specialization), the provision of relief workers, and rest pauses are often helpful. The basic, most important, principle is "not to do—not to repeat" possibly injurious motions and exertions and to avoid unsuitable postures.

As a general rule, tools and tasks should be designed so that they can be handled without causing wrist deviations. The wrist should not be severely flexed, extended, or laterally pivoted, but should, in general, remain aligned with the forearm, which itself should not be twisted (should be neither pronated nor supinated). The elbow angle should be varied about a median 90 degrees. The upper arms usually should hang down along the sides of the body. The head should be held fairly erect. The trunk should be mostly upright when standing; when sitting for long periods, leaning on a tall, well-shaped backrest is desirable. Severe twisting of the trunk should not be required at work. There should be enough room for the legs and feet to allow standing or sitting comfortably. It is important that the postures of the body segments and of the whole body be varied often during the working time.

Jobs should be analyzed for their movement, force–torque, and posture requirements. This can best be done by using the well-established industrial engineering procedure of "motion and time study" (Konz and Johnston 2000). Each element of the work should be screened for factors that can contribute to OD, and any that are found should be eliminated by human-factors engineering. Table 8–8 (adapted from Kroemer 1989b) provides an overview of generic ergonomic measures to fit the job to the person.

The avoidance of OD, whether by appropriate planning of new work or by redesigning an existing workstation, follows one simple generic rule: Let the operator perform "natural" activities (ie, those for which the human body is suited). Avoid highly repetitive activities and those in which straining forces or torques must be exerted or in which an awkward posture must be maintained over a prolonged time. The opposite way—that is, selecting persons who seem to be especially able to perform work that most people cannot do or letting several people work at the same workstation alternately in order to distribute the overload—is basically an inappropriate measure that should be applied only as a stopgap until a good solution can be found.

Fit the job to the person; do not attempt to fit the person to the job.

In addition to keeping the number of repetitions and the amount of energy (force or torque) small, the following posture-related measures should be considered:

1. Provide a chair with a headrest, so that one can relax the neck and trunk muscles at least temporarily;
2. Provide an armrest, possibly cushioned, so that the weight of the arms will not be carried by muscles crossing the shoulders and elbow joints;
3. Provide flat, possibly cushioned surfaces on which the forearms may rest while the fingers work;
4. Provide a wrist rest for people operating traditional keyboards, so that the wrist does not drop below the level of the keys.
5. Round, curve, and pad all edges that otherwise might be point sources of pressure.
6. Select jigs and fixtures to hold workpieces in place, so that the operator does not have to hold the piece.

TABLE 8–8. Ergonomic Measures to Avoid Common Overuse Disorders

<table>
<tr><th>Disorder</th><th>Avoid in general</th><th>Avoid in particular</th><th>Do</th><th>Design</th></tr>
<tr><td>Carpal tunnel syndrome</td><td>Rapid, often repeated, finger movements, wrist deviation</td><td>Dorsal and palmar flexion, pinch grip, vibrations between 10 and 60 Hz</td><td rowspan="13">Use large muscles, but infrequently and for short durations
—
Let wrists be in the line with the forearm
—
Let shoulder and upper arm be relaxed
—
Let forearms be horizontal or more declined</td><td rowspan="14">the work object properly
—
the job task properly
—
hand tools properly ("bend tool, not the wrist")
—
round corners, pad
—
placement of work object properly</td></tr>
<tr><td>Cubital tunnel syndrome</td><td colspan="2">Resting forearm on sharp edge or hard surface</td></tr>
<tr><td>DeQuervain's syndrome</td><td colspan="2">Combined forceful gripping and hard twisting</td></tr>
<tr><td>Epicondylitis</td><td>"Bad tennis backhand"</td><td>Dorsiflexion, pronation</td></tr>
<tr><td>Pronator syndrome</td><td>Forearm pronation</td><td>Rapid and forceful pronation, strong elbow and wrist flexion</td></tr>
<tr><td>Shoulder tendonitis, rotator cuff syndrome</td><td>Arm elevation</td><td>Arm abduction, elbow elevation</td></tr>
<tr><td>Tendonitis</td><td>Often repeated movements, particularly while exerting a force; hard surface in contact with skin; vibrations</td><td>Frequent motions of digits, wrists, forearm, shoulder</td></tr>
<tr><td>Tendosynovitis, DeQuervain's syndrome, ganglion</td><td>Finger flexion, wrist deviation</td><td>Ulnar deviation, dorsal and palmar flexion, radial deviation with firm grip</td></tr>
<tr><td>Thoracic outlet syndrome</td><td>Arm elevation, carrying</td><td>Shoulder flexion, arm hyperextension</td></tr>
<tr><td>Trigger finger or thumb</td><td>Digit flexion</td><td>Flexion of distal phalanx alone</td></tr>
<tr><td>Ulnar artery aneurism</td><td colspan="2">Pounding and pushing with the heel of the hand</td></tr>
<tr><td>Ulnar nerve entrapment</td><td>Wrist flexion and extension</td><td>Wrist flexion and extension, pressure on hypothenar eminence</td></tr>
<tr><td>White finger, vibration syndrome</td><td>Vibrations, tight grip, cold</td><td>Vibrations between 40 and 125 Hz</td></tr>
<tr><td>Neck tension syndrome</td><td>Static head posture</td><td>Prolonged static head/neck posture</td><td>Alternate head and neck postures</td></tr>
</table>

Source: Adapted from Kroemer 1989b.

7. Select and place jigs and fixtures so that the operator can easily access the workpiece without contorting the hand, arm, neck, or back.
8. Select and place bins and containers so that the operator can reach into them with least possible bending and twisting of the hand, wrist, arm, neck, and trunk.
9. Select tools whose handles distribute pressure evenly over large surfaces of the operator's digits and palm;
10. Select hand tools and working procedures that do not require pinching grips.
11. Select the lightest possible hand tools.
12. Select hand tools that are properly angled so that the wrist will not be bent;
13. Select hand tools that do not require the operator to apply a twisting torque.
14. Select hand tools whose handles are shaped so that the operator does not have to apply much grasping force to keep the tool in place or to press it against a workpiece;
15. Avoid tools with sharp edges, fluted surfaces, or other prominences that press into tissues of the operator's hand.
16. Suspend or otherwise hold tools in place so that the operator does not have to do so for extended periods.
17. Select tools that do not transmit vibrations to the operator's hand.
18. If the hand tool must vibrate, have energy-absorbing or energy-dampening material between the handle and the hand. (Yet, the resulting handle diameter should not become too large.)
19. Make sure that the operator's hand does not become undercooled, which may be a problem particularly with pneumatic equipment.
20. Select any gloves that are worn to be of the proper size, texture, and thickness.

☛☛☛ *Four decades ago, Peres wrote the following:*

"It has been fairly well established, by experimental research overseas and our own experience in local industry, that the continuous use of the same body movement and sets of muscles responsible for that movement during the normal working shift (not withstanding the presence of rest breaks), can lead to the onset initially of fatigue, and ultimately of immediate or cumulative muscular strain in the local body area . . .

"It is sometimes difficult to see why experienced people, after working satisfactorily for, say, 15 years at a given job, suddenly develop pains and strains. In some cases these are due to degenerative arthritic changes and/or traumatic injury of the bones of the wrist or other joints involved. In other cases, the cause seems to be compression of a nerve in the particular vicinity, as for example, compression of the median nerve in carpal tunnel syndrome. However, it may well be that many more are due to cumulative muscle strain arising from wrong methods of working" (Peres 1961, pp. 1, 11). ☚☚☚

RESEARCH NEEDS

Our current knowledge of about the biomechanical relationships between activities (on the job, in daily living, and during leisure) and ODs is limited chiefly to exertions of fairly large forces, high exertion frequencies, and certain body postures that are maintained over a long period. Yet, even for those gross muscular activities, the exact causal relationships to ODs are not totally clear

(Bernard 1997). Some causal factors are not well defined, and for most, the critical threshold values are not known.

Forcefulness

For example, the force or torque requirement of a task may be measured statically (isometrically) or in dynamic terms. Silverstein's (1985) definition (see earlier) is apparently applicable to static exertions, while many activities are in fact dynamic in nature. Some ODs are explicitly related to motion (ie, to dynamics, as indicated by the term "repetitive motion injury"). The exertion of energy under static conditions establishes very different body strains than under dynamic conditions (Kroemer et al. 1990, 1997). Unfortunately, our current knowledge base seems to be largely dependent on the assumption of isometric muscle efforts (ie, on a static condition).

Repetitiveness

Another major problem, both in concept and in application, is the frequency of activities related to ODs. In numerically describing the "frequency" of an activity, one presumes that the actions occur at regular intervals during the recording time. Yet, in reality, this is commonly not the case over a day's working time: Activities may bunch together in some periods and occur seldom during others. It is not known how an uneven distribution of activities over the workday may be related to the occurrence of ODs.

Shop activities, usually with exertion of muscle strength, have been linked to ODs. In the office, the operation of a keyboard or mouse requires rather small energies to be exerted by the fingers per activation, but the number of such actions per hour is often high—in some cases, up to 20 thousand. This brings up the unresearched problem of the interrelation between energy (force and displacement) and repetitiveness with respect to ODs. It is likely that such interrelation is rather complex, and certainly it includes factors beyond (static) force, frequency, and body posture (Latko et al. 1997; Marras and Schoenmarklin1991; Moore et al. 1991; Occhipinti and Colombini 1998). Clearly, much research must be conducted to identify the components of activities that may lead to ODs and to understand why and how these physical events, singly or combined, overload body structures and tissues. When these relationships are understood, it should be possible to establish exact thresholds or doses for factors such as posture, force or torque, displacement, and their rates of occurrence and duration that separate beneficial from harmful conditions.

SUMMARY

The human body is the traditional measure for sizing hand tools, equipment, and workstations. (See Table 8–9. One may assume different postures at work; sitting is generally preferred, and one can apply forces with the feet better while seated, but walking and standing allow us to cover more space and exert larger hand forces.

The hands are our primary means of performing work; they operate with the finest control directly in front of the body, at belly to chest height. Hand tools are often needed when finger manipulation alone is insufficient, but they should be designed so that they are helpful for the intended purpose, not stressful or even potentially damaging to the body. Foot operation of controls is more forceful, but frequent foot operation should be required only if the operator sits.

Overuse disorders are caused by often-repeated activities, particularly if these require extensive body energy, force, and torque; "unnatural" postures, especially in the wrists, arms, shoulders, neck, and back; and vibrations, impacts, and the cold. Although in many cases exact injury mechanisms for combined stresses are not yet well understood, rules and recommendations for the design of proper equipment and for its use are at hand.

TABLE 8–9. Folk Norms of Measurement.

Inch	Breadth of thumb; length of distal phalanx of little finger
Phalanx	Length of distal phalanx of thumb; length of middle phalanx of middle finger (two inches)
Hand	Width of palm across knuckles, length of index finger (two phalanges)
Span (of hand)	Distance between tips of spread thumb and index finger (two hands, four phalanges)
Foot	Length of foot (three hands, six phalanges)
Ell	Length from elbow to tip of extended middle finger (three spans, six hands)
Step	Distance covered by one step (four spans, 16 phalanges)
Fathom	Distance between tips of fingers of the hands with arms extended laterally ("span akimbo") (three steps)

Source: Reprinted with permission from Rudolph J. Drillis,"Folk Norms and Biomechanics," *Human Factors,* Vol. 5, 427–441, 1963.

CHALLENGES

Which are the structures that keep the spinal column in balance?

What are the problems associated with drawing a person's attention to perceived working conditions in the course of an interview.

What changes would be needed in conventional lathe design to allow the operator to sit?

What makes work in tight, confined spaces so difficult?

Why is it difficult to exert large forces with the foot when one is standing?

What are the effects of using the trunk muscles when sitting without a backrest?

How can one's comfort diminish with the length of time one sits?

Occasionally, one hears the argument that sitting may be more conducive to falling asleep on the job than standing is. Is this true? Why or why not?

Can one sit too comfortably?

Come up with some alternative design solutions for the foot controls in an automobile.

Come up with some alternative design solutions for the hand-control functions of flight direction and engine speed in an airplane.

What are the disadvantages of contouring hand tools to fit the hand closely?

Describe some tests for assessing the usability of gloves.

Describe some means of measuring the pressure distribution of surfaces that "form-fit" the human body, such as a hand tool or the surface of a seat?

Often, force is applied to an object not in a continuous way, but in steps, such as in the force required to set an object into motion and then the force required to keep it in motion. How could such force exertions be measured?

Consider the statement that overexertion injuries should not occur in the normal use of the body. Is this statement true? Why or why not?

What are the advantages of being aware of signs of body strain related to repetitive work? Are there disadvantages as well?

Is "social awareness" of problems at work detrimental or advantageous to productivity?

Which biomechanical models could be applied to explain cumulative trauma injuries to the hand–arm system as a result of manipulations?

Describe specialized job analyses to check for potential overuse disorders.

Under what conditions might it be permissible to select persons to do difficult jobs, instead of fitting the job better to human capabilities?

Discuss the possible interactions among forcefulness, repetitiveness, and posture as a cause of cumulative trauma.

Chapter 9

The Office (Computer) Workstation

OVERVIEW

The modern office, equipped with computers, has little resemblance to the rooms in which male clerks handwrote entries in ledgers or penned letters a century and longer ago. By the middle of the 20th century, clerks had changed from standing at work to sitting, and most office employees were females. The idea that "healthy sitting is sitting upright" prevailed, and office furniture was designed for that body position, until it became obvious that people in modern offices sit any way they like—apparently without bad health consequences.

Different work tasks, the use of computers, and changed attitudes give reasons to rethink the old recommendations for office furniture design. Furniture should accommodate a wide range of body sizes, body postures, and activities; should further the performance of the task at hand, facilitate vision, and allow interaction with coworkers; and should improve people's well-being and make them feel well in their work environment.

INTRODUCTION

Three hundred years ago, Ramazzini discussed why people who have to work standing (as opposed to walking) become so fatigued. His explanation was that the same muscles have to be kept tensed to maintain the upright posture. He also stated that workers who sit still, stooped, looking down at their work (such as tailors) often become round shouldered and suffer from numbness in their legs, lameness, and sciatica. Believing that "all sedentary workers suffer from lumbago," Ramazzini advised that workers not stand or sit still, but move the body and "take physical exercise, at any rate on holidays" (Wright's translation of 1993; pp. 180–185).

In the offices of the late 1800s and the early 1900s, when the clerks were male, it was common to stand while working. Then the concept changed: It is now accepted that one should sit in the office. Yet, low back pain and musculoskeletal discomfort, often together with eyestrain, constitute the majority of subjective complaints and objective symptoms of computer operators.

IS THERE A NORMAL, HEALTHY, IDEAL POSTURE?

☛☛☛ *Grieco (1986) believes that* Homo erectus *was, and* Homo sapiens *is, biomechanically adapted to moving around, but not for standing still or sitting still, and that the human spinal column, suitable for a body in four-legged motion, has not had enough time to adapt to the upright position used in bipedal locomotion. Grieco also fears that the current rapid transformation to* Homo sedens *will lead to serious maladaptation of the spine.* ☚☚☚

THEORIES OF "HEALTHY" STANDING

In the 19th century, body posture was of great concern to physiologists and orthopedists, as evidenced in a review by Zacharkow (1988). Von Meyer speculated in 1873 about the biomechanical consequences of relaxed standing in contrast to the stiff military "at attention" position. In 1889, observing men standing easy and unconstrained, head high and looking straight forward, Staffel classified standing postures according to their spinal curvatures. His "normal" posture is characterized by a straight spine in the frontal or rear view; when the spine is seen from the side, a plumb line from the top of the cranium passes through vertical vertebrae, shoulder joint, middle between the back and chest contours, lumbar vertebrae, and trochanter of the femur, just behind the center of the hip joint and slightly in front of the center of the knee and the ankle joints. This upright or straight posture appeared balanced and healthy to Staffel. Farmers and laborers whose work required excessive spine loading and who maintained a bending posture often had back curvatures that diverted from "normal": flat, overly lordotic, and kyphotic or scoliotic. Staffel and his contemporary orthopedic physicians found these nonnormal postures unhealthy and concluded that they had to be avoided, especially at a young age. Subsequent medical and popular literature abounded in speculative wordy descriptions of human posture until, in 1946, a numerical assessment using the postural landmarks of the spine and pelvis was published by Appleton (Bonne 1969).

The "normal" posture as recommended by Staffel and his 19th-century contemporaries and has been promoted ever since by physicians, orthopedists, physical therapists, mothers, teachers, and military superiors. Even today, that "upright" or "straight" standing posture with slight lordoses (forward bends) in the lumbar and cervical spines and a light kyphosis (backward bend) in the thoracic spine is stereotypically considered "good and proper" and is often called healthy, balanced, or neutral (Merrill 1995). For a discussion of the "neutral" posture, see Chapter 8. But that posture is certainly impractical for working, since we do move about—and should do so—instead of standing stiff and still.

THEORIES OF "HEALTHY" SITTING

Sitting Upright?

Concurrent with his treatise on standing postures, in 1884 Staffel published his theories about "hygienic" sitting postures. He recommended an erect posture of the trunk, neck, and head, with normal slight lordoses in the lumbar and cervical areas and a light kyphosis in the thoracic spine, similar to the desired back posture when standing erect. Staffel and his colleagues were particularly concerned about the postural health of children, and therefore, they generally agreed that school seats and desks should be designed, and the children exhorted, to maintain that "upright" back, neck, and head posture. A great number of design proposals for school furniture, including seats and seat–desk combinations (Zacharkow 1988), believed suitable to promote that posture, were published from the 1880s on. In fact, the same sitting posture, with the thighs horizontal and lower legs vertical, was also advised for adults, especially those working in the seated position in offices. Thus, office furniture was endorsed that supposedly made people sit in that erect posture—but only the lowly employees: Into the 1970s, managers habitually enjoyed an ample armchair with a high back and comfortable upholstery (called a "Chef-Sessel"), while the secretary sat on a small, hard chair with a miserable little board for a backrest.

The simple concept that sitting upright, with the thighs horizontal and lower legs vertical, means "sitting healthily" endured for a surprisingly long time; even the ANSI/HFS Standard 100 of 1988 used this posture to prescribe office furniture.

☛☛☛ *For at least a century, it was a common belief that standing and sitting with a straight back is physically desirable and socially proper for pupils and adults alike. Of course, there is nothing wrong with voluntarily sitting or standing upright, but it is false to require that an erect back be maintained for long periods of time, such as during work while sitting in the office. Apparently, the human body is adapted to change—to moving about. Sitting or standing still for extended periods is uncomfortable and leads to compression of tissue, a reduction in metabolism, a deficiency in blood circulation, and an accumulation of extracellular fluid in the lower legs.* ☚☚☚

Maintaining lumbar lordosis?

The 19th-century concept of the proper sitting posture emphasized maintaining a "normal" lumbar lordosis, similar to that of a healthy person standing upright.

When one sits down on a hard, flat surface, without a backrest, the ischial tuberosities (inferior protuberances of the pelvic bones) act as fulcra around which the pelvic girdle rotates under the weight of the upper body. Since the bones of the pelvic girdle are linked by connective tissue to the thighs and the lower trunk, rotation of the pelvis affects the posture of the lower spinal column, particularly in the lumbar region. If the rotation of the pelvis is rearward, the normal lordosis of the lumbar spine is flattened. (See Figure 9–1.) This was deemed highly undesirable by orthopedists and physiologists. Hence, avoiding the backward rotation of the pelvis is a main issue of many theories of seat design.

Given the tissue connections between pelvis and thigh—particularly the effects of muscles spanning the hip joint or even both the knee and the hip joints (eg, the hamstrings, quadriceps, rectus femoris, sartorius, tensor fasciae latae, and psoas major), the actual hip and knee angles also affect the location of the pelvis and hence the curvature of the lumbar spine. At a large hip angle, a forward rotation of the pelvis on the ischial tuberosities is likely, accompanied by lumbar lordosis. (These actions on the lumbar spine take place if associated muscles are relaxed; muscle activities or changes in trunk tilt can counter the effects.)

Accordingly, in 1884, Staffel proposed a forward-declining seat surface to open up the hip angle and thereby bring about lordosis in the lumbar area. This concept was reinforced by Keegan in 1952 and led to a seat pan design with an elevated rear edge that was popular in Europe in the 1960s. Since then, Mandal (1975, 1982) and Congleton et al. (1985) have again

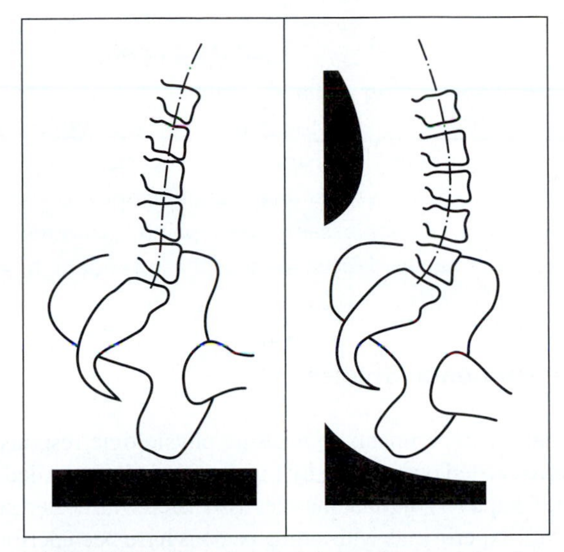

Figure 9–1. Positions of the pelvis and lumbar spine on a flat and on an inclined seat surface (from Kroemer and Robinette 1968).

promoted the idea that the whole seat surface should slope foreward and downward. To prevent the buttocks from sliding down the forward-declined seat, the surface of the seat may be shaped to fit the human underside (Congleton), or one may counteract the downward–forward thrust either by bearing down on the feet (Mandal) or by propping the knees or upper shins on special pads. One may call this posture "semisitting" or "semikneeling." Figure 9–2 gives an example.

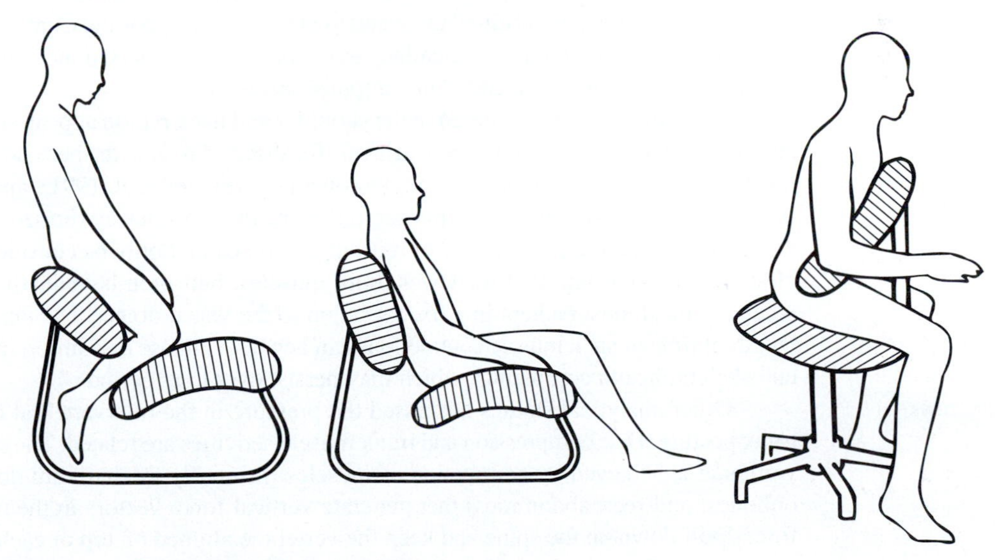

Figure 9–2. Modes of sitting.

SEMISITTING

A major advantage of semisitting is that the trunk is mobile, particularly if there is no backrest. If there is no knee pad in the way, it is easy to get down on the support and to get up from it: therefore, semisitting is suitable to relieve the worker from continuous standing. (See Chapter 8.)

Among the disadvantages of semisitting is the tendency to slide off the support surface, counteracted either by fatiguing leg thrust or by unpleasant or even painful pressure against the shin pads. In the office, where one tends to sit rather low, it can be difficult to move the legs out of the confined space between pads and seat.

Experimental Studies

Muscle tension

Analytic experiments to measure physiologic responses of the human body to certain postures were carried out in 1951 by Lundervold, who recorded and interpreted electromyograms (EMGs; see Chapter 1) of trunk muscles associated with seated positions. Based on these early studies, many EMG experiments with sitting persons have been performed, especially by Andersson and Oertengren (1974; see the summary by Chaffin et al. 1999). EMGs show various activities of the five major trunk muscle pairs (erector spinae, latissimus dorsi, internal and external oblique, and rectus abdominus) exerting pull forces along the length at the trunk and stabilizing the spinal column. Variations in postures, external loads, and backrests or other seat features influence these muscular activities. Muscular strains, especially in the low back area, can be substantial while one is leaning over the desk to lift something with extended arms. However, in regular sitting, the magnitude of the tension of muscles in the trunk is rather low—typically, well below one-tenth of the muscles' maximal contraction capabilities. How to interpret these weak EMG signals is controversial: Recording them may not be very reliable, and one should not necessarily assume that little muscle use (shown as flat EMG signals) is preferable over more extensive use (Wiker et al. 1989).

The idea of increasing muscular activity when sitting has been promoted to maintain suitable muscle use and training, to facilitate electrolyte and fluid balance for intervertebral disc metabolism, to improve blood circulation, and to prevent blood pooling. Continual muscular activities or bursts of activities (sometimes called "dynamic sitting") can be done voluntarily or may be encouraged by messages embedded in computer software.

Muscular tensions observed in the shoulder and neck region appear, in contrast to those observed in the lower trunk, to be more critical. Tension and pain in the neck area are among the most mentioned health complaints of computer operators (Sauter et al. 1991), apparently more related to vertical tilt ("nodding") in the midsagittal plane than to sideway (horizontal) motions (Collins et al. 1990). The relative intensity of muscle tension cannot only be considerably higher than the 10 or less percent reported for lower trunk muscles, but often is maintained over long periods when the head must be kept in a fixed relation to the visual object. The intensity, frequency, and length of time of such muscle contractions can generate intense discomfort, pain, and related musculoskeletal health complaints, which may persist over long periods.

Disc pressure

Other analytical studies addressed the pressure in the intervertebral discs as a function of trunk posture. Disc compression and trunk muscle activities are related: The stability of the stacked vertebrae is achieved by contraction of muscles (primarily the latissimi dorsi, erectores spinae, obliques, and recti abdominus) that generate vertical force vectors in the trunk. Together, these forces pull down on the spine and keep the vertebrae aligned on top of each other. Each vertebra rests upon its lower one, cushioned by the spinal disc between their main bodies and also supported laterally posteriorly in the two facet joints of the articulation processes. (See Chapters 1 and 2.)

Since the downward pull of the muscles generates disc and facet joint strain (in response to upper body weight and external forces), one should expect close relationships between trunk muscle activities and disc pressures. In the 1970s, Andersson and Oertengren (1974) and Andersson et al. (1974) inserted pressure transducers into the L3/L4 spinal disc of four subjects sitting on an experimental stool (see Figure 9–3) and of three female volunteers sitting on office chairs (Figure 9–4; for a compilation of the findings and their review, see Chaffin and Andersson 1984.)

All these experiments showed that the amount of intradisc force in the lumbar region was dependent upon the trunk posture and body support. When the subject was standing at ease, the forces in the lumbar spine were in the neighborhood of 330 newtons, as Figure 9–3 shows. The magnitude increased by about 100 N when the subject sat on a stool without a backrest. It made little difference if one sat erect with the arms hanging or relaxed with the lower arms on the thighs. (Sitting relaxed, but letting the arms hang down, increased the internal force to nearly 500 N.) Thus, sitting down produced an increase in spinal compression force in the lumbar region compared with standing, but the differences among sitting postures were not very pronounced.

About the same force values were measured in persons sitting on an office chair with a small lumbar support. (See Figure 9–4.) Sitting with the arms hanging, writing with the arms

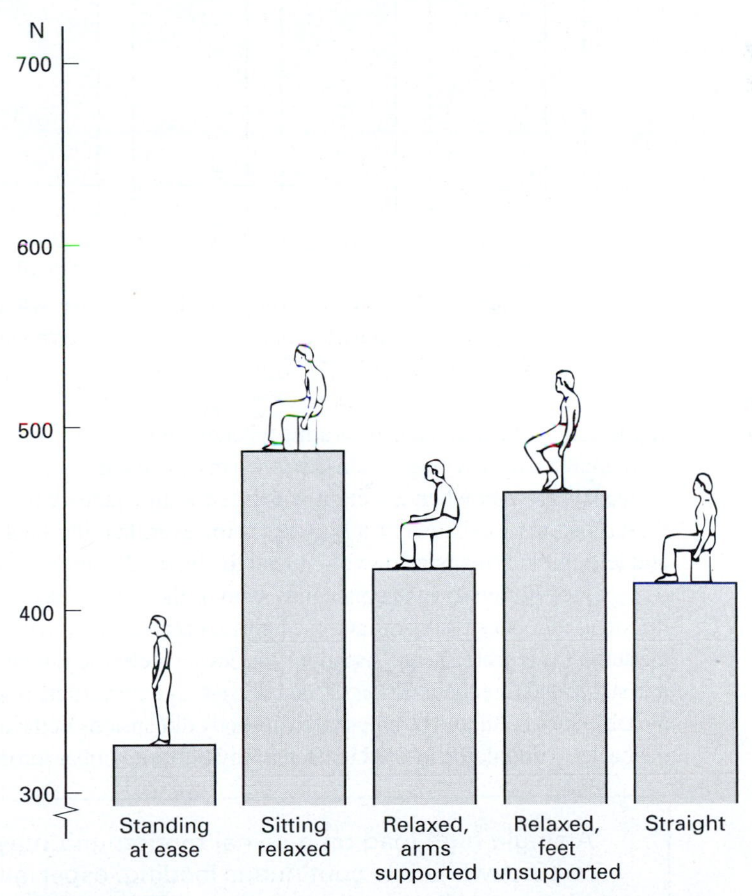

Figure 9–3. Forces in the third lumbar disc when standing, or when sitting on a stool without a backrest (with permission from Chaffin and Andersson 1984).

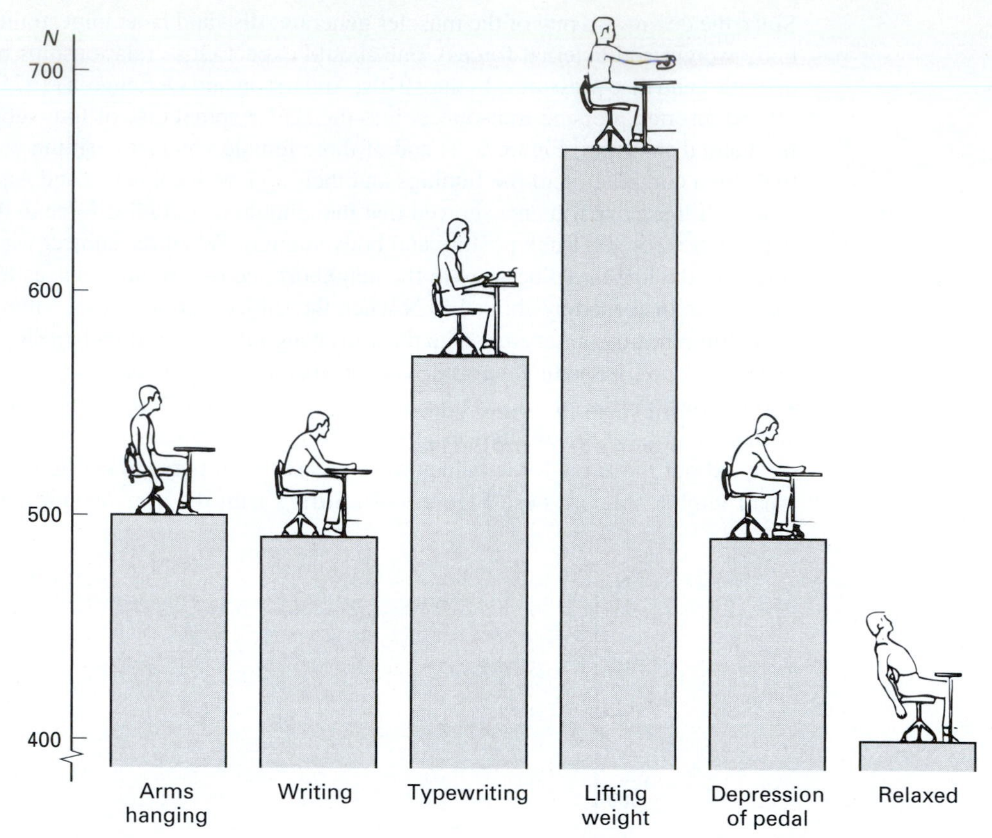

Figure 9–4. Forces in the third lumbar disc when sitting on an office chair with a small lumbar backrest (with permission from Chaffin and Andersson 1984).

supported on the table, and activating a pedal led to forces of around 500 N. The spinal forces also were increased by typing, when the forearms and hands were lifted to keyboard height. (A further increase was seen when a weight was lifted in the hands with forward-extended arms.) None of these postures made use of a backrest. However, leaning back decidedly over a small backrest and letting the arms hang down reduced the internal compression forces to approximately 400 N.

Since Andersson and colleagues' work is the only known analytic study of this kind, and given the small number of subjects and the apparent scatter of the data, the entire concluding statement of the authors is repeated here: "Assuming that low myoelectric activity and low disc pressure are desirable, it is suggested that the position of the backrest and the vertical distance between the seat surface and the table surface should be adjusted to the body dimensions of the occupant. It is further suggested that the backrest should be located in such a way that the lumbar lordosis is preserved" (p. 120).

"A single high load on a spinal motion unit may produce an end plate fracture, while low continuous loading, especially in a flexed spinal position, may lead to disc herniation. Thus, sitting in a 'low demand' situation may produce injury." Aspects of soft tissue responses "explain apparently paradoxical findings that both high loads and low loads (as in sitting) may be associated with low back pain and that high demand manual tasks and low demand manual tasks (such as keyboard work) may lead to upper limb disorders." (Richard Wells, quoted on pages 47–48 by National Research Council 1999).

Back support

The analytic studies just discussed indicate the importance of supporting the back by leaning it on a rearward-declined backrest and by maintaining lumbar lordosis. Figure 9–5 shows the effect of back support: When the backrest is upright, it cannot support any body weight, and disc forces between about 350 and 660 N may arise. Declining the straight backrest behind the vertical brings about decreases in internal force, because part of the upper body weight is now transmitted to the backrest and hence does not rest on the spinal column. An even more pronounced effect is brought about by making the backrest protrude toward the lumbar cavity. A protrusion of 5 centimeters nearly halves the internal disc forces from the values associated with the flat backrest; protrusions of 4 to 1 cm in the lumbar region bring about proportionally smaller effects.

Andersson et al. summarized the available findings in 1986 and concluded that "In a well-designed chair the disc pressure is lower than when standing" (p. 1113). Relaxed leaning against a declined backrest is the least stressful sitting posture. This posture is often freely chosen by persons working in the office if there is a suitable backrest available: "An impression which many observers have already perceived when visiting offices or workshops with VDT workstations [is that] most of the operators do not maintain an upright trunk posture. . . . In fact, the great majority of the operators lean backwards even if the chairs are not suitable for such a posture" (Grandjean et al. 1984, pp. 100–101).

Experiments on the biomechanical effects of sitting postures yield three important findings. The first is that sitting without the use of a backrest or armrests may increase disc pressure over standing by one-third to one-half. The second is that there are no dramatic disc-pressure differences between sitting straight, sitting relaxed, and sitting with supported arms if there is no backrest or only a small lumbar board. The third finding is that leaning on a well-designed backrest can bring about disc pressures that are as low as, or even lower than, those experienced by a standing person.

There Is No *One* Healthy Posture

Neither theories nor practical experiences endorse the idea of a single proper, healthy, comfortable, efficient sitting posture. Thus, the traditional postulate that everybody should sit upright and that furniture should be designed to end that is mistaken. Instead, there is general agreement that many postures may be comfortable (healthy, suitable, efficient, etc.) for short periods, depending on one's body, preferences, and work activities. (See Chapter 8.)

Consequently, furniture should allow body movements and various postures. For variation, furniture should permit easy adjustment in its main features, such as the seat height and angle, backrest position, and knee pads and footrests; and the computer workstation should allow easy variations in the location of the input devices and the display. Motion, change, variation, and adjustment to fit the individual are central to his or her well-being.

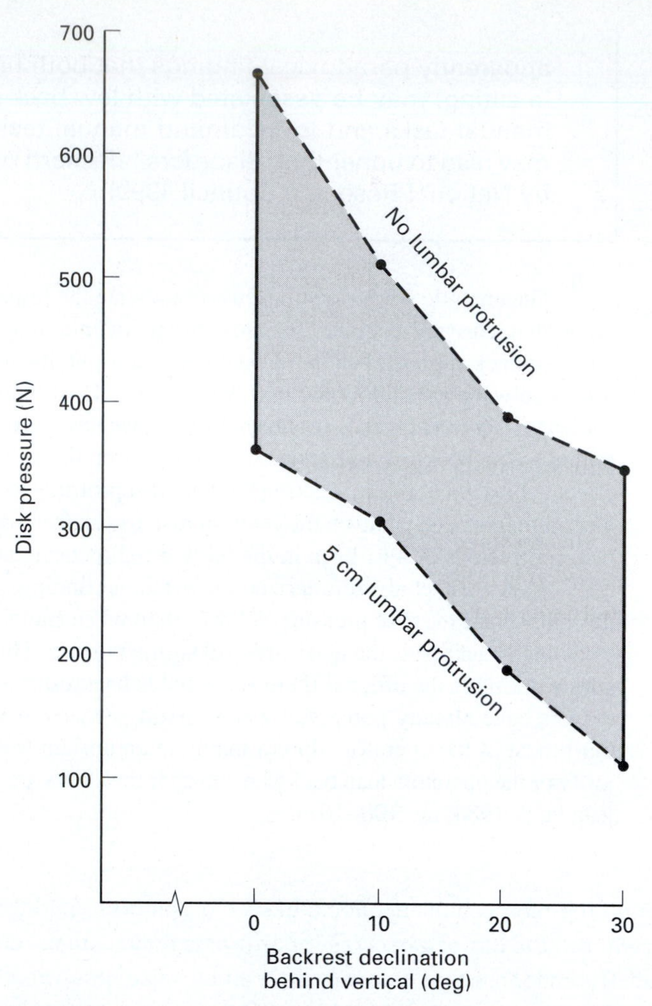

Figure 9–5. Disc forces (L3/4) depending on backrest angle and lumbar pad size. SOURCE: Adapted from Chaffin, D. B. and Anderson G. J. B., *Occupational Biomechanics.* Copyright 1984 by John Wiley & Sons, Inc. Reprinted by permission of John Wiley & Sons, Inc.

"FREE POSTURING"

Bendix et al. (1996) reported that, while reading, persons often assumed a kyphotic lumbar curve even when sitting on a chair with a lumbar pad that should have produced a lordosis. Apparently, people sit any way they want, regardless of how experts think they should sit.

Allowing persons freely to select their posture has led in two instances to surprisingly similar results. In 1962, Lehmann showed the contours of five persons "relaxing" under water. Sixteen years later, relaxed body postures were observed in astronauts and reported by NASA (1978). The similarity between the postures under water and in space is remarkable. (See Figure 9–6.) One

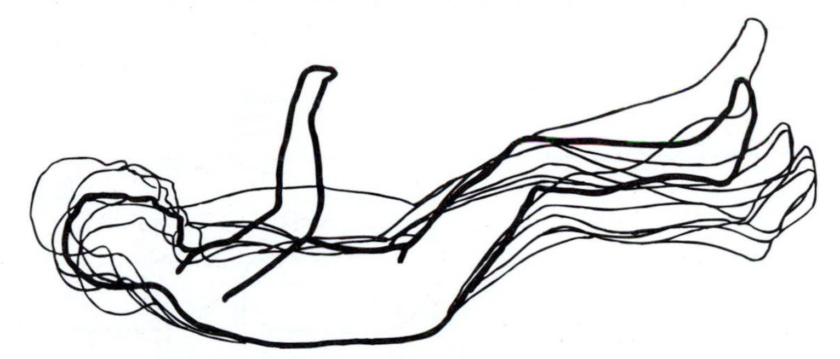

Figure 9–6. Relaxed underwater body postures (with permission from Lehmann 1962), superimposed on the relaxed posture in weightlessness (from NASA/Webb 1978).

might assume that, in both cases, the sum of all tissue torques around body joints was nulled. Coincidentally or not, the shape of "easy chairs" is quite similar to the contours shown in both figures. Some "executive" computer workstations, sketched in Figure 9–7, use similar support shapes.

If any label can be applied to current theories about proper sitting, it may be "free posturing," as sketched in Figure 9–8.

The "free-posturing" design principle has these basic ideas:

- Allow the operator to freely assume a variety of sitting (or standing) postures, adjust the workstation to her or his preferences, and even get up and move about.
- Design for a variety of user dimensions and user preferences.
- Use new technologies as soon as possible at the workstation. For example, radically new keyboards and input devices, including voice recognition, may be available soon, and display technologies and the placement of displays are undergoing rapid changes.

☛☛☛ *"Egyptian tomb reliefs illustrate clearly that, even at that time, dignified man had to sit with back straight, thighs horizontal, and lower legs vertical. How was it, in spite of all the evidence against it being either comfortable or natural, that this stilted posture came to be accepted as standard, whether for sitting on a throne, for dining, for contemplation in the privacy of the boudoir, or for working in an office?" (Editorial introduction to March 1986 issue of* Ergonomics.*)* ☚☚☚

Asking the User

Shape of the backrest

A desired lumbar lordosis could be achieved simply by opening the hip angle to more than 90 degrees. Sitting on a tall, forward-declined seat pan with declining thighs rotates the pelvis forward and bends the lumbar spine into lordosis. This kind of sitting and seat has been advocated repeatedly (eg, by Congleton et al. 1985 and Mandal 1982), but has not received widespread acceptance, regardless of whether a backrest is attached to the seat. There is reason to believe that the discomfort associated with the seat stems mainly from the pronounced declination of the pan

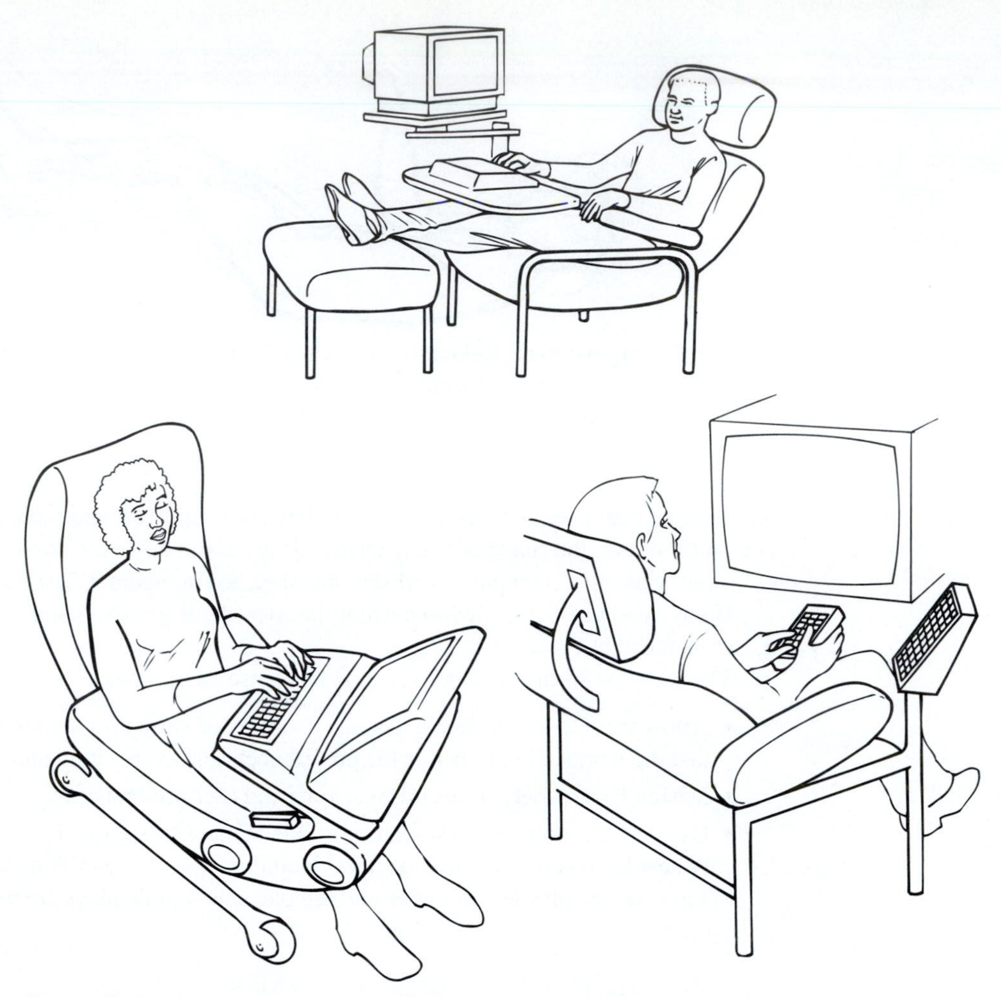

Figure 9–7. Body postures in "executive" computer workstations.

and the ensuing semisitting, which make it necessary to bear down on the feet or push against shin pads.

A well-designed backrest alone could bring about lordosis of the lumbar spine by pushing that section of the back forward. Old wooden school benches simply had a horizontal wood slat at lumbar that forced the seated pupil to bend the lower back forward to avoid painful contact with the slat; hence, the child sat "up," to the satisfaction of the teacher. There are more subtle and agreeable ways, however, to promote lumbar concavity—for example, by a fixed lumbar pad (Akerblom 1948), a portable pad (Vink et al. 1994), inflatable lumbar cushions incorporated into the seat back of car and airplane seats, and adjustable lumbar supports in some office chairs.

Shape of the whole chair

Of course, one can shape the total backrest to support the back fittingly: Apparently independently of each other, Ridder (1959) in the United States and Grandjean and his coworkers (1963) in Switzerland found rather similar backrest shapes generally accepted. In essence, these shapes

Figure 9–8. "Free posturing."

follow the curvature of the rear side of the human body: concave at the bottom, to accept the buttocks; convex slightly above, to fill in the lumbar lordosis; then rising nearly straight, but declined backward to support the thoracic area; and then convex again at the top, to follow the neck lordosis. Combined with a suitably formed and upholstered seat pan, this shape has been used successfully for seats in automobiles, aircraft, and passenger trains and for easy chairs. In the traditional office, these "first-class" shapes were enjoyed by the boss, while lowly employees had to use simpler designs. So-called secretarial chairs had a small, often hard-surfaced seat pan and a slightly curved support for the low back; the more recent "task chair" is an improved version.

Comfort vs. discomfort

The concept of comfort, as related to sitting, was elusive as long as it was defined, simply and conveniently, but falsely, as the absence of discomfort. Helander and Zhang showed in 1997 that, in reality, these two aspects are not extremes of one single scale. Instead, there are two scales, one for comfort and the other for discomfort, that are not even parallel, but partly overlap.

Discomfort is expressed in such terms as feeling stiff, strained, cramped, tingly, numb, unsupported, fatigued, restless, sore, and in pain. Some of these attributes can be explained in terms of circulatory, metabolic, or mechanical events in the body; others go beyond such physiological and biomechanical phenomena.

Users can rather easily describe design features that result in feelings of discomfort, such as chairs in wrong sizes, that are too high or too low, or with hard surfaces or edges; but avoiding these mistakes does not by itself make a chair comfortable.

We associate *comfort* in a sitting condition in terms of well-being if we feel supported, safe, pleased, and content; in terms of biomechanics if we are relaxed and restful; and in terms of felt sensation if we are under the impression of experiencing warmth, softness, plushness, and spaciousness. Even esthetics plays a role if we like the appearance, color, or ambience of a piece of furniture.

Upholstery, for example, strongly contributes to the feeling of comfort because it "breathes" by letting heat and humidity escape and supports the body by distributing pressure along the contact area so that it is neither too soft nor too stiff. However, exactly what feels comfortable depends very much on the person, individual habits, on the environment and the task, and on the passage of time.

Helander and Zhang characterized discomfort and comfort separately with respect to sitting in a chair by means of seven respective statements expressing one's feelings or impressions about both the chair and its effects on the body or mind. Rated in a series of nine steps from "not at all" to "extremely," the statements about discomfort were as follows:

1. I have sore muscles.
2. I have heavy legs.
3. I feel uneven pressure.
4. I feel stiff.
5. I feel restless.
6. I feel tired.
7. I feel uncomfortable.

The following statements, also rated in terms of the nine steps, were about comfort:

1. I feel relaxed.
2. I feel refreshed.
3. The chair feels soft.
4. The chair is spacious.
5. The chair looks nice.
6. I like the chair.
7. I feel comfortable.

Unless a chair was truly unsuitable, Helander and Zhang found it difficult to rank the chair by attributes of discomfort, because the body is surprisingly adaptive (except when the sitter has a bad back). By contrasts, the comfort descriptors proved sensitive and discriminating for ranking chairs in terms of preference. Helander and Zhang's subjects said it was easy to rank chairs in terms of their *overall* comfort or discomfort (statement 7 in each list) after being exposed to the preceding more detailed descriptors. It is interesting to note that the rankings of chairs were established early during the trials and did not change much with the length of time the subject sat. Still, it is not clear whether a few minutes of sitting on chairs is sufficient to assess them or whether it takes longer trial periods.

ERGONOMIC DESIGN OF THE OFFICE WORKSTATION

Whether an ergonomically designed office workstation is successful depends on several interrelated properties, sketched in Figure 9–9. The work tasks, the work postures, and the work activities all interact and are influenced by the workstation conditions, including furniture and other equipment, and by the environment. All of these must "fit" the person. Work postures and activities determine, to a large extent, the person's physical well-being and work output. Of course, job demands, content, control, and other social and organizational factors also influence the individual's feelings, attitudes, and performance.

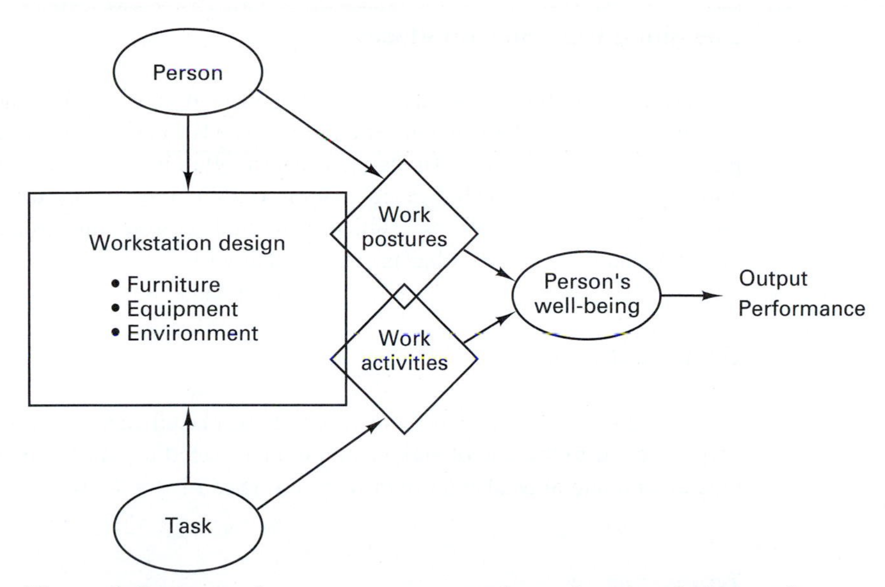

Figure 9–9. Interactions among person, task, workstation design, and performance.

For the layout of the workstation and the work task, it is useful to think of three main links between a person and the job:

1. The first is the *visual interface:* One must look at, among other things, the keyboard, the computer screen, the printed output, and source documents.
2. The second is *manipulation:* The hands operate keys, a mouse, or other input devices, manipulate a pen and paper, and hold a telephone. Occasionally, the feet operate controls, as, for example, in starting and stopping a dictation machine.

The intensities of the visual and motor requirements depend on the specific job. (See Table 9–1.)

3. The third is *body support:* The body is linked to the seat at the undersides of the thighs and buttocks and at the back with the backrest. Armrests or wrist rests may be other support links.

TABLE 9–1. Linking the Human with the Computer in Different Work Tasks

Task	*Visual requirements*	*Motor input requirements*	*Continuity of requirements*
Data entry	High (source and screen)	High (keyboard)	Few interruptions
Data acquisition	High (screen)	Medium (keyboard)	Varies
Word processing	High to medium (source, screen, and keyboard)	High (keyboard)	Few interruptions
Interactive communication	Medium (screen)	Medium	Varies
CAD	High (screen and source)	Low	Frequent interruptions

SOURCE: Modified from Kroemer 1988a.

Designing the Visual Interface

In conventional offices, the paper for writing or reading is usually placed on the regular working surface, roughly at elbow height. If an object needs to be looked at more closely, it is lifted to a proper relation to the eyes. An inclined surface often has been recommended for easier reading, making an angle with the line of sight closer to 90 degrees, but the disadvantage of the sloped surface is that work materials may slide or roll off. Therefore, one usually has a horizontal table or desk on which the various objects are placed at will.

Document holder

A source document from which one reads can be placed on a holder that puts the document about perpendicular to the line of sight. But a holder placed too far to one side causes a twisted body posture and lateral head and eye movements. (See Figure 9–10.)

Typewriter

When a typewriter is used, there are several visual targets. One is the "printing" area—the platen on a conventional typewriter and the display area on an electronic typewriter. This area is usually only a few centimeters above the keyboard and is located fairly well with respect to the eyes in terms of distance, direction, and preferred line-of-sight angle. (See Chapter 4.)

Figure 9–10. "The computer user syndrome" (with permission from Grant 1990).

Computer

With the large increase in the number of keys from the typewriter (fewer than 50) to most computer keyboards (often more than 100), many operators have reverted to scanning the keys, whereas previously typists were able to do their job without looking at them. Thus, unfortunately, most computer keyboards have become rather large, nearly horizontal visual target areas. In addition, one must look at the display area of the computer monitor, which is correctly placed in front of the operator at about right angles to the line of sight. The third visual target is often some sort of source document, in some cases fairly large, such as a drawing used in computer-aided design.

Placing the visual targets

The problem of placing all these displays is mostly one of available space within the center of the person's field of view. Most people prefer to look downward at close visual targets at angles between 20 and 60° below the ear–eye plane, as discussed in Chapter 4. (This inclination is up to 45 degrees below the horizon when one sits "upright.") The natural way of focusing at a near target with the least effort is to incline the head slightly forward and to rotate the eyeballs downward (not holding the trunk and neck erect and looking straight ahead at the screen). Mistakenly putting the monitor up high behind the keyboard (often by placing it on the computer itself or on a so-called monitor stand) is rather uncomfortable for most viewers. Instead, the display should be placed immediately behind the keyboard, with the lower edge of the screen as near as possible to the rear section of the keyboard. The source document should be placed next to the display. (See Figure 9–11.) More refined solutions may incorporate all three visual areas in one setup; one such solution is shown in Figure 9–12.

Selection criteria for a good display are discussed in Standard ANSI/HFES 100/2000 and in Chapter 10 of this book.

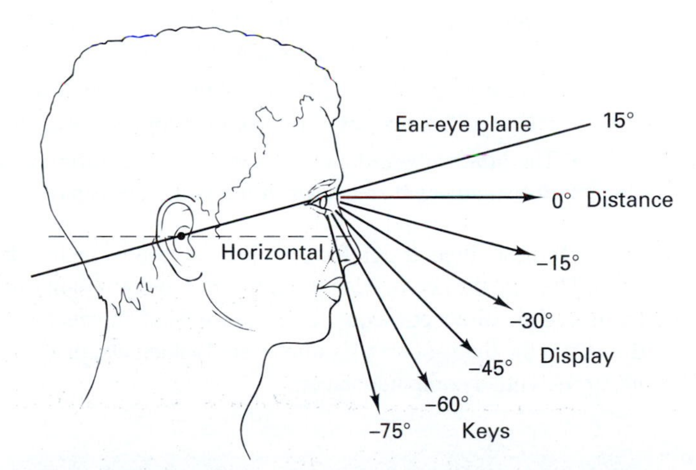

Figure 9–11. Suggested angles of the line of sight to observe keyboard and display with current computer workplace technology. (Note that this is a compromise solution for imperfect components.)

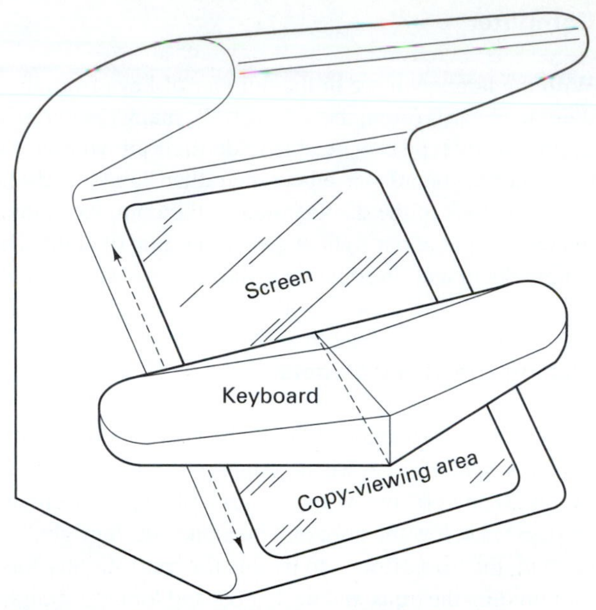

Figure 9–12. Proposed combination of screen, source display, and keyboard (with permission from Grant 1990).

Proper Office Lighting

For the engineer, the most important design factors are illumination, luminance, luminous contrast, and how they are distributed. (See Chapter 4.)

- ***Illumination*** is the amount of lighting falling on a surface. The light may come from the sun or from luminaires (lamps).
- ***Luminance*** is the amount of light reflected or emitted from a surface. Light may be reflected from ceiling, wall, or tabletop, or it may be emitted from a VDT screen.
- The ***luminous*** *contrast ratio* describes the difference between the luminance values of two adjacent areas, assuming that there is a defined boundary between them.

Of these three design factors, illumination is the best known, though least useful, phenomenon. The sun, the sky, or a lamp send visible energy (light) into the surroundings. Yet, what is emitted is of little direct consequence, because what "meets the eyes" is luminance (unless one stares directly at the light source). Examples are looking at a piece of paper, seeing the surroundings about us, or viewing a computer screen.

Workplace illumination shall be selected not to degrade the image quality of the computer display. On light-emitting displays, such as CRTs, the ambient illuminance should be between 200 and 500 lux, while 300 to 750 lux are suitable with flat-panel or other light-reflective displays (ANSI/HFES 100-2000).

Luminance counts

The quantity that determines human vision is the luminance impinging on the retina. Luminance results from the light energy emitted or reflected from the visual target and from its visual surroundings in our field of view. In rooms, most light energy that reaches the eyes is that reflected from surfaces—mostly walls and ceilings. Excessive luminance often glare.

Glare

Unless we look directly into a light source (eg, the sun or a lamp), we experience glare indirectly as light reflected from a "shiny" object. Polished surfaces reflect incoming light at the same angle at which it was received. Mirrors have a background coating that absorbs little light and reflects nearly all of the incoming energy. The *reflectance* of a surface (generally expressed in percent) indicates the portion of incident light that is reflected by the surface. To avoid specular (ie, "directed") glare, surfaces can be made matte by giving them a rough surface that reflects incoming light in various directions.

Contrast

The larger the luminous contrast between two (reflecting or emitting) areas and the better defined their common boundary, the easier it is for the eye to distinguish them.

Room Surfaces

The distribution of light within a room depends on the location of light sources (luminaires) and the direction of light flow from them, as well as on the reflectances of the ceiling, walls, and other surfaces. As a rule, the reflectances and colors of the room surfaces should be chosen so that there is a continuous decrease in reflectance from the ceiling to the floor. The ceiling reflectance should be about 80 to 90 percent—accomplished for instance, by white paint. The walls should have reflectances of 40 to 60 percent, which corresponds to bright beige, yellow, or green. The floors should have a reflectance of approximately 20 to 40 percent, such as by medium blue-green or brown-beige colors. The surfaces of furniture and equipment should reflect at 25 to 45 percent.

Avoiding Glare from the Computer Screen

One speaks of "indirect" (or reflected) glare at the computer workstation if a light source is reflected from the monitor screen into the eyes, in much the same way as a window, a lamp, or a white shirt is reflected in a mirror. The reflected bright spot or surface on the screen reduces the contrast between background and displayed characters (eg, lines, letters, and numbers). The attempt to discern characters can lead to "eye fatigue" (a rather undefined, but descriptive and popular term). A similar irritating condition called "direct" glare is caused by an intensive light source (the sun, light coming from a window, or a bright lamp) that shines directly into the eyes of the computer operator, generating high illumination of the retina, making it difficult for the rods and cones to discern contrasts between characters and their background on the screen.

Two other glare conditions, less well defined, but still common, are *washout* or *veiling* of the contrast on the screen and *stray glare.* Washout is caused by high ambient illumination (a form of indirect glare); stray glare is caused by reflection from shiny surfaces in the field of vision (also a form of indirect glare).

The problem of glare coming from a computer screen is completely resolved if no sources of high luminance are visible to the human operator. This is achieved by

- not placing any bright objects within the operator's field of view (ie, within his or her cone of vision) and
- positioning or angling the screen so that any existing bright objects are outside the operator's cone of vision.

Windows

Windows, liked by so many people, often generate difficult lighting conditions for the designer. On a bright day, they are sources of high intensity and strongly directed light, able to generate both direct and indirect glare. Furthermore, when hot, window panes radiate heat into the room; when cold, they draw warmth from the room. (See Chapter 5.)

Several means of controlling the light coming from a window can be employed:

- The window panes may be colored dark to reduce the amount of illumination coming through.
- Horizontal or vertical louvers ("miniblinds") can keep the light out. Vertical blinds are best when the sun is low, horizontal blinds when the sun is high. Properly adjusted, blinds can screen out the direct rays of the sun, but still allow people to look out through the window. Also, blinds can be removed if not needed—for example, when there is no sunshine. Light-colored louvers are advantageous at night, when the darkness outside would turn the windows into dark surfaces reflecting on the inside like mirrors.
- Curtains can be used to hinder direct sunlight from entering the room. Light-colored curtains absorb little energy from sunlight and hence do not heat up and then send warmth into the room, but they are highly luminous and thus may generate a source of high light intensity. Therefore, with respect to light control, dark-colored curtains are preferred.

PLACING LAMPS

Figure 9–13 shows proper illumination arrangements, as well as the visibility problems that can be generated by imperfect lamp positions. A light source positioned behind the operator (A) can generate specular glare on the computer screen. A luminaire located above the operator (B) may wash out the contrast on the screen by creating a veiling luminance and may cause reflections from the table and keyboard. A light in front of the operator (C) may cause direct glare by shining into the eyes. The figure also shows the seated operator from behind; a luminaire above the person, as in (B), may cause reflections from a shiny workstation surface. But placing lamps to the sides, as in (D) and (E), do not cause any problems for this operator.

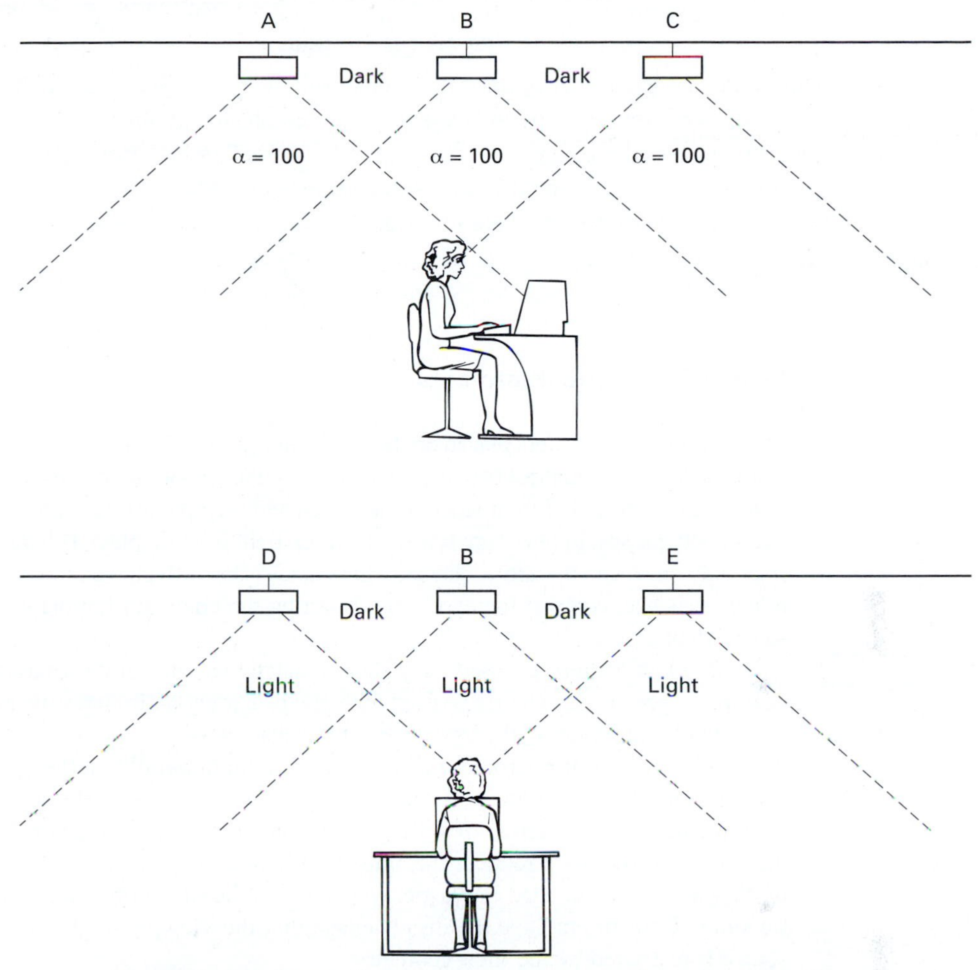

Figure 9–13. Luminaire A may generate specular reflections, B may wash out screen contrast, and C may cause direct glare. Luminaire B, positioned above the operator, may cause reflections from the workstation surface. Locations D and E are more suitable for the operator (adapted with permission from Helander 1982).

INDIRECT LIGHTING

The positions of light sources may be lowered from near the celing and some, most, or even all of the emitted luminous flux directed upward. In this case, the workstation may be illuminated partly by direct light from the luminaire and partly, mostly, or entirely by indirect light reflected from the ceiling. This arrangement has two advantages:

1. Moved closer to the workstation, the luminaire generates defined areas of direct illumination.
2. The illumination reflected from the ceiling does not create specular reflections.

TASK LIGHTING

One may illuminate only certain objects in the office with high intensity, such as the source document. This configuration allows easier reading while keeping the overall illumination low. For such spot lighting, often called task lighting, a couple of arrangements can be used: The task luminaire may be mounted in the ceiling and shine a spotlight at the target; or, more commonly, a lamp is placed near the source document and directed at it. Care should be taken that the light source and the lighted surface do not generate direct or indirect glare for the operator or anyone else nearby.

Screen Filters and Treatments

The purpose of both filters placed on the screen and "antiglare coatings" is to control reflections that would decrease the contrast between characters and the rest of the screen. Ambient illumination passes through the filter before it reaches the screen and then passes through the filter again on its way out. In both passes, light energy is lost. However, light from displayed characters passes through the filter only once, and therefore, character luminance is less affected than screen-reflected ambient luminance; hence, contrast increases. Yet, lowering the character luminance decreases visibility to some extent.

Filters

If colored filters are used, they should match the color of the characters on the screen. For example, a green filter should be used with green characters. In that way, luminous contrast is enhanced without reducing the luminance of the characters.

Microlouver or micromesh filters allow only light parallel to the openings to pass through, thus preventing much of the ambient illumination from getting to the screen. This improves contrast, owing to the reduction of veiling reflection. One disadvantage of micromesh filters is that they collect dirt, which reduces the amount of light that can pass through the filter. Microlouver filters are often embedded in plastic, which solves the dirt problem, but may create specular reflections. Both filters have the disadvantage that the viewing angle of the operator toward the screen is restricted by the filter geometry.

Coatings

Antireflection coating of the screen surface is occasionally used. As with the coating of camera lenses, a quarter-wavelength coating is commonly employed, which reduces specular reflections from the surface of the filter. The main disadvantages of a coating are that it can be degraded by fingerprints, it loses effectiveness with age, and it is fairly expensive.

Specular reflections can also be reduced by making the screen matte—for example, by etching. This eliminates the shiny mirrorlike surface by diffusing light. Unfortunately, it also reduces the contrast between the characters and the background and makes the edges of the characters less defined.

Hoods, sunglasses, dividers

A hood that protrudes from the top and the sides of the screen toward the operator can shield the screen from ambient illumination. However, it is often difficult to position such a hood without casting a shadow on the screen from the edge of the hood. Also, a hood may make the operator feel as if he or she were looking into a tunnel.

Some computer operators wear sunglasses at work. Sunglasses do not increase contrast, but instead reduce the luminous level of both the characters and the background alike. Visibility is therefore diminished.

Sometimes it is easy to place a shield or divider between the light source and the display. (See Figure 9–14.) A piece of dividing wall can give the feeling of a small private office.

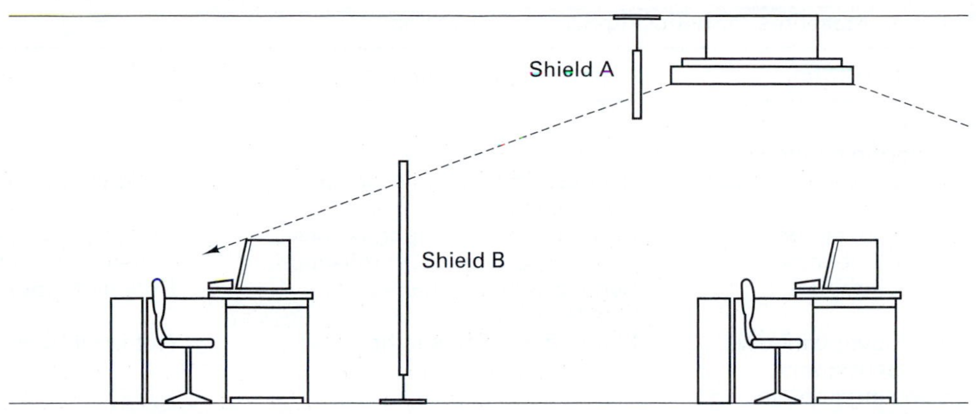

Figure 9–14. Shields used to keep light from shining directly onto a CDT (adapted with permission from Helander 1982).

Design for "No Bright Spots"

It is more effective to plan the layout of an office together with its illumination with an eye toward avoiding glare rather than counteracting existing glare conditions. The best solution is to install the proper light sources, because this eliminates direct and indirect glare at their points of origin. The second-best solution is to intercept the flow of light from a source to the eyes of the operator or to the screen, but this approach is usually more costly and less effective than installing the proper light sources. The third and least desirable solution is to apply means at the display surface to reduce reflections and improve contrast, because most such treatments result in diminished visibility owing to the loss of light energy. Table 9–2 summarizes the various measures and their advantages and disadvantages.

Lenses to Correct Vision Defects

Computer operators who experience vision difficulties and eyestrain often learn, upon taking a vision test, that they have eye deficiencies. To compensate, they need artificial lenses—eyeglasses or contact lenses. But using corrective lenses incorrectly can generate new problems, particularly in the case of so-called reading glasses. These are ground for a viewing distance of about 40 centimeters and a downward tilt of the line of sight, but many visual targets in the computer area are placed further away, including the screen, which, unfortunately, is too often placed well behind and above the keyboard. If such a visual target is beyond the focusing distance, one is tempted either to squint the eyes while trying to focus or to move the head forward to bring it closer to the correct focusing distance. The first attempt may lead to "eye fatigue," the second to improper neck posture and muscle tension. The effect is even more pronounced if one wears bifocals or trifocals, where the lowest section is meant for reading. In this case, one is likely to tilt the head severely backward in order to get the display on the screen on the "line of sight," which the reading lens of the glasses determines to be downward with respect to the head. The ensuing strong backwards tilt of the head requires muscular tension and often results in a headache. The harmful head and neck bends are sketched in Figure 9–15. The solution is to bring the display down, by placing it low behind the keyboard, and to wear corrective lenses in which the full surface is shaped for the correct viewing distance, even if this blurs the impression of objects further away. Orchestra musicians, for example, habitually have their glasses

TABLE 9–2. Measures for Reducing Screen Reflections

Intervention	*Advantages*	*Disadvantages*
At the source:		
1. Covering windows		
(a) Dark film applied	Reduces veiling and specular reflections	Difficult to look out
(b) Louvers or miniblinds	Exclude direct sunlight, reduce veiling and specular reflections	Must be readjusted in order to keep out light rays or to look out
(c) Curtains	Reduce veiling and specular reflections	Difficult to look out
(d) Cover windows permanently	Eliminates outside light	Not liked by most employees
2. Selecting and placing luminaries		
(a) Control of location and direction of illumination	Reduces veiling reflections, may eliminate specular reflections	None
(b) Indirect lighting	Reduces specular reflections, allows larger number of workstations per square unit	None (but used more electric energy)
(c) Task illumination	Increases visibility of source document	None (if properly arranged)
Between source and workstation:		
3. Shield or screen between luminaires or windows and workstation	Reduces direct light flux and reflections	Might create the impression of an isolated workplace
At the workstation:		
4. Readjust or relocate entire workstation	Eliminates all reflections	May need more space, alter office design
5. Tilt screen	Eliminates specular reflection	May force operator to assume awkward posture
6. Use neutral density (gray) filter	Reduces veiling reflection, increases character contrast	Reduces character luminance
7. Use color filter (same color as characters)	Reduces veiling reflection, increases character contrast	Decreases character luminance
8. Use micromesh, microlouver filter	Reduces veiling reflection, increases character contrast	Operator must look directly onto the screen; embedded filter may get dirty
9. Use polaroid filter	Reduces veiling reflection, increases character contrast,	Decreases character luminance
10. Use quarter-wavelength reflection coating	Eliminates specular reflection	Expensive, difficult to maintain
11. Put matte finish on screen surface	Decreases specular reflection	Increases character edge spread (fuzziness), increases veiling reflection
12. Use screen hood	Reduces veiling and specular reflections	Difficult to avoid shadow on the screen; feeling of watching television
At the operator:		
13. Wear tinted glasses	None	Reduces visibility

SOURCE: Adapted from Helander 1982.

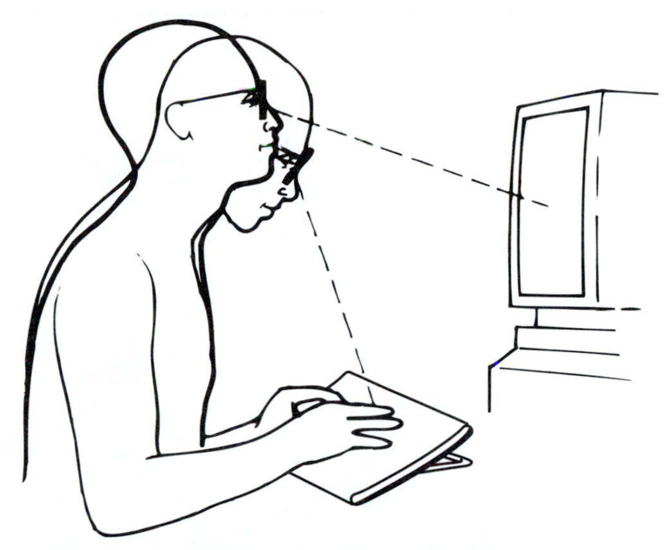

Figure 9–15. Excessive head and neck postures assumed when looking through "reading glasses" at a monitor or keyboard.

ground to a distance of about one meter so that they can focus on their sheet music. Figure 9–16 shows that they still have a dilemma: They can see the notes clearly, but not the conductor.

Designing the Motor Interface

With current technology, the operator transmits most of the data to the computer by hand. The common interface is the keyboard, often accompanied by other input devices, such as a mouse, trackball, joystick, or light pen. Their design (discussed in some detail in Chapter 10) codetermines the workload and motor activities of the operator, as well as the layout of the computer workstation.

Keyboard

Keyboard design

Unfortunately, the keyboard of the old typewriter is still used, largely unchanged, as the major input device for computers. The conventional QWERTY keyboard has several unergonomic features, such as letters that frequently follow each other in common text (eg, in English, *t* and *h*) spaced apart on the keyboard. Most likely, this was originally done so that the keys would not entangle if struck in rapid sequence. Another characteristic is that the "columns" of QWERTY keys run diagonally from left to right, which was also necessary on early typewriters due to mechanical constraints. With the QWERTY columns crooked, the rows of keys are straight, while the fingertips are not naturally aligned side by side. The keyboard must be operated with pronated hands ("thumbs down"), owing to the horizontal arrangement of the rows of keys. Furthermore, most desktop keyboards have excessively many keys (usually 101 or more), of which one must be correctly selected so that the desired character will be produced. Cognitively, this requires a difficult

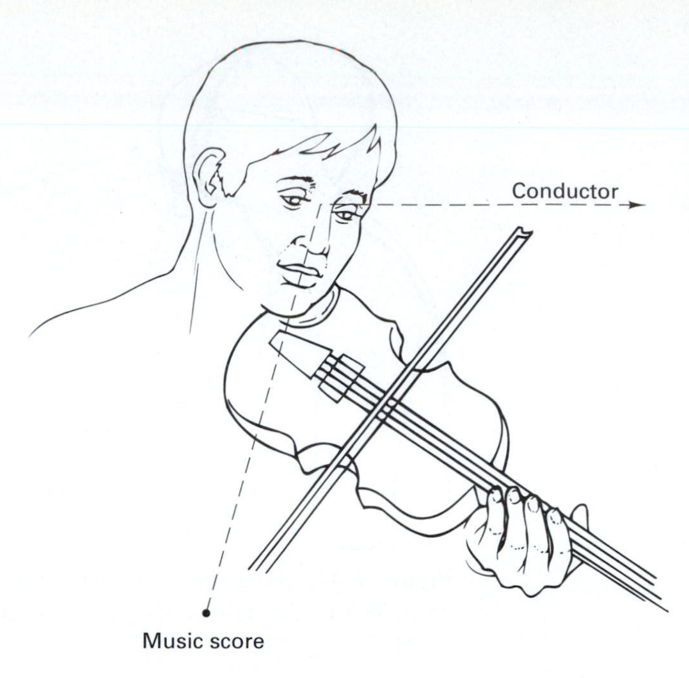

Figure 9–16. "The musician's malady" (with permission from Grant 1990).

multichoice decision; motorically, it requires the complex use of muscles to move the finger to the proper key.

ODs of keyboarders

The overuse disorders discussed in Chapter 8, often believed to occur mainly with repetitive tasks that require much physical effort, are common in keyboard operators. Causal or contributing factors are the frequency of operation of the keyboard, combined with awkward forearm and wrist postures (especially pronation and lateral deviation), but the measurement of specific tissue strains is difficult, and relatedly, establishing thresholds of strain doses is a still unresolved task (Brown et al. 1985; Hagberg and Rempel, 1997; Keir et al. 1996; Martin et al. 1998; National Research Council, 1999; Tittiranonda et al. 1999).

Keyboard improvements

The old mechanical keyboards on typewriters had strong key resistances and required large key displacements. Hence, it was suspected that weak fingers, particularly the little ones, were overworked. Accordingly, many recommendations for improving the traditional keyboard were proposed. (See, for example, Kroemer 1972b, 1997; Alden, Daniels, and Kanarick 1972; and Noyes 1983a.) Among the suggestions were relocating the letters on the keyboard and providing new geometries for the keyboard, such as various curved arrangements of the keys. It was also suggested that the keyboard be divided into one half for the left hand and one for the right, arranged so that the center sections are higher than the outsides, thus avoiding the pronation of the hand required on the flat keyboard (Heidner 1915, Klockenberg 1926). Another idea was to activate two or more keys simultaneously to generate one character (called chording; see Noyes 1983b; Gopher and Raij 1988; Keller et al. 1991). A new idea is to use chording with ternary instead of the usually binary keys (Kroemer 1991b). Examples of several different keyboards are given in Figure 9–17; further information on keyboards and on other entry devices is presented in Chapter 10.

Small keyboards can be placed nearly anywhere at the user's convenience, such as on one's lap, or they may be incorporated in an armrest, in a glove, or even in the shell of a space suit. New developments may radically change the nature and appearance of keyboards and hence free body and hand from forced postures.

ANSI/HFES 100(2000) presents recommendations for specific keyboard design features, such as key spacing, size, actuation force, and displacement. With the rapid development of computer input technology, today's guidelines may be outdated tomorrow. (See Chapter 10.)

Other input devices

The design of the keyboard or any other manipulated input device (eg, a mouse or a puck) determines hand and arm posture. The essentially horizontal surface of conventional keyboards and of most mice requires that the palm also be kept approximately horizontal; this is uncomfortable, because such strong pronation in the forearm is near the extreme of what is anatomically possible, with the upper arm hanging and the forearm extended forward. (Typists have long been observed to lift the elbows to ease the pronation, with resultant muscle tension about the shoulder.) A more convenient angle of forearm and wrist rotation is achieved by a sideways-down tilt of the mouse surface in contact with the palm; splitting a keyboard and tilting the sections down to the side was proposed as early as 1915 by Heidner and, in more detail, in 1926 by Klockenberg.

Of course, the location of input devices of any kind that are operated by hand (or by foot) determines the position of the hand (or foot) and, through corresponding arm (leg) position, the posture of the trunk. As a rule, the best position for the hand is in front of the body, at elbow height. If the mouse, puck, stick, or other input instrument is used with a keyboard, all the input devices should all be placed together; often, the keyboard housing contains a mouse, trackball, trackpad, or nipple.

Supporting the arm and the hand in a suitable fashion relieves shoulder and back muscles (Aaras et al. 1998a,b; Farris et al. 1998; Garcia et al. 1998). Depending on the task and conditions, various kinds of arm and hand rests can be employed during the work, or at least at breaks during extended work. Breaks are helpful, but by themselves are not effective in overcoming unsuitable equipment and workloads (Balci et al. 1998; Neuffer et al. 1997).

ANSI/HFES 100 recommends overall design dimensions for various input devices, including mice, pucks, trackballs, joysticks, styli, light pens, tablets, overlays, and touch-sensitive panels.

> In considering the details of ANSI/HFES 100, one should keep in mind that this standard merely specifies "acceptable" applications based "on accepted human factors engineering research and experience" and "does not apply to operator health considerations or work practices." The standard recognizes that alternative computer workstation technologies are acceptable and that the development and use of novel solutions is not to be impeded.

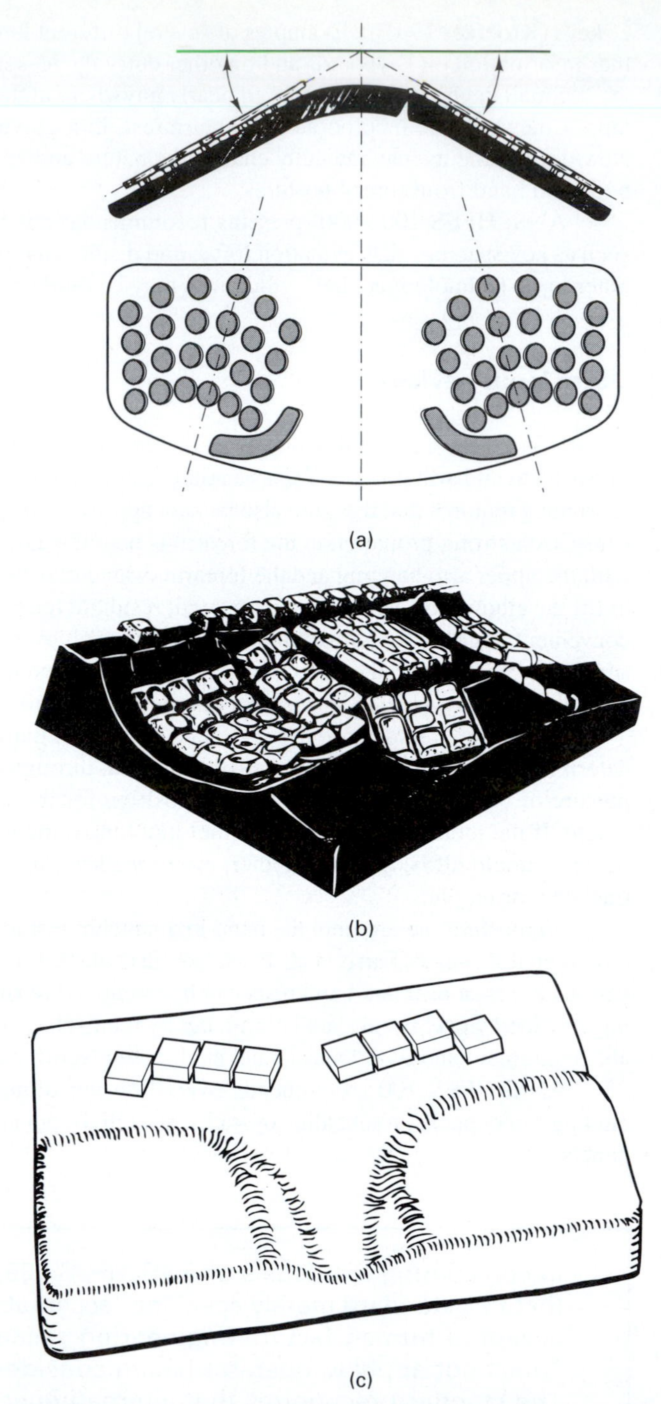

(a)

(b)

(c)

Figure 9–17. Examples of keyboards: (a) K-keyboard; (b) Kinesis™ and (c) the ternary chord keyboard.

Designing the Sit-Down Workstation

One of the first steps in designing office furniture is to establish its main clearance and external dimensions; these properties derive essentially from the body dimensions and work tasks of the people in the office. The common design procedure is to start from the floor, upon which one stacks the heights of chairs, supports for input devices (such as the keyboard, mouse pad, etc.), tables and desks. Main vertical anthropometric inputs used to determine the height clearances are the lower leg (popliteal and knee) heights, the thigh thickness, and the elbow, shoulder, and eye heights. Furniture depths and widths are selected mostly to fit horizontal body dimensions (especially popliteal and knee depths, the hip breadth, and reach capabilities), as well as work task and equipment space needs. (See Chapter 1 and ANSI/HFES 100).

Three design strategie

Of course, there are many possible design approaches, but basically, three alternative strategies determine the height of equipment. The first strategy assumes that all components are adjustable in height. To fit the user, the adjustment range for the seat height is established first; then heights for equipment supports (primarily the keyboard, display, and other working surfaces) are calculated. The second strategy assumes that the height of the major work surface (the table height in traditional offices) must be fixed, but that seat and display heights are adjustable. The third strategy presumes a fixed seat height, but support and display heights are adjustable.

Typical results for conventional furniture are listed in Table 9–3. The figures include height adjustments that fit about 90 percent of the U.S. civilian population, excluding only females who are smaller than the 5th percentile in relevant dimensions and males who are larger than 95th percentile in relevant dimensions. (The actual body dimensions will be different for various user groups, but the principal results are similar.)

First strategy: The seat height is determined from the popliteal height, adding 2 centimeters for heels. This results in a range of seat heights above the floor adjustable from about 37 to 50 cm. Then, the thigh thickness is added to calculate the necessary clearance height underneath the support structure; also, adding 2 cm for the thickness of the support structure results in support surface heights of 53 cm to 70 centimeters. The next step is to determine the eye height above the seat pan, considering the actual trunk and head posture. From this number, the center height of the display is determined, using values for the preferred viewing distance and the preferred angle of sight (Kroemer 1985c; Hill and Kroemer 1986.) Accordingly, the height of the center of the computer display should be between 93 and 122 centimeters above the floor. Note that a footrest is not needed in this design approach.

TABLE 9–3. Adjustment Ranges for VDT Workstation Heights, in cm Above the Floor

	Height above floor		
Item	*First strategy*	*Second strategy*	*Third strategy*
	All adjustable	Support for keyboard fixed, all other adjustable	Seat fixed, all other adjustable
Seat Pan	37 to 50	50 to 55	Fixed at 50
Support surface for keyboard or other data entry device	53 to 70	Fixed at 70	65 to 70
Center of display	93 to 122	106 to 127	106 to 122
Footrest	Not needed	0 to 18	0 to 13

Second strategy: The support for the keyboard (the "table") is held fixed at 70 centimeters above the floor, so that even the tallest users fit underneath. The seat heights are adjusted below this height according to thigh thickness. Consequently, many persons need footrests. Also, the display must be arranged slightly higher than in the previous strategy, where by most people used lower seats and keyboard heights.

Third strategy: The seat height is fixed at 50 centimeters above the floor to accommodate even persons with long legs; persons with shorter legs need to use footrests. Following the same logic as before results in support surfaces for the keyboard and for the display at intermediate height ranges.

Advantages and disad-vantes

Each of the foregoing strategies brings about design solutions with specific advantages and disadvantages.

1. The first strategy, requiring complete adjustability, easily accommodates all persons and does not require footrests. However, the adjustment ranges for the seat pan height, support surface height, and display height are the largest of all the strategies.
2. The second approach allows the height of the support surface (the "table height") to be kept constant for all workstations, but requires the tallest seat height, yet with relatively little adjustment needed. Footrests are often necessary, and to considerable heights. The display still needs to be adjusted up and down, although slightly less than in the first design strategy.
3. The third design approach allows chairs of constant height to be used, but requires the widespread use of footrests, which, however, need not be as high as in the second design strategy. Of course, the support surfaces for the keyboard and display must be independently adjustable, but the needed adjustment ranges are the narrowest of all approaches.

Design Furniture for Change

Maintaining a given posture, even if it appears comfortable in the beginning, becomes stressful with time. Changes in posture are necessary, best combined with a brief period of physical activity. Hence, the computer workstation should be designed to be easily adjustable so that the operator can move between a variety of postures. This, obviously, requires a chair that is comfortable in many positions.

All components of the workstation must fit each other, and each must suit the operator. Figure 9–18 sketches various adjustment features that allow matching the seat height with the support height of the input devices or table (possibly while using a footrest) and positioning the monitor.

The furniture at the computer workstation consists primarily of the seat and the supports for the data-entry device, the display, and the working surface. All of these should be independently adjustable. Recommended height adjustment ranges and other workstation characteristics are shown in Table 9–4 (Kroemer 1988; Tougas and Nordin 1987).

TABLE 9–4. Height Adjustment Ranges Above the Floor, in cm, for Computer Workstations

Seat pan	37 to 50
Keyboard support	60 to 70
Screen center	60 to 130
Work surface	60 to 70

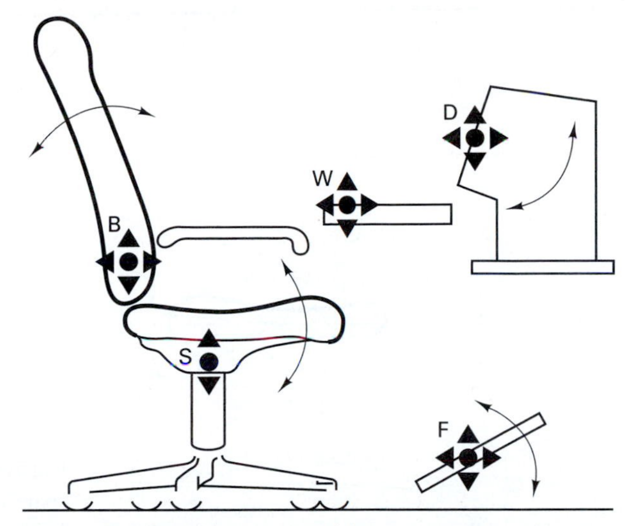

Figure 9–18. Adjustments of the components of a computer workstation.

Seat Pan

Stools for semisitting may be acceptable or comfortable for some individuals, but should not be generally prescribed, because most people are used to more conventional seats. The surface of the seat pan must support the weight of the upper body comfortably and securely. Hard surfaces generate pressure points, which should be avoided by providing suitable upholstery, cushions, or other surfaces that can elastically or plastically adjust to the body's contours.

The only inherent limitation to the size of the seat pan is that it should be short enough that the front edge does not press into the sensitive tissues near the knee. The height of the seat pan must be widely adjustable, preferably down to about 37 cm and up to 58 cm, or at least to 50 or 51 cm, to accommodate persons with short and long lower legs. The adjustment must be easily accomplished while one is sitting on the chair.

Usually, the seat pan is essentially flat, between 38 and 42 cm deep and at least 45 cm wide. A well-rounded front edge is mandatory. Often, the side and rear borders of the pan are slightly higher than the central parts of the surface, an effect usually achieved by making the inner sections more compressible. Figures 9–19 and 9–20 illustrate the major dimensions of the seat pan and backrest.

In the side view, the seat pan is often essentially horizontal, but tilting it slightly backward or forward is usually perceived as comfortable and desirable. Seat pans that are higher in their rear portion and lower at the front facilitate "opening the hip angle." Sitting on a distinctly forward-declined seat surface is comfortable for some persons.

Backrest

The backrest serves two purposes: to carry some of the weight of the upper trunk, arms, and head and to allow muscles to relax. Both purposes can be fulfilled only when the trunk reclines on the backrest.

The backrest should be as large as can be accommodated at the workplace: This means up to 85 cm high and at least 30 cm wide. The backrest should provide support from the head and

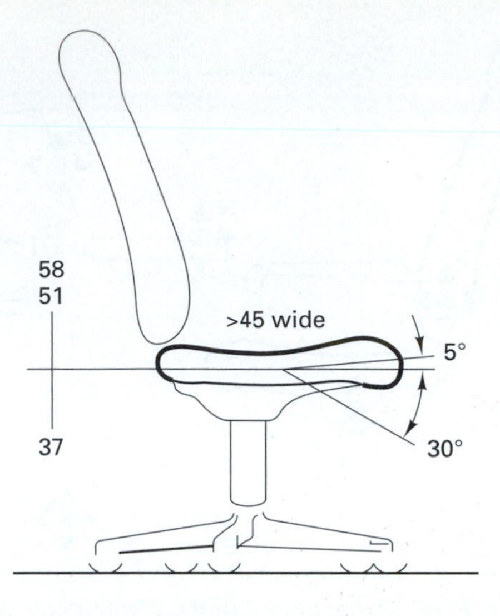

Figure 9–19. Essential dimensions of the seat.

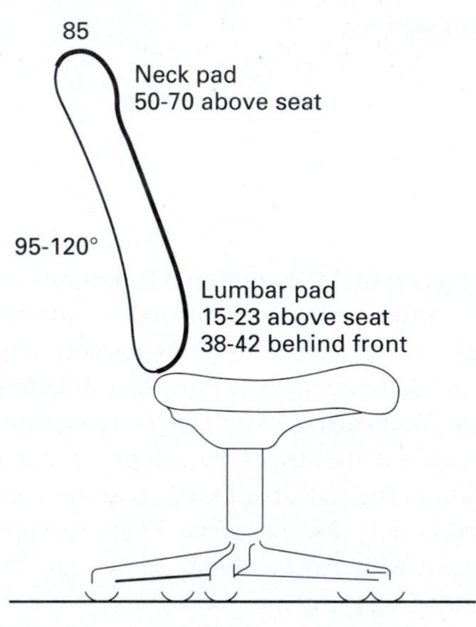

Figure 9–20. Essential dimensions of the backrest.

neck on down to the lumbar region. For this purpose, in side view it is usually shaped to follow the contours of the back, specifically in the lumbar and the neck regions. An adjustable pad for supporting the lumbar lordosis (eg, an inflatable cushion) is appreciated by many users. The lumbar pad should be adjustable from 15 to 23 cm, a cervical pad from 50 to 70 cm, above the seat surface.

The angle of the backrest must be easily adjustable while the person is seated. It should range from slightly behind upright (95° from the horizontal) to 30° behind vertical (120° from the horizontal), with further declination for rest and relaxation desirable. Whether the angle of the back of the seat should be mechanically linked to the seat pan angle is apparently a matter of personal preference. (Note that seat stability must be guaranteed, even if the backrest is strongly declined.)

Armrest

Armrests allow supporting the weight of the hands, arms, and even portions of the upper trunk. Thus, armrests are useful, even if only for short periods. They must be well located, with a suitable load-bearing surface. Adjustability in height, width, and, possibly, direction is desirable.

However, armrests can also hinder moving the arm, pulling the seat toward the workstation, or getting in and out of the seat. In these cases, having short armrests or even no armrests at all is appropriate.

Footrest

If the seat pan cannot be sufficiently lowered for the seated person, then a footrest is needed. Hence, the presence of footrests usually indicates a deficient workplace design.

A footrest should be high enough that the sitting person has the thighs nearly horizontal. It should not consist of a single bar or another very small surface, because this severely limits the ability of the sitting person to change the posture of the legs. Instead, the footrest should provide a support surface that is about as large as the total leg room available in the normal work position.

Work surface and keyboard support

The table or other work surface should be adjustable in height between about 60 and 70 cm, and even a bit higher for very tall persons, to permit proper hand–arm and eye locations. Often, a keyboard or other input device is placed on the work surface. Putting the keyboard on a tray lower than the work surface can be useful, especially if the table is a bit high for a person, but it also may reduce the clearance height for the knees. The tray should be large enough to allow placement of the mouse pad or trackball as well, unless these are built into the keyboard.

Monitor support

It is advantageous to use a separate support for the computer monitor, so that the display can be adjusted in height independently from the table or work surface; the position of the screen greatly affects the body position of the computer operator (Ankrum and Nemeth 1995; Bauer and Wittig 1998; Dowell et al. 1997; Turville et al. 1998; Villanueva et al. 1996). Height adjustment of the display is facilitated by a spring-supported or motor-driven suspension system of the support surface. As a rule, the screen should be at a proper viewing distance, usually about 50 cm from the eyes. The all-too-common practice of putting the monitor on the CPU box and possibly also on a stem to adjust the angle of viewing lifts the screen much too high for most users, who, as a consequence, tilt their head back and then may suffer from neck and back problems. Instead, the monitor should be located low and behind the keyboard, so that one looks down at it.

Designing the Stand-Up Workstation

One way to change the working posture is to allow the computer operator, at his or her own choosing, to work standing up for some period of time. Stand-up workstations can often use a spare computer in the office, to which work activities can be switched from the sit-down workstation for a while; or one may stand while reading, writing, or telephoning. Stand-up workstations should be adjustable to have the input device at approximately elbow height (e.g., between 90 and 120 centimters) when the operator is standing. As in the sit-down workstation, the display should be located close to the other visual targets, such as the source document and the keyboard. If the work surface is used for reading or writing, it may be inclined slightly. A footrest at about two-thirds knee height (approximately 30 centimeters) is often welcomed, so that the person can prop one foot up on it temporarily. This resting posture brings about changes in rotation of the pelvis and curvature of the spine.

Data-Entry Devices

The data-entry device at a workstation is often a keyboard. The usual binary "tapping" keyboards have an essentially flat surface across their 101-plus key tops, together with various other ergonomically suspect design features (as mentioned earlier in this chapter, as well as in Chapter 10). Other keyboard designs have been recommended, particularly those that comply with the natural rotation of the wrist (actually, of the forearm) and the movement of the fingertips. Arguments against human-engineered keys were that their posture and health advantages were unproven, that they require prolonged relearning, and that throughput was worse, or at least not better, than with the old-fashioned keyboard (Lewis et al. 1997). These arguments have proven to be exaggerated or even invalid; one quickly gets used to properly selected ergonomically designed keyboards, and they are advantageous (Marklin et al. 1997, 1999; Simoneau et al. 1999: Smith et al. 1998; Swanson et al. 1997).

Like the keyboard, other entry devices are stationary, such as joysticks and trackballs. Among the movable input devices are the mouse, which is rolled on a horizontal surface, and the light pen, which is pointed by hand to targets on the screen. Each of these may be appropriate for given tasks. (For more information, see Chapter 10 and ANSI/HFES 100.)

Display

The screen of the display unit, particularly its optical quality, is of major ergonomic importance. In general, the screen should provide a stable image (that neither flickers nor jitters), showing characters of good contrast against the background. Dark characters on a light background are often preferred. The characters and symbols displayed should be in a clean font with clear edge definition, of appropriate size, and suitably spaced.

The luminance level (brightness) and contrast should be adjustable by the user. The use of more than two colors should be considered cautiously, since some colors pose visual problems, and because a large variety of colors displayed on the same screen may be more irritating than useful. (For more information, see Table 9–5, ANSI/HFES 100, and Chapter 10.)

TABLE 9–5. Recommendations for Characters on Displays.

Character height	minimum:	16 minutes of arc
	preferred:	20 to 22 minutes
	maximum:	45 minutes
Character height/width	minimum:	1/0.5
	preferred:	1/0.7 to 1/0.9
Character stroke width	minimum:	1/12 of height
Spacing between characters	minimum:	10% of height
Spacing between words	minimum:	one width
Spacing between lines	minimum:	double the stroke width, or 15% of height (whichever is greater)

SOURCE: Excerpted from ANSI 100, Human Factors Society, 1988.

Environment

Office layout

The office layout should follow sound architectural design principles, providing enough room and privacy to the individual. Personal preferences and job attributes may suggest either a separate room or cubicle or an open layout ("office landscape") to facilitate communication with coworkers.

Static electricity in the office

Static electricity may be a problem, not only because it can be irritating and may cause skin rashes in some operators, but also because it attracts dust to the screen. Proper material especially of carpets, grounding, and increased humidity of the room air can help solve the problem.

Light in the office

The illumination level at the computer workstation is lower than that required in ordinary offices. The reasons are that reflections on the screen (glare) must be avoided and that the screen is itself a source of light which must be frequently viewed by the operator. Therefore, the general room illumination should be between 300 and 700 lx, with the lower levels appropriate when the hard copy (paper source document) used is of high quality (good contrast). If it is difficult to read the source document at low-level illumination, a lamp may be used that shines exclusively upon the copy. The distribution of the illumination should be fairly constant throughout (with the exception of the spotlighted area) and should be either diffuse or directed so that there are no reflections on the screen. Wall and ceiling colors affect the absorption and reflection of light, as has been discussed, but otherwise are a matter of personal preference.

Sound in the office

Noise is usually not a major concern, since most computers operate at fairly low sound levels. However, a fan or printer may need attention. Background noise and interference from other workstations or equipment must be considered. The general recommendation is to keep the sound levels as low as possible, such as 60 dBA.

Climate in the office

Temperature can be a problem at some computer workstations where the equipment emits heat. The same requirements as in other offices apply, meaning that the effective temperature should be in the low 20°C region, the relative humidity around 50 percent, and the air movement low, such as 0.5 m/s. Small deviations from these values should have no effects on the production or well-being of the office workers, but complaints about the office being too hot or too cold are frequent in the United States.

Changes through Technical Developments

Some tasks currently performed with computers, such as the prolonged, simple input of numbers or of words, probably should not be imposed on humans. Not only are such tasks boring, but the repetitive finger movements they require may be a source of cumulative trauma disorders, such as tendonitis in the hands or arms and carpal tunnel syndrome. (See Chapter 8.) Repetitive entry should be automated, by machine recognition of characters, by scanning, or by voice input. Feedback from the computer can also be given to the operator through acoustical means, rather than just by being displayed on the screen.

Clever software offers possibilities for facilitating the task of the computer operator. Among such software are automated programs for grammar and spelling, programs for stringing characters together, and algorithms that check and indicate outliers in data, unusual events, repetitive occurrences. Certainly, a wide variety of opportunities exist to improve and facilitate the work of the computer operator.

Job Content and Work Organization

Many persons are proud to be autonomous in performing their work, to take responsibility for its quality and quantity, and to control its timing. Most prefer to perform larger tasks from beginning to end, instead of simply doing specialized tidbits. The ability to receive direct feedback about one's work, preferably through a daily review supplemented by constructive comments from the supervisor, contributes significantly to the feeling of achievement and satisfaction. Within the limitations set by the requirement that certain work needs to be done, the operator should be free to distribute the workload, both in amount and pace, according to his or her own preferences and needs. Communicating with colleagues and maintaining social relations are essential, although the intensity of each varies with the individual. Isolating people or submitting them to cold, formal relationships is usually detrimental to their well-being and performance.

The organization of one's working time—particularly with regard to changes in work periods and rest pauses—is important for carrying out many computer tasks. Most people are bored by repetitive, monotonous, and continuous work; instead, varying tasks of different lengths should be done, so that the computer operator has occasion and reason to shift from one task to another, to move away from the computer for a while, to do something else, like getting or taking away materials (see Figure 9–21), or simply to take a break. The "recovery value" of many short rest pauses is larger then that of a few long breaks. Certainly, long periods of sitting still, especially in awkward postures such as those shown in Figure 9-22, must be avoided. Psychosocial, musculoskeletal, and visual stress and strains can all be reduced, and even avoided, after one identifies improper versus suitable conditions and introduces appropriate behaviors, best done by "participatory ergonomics" (Aaras et al. 1998; Lu and Aghazadeh 1998; Nelson and Silverstein, 1998; Vink and Kompier 1997).

Flexibility in the attitude of management and in one's work organization gives the individual latitude in the way a task is performed, the manner in which furniture is used, and how the workplace is arranged. Indeed, providing freedom for individual variations from the conventional norm acknowledges that persons in the office differ in their physiques and work preferences.

The human is the most important component of any system, since she or he drives the output. Hence, the human must be accommodated first: The design of the workplace components should fit all operators and allow many idiosyncratic variations in working posture. The myth of "one healthy upright posture, good for everybody, anytime" must be abolished, both as a guide for posture and as a design template.

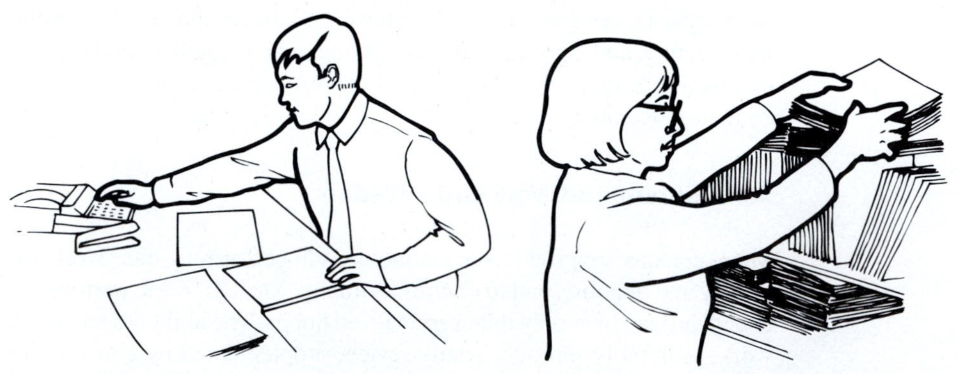

Figure 9–21. Occasional (but not habitual) reaching and stretching are desirable (courtesy of Herman Miller).

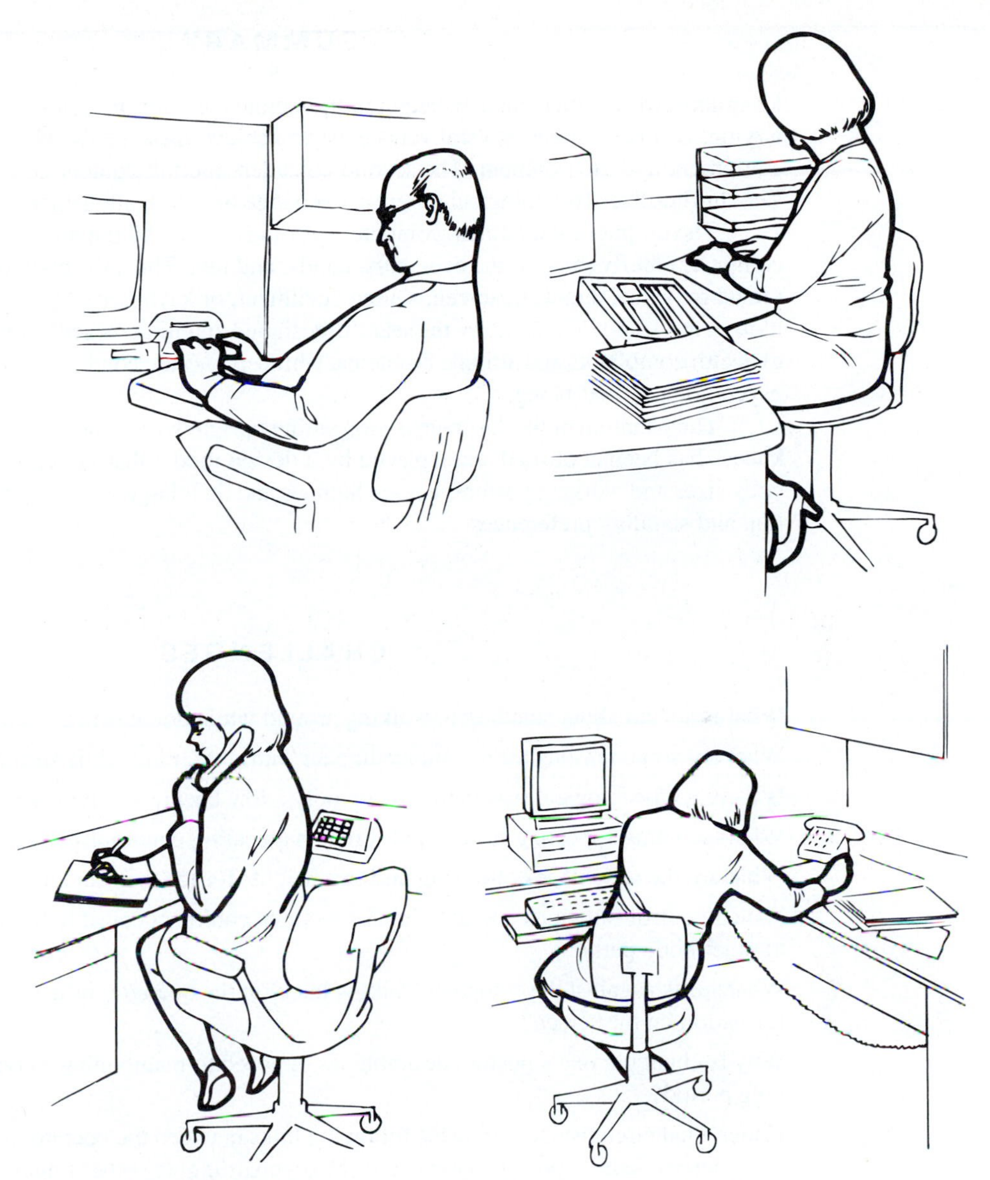

Figure 9–22. Contorted postures can lead to overuse disorders (courtesy of Herman Miller).

SUMMARY

Long-term work with computers requires ergonomic measures to assure that the job is healthy, satisfying, and productive. Several ways exist to achieve these goals. The first concerns the work equipment and environment. The second considers the job content and the work organization. The third utilizes technological progress to change the work altogether.

The output of the human–computer system is driven by the human, who interacts with the computer chiefly through the eyes, ears, hands, and feet. Thus, the body posture is largely determined by visual targets (a screen, source document, or keyboard), by input devices (a keyboard, mouse, etc.), and, of course, by the seat. Ill-designed and ill-arranged computer workstations lead to health complaints and attitude problems, while ergonomic conditions further both physical and psychological well-being.

The phantom of the "average person sitting upright with right angles at the elbows, hips, and knees" has been abolished and replaced by a design model that incorporates the actual range of body sizes and working postures among humans and their large variations reflecting individual sitting and standing preferences.

CHALLENGES

What is so bad about standing or walking around while doing office work?

What are some reasons for recommending an "upright" trunk while sitting?

Is there a physiological reason to provide only a low backrest to typists or word processors?

What are some appealing factors and some unappealing conditions associated with office work?

What are the advantages of having uniform heights for (a) seats and (b) work surfaces?

Which control functions and data-input functions could conceivably be shifted from the fingers to other body parts?

What speaks against the computer "talking back" at the operator, instead of simply displaying information on the screen?

Why is changing one's posture desirable, as opposed to maintaining a comfortable posture for a long period?

Under what circumstances may the following links between the operator and the computer be the main determiners of posture: (a) vision; (b) keyboarding; (c) other finger-control interactions?

What are the main arguments pro and con individual versus multiperson offices?

Should one strive to have windows in the office, in spite of the difficulties of controlling illumination?

Should eye examinations and the provision of suitable corrective lenses be of more concern to the worker or to the employer?

Does "participatory ergonomics" work better than expert interventions?

Chapter 10

Selection, Design, and Arrangement of Controls and Displays

OVERVIEW

"Knobs and dials" were researched extensively in the 1940s and 1950s, thus, well-proven design recommendations are available for traditional controls and displays. Unfortunately, they rely more on everyday experience than on known psychological rules and principles.

Advances in technology, as well as user complaints and preferences have brought up new issues, related particularly to computer data entry and the display of information. Also, labeling and warnings have become ergonomically and legally important.

INTRODUCTION

Controls (called activators in ISO standards), transmit inputs to a piece of equipment. They are usually operated by hand or foot. The results of the control inputs are shown to the operator either in terms of displays or indicators or by the ensuing actions of the machine.

The 1940s and 1950s are often called the "knobs and dials era" in human-factors engineering, because much research was performed on controls and displays. Thus, this topic is rather well researched, and summaries of earlier findings were compiled by Van Cott and Kinkade (1972), Woodson (1981), and McCormick and Sanders (1982). Military and industry standards (for example, MIL STD 1472 and HDBK 759; SAE J 1138, 1139, and 1048; and HFS/ANSI 100) established detailed design guidelines. These matter-of-fact practical recommendations pertain to controls and displays used in well-established designs and Western stereotypes. Yet, overriding general laws based on human motion or energy principles, or on perception and sensory processes, are not usually known. Thus, the current rules for selection and design are likely to change with new kinds of controls and displays, as well as new circumstances and applications.

CONTROLS

One distinguishes the following control actions:

1. Activate or shut down equipment, such as with an on–off key lock.
2. Make a "discrete setting," such as a separate or distinct adjustment (eg, selecting a TV channel or entering data via a computer keyboard).
3. Make a "quantitative setting," such as a temperature on a thermostat. (This is a special case of a discrete setting.)
4. Apply "continuous control," such as in steering an automobile.

CONTROL SELECTION

Controls are selected for their functionality (usefulness) in accordance with the following guidelines:

1. The type of control shall be compatible with stereotypical or common expectations. For example: use a push button or a toggle switch, not a rotary knob, to turn on a light.
2. Size and motion characteristics of controls shall be compatible with stereotypical experience and past practice. For example, have a fairly large steering wheel, not a small rotary control, for (two-handed) operation in an automobile.
3. The direction of operation of a control shall be compatible with stereotypical or common expectations. For example, an *on* control is pushed or pulled, not turned to the left.
4. Operations requiring fine control and small forces shall be done with the hands, while gross adjustments and large forces are usually exerted with the feet.
5. The control shall be safe in that it will not be operated inadvertently or operated in false or excessive ways.

There are few "natural rules" for the selection and design of controls. One is that hand-operated controls are expected to be used for fine control movements, whereas foot-operated controls are usually reserved for large force inputs and gross control. Yet, consider the pedal arrangements in modern automobiles, where vital and finely controlled operations, nowadays requiring fairly little force, are performed with the feet on the gas and brake pedals. Furthermore, the movement of the foot from the accelerator to the brake is very complex. Probably no topic in ergonomics has received more treatment in course projects and master's theses than this against-all-the-rules, but commonly used, arrangement.

COMPATIBILITY OF CONTROL–MACHINE MOVEMENT

Controls shall be selected so that the direction of their movement is compatible with the response movement of the controlled machine, be it a vehicle, equipment, a component, or an accessory. Table 10–1 lists such compatible movements.

The concept of compatibility—the relationship between control and display—arises out of a context or situation in which an association appears manifest or intrinsic (eg, locating a control next to its related display). Yet, many, if not all, relationships depend on what one has learned—what is commonly used in one's culture. In the Western world, red is conceived to mean "danger" or "stop" and green to mean "safe" or to "go." Such a relationship is called a population stereotype, probably learned during early childhood. A population stereotype may differ from one user group to another. For example, in Europe a switch is often toggled downward to turn on a light, while in the United States the switch is pushed up. Outside the Western world, one may encounter quite different stereotypical expectations and uses. For example, in China, some longtime conventions exist that are different from those in the United States (Courtney 1994).

CONTROL ACTUATION FORCE OR TORQUE

The force or torque applied by an operator to actuate a control shall be kept as low as feasible, particularly if the control must be operated often. If jerks or vibrations arise, it is usually better to stabilize the operator than to increase the resistance of the control in order to prevent uncontrolled or inadvertent activation.

Operator-applied force or torque depends on the control's mechanical built-in resistance, which can be of five types:

1. *Intertial.* A function of mass and acceleration (according to Newton's second law), intertia resists changes in position (movement) and helps to maintain a steady motion.
2. *Stiction.* A static (resting) friction that must be overcome to initiate motion, stiction helps avoid accidental movement, but provides little feedback.
3. *Coulomb friction.* A dynamic friction that must be overcome to maintain movement, coulomb friction provides feedback about both the presence and direction of motion of the control.
4. *Elastic.* A spring-type linear or nonlinear resistance to displacement, elastic resistance provides feedback about the direction and amount of motion of the control and is often employed in "dead person" control because it re-sets toward an initial position.
5. *Viscous damping.* A resistance whose magnitude depends on velocity, viscous damping helps to stabilize movement of the control.

Many design recommendations for rotational controls list values of the tangential force at the point of application, instead of the torque, because that is usually the most practical information.

TABLE 10–1. Control Movements and Expected Effects. 1 means most preferred, 2 less preferred.

	Direction of control movement											
Effect	*Up*	*Right*	*Forward*	*Clockwise*	*Press*, Squeeze*	*Down*	*Left*	*Rearward*	*Back*	*Counter–clockwise*	*Pull**	*Push***
On	1	1	1	1	2	—	—	—	—	—	1	—
Off	—	—	—	—	—	1	2	2	—	1	—	2
Right	—	1	—	2	—	—	—	—	—	—	—	—
Left	—	—	—	—	—	—	1	—	2	—	—	—
Raise	1	—	—	—	—	—	—	2	—	—	—	—
Lower	—	—	2	—	—	1	—	—	—	—	—	—
Retract	2	—	—	—	—	—	—	1	—	—	2	—
Extend	—	—	1	—	—	2	—	—	—	—	—	2
Increase	2	2	1	2	—	—	—	—	—	—	—	
Decrease	—	—	—	—	—	2	2	1	—	2	—	—
Open Valve	—	—	—	—	—	—	—	—	—	1	—	—
Close Valve	—	—	—	1	—	—	—	—	—	—	—	—

Source: Modified from Kroemer 1988b.

*With trigger-type control.

**With push–pull switch.

CONTROL–EFFECT RELATIONSHIPS

The relationships between the action of the control and the resulting effect shall be made apparent through common sense, habitual use, similarity, proximity and grouping, coding, labeling, and other suitable techniques, discussed later. Certain types of control are preferred for specific applications. Tables 10–1, 10–2, and 10–3 help with the selection of hand controls.

CONTINUOUS VERSUS DETENT CONTROLS

Continuous controls shall be selected if the operation that is to be controlled can be anywhere within the adjustment range of the control with no need to be set in any given position. But if the operation must be in discrete steps, these shall be marked and secured by "detents" or "stops" in which the control comes to rest.

STANDARD PRACTICES

Unless other solutions can be demonstrated to be better, the following rules apply to common equipment:

1. One-dimensional steering is by a steering wheel.
2. Two-dimensional steering is by a joystick or by combining levers, wheel, and pedals.
3. Primary vehicle braking is by pedal.
4. Primary vehicle acceleration is by a pedal or lever.
5. Selection of a transmission gear is by lever or by a switch with legends as indicators.
6. Valves are operated by round knobs, handwheels, or T-handles.
7. Selection of one (of two or more) operating modes can be by toggle switch, push button, bar knob, rocker switch, lever, or legend switch.

Table 10–3 protects an overview of controls suitable for various operational requirements.

TABLE 10–2. Control–Effect Relations of Common Hand Controls. 1 means most preferred, 3 least preferred.

Effect	*Key lock*	*Toggle switch*	*Push button*	*Bar knob*	*Round knob*	*Thumbwheel Discrete*	*Thumbwheel Continuous*	*Crank*	*Rocker switch*	*Lever*	*Joystick or ball*	*Legend switch*	*Slide**
Select On–Off	1	1	1	3	—	—	—	—	1	—	—	1	1
Select On–Standby–Off	–	2	1	1	—	—	—	—	—	1	—	1	1
Select Off-Mode 1–Mode 2	—	3	2	1	—	—	—	—	—	1	—	1	1
Select one of several related functions	—	2	1	—	—	—	—	—	2	—	—	—	3
Select one of three or more discrete alternatives	—	—	—	1	—	—	—	—	—	—	—	—	1
Select operating condition	—	1	1	2	—	—	—	—	1	1	—	1	2
Engage or disengage	—	—	—	—	—	—	—	—	—	1	—	—	—
Select one of mutually exclusive functions	—	—	1	—	—	—	—	—	—	—	—	1	—
Set value on scale	—	—	—	—	1	—	2	3	—	3	3	—	1
Select value in discrete steps	—	—	1	1	—	1	—	—	—	—	—	—	1

Source: Modified from Kroemer 1988b.

*Estimated, no experiments known.

TABLE 10–3. Selection Of Controls

Controls Suitable for Small Operating Force	
Two discrete positions	Key lock, hand operated Toggle switch, hand operated Push button, hand operated Rocker switch, hand operated Legend switch, hand operated Bar knob, hand operated Slide, hand operated Push–pull switch, hand operated
Three discrete positions	Toggle switch, hand operated Bar knob, hand operated Legend switch, hand operated Slide, hand operated
Four to 24 discrete positions, or continuous operation	Bar knob, hand operated Round knob, hand operated Joystick, hand operated Continuous thumbwheel, hand operated Crank, hand operated Lever, hand operated Slide, hand operated Trackball, hand operated Mouse, hand operated Light pen, hand operated Crank, hand operated
Continuous slewing, fine adjustments	Round knob, hand operated Track ball, hand operated
Controls Suitable for Large Operating Force	
Two discrete positions	Push button, foot operated Push button, hand operated Detent lever, hand operated
Three to 24 discrete positions	Detent lever, hand operated Bar knob, hand operated
Continuous operation	Handwheel, hand operated Lever, hand operated Joystick, hand operated Crank, hand operated Pedal, foot operated

Source: Modified from Kroemer 1988b.

ARRANGEMENT AND GROUPING OF CONTROLS

Several operational rules govern the arrangement and grouping of controls.

LOCATE FOR EASE OF OPERATION

Controls shall be oriented with respect to the operator. If the operator has different positions (such as in driving and operating a backhoe), the controls and control panels shall move with the operator, so that in each position their arrangement and operation are the same for the operator.

PRIMARY CONTROLS FIRST.

The most important and most frequently used controls shall have the best positions with respect to ease of operation and reach.

GROUP RELATED CONTROLS TOGETHER.

Controls that have sequential relations, that are related to a particular function, or that are operated together shall be arranged into functional groups (together with their associated displays). Within each functional group, controls and displays shall be arranged according to their operational importance and sequence.

ARRANGE FOR SEQUENTIAL OPERATION.

If controls are operated in a given pattern, they shall be arranged to facilitate that pattern. The common arrangements are left to right (preferred) and top to bottom, as in reading material in the Western world.

BE CONSISTENT.

The arrangement of functionally identical or similar controls shall be the same from panel to panel.

DEAD-MAN CONTROL.

If the operator becomes incapacitated and either lets go of a control or continues to hold onto it, a "dead-man control" design shall be utilized that either returns the system to a noncritical state or shuts it down.

GUARD AGAINST ACCIDENTAL ACTIVATION.

Numerous ways to guard controls against inadvertent activation may be applied, such as putting mechanical shields at or around them or requiring critical forces or torques. (See later.) Note that most of these methods will reduce the speed of operation.

PACK TIGHTLY, BUT DO NOT CROWD.

Often, it is necessary to place a large number of controls in a limited space. Table 10–4 indicates the minimal separation distances for various types of controls.

TABLE 10–4. Minimal Separation Distances (in mm) for Hand Controls*

	Key lock	*Bar knob*	*Detent thumbwheel*	*Push button*	*Legend switch*	*Toggle switch*	*Rocker switch*	*Knob*	*Slide switch*
Keylock	25	19	13	13	25	19	19	19	19
Bar knob	19	25	19	13	50	19	13	25	13
Detent Thumbwheel	13	19	13	13	38	13	13	19	13
Pushbutton	13	13	13	13	50	13	13	13	13
Legend switch	25	50	38	50	50	38	38	50	38
Toggle switch	19	19	13	13	38	19	19	19	19
Rocker switch	19	13	13	13	38	19	13	13	13
Knob	19	25	19	13	50	19	13	25	13
Slide switch	19	13	13	13	38	19	13	13	13

*The given values are measured edge to edge with the controls in their closest positions. For arrays of controls and operation with gloves, larger distances are recommended in some cases. (See Boff and Lincoln 1988.)

CONTROL DESIGN

The following descriptions, together with Tables 10–5 through 10–19 and Figures 10–1 through 10–13, present design guidance for various detent and continuous controls. The first set, keylocks through rocket switch, consists of detent controls.

KEY LOCK

Key locks (also called key-operated switches) are used to prevent unauthorized operation of a machine. Key locks usually set into *on* and *off* positions that are always distinct.

Design recommendations are given in Figure 10–1 and Table 10–5. Other recommendations are as follows:

1. Keys with teeth on both edges (preferred) should fit the lock with either side up. Keys with a single row of teeth should be inserted into the lock with the teeth pointing up.
2. Operators should normally not be able to remove the key from the lock unless the key is turned *off.*
3. The *on* and *off* positions should be labeled.

BAR KNOB

Detent bar knobs (also called rotary selector switches) should be used for discrete functions when two or more detented positions are required.

Knobs shall be bar shaped with parallel sides, and the index end shall be tapered to a point.

Design recommendations are illustrated in Figure 10–2 and listed in Table 10–6.

DETENT THUMBWHEEL

Detent (or: discrete) thumbwheels for discrete settings may be used if the function requires a compact input device for discrete steps.

Design recommendations are illustrated in Figure 10–3 and listed in Table 10–7.

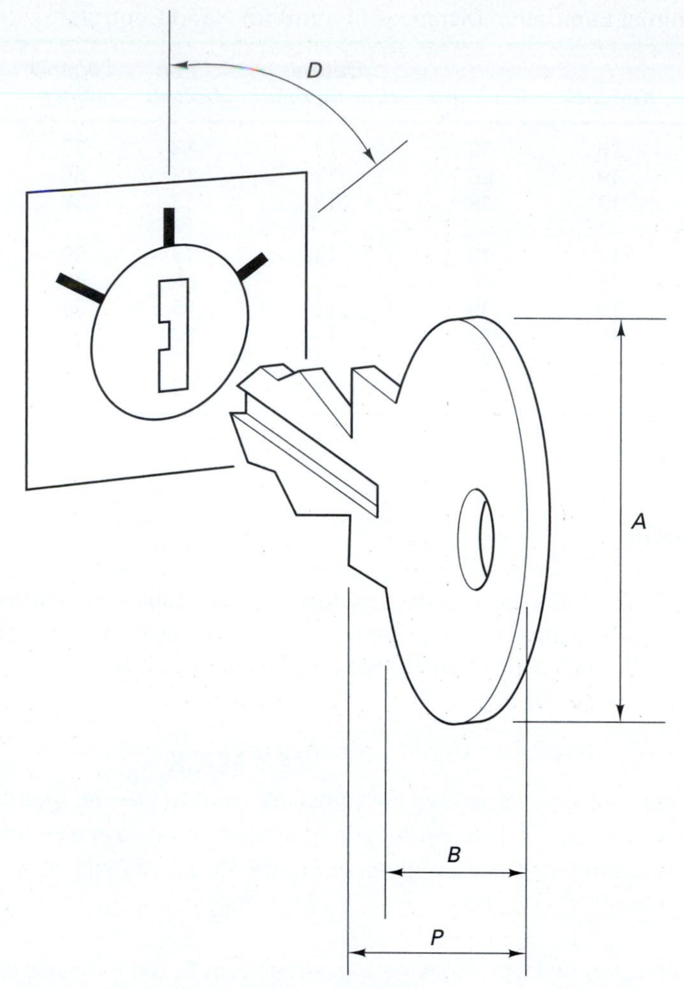

Figure 10–1. Key lock.

TABLE 10–5. Dimensions of a Keylock

	A Height (mm)	B Width (mm)	P Protrusion (mm)	D Displacement (degrees)	S* Separation (mm)	R† Resistance (Nm)
Minimum	13	13	20	30	25	0.1
Preferred	—	—	—	—	—	—
Maximum	75	38	—	90	25	0.7

Source: Modified from MIL–HDBK 759, MIL-STD-1472F.

Note: Letters correspond to measurements illustrated in Figure 10–1.

*Between closest edges of two adjacent keys.

†Control should snap into detent position and not be able to stop between detents.

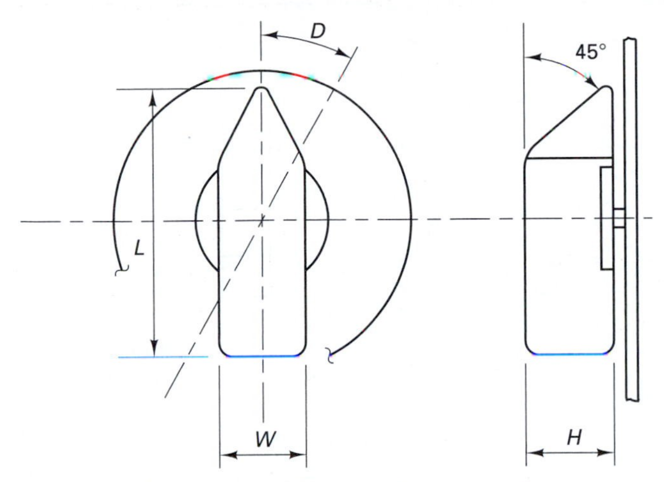

Figure 10–2. Bar knob. (Rotary selector switch)

TABLE 10–6. Dimensions of a Bar Knob

	L Length (mm)	W Width (mm)	H Height (mm)	r† Resistance (Nm)	D Displacement (degrees)	S, Separation: One hand, random operation (mm)	S, Separation: Two hands, simultaneous operation (mm)
Minimum	25 38*	13	16	0.1	15 30**	25 38*	75 100*
Preferred	—	—	—	—	—	50	125
Maximum	100	25	75	0.7	40 90**	— 63*	150 175*

Source: Modified from MIL–HDBK 759, MIL-STD-1472F.

Note: Letters correspond to measurements illustrated in Figure 10–2.

*If operator wears gloves.

†High resistance with large bar knob only. Control should snap into detent position and not be able to stop between detents.

**For blind positioning.

PUSH BUTTON

Pushbuttons should be employed for single switching between two conditions, for the entry of a discrete control order, or for the release of a locking system (eg, a parking brake). Push buttons can be used for momentary contact or for sustained contact, with or without detent.

Design recommendations are given in Figure 10–4 and listed in Table 10–8. Other recommendations are as follows:

The push button surface should normally be concave (indented) to fit the finger; or the surface may be convex for operation with the palm of the hand. When either shape is impractical, the surface shall provide high frictional resistance to prevent slipping. A positive indication of control activation shall be provided, for example, by snap feel, an audible click, or integral light.

For pushbuttons used on keyboards, see Chapter 9 and later in this chapter.

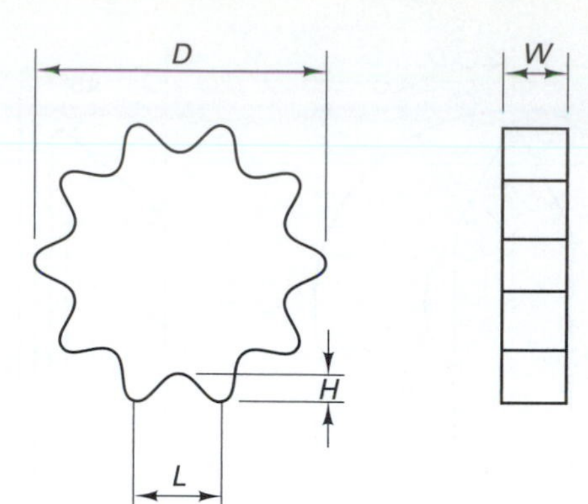

Figure 10–3. Discrete thumbwheel. (Detent thumbwheel)

TABLE 10–7. Dimensions of a Discrete Thumbwheel

	D *Diameter* *(mm)*	*W* *Width* *(mm)*	*L* *Through distance* *(mm)*	*H* *Through depth* *(mm)*	*S* *Separation, side by side* *(mm)*	*R** *Resistance* *(N)*
Minimum	29	3	11	3	10	2
Preferred	—	—	—	—	—	—
Maximum	75	—	19	6	—	6

Source: Modified from MIL–HDBK 759, MIL-STD-1472F.
Note: Letters correspond to measurements illustrated in Figure 10–3.
*Control should snap into detent position and not be able to stop between detents.

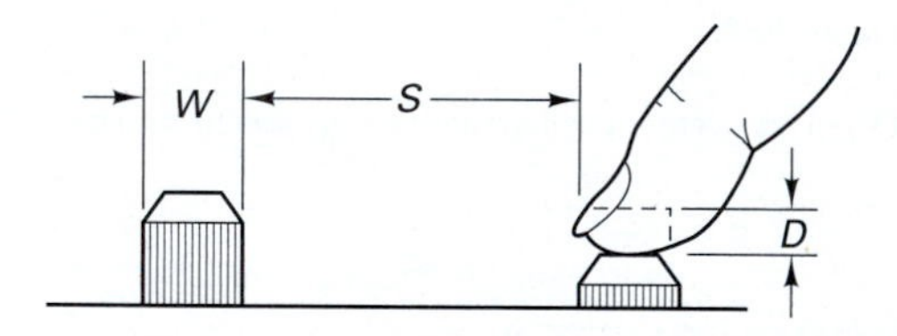

Figure 10–4. Push button. (Operated by finger, thumb, or palm)

PUSH–PULL SWITCH

Push–pull controls have been used for discrete settings, commonly *on* and *off;* intermediate positions have been occasionally employed (eg, for the air–gasoline mixture of a combustion engine). Push–pull controls generally have a round flange under which to hook the fingers. Their diameter should be not less than 19 mm, protruding at least 25 cm from the mounting surface. The separation between adjacent controls should be at least 35 mm. There should be at least 13 mm displacement between the settings (MIL STD 1472F).

TABLE 10–8. Dimensions of a Pushbutton

Operation	*W, Width of square or diameter*			*R, Resistance*				*D Displacement (mm)*		*S, Separation*			
										Single finger			
	Fingertip (mm)	*Thumb (mm)*	*Palm of hand (mm)*	*Single finger (N)*	*Other[†] fingers (N)*	*Thumb (N)*	*Palm of hand (N)*	*Finger-tip*	*Thumb or palm*	*Single operation (mm)*	*Sequential operation (mm)*	*Different fingers (mm)*	*Palm or thumb (mm)*
Minimum	10	19	40	3	1	3	3	2	3	13	6	6	25
	19*	25*	50*							25*			
Preferred	—	—	—	—	—	—	—	—	—	50	13	13	150
Maximum	25	25	70	11	6	23	23	6	40	—	—	—	—

SOURCE: Modified from MIL–HDBK 759, MIL-STD-1472F.

Note: Letters correspond to measurements illustrated in Figure 10–4.

*If operator wears gloves.

**Depressed button shall stick out at least 2.5 mm.

[†]Actuated at the same time.

LEGEND SWITCH

Detent legend switches are particularly suited to display qualitative information on the status of equipment that requires the operator's attention and action.

Design recommendations are given in Figure 10–5 and listed in Table 10–9. Legend switches should be located within a 30° cone along the operator's line of sight.

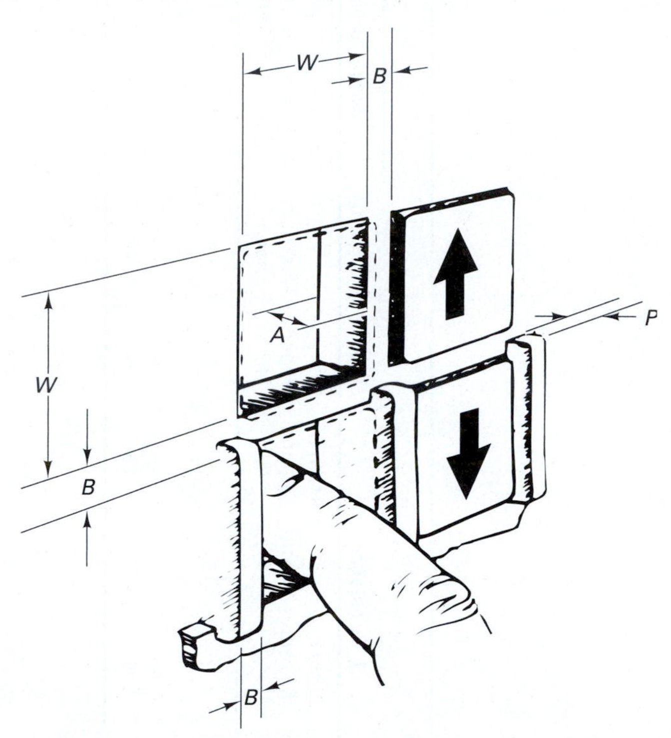

Figure 10–5. Legend switch. (Modified from MIL-HDBK 759.)

TABLE 10–9. Dimensions of a Legend Switch

	W, Width of square or diameter (mm)	*A Displacement (mm)*	*B Barrier width (mm)*	*P Barrier depth (mm)*	*R Resistance (N)*
Minimum	25*	3	3	5	3
Preferred	—	—	—	—	—
Maximum	38*	6	6	6	17

Source: Modified from MIL—HDBK 759, MIL-STD-1472F.

Note: Letters *W, A, B,* and *P* correspond to measurements illustrated in Figure 10–5.

*If operator wears gloves

TOGGLE SWITCH

Detent toggle switches may be used if two discrete positions are required. Toggle switches with three positions shall be used only where the use of a bar knob, legend switch, array of push buttons, etc., is not feasible.

Design recommendations are given in Figure 10–6 and listed in Table 10–10. Toggle switches should be oriented so that the handle moves in a vertical plane, with *off* in the down position. Horizontal actuation shall be employed only if compatibility with the controlled function or the location of equipment is desired.

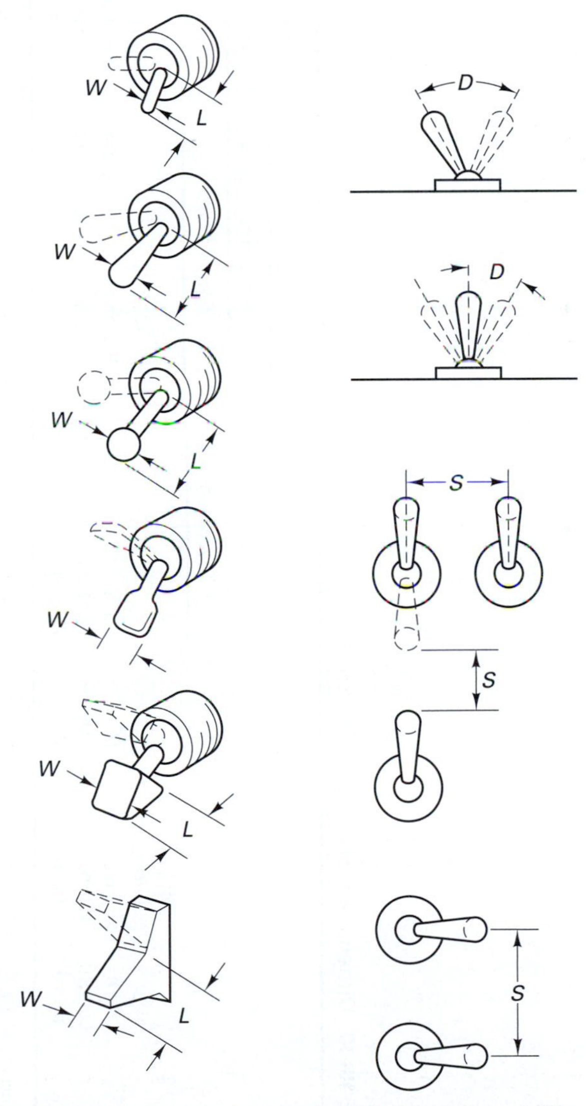

Figure 10–6. Toggle switch. (Modified from MIL-HDBK 759.)

TABLE 10–10. Dimensions of a Toggle Switch

				D, Displacement		*S, Separation*				
						Horizontal array, vertical operation				
						Single finger				
	L Arm Length (mm)	*W, Tip width or diameter (mm)*	*R*† *Resistance (N)*	*2 Positions (degrees)*	*3 Positions (degrees)*	*Random operation (mm)*	*Sequential operation (mm)*	*Several fingers, simultaneous operation (mm)*	*Vertical array, horizontal operation (mm)*	*Toward each other, tip to tip (mm)*
Minimum	13 38*	3	3	30	17	19	13	16 32*	25 38*	25
Preferred	—	—	—	—	25	50	25	19	—	—
Maximum	50	25	11	80	40	—	—	—	—	—

Source: Modified from MIL-HDBK 759, MIL-STD-1472F.

Note: Letters correspont to measurements illustrated in Figure 10–6.

*If operator wears gloves.

†Control should snap into detent position and not be able to stop between detents.

ROCKER SWITCH

Detent rocker switches may be used if two discrete positions are required. Rocker switches protrude less from the panel than do toggle switches.

Design recommendations are given in Figure 10–7 and listed in Table 10–11. Rocker switches should be oriented so that the handle moves in a vertical plane, with *off* in the down position. Horizontal actuation shall be employed only if compatibility with the controlled function or the location of equipment is desired.

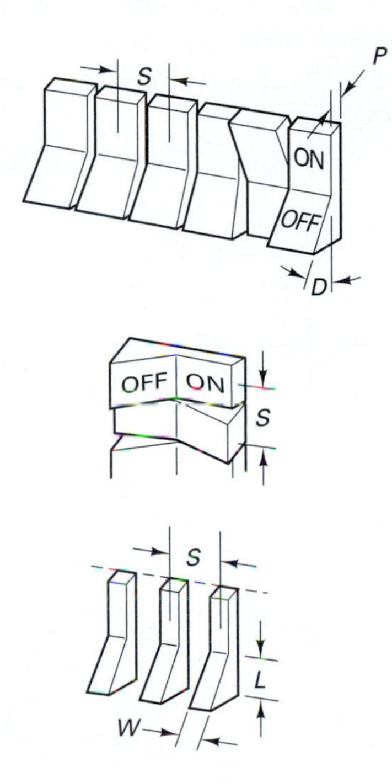

Figure 10–7. Rocker switch. The narrow switch (bottom) is especially desirable for tactile definition when gloves are worn. (Modified from MIL-HDBK 759.)

TABLE 10–11. Dimensions of a Rocker Switch

	W Width (mm)	*L* Length (mm)	*D* Displacement (degrees)	*P* Protrusion, depressed (mm)	*S* Separation, center to center (mm)	*R* Resistance (N)
Minimum	6	13	30	2.5	19 32*	2.8
Preferred	—	—	—	—	—	—
Maximum	—	—	—	—	—	11

Source: Modified from MIL–HDBK 759, MIL-STD-1472F.

Note: Letters *D, L, S,* and *W* correspond to measurements illustrated in Figure 10–7.

* If operator wears gloves.

The following set of controls (knob through slide) is for continuous operations:

KNOB

Continuous knobs (also called round knobs or rotary controls) should be used when little force is required and when precise adjustments of a continuous variable are required. If positions must be distinguished, an index line on the knob should point to markers on the panel.

Design recommendations are given in Figure 10–8 and listed in Table 10–12. Within the range specified in the table, the knob size is relatively unimportant if the resistance is low and the knob can be easily grasped and manipulated. The smallest diameter is 10 mm, the longest 100 mm.

When panel space is extremely limited, knobs should be small and should have resistances as low as possible without permitting the setting to be changed by vibration or by inadvertent touching.

Unless otherwise specified, control knob styles best conform to the guidelines established in military standards (eg, MIL-STD 1472 or MIL-HDBK 759).

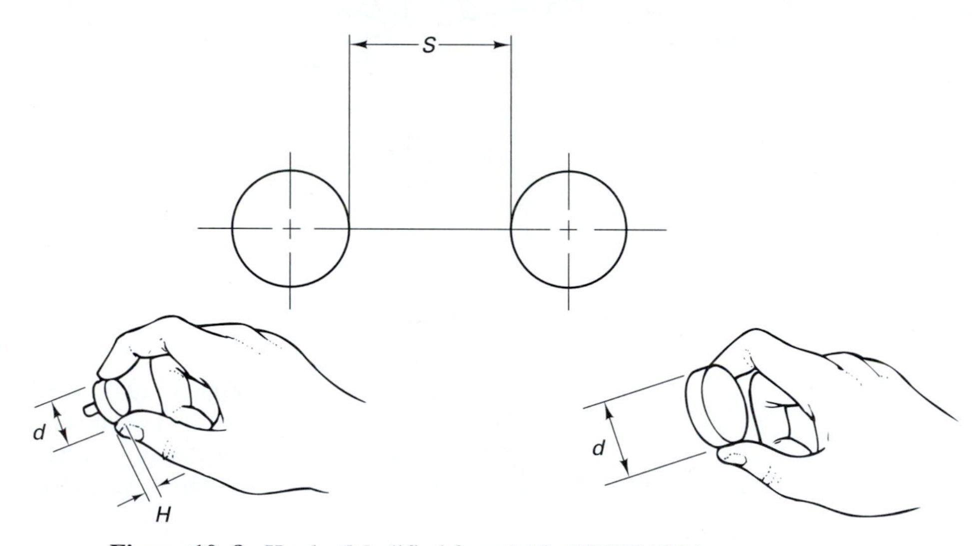

Figure 10–8. Knob. (Modified from MIL-HDBK 759.)

TABLE 10–12. Dimensions of a Knob

	H	*D, Diameter*		*T, Torque*		*S, Separation*	
	Height (mm)	*Fingertip grip (mm)*	*Thumb and finger grasp (mm)*	*Up to 25 mm in diameter (N m)*	*Over 25 mm in diameter (N m)*	*One hand (mm)*	*Two hands simultaneously (mm)*
Minimum	13	10	25	—	—	25	50
Preferred	—	—	—	—	—	50	125
Maximum	25	100	75	0.03	0.04	—	—

Source: Modified from MIL–HDBK 759.

Note: Letters correspond to measurements illustrated in Figure 10–8.

CRANK

Continuous cranks should be used primarily if the control must be rotated many times. For tasks involving large slewing movements or small, fine adjustments, a crank handle may be mounted on a knob or handwheel.

The crank handle shall be designed so that it turns freely around its shaft, especially if the whole hand grasps the handle.

Design recommendations are given in Figure 10–9 and listed in Table 10–13.

HANDWHEEL

Handwheels that are designed for continuous nominal two-handed operation should be used when the breakout or rotation forces are too large to be overcome with a one-hand control—provided that two hands are available for the task.

Knurling or indentation shall be built into a handwheel to facilitate the operator's grasp.

When large displacements must be made rapidly, a spinning handle (a so-called spinner crank) may be attached to the handwheel when doing so is not overruled by safety considerations.

Design recommendations are given in Figure 10–10 and listed in Table 10–14.

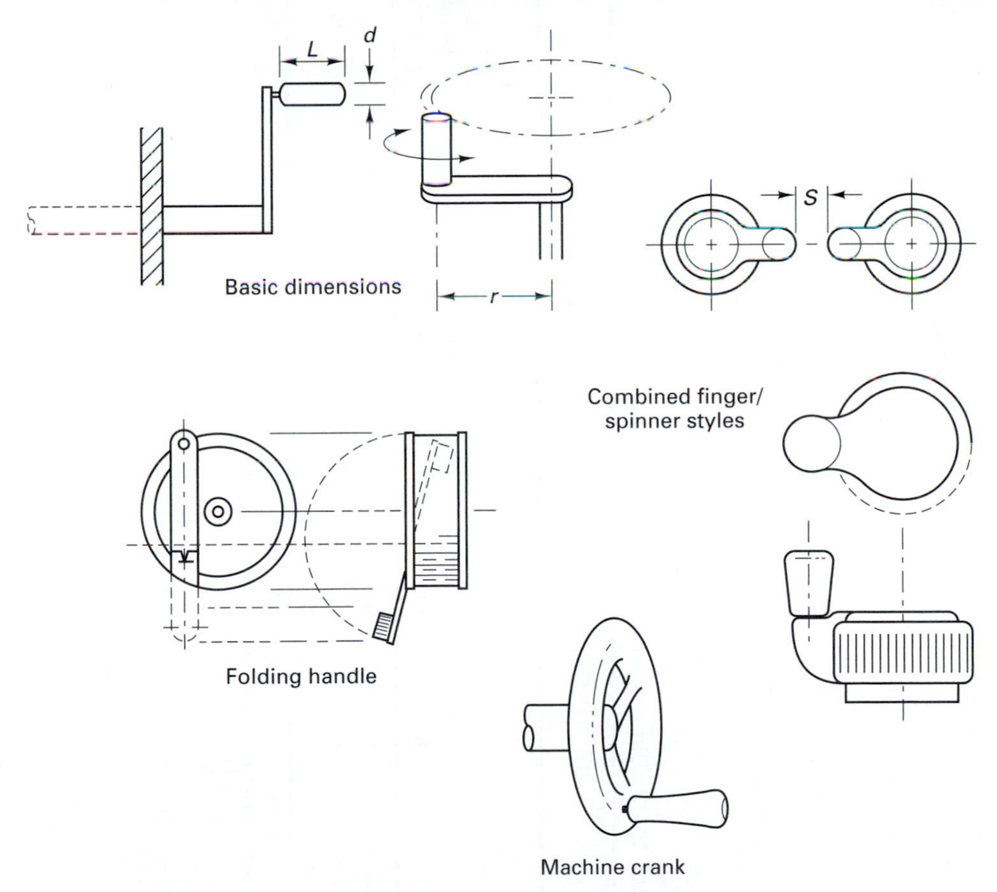

Figure 10–9. Crank. (Modified from MIL-HDBK 759.)

TABLE 10–13. Dimensions of a Crank

	Operated by finger and wrist movement (Resistance below 22 N)					*Operated by arm movement (Resistance above 22 N)*				
			r, Turning radius					*r, Turning radius*		
	L Length (mm)	*d Diameter (mm)*	*Below 100 RPM (mm)*	*Above 100 RPM (mm)*	*S Separation (mm)*	*L Length (mm)*	*d Diameter (mm)*	*Below 100 RPM (mm)*	*Above 100 RPM (mm)*	*S Separation (mm)*
Minimum	25	10	38	13	75	75	25	190	125	75
Preferred	38	13	75	58	—	95	25	—	—	—
Maximum	75	16	125	115	—	—	38	510	230	—

Source: Modified from MIL–HDBK 759, MIL-STD-1472F.

Note: Letters correspond to measurements illustrated in 10–9.

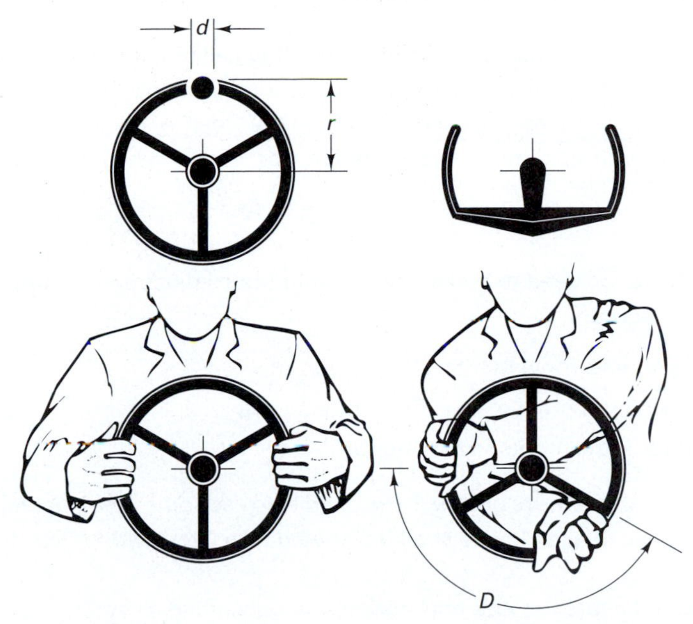

Figure 10–10. Handwheel. (Modified from MIL-HDBK 759.)

TABLE 10–14. Dimensions of a Handwheel

	r, Wheel radius					
	With power steering (mm)	*Without power steering (mm)*	*d Rim diameter (mm)*	*Tilt from vertical (degrees)*	*R Resistance (N)*	*D Displacement, both hands on wheel (degrees)*
Minimum	175	200	19	30 Light vehicle	20	—
Preferred	—	—	—	—	—	—
Maximum	200	255	32	45 Heavy vehicle	220	120

Source: Modified from MIL–HDBK 759.
Note: Letters correspond to measurements illustrated in 10–10.

LEVER

Levers may be used when a large force or displacement is required for continuous adjustments or when multidimensional movements are required. There are two kinds of levers, often called joysticks (or simply sticks):

1. The *force joystick* does not move (and hence is isometric) but transmits control inputs according to the force applied to it. The device is especially suitable when a return to the center of the system is desired after each control input.

2. The *displacement joystick* (often misleadingy called isotonic) moves and transmits control inputs according to its spatial position, direction of movement, or speed. Elastic resistance that increases with displacement may be used to improve the "feel" of the stick. The device is appropriate when control in two or three dimensions is needed, particularly when accuracy is more important than speed.

Design recommendations are given in Figure 10–11 and listed in Table 10–15. Other recommendations are as follows:

When levers are used to make fine or continuous adjustments, support shall be provided for the appropriate limb segment:

- For large hand movements: elbow support;
- For small hand movements: forearm support;
- For finger movements: wrist support.

When several levers are grouped in proximity to each other, the lever handles shall be coded. (See later).

When practicable, all levers shall be labeled (see later) with regard to their function and direction of motion.

High-Force Levers may be appropriate for occasional or emergency use. They shall be designed to be either pulled up or pulled back toward the shoulder, with an elbow angle of 150° (±30°). The force required for the operation of these levers shall not exceed 190 N. The diameter of the handle shall be from 25 to 38 mm, and the length of the lever shall be at least 100 mm. Displacement should not exceed 125 mm. The clearance behind the handle and at its sides shall be at least 65 mm along the entire path of the handle. The lever may have a thumb button or a clip-type release.

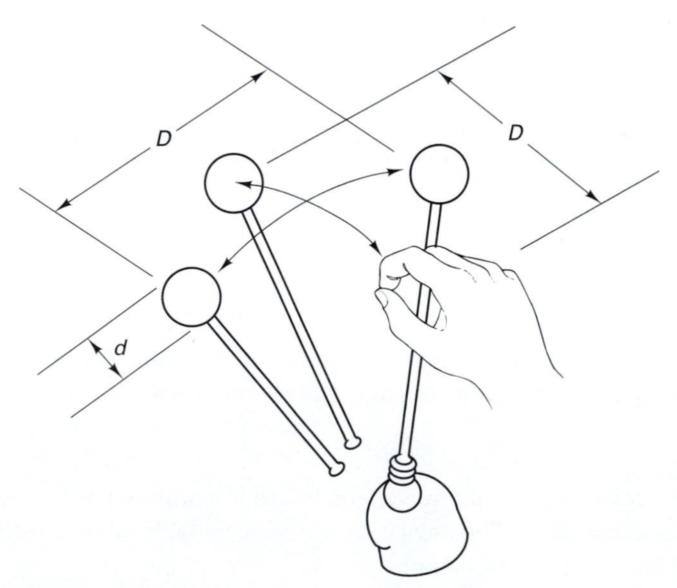

Figure 10–11. Lever. (Joystick, stick)

CONTINUOUS THUMBWHEEL

Thumbwheels for continuous adjustments may be used as an alternative to round knobs if the compactness of the thumbwheel is beneficial.

Design recommendations are given in Figure 10–12 and Table 10–16.

SLIDE

Slide switches are used to make continuous settings—for example, in music mix-and-control stations. Design recommendations, by Cushman and Rosenberg (1991) and from MIL-STD 1472D, are compiled in Figure 10–13 and Table 10–17.

TABLE 10–15. Dimensions of a Lever

	d, Diameter		*R, Resistance*				*D, Displacement*		*S, Separation*	
			Fore–aft		*Left–right*					
	Finger grip (mm)	*Hand grip (mm)*	*One hand (N)*	*Two hands (N)*	*One hand (N)*	*Two hands (N)*	*Fore–aft (mm)*	*Left–right (mm)*	*One hand (mm)*	*Two hands (mm)*
Minimum	13	32	9	9	9	9	—	—	50*	75
Preferred	—	—	—	—	—	—	—	—	100	125
Maximum	38	75	135	220	90	135	360	970		

Source: Modified from MIL-HDBK 759.

Note: Letters *d* and *D* correspond to measurements illustrated in Figure 10–11.

*About 25 mm if one hand usually operates two adjacent levers simultaneously.

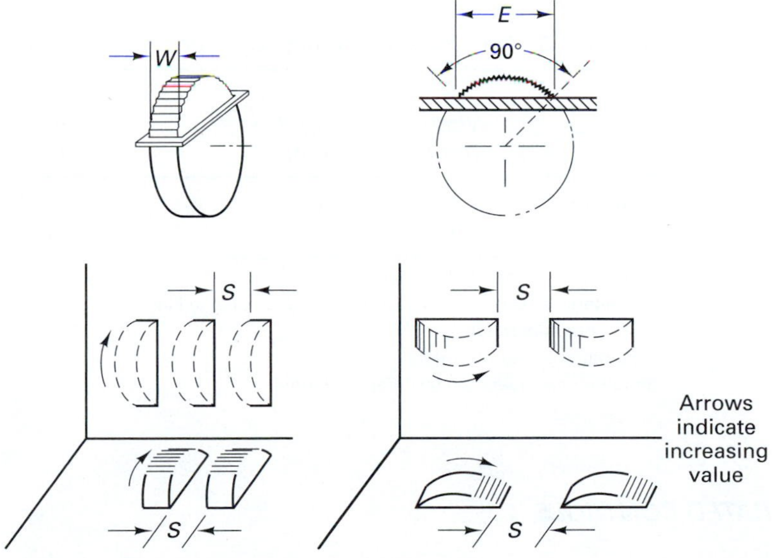

Figure 10–12. Continuous thumbwheel. (Modified from MIL-HDBK 759.)

TABLE 10–16. Design Recommendations for a Continuous Thumbwheel

	E Rim exposure (mm)	*W* Width (mm)	*S, Separation* Side–by–side (mm)	*S, Separation* Head–to–foot (mm)	*R* Resistance (N)
Minimum	—	—	25 38*	50 75*	—
Preferred	25	3	—	—	—
Maximum	100	23	—	—	3†

Source: Modified from MIL–STD 759, MIL-STD-1472F.
Note: Letters correspond to measurements illustrated in Figure 10–12.
*If operator wears gloves.
†To minimize danger of inadvertent operation.

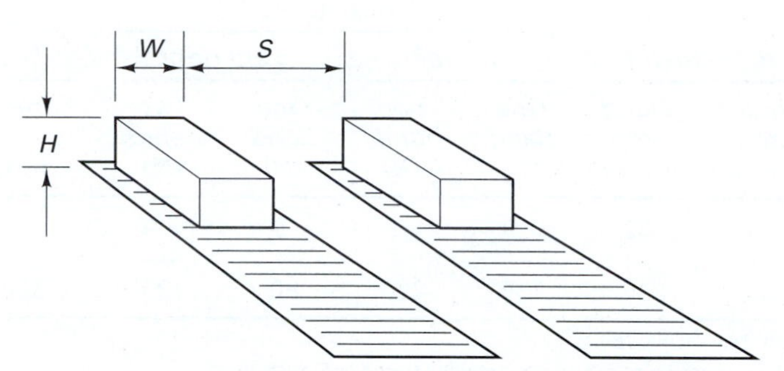

Figure 10–13. Slide. (Modified from MIL-HDBK 759.)

TABLE 10–17. Dimensions of a Slide Operated by Bare Finger

	W Width (mm)	*H* Height (mm)		*S* Separation (mm)		*R* Resistance (N)
Minimum	6	6	19*	13†	16**	3
Maximum	25	—	50*	25†	19**	11

Source: Modified from MIL–STD 1472D.
Note: Letters correspond to measurements illustrated in Figure 10–13.
*Single-finger operation.
†Single-finger sequential operation.
**Simultaneous operation by different fingers.

FOOT-OPERATED CONTROLS

As mentioned in Chapter 8, long-duration foot operation of controls should be required only from a seated person.

Foot-operated switches may be used if only two discrete conditions (such as *on* and *off*) need to be set. Recommended design parameters are listed in Table 10–18.

Foot operation can be more forceful than hand operation, but is less precise and less suitable for continuous adjustments. Nevertheless, in vehicles and cranes (but seldom in other applications), pedals are used for continuous adjustments, such as to control an automobile's speed. Recommended design dimensions are listed in Table 10–19.

TABLE 10–18. Dimensions of a Foot–Operated Switch for Two Discrete Positions

	W Width or diameter (mm)	*D* Displacement (mm)				*F* Resistance (N)	
		Operation with regular shoe	*Operation with heavy boot*	*Operation by ankle flexion*	*Operation by whole leg movement*	*Foot does not rest on control*	*Foot rests on control*
Minimum	13	13	25	25	25	18	45
Maximum	—	65	65	65	100	90	90

Source: Modified from MIL–STD 1472D.

REMOTE-CONTROL UNITS

Remote-control units, usually simply called "remotes," are small handheld control panels that are manipulated at some distance from a computer, with which they communicate by cable or radiation.

Remote-control units used with TV and radio equipment in the home got so difficult to understand and operate in the 1980s, that they became almost proverbial for useless gadgetry. In 1991, a simple and inexpensive remote control unit "to replace all remotes" came on the market and was an immediate commercial success.

In the robotics industry, remote-control units (frequently called "teach pendants") are used by a technician, often to specify the point in space to which the robot effector (its tool or gripper) must be moved to operate on a workpiece.

Many human-engineering issues are pertinent. One is the need of the human operator to see, at least initially, the exact location of the robot effector in relation to the workpiece. This may be difficult to do (for instance, because of lack of illumination), and it may be dangerous for the operator, who must step into the operating area of the machine. A second ergonomic concern is the proper design and operation of the controls. Joysticks are preferable in principle, because they allow continuous control of movement in a plane or in space; push buttons, the more common solution, permit control only in a linear or angular

TABLE 10–19. Dimensions of a Pedal for Continuous Adjustments (e.g., Accelerator or Brake)

	H		*D, Displacement (mm)*				*R, Resistance (N)*				*Edge-to-edge separation, S*	
	Height or depth (mm)	*W Width (mm)*	*Operation with regular shoe*	*Operation with heavy boot*	*Operation by ankle flexion*	*Operation by whole leg movement*	*Foot does not rest on pedal*	*Foot rests on pedal*	*Operation by ankle flexion*	*Operation by whole leg movement*	*Random operation by one foot*	*Sequential operation by one foot*
Minimum	25	75	13	25	25	25	18	45	—	45	100	50
Maximum	—	—	65	65	65	180	90	90	45	800	150	100

Source: Modified from MIL–STD 1472F.

direction. Yet, push buttons are easier to protect from inadvertent operation and from damage when the unit is dropped. The proper arrangements of sticks or buttons in arrays on the surface of the remote is similar to the design aspects discussed in this chapter. Particular attention must be paid to the ability to immediately stop the robot in emergencies and to move it away safely from a given position, even by an inexperienced operator.

CODING

There are numerous ways to help identify hand-operated controls and how to operate them, to indicate the effects of their operation, and to show their status. It is important that, throughout the system, uniform coding principles be employed. Coding of foot-operated controls is difficult, and no comprehensive rules are known.

The major coding means are as follows.

LOCATION.

Controls associated with similar functions shall be in the same relative location from panel to panel.

SHAPE.

Shaping controls to distinguish them can appeal to both the senses of sight and touch. Sharp edges shall be avoided. Various shapes and surface textures have been investigated for diverse uses. Figures 10–14 through 10–20 show examples.

SIZE.

Up to three different sizes of controls can be used for discriminating by size. Controls that have the same function on different items or equipment shall have the same size (and shape).

MODE OF OPERATION.

One can distinguish controls by their different manners of operation, such as pushing, turning, and sliding. If the operator is not familiar with the control, a false manner of operation may be tried first, which is likely to increase operation time.

LABELING.

While proper labeling (see later) is a secure means of identifying controls, it will work only if the labels are in fact read and understood by the operator. The label must be placed so that it can be read easily, is well illuminated, and is not covered. Yet, labels take time to read. Transilluminated (back-lighted) labels, possibly incorporated into the control, are often advantageous.

COLOR.

Most controls are either black (number 17038, 27038, or 37038 in FED-STD 595) or gray (26231 or 36231). The following colors may also be selected: red (11105, 21105, 31105, 14187); green (14187); orange-yellow (13538, 23538, 33538); and white (17875, 27875, 37875). Use blue (1523) only if an additional color is absolutely necessary. Note that the use of color requires sufficient luminance of the colored surface. (MIL-STD-1472F.)

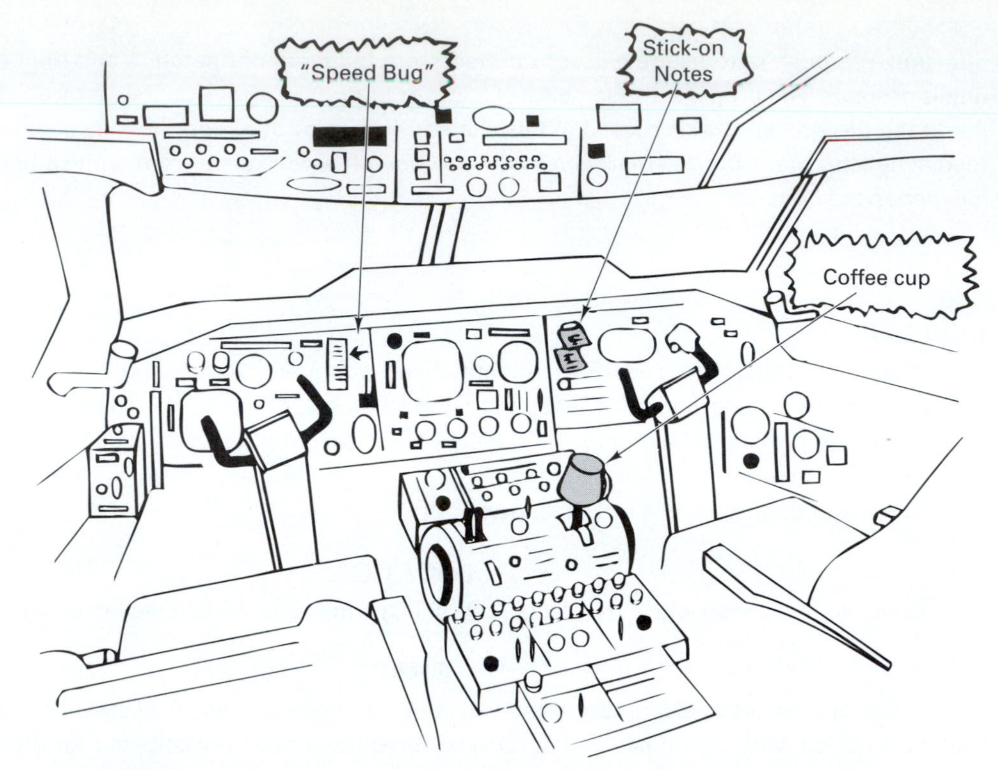

Figure 10–14. Informal coding used by aircraft crews (with permission from Norman 1991).

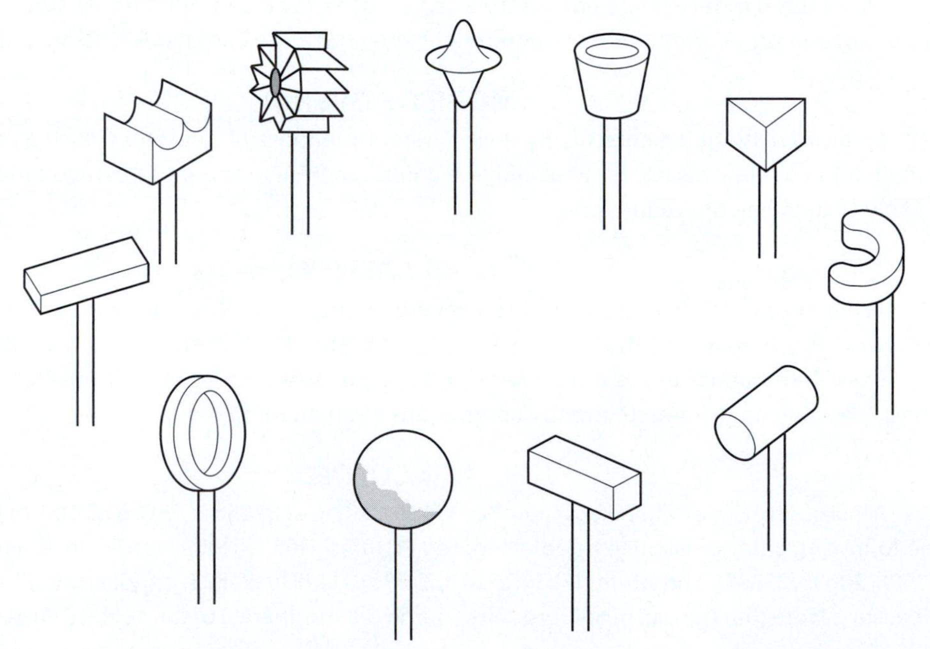

Figure 10–15. Examples of shape-coded aircraft controls. (Jenkins 1953.)

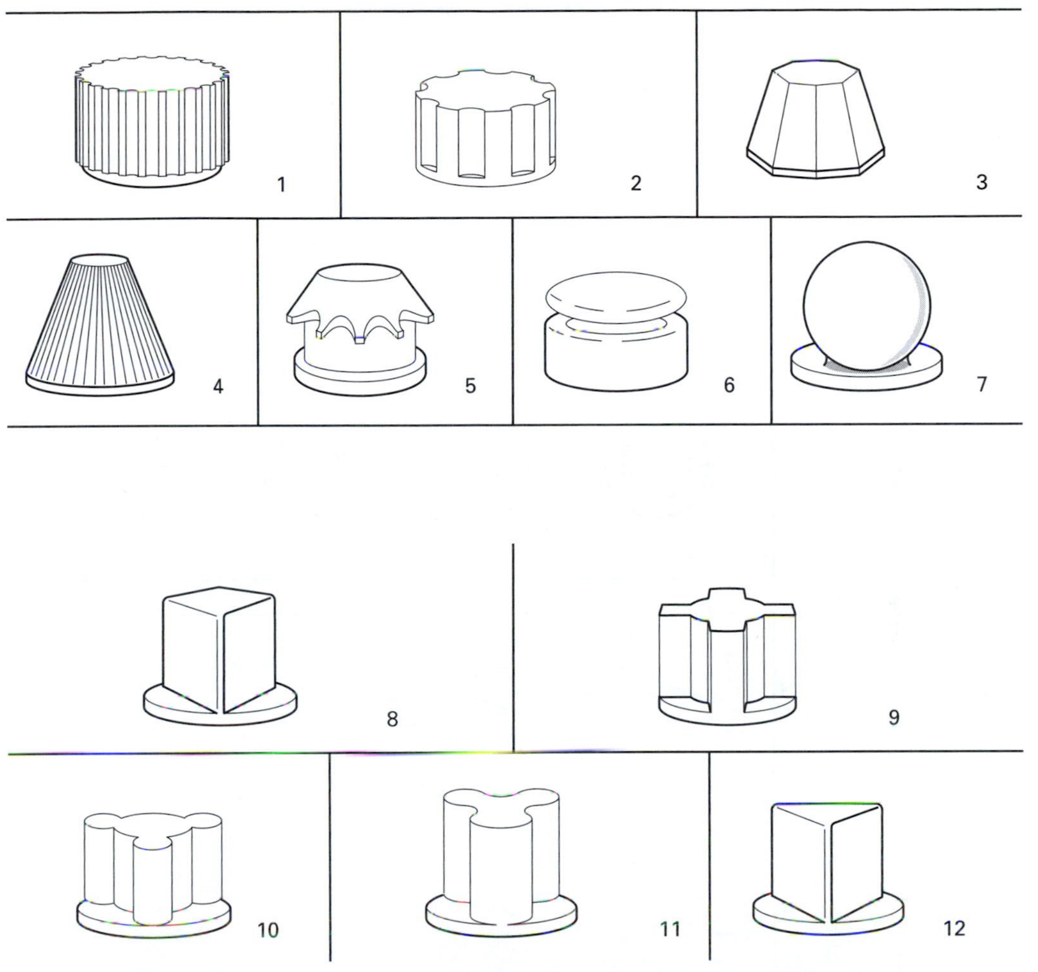

Figure 10–16. Examples of shape-coded knobs of approximately 2.5 cm diameter and 2 cm height. (Hunt and Craig 1954.) Numbers 1 through 7 are suitable for full rotation (but do not combine 1 with 2, 3 with 4, or 6 with 7). Numbers 8 through 12 are suitable for partial rotation. Recommended combinations are 8 with 9, 10 with 11, and 9 with 12 (but do not combine 8 with 12, 9 with 10 or 11, or 11 with 3 or 4).

REDUNDANCY.

Often, coding methods such as location, size, and shape, or color and labeling, can be combined. Combining codes has advantages. For example, the combination of codes can generate a new set of codings. Or, through offering multiple ways to achieve the same kind of feedback, it provides redundancy. For instance, if there is no chance to look at a control, one can still feel it; also, one knows that on a traffic light, the top signal is red, the bottom green.

The various types of codings have certain advantages and disadvantages, as listed in Table 10–20. Table 10–21 indicates the largest number of coding stimuli that can be used together.

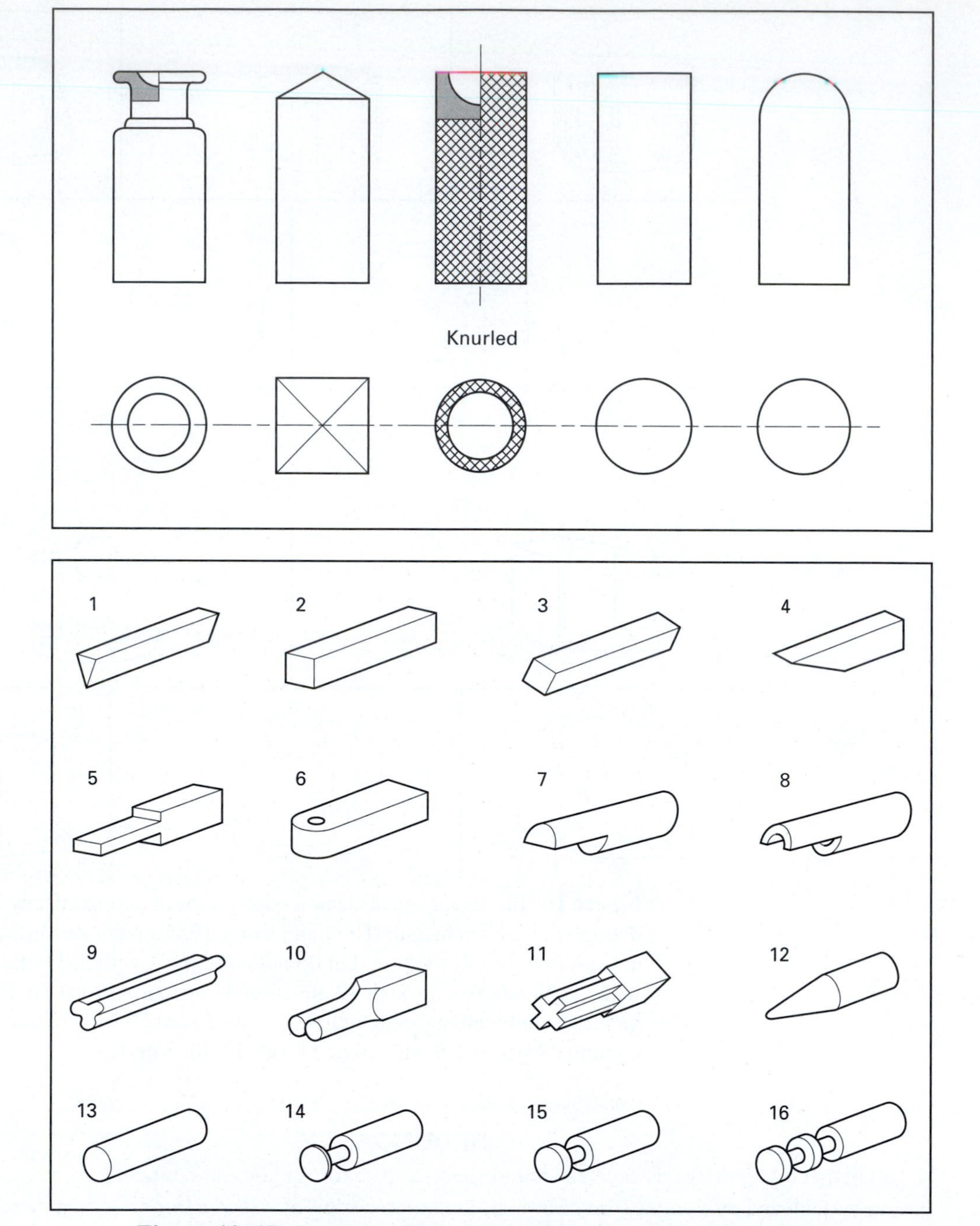

Figure 10–17. Examples of shape-coded toggle switches (top: Stockbridge 1957; bottom: Green and Anderson 1955) of approximately 1 cm diameter and 2.2 cm length.

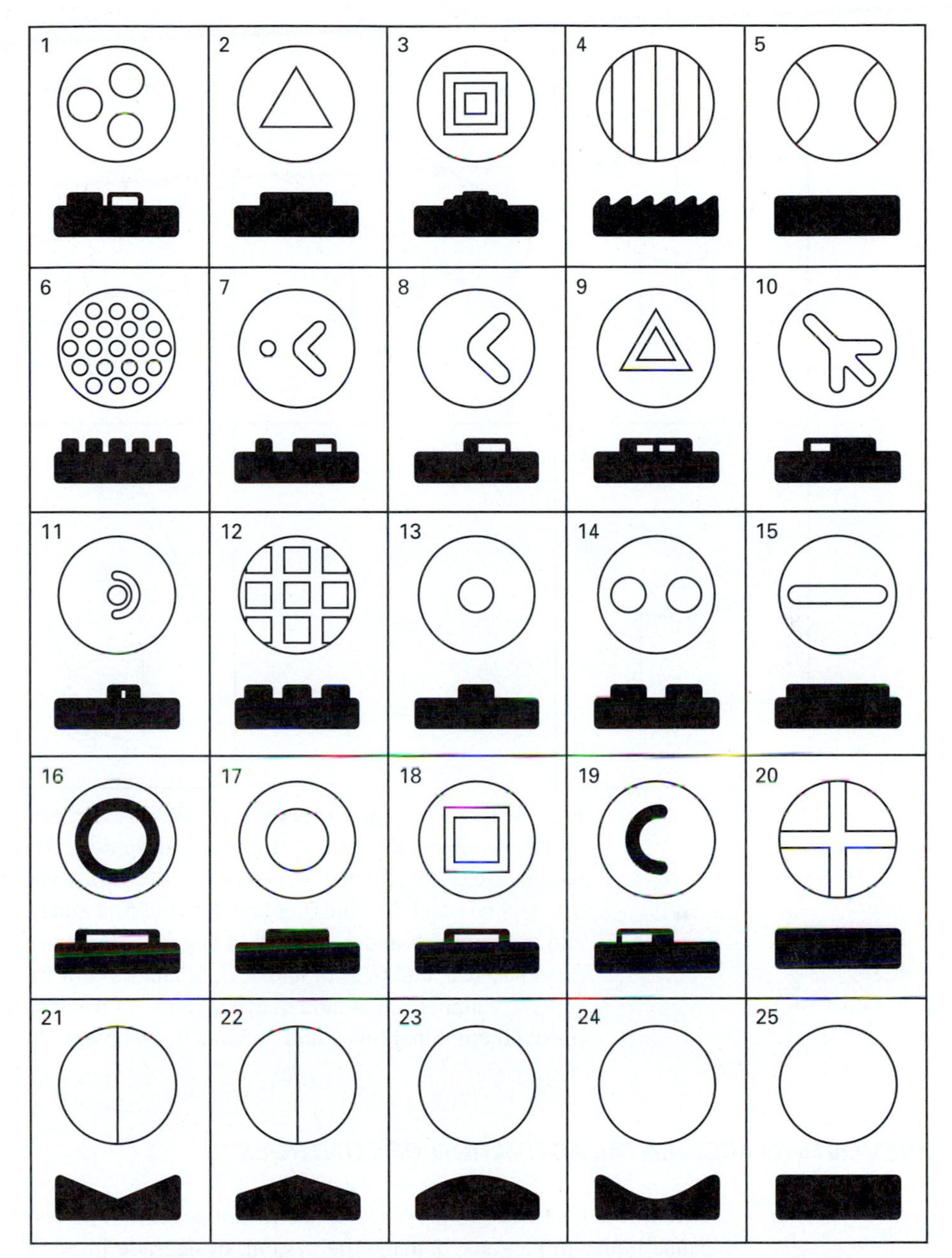

Figure 10–18. Examples of push buttons shape coded for tactile discrimination (with permission from Moore 1974). Shapes 1, 4, 21, 22, 23, and 24 are best discriminable by bare-handed touch alone, but all shapes were confused on occasion.

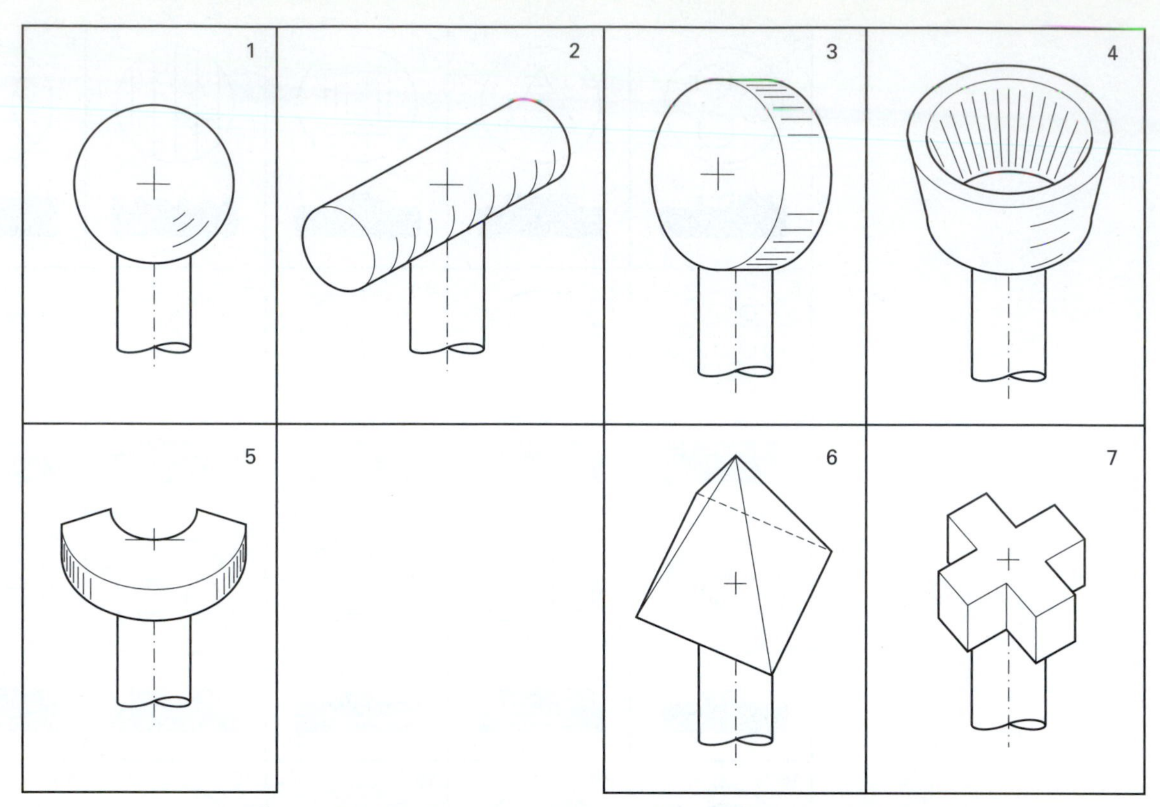

Figure 10–19. Shapes proposed (by K.H.E. Kroemer in 1980) for use on finger-operated controls of mining equipment. The following combinations are recommended for concurrent use: Two handles—1 and 2, 1 and 5, 1 and 6, 1 and 7, 2 and 3, 2 and 4, 2 and 5, 2 and 6, 2 and 7, 3 and 5, 3 and 6, 3 and 7, 4 and 5, 4 and 6, 4 and 7. Three handles—1, 2, and 6; 1, 2, and 7; 2, 3, and 6; 2, 3, and 7; 3, 5, and 6. Four handles—1, 2, 3, and 6; 1, 2, 3, and 7; 2, 3, 4, and 6; 2, 3, 4, and 7. Five handles—1, 2, 4, 5, and 6. Avoid combinations 1 and 3, 3 and 4, and 5 and 7.

PREVENTING ACCIDENTAL ACTIVATION OF CONTROLS

Often, it is necessary to prevent the accidental activation of controls, particularly if that might cause injury to persons, damage the system, or degrade important system functions. There are various means of preventing accidental activation, some of which may be combined:

1. Locate and orient the control so that the operator is unlikely to strike it or move it accidentally in the normal sequence of operations.
2. Recess, shield, or surround the control by physical barriers.

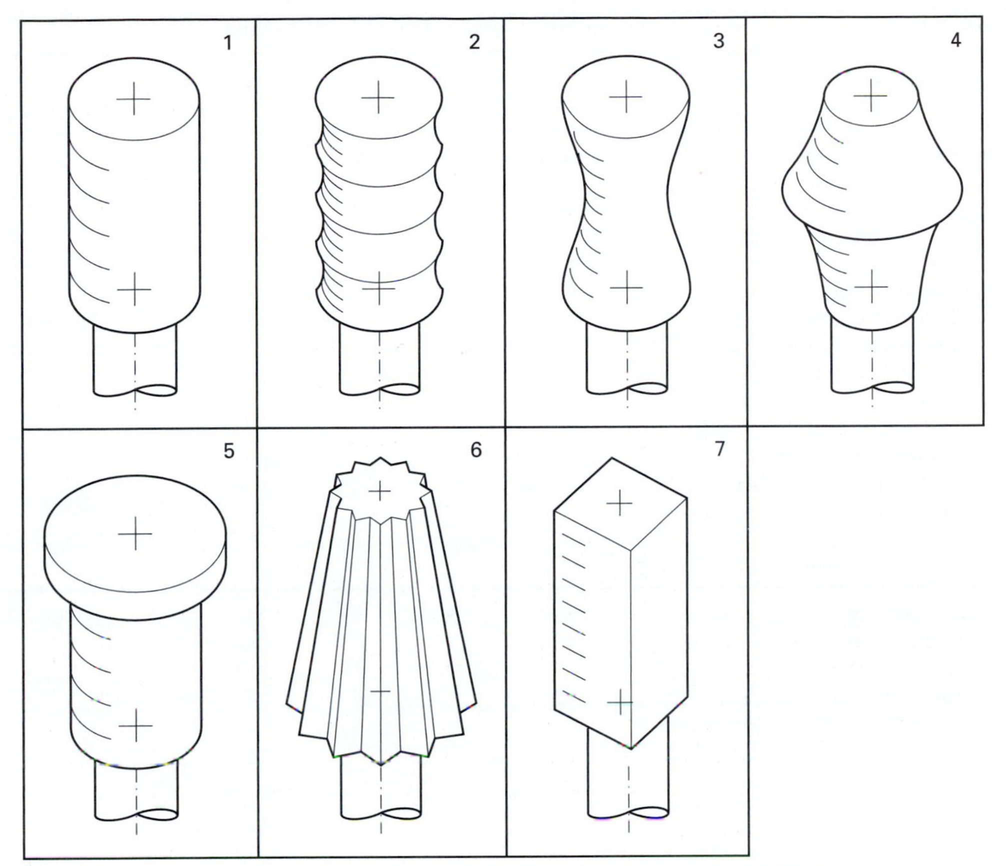

Figure 10–20. Shapes proposed (by K.H.E. Kroemer in 1980) for use on lever handles on mining equipment. The following combinations are recommended for concurrent use: Two handles—1 and 5, 1 and 6, 1 and 7, 2 and 6, 2 and 7, 3 and 4, 3 and 6, 3 and 7, 4 and 5, 4 and 6, 4 and 7, 5 and 6, 5 and 7. Three handles—1, 2, and 6; 1, 2, and 7. Four handles: 1, 2, 4, 5; 1, 2, 4, and 6; 1, 2, 4, and 7. Five handles: 1, 2, 3, 4, and 6; 1, 2, 3, 4, and 7. Avoid combinations 2 and 5, 3 and 5, and 6 and 7.

3. Cover or guard the control.
4. Provide interlocks between controls so that either the prior operation of a related control is required or an extra movement is necessary to operate the control.
5. Provide extra resistance (viscous or coulomb friction, spring-loading, or inertia) so that an unusual effort is required to actuate the control.
6. Provide a locking mechanism, so that the control cannot pass through a critical position without delay.

Note that these means usually slow down the operation of the system, which may be detrimental in an emergency.

TABLE 10–20. Advantages and Disadvantages of Coding Techniques

Advantages	*Location*	*Shape*	*Size*	*Mode of Operation*	*Labeling*	*Color*
Improves visual identification	X	X	X		X	X
Improves nonvisual identification (tactual and kinesthetic)	X	X	X	X		
Helps standardization	X	X	X	X	X	X
Aids identification under low levels of illumination and colored lighting	X	X	X	X	(When transilluminated)	(When transilluminated)
May aid in identifying control position (setting)		X		X	X	
Requires little (if any) training; is not subject to forgetting					X	
Disadvantages						
May require extra space	X	X	X	X	X	
Affects manipulation of the control (ease of use)	X	X	X	X		
Limited in number of available coding categories	X	X	X	X		X
May be less effective if operator wears gloves		X	X	X		
Controls must be viewed (ie, must be within visual areas and with adequate illumination present)					X	X

Source: Modified from Kroemer 1988b.

Note: X indicates presence of advantage or disadvantage..

TABLE 10–21. Maximal Number of Stimuli for Coding

Visual stimuli	*Max. No.*	
Light intensity (brightness)	2	
Color of surfaces	9	5* (surfaces and lights combined)
Color of lights (lamps)	3	
Flash rates of lights	2	
Size	3	10*
Shape	5	
Auditory stimuli		
Frequency	4	
Intensity (loudness)	3	
Duration	2	

Source: Adapted from information compiled by Cushman and Rosenberg 1991.

*According to MIL-STD-1472F.

COMPUTER INPUT DEVICES

Within just a few decades, the computer has changed the way in which "office work" is done, how information is distributed, and how work systems are designed and run. Given that the computer is such a dynamic tool, however, it is surprising that most of our inputs still are by old-fashioned binary *(on–off)* push buttons.

KEYS

Keys on conventional keyboards are of the nondetent push button type, commonly arranged side by side in straight rows and straight columns (but in curved and slanted columns on the QWERTY keyboard; see shortly.)

NUMERICAL KEYPADS

Two different kinds of numerical key sets are widely used. One, usually associated with telephones, is arranged thus:

1 2 3
4 5 6
7 8 9
0

The so-called calculator key set is arranged like this:

7 8 9
4 5 6
1 2 3
0

A person's keying performance may be more accurate and slightly faster with the telephone key set than with the calculator arrangement, but the differences are slight. The most important feature appears to be the placement of the zero key below the three rows of the other keys. (Conrad and Hull 1968; Marteniuk et al. 1996).

Computer Keyboards

Sholes's Keyboard

C. Latham Sholes successfully designed, produced, and marketed the keyboard originally meant for typewriters and now employed, with additional keys, on computers. His "type-writing machine" (U.S. Patent 207,559 of August 27, 1878) shows a keyboard with four straight rows of 11 round keys each, as illustrated in Figure 10–21. Starting from the left, the keys on the third row from the operator are labeled QWERTY. This arrangement of six keys is used nowadays as a short label for the arrangement of all the letter ("alpha" or "alphabetic") keys; it is remarkable that the current QWERTY keyboard is still arranged in a fashion similar to Sholes's. How it came to that arrangement is no longer known.

Figure 10–21. The 1878 Sholes keyboard.

The origin of the QWERTY key set

Sholes made 14 specific technical claims on page 4 of his 1878 U.S. Patent No. 207,559 of a "type-writing machine," but none mentions key selection or layout. His Figure 1 shows a frontal view of four horizontal rows of keys, staggered so that the row farthest from the operator is the highest; Figure 2 is a top view with four straight rows, on which the third row of the keyboard starts, from the left, with the letters QWERTY. Each of the four rows of keys consists of 11 keys, for a total of 44.

Previous 1878 patents by Sholes (207,558, 207,557, and 200,351 with Glidden; 199,382 by himself) show keyboards of three rows with seven keys each; the keys are not labeled with letters. His earlier patent 182,511 of 1876 (with Schwalbach) also shows a three-row keyboard with 11 keys in the two near rows and 10 keys in the far row; the keys are not labeled with letters. Sholes's earlier patents 79,868 and 79,265 of 1868 (with Glidden and Soulé) show two rows of keys "similar to the keys of a piano or melodeon" (79,868, p. 2). Neither of these patents explains the layout of the keys, but in patent 79,868 the keys are lettered in numerical and alphabetic order along the two rows.

Lacking any statements by Sholes or his contemporaries in the patents, one observes that the QWERTY layout shows some remnants of an alphabetic arrangement, from which one can surmise similarities to the arrangement of the printer's "type case" in which pieces were assorted according to convenience of use and not according to the alphabet. Both Sholes and James Densmore, with whom he worked in the 1870s, were printers by trade.

Another reason for the arrangement of the letters and keys was probably the intent to avoid the tendency of the type bars to collide and stick together when neighboring ones were activated in quick sequence. It is likely that this led to a separation of certain bars and keys on the keyboard; yet, there is no contemporaneous evidence for any of this.

Numerous earlier attempts to design "writing machines" are known. The Herkimer County Historical Society (1923) and Martin (1949) list about 100 proposals. They show a great variety of keyboards, some similar to the extended black and white keys on the clavier, some in double or triple rows of buttonlike keys, and some with keys arranged in concave or convex circular sections. In most cases, the mapping of keys to certain letters, numerals, or other signs is not shown or not clear. Thus, inventions that precede those by Sholes and his coworkers do not provide any obvious insights regarding the origins of the QWERTY keyboard.

The computer keyboard

The 1878 QWERTY layout was quickly adopted, with some modifications, internationally for typewriters. For computer use, keys were added, mostly "function" keys, to the left, behind, and especially to the right of the original QWERTY set. With an additional numeric keypad and a cursor-control keypad (commonly on the right side), in the 1980s the total number of keys was customarily just over 100—in some case about 125—as compared to the 45 to 50 keys on the old typewriter. The 100 or so keys require large finger reaches to those on the edge of the keyboard, and identifying a special key often becomes a visual search task, making the typists' "touch typing" (without looking at the keys) a largely lost skill.

On the customary 1980 and 1990s computer keyboards, all the keys were arranged side by side in straight rows, which the fingertips are not. The columns of keys followed two different design rules: On the QWERTY set, the columns were bent and ran slanted to the left, as seen by the operator; but on the newly added key sets, the columns were straight. Apparently, neither the original QWERTY keyboard nor its successors were "human engineered."

Redesigning the keyboard

The mapping from keys to characters has some serious ramifications. For example, letters that frequently follow each other in English text, such as *t* and *h,* are spaced apart on the keyboard. Since the early 1900s, many proposals were published (reviewed, eg, by Alden et al. 1972; Gilad and Pollatsheck 1986; Gopher and Raij 1988; Hirsch 1970; Kroemer 1964, 1972, 1997, 1998; Lewis et al. 1997; Lithrick 1981; Michaels 1971; Norman and Fisher 1982; Noyes 1983a; and Seibel 1972) to make keying operations easier and to improve typing. Included among the proposals were suggestions to relocate keys within the standard key-set layout and to change the entire keyboard layout—for example, by breaking up the columns and rows of keys or by dividing the keyboard into separate (and tilted) sections, as in the German patent 1,255,117 (Max-Planck-Gesellschaft 1968). A terminology for describing major design features of keyboards is given in Figure 10–22.

From mechanics to electronics

The mechanics of the old typewriter made changes difficult until new technologies appeared in the 1960s. On the mechanical typewriter, the top of the key served as the energy input from the finger into a lever system that ended in the final type bar striking the ribbon to the paper. Later, in the so-called electric typewriter, auxiliary energy input reduced the work required of the typist's finger. Then the keys became true switches for electric or electronic circuitry, and the physical effort needed from the finger was very much reduced. This changed the dynamic work (force–displacement integral) required from the key operator's fingers profoundly, but the measurement procedures commonly used into the 1990s still relied on static force application to depress the key. This is not indicative of the keyboard's actual operation (Olacsi and Beaton 1997), which varies with key designs and among operators (Gerard et al. 1996; Rempel et al. 1997, 1999; Serina and Rempel 1996). Using the new switch technology, for continuous keyboarding operators prefer keys that require a (static) force of less than 0.6 N with a snapback (wherein activation of the key is felt by a reduced key resistance) within a total displacement of less than 4 mm, as per ANSI standards (ANSI/HFES 100 of 1988 and 2000). Yet, seemingly minute differences in key switch force during displacement can make a major difference in the operator's biomechanics and effort (Dennerlein et al. 1998; Rempel et al. 1999).

New keys

Instead of simply improving conventional key and keypad designs, radically new solutions have been proposed that include the use of chording for generating letters, numerals, or words, instead of single-key activations. This approach has been used, for example, by court reporters for on-the-spot recording of verbal discussions. It also has been used to sort mail. While traditional keys are tapped down for activation and are binary (ie, have the only two states, *on* and *off,*), one could use three-position (ternary) keys that are rocked back and forth. With this technique, one places the fingertips on multidirectional switches that can be tapped, rocked, and moved sideways. Finger movements may be registered by means of instrumented gloves (Kroemer 1997; Lewis et al. 1997). Of course, voice recognition provides input to the computer that requires neither finger nor switch movement.

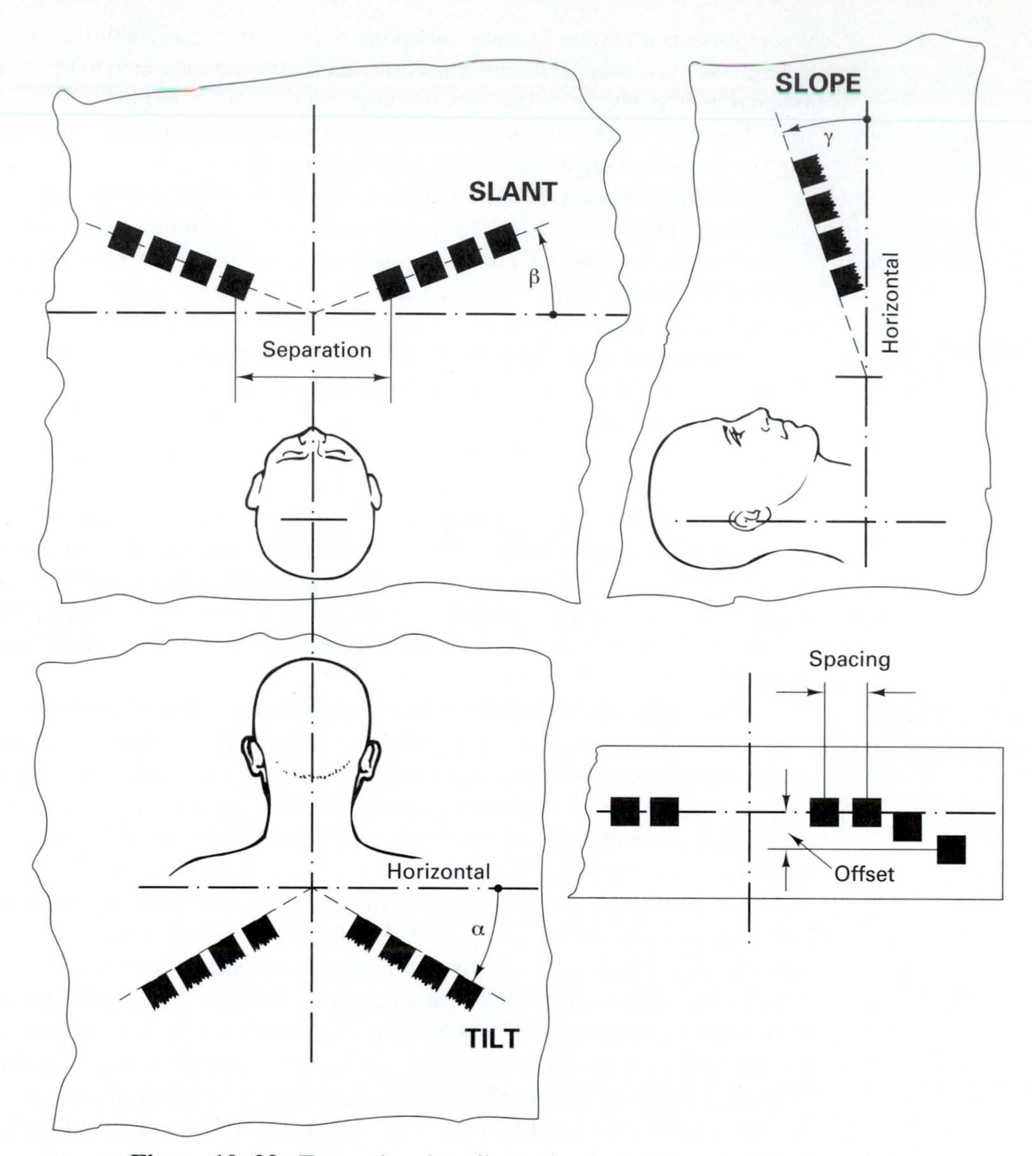

Figure 10–22. Terms that describe major design factors of keyboards.

Alternative keyboards

In 1915, Heidner had already recognized that the "flat" keyboard with its horizontal rows of keys required the typist to rotate the forearms into extreme pronation, a straining posture difficult and often painful to maintain. Consequently, Heidner and later Klockenberg (1926) proposed to break the QWERTY keyboard into two halves and to decline the sections laterally to facilitate the work of the key operator. Splitting the keyboard, angling the sections in slant, tilt, and slope (see Figure 10–22), and relocating the keys from the old flat rows and columns along curved and dished shapes that follow the natural structure and mobility of hands and its digits were features of numerous designs and patents since 1967 (Kroemer 1997, 1998: Lewis et al. 1997). By the end of the century, such human-engineered alternative keyboards had become strong competition to conventional keyboards, with users of the new keyboards hoping to avoid musculoskeletal overuse disorders (see Chapter 8) while maintaining or even improving their keying performance (Marklin et al. 1997; National Research Council, 1998, 1999; NIOSH 1997; Smith et al. 1998; Swanson et al. 1997).

Other input devices

In addition to the keyboard, a variety of other input devices has been used, including the following:

1. *Mouse.* A palm-sized, hand-contoured block with one or more finger-operated buttons. The mouse is slid on a surface (mouse pad), commonly to move a cursor.
2. *Puck.* Similar in shape to a mouse, but typically has a reticular window and is used on a digitizing surface (tablet).
3. *Trackball.* Consists of a ball mounted in an enclosure. The protruding surface of the ball is rotated by the palm or the digits, usually to move a cursor.
4. *Joystick.* A short lever operated by a fingertip (often as a "nipple" among the keys of a keyboard) or, if larger, by the hand, typically to move a cursor for pointing or tracking.
5. *Stylus.* A pencil-shaped, handheld device frequently used to select objects, for freehand drawing, or to move a cursor, often on a tabletop digitizing surface.
6. *Light pen.* Similar to a stylus, but commonly used on a CRT display surface.
7. *Tablet.* A flat, slatelike panel over which a stylus or puck is slid, usually to move a cursor or select objects.
8. *Overlay.* An opaque sheet superimposed on a tablet system and providing graphic capabilities.
9. *Touch screen (touch-sensitive panel).* An empty frame or transparent overlay, mounted over the display screen, that locates the position of a finger or pointing device. A touch screen is used to select, move, or draw, objects.

The current ANSI/HFES 100 standard provides design recommendations for regular keys and keyboards, as well as for other input devices. The typical uses of all of these are to input a single bit of information (eg, one character or number) or to move a cursor, manipulate objects on the screen, insert or retrieve information by pointing, or digitize information; but the tasks change with new technology and software. Therefore, any lists of recommendations, compilations, including the ANSI standard, become outdated quickly, and the designer must follow the current literature closely. See, for example, Adams et al. 1993; Bohan et al. 1998: Bringelson et al. 1998; Fernstroem and Ericson 1997: Harvey and Peper, 1997; Huffman and Lehman, 1997; and Rempel et al. 1999.

NEW DESIGNS

Traditionally, computer data have been entered by mechanical interaction between the operator's fingers and such devices as a keyboard, mouse, trackball, or light pen. Yet, there are many other means of generating inputs. Voice recognition is one fairly well-known method, but others can be employed that utilize, for example,

1. The hands and fingers, for pointing, gestures, sign language, or tapping;
2. the arms, for gestures, making signs, or moving or pressing control devices;
3. the feet, for motions and gestures, or moving or pressing devices.
4. the legs, for gestures or moving or pressing devices;
5. the torso, including the shoulders, for positioning and pressing;
6. the head, for positioning and pressing;
7. the mouth, for lip movements, use of the tongue, or breathing, such as through a tube;
8. the face, for grimaces and other facial expressions;
9. the eyes, for tracking.

Combinations and interactions of these different inputs also can be used, similar to the way we utilize them to communicate face to face, to convey information, and to express our meaning and mood. Of course, the input method selected must be clearly distinguishable from environmental clutter or other "loose energy" that interferes with the pickup from the sensor(s).The ability of a sensor to detect the input signals depends on the type and intensity of the signal generated. For example, it may be quite difficult to distinguish between different facial expressions, whereas it is much easier to register the displacement of, or pressure on, a transmitter.Thus, the use of other-than-conventional input methods depends on the state of technology, which includes the tolerance of the system to either missed or misinterpreted input signals.

Pointing with the finger or hand, for example, is an attractive means of registering input, for a number of reasons. Pointing is a natural, easily learned and controlled activity—a common means of conveying information. Pointing is also a dynamic activity: Motion (or position) could be sensed either from the digit or hand directly or by means of markers such as reflective surfaces, attached, for example, to rings worn on the fingers. Sensors are able to track position, direction, and movement; thus, pointing could be observed fairly easily and could be discrete or continuous and in one, two, or three dimensions. However, there is the possibility of missed inputs, misinterpreted inputs, unintended (false) inputs, and fatigue or overuse effects when many pointing actions are executed quickly and repeatedly.

Such ideas are not far fetched, either in terms of sensor technology or in terms of human input activities; most of the input devices just listed are not older than a decade or two. Even such simple measures as returning to a small set of keys, as on the old QWERTY keyboard, and resting the wrists on suitable pads during frequent breaks might alleviate many operators' cumulative trauma disorders discussed in Chapter 8.

The example of sign language used by deaf persons shows that it is often unnecessary to dissect information into letters, as we customarily do while communicating with computers through single-character keyboards; most inputs could be by other means and in chunks.

DISPLAYS

Displays provide the operator with information about the status of equipment. Displays are either visual (often a light, scale, counter, CRT, or flat panel), auditory (eg, a bell, horn, beep, or recorded voice), or tactile (such as shaped knobs or Braille writing). Labels and instructions or warnings are special kinds of displays.

The following are the "four cardinal rules" for displays:

1. Display only that information which is essential for performing the job adequately.
2. Display information only as accurately as is required for the operator's decisions and control actions.
3. Present information in the most direct, simple, understandable, and usable form possible.
4. Present information in such a way that failure or malfunction of the display itself will be immediately obvious.

SELECTING THE DISPLAY

Selecting either an auditory or a visual display depends on its purpose. The objective may be to provide

1. Status information, about the current state of the system, the text input into a word processor, etc.
2. Historical information, about the past state of the system, such as the course run by a ship.
3. Predictive information, such as the future position of a ship, given certain steering settings.
4. Instructional information, telling the operator what to do and how to do it.
5. Commanding information, giving directions or orders for a required action.

What kind of display to select depends on the existing conditions and circumstances of its use:

1. An auditory display is appropriate if the environment must be kept dark, the operator must move around, and the message is short, is simple, requires immediate attention, or deals with events in time.
2. A visual display is appropriate if the environment is noisy, the operator stays in place, and the message is long, is complex, will be referred to later, or deals with spatial location.

VISUAL DISPLAYS

There are three basic types of visual displays:

1. The *check display* indicates whether or not a given condition exists. (For example, a green light indicates normal functioning.)
2. The *qualitative display* indicates that status, approximate value, or trend of a changing variable. (For example, a pointer indicates the "normal" range of a variable.)
3. The *quantitative display* indicates an exact numerical value that must be read (eg, off a clock) or shows exact information that must be ascertained (eg, the operator's location on a map).

CHECK DISPLAYS

LIGHT, OR COLOR, SIGNALS

Signals transmitted by light (color) are often used to indicate the status of a system (such as *on* or *off*) or to alert the operator that the system or a portion thereof is inoperative and that special action must be taken. Common light (color) coding systems are as follows (see Table 10–22):

- A *white* signal has no correct–incorrect implications, but may indicate that certain functions are on.
- A *green* signal indicates that equipment which is being monitored is in satisfactory condition and that it is all right to proceed. For example, a green display may provide such information as "go ahead," "in tolerance," "ready," or "power on."

TABLE 10–22. CODING OF INDICATOR LIGHTS

	Color			
Size/Type	*Red*	*Yellow*	*Green*	*White*
13 mm diameter or smaller; steady	Malfunction; action stopped; failure; stop	Delay; check; recheck action	Go ahead; in tolerance; acceptable; ready	Function or physical position; action in progress
25 mm diameter or larger; steady	Master summation (system or subsystem)	Extreme caution (impending danger)	Master summation (system or subsystem)	—
25 mm diameter or larger; flashing	Emergency (impending personnel or equipment disaster)	—	—	—

Source: Modified from MIL–STD-1472F.

- A *yellow* signal advises that a marginal condition exists and that alertness is needed, that caution be exercised, that checking is necessary, or that an unexpected delay exists.
- A *red* signal alerts the operator that the system or a portion thereof is inoperative and that a successful operation is not possible until appropriate corrective or overriding action has been taken or must be taken. Examples of the use of a red-light signal are to provide information about malfunctions, failures, errors, and so on.
- A *flashing red* signal denotes an emergency that requires immediate action to avert impending personal injury, damage to equipment, or some other serious consequence.
- *Blue* has no special signal meaning so far, but, together with a flashing red light is used in the United States on police and other emergency vehicles.

Firefighters have traditionally painted their trucks "fire red" to make them visible, but lime yellow may get better attention in traffic and at night and reduce the high rate of traffic accidents of firefighters (Solomon and King 1997).

Emergency Signals

An emergency alert is best indicated by an auditory warning signal accompanied by a flashing light. Within rooms or cabs, an emergency light should be within 30° of the operator's normal line of sight and should be larger than general status indicators. The luminance contrast C with the immediate background (see Chapter 4) should be at least 3 to 1. The flash rate should be between three and five pulses per second, with *on* time about equal to *off* time. If the flasher device should fail, the light should remain *on* steadily; warning indicators must never turn off merely because a flasher fails. A text message (such as "DANGER—STOP") should be used if feasible. The operator may acknowledge the emergency by turning off either the light or sound signal.

The environment may be very "busy" in terms of noise or lights that can mask warning signals; examples are automobile traffic, construction sites, and the interior of trucks or other vehicles. Operators of noisy equipment may have difficulties hearing the horn of a train (Robinson et al. 1997; Seshagiri 1998), and construction workers may not notice the backup warning of trucks, especially if they wear certain types of devices to protect their hearing. These kinds of situations lead to special considerations in the design of the warning systems and hearing protection devices. (See Chapter 4; see also van den Heever and Roets 1996; Casali and Wright 1995; Haas and Casali 1995; Robinson and Casali 1995; and Seshagiri and Stewart 1992.)

Qualitative and Quantitative displays

More complex displays provide information that is either of a qualitative kind (eg, cold–normal–hot) or that actually presents exact quantitative data (eg, in degrees). Traditionally, four types of displays are distinguished:

1. A moving pointer (and fixed scale).
2. A moving scale (and fixed pointer).
3. Counters.
4. A pictorial. (conic display).

All these displays used to be mechanical, but nowadays they can be generated electronically. Table 10–23 lists the four kinds of displays and their relative advantages and disadvantages.

To ease the (cognitive) load of the operators, it may be advisable to reduce the information content of a display. For example, a quantitative temperature indicator may be reduced to a qualitative one by changing a numerical display (of the temperature, in degrees) to indicate simply "too cold," "acceptable," or "too hot."

For a quantitative display, it is usually preferable to use a moving pointer over a fixed scale.

SCALES AND POINTERS

The scale may be straight (either horizontally or vertically), curved, or circular. Scales should be simple and uncluttered; graduations and numbers should be laid out such that correct readings can be taken quickly. Numerals should be located outside the scale markings so that they are not obscured by the pointer. On the other side of the scale, the pointer should end with its tip directly at the markings. Figure 10–23 provides related information.

The scale should show only such fine divisions as the operator must read. All major marks should be numbered. Progressions of either 1, 5, or 10 units between major marks are best. The largest admissible number of unlabeled minor graduations between major marks is nine, but only with a pronounced graduation at the middle one. Numbers should increase from left to right, bottom to top, and clockwise. Recommended dimensions for the scale markers are presented in Figure 10–24. The dimensions shown there are suitable even for low illumination.

CODING

The information that is displayed may be coded by the following means:

- Shape: straight or circular.
- Shades: black and white, or gray.
- Lines and crosshatched patterns.
- Figures, pictures, or, icons: symbols presented at various levels of abstractions (eg, the outline of an airplane against the horizon, to indicate the flying angle)
- Colors: See Chapter 4 for basics.
- Alphanumerics: letters, numbers, words, and abbreviations.

"Consider the digital watch and digital speedometer. Although both are more accurate and easier to read than their analog counterparts, many people . . . prefer products with analog displays. One reason is that analog displays also provide information about deviations from reference values. For example, a watch with an analog display also shows the user how many minutes remain before a specific time. Similarly a speedometer with an analog display shows the driver of a car both the car's speed and how far it is above (or below) the posted speed limit" (Cushman and Rosenberg 1991, pp. 92–93).

TABLE 10–23. CHARACTERISTICS OF DISPLAYS

Use	*Moving Pointer*	*Moving Scale*	*Counters*	*Pictorial Displays*
Quantitative information	*Fair* Difficult to read while pointer is in motion	*Fair* Difficult to read while scale is in motion	*Good* Minimum time and error for exact numerical value, but difficult to read when moving	*Fair* Direction of motion–scale relations sometimes conflict, causing ambiguity in interpretation
Qualitative information	*Good* Location of pointer easy; numbers and scale need not be read; changes in position easily detected	*Poor* Difficult to judge direction and magnitude of deviation without reading numbers and scale	*Poor* Numbers must be read; changes in position not easily detected	*Fair* Easily associated with real–world situation
Setting	*Good* Simple and direct relation of motion of pointer to motion of setting knob; position change aids monitoring	*Fair* Relation to motion of setting knob may be ambiguous; no change in position of pointer to aid change monitoring; not readable during rapid setting	*Good* Most accurate monitoring of numerical setting; relation to motion of setting knob less direct than for moving pointer; not readable during rapid setting	*Fair* Control–display relationship easy to observe
Tracking	*Good* Pointer position readily controlled and monitored; simplest relation to manual control motion	*Fair* No changes in position to aid monitoring; relation to control motion somewhat ambiguous	*Poor* No gross changes in position to aid monitoring	*Fair* Control–display relationship easy to observe
Difference estimation	*Good* Easy to calculate positively or negatively by scanning scale	*Fair* Subject to reversal errors	*Poor* Requires mental calculation	*Good* Easy to calculate either quantitatively or qualitatively by visual inspection
General	Requires largest exposed and illuminated area on panel; scale length limited unless multiple pointers used	Saves panel space; only small section of scale need be exposed and illuminated; allows long scale	Most economical use of space and illumination; scale length limited only by available number of digit positions	Pictures and symbols need to be carefully designed and pretested

Source: Modified from MIL–HDBK 759, MIL-STD-1472F.

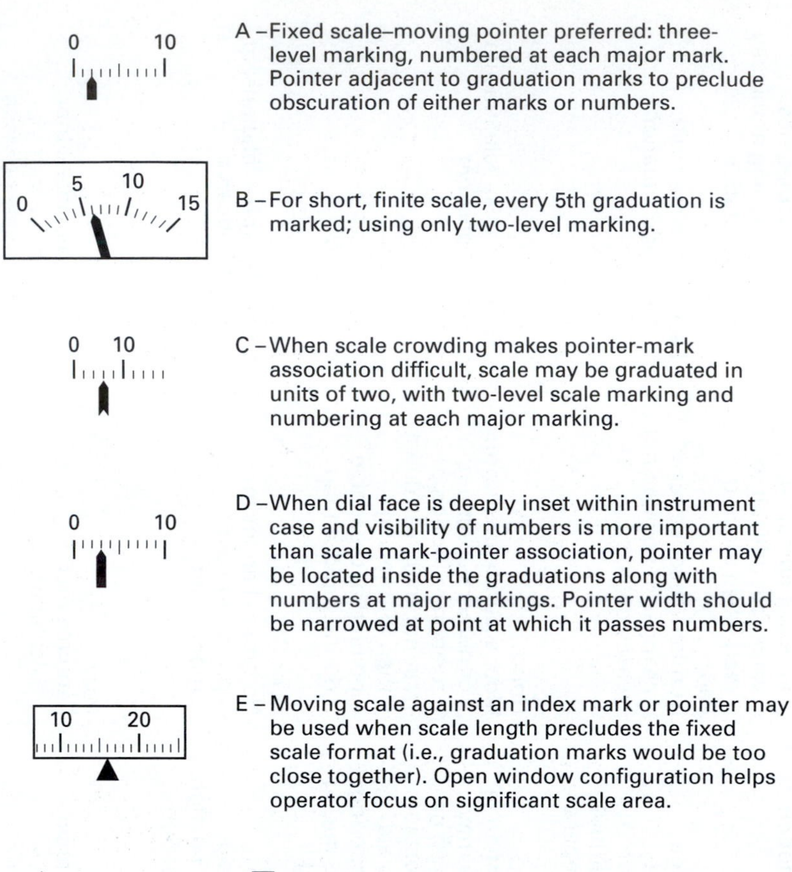

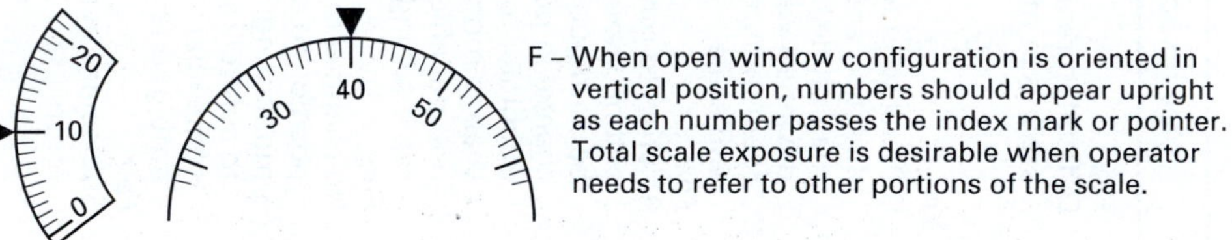

Figure 10–23. Scale graduation, pointer position, and scale numbering alternatives. (Modified from MIL-HDBK 759.)

ELECTRONIC DISPLAYS

Beginning in the 1980s, mechanical displays (such as actual pointers moving over printed scales) were increasingly replaced by electronic displays, either solid-state devices such as light-emitting diodes or displays with computer-generated images. In many cases, electronically generated displays were overly complex and colorful, fuzzy, and hard to read. Often, they required exact focusing and close attention, which distracted the operator from the main task. A typical bad example is road map displays in cars (Wierwille 1992). In these cases, the first three of the "four cardinal rules" listed earlier were often violated. Furthermore, many electronic pointers, markings, and alphanumerics did not comply with established human-engineering guidelines, especially when the displays were generated by line segments, scan lines, or dot matrices.

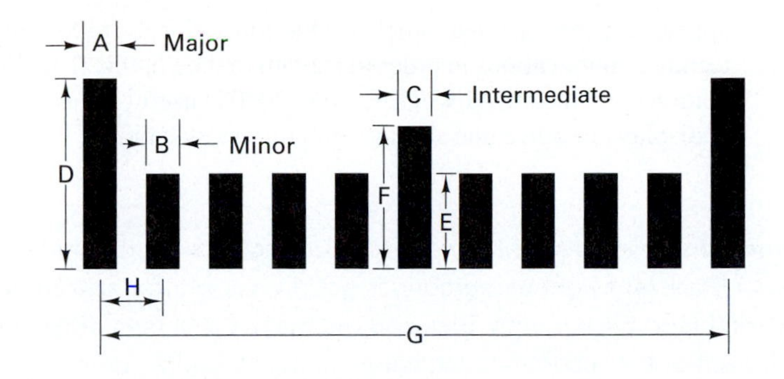

Minimum scale dimensions suitable even for low illumination

Dimension (in mm)	Viewing distance (in mm) 710	910	1,525
A (Major index width)	0.89	1.14	1.90
B (Minor index width)	0.64	0.81	1.37
C (Intermediate index width)	0.76	0.99	1.63
D (Major index height)	5.59	7.19	12.00
E (Minor index height)	2.54	3.28	5.44
F (Intermediate index height)	4.06	5.23	8.71
G (Major index separation between midpoints)	17.80	22.90	38.00
H (Minor index separation between midpoints)	1.78	2.29	.381

Figure 10–24. Scale marker dimensions. (MIL-HDBK 759.)

Electronic displays come with computers in offices, in cockpits of airplanes and automobiles, in cameras, and, increasingly, with many new appliances. They must meet basic human vision requirements discussed in Chapter 4 although some users are willing to tolerate imperfections of developing technology. Among the most critical aspects are the following:

- Geometry: size, viewing angle, and distance.
- Spatial qualities: imaging capacity, raster modulation, fill factor, pixel faults, linearity, and distortion.
- Temporal qualities: image formation time, jitter and flicker.
- Photometry and colorimetry: illuminance and display luminance, contrast, color, color differences, size of colored objects, image polarity.

Innovations in display techniques develop rapidly, together with new applications under often new circumstances. Therefore, printed statements that are current and pertinent when they appear become obsolete quickly. One must closely follow the latest information appearing in technical publications in order to stay up to date in this field. With this note of caution, the information contained in AINS/HFES 100-2000 is useful for assessing, selecting, and designing visual displays in office and similar computer workstations.

Monochrome displays have only one color, preferably near the middle of the color spectrum. (See Chapter 4.) Black on a light background is good, as is white ("achromatic"), green, yellow, amber, or orange on a dark (black) background. Blue and red make the screen difficult to read. While measured performance with each of the recommended colors appears similar, personal preferences exist that make it advisable to provide a set with switchable colors (Cushman and Rosenberg 1991). Note that a screen with a dark background is more likely than one with a light background to generate glare in terms of reflecting external light sources.

Chromatic displays have several colors. The colors must contrast sharply with the background and should be easily discriminated. It is best to display not more than four colors simultaneously; more may be used if necessary, if the user is experienced, and if the stimuli subtend at least 45′ of visual angle (Cushman and Rosenberg 1991; Kinney and Huey 1990).

The following phenomena have relevance to displaying visual information:

- *Albney effect.* Desaturating a colored light (by adding white light) may introduce a shift in hue.
- *Assimilation.* A background color may appear to be blended with the color of an overlying structure (as with, eg, alphanumeric characters). The effect is the opposite of color contrast.
- *Bezold–Brucke effect.* Changing the luminance of a colored light is usually accompanied by a change in perceived hue.
- *Chromostereopis.* Highly saturated reds and blues may appear to be located in different depth planes (ie, in front of or behind the display plane). The phenomenon may induce visual fatigue and a feeling of nausea or dizziness.
- *Color adaptation.* Prolonged viewing of a given color reduces an observer's subsequent sensitivity to that color. As a consequence, the hues of other stimuli may appear to be shifted.
- *Color "blindness."* About 10 percent of all Western males, but only about 1 percent of females, have hereditary color deficiencies, mostly such that they see colors (especially reds and greens) differently or less vividly.
- *Color contrast.* The hue of an object (eg, a displayed symbol) is shifted toward the complementary color of the surrounding background. The effect is the opposite of assimilation.
- *Desaturation.* Desaturating highly saturated colors may reduce the effects of adaptation, assimilation, chromostereopsis, and color contrast but increase the Albney effect.
- *Liebmann effect.* Removing luminance contrast from a color image may produce subjectively fuzzy edges.

- *Receptor distribution.* The distribution of cones (color-sensitive receptors) on the retina is uneven. Most sensitivity to red and green (particularly under high illumination) is near the fovea, at the periphery is more sensitivity to blue and yellow occurs. (See Chapter 4.)

Although many variables affect (either singly or combined, through interacting with each other) the usability of complex color displays, the following guidelines apply to the use of color in displays (adapted from Cushman and Rosenberg 1991):

- Limit the number of colors in a display to four if users are inexperienced or if use of the display is infrequent.
- No more than seven colors should ever be used.
- The particular colors chosen should be widely separated from one another in wavelength, in order to maximize discriminability. Colors that differ from one another only with respect to the amount of one primary color (eg, different shades of orange) should not be used.

Suggested combinations are:

1. green, yellow, orange, red, and white;
2. blue, cyan, green, yellow, and white; and
3. cyan, green, yellow, orange, and white.

Avoid

1. reds with blues;
2. reds with cyans; and
3. magentas with blues.

In general, avoid displaying several highly saturated, spectrally extreme colors at the same time. Some other guidelines are as follows:

- The color of alphanumeric characters must contrast sharply with that of the background.
- Blue (preferably desaturated) is a good color for backgrounds and large shapes. Blue should not be used for text, thin lines, or small shapes.
- Red and green should not be used for small symbols and small shapes in peripheral areas of large displays.
- Using opponent colors (eg, red and green or yellow and blue) adjacent to one another or in an object–background relationship is sometimes beneficial and sometimes detrimental. No general guidelines can be given.

- Use shape together with color as a redundant cue (eg, make all yellow symbols triangles, all green symbols circles, and all red symbols squares). Such redundant coding makes the display much more acceptable to users who have color deficiencies.
- With increasing numbers of colors, increase the sizes of the objects coded with those colors.

LOCATION AND ARRANGEMENT OF DISPLAYS

The following guidelines apply to the location and arrangement of displays:

- Orient displays within the normal viewing area of the operator, with their surfaces perpendicular to the line of sight. The more critical the display, the more centered it should be within the operator's central cone of sight (Chapter 9 discusses locating the monitor in computer offices.)
- Avoid glare. (See Chapters 4 and 9.)
- Arrange displays so that the operator can locate and identify them easily without unnecessary searching.
- Group displays functionally or sequentially so that the operator can use them easily.
- Make sure that all displays are properly illuminated or luminant, coded, and labeled according to their function.

A group of instruments with pointers can be arranged so that all the pointers are aligned under normal conditions. Then, if one of the pointers deviates from the normal direction, its displacement from the aligned configuration is particularly obvious. Figure 10—25 gives examples of such arrangements.

CONTROL–DISPLAY ASSIGNMENTS

When the results of control settings are shown on a display, the controls and the display should be located close to each other to reflect their relations. This allows convenient, fast setting of the correct control. The assignment is clear when the control is directly below or to the right of the display. Other expected spatial relationships exist, but the stereotypes are often not strong and may depend on the user's background and culture. Figure 10–26 presents examples of suitable arrangements.

A frequent course assignment for students of ergonomics is to design a stove top in which the controls for four burners are located so that their relations are unambiguous. (See, for example, Hsu and Peng 1993). No solution has ever been found that combines a satisfactory assignment, safety of operation, and easy cleaning of the stove top.

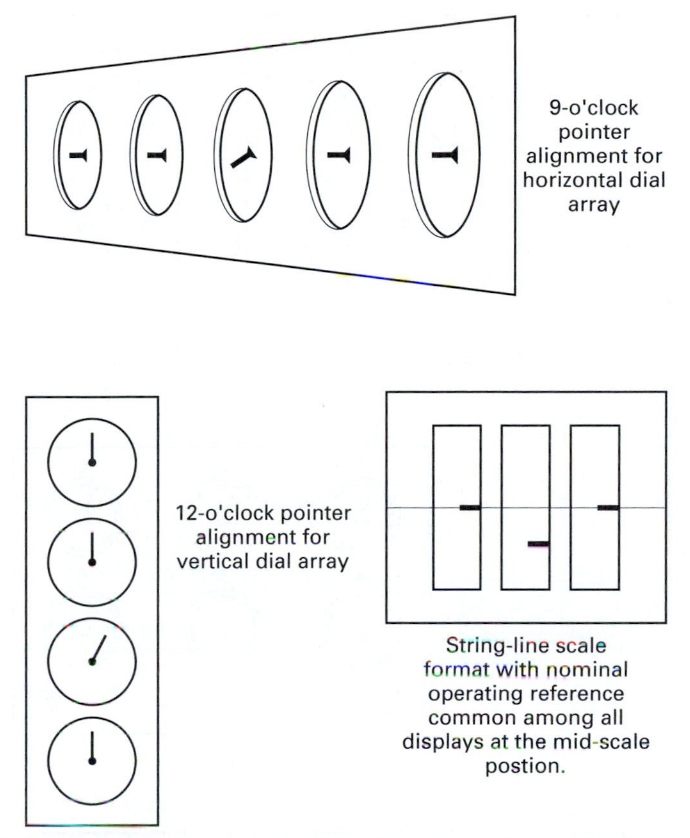

Figure 10–25. Aligned pointers for rapid reading. (Modified from MIL-HDBK 759.)

Relationships among expected movements of controls and displays are influenced by the types of controls and displays. When both are congruous (eg, when both are linear or rotary), the stereotype is that they move in corresponding directions (eg, both up or both clockwise). When the movements are incongruous, their preferred relationship can be taken from Figure 10–27. The following rules apply generally:

1. Gear-slide ("Warrick's") rule: A display (pointer) is expected to move in the same direction as does the side of the control close to ("geared with") the display.
2. Clockwise for increase: Turning the control clockwise should cause an increase in the displayed value.

Control–Display Ratio

The control–display (C/D) ratio describes how much a control must be moved to adjust a display. The ratio is like a gear ratio: If much control movement produces only a small display motion, one speaks of a high C/D ratio and of the control as having low sensitivity. The converse ratio, D/C, is often called the gain, an expression that resembles a transfer function.

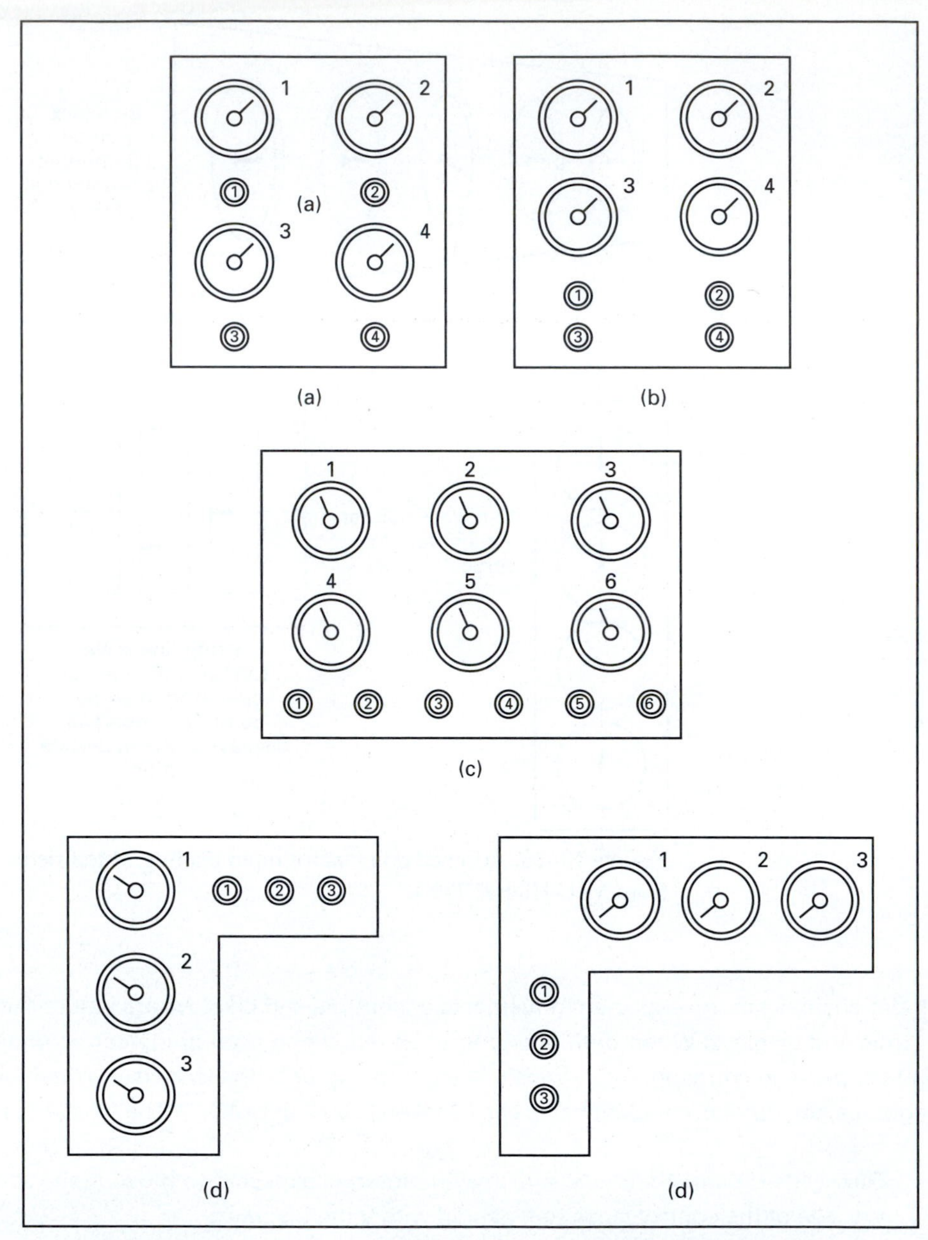

Figure 10–26. Difficult control–display relationships. (Adapted from MIL-HDBK 759.)

Usually, two distinct movements are involved in making a setting: first a fast motion to an approximate location and then a fine adjustment to the exact setting. The optimal C/D ratio is that which minimizes the sum of the two movements. For continuous rotary controls, the C/D ratio is usually 0.08 to 0.3; for joysticks, it is 2.5 to 4. However, the most suitable ratios depend much on the given circumstance and must be determined for each application (Boff and Lincoln 1988; Arnaut and Greenstein 1990).

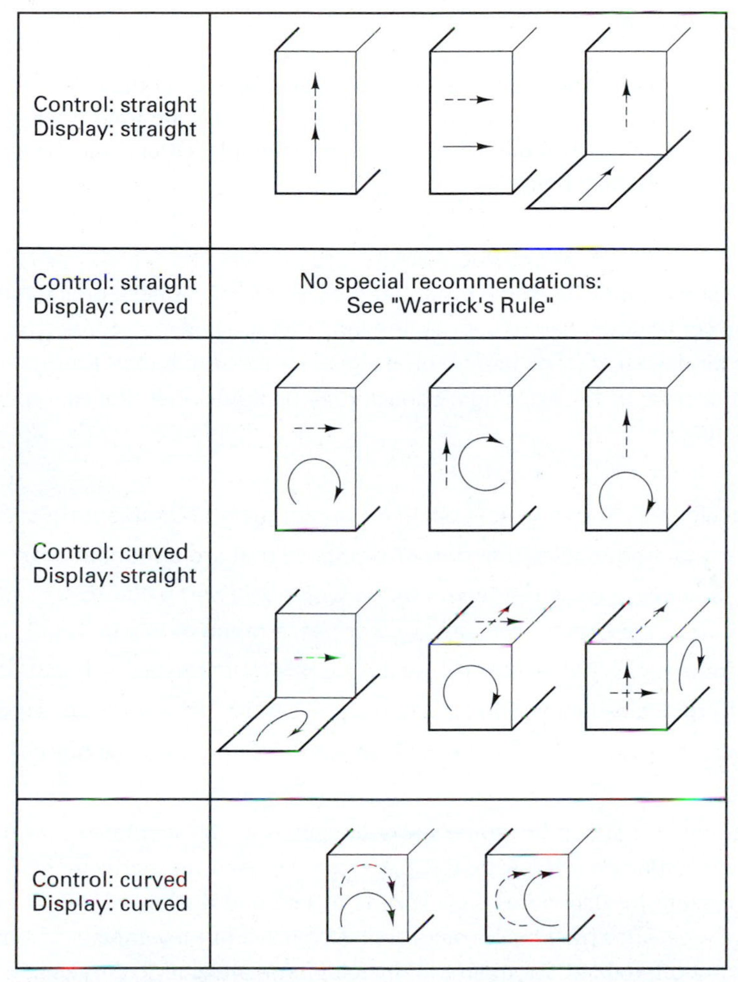

Figure 10–27. Compatible control–display directions. (Modified from Loveless 1962.)

DISCON: DISPLAY AND CONTROL IN ONE

New technology blurs the traditional differentiation between control and display, as, for instance, when one can simply drag or change an object shown on a computer monitor while it is on the screen. Another, much more complex, example is the U.S. Air Force's Visually Coupled Targeting and Acquisition System (VCATS), which combines the information about a target that the pilot sees in a helmet-mounted display with the aiming of a weapon (eg, a missile), all done automatically by an onboard computer (Rastikis 1998). A possible scenario for future DisCons is that sensors track the line of sight from the eyes to an object and the person logs that information into a computer by saying "get" and changes the object's setting by word or some other action of the body.

AUDITORY DISPLAYS

As mentioned earlier, auditory signals are better than visual displays when the message must attract attention. Therefore, auditory displays are used predominantly as warning devices, especially when the message is short or simple. Often, auditory displays are utilized together with a flashing light.

Auditory signals may be single tones, sounds (mixture of tones), or spoken messages. Tones and sounds may be continuous, periodic, or at uneven timings. They may come from horns, bells, sirens, whistles, buzzers, or loudspeakers. The use of tonal signals is recommended for qualitative information, such as indications of status, or for warnings; speech may be appropriate for all types of messages. Following are some guidelines:

1. Tonal signals should be at least 10 dB louder than the ambient noise. (See Chapter 4.)
2. Use signal frequencies that differ from those that are dominant in the background noise.
3. The signal frequency range should be within 200 and 5,000 Hz; the best range is between 500 and 1,500 Hz. If the signal undulates or warbles, a range of 500 to 1,000 Hz is advantageous.
4. Use frequencies below 1000 Hz when signals must travel long distances (eg, more than 300 m).
5. Use frequencies below 500 Hz when signals must "bend around" or pass through sound barriers.
6. Buzzers may have frequencies as low as 150 Hz and horns as high as 4,000 Hz.

The tonal signal can be made more conspicuous by increasing its intensity, by interrupting it repeatedly, or by changing its frequency. For example, one may increase the frequency from 700 to 1,700 Hz in 0.85 s, be silent for 0.15 s (for a cycle time of 1 s), and then start over.

Verbal messages may be prerecorded or digitized or may consist of synthesized speech. The first two techniques are often used, for example, by telephone answering services or in "talking products" and are characterized by good intelligibility and natural sound. Synthesized speech usually consists of compositions of phonemes; this does not sound natural, but is effective in converting written text to speech and may sound startling, thus attracting attention to the message.

LABELS AND WARNINGS

Labels

Ideally, no label should be required on any piece of equipment or control to explain its use. Often, however, it is necessary to use labels so that one may locate, identify, read, or manipulate controls, displays, or other equipment. Labeling must be done so that the information is provided accurately and rapidly. For this, the following guidelines apply:

1. *Orientation.* A label and the information printed on it shall be oriented horizontally so that it can be read quickly and easily. (Note that this guideline applies only if the operator is used to reading horizontally, as in Western countries.)
2. *Location.* A label shall be placed on or near the item that it identifies.
3. *Standardization.* Labels shall be placed consistently throughout the equipment and system. A label shall primarily describe the function of the labeled item. ("What does it do?")
4. *Abbreviations.* Common abbreviations may be used. If a new abbreviation is necessary, its meaning shall be obvious to the reader. The same abbreviation shall be used for all tenses and for the singular and plural forms of a word. Capital letters shall be used, periods normally omitted.
5. *Brevity.* The inscription on the label shall be as concise as possible without distorting the intended meaning or information. The text shall be unambiguous, redundancy minimized.
6. *Familiarity.* If possible, words that are familiar to the operator shall be chosen.
7. *Visibility and Legibility.* The operator shall be able to be read the label easily and accurately at the anticipated actual reading distances, at the anticipated worst illumination level, and within the anticipated vibration and motion in the environment. Some considerations are the important contrast between the lettering and its background; the height, width, stroke width, spacing, and style of the letters; and the specular reflection of the background, cover, or other components.
8. *Font and Size.* Typography (the style, font, arrangement, and appearance of the written word) determines the legibility of a message. The font (typeface) should be simple, bold, and vertical; good fonts in these respects are Futura, Helvetica, Namel, Tempo, and Vega.

Note that most electronically generated fonts (such as by an LED, an LCD, or a matrix) are inferior to printed fonts; thus, special attention must be paid to make them as legible as possible.

The recommended height of characters depends on the viewing distance:

For a viewing distance of 35 cm, the suggested height is 22 mm.

For a viewing distance 70 cm, the suggested height is 50 mm.

For a viewing distance 1 m, the suggested height is 70 mm.

For a viewing distance 1.5 m, the suggested height is at least 1 cm.

The *ratio of strokewidth to character height* should be between 1:8 and to 1:6 for black letters on a white background and between 1:10 and 1:8 for white letters on a black background.

The *ratio of character width to character height* should be about 3:5.

The *space between letters* should be at least one stroke width.

The *space between words* should be at least one character width.

For continuous text, one should mix upper- and lowercase letters. (For labels, use uppercase letters only).

WARNINGS

Ideally, all devices should be safe to use. This is, in safety engineering, the first priority. In reality, complete safety cannot often be achieved through design alone. Then the next priority is to remove the dangerous interface; for example, eliminate all railroad crossings. If that is impossible, then step down to the next level, guarding. If a hazard still remains, one must warn users of the dangers associated with the use of the product and provide instructions for its safe use to prevent injury or damage (Christensen 1993; Hammer and Price 2000; Kitzes 1996; Lehto and Salvendy 1995).

Reasons to Warn

One must warn of a known or knowable potential danger of injury from the normal use of a product:

1. The more serious the potential injury, the greater is the duty to warn.
2. The less obvious the danger, the greater is the duty to warn.
3. The more insidious the onset of injury, the greater is the duty to warn.
4. The larger the number of people at risk. the greater is the duty to warn.

What to Warn About

People must be warned about various concerns, such as:

1. The product must be used properly.
2. Improper or excessive use may cause serious injury.
3. Certain people are at particularly elevated risk and must take extra care.
4. Seek medical attention immediately if any symptoms appear.

Whom to Warn

A product manufacturer has the duty to warn:

1. Potential users of the product, so that they can learn to protect themselves.
2. Potential customers (purchasers of the product) whose personnel might be at risk. In this way, work practices and the workstation (where the product is used) may be modified.
3. Sales, marketing, and service persons who come in contact with users and customers.
4. The general public.

How to Warn

People may be warned by:

1. Warnings on the product itself. (Active warnings are better than passive ones; see shortly).
2. Instructions in a product manual or directions for using the product.
3. Promotional literature and advertising.
4. Sale and service staff's instructions to customers and users.

Active Versus Passive Warnings

Active warnings are preferable to passive warnings. Active warnings usually consist of a sensor that notices an inappropriate use and of a device that alerts the human to an impending danger. Yet, in most cases, passive warnings are used, usually consisting of a label attached to the product and of instructions for safe use of the product in the user manual. Such passive warnings rely completely on the human to recognize an existing or potential dangerous situation, to remember the warning, and to behave prudently.

Factors that influence the effectiveness of product warning information have been compiled by Cushman and Rosenberg (1991) and are listed in Table 10–24.

TABLE 10–24. Factors That Influence the Effectiveness of Warning on Products

	Effectiveness of Warning	
Situation	*Low*[1]	*High*[2]
User is familiar with product	X	
User has never used product before		X
High accident rate associated with product		X
Probability of an accident is low	X	
Consequences of an accident are likely to be severe		X
User is in a hurry	X	
User is poorly motivated	X	
User is fatigued or intoxicated	X	
User has previously been injured by product		X
User knows good safety practices		X
Warning label is adjacent to the hazard		X
Warning label is very legible and easy to understand		X
Active warnings alert user only when some action is necessary		X
Active warnings frequently give false alarms	X	
Product is covered with warning labels that seem inappropriate	X	
Warning contains only essential information		X
Source of information on warning is credible		X

Source: Reprinted with permission from Cushman and Rosenberg 1991.
[1]Low probability of behavioral change.
[2]High probability of behavioral change.

Design of Warnings

Labels and signs for passive warnings must be carefully designed by following the most recent government laws and regulations, recognized national and international standards, and the best applicable human-engineering information. Most warnings are in writing, but auditory warnings (see earlier and Chapter 4) or tactile means (Chapter 4) may be employed and are often advantageous (Bogner et al. 1998; Edworthy and Stanton 1995; Gilliland and Schlegel 1994; Momtahan et al. 1993).

Visual warnings

Warning labels and placards may contain text, graphics, and pictures, often redundantly. Graphics—particularly pictures, pictograms and icons—can be used by persons with different cultural and language backgrounds if the depictions are selected carefully. (For recent overviews, see, eg, Edworthy and Adams 1996; Parsons et al. 1999.) One must remember, however, that designing of a "safe" product is far preferable to applying warnings to an inferior product. Furthermore, users of different ages and experience and users of different national and educational backgrounds may have rather different perceptions of dangers and warnings.

Symbols and Icons

Symbols or icons are simplified drawings of objects or abstract sign, meant to identify an object, warn of a hazard, or indicate an action. They are common in public spaces, automobiles, computer displays, and maintenance manuals. The Society of Automotive Engineers (SAE) and the International Standardization Organization (ISO), for example, have developed extensive sets of symbols and established guidelines for developing new ones. Some of the symbols for use in vehicles, construction machinery, cranes, and airport handling equipment are reproduced in Figure 10–28. Note that both abstract symbols (eg, a line for *on*) and simplified pictorials may require learning or the viewer's familiarity with the object. (Does an hourglass denote hours elapsed?) If one is developing new symbols, the cultural and educational background of the viewer must be carefully considered: Many symbols have ancient roots, and many invoke unexpected and unwanted reactions (Fruitiger 1989).

The following guidelines have been adapted from those of ISO Technical Committee 145 (dated October 29, 1987):

- Symbols should be simple, clear, distinct, and logical to enhance their recognition and reproduction.
- Symbols should incorporate basic elements that can be used alone or combined as necessary into a logical symbolic language which, if not immediately obvious, is at least readily learned.
- Graphical clarity should prevail in disputes over logical consistency, because no symbol is recognizable, no matter how logical, if it cannot be distinguished from other symbols.
- A minimum of detail should be included; only details that enhance the recognition of the symbol or what it denotes should be allowed, even if other details are accurate renditions of the machine or equipment.

SUMMARY

Although psychological rules on how humans best perceive displayed information and operate controls are generally absent, many well-proven design recommendations apply to traditional controls and dial-type displays. With the widespread use of the computer, new input devices have emerged, such as the mouse, trackball, and touch screen. The classic typing keyboard has mushroomed; a redesign is urgently needed. Probably other keyboard designs and even new input methods such as voice recognition should replace the monstrous arrangement of more than 100 keys that apparently overloads many users.

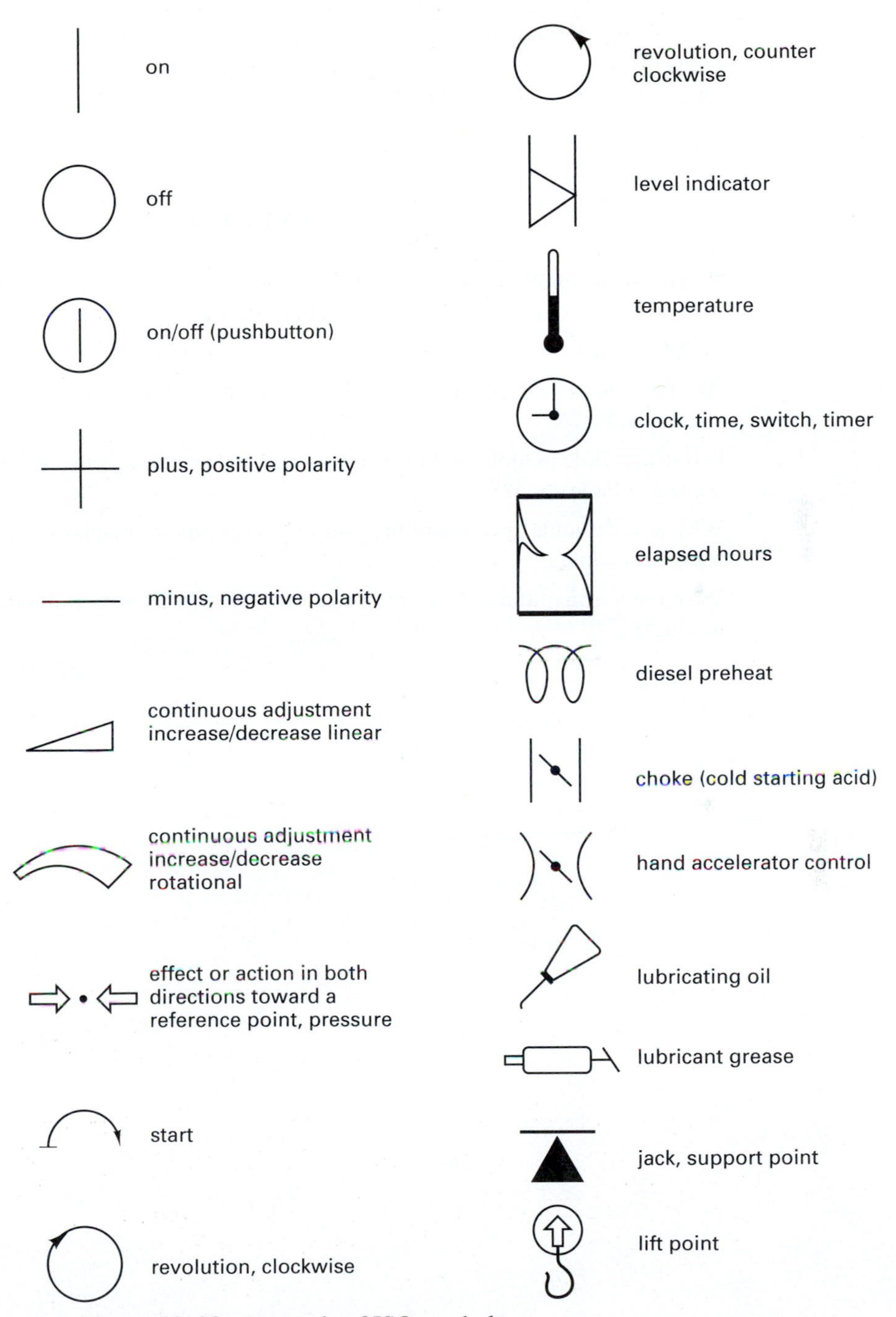

Figure 10–28. A sample of ISO symbols.

Display technology also has changed. Electronic displays, many of which often caused eye aches and headaches, can now be made in a quality similar to hard copy. The use of color, if properly done, makes the transmission of visual information interesting and easy.

Warning signals, instructions on how to use equipment, and warnings to avoid their misuse and of the danger they present are important, both ergonomically and juristically.

Computer-generated messages and displays were rather deficient when they first appeared, but better human-engineering and advancing technology provide novel solutions, including the interpretation of displays and controls.

CHALLENGES

What are the implications of future developments in control and display technology, if so few general rules exist that explain the current usage of these devices?

Should one try to generate the same "population stereotypes" throughout the world?

Are there better means of controlling the direction of an automobile than by the conventional steering wheel?

Is it worthwhile to look for different keyboard and key arrangements if voice activation might become available soon?

Why is it difficult to compare the usability of computer displays with that of information displayed on paper?

Why must a manufacturer warn about risks associated with the wrong or excessive use of a product?

Chapter 11

"Handling" Loads

OVERVIEW

We all "handle" loads daily. We lift, hold, carry, push, pull, and lower objects while moving, packing, and storing them. The material may be soft or solid, bulky or small, or smooth or with corners and edges; it may come as bags, boxes, or containers, with or without handles. We may handle objects occasionally or repeatedly, during leisure activities or as part of our job. On the job, the ergonomic design of materials, containers, and workstations can help to avoid overexertions and injuries, as should instructions and training on how to "lift properly." Still, For some jobs, it may be necessary to try do select only those persons who are physically capable of strenuous work.

INTROUCTION: STRAINS ASSOCIATED WITH HANDLING LOADS

Manipulating even lightweight and small objects can strain us because we have to stretch, move, bend, or straighten out body parts. Heavy loads pose an additional strain on the body, owing to their mass or bulk or their lack of handles.

Handling materials is among the most frequent and the most severe causes of injury all over the world, with strains in the low back area predominating (Hashemi and Dempsey 1997; Marras et al. 1995). The direct and indirect cost are enormous, and the human suffering associated with, for example, low back injuries is immeasurable.

Exerting force and energy in lifting an object with one or both hands strains the hands, arms, shoulders, trunk, and, since one usually stands, also the legs. The same parts of the musculoskeletal system are under stress in lowering an object and in pushing or pulling it, but the directions and magnitudes of the external and internal force and torque vectors are different.

The primary area of physiological and biomechanical concern has been the low back, particularly the discs of the lumbar spine. Thus, the operative words in the literature are "low back pain related to lifting." Yet, when one considers all the musculoskeletal structures within the trunk (see Chapter 2), a variety of elements may be strained, either singly or combined, including, of course, the spinal vertebrae (primarily their disc or facet joints), but also connective tissue such as ligaments and cartilage, as well as muscles with their tendons. All of these elements may be beset with insults, sprains, or trauma. Chiefly, the compression strain of discs and vertebrae has been studied, but tension is the primary loading of elastic elements such as muscles and connective tissues. Tension strains can be in form of linear elongation, bending movements, or twisting torque. All structures are subject to shear.

The loading of the body may come from activities done on external objects, such as lifting, lowering, pushing, pulling, carrying them; or holding; thus, the strains may be dynamic or static, of fast or slow onset, and of short or long duration. They may be single or several events; if the same or similar strains reappear, the repetition may be at regular or irregular intervals, and the strains may be of similar or dissimilar magnitudes.

Body structures are also strained just by moving one's own body (according to Newton's second law, ie, force = mass × acceleration) or simply by maintaining a posture through muscle tension without any external load. Longitudinal contraction of the trunk muscles compresses the spinal column, which is (possibly with some help from intraabdominal pressure; Chapter 2) the only solid load-bearing structure of the trunk.

Such musculoskeletal strain may be experienced in sports, most obviously in weight lifting. Other sources of stresses are leisure and occupational activities, the latter often labeled "manual material(s) handling" (MMH).

☛☛☛ *Since the Latin word "manus" means hand, the term "manual material handling" is redundant.* ☚☚☚

In the course of material handling, one may exert energy intentionally toward an outside object, or the body may be subjected to unexpected energies, such as in catching a falling object or by accidents (e.g., slipping and falling). Aside from a focus on sports, the literature has dwelt primarily on three types of material handling: lifting—that is, moving an object "by hand" from a lower position to a higher one; the opposite action, lowering; and pushing and pulling, carrying, and holding. This chapter summarizes the present knowledge in these areas. Based on this knowledge, ergonomic intervention to prevent overexertions of material handlers are examined.

ASSESSING BODY CAPABILITIES RELATED TO MATERIAL HANDLING

Handling material requires exerting energy or force to lift, lower, push, pull, carry, or hold objects—that is, to move objects or keep them from moving. The energy needed to do these tasks must be generated within the body and exerted in terms of force or torque over time to the outside object. In the past, research was mostly directed at cardiocirculatory loading and energy generation within the body over an extended period of work (see Chapter 2) or was concerned with forces applied to the object being handled, mostly during one episode of lifting or just a few repeated episodes. (See Chapter 1.)

A person's ability to lift (lower, push, pull, or carry) material over hours in activities that involve the whole body (or large segments thereof) is likely to be limited by his or her metabolic and circulatory capabilities. Given the inefficiency of the body in terms of generating energy, moving the body for lifting (lowering, etc.) objects taxes its abilities, usually to such an extent that only a fairly small external load may be moved, because so much energy is spent on moving body parts. This fact became obvious in the development of the 1981 NIOSH *Lifting Guide:* MMH activities that are performed several times a minute, over hours, strain mostly metabolic and circulatory functions.

However, if a very strong force, such as in lifting a heavy object, must be exerted just once or occasionally, then, indeed, the ability to generate that force once is the limiting factor, as it strains musculoskeletal components of the body. This experience was apparently the reason that, in the past, guidelines were used which relied on stating maximal weights (ILO 1988; Straker et al. 1996). Of course, simply setting a weight (mass) limit for objects to be handled is neither reasonable nor prudent, because one might exert a large force even on a fairly small mass if much acceleration is applied, as per Newton's second law.

Most material-handling work is somewhere between the two extremes of doing single maximal efforts and handling light loads over long periods. Thus, assessment of human abilities to move material has been done primarily by biomechanical and psychophysical methods, with some use of metabolic and circulatory methods, as appropriate (Ayoub and Dempsey 1999; Capodaglio et al. 1995, 1997; Craig et al. 1998; McGill et al. 1995, 1996; Mirka and Baker 1996; NIOSH, 1981, 1991; Straker et al. 1997). Psychophysics bridges the range between assessments of metabolic capabilities and of one-time muscle strength.

Psychophysical Measurements

The psychophysical approach relies on the assumption that the human can sense and integrate the perception of strain on all body functions and capabilities, be they metabolic, circulatory, muscular, or connective tissue related, or on the bony structures. (See Chapter 7.) Judging the perceived strain contains an overall assessment of acceptability, suitability, and willingness to perform a stressful task. And, indeed, while we may have great difficulty in singling out the functions of the body upon which to base our judgments, we make judgments throughout our life. Psychophysical methods have become an important part of research regarding human material-handling capabilities (Ayoub and Dempsey 1999; Ciriello et al. 1993; Genaidy et al. 1990; Karwowski, 1991; Kuorinka and Forcier 1995; Nordin et al. 1997; Snook 1978; Snook and Cierello 1991; Wiktorin et al. 1996).

Biomechanical Measurements

Spinal Compression

Biomechanical methods for evaluating strains and capabilities of the body have become widely used in the last few decades. The initial research interest concerned mainly the spinal column, particularly with respect to the responses of the vertebrae and the vertebral discs to compression (Kazarian and Graves 1977; Schultz and Andersson 1981). The calculation of compression strain is complicated by the fact that the human spine is not a straight column, but is anterior–posteriorly curved even when "erect," and that this curvature changes with different trunk postures. Diurnal variations and daily activities modify the mechanical parameters of the spinal unit. The strains (measured, modeled, and calculated) in the spinal column were initially treated as static (Chaffin 1981; Chaffin and Andersson 1984; Jaeger and Luttmann 1986; NIOSH 1981). Dynamical considerations entered into the models only slowly, because of the complexity of calculating the strains involved.

Involvement of the Torso Muscles

Early models were descriptive, mathematically expressing observed behavior, and did not include much analytical understanding of the sources and existing amounts of body strain; but of late, analysis and modeling of the strains in the torso have made great strides, with particular interest directed at the activation of the various muscles within the trunk in movement, such as lifting (Ayoub 1982; Asfour et al. 1988; Bush-Joseph et al. 1988; Gagnon and Smyth 1992; Marras et al. 1995; Marras and Rangarajulu 1987; Nussbaum and Chaffin 1996; Sommerich and Marras 1992).

The muscles that develop force vectors that run between the inferior and the superior parts of the trunk (in essence, they try to "pull the shoulders onto the hips") are the right and left latissimus dorsi, right and left erector spinae, right and left internal and external obliques, and right and left rectus abdominus (Schultz and Andersson 1981). Their locations are sketched in Figure 11–1.

Marras and Reilly (1988) inserted wire electrodes into each of these muscle groups and measured their activities via EMG during controlled conditions of trunk motion. In this manner, they were able to identify both which muscles were involved and the sequence and intensity with which those muscles were called to action. As might be expected, muscle recruitment is quite different in quality and quantity when the muscles hold the trunk in a static position, as opposed to when they move the torso.

Depending on the lifting task—particularly the height, distance from the body, speed, and sideways displacement (asymmetry) of the object lifted—different sequences and magnitudes of muscle activation were recorded. Figure 11–2 shows the events that occurred in an isovelocity trunk motion from a forward bent posture to upright—a 67° motion at half the subjectively possible speed. In this case, the latissimus dorsi, erector spinae, and external oblique muscles were activated first, together with a buildup of intraabdominal pressure, which was the last effect to subside after the motion.

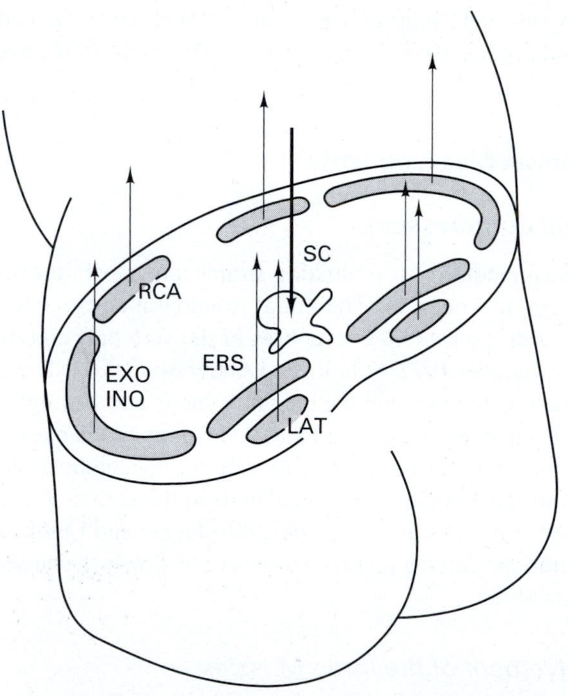

Figure 11–1. Sketch indicating the pull forces generated by the trunk muscles (RCA, rectus abdominus; EXO, INO, external and internal obliques; ERS, erector spinae; LAT, latissimus dorsi) resulting in spinal compression (SC). (Modified from Schulz and Andersson 1981.)

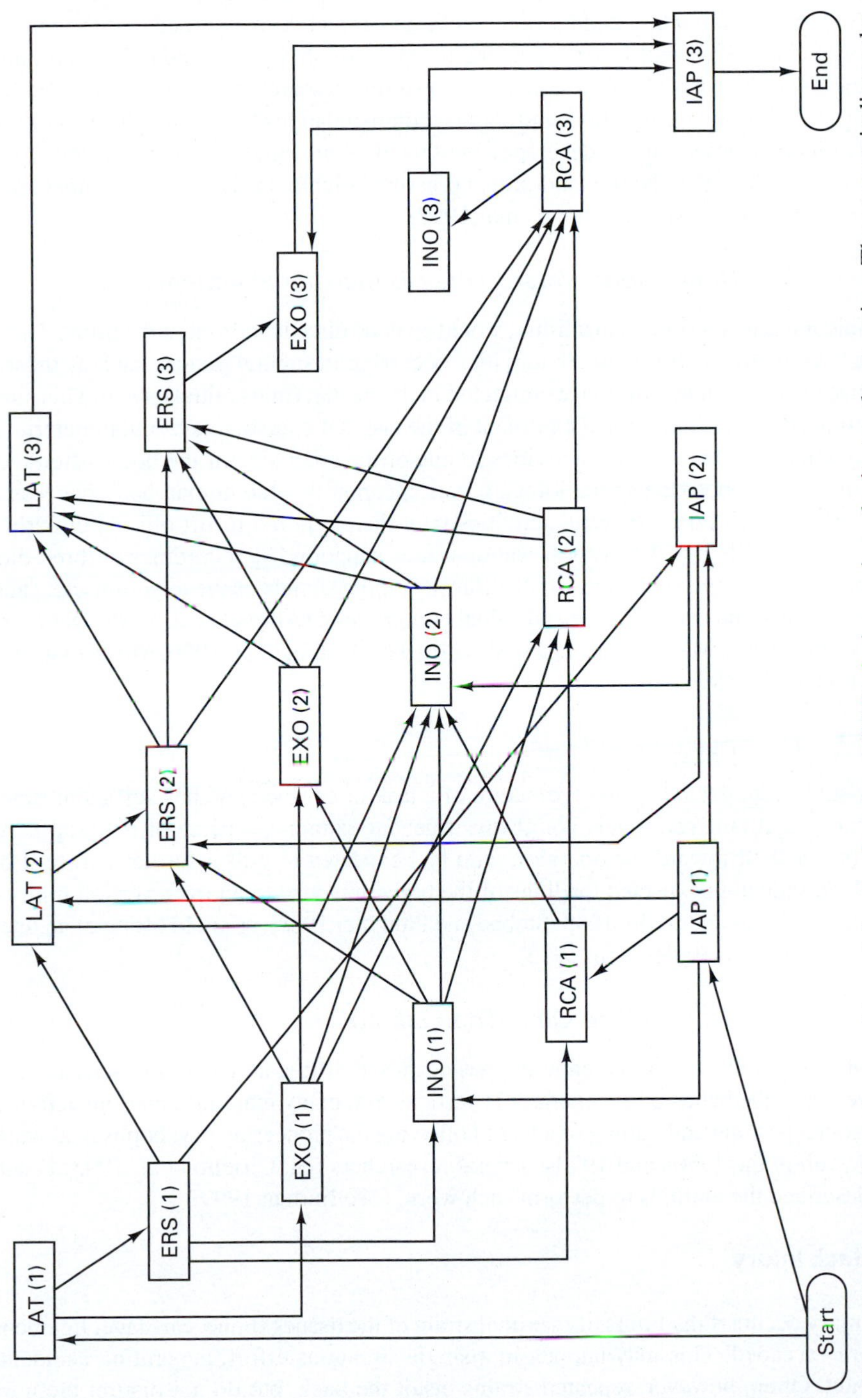

Figure 11–2. Muscle tensions and intraabdominal pressure (IAP) during an isovelocity trunk extension. The arrows indicate the start (1), peak (2), and end (3) of activities. (Courtesy of Prof. W. S. Marras, Ohio State University.)

Intraabdominal Pressure

The just mentioned studies also showed that the intraabdominal pressure (IAP) changes very much with the trunk loading and motion conditions. It had been assumed that the pressure column between the inferior and superior diaphragms of the trunk cavity would relieve the spinal column from having to transmit all of the compression strain. Researchers estimated that the IAP would carry up to 15 percent of the strain, and some recommendations for suitable lifting conditions had largely relied on observing the development of the pressure buildup (Freivalds 1989). Yet, new research findings show that the intraabdominal pressure column does *not* reduce spinal compression loading (Marras and Mirka 1996; Kumar 1997).

Two-Dimensional Versus Three-Dimensional Activities

Initial research assumed that lifting would be done directly in front of the trunk. This has been called a "symmetrical" lift, with all activities occurring in sagittal planes, such as those of the left and right hand, with mirror images reflected in the medial (midsagittal) plane. This simplistic concept allowed a two-dimensional treatment of the task. Of course, much actual material handling is not symmetrical, but involves activities to one or the other side of the body, often requiring a twisting motion about the spinal joints. It was apparent that the human body has less of an ability to perform such nonsymmetric activities than than required to lift objects directly in front of the body (NIOSH 1981). This observation is now supported by recordings of three-dimensional lumbar motions, by evaluations of electromyographic signals of the torso muscles, and by models of distributed muscle activities and spinal compression (Allread et al. 1996; Davis and Mirka 1997; Hughes 1995; Marras et al. 1993; McGill 1999; McGill et al. 1996; Mirka et al. 1997; Nussbaum and Chaffin 1996).

Unexpected Loadings

Usually, the person knows the nature of a task in advance and has sufficient time to prepare for carrying it out. Yet, this is not always true; for example, a piece of material may suddenly fall from a shelf toward the operator, who had expected to pull it further before it began to move. Such sudden unexpected loadings of the body may lead to an overexertion injury (Gagnon et al. 1995; Marras and Reilly 1988; Stobbe and Plummer 1988), as could lifting of an unexpectedly light or heavy load (Butler et al. 1993).

Material Handling Other Than Lifting

Most research in material handling has dealt with lifting and lowering activities, where the force vector in the hands of the operator is vertical. Yet, many material-handling activities involve horizontal pushing and pulling of a load. Following the pioneering psychophysical work of Snook and Ayoub in the 1960s and 1970s, several researchers (eg, Ciriello et al. 1993; Daams, 1993) have described the abilities to perform such work. (See Imrhan 1999.)

Back Injury

Injury occurs if the limits of maximal strain of the tissues (bone, cartilage, ligaments, or muscles) are exceeded. This may happen in a single strenuous effort, generating accidental trauma and pain. Often, however, repeated strains insult the back, but do not disrupt the normal pattern of work until, finally, the onset of pain signals that the accumulated loadings have added up to a cumulative overloading.

Pain

Low back pain, often abbreviated LBP, is the most common general indicator of overexertion of the body; it is frequently a sign of improper work design or practice.

Neither the facets of the apophyseal joints nor the intervertebral discs seem to have pain-sensitive nerves. Thus, the three load-bearing elements (the two facet joints and one disc) of each spinal unit can be injured without any sensation of pain (Jensen et al. 1994). If the discs or facet joints are repeatedly injured, degenerative changes may set in. Radiography has shown disc or joint degeneration to be as frequent in persons who have low back pain as it is in those who do not suffer from it. To complicate the case even further, mechanical derangements of the intervertebral joint, such as a decrease in disc height or a change in the positions of the components of the facet joints, may produce clinical symptoms that appear days after the acute phase of the injury is over (Andersson 1991).

The actual reasons for back pain are quite often not clear. Different individual factors have been associated with the likelihood of an incidence of a back injury and with its severity, particularly in its lumbar section. Among these factors are degenerative disc disease, congenital anomalies, spondylolisthesis, differences in leg length, and previous injuries. Whether and to what extent these conditions contribute to low back pain, and how they can be diagnosed and treated, has often been a topic of heated discussions, particularly between physicians and chiropractors. While it was long believed that a large portion of all chronic low back pain is discogenic, muscular or ligament tissue problems near the spine may be just as frequent and important (Andersson 1991, 1999; Snook 1988, 1991).

A major difficulty in recognizing and analyzing the cause of a back injury is that it may happen without generating any pain. Often, low back pain cannot be traced to one specific incident of overexertion.

Return to Work

Moderate activity increases the nutrition of the spinal discs, mostly by passive diffusion, aided by the pumping action of compressing and releasing the disc tissue, and by bending and unbending. Rest and inactivity inhibit the healing of strained disc tissue. After one rises from bed in the morning, bending strains the fluid-filled and flexible disc tissue about three times more than later during the day (Nachemson 1989).

About every second person returns to work within one week after an incidence of low back pain. Only a few patients with unrelenting low back pain eventually need surgery. With increasing duration of the absence from work, the successful return of the worker becomes less likely; yet altogether, about 9 out of 10 LBP sufferers eventually return to work, some to medically recommended restricted work with reduced strain on the back. The likelihood of a successful return in patients above 45 years old is only about half that of persons younger than 24 years. Youthful workers have more frequent, but less severe, back disorders than do older workers. The highest incidence for females is between the ages of 24 and 34, for males between 20 and 24 years. Altogether, males and females are afflicted with low back pain about equally often; however, since fewer female workers than males are employed in the United States, more compensation cases are reported for males than for females (Andersson 1991). Chronic LBP lasting three months or longer often becomes disassociated from its physical basis and is instead associated with emotional distress (Snook 1991).

Snook (1988) listed the following obstacles to returning to work:

1. Associated with the worker:
   - Malingering
   - Psychological disability
   - Illness behavior
2. Associated with management:
   - Lack of follow-up or encouragement
   - Requirement to be 100 percent rehabilitated
   - No modification of the work
3. Associated with the union:
   - Rigid work rules
   - Referrals to "friendly" physicians
   - Referrals to lawyers
4. Associated with the practitioner:
   - Inappropriate treatment
   - Ineffective treatment
   - Prolonged disability
5. Associated with the lawyer:
   - Lump-sum settlements instead of rehabilitation.

PERSONNEL TRAINING

There are three major ergonomic approaches for safer and more efficient manual material handling: personnel training, personnel selection, and job design. The first two fit the person to the job, while the third fits the job to the person. For each, knowledge of the operator's capabilities and limitations is necessary. Thus, the research and knowledge base overlaps in all procedures.

Since the lifting of loads is associated with a large percentage of all back injuries, training in "safe lifting" procedures has been advocated and conducted for decades. Studies indicate that approximately one-half of compensable low back injuries are associated with lifting. Industrial experience has also identified lifting as a major cause of back injury. Hence, "training how to lift safely" has been targeted at persons performing manual material-handling activities in industry, as well as at groups of industries and jobs.

Training is expected to reduce both the frequency and the severity of injuries, to develop specific material-handling skills, to further the worker's awareness and self-responsibility, and to improve specific physical fitness characteristics. Various instructional styles and media have been used. Training, either generic or customized, has been done in a single session or in several sessions, at various times during employment, and at the job site, in classrooms, or in outside workshops. Participants in these training efforts have been selected at random or chosen according to their risk, previous injuries, age, etc.; either they were volunteers or they included all employees in an establishment. A large variety of experimental variables (independent, controlled, and dependent) were used, with various experimental designs and treatments.

Studies on the outcome of training and the results of intervention reported in the literature are scientifically deficient when they do not follow an experimental design that allows the outcome of the experiment to be assessed for reliability and validity. Granted that it is difficult (because of the interference with work, the time needed, and the expense) to conduct a field experiment in which one varies only the independent variable and excludes confounding variables and uncontrollable interferences, still, including a control group in the experimental design (long advised, for example, by Cook and Campbell in 1979 and by Smith in 1976) is often feasible and allows the outcome of the experimental training to be evaluated.

The large variety of past training approaches and their results has been reviewed (Kroemer 1992b; see also the 1994 edition of this book); therefore, a short account of the earlier findings, together with new results, is both sufficient and revealing.

Training in Proper Lifting Techniques

Instructions on how to lift are meant to affect a person's lifting behavior and to reduce the likelihood of an overexertion strain or injury. In the laboratory, it has been shown that proper training regimes can increase one's ability to lift. Sharp and Legg (1988) reported that after four weeks of training, initially inexperienced lifters increased their work output significantly while maintaining their energy expenditures. The improvement was attributed to increased skill levels attained through improved neuromuscular coordination and to possible increases in muscular endurance. Genaidy et al. (1990) used six weeks of training and also found that muscular endurance, muscular strength, and cardiovascular endurance were improved. Yet, no tightly controlled experiments with large numbers of industrial material handlers have been performed, and the validity of laboratory findings for industrial environments has not been established.

Many lifting instructions include the admonition "keep the back straight." However, what that means is that one should maintain the "natural curvature" of the "standing" spine—in particular, its lumbar lordosis. This imprecise use of words has led to much confusion.

From the 1940s on, the method advocated was the straight-back/bent-knees lift (squat lift), whereby workers lowered themselves to the load by bending the knees and then lifted the load by using the leg muscles. Yet, biomechanical and physiological research has shown that the leg muscles used in this lift do not always have the needed strength; also, awkward and stressful postures may be assumed if one tries to enforce the technique under unsuitable circumstances (eg, when the object is bulky). Hence, the straight-back, bent-knees action evolved into the so-called kinetic lift, in which the back is kept mostly straight while the knees are unbent; feet, chin, arm, hand, and torso positions are prescribed. Another variant was the free-style lift, which appeared to be better for some workers than the straight-back, bent-knees technique. In some situations, the stoop lift (with a strong bend at the waist and straight knees) may be superior to the squat (bent knees and a flat back) posture that is usually advocated. Sedgwick and Gormly (1998) discussed the suitability of leg and stoop lifts and believed that an intermediate "semisquat" lift was most versatile. The contradictory findings confirm Jones's 1972 statement that no single lifting method is best for all situations. Therefore, the training of proper lifting techniques is an area of confusion: What method should be taught?

Note that even the terminology is somewhat confusing, as it only describes the posture at the beginning of the lift. For a remedy, Burgess-Limerick and Abernathy (1997) proposed definitions based on body joint angles.

Unsuccessful Training

Reviews by Brown (1972, 1975) and Yu et al. (1984) detected no significant reductions in back injury that could be attributed to education over four decades. Snook et al. (1980) compared the number of back injuries in companies that conducted training programs in "safe lifting" with the number of injuries in companies that did not have such programs and found no significant difference. Several studies on nurses did not find any effects on the incidence of low back injury after receiving repeated instruction on lifting procedures; in fact, the principles taught were seldom used. Although neither the quality nor the content of the programs was investigated, the general conclusion was that training was not an effective preventive program for low back injuries.

Scholey and Hair (1989) reported that 212 physical therapists involved in back care education had the same incidence, prevalence, and recurrence of back pain as a carefully matched control group.

Successful Training

In 1981, Hayne suggested that the three essential components of a training program are "knowledge, instruction, and practice," but he neither provided sufficient information on the contents of such a program nor indicated how to evaluate its effectiveness reliably. Davies (1978) reported decreases in the incidence of back injury after three carefully designed and properly carried-out training programs; unfortunately, he did not describe the specific criteria for "carefully designed" and "properly carried out." Miller (1977) reported success in decreasing the frequency of back injuries by using a five-minute slide-and-cassette program, a film, and posters to emphasize the theme "When you lift, bend your knees." No cohort, however, was observed. Hall (1973) advocated "clean lifting" (a form of the straight-back, bent knees posture), based on his own personal experience with a bad back, but only presented anecdotal evaluative information.

Most training targets the individual worker or groups of workers. But one can also educate supervisors, health and safety professionals, and management personnel regarding their awareness of MMH problems, of ergonomic job design principles, and of how to respond to low back pain and injury once it has occurred. Such supervisor and manager training is probably of great importance for an effective injury prevention program, but there are no actual data to support that plausible program.

Statistically, back pain has been related strongly to personal dissatisfaction with, and negative attitude to one's job. This finding generates important questions regarding psychosociological aspects on and off the job. So far, these questions have found few answers.

Back Schools

The back-school concept is often traced to Fahrini (1975), who suggested education as a conservative treatment for patients with low back pain, as opposed to radical treatment by surgery. He began using this treatment as early as 1958; other schools have since adopted the approach and in the 1980s it became popular for rehabilitating back-injured patients with health-care providers, physicians, nurses, and physical therapists. This training approach emphasizes knowledge, awareness, and changes in attitude by instructing the individual in anatomy, biomechanics, and injuries of the spine. The overall goal is to encourage the person to take responsibility for his or her own health by means of proper nutrition, physical fitness, and awareness of the effects of posture and movement on the back. Such programs may also provide vocational guidance and information on stress management—indeed, even on drug use and abuse.

Controlled research concerning back schools has been limited, but back-school patients usually express an increased understanding of their own back problem and a feeling of better control of pain. A rather careful study on the effectiveness of back schools was conducted by Bergquist-Ullman and Larsson (1977): Low-back-pain patients from a plant in Sweden were randomly assigned to one of three treatments: back school, physical therapy, and placebo (short-wave diathermy). The results indicated that back-school and physical therapy treatments were equally effective in reducing the number of days needed for pain relief, achieved by both treatments in fewer days than with the placebo. Of the equally effective back school and therapy, the authors preferred back school as more economical, because patients are treated in groups rather than singly.

Typical industrial case studies involve a training program, together with the institution of new safety rules and job redesign and with a campaign of posters, booklets, and cards given out with paychecks. Often, reported results indicate large reductions in compensation costs, together with fewer lost workdays per injury—but, as an unfortunate rule, a control group is not used. One exception is a study of 3,424 employees of the Boeing Company: No significant differences were found, either in the occurrence of back pain or in the number of lost workdays, between (healthy) employees who attended back school and a control group that did not.

One may even see an increase in the number of cases of reported pain, but this can be attributed to a change in management attitude and a willingness on the part of employees to report back problems early—a positive step toward avoiding more serious injury.

Morris (1984) stated key factors for motivating compliance in efforts to prevent back injuries:

- Recognizing the problem and the employee's personal role in it.
- Gaining knowledge about the spine.
- Understanding the relevance of the problem by seeing the working environment on slides or videotapes.
- Being able to alter one's behavior by learning proper techniques of material handling.
- Practicing the new techniques.
- Following up the program at the workplace, as well as training new employees and using the program in safety meetings.

The concept of the postinjury back school, with its emphasis on changes in attitude and awareness to encourage compliance with proper procedures at work, is an appealing approach that goes beyond the traditional teaching of "safe lifting." However, one can hardly draw general conclusions from the case studies mentioned. Clearly, in many instances the management is happy with the results (especially when they are published) and convinced of the program's effectiveness, even if its success may in reality be a "Hawthorne effect"—the positive result of "treating" persons with actions that themselves may be ineffective, but that people react positively to because of the show of concern and interest (Roethlisberger and Dickson 1939.)

The Hawthorne effect: "bread and butter" for consultants

Consultants called by management to improve (sufficiently bad) conditions of manual material activities will be successful if the improvement strategies are well intended and well directed by involving the material handlers. If, after the campaign, the level of performance sags, as is to be expected, management is likely to call upon the same consultants again because "they were successful previously."

Only a few reports on the effectiveness of training mention control groups; according to these reports, back-school therapy was successfully applied to patients with back pain (Bergquist-Ullman and Larsson 1977; Carlton 1987; Lankhorst et al. 1983; Moffett et al. 1986). More recently, Daltrov et al. (1997) conducted a controlled study with nearly 4,000 postal employees. Experienced physical therapists trained 2,534 workers and 134 supervisors. Units consisting of workers and their supervisors participated in a two-session back school (three hours of training), followed by three to four reinforcement sessions over the next five years. Persons from either the intervention or the control groups who were injured during that period (75 cases) were randomized again to either receive or not receive training after returning to work. During the observation period of more than five years, 360 persons reported back injuries; the rate was 21.2 injuries per 1,000 worker-years of risk. The median time off work was 14 days, the median cost $204. A comparison of the intervention and control groups showed no effects of the education program on the rate of low back injuries, the cost per injury, the time off work per injury, or the rate of repeat injuries, but the knowledge of safe behavior was increased among the trainees.

Fitness and Flexibility Training

Another approach to training workers in the prevention of low back injury is that of physical fitness training. Material handling is physical work, and it is reasonable to assume that many aspects of physical fitness, especially musculoskeletal strength, aerobic capacity, and flexibility may be associated with the ability to perform MMH tasks without injury. Exercise has been used in the treatment of back injury for many years, although its exact role and effectiveness are not completely understood.

Despite the well-known practices of Chinese and Japanese workers doing their exercises before work, and of athletes training for better performance, the use of fitness training as a preventive program is rare in North America.

Musculoskeletal strength is one of the aspects of physical fitness that is generally believed to be related to back injury.

Experience indicates that musculoskeletal injuries occur up to three times more frequently in weaker workers than in stronger workers (Chaffin et al. 1978), but there is also the unexpected finding that "strong" persons may be injured more often than their weaker colleagues (Battie et al. 1989). Although the concept of strength training within an industrial environment as a means of preventing injury is occasionally mentioned, no major research literature on this topic seems to exist.

Flexibility, particularly of the trunk, appears to be needed for bending and lifting activities that are part of material-handling tasks. Locke (1983) suggested that a series of 25 stretching exercises be used as a means of warming up and as a starting point for fitness improvement. Yet, Nordin (1991) stated that flexibility measures have been found to be poor predictors of back problems.

Chenoweth (1983a, 1983b) reported that an industrial fitness program was highly successful with participants who volunteered compared with their control-group worker collegues. Unfortunately, the control group seems not to have been selected from persons who volunteered to participate in the program.

An empirical study investigated fitness and low back injury in 1,652 Los Angeles County firefighters to determine the relationships between five strength and fitness measures with the occurrence of back injury over a three-year period (Cady et al. 1979a, 1979b). Individuals were rated to be on one of three levels of fitness (high, middle, or low), based on measurements of flexibility, isometric lifting strength, recovery heart rate, blood pressure, and endurance. The results show that the fittest firefighters had the lowest percentage of back injuries and the least fit group the largest percentage; but the fittest group also had the most severe injuries. This study has often been used to suggest that physical fitness may help to prevent back injuries, but Nordin (1991) cited two more recent longitudinal studies that did not show any association between fitness level (measured via maximal oxygen uptake) and reported back pain.

Regaining and improving fitness, including flexibility, while recovering from a back injury or other disability has always been of concern to patients and their health caretakers. Parts of the back-school concept have been incorporated into "work hardening," wherein specific body abilities deemed necessary to perform a job are improved through purposeful exercises. Fitness training for preventing back injury is viewed with great interest; however, as in the other two training approaches discussed, evidence is insufficient at this time to allow making sound judgments on the effectiveness of the approach in general or on specific programs.

Training: What and How?

Content

The basic question has not yet been answered: "What to teach?" Certainly, the content of a training course is dependent on the aims of the course. Previous efforts have generally been in three areas, as mentioned earlier:

1. Training of specific lifting techniques (ie, improvement of skills).
2. Teaching the biomechanics and the awareness of, and self-responsibility for, back injuries, thereby changing attitudes.
3. Making the body physically fit so that it is less susceptible to injury.

Although the aim—preventing injury—is the same in each case, the methods of how to achieve that aim are quite different. The traditional approach of training a specific lifting technique alone does not appear effective, mostly because there is no one technique that is appropriate for all lifts. Most courses are therefore considered unrealistic and centered too much on protecting the back, as Sedgwick and Gormly (1998) reported from consensus meetings with over 900 Australian health professionals. Yet, they proposed to teach a "semisquat" lifting method throughout the meetings.

The method of preventing injury by increasing one's knowledge of the body and promoting changes in attitude so that workers feel responsible for their bodies has a basic, almost simplistic, appeal and should be quite applicable. However, exactly what should be taught, and how, is still open. How much knowledge is needed of kinesiology, biomechanics, and physiology? What method is most effective? Barker and Atha (1994) reported that written guidelines, such as those commonly handed to untrained industry personnel, worsened lifting performance.

Another key to awareness and changing one's attitude may be the attention paid to material-handling problems by managers, supervisors, or training instructors: Making employees aware that management is concerned about its workers is an underlying theme in the training received from back schools. (Perhaps this is the only element necessary if a Hawthorne effect is deemed successful.)

Having a physically fit body is an intuitively and rationally convincing property for injury-free load handling. Yet, research regarding the role of physical training as a preventive method is still sparse. For example, would mobility, strength, or endurance training of workers effectively reduce back injury? Are Asian-style or warm-up exercises before the work shift beneficial?

Beyond these approach-specific research questions are the concomitant problems of implementing two or more approaches in training. For example, how exactly does physical fitness interact with changes in attitude? In many cases, the programs presented in the literature contain elements of knowledge and attitude change, physical exercises, and specific lifting techniques. It is not at all clear whether this combined approach is the best, or even effective at all.

Criteria

Another major and basic question is how to judge the effectiveness of any given training. The most commonly used methods rely on "objective" data derived from company records on cost and productivity and from medical records to compare quantitative measures before and after training. The use of company loss data is most common, but sometimes the exact meaning of the numbers is not clear, particularly if other actions take place during the period of data recording that may have had effects on the loss statistics. Apparently, there is not enough standardization among companies in the definition of terms, or in the actual derivation of the statistics, to allow reliable industrywide comparisons. "Subjective" data result from asking trainees or managers questions such as "Was training worthwhile?" The value of such judgments is uncertain.

Retention and Refreshment

It would not make sense to do long-term evaluations of training if the students never learned the material in the first place. So how can we measure whether the workers have actually absorbed what was presented in training sessions? Some test or demonstration as a criterion of learning directly following the training could be used as an evaluative tool, but what measure to use and how well one must do on the measure to be considered trained (or having a changed attitude) has not been discussed in the literature.

Another unanswered question concerns the retention of information by the trainees after training has been completed. Was there sufficient original learning? Retention may have been good immediately following instruction, but why, then, is there usually an increase in injuries (or whatever measure is used) as time passes? Is this a reflection of good training that is forgotten over time, or was the training not "good enough to last" in the first place? To increase a person's retention of information is a traditional concern. Can interference be reduced? Is the transfer from the learning environment to the work environment easily made? Refresher courses (of some kind) probably should be offered, yet the time intervals between instances of training need to be established.

Instructional Style and Media

Once it has been determined what should be taught, attention must be paid to the formation of the course itself. It seems from the literature that many courses are taught in a lecture format, with techniques practiced at some time during the session. Films and videotapes, audiotapes, posters, and cards in paycheck envelopes have also been used. Programmed instruction, computer-aided instruction, and interactive video are at hand. The relative effectiveness of these methods has not been determined.

Where should the sessions be held? Would a classroom with desks and chairs be more appropriate than a lecture hall? The work site appears most appropriate, but it may not be suitable for instructional purposes. Is it best to train employees who work together as a group, or should the group be split up? Of course, in any situation, there are practical constraints that might limit one's ability to implement training ideals, but currently information is not available on which to base sound judgment.

Training "customized" for certain industries or jobs

Most industrial back injuries have not been associated with objective pathological findings, and about every second episode of back pain cannot be linked to a specific incident. A variety of actions and events at work have been associated with low back injuries. In the mining industry, overexertion, slips, trips, falls, and jolts in vehicles are the most frequently mentioned events (Bobick and Gutman 1989). Back injuries have been directly associated with lifting (37 to 49 percent of the cases), pulling (9 to 16 percent), pushing (6 to 9 percent), carrying (5 to 8 percent), lowering (4 to 7 percent), bending (12 to 14 percent), and twisting (9 to 18 percent); the percentages vary considerably among industries and occupations. The industries with the greatest incidence of compensation claims for back injuries are construction and mining. Occupations with a high incidence of low back pain are garbage collector, truck driver, nurse, and, of course, material handler.

Load-handling specifics differ much among industries and jobs, as well as within one industry or profession. They depend on the specific task, on available handling aids and equipment, and on many other conditions. Therefore, a question to ask is "How are training recommendations applicable across settings?" Even the group-related characteristics of material handlers in different industries might be important in designing a training program; for example, hospital workers might have higher educational skills than heavy-industry workers. Women employees may be predominant in a given industry or occupation, which might then influence MMH training, because, on the whole, women are about two-thirds as strong as men. It is not known how personal or task-specific characteristics should influence training.

In the United States, about 2 of every 100 employees report a back injury each year. This figure poses another problem regarding the effectiveness and cost of back-care instructions. Of the actually reported injuries, about every tenth is serious, yet these few serious injuries are responsible by far for the largest portion of the total cost. Hence, if one wanted to specifically prevent these serious injuries, 2 of every 1,000 employees would be the target sample, while all one thousand must be in the educational program. Even if one wanted to address all persons who might suffer from any kind of back problem—about 20 out of 1,000—this is still a rather expensive approach that may not appear cost effective to the administrator.

Ethical Considerations

Workers get injured and experience pain; so who shall receive training, everyone or just individuals at high risk? In conducting research, a no-treatment control group is often used. Yet, what are the responsibilities of the researcher in choosing whether to train or not train people in ways that might help prevent back injury? In the case of fitness training for preventing injury, there is the potential for injury from the training itself. Therefore, should the training be mandatory or voluntary? If it is voluntary, will the training be given during working hours or on the worker's own time? Some individuals might feel that mandatory fitness training and nutrition guidelines, for example, decree a change in lifestyle that they may not want. How far can employers go?

☛☛☛ *The willingness to exercise was investigated by questioning 444 employees of a Canadian power company who had agreed to take a fitness test. The prevalent determinants, on or off the job, are apparently the following:*

- *The already existing habit of exercising: Those who exercise are likely to continue; those who don't exercise probably won't even start to.*

- *The perceived obstacles to exercising: If exercise is difficult to arrange or if it is difficult to succeed in exercising, one is likely not to try or to drop out.*
- *The attitude: if exercise is considered positive and worthwhile, it is likely to be pursued.*

Unfortunately, this disappointing result appears to hold for all population subgroups tested so far (Godin and Gionet 1991).

Summary of Review of Training

The review of training we have just presented covered about 100 published sources. Yet, the available literature remains spotty: It neither addresses all topics of interest nor indicates many clear results. The first deficiency relates to the multifaceted magnitude of the problem, the second to deficiencies in the experimental design and control of training approaches.

Jensen (1985) developed a block diagram that demonstrates the sequence of

- needs assessment,
- development or selection of a suitable training program,
- pretraining assessment,
- doing the training, and
- posttraining assessment,
- followed by improvements and possible repetition of the program.

This systematic approach, shown in Figure 11–3, helps to evaluate the reliability, significance, and validity of training in safe manual material handling.

Given the scarcity of information, hardly any training guidelines are well supported by controlled research. This leaves much room for speculation, guesswork, and charlatanry regarding the "best" way to train people for the prevention of back injuries related to MMH. This state of affairs is deplorable and needs to be remedied, since common sense indicates that training should be successful.

The issue of training for the prevention of back injuries in load handling still is confused, at best. Some—possibly most—training approaches are not effective in preventing injury, or their effects may be so uncertain and inconsistent that money and effort expended on training programs might be better spent on ergonomic job design. "In spite of more than 50 years of concerted effort to diminish task demand, the incidence of compensable back injuries has not wavered. . . . Rather than pursuing the 'right way to lift,' the more reasonable and humane quest might be for workplaces that are comfortable when we are well and accommodating when we are ill" (Hadler 1997, p. 935).

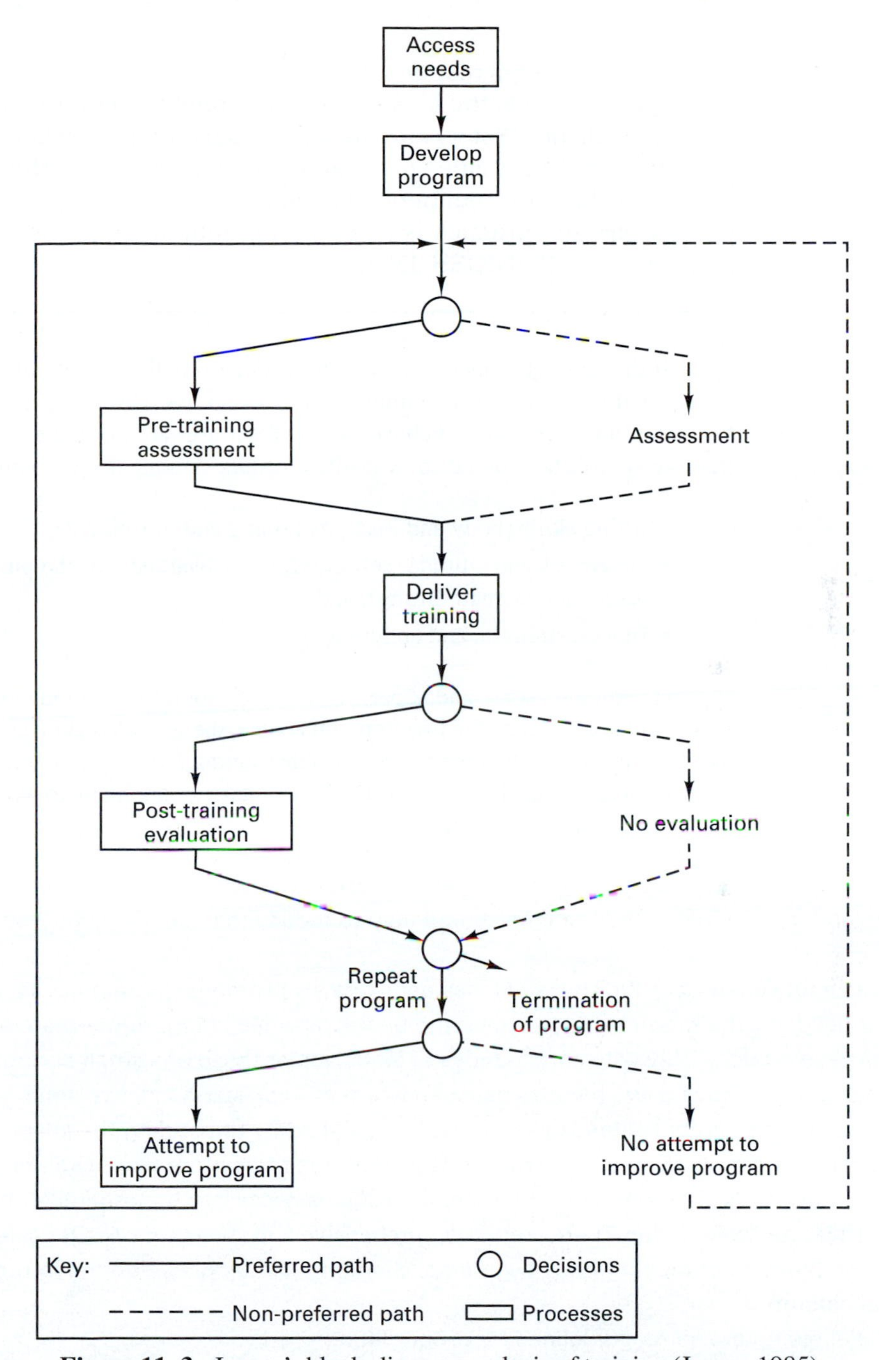

Figure 11–3. Jensen's block-diagram analysis of training (Jensen 1985).

> The legal responsibility of employers to provide training cannot be ignored; thus, the idea to abandon training in material handling appears unrealistic. "Yet so long as it is a legal duty [in the United States] for employers to provide such training or for as long as the employer is liable to a claim of negligence for failing to train workers in safe methods of MMH, the practice is likely to continue despite the lack of evidence to support it" (NIOSH 1981, p. 99).

If the job requirements are stressful, "doctoring the symptoms," via such techniques as behavior modification, will not eliminate the inherent risk. Designing a safe job is fundamentally better than training people to behave safely. Yet it appears plausible to expect that at least certain training approaches should show positive results. Among these, training for

- lifting skills (body and load positioning and movement),
- awareness and attitude (self-control and awareness of the physics and biomechanics associated with lifting), and
- fitness, strength, and endurance

appeal to common sense and appear theoretically sound, even though none of them has yet proven successful according to the literature. Of course, the so-called Hawthorne effect is likely to reduce incidents immediately after any reasonable training (which is, of course, a positive and desirable result), but after a fairly short time the injury statistics are likely to worsen again—which provides a reason to continue training.

Several general recommendations should help to reduce the risk of overexertion injuries when lifting:

Leg lift versus back lift. The leg lift has been heavily promoted in training, as opposed to the back lift. It is, indeed, normally better to straighten the bent legs while lifting, rather than unbending the back. But leg lifts can be done only with certain types of loads: either those of a small enough size that fits between the legs or those having two handles between which one can stand (eg, two small suitcases instead of one big case). Large and bulky loads cannot usually be lifted by unbending the knees without contorting the body; if one attempts to do so, one may in fact stress the torso more than when one is also unbending the back slightly. Hence, proper task and material design is necessary to permit leg lifts.

Rules for "safe" lifting. There are no comprehensive and sure-shot rules for safe lifting, which is a complex combination of moving body segments, changing joint angles, tightening muscles, and loading the spinal column.

Following are some guidelines for proper lifting:

1. *Eliminate manual lifting (and lowering) from the task and workplace.* If lifting needs to be done by a person, perform it between knuckle and shoulder height.
2. *Be in good physical shape.* If you are not used to lifting and to vigorous exercise, do not attempt to perform difficult lifting or lowering tasks.

3. *Think before acting.* Place material conveniently. Make sure that sufficient space is cleared. Have handling aids available.
4. *Get a good grip on the load.* Test the weight before trying to move it. If it is too bulky or heavy, get a mechanical lifting aid, or somebody else to help, or both.
5. *Get the load close to the body.* Place the feet close to the load. Stand in a stable position, and have the feet point in the direction of movement.
6. *Involve primarily straightening of the legs in lifting.*

Following are some things to avoid:

1. Do *not* twist the back or bend sideways.
2. Do *not* lift, lower, push, or pull awkwardly.
3. Do *not* hesitate to get help, either mechanical or by another person.
4. Do *not* lift or lower the object with the arms extended.
5. Do *not* continue heaving when the load is too heavy.

The foregoing rules and recommendations not only should be part of a worker training program, but also should tell the engineer and manager how to design the job. For example, the loads should be

- of proper size to be lifted by straightening the knees,
- placed at the proper height to be handled in front of the trunk, and
- of the proper form and shape so that one can get a good grip.

Obviously, the topic of training in MMH techniques to prevent injury, particularly to the back, has not been studied in a cohesive or systematic manner. The overall impression is that many health and safety professionals, industrial engineers, and managers have accepted the general ideas of training workers in proper lifting and handling techniques and of improving their awareness and attitude, but have not yet clearly determined what content and which media to use or made evaluations of the long-term effectiveness and worth of the various training approaches. Much applied research needs to be done until training theory and practice are integrated and successful.

PERSONNEL SELECTION BY PHYSICAL TESTING

While training is one approach to "fit the person to the job," the other is the selection of suitable persons—that is, the screening of individuals with the purpose of placing those on strenuous jobs who can do them safely. This screening may be done either before the person is employed, before he or she is placed in a new job, or on the occasion of a routine examinations the person takes during employment.

A basic premise of selecting personnel by physical characteristics is that the risk of overexertion injury from manual material handling decreases as the handler's capability to perform such activity increases. This means that a test should be designed that allows to judge the match between a person's capabilities for load handling and the actual load-handling demands of the job. Accordingly, the matching process requires that one know, quantitatively, both the job requirements and the related capabilities of the individual. (Of course, if the job requirements are excessive, they should be lowered before any matching is attempted.)

Limitations in Capability

The human body must maintain a balance between the external demands of the work and related internal capacities. The body is an energy "factory" that converts chemical energy derived from nutrients into externally useful physical energy. The final stages of this metabolic process take place at the skeletal muscles, which need oxygen transported from the lungs by the blood. The blood flow also removes by-products generated in the conversion of energy, such as carbon dioxide, water, and heat, which are dissipated in the lungs, where oxygen is absorbed into the blood. Heat and water are also dispelled through the skin (as sweat). The blood circulation is powered by the heart. (See the first chapters of this book for details.)

Thus, the pulmonary system, the circulatory system, and the metabolic system establish *central* limitations on a person's capability to perform strenuous work.

The person's capability may be limited also by his or her muscular strength, by the ability to move the body joints, and, in material handling, often by the responses to stress of the spinal column and its supporting connective tissues. As discussed earlier, these are *local* limitations on the force or work that a person can exert.

Hence, both central and local limitations determine a person's capability to perform. While the person is handling material, the force or torque exerted with the hands must be transmitted through the body—that is, via the wrists, elbows, shoulders, trunk, hips, knees, ankles, and feet—to the floor. In this chain of force vectors, the weakest link determines the capability of the whole body to do the job. If muscles are weak, or if they have to pull under mechanical disadvantages, the available handling force is reduced. Often, the lumbar section of the spinal column is the weakest link in the kinematic chain: Muscular or ligament strain or painful displacements of the vertebrae or the intervertebral discs may limit a person's ability to handle materials.

Assessment Methods

Various methods have been developed to assess an individual's capabilities for performing specified handling tasks (Himmelstein and Andersson 1988). Such assessments rest on a fundament of (by nature, general) epidemiological data or more specific etiological information.

The medical (physical) examination primarily identifies persons with physical impairments or diseases. Since X-ray examinations are now used sparingly, the examination screens out the medically unfit on the basis of their physiological and orthopaedic traits. Unless specific job requirements are known to the physician or nurse performing the testing, he or she must evaluate the person's capabilities against generic job demands. Hence, the physical examination is often not a specific match of the person's capabilities (actually, limitations of capability) with the job demands.

The physiological examination usually identifies individual limitations in central capabilities (eg, of pulmonary, circulatory, or metabolic functions). This examination provides essential criteria if these functions are indeed highly taxed by the material-handling job, such as in frequent movements with heavy loads and much involvement of the body. However, this is now seldom the case in technically developed industries.

The biomechanical examination addresses mechanical functions of the body, primarily of the musculoskeletal type (eg, load-bearing capacities of the spine and muscle strength exertable in certain postures or motions). Biomechanical methods rely on explicit models of such body functions, which is both their strength and their weakness: The results are only as good (reliable and valid) as the underlying models.

The psychophysical examination addresses all (local or central) functions strained in the test; hence, it may include all or many of the systems checked via medical, physiological, or biomechanical methods. The psychophysical examination filters the strain experienced through the sensation of the subject who rates the perceived exertion. In tests of maximal voluntary exertions, the subject decides how much strain is acceptable under the given conditions (eg, what the maximal weight is that he or she is willing to lift).

The physician is expected to perform essentially all of these investigations during the medical examination, with the additional task of relating them to general and specific (but often unspecified) job demands, to determine whether a proper match exists.

Techniques Using "Strength" for Screening

Several screening techniques exist that rely on assessing a person's "strength" in terms of the force or torque the person exerts on an instrument or the weight (mass) that the person can lift. These outcomes should serve to select persons who are able to perform defined material-handling activities with no risk or an acceptably small risk of overexertion injuries—that is, those whose capabilities match the job demands, within a margin of safety.

In the past, managers, foremen, or physicians had to rely on many intuitive, experience-based guesses; but now, a number of preemployment or placement tests have been developed on the basis of the models and methods just discussed (Rice 1999). Primarily, these tests differ in the techniques used to generate the external stresses that strain the local and central function capabilities to be measured.

STATIC TESTS

Static techniques require the subject to exert isometric muscle strength against an external measuring instrument. Since muscle length does not change, there is no displacement of any body segments involved; hence, there are no time derivatives of displacement either. This establishes a mechanically and physiologically simple case that allows straightforward measurement of the person's isometric muscle strength capability. The 1974 standardized procedure (Caldwell et al. 1974) has been widely accepted (with some minor modifications) for such static strength measurement, which in turn has become part of well-established screening techniques claimed to be effective (Ayoub et al. 1984; Anderson and Catterall 1987; Laughery et al. 1988; Chaffin 1981; Kumar et al. 1988; NIOSH 1981).

DYNAMIC TESTS

Dynamic techniques appear more relevant to actual material-handling activities (Ayoub and Mital 1989; Stevenson et al. 1989), but they are also more complex because of the large number of possible displacements and of their time derivatives, viz., velocity, acceleration, and jerk. (See Chapter 1). Hence, most current dynamic-testing techniques employ either one of two ways to generate dynamic stresses.

In the **isokinematic** (or isovelocity—often falsely called isokinetic) technique, the subject moves the limb (or trunk) at a constant angular velocity about a specific joint (usually the knee, hip, shoulder, or elbow) while exerting maximal voluntary torque. Several kinds of test equipment are on the market. They are so designed that the exerted torque can be monitored continuously throughout the angular displacement. Since this equipment is rather costly, simplified versions are available in which a handle is moved at constant speed while angular joint speeds are not specified. (Note that in the initial and final portions of the test the limb speed is not constant and that, in fact, muscle movement is probably not constant throughout the test. (See Chapter 1.)

The **isoinertial** technique requires the subject to move a constant mass (weight) between two defined points. The maximal load that can be lifted (or lowered, carried, or held) by the subject is the measure of that person's capability, while the forces or torques actually exerted depend (according to Newton's second law) on mass and acceleration. This rather realistic technique is part of many tests that are currently employed.

Pros and Cons of Screening Techniques

Ease

The main advantage of the static (isometric) techniques is their conceptual simplicity: Putting a person into a few "frozen" positions and measuring the force that she or he can generate in a given direction over a period of just a few seconds is indeed easily understood. Furthermore, since no displacement takes place (muscle lengths and joint locations do not change), the physical conditions are very simple, since all time derivatives of displacement are zero. This allows the use of rather simple instruments and permits relatively easy control of the experimental conditions. It is for these and other reasons that static strength measurement was initially suggested by Chaffin et al. (1978) and promoted by NIOSH in 1981 as a first analytical assessment of individual capabilities. Static strength testing was a significant step forward from the earlier simplistic assumption of single "safe" loads that children, women, and men supposedly could lift without danger of overexertion, but it has been shown to have rather low predictive power for dynamic tasks.

This unsatisfactory relationship between static testing and a person's dynamic performance capability triggered researchers to develop more suitable dynamic measuring techniques (eg, by Ayoub 1982; and Kroemer 1982). The apparent advantage of dynamic measuring is its resemblance to actual material handling, which is usually done in motion, not motionless. The testing difficulty lies in the fact that body members can be moved in a variety of dynamic conditions, depending, for example, on the actual path of motion, the time consumed, and the accelerations and decelerations applied to the different masses involved (Hsiang and McGorry 1997; Mirka et al. 1998). Overcoming this difficulty by keeping the first derivative of displacement (velocity) constant is characteristic of isokinematic (or isovelocity, not isokinetic) measurement techniques. Their main attraction lies in the ability to measure the force or torque that is exerted while controlling the speed with which the person moves. The main disadvantage of these techniques is that they keep the speed constant, which is not the way in which objects are commonly handled. This may explain why some isokinematic test results have been found of little value for selecting employees (Burgdorf et al. 1995; Dueker et al. 1994).

Still left for consideration is the apparently oldest and most practical test of lifting capability: actually to lift loads, starting with small ones and increasing them until the largest one that can be lifted is found. This incremental isoinertial test is easily understood, executed, and controlled;

is safe, reliable, and valid; and has been used to test millions of U.S. military recruits, male and female (McDaniel et al. 1983; Stevenson et al. 1996; Teves et al. 1985). A strategy of increasing or decreasing the loads to determine a person's lifting capacity to the nearest 5 lb is shown in Figure 11–4. The equipment is inexpensive and robust. Such an isoinertial test is an efficient, reliable, and realistic approach to assessing an individual's capability to lift.

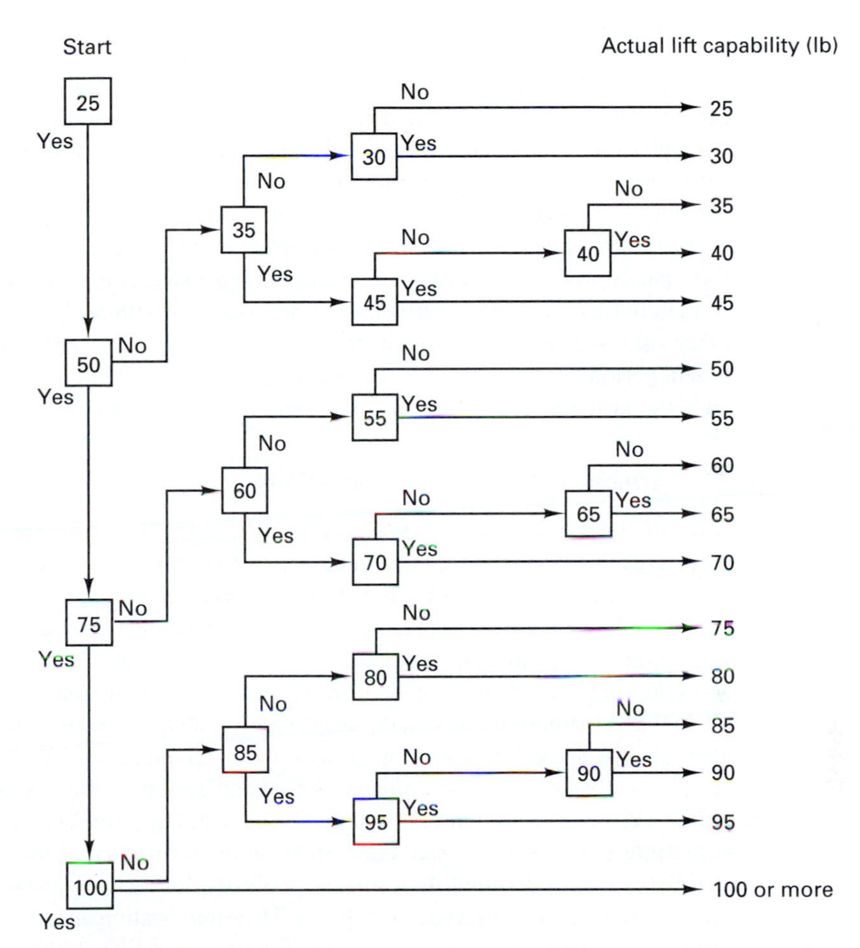

Figure 11–4. Sequence for determining actual lifting capability used in "LIFTEST" (Kroemer 1982).

Realism and Safety

In Chapter 1, we saw that all tests with maximal voluntary exertions depend on the subject's will to achieve a maximal performance. The outcomes of all of the testing procedures mentioned in the previous section are strongly, yet often imperceptibly, affected by the testee's evaluation of the body strain she or he experiences and the associated motivation to exert more or less effort. All these tests are "psychophysical and subjective," regardless of whether they are static or dynamic. (See Chapter 7.) Waikar et al. (1991) found that the subjectively acceptable weight of an object to be lifted was below that predicted by a biomechanical model. The reason may be that the model used was unrealistic or that, indeed, the subjective assessment of one's own body strain is the best (safest) method to determine what is suitable and what is not. This thought lends some credence to the psychophysical assessment techniques used prominently by Snook and coworkers (1978, 1991), the results of which are presented later in the chapter. These researchers gave their subjects monetary incentives to work as hard as they could, but instructed them to do so without becoming unusually tired, weakened, overheated, or out of breath.

Validity

A major problem in all current techniques for testing a person's lifting strength is the predictive power of the tests with regard to real-life, true working conditions. Given that static (isometric) strength testing can no longer be considered a sufficient measure of an individual's ability to perform most lifting tasks, isoinertial incremental strength testing appears to be a better procedure for selecting personnel. Still, while the face validity of isoinertial testing is high, rigorous assessments of its validity have not yet been performed (Fernandez et al. 1991; Lee and Chen 1996). Fortunately, to do so is straightforward and inexpensive. A typical scenario includes a large group of material handlers whose capabilities are assessed with an isoinertial test and, possibly, with other tests. While no job assignments are made, the workers' performance is observed over a sufficient period of time, say, one year. Validity (fidelity) is demonstrated if those persons who tested better also handle loads better (ie, have fewer overexertions and injuries).

Ethical and Legal Considerations

A major concern in all tests is to ensure that the individuals who are tested are not treated unfairly as a result of the testing. In essence, this is the problem of balancing the needs of the system (as perceived by the employer) and of individuals (on whom the system finally depends). The test criteria to be met must be representative of the actual and necessary requirements on the job. The testing procedures must be valid and reliable. A further problem is that both the criteria and the procedures are based on group data and do not take into account individual variations (Webb and Tack 1988). Although the specific legal requirements are different from country to country, in general, tests should not discriminate against any given group. Yet, for example, as a group, men are physically stronger than women (Astrand and Rodahl 1986; NASA/Webb 1978), and younger employees have been found to have more frequent, but often lighter, injuries than older employees (Bigos et al. 1986). Correlations between the occurrence of low back pain and body weight, static strength, or postural deficiencies (scoliosis, lordosis, kyphosis, or an uneven leg length), unless extreme, are low (Andersson 1991). Therefore, testing an individual to match her or his personal capabilities with specified job requirements is preferred over making group assessments.

ERGONOMIC DESIGN FOR LOAD HANDLING

How the job is designed determines the stress imposed by the work on the human material handler. The size of the object that is handled, its weight, whether it has handles or not, the layout of the task, the kinds of body motions to be performed, the body forces and torques to be exerted, the frequency of these efforts, the organization of work and rest periods, and other engineering and managerial aspects determine whether a job is well designed and whether it is safe, efficient, and agreeable to the operator.

The following text discusses the facility layout, the environment, and the equipment pertaining to handling loads. (See Kroemer 1997 for a more extensive treatment.)

FACILITY LAYOUT

The layout of the overall work facility contributes to the safe and efficient transfer of materials. Organizing the flow of material in general and designing it carefully in detail determines the involvement of people and how they must handle the material.

In the real world, one encounters either one of two situations: A facility exists, it must be used, and the building and its interior layout cannot be changed significantly. By small changes, one must make the best use of what there is. In the second case, one can plan and design a new facility and its details according to the process at hand, to suit the workers and their work easy and safe.

The opportunity to "do it right at the planning stage" is most desirable, allowing the closest approximation to the optimal solution. Striving for the best (possible) solution is also the purpose for modifying a given facility. Therefore, we shall use the ideal case as a guide, even when only modifications may be possible.

It is the purpose in facility layout, or facility improvement, to select the most economical, safest, and efficient design of buildings, departments, and workstations. Of course, specific details depend on the overall process.

A facility with a well-planned material flow has a few short transportation lines. Reducing the movement of material through the proper facility layout can lower the cost of transporting materials considerably, which usually accounts for 30 to 75 percent of the total operating cost and is even higher in some instances. Unfortunately, moving materials adds no value to the product and is dangerous. Hence, reducing material handling is both a major safety consideration and a cost factor.

PROCESS VS. PRODUCT LAYOUT

There are two major design strategies: process layout and product layout. In the first, all machines or processes of the same type are grouped together, such as all heat treatment in one room, all production machines in one section, and all assembly work in one division. Figures 11–5 and 11–6 are examples of process layouts.

There is a major advantage to this design strategy: Quite different products or parts may flow through the same workstations, keeping machines busy. But much floor space is needed, and there are no fixed flow paths. Also, process layout requires relatively much material handling. It is worthwhile to study Figures 11–5 and 11–6 carefully to determine what improvements should be made in each case to improve the conditions depicted.

In product layout, all machines, processes, and activities needed for work on the same product are grouped together. This results in short throughput lines, and relatively little floor space is needed. However, the layout suits only the specific product, and a breakdown of any single machine or of special transport equipment may stop everything. Still, overall, product layout is advantageous for material handling because routes of material flow can be predetermined and planned well in advance.

FLOWCHARTS AND FLOW DIAGRAMS

Figures 11–5 and 11–6 show how easy it is to describe events and activities with simple sketches, symbols, and words.

The *flow diagram* is a picture or a sketch of the activities and events one is interested in. It indicates their sequence and where they take place.

The *flowchart* is a listing or table of the same activities and events depicted in a flow diagram. It can easily be put together and stored in a computer. Indicated on the chart are the durations of the activities

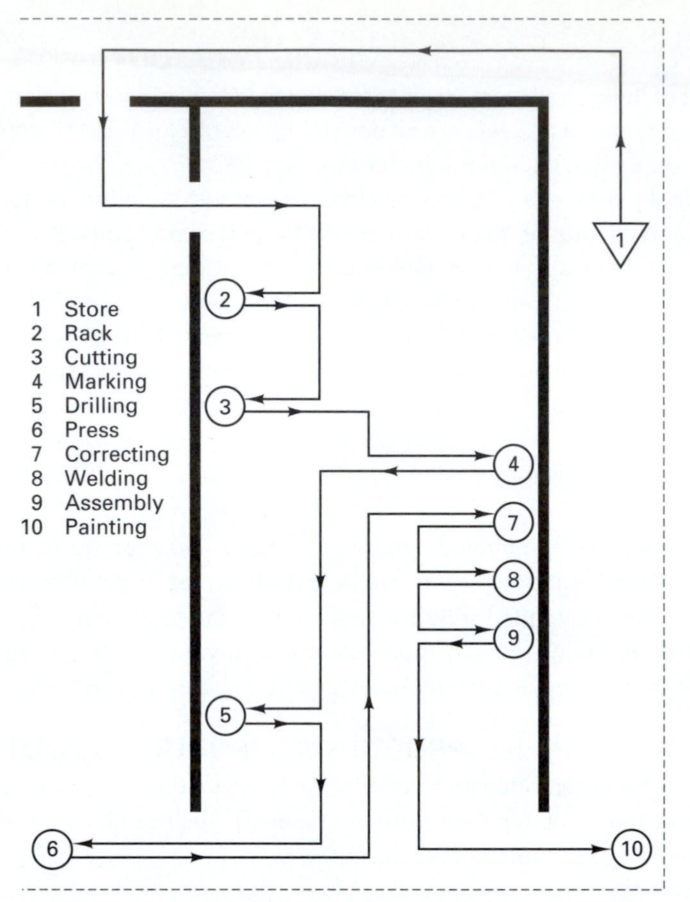

Figure 11–5. Flow diagram of bus seat production. (Adapted from ILO 1974.)

and events, as well as detailed information on related facts or conditions. The events depicted in Figure 11–6 are charted in Figure 11–7.

When flow diagrams or flowcharts indicate activities performed on a given material, the diagram or chart is called a "material (-centered)" diagram or chart. When it shows the activities of one person, it is called an "operator (-centered)" diagram or chart.

Making a flowchart or setting up a flow diagram is simple, straightforward task that can be done even by a layperson. The chart and the diagram help one to understand what is done, when it is done, how long it takes, what is involved, and which hazards are present. Many industrial engineering texts explain in detail how to make flowcharts and flow diagrams, and how to interpret them in order to initiate ergonomic improvements in the work environment.

An application of flow diagrams to a receiving facility is shown in Figure 11–6. Even a cursory look shows many workstations, many long transports, and many delays—altogether, a waste of time, space, and

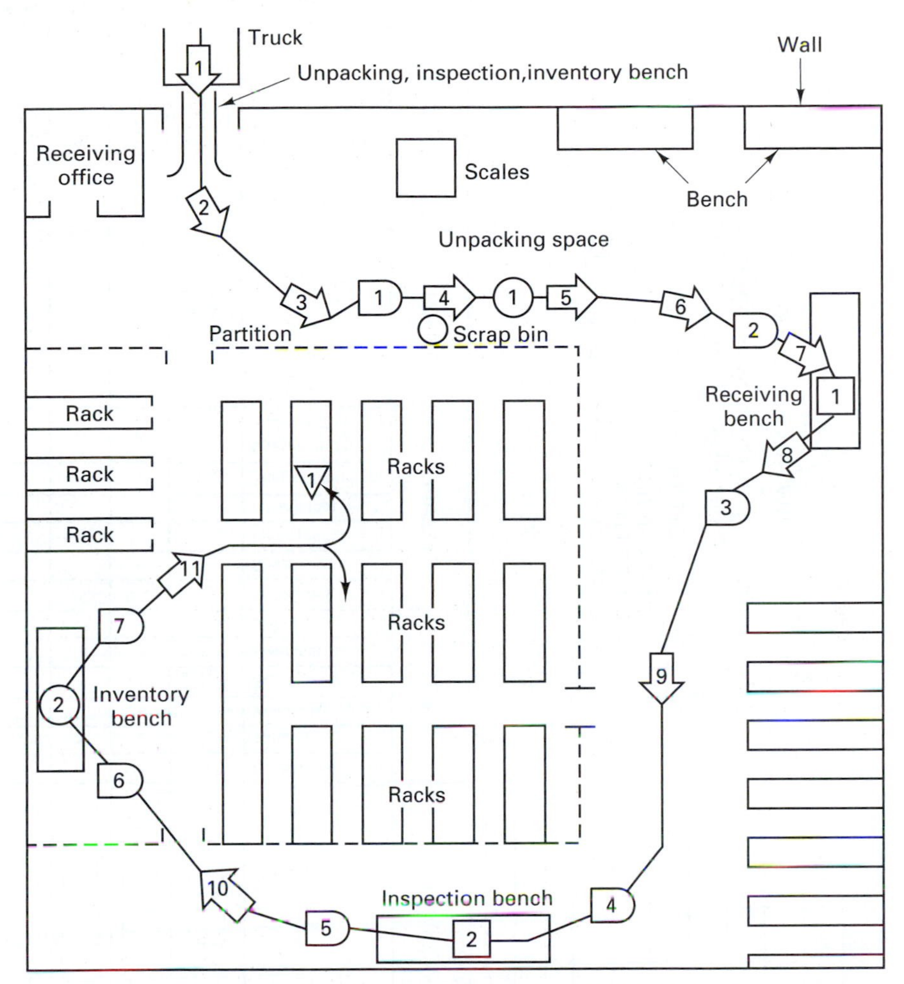

Figure 11–6. Flow diagram of receiving, inspection, taking inventory, and storage: original setup. (Adapted from ILO 1974.)

personnel. Also shown are many possible sources of injuries caused by the movement of materials. Figure 11–7 details all the actions that are necessary to perform the work under the original conditions. It reveals that, in order to carry out just two operations and two inspections, 11 transports are needed. Seven delays are encountered, the material travels 61 meters, mostly by human hand, and it takes nearly 2½ hours to store the material. Twenty-six hazards are present.

Figure 11–8 shows a simple solution. Unpacking, inspection, and inventory taking are all combined in one location, close to the entry. The partition is largely removed, and, through the use of a conveyor, the transport lines have become short and simple. Listing the activities of the ergonomically designed work in a newflow chart indicates that both the time needed to place the material in storage and the distance traveled by the material have been cut in half; and the human labor has become much less intensive and dangerous.

SUMMARY

	PRESENT		PROPOSED		DIFFERENCE	
	No.	Time	No.	Time	No.	Time
○ Operations	2	8				
⇨ Transportations	11	26				
□ Inspections	2	35				
D Delays	7	85				
▽ Storages	1	2				
Distance travelled	61 m		m		m	
Number of hazards: High	0					
Number of hazards: Medium	8					
Number of hazards: Low	8					

FLOW PROCESS CHART

No. 1
Page 1 of 2

JOB Receive, check, inspect, inventorize and storage of parts received in cartons.

☐ OPERATOR OR ☒ MATERIAL

CHART BEGINS 9:15 a.m.
CHART ENDS 11:31 a.m.
CHARTED BY KHEK Date 10/16/89

METHOD PRESENT OR PROPOSED	ACTIVITY					FACTS						HAZARDS						COMMENTS	CONTROL ACTION
Describe in Detail Each Shop	Operation	Transport	Inspection	Delay	Storage	Distance m	Time min	Weight kg	Size m.m.m	Freq/shift	# of people	Falling mat	Sharp edges	Pinch points	Hazard mat	Manual handling	Overall rating High/med/low	Identify Important Aspects, Particularly Hazard Potentials	Eliminate Combine Simplify Personal Protection etc.
1 Carton lifted from truck; placed on inclined roller conveyor		1				2	2	75	0.7 0.7 0.5	4 0 0	2	√				√	L	Back injury possible	Change
2 Slid on conveyor		2				6	1	•	•	•	2	√					L		Combine
3 Stacked on floor		3				6	4	•	•	•	2			√			L		Eliminate
4 Await unpacking				1		–	32										/		Eliminate
5 Unstacked		4				1	2	•	•	•	2	√		√			M		Eliminate
6 Lid removed, shipment papers removed	1					–	3	•	•	•	2		√	√			M		Change
7 Place on hand truck		5				1	1	•	•	•	2	√		√			M	Back injury possible	Eliminate

Figure 11–7. Flowchart of the setup shown in Figure 11–6.

8	Trucked to receiving bench	⇨ 6	9	4	▪	▪	▪	1	√					L		Combine
9	Await removal from truck	D 2	–	10										/		Eliminate
10	Placed on bench	⇨ 7	1	1	▪	▪	▪	2			√			L	Back injury possible	Eliminate
11	Parts taken from carton, checked, replaced	□ 1	–	15	▪	▪	▪	1		√	√			M		Simplify
12	Carton placed on hand truck	⇨ 8	1	1	▪	▪	▪	2	√		√			M	Back injury possible	Eliminate
13	Await transport	D 3	–	5										/		Eliminate
14	Trucked to inspection bench	⇨ 9	17	6	▪	▪	▪		√					L		Eliminate
15	Await inspection	D 4	–	10										/		Eliminate
16	Parts removed, inspected for function, replaced	□ 2	–	20	▪	▪	▪	3		√	√			M		Simplify
17	Await transport	D 5	–	5										/		Eliminate
18	Trucked to inventory bench	⇨ 10	9	5	▪	▪	▪	1	√					L		Eliminate
19	Await inventorizing	D 6	–	15										/		Eliminate
20	Inventized, removed, numbered, replaced	○ 2	–	5	▪	▪	▪	1	√	√	√			M		Change
21	Await transport	D 7	–	8										/		Eliminate
22	Transport to storage rack	⇨ 11	7	4	▪	▪	▪	2	√					L		Change
23	Carton stored	▽ 1	1	2	▪	▪	▪	2	√		√			M		Personnel protection equipment
24		○ ⇨ □ D ▽														
25		○ ⇨ □ D ▽														
26		○ ⇨ □ D ▽														

Figure 11–7. *(cont.)*

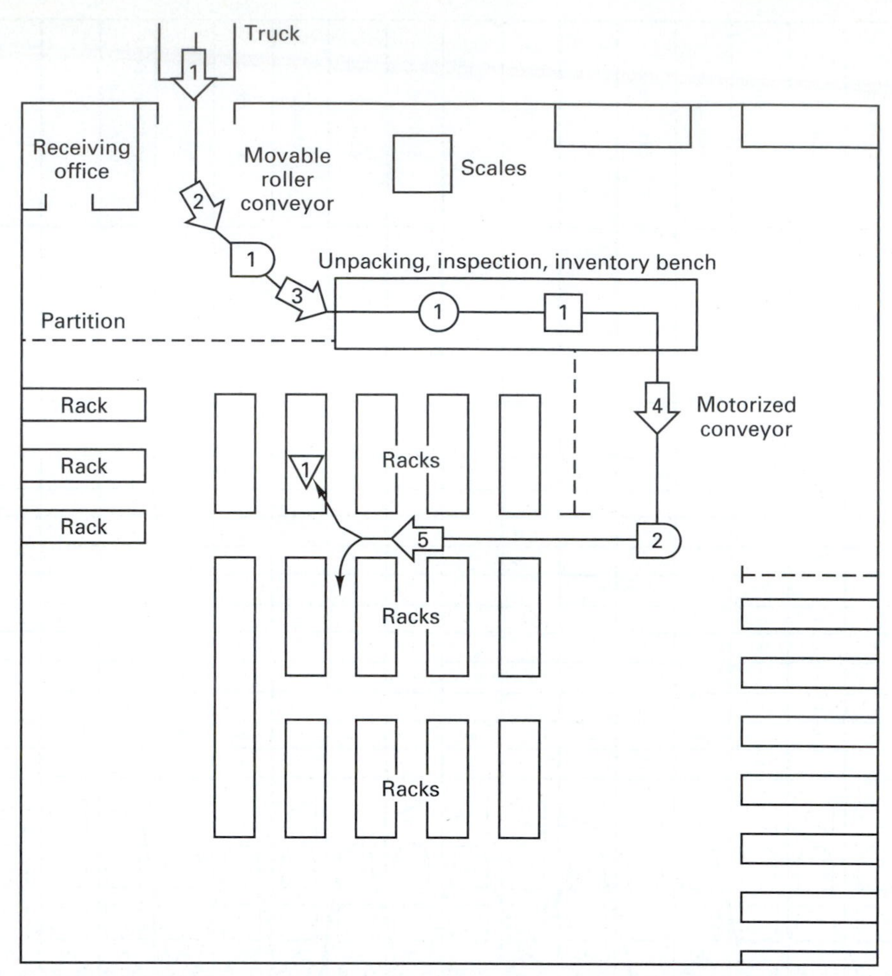

Figure 11–8. Flow diagram of improved setup. (Adapted from ILO 1974.)

THE WORK ENVIRONMENT

The work environment can be made to contribute to safe manual material activities if it is well designed and maintained.

The **visual environment** should be well lit, clean, and uncluttered, allowing good depth perception and discrimination of visual details, differences in contrast, and colors.

The **thermal environment** should be within zones that are comfortable for the physical work, usually in the range of about 18 to 22 degrees Celsius. Thermal stress resulting from conditions that are too hot or too cold may contribute to safety problems in handling materials.

The **acoustical environment** should be agreeable, with sound levels preferably below 75 dBA. Warning sounds and signals indicating unusual conditions should be clearly perceptible by the operator. High levels of noise can contribute to an overall strain on the operator and hence affect the safety of material handling.

(For more information on environmental conditions, see Chapters 4 and 5.)

Good **housekeeping** helps to avoid injuries. Safe gripping of the shoes on the floor, and good support from the chair when the worker is seated, are necessary conditions for safe material handling. Poor coupling with the floor can result in slipping, tripping, or misstepping. Floor surfaces should be kept clean to provide a good coefficient of friction with the shoes (Myung and Smith 1997). Clutter, loose objects on the floor, dirt, spills, and the like can reduce friction and cause the worker to slip and fall. (See Figure 11–9.)

WORK EQUIPMENT

Equipment at the workplace either may assist the material handler or may do the actual transportation. (See Figures 11–10 through 11–12.)

Equipment for assistance includes

- Lift tables, hoists, and cranes;
- Ball transfer tables and turntables; and
- Loading and unloading devices.

Transportation equipment includes

- Nonpowered dollies, walkies, and trucks;
- Powered dollies ("walkies,") rider trucks, and tractors;
- Conveyors and trolleys; and
- Overhead and mobile cranes.

Obviously, several of these pieces of equipment, such as hoists, conveyors, and trucks, can be used both at the workplace and for in-process movement.

Finally, there is a group of material-moving equipment that is used primarily at receiving stations and in warehouses. This group includes

- Stackers,
- Reach trucks,
- Lift trucks,
- Cranes, and
- Automated storage and retrieval systems.

WILL THE EQUIPMENT BE USED?

All of the equipment just mentioned can take over the requirements of holding, carrying, pushing, pulling, lowering, and lifting of materials that would otherwise be performed manually by a person. However, whether these tasks will indeed be done by machines depends not only on economical considerations, but also on the layout and organization of the work itself. For example, will an operator use a hoist to lift material that is to be fed into a machine if doing so is time consuming or awkward? Will a lift table be installed next to an assembly workstation if it means removing or relocating another workstation in order to make sufficient room?

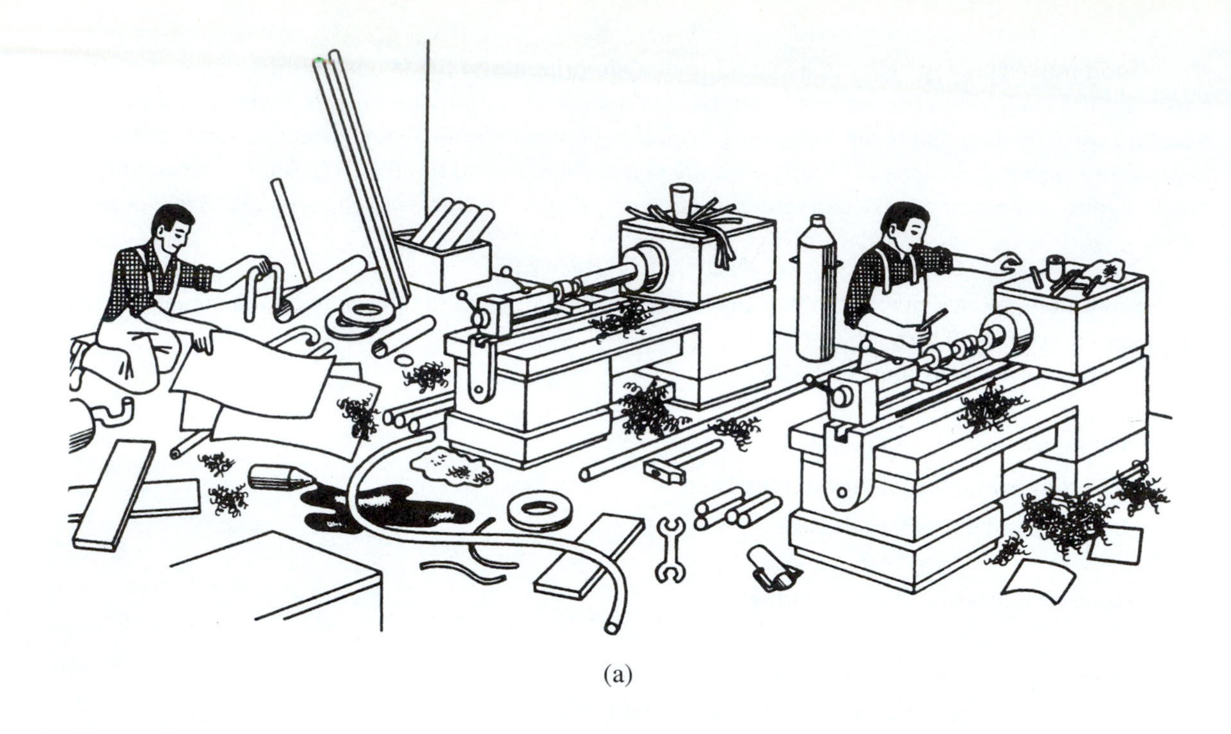

(a)

(b)

Figure 11–9. (a) Bad and (b) good housekeeping. (Courtesy of ILO 1988.)

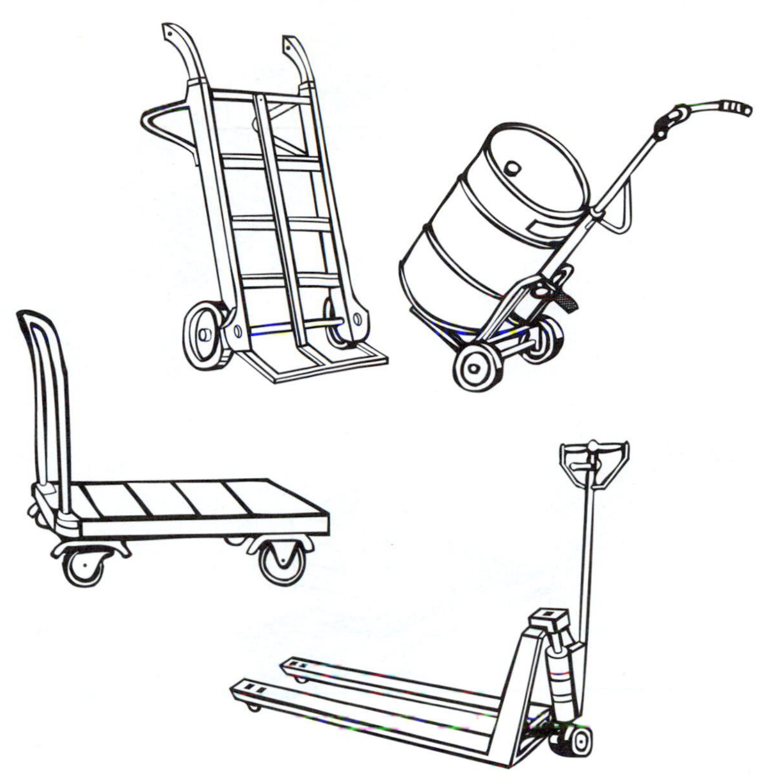

Figure 11–10. Simple carts and dollies for rolling materials instead of carrying them.

Obviously, the layout of the facility, as well as the design of the workplace must be suitable for the use of equipment. Furthermore, the operator must be trained that it is worthwhile to go through the effort of using a hoist instead of heaving the material by hand and must be convinced that this behavior is not only tolerated, but in fact encouraged, by management.

ERGONOMIC DESIGN OF EQUIPMENT

Equipment not only must be able to perform the material-moving job, but also must "fit" the human operator. Ease of use must be considered together with the safety of personnel working with or along the equipment. Unfortunately, some material-moving equipment, such as cranes, hoists, conveyors, and powered and hand trucks, show an alarming neglect of human factors and safety principles in their design. The actual use of hand trucks depends on the following characteristics associated with the trucks (Mack et al. 1995; Young et al. 1997):

1. The human force required to operate the trucks.
2. The stability of the truck.
3. The ease of steering the vehicle.
4. The design and location of the handle.

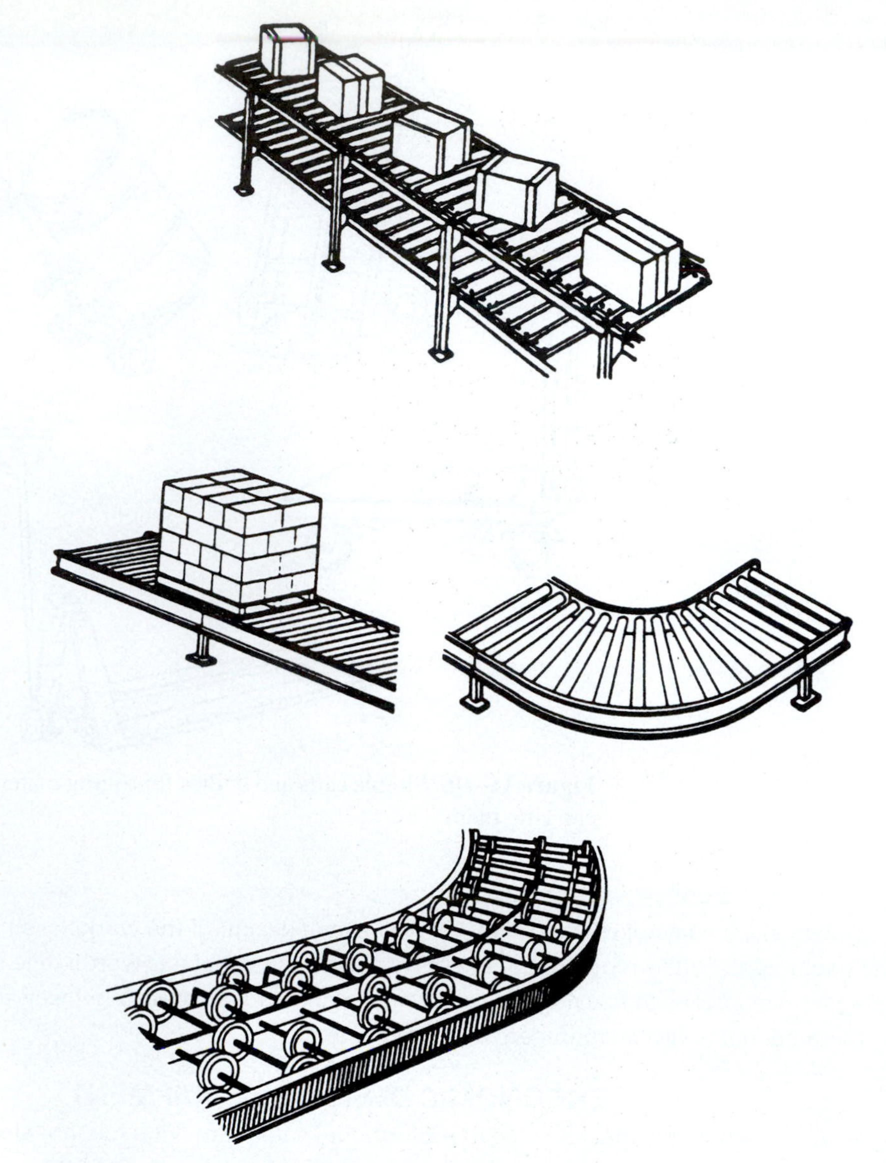

Figure 11–11. Conveyors on which one can push objects easily.

5. The ease of starting and stopping the truck.
6. The ease of loading and unloading.
7. The security of the load.

Older forklifts provide the overall worst example of bad human engineering: When lifting, carrying, and lowering a load, it obstructs the operator's view. Furthermore, the driver often has little space in which to sit and is subjected to vibrations and impacts transmitted from the rolling wheels. (See Figure 11–13.)

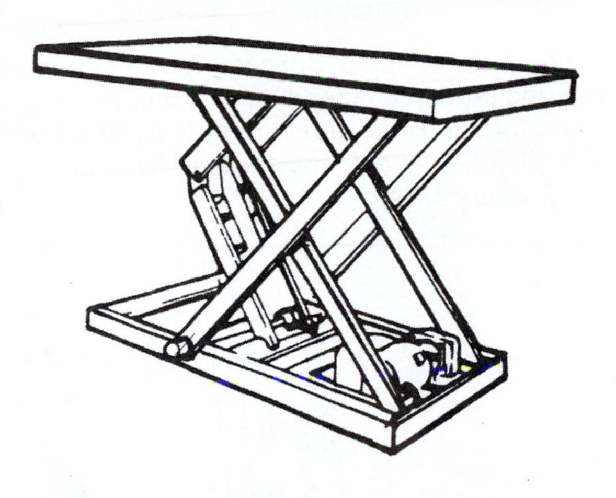

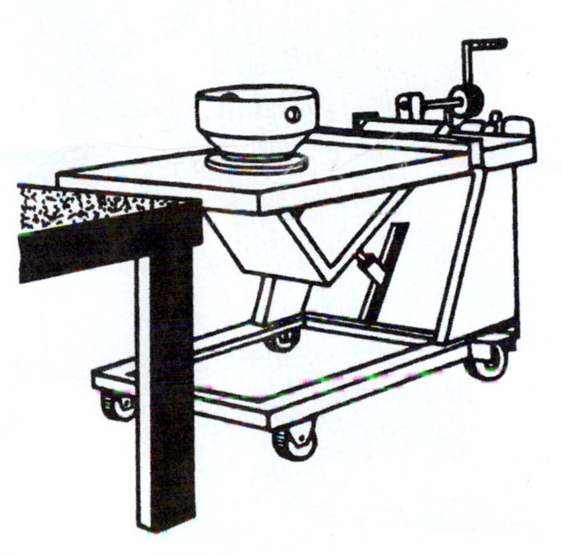

Figure 11–12. Lift tables.

STILL SOUNDS FAMILIAR, UNFORTUNATELY!

"I was shocked, dismayed, and perturbed. Recently I attended a regional industrial exhibition that had an emphasis on materials-handling equipment. I intentionally went around looking for bad or lacking human engineering. I found plenty . . inadequate labels; wrong-size controls; lack of shape, position, color coding; controls that could be inadvertently actuated; absence of guard rails; unintelligible instructions; slippery surfaces; impossible reach requirements; sharp edges; unguarded pinch-points; extreme strength requirements; lack of guards; spaces needed for maintenance too small for the human hand; poorly located emergency switches; and so on, and so on! . . Spacecraft and supersonic aircraft and missile monitoring equipment need human engineering; so, too, do hydraulic hoists and forklift trucks and conveyor systems and ladders." (Excerpts from R. B. Sleight's letter to the editor of the *Human Factors Society Bulletin,* p. 7, January 1984.)

Figure 11–13. Typical forklift truck. While the truck has a cage to protect the driver, the lift mechanism makes it very difficult to see what is ahead, even if there is no load.

Research in ergonomics and human factors research work has provided information on design features that are needed to fit equipment to the human—in particular, in order to provide safe and efficient working conditions. Much of this information was originally developed for military applications, but has found its way into industrial settings as well.

Overall dimensions and space requirements can be derived from human body and reach dimensions, as demonstrated in Chapter 8. The purposeful application of such information ensures, for example, that a driver compartment of a lift truck accommodates all sizes of drivers or that operators do not strain themselves in trying to reach an object hanging from an overhead conveyor or in grasping material in the far bottom corner of a transport bin. Of course, one should never design for the "average person," because this ghost does not really exist. Instead, one must design for body size ranges, such as from the 5th to the 95th percentile. (See Chapter 1.)

Handles on containers or tools should be so designed and oriented that the hand and forearm of the operator are aligned. The operator should not be forced to work with a bent wrist. (Recall the stricture to "Bend the handle, not the wrist.") Carpal tunnel syndrome and other cumulative trauma problems are less likely to occur if a person can work with a straight wrist. (See Chapter 8.)

Handles should be of such shape and material that squeezing forces are distributed over the largest possible area of the palm. Handle diameters should be in the range of 2½ to 5 centimeters. The surfaces of handles should be slip resistant. Coupling with the hands is best effected by protruding handles or by gripping-notch types, while handhold cutouts or drawer-pull types are acceptable. If they are on a boxlike object (of about 40 × 40 × 40 cm, weighing between 9 and 13 kg), one should arrange the location of handles according to the scheme shown in Figure 11–14: The right side of the box is divided into nine areas, with area 5 in the center. (The same numbering system is on the left side of the box, with area 1 again on top and close to the body.) The best, or at least suitable, handle locations are listed in Table 11–1. The worst case is not to have any handles, which makes the material handler first tilt the box and then move it (Authier et al. 1994; Gagnon 1997; Mirka et al. 1998).

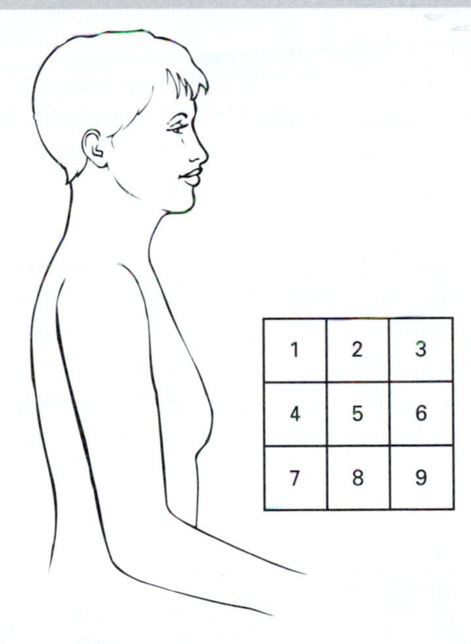

Figure 11–14. Scheme describing locations of handles in a box-type container. Each side is divided into nine regions, with number 1 on top and closest to the body.

TABLE 11–1. Preferred Locations of Handles (See Figure 11–14) on 40 × 40 × 40 CM Boxes

	For lifting and lowering in front of the body	*For work with sideways twisting of the body*
Best	2/2	6/8, 8/8 for heavy loads
Acceptable	3/8, 6/8	3/8, 2/2

Source: From Drury et al. 1989.

Handles or cutouts may be angled (against the horizontal), 2/2 at about 83 degrees, 3/8 at 76 degrees, 6/8 at 60 degrees, and 8/8 at 50 degrees.

Containers and trays must be designed to have the proper weight, size, balance, and coupling with the human hand. The container should be as light as possible, to add little to the load of the material. The size of the container and the arrangement of its handles should be such that the center of the loaded tray is close to the body. The tray should be well balanced, with its weight centered between, but below, the handholds. Big, heavy, pliable bags are usually more difficult to handle than boxes or trays.

Written instructions and labels are not necessary if equipment and its operation are designed to be "perfectly obvious." However, when instructions are needed, labeling should be done according to these rules (see also Chapter 10):

1. Write the instructions in the simplest, most direct manner possible.
2. Give only the information that is needed.
3. Describe clearly the required action. Never mix different instruction categories, such as operation, maintenance, and warning.
4. Use familiar words.
5. Be brief, but not ambiguous.
6. Locate labels in a consistent manner.
7. Make words read horizontally, not vertically.
8. Make the label a color that contrasts with the equipment it is on.

ERGONOMIC DESIGN OF THE WORKSTATION AND WORK TASK

The most effective and efficient way to reduce the number and severity material-handling injuries is to design equipment ergonomically, so that job demands are matched to human capabilities.

Designing the job to fit the human can take several approaches. The most radical solution is to "design out" manual material movement by assigning it to machines: If no people are involved, no people are at risk. If people must be involved, the weight and size of the load should be kept small best accompanied by ergonomic design of the work task—that is, by selecting the proper type of material-handling movements (eg, a horizontal push instead of a vertical lift) and their frequency of occurrence. The location of the object with respect to the body is very important; it is best between hip and shoulder height, directly in front of the body so as to avoid twisting or bending the trunk. (See Figures 11–15 and 11–16.) The object itself is

Figure 11–15. Store material at the proper height. (Modified from ILO 1988.)

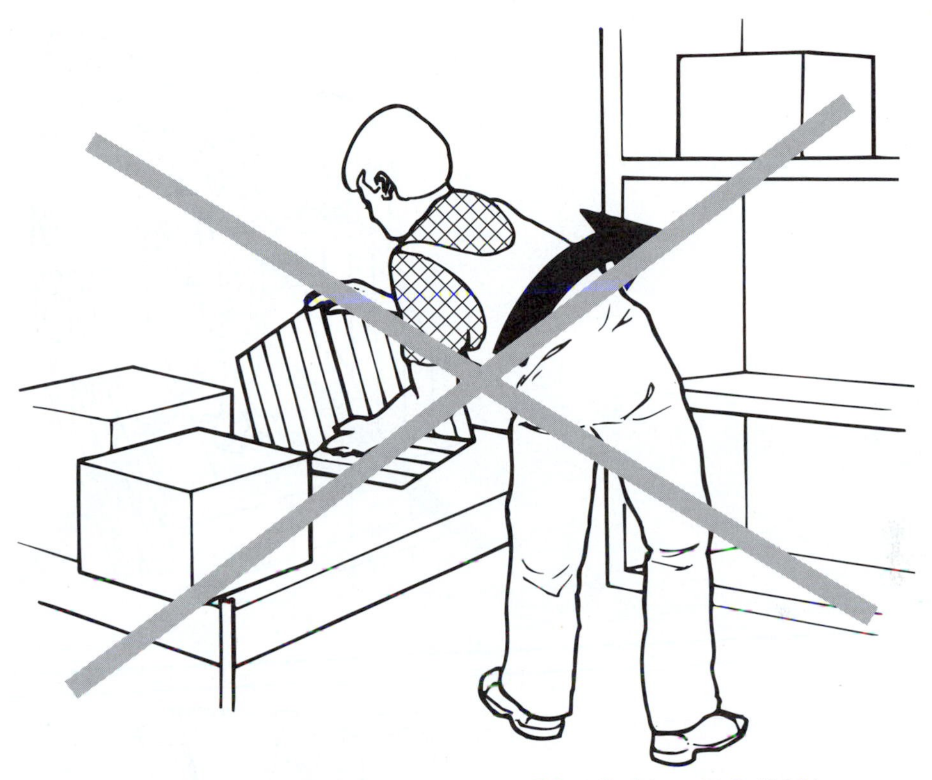

Figure 11–16. Avoid twisting the trunk. (Modified from ILO 1988.)

important, of course, regarding its bulk, its pliability (eg, a firm box vs. a pliable bag), and its ability to be grasped securely (through its shape or with handles). Naturally, the workplace itself must be well designed and maintained, with the proper working height, material provided in containers from which it can be removed easily (see Figure 11–17), a nonslip floor, and, in general, a clean, quiet, orderly, suitable environment. Always keep in mind that all ergonomic conditions contribute to avoiding stress, physical overexertions, and accidents.

Permissible Loads for Handling

Only a few decades ago, it was believed that certain weights could be lifted safely by men, women, or children (International Labor Office 1988). This simplistic idea is not valid. Through epidemiological, medical, physiological, biomechanical, and administrative approaches, new knowledge on human capabilities for manual material handling has been gained. The individual's capabilities depend on such the variables as frequency, location, direction, and other details of the MMH activities. Two major sets of recommendations are applicable, but the reader is cautioned that they are based on assumptions and approaches that still need refinement and further evaluation (Fredericks et al. 1998; Hidalgo et al. 1995, 1997; Lavender et al. 1997; Potvin 1997; Waters and Putz-Anderson 1998, 1999).

Figure 11–17. Deliver and store containers at the proper working height. Note that even when the container is propped as shown, the worker still has to bend the back to reach a far object.

LIMITS FOR LIFTING AND LOWERING

In 1981, a panel of experts prepared a *Work Practices Guide for Manual Lifting* for the U.S. National Institute for Occupational Safety and Health (NIOSH 1981). For the first time, there was a document that contained distinct recommendations for acceptable masses to be lifted that differed from the previous assumptions that one could establish one given weight that would be "safe" to lift.

In the 1981 guide, two different threshold curves were established. The lower, called the *action limit* (AL) was thought to be safe for 99 percent of working men and 75 percent of working women in the United States. The AL values were dependent on the starting height of the load, the length of its upward path, its distance in front of the body, and the frequency of lifting. If the actual weight was above the AL value, engineering or managerial controls had to be applied to bring the load value down to the acceptable limit. However, under no circumstances was a lifting task allowed if the load was three times larger than the AL values. This threshold was called the *maximum permissible load* (MPL).

A decade later, NIOSH revised the technique for assessing overexertion hazards of manual lifting (Putz-Anderson and Waters 1991). Unlike the old document, the new one contains two separate weight limits, but has only one *recommended weight limit* (RWL). This is the key concept in the 1991 guide; it represents the maximal weight of a load that may be lifted or lowered under the best possible conditions by about 90 percent of American industrial workers, male or female, physically fit, and accustomed to physical labor.

The 1991 equation used to calculate the RWL resembles the 1981 formula for calculating the AL, but includes new multipliers to reflect asymmetry and the quality of hand–load coupling. The 1991 equation allows a maximum *Load Constant* (LC) (permissible only under the most favorable circumstances) with a value of 23 kg (51 lb), which may not be exceeded under any circumstances. This is quite a reduction from the maximal 40 kg in the 1981 NIOSH guidelines.

The following statements apply to the 1991 guidelines:

1. The equation does *not* include safety factors for such conditions as unexpectedly heavy loads, slips, or falls or for temperatures outside the range from 19°C (66°F) to 26°C (79°F) and for humidity outside the range from 35 to 65 percent.
2. The equation does *not* apply to one-handed tasks performed while the person is seated or kneeling or to tasks carried out in a constrained workspace.
3. The equation assumes that other manual handling activities and body motions requiring a high expenditure of energy, such as that expended in pushing, pulling, carrying, walking, climbing, or holding an object, are less than 20 percent of the total work activity for the work shift (Waters 1991).
4. The equation assumes that the shoe–floor-surface coupling provides a coefficient of static friction of at least 0.4 between the sole of the shoe and the standing surface.
5. The equation may be applied to lifting or lowering tasks (ie, acts of manually grasping and moving an object of definable size, without mechanical aids, to a different height) under the following circumstances:
   - The duration of the task is normally between two and four seconds. The load is grasped with both hands.
   - The motion is smooth and continuous.

- The posture is unrestricted. (See earlier.)
- Foot traction is adequate. (See earlier.)
- The temperature and humidity are moderate. (See earlier.)
- The horizontal distance between the two hands is no more than 65 cm (25 in).

The **Lifting Index (LI)** is calculated as LI = *L*/RWL, with *L* the actual load. If

- LI is at or below 1, no action must be taken.
- LI exceeds 1, the job must be ergonomically redesigned.

The **Recommended Weight Limit (RWL)** is calculated as (Waters and Putz-Anderson 1999):

$$RWL = LC \times HM \times VM \times DM \times AM \times FM \times CM,$$

where **LC** is the load constant of 23 kg (51 lb);

HM is the horizontal multiplier, where *H* is the horizontal location (distance) of the hands from the midpoint between the ankles at the start and the end points of the lift or lower (see table that follows);

VM is the vertical multiplier, where *V* is the vertical location (height) of the hands above the floor at the start and end points of the effort (see table that follows);

DM is the distance multiplier," where *D* is the vertical travel distance from the start to the end points (see table that follows);

AM is the asymmetry multiplier, where *A* is the angle of asymmetry [ie, the angular displacement of the load from the medial (midsagittal) plane that forces the operator to twist the body], measured at the start and end points of the effort and projected onto the floor (see table that follows);

FM is the frequency multiplier, where *F* is the frequency of the lifting or lowering, expressed in lifts or lowers per minute, and depending on the duration of the task (see table that follows); and

CM is the coupling multiplier, where *C* indicates the quality of the coupling between hand and load (see table that follows).

Each of HM, VM, DM, AM, FM, and CM can assume a value between zero and unity.

The following values are entered into the equation for RWL:

	METRIC	U.S. CUSTOMARY
LC	23 kg	51 lb
HM	$25/H$	$10/H$
VM	$1 - (0.003 \mid V - 75 \mid)$	$1 - (0.0075 \mid V - 30 \mid)$
DM	$0.82 + (4.5/D)$	$0.82 + (1.8/D)$
AM	$1 - (0.0032A)$	$1 - (0.0032A)$

FM (See listing that follows.)

CM (See listing that follows.)

The variables themselves can have the following values:

H is between 25 cm (10 in) and 63 cm (25 in). Although objects can be carried or held closer than 25 cm in front of the ankles, most objects that are closer cannot be lifted or lowered without encountering interference from the abdomen. Objects farther away than 63 cm (25 in) cannot be reached and cannot be lifted or lowered without losing one's balance, particularly when the lift is asymmetrical and the operator is small.

V is set to be between zero and 175 cm (70 in) because few people can lift higher.

D is equal to $V_{end} - V_{start}$ for lifting; for a lowering task, *D* equals $V_{start} - V_{end}$.

A is set to be between zero and 135°.

F is between one lift or lowering every five minutes (over a working time of eight hours) to 15 lifts or lowers every minute (over a time of one hour or less), depending on the vertical location *V* of the object. Table 11–2 lists the frequency multipliers, FM.

C is between 1.00 ("good") and 0.90 ("poor"). The effectiveness of the coupling may vary as the object is being lifted or lowered; indeed, a "good" coupling can quickly become "poor." Three categories are defined in detail in the NIOSH publication and result in the following listing of values for the coupling multiplier CM:

COUPLINGS	$V < 75$ cm (30 in)	$V \geq 75$ cm (30 in)
Good	1.00	1.00
Fair	0.95	1.00
Poor	0.90	0.90

LIMITS FOR LIFTING, LOWERING, PUSHING, PULLING, AND CARRYING

Snook and Ciriello published extensive tables of loads and forces found acceptable by male and female workers for continuous manual material-handling jobs; with their jobs not limited to lifting and lowering, as in the 1991 *NIOSH Guide,* but also pertaining to pushing, pulling, and carrying. (The original data sets were published by Snook in 1978, first updated in 1983 by Ciriello and Snook, revised in 1991 by Snook and Ciriello, and again updated by Ciriello, Snook, and Hughes in 1993.)

TABLE 11–2. Frequency Multipliers FM

Frequency, lifts/min	Work duration (continuous) ≤ 8 hr		≤ 2 hr		≤ 1 hr	
	V < 75*(cm)	*V* ≥ 75(cm)	*V* < 75 (cm)	*V* ≥ 75 (cm)	*V* < 75 (cm)	*V* ≥ 75 (cm)
0.2	0.85	0.85	0.95	0.95	1.00	1.00
0.5	0.81	0.81	0.92	0.92	0.97	0.97
1	0.75	0.75	0.88	0.88	0.94	0.94
2	0.65	0.65	0.84	0.84	0.91	0.91
3	0.55	0.55	0.79	0.79	0.88	0.88
4	0.45	0.45	0.72	0.72	0.84	0.84
5	0.35	0.35	0.60	0.60	0.80	0.80
6	0.27	0.27	0.50	0.50	0.75	0.75
7	0.22	0.22	0.42	0.42	0.70	0.70
8	0.18	0.18	0.35	0.35	0.60	0.60
9	0	0.15	0.30	0.30	0.52	0.52
10	0	0.13	0.26	0.26	0.45	0.45
11	0	0	0	0.23	0.41	0.41
12	0	0	0	0.21	0.37	0.37
13	0	0	0	0	0	0.34
14	0	0	0	0	0	0.31
15	0	0	0	0	0	0.28
>15	0	0	0	0	0	0

Source: From Putz-Anderson and Waters 1991.

The following prerequisites apply:

- Two-handed symmetrical material handling in the medial (midsagittal) plane (ie, directly in front of the body); yet, a light body twist may occur during lifting or lowering.
- Moderate width of the load, such as 75 cm or less.
- Good couplings of hands with handles and shoes with the floor.
- Unrestricted working postures.
- A favorable physical environment, such as about 21°C and a relative humidity of 45 percent.
- Only minimal other physical work activities.
- Material handlers who are physically fit and accustomed to labor.

The format of Snook and Ciriello's recommendations is different from that of the NIOSH guidelines. The NIOSH values are unisex, while Snook and Ciriello's data are separated for female and males. Also, their data are grouped with respect to the percentage of the worker population to whom the values are acceptable, whereas the NIOSH data are not. In the excerpts reprinted here, only those values are listed that are said to be acceptable to 50 percent, 75 percent, or 90 percent of the worker population. (See the original tables for more information.)

Lifts and powers are subdivided into three different height areas:

- floor to knuckle height,
- between knuckle and shoulder heights, and
- shoulder to overhead reach heights.

Tables 11–3 through 11–7 show, in abbreviated form, Snook's and Ciriello's 1991 recommendations for suitable loads and forces in lifting, lowering, pushing, pulling, and carrying. Their original tables should be consulted for more information.

It is of interest to note that, similar to the NIOSH recommendations, the data in Snook and Ciriello's (1991) study also indicate that the absence of handles reduces the loads that people are willing to lift and lower by an average of about 15 percent. If the objects become so wide or so deep as to be difficult to grasp, the lifting and lowering values are again considerably reduced. If several of the material-handling activities occur together, the most strenuous task establishes the handling limit; the tables give the lowest percentage acceptance of a given set of parameters for the most limiting task condition.

If actual loads or forces exceed the values shown in the table, engineering or administrative controls should be applied. Snook believes that industrial back injuries could be reduced by about one-third if loads that lie above the values acceptable to 75 percent of material handlers could be eliminated.

COMPARING THE RECOMMENDATIONS

The recommendations by NIOSH and by Snook and Ciriello indicate material-handling conditions that are deemed suitable, or unsuitable for workers in the United States.

For new systems, the data are planning guides for material-handling conditions that either can be performed by persons or should be assigned to machines. To evaluate existing material-handling systems, one can compare existing job requirements with the data in the tables in order to seek out those task demands that are likely to exceed human capabilities. Then the working conditions should be changed by engineering intervention (ie automation, mechanization, or a lowering of the demands) managerial intervention (ie, selecting and training workers and rest pauses or job rotation). Clearly, the engineering intervention is preferable, because it eliminates the source of the risk.

In general, one finds the 1981 NIOSH action limits for lifting to be quite similar to the lift data recommended by Snook in 1978 for 90th-percentile male workers. (This is not surprising, because Snook's lift data were part of the basic information used to develop the 1981 NIOSH guidelines.) The 1991 NIOSH recommendations are also, in principle, similar to many of the 1991 Snook and Ciriello data on lifting and lowering. In some cases, however, particularly at extreme working conditions and for very frequent activities, discrepancies between the sets of recommendations are substantial.

Furthermore, as discussed earlier in this chapter, assessments of "suitable" efforts in manual material handling have been developed using four different sets of criteria:

1. Physiological, mostly metabolic and circulatory strains.
2. Psychophysical, mostly subjective assessments of what one is willing to do.
3. Biomechanical, mostly assessing compression values in the lumbar area of the back, usually based on calculations from models.
4. Intraabdominal pressure (IAP), not much used anymore.

TABLE 11–3. Maximal Acceptable Lifting Weights (kg)—see the complete tables by Ciriello et al. 1993.

			Floor level to knuckle height One lift every								*Knuckle height to shoulder height One lift every*								*Shoulder height to overhead reach One lift every*							
Width (a)	*Distance (b)*	*Percent (c)*	*5 sec*	*9 sec*	*14 sec*	*1 min*	*2 min*	*5 min*	*30 min*	*8 hr*	*5 sec*	*9 sec*	*14 sec*	*1 min*	*2 min*	*5 min*	*30 min*	*8 hr*	*5 sec*	*9 sec*	*14 sec*	*1 min*	*2 min*	*5 min*	*30 min*	*8 hr*
	Males																									
		90	9	10	12	16	18	20	20	24	9	12	14	17	17	18	20	22	8	11	13	16	16	17	18	20
34	51	75	12	15	18	23	26	28	29	34	12	16	18	22	23	23	26	29	11	14	17	21	21	22	24	26
		50	17	20	24	31	35	38	39	46	15	20	23	28	29	30	33	36	14	18	21	26	27	28	31	34
	Females																									
		90	7	9	9	11	12	12	13	18	8	8	9	10	11	11	12	14	7	7	8	9	10	11	12	12
34	51	75	9	11	12	14	15	15	16	22	9	10	11	12	13	13	14	17	8	8	9	11	11	11	12	14
		50	11	13	14	16	18	18	20	27	10	11	13	14	15	15	17	19	9	10	11	12	13	13	14	17

(a) Handles in front of the operator (cm).
(b) Vertical distance of lifting (cm).
(c) Acceptable to 50, 75, or 90 percent of industrial workers.

Conversion
1 kg = 2.2 lb
1 cm = 0.4 in

TABLE 11–4. Maximal Acceptable Lowering Weights (kg)—see the complete tables by Ciriello et al. 1993.

			Knuckle height to floor level One lowering every								*Shoulder height to knuckle height One lowering every*								*Overhead reach to shoulder height One lowering every*							
Width (a)	*Distance (b)*	*Percent (c)*	*5 sec*	*9 sec*	*14 sec*	*1 min*	*2 min*	*5 min*	*30 min*	*8 hr*	*5 sec*	*9 sec*	*14 sec*	*1 min*	*2 min*	*5 min*	*30 min*	*8 hr*	*5 sec*	*9 sec*	*14 sec*	*1 min*	*2 min*	*5 min*	*30 min*	*8 hr*
	Males																									
		90	10	13	14	17	20	22	22	29	11	13	15	17	20	20	20	24	9	10	12	14	16	16	16	20
34	51	75	14	18	20	25	28	30	32	40	15	18	21	23	27	27	27	33	12	14	17	19	22	22	22	27
		50	19	24	26	33	37	40	42	53	20	23	27	30	35	35	35	43	16	19	22	24	28	28	28	35
	Females																									
		90	7	9	9	11	12	13	14	18	8	9	9	10	11	12	12	15	7	8	8	8	10	11	11	13
34	51	75	9	11	11	13	15	16	17	22	9	11	11	12	14	15	15	19	8	9	10	10	12	13	13	16
		50	10	13	14	16	18	19	20	27	11	13	13	14	16	18	18	22	10	11	11	12	14	15	15	19

(a) Handles in front of the operator (cm).
(b) Vertical distance of lowering (cm).
(c) Acceptable to 50, 75, or 90 percent of industrial workers.

Conversion
1 kg = 2.2 lb
1 cm = 0.4 in

TABLE 11–5. Maximal Acceptable Push Forces (N)—see the complete tables by Ciriello et al. 1993.

		One 2.1-meter push every										One 30.5 m push every					
	Height (a)	Percent (b)	6 sec	12 sec	1 min	2 min	5 min	30 min	8 hr		Height (a)	Percent (b)	1 min	2 min	5 min	30 min	8 hr
Males	95	90	206	235	255	255	275	275	334	INITIAL PUSH FORCES	95	90	167	186	216	216	265
		75	275	304	334	324	353	353	432			75	206	235	275	275	343
		50	334	373	422	422	442	442	530			50	265	294	343	343	432
Females	89	90	137	147	167	177	196	306	216		89	90	118	137	147	157	177
		75	167	177	296	216	235	245	265			75	147	157	177	186	206
		50	196	216	245	255	285	294	314			50	177	196	206	226	255
Males	95	90	98	128	159	167	186	186	226	SUSTAINED PUSH FORCES	95	90	79	98	118	128	157
		75	137	177	216	216	245	255	304			75	108	128	157	177	206
		50	177	226	225	285	324	335	392			50	147	167	196	226	265
Females	89	90	59	69	88	88	98	108	128		89	90	49	59	59	69	88
		75	79	106	128	128	147	157	186			75	79	88	88	98	128
		50	168	147	177	177	196	206	255			50	98	118	118	128	167

(a) Vertical distance from floor to hands (cm).
(b) Acceptable to 50, 75, or 90 percent of industrial workers.

Conversion
1 $kg_f = 2.2\ lb_f = 9.81$ N
1 cm = 0.4 in

TABLE 11–6. Maximal Acceptable Pull Forces (N)—see the complete tables by Ciriello et al. 1993.

	Height (a)	Percent (b)	One 2.1-meter pull every 6 sec	12 sec	1 min	2 min	5 min	30 min	8 hr
			INITIAL PULL FORCES						
Males	95	90	186	216	245	245	265	265	314
		75	226	265	304	304	314	324	383
		50	275	314	353	353	383	383	461
Females	89	90	137	157	177	186	206	216	226
		75	157	186	206	216	245	255	265
		50	186	226	245	255	285	294	314
			SUSTAINED PULL FORCES						
Males	95	90	98	128	157	167	186	196	235
		75	128	167	206	216	245	255	294
		50	157	206	255	265	304	314	363
Females	89	90	59	88	98	98	108	118	137
		75	79	118	128	128	147	157	196
		50	98	147	157	167	186	196	245

(a) Vertical distance from floor to hands (cm).
(b) Acceptable to 50, 75, or 90 percent of industrial workers.

Conversion
1 kg_f = 2.2 lb_f = 9.8 N
1 cm = 0.4 in

TABLE 11–7. Maximal Acceptable Carry Weights (kg)—see the complete tables by Ciriello et al. 1993.

	Height (a)	*Percent (b)*	*One 2.1-meter carry every 6 sec*	*12 sec*	*1 min*	*2 min*	*5 min*	*30 min*	*8 hr*
		90	13	17	21	21	23	26	31
Males	79	75	18	23	28	29	32	36	42
		50	23	30	37	37	41	46	54
		90	13	14	16	16	16	16	22
Females	72	75	15	17	18	18	19	19	25
		50	17	19	21	21	22	22	29

(a) Vertical distance from floor to hands (cm).
(b) Acceptable to 50, 75, or 90 percent of industrial workers.

Conversion
1 kg = 2.2 lb
1 cm = 0.4 in

Note that the 1991 NIOSH recommendations are based on an amalgam of all of these criteria, plus epidemiologic inputs. Comparisons of the results are very difficult because of different underlying concepts, conditions, procedures, criteria, and units. In spite of these difficulties, Garg and Ayoub (1980) found that recommendations resulting from the physiological, psychophysical, and biomechanical approaches were not in general agreement. Specifically, the recommended lift weights based on the psychophysical criteria were lower than those based on the biomechanical approach (this was also was found to be the case in 1991 by Waikar et al.), and the lift loads based on the physiological criteria were lower than those based on the psychophysical criteria, for greater lift frequencies.

Dempsey (1999) compared physiologically based recommendations with Snook et al.'s psychophysically based data and found that latter the approach resulted in recommendations that exceeded physiological limits for activities performed with high frequency.

Nicholson (1989) compared lifting recommendations derived from psychophysical, biomechanical, and IAP criteria. The IAP recommendations were somewhat higher at working heights above the shoulder, but often much lower for work below shoulder height, than the psychophysical recommendations. The latter compared reasonably well with the biomechanical recommendations, but the IAP recommendations were only about half the biomechanical values. Nicholson also compared pushing and pulling recommendations: For pushing, the psychophysical values were in good agreement with IAP recommendations, but for pulling, the IAP data were considerably higher than the psychophysically based recommendations.

Potvin and Bent (1997) compared1991 NIOSH recommendations with those of Snook at al. with respect to the width of lifted objects and found that the NIOSH data should be corrected to agree with the psychophysically based findings.

Effects of Multiple Work Activities

Straker et al. (1996) investigated the effects of combining the activities of lifting, lowering, pushing, pulling, and carrying into one work task; the resulting acceptable load limits were quite different for the separate tasks compared with several of their combinations.

Obviously, the various recommendations overlap considerably, but different approaches may lead to differing results. In the case of conflicting guidelines, it is prudent to use the recommendation that protects the human material handler best.

Effects of Constrained Space and Postures

So far in this chapter, we have assumed that the material handler was free to adopt any body postures suitable for the job. Yet, under certain conditions, such as in underground mines or in aircraft cargo holds, only limited room is available. A stooped, kneeling, bent, sitting, supine, prone or otherwise restricted body posture reduces the ability to handle objects, often severely, in terms of forcefulness, direction, and distance.

Use of Lifting Belts

The intraabdominal pressure was believed to help support the curvature of the spinal column and to relieve it from some of the compression force generated by the outside load and by the body weight in lifting or lowering loads. (See Chapter 1 and Figure 11–1.) Placing an external wrapping around the abdominal region may make the walls of this internal pressure column stiffer. For example, porters and workers in Nepal traditionally wear a cloth called a *patuka* wound around the waist (Shah 1993), while weight lifters commonly use fairly stiff, wide, contoured belts. It has been advocated that people who do heavy manual material handling should also wear such abdominal belts (called variously back belts, lift belts, back braces, or back supports). A large number of studies has been performed, summarized, and reviewed, for example, by NIOSH (1994) and, more recently, by Lavender et al. (1998), McGill (1999), and Thoumier et al. (1998). The conclusions neither summarily favor nor disfavor the wearing of support belts in industrial jobs.

- Certain material handlers, especially persons who have suffered a back injury, may benefit from a suitable belt.
- Those who wish to wear a belt should be screened for cardiovascular risk, which may be increased by the pressure from the belt.
- Belt wearers should receive an education such as that provided in in back school, because the presence of the belt may provide a false sense of security.
- Belts should not be considered for long-term use.
- Belts are not a substitute for the ergonomic design of the work task, the workplace, and work equipment.

Altogether, the use of lifting belts for industrial material handling does not seem to be an effective way of preventing overexertion injuries: Even competitive weight lifters suffer back injuries.

ERGONOMIC RULES FOR INDUSTRIAL MANUAL HANDLING TASKS, PARTICULARLY LIFTING OF LOADS

Although general rules may not cover all specific cases and there are always the proverbial exceptions, the following guidelines are useful:

Ground Rule: Eliminate the manual task if at all possible. (No exposure, no injury.)

Rule 1: Reduce the sizes, weights, and forces involved.

Rule 2: Provide good handholds.

Rule 3: Keep the object close to the body, always in front; don't twist.

Rule 4: For lifting, keep the trunk up and the knees bent.

Rule 5: Minimize the distance through which the object must be moved.

Rule 6: Move horizontally, not vertically: Convert lifting and lowering to pushing, pulling, or carrying. (Provide material at the proper working height.)

Rule 7: Plan all movements, and make them smooth.

Rule 8: Don't lift (lower) anything that must be lowered (lifted) later.

If material-handling tasks cannot be avoided altogether, take measures to facilitate the task and make it less hazardous and strenuous. (See Figures 11–18, 11–19 and 11–20.)

The following guidelines address first the (most hazardous) tasks of lifting and lowering, then carrying and holding, and finally pushing and pulling.

ELIMINATE THE NEED TO LIFT OR LOWER AN OBJECT MANUALLY

- by supplying the material at the working height, perhaps by raising (or lowering) the operator or the work area.

If lifting or lowering cannot be eliminated, use, for example,

- a lift table, lift platform, elevated pallet, or lift truck;
- a crane, hoist, or elevating conveyor; or
- a work dispenser, gravity dump, gravity chute, etc.

Figure 11–18. It may be easier to have somebody help in handling an object than try to do it alone. (Modified from ILO 1988.)

ELIMINATE THE NEED TO CARRY CONVERTING THE TASK TO PUSHING OR PULLING

- with the use of a conveyor, cart, dolly, or truck, with tables or slides between workstations, and by rearranging of workplace; and
- by increasing the "unit weight" so that it must be handled mechanically (eg, via palletized loads).

If carrying cannot be eliminated, reduce the weight that must be handled by

- using lighter material for the object,
- making the container smaller,
- making the container lighter,
- making the object smaller,
- assigning two workers to the job, and
- reducing the carrying distance by
- changing the layout of the workplace or
- getting the operation closer to the previous or next operation with the use of conveyors.

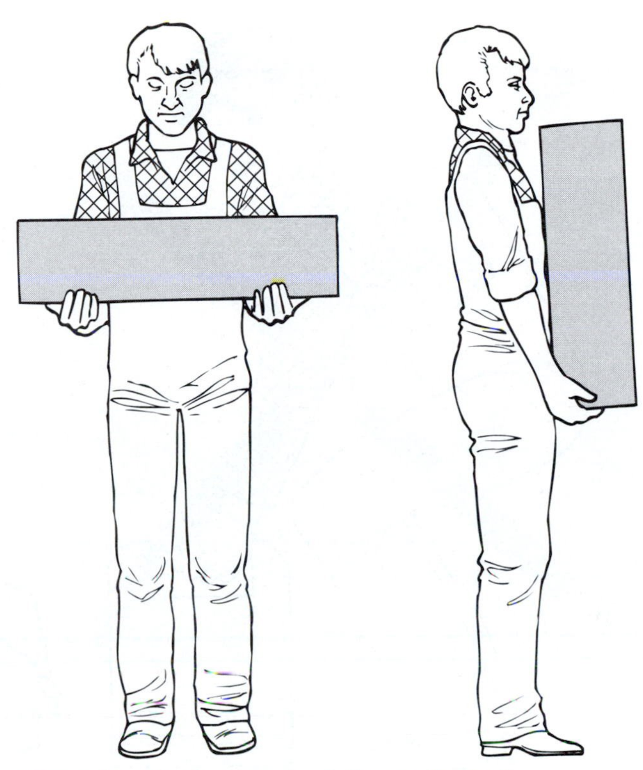

Figure 11–19. Keep the load close to the body.

ELIMINATE THE NEED FOR MANUAL HOLDING

by using automatic feed and unload systems, jigs, fixtures, support stands, and tables.

If manual holding cannot be eliminated,

- reduce the weight of the object to be held,
- reduce the time the object must be held, or
- hold the object in front of the trunk and as close to the trunk as possible.

ELIMINATE THE NEED FOR PUSHING OR PULLING

by using a conveyor, lift truck, powered truck, slide, chute, etc.

If pushing or pulling cannot be eliminated, reduce the force required by

- reducting the weight or size of the load,
- using ramps, a conveyor, a dolly or truck, wheels and casters, or air bearings, together with good maintenance of the equipment and floor surface.

Figure 11–21 summarizes these ergonomic rules and their results.

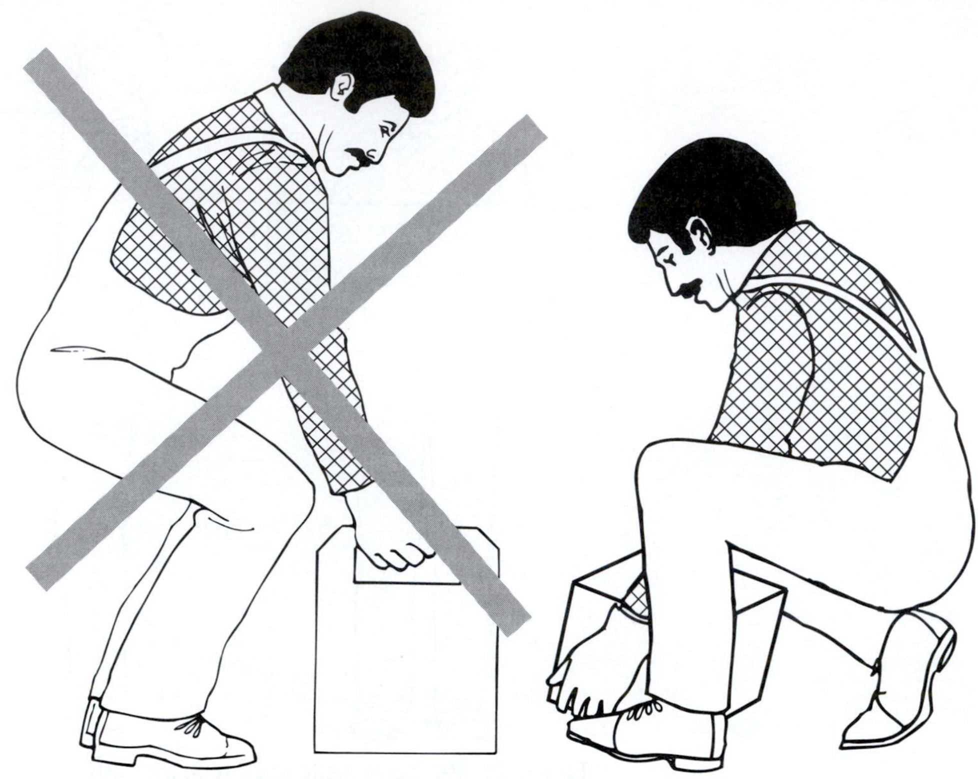

Figure 11–20. If lifting from the floor, keep the load between the legs, never in front of the knees.

ACTIVITIES OTHER THAN INDUSTRIAL LIFTING, LOWERING, AND CARRYING

Past research has dealt predominantly with the lifting of loads, often considering only the ability to generate an upward-directed static force. Aside from static (isometric) horizontal push or pull forces (see Chapter 8) and push–pull and carrying forces in material handling, little information is available.

Carrying Techniques

Snook and his colleagues' data on industrial carrying assume that the load is carried by the hands, but there are other techniques of carrying loads that are hardly ever mentioned in the literature, yet are commonly used. One is to carry loads on or in the arms, often by clasping the load and pressing it against the trunk. This everyday technique is not easily described in biomechanical terms, and the associated muscular effort and expenditure of energy depend very much on the actual manner of clasping and on the load carried. The other common, but seldom described, technique is to support a load on the hips, such as a mother does when carrying her small child.

Obviously, loads can be carried in many different ways. The technique that is most appropriate depends on a number of variables: the amount (weight) of load, its shape and size, its rigid-

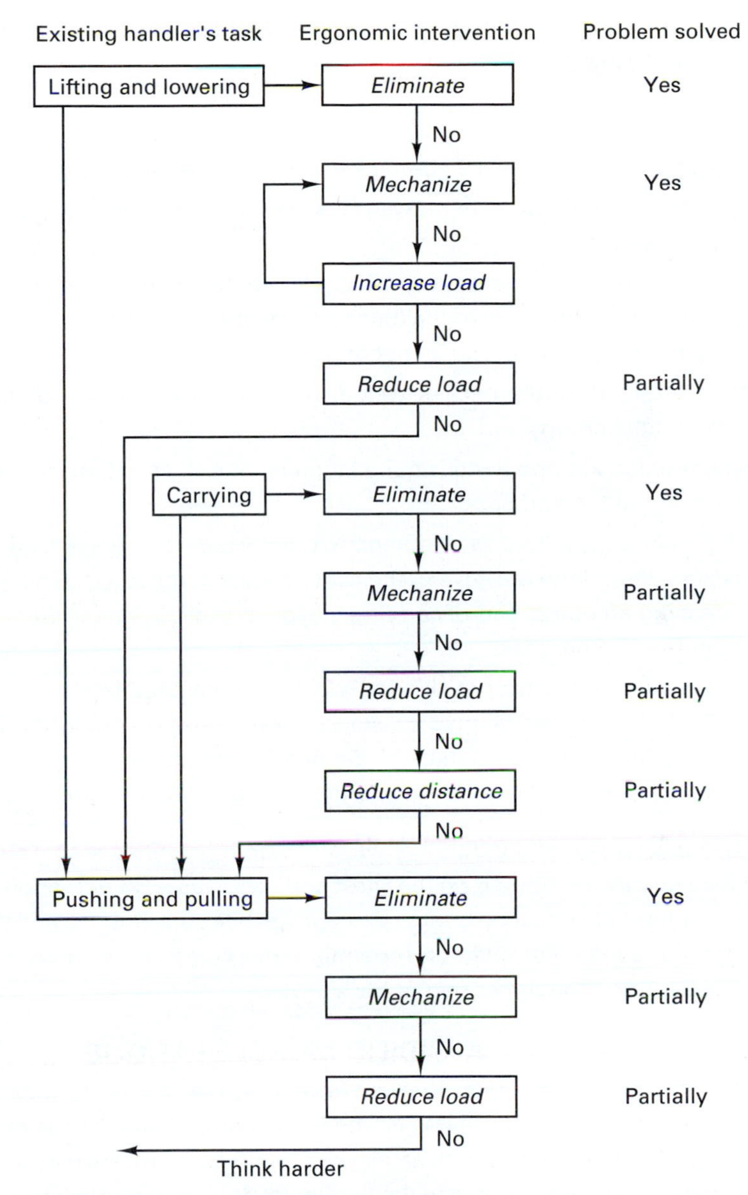

Figure 11–21. Ergonomic interventions.

ity or pliability, the provision of handholds or other means of attachment, and the bulkiness or compactness of the load. What is the best technique also depends on the distance the load is carried; on whether the path is straight or curved, flat or inclined, and with or without obstacles; and on whether one can walk freely or must duck (because of limited space) or hide, such as a soldier who does not want to be detected. Many of these aspects have not been formally investigated, but some have been the object of studies (Ayoub and Smith 1988; Datta and Ramanathlan 1971; Haisman 1988; Holewijn and Lotens 1992; Johnson et al. 1998; Kirk and Schneider 1990; Knapick et al.

1996; Legg et al. 1997; Mello et al. 1988). The conclusions of these studies may be summarized as follows:

- In general, the load should be carried close to the midaxis of the body, near waist height. The further away from there, either toward the feet or toward one side, the more demanding the carrying becomes.
- Carrying a medium load, say, 25 to 30 kg, distributed on the chest and back is the least energy-consuming method of carrying (but a load on the front and sides must be kept small in size and very close to the body, so as not to hinder movements).
- Carrying the load on the back or well distributed across both shoulders and the neck also costs fairly little energy.
- Carrying the load in one hand is quite fatiguing and stressful, particularly for the muscles of the hand, shoulder, and back.
- Although carrying a load in one hand is biomechanically and physiologically disadvantageous, it is often done because of the convenience of quickly grasping and securely holding and releasing an object and of carrying it over short distances while the other hand remains free to perform other tasks.
- Carrying a load of the proper size and weight on the head is also suitable if one is used to doing so. While the load is quite a distance away from the center of body mass, it is vertically above it. Thus, carrying a load on the head requires much balancing skill and a healthy spine, but does not demand much energy beyond that needed to move the body.

Table 11–8 is a survey of the different ways of carrying a 30-kg load. Of the conditions investigated, carrying the load evenly distributed on the chest and back was least demanding. For very small loads, the location probably has little effect on energy demands, while convenience and the availability of space may be the determining factors. For example, one might put items into pockets along the thighs, around the waist, on the chest, or on the upper part of the sleeves.

AVOIDING MUSCLE FATIGUE

Anecdotal information indicates that carrying heavy loads (such as logs of wood or suitcases) in the hands is fatiguing for the muscles that keep the fingers closed and for the muscles that cross the shoulders. Attempting to stabilize a load with an elevated hand and arm, such as a load carried on the head, on the back, or on the shoulders, can cause the person to stop carrying the load, due to local muscle fatigue (not necessarily to exhaustion in general). Fatigue may also occur when the load is placed asymmetrically, such as when it is carried in one hand, on one shoulder, or on the back or chest alone. In these cases, the one-sided loading must be counteracted by effort of muscles, which are likely to become fatigued if they are pulled into action for long periods of time, especially at high percentages of their total strength.

Another major problem associated with carrying a load is that of local discomfort, often brought about by ischemia (ie, lack of blood supply due to compression of tissues). This happens, for example, when a small hand grip cuts into the tissues of the hand or when a narrow strap across the shoulder compresses the tissue.

TABLE 11–8. Techniques For Carrying Loads of Approximately 30 kg

	Estimated energy expenditure for carrying 30 kg on straight flat path (kcal/min)	*Estimated muscular fatigue*	*Local pressure and ischemia*	*Stability of loaded body*	*Special aspects*	
In one hand	?	Very high	Very high	Very poor	Load easily manipulated and released	Suitable for quick pickup and release; for short-term carriage even of heavy loads.
In both hands, equal weights	Very high, about 7	High	High	Poor		
Clasped between arms and trunk	?	?	?	?	Compromise between hand and trunk use	
On head, supported with one hand	Fairly low, about 5	(High if hand guidance needed)	?	Very poor	May free hand(s); strongly limits body mobility; determines posture; pad is needed	If accustomed to this technique, suitable for heavy and bulky loads.
On neck often with sherpa-type, strap around forehead	Medium, about 5.5,	?	?	Poor	May free hand(s); affects posture	
On one shoulder	?	High	Very high	Very poor	May free hand; strongly affects posture	Suitable for short-term transport of heavy and bulky loads.
Across both shoulders by yoke, held with one hand	High about 6.2.	?	High	Poor	May free hand(s); affects posture	Suitable for bulky and heavy loads; pads and means of attachment must be carefully provided.
On back	Medium, 5.3, backpack; 5.9 for bag held in place with hands	Low	?	Poor	Usually frees hands; forces forward trunk bend; skin-cooling problem	Suitable for large loads and long-time carriage. Packaging must be done carefully, attachment means shall not generate areas of high pressure on body.
On chest	?	Low	?	Poor	Frees hands; easy hand access; reduces trunk mobility; skin-cooling problem	Highly advantageous for several small loads that must be accessible.
Distributed on chest and back	4.8, lowest	Lowest	?	Good	Frees hands; may reduce trunk mobility; skin-cooling problem	Highly advantageous for loads that can be divided or distributed; suitable for long-durations.
At waist, on buttocks	?	Low	?	Very good	Frees hands; may reduce trunk mobility	Around waist for smaller items, distributed in pockets or by special attachments; superior surface of buttocks often used to partially support backpacks.
On hip	?	Low	?	Very good	Frees hands; may affect mobility	Often used to prop up large loads temporarily.
On legs	?	High	?	Good	Easily reached with hands; may affect walking	Requires pockets in garments or special attachments.
On foot	Highest	Highest	?	Poor	Usually not useful	

In general, carrying a load is an inefficient method of doing external work. Indeed, if the load is very small, the energy required to move the body is high in relation to the energy spent to carry the object. This phenomenon is well explained by the low energy efficiency of the body, meaning that large amounts of energies must be input into the body in order to generate a fairly small amount of external work. (See Chapter 2.) At the other extreme, carrying heavy loads (with the possible exception of skillfully balancing a load on the head) is a highly fatiguing effort for the muscles employed. Thus, given the low energy efficiency of these muscles, the human body is, in general, not a good means of transport for moving loads over large distances and long periods of time. Yet, the human's ability to perform finely controlled and carefully executed actions allows the manipulation of precious and fragile loads.

Some general principles that apply to carrying loads are as follows:

- Use large muscle groups to avoid fatigue.
- Keep the load close to the body, for the sake of stability and minimal effort.
- Keep the load close to the center of mass of the body to maintain a low center of mass.
- Avoid pressure points and other highly concentrated loads on body tissues.
- Ensure adequate freedom of movement for the body in general and the arms and trunk specifically.

WALKING ON DIFFERENT TERRAIN

The technique that is most suitable for carrying a load also depends on the terrain to be covered. Walking on a smooth, solid surface is the least energy demanding. (See Chapter 2, Table 2-2.) The effort increases as one walks on a dirt road, through light brush, on hard-packed snow, through heavy brush, or on a swampy bog. Walking in loose sand demands about double the energy that walking on smooth blacktop does. In soft snow of approximately 20 cm depth, the energy demand is about three times that of walking on blacktop, and it is about four times when the snow is 35 cm deep. Different kinds of snowshoes can be helpful (Duggan and Haisman 1992; Knapik et al. 1997; Pandolf et al. 1996, 1997).

Moving Patients

Working as a nurse is an occupation with a high risk of lower back injuries. Nurses and their aides must often lift and move "precious loads"—patients. Of course, these individuals have asymmetrical shapes, varying sizes and weights, and no handholds. They are often difficult to move because they "go limp," they may not want to be moved, or they may be uncooperative because they are in pain or uncomfortable. Nevertheless, nurses and their aides must move them, primarily in, from, or to beds. While moving their patients, nurses may have to assume awkward positions (Winkelmolen et al. 1994). Often, one nurse alone cannot move the patient, because doing so would generate a high risk of overexertion and, consequently, back injury. Therefore, it is often advisable to call helpers; yet, the ability of teams of persons to lift is not simply the sum of each team member's individual lifting capacity, but considerably less.

These problems have given rise to repeated attempts to redesign hospital beds, especially to provide a means of raising, lowering, or tilting the beds, and of lifting and transferring patients. The solution is to use equipment that is, in principle, similar to some of the material-handling devices discussed earlier in the chapter, but used under different conditions and on an extraordinarily precious, sensitive, and fragile human beings as load. So far, neither the equipment nor the procedures have been generally accepted (Le Bon and Forrester 1997). External lifting aids of the hoist variety have not become popular because of their awkwardness, the need to move them from bed to bed, and their unpleasant appearance and high price.

Gurneys, on the other hand, are more amenable to ergonomic design, because they are used not for patients' beds, but rather as a means of short-term transportation. Thus, gurneys are often both adjustable in height and narrow, traits that facilitate handling patients. Among the ideas put forth to help move the patient laterally on a gurney is to do so by smooth gliding instead of lifting. Gliding can be facilitated by having a fairly slippery gurney surface (possibly by employing rollers, as on conveyors) and by using some kind of a stiff bedsheet with handles.

HELPING EACH OTHER

If a load appears too heavy or large to be handled alone, one often calls another person to help. (See Figure 11–18.) This approach is quite common in hospitals, where nurses must lift and move patients (Johnson and Lewis 1989; McMulkin and Sivasubramanian 1998; Sharp et al. 1993; Knibbe and Friele 1996). How much can be gained by helping each other? A second person might help to stabilize and balance a load that otherwise would be difficult to handle, but the strengths of two persons do not simply add up, partly because the timing of their efforts is not exactly the same, particularly if the load is difficult to grasp and handle. Even under conditions of good coordination and suitable placement of the hands and feet, at best 90 percent, but usually only about 80 percent, of the sum of the lifting strengths of two or three persons can be applied in isometric exertion. If the effort is dynamic instead of static, two persons together can generate only about two-thirds of their combined single strengths, and the fraction is even less if three persons cooperate (Karwowski 1988; Karwowski and Pongpatanasuegsa 1988; Oriet and Dutta 1989).

SUMMARY

Avoiding unnecessary strains, overexertions, and injuries in manual material movement is necessary for both ethical and economic reasons. Basically, two major approaches exist: to fit the person to the job through training or selection and to fit the job to the person through ergonomic design.

Training for "safe" manual material handling has been attempted in many different ways. Training relies on the assumption that there are safe procedures that can be identified, taught, and followed. Unfortunately, no single material-handling technique, nor even any one training procedure, has yet been proven successful and lasting. Nevertheless, many claim (often short-lived) improvement. Study design and evaluation techniques are at hand to assess which training techniques are successful.

Personnel selection relies on the assumption that a stronger worker would be less susceptible to overexertion than a weaker colleague. While static strength measurements have been fairly successful in assessing individual capabilities during the last decades, the actual job demands of material handling generally are better reflected in dynamic test exertions. Such dynamic strength tests have been developed recently and applied to large numbers of military personnel. The techniques used are now available for application in industry.

Ergonomic design of the load, task, equipment, and workstation, including the work environment and the total facility, appears to be the most successful approach. Human-engineering solutions can avoid or remove many causes of overexertion injuries, thus generating fundamentally safer, more efficient, and more agreeable working conditions. This is because ergonomic design does not just "doctor the symptoms," as worker selection and training do. Thus, even if (as Snook cautioned) two-thirds of all incidents may be unavoidable, reducing suffering and expenses by one-third would be a major success of ergonomic design.

Developing successful personnel selection and training procedures and combining them with ergonomic design of the task, workplace, and equipment should help to get the problem of overexertion and injuries to material handlers under control. In fact, the combination of all three approaches should provide the highest probability of success and the best efficiency; this comprehensive approach has been reported to be successful in health care facilities, where various and difficult manipulations have to be done on precious loads.

Both workers and managers are important players: Each cannot do without the other. While material handlers are the direct recipients of proper ergonomic measures, the manager must fully understand and support them. Physical work, such as moving materials manually, is accompanied by physical exertion of the body, the expenditure of energy, the generation of forces, and accompanying fatigue and aches. Low-back-pain symptoms, for example, are likely to appear in nearly everybody's life, whether one works in the shop, in an office, or at home. When workers experience low back pain during work, management should not allow adverse situations to develop that are likely to result in prolonged disability. Instead, understanding and acceptance of low-back problems, early interventions, good follow-up and communication, and programs that encourage workers to return to work early may prevent, alleviate, or shorten the duration of disability.

CHALLENGES

What determines the time-dependent forces and impulses that must be applied to a load? How do these forces and impulses strain the body?

Why is the low back so frequently overexerted?

What are the relations between force and torque in the spinal column?

What relationships exist among compressive forces, shear forces, bending torques, and twisting torques? How are these transmitted by joints, bones, and discs and by the ligaments and muscles that attach to vertebrae?

What role does the development of skills (experience) play in load handling? Which specific traits can be assessed using the biomechanical approach?

Under what conditions would physiological functions (such as metabolism and muscular efforts) establish limitations for a load handler's capacity for handling loads?

Explain the mechanism whereby contracting the longitudinal trunk muscles loads the spinal column?

How does "asymmetrical lifting" influence the use of different muscle groups and the loading of the spinal column?

What is the role of intraabdominal pressure in lifting tasks and in pushing and pulling, as well as carrying?

What effects might the use of "lifting belts" have on one's attitude toward, and ability in, material handling?

Would "staying in practice" keep an aging worker safe from a lifting injury?

What can be done to speed up injured workers return to their jobs?

What needs to be done to make training for safe material handling more effective?

Which lifting techniques are appropriate for all tasks and conditions?

How can the success of training be assessed?

Is "work hardening" a promising approach?

How much intrusion into one's lifestyle should a person accept to become "fit for material handling on the job"?

What are the responsibilities of the employer and the employee with regard to promoting noninjurious material handling on the job?

Should training for proper material handling be done during working hours?

Should an employer provide exercise facilities for employees?

Is there anything wrong with use of the Hawthorne effect to reduce overexertion injuries?

Why is there not one safe load?

What means are appropriate for selecting persons who are capable of handling heavy loads or handling loads frequently?

Which medical examination techniques can be employed to select persons healthy enough to become material handlers?

How well can static strength testing predict a person's ability to handle loads?

What are the major ways in which ergonomic design of the work process can reduce manual labor?

Should one introduce new symbols into flowcharts to identify details of load handling?

How can one redesign the forklift truck to provide better vision and riding conditions for the operator?

What major procedural differences underlie the recommendations by Snook and by NIOSH?

Which loads are particularly suitable to being carried close to the chest or on the back?

What ergonomic means can be devised to facilitate the moving of patients in hospitals?

Chapter 12

Designing for Special Populations

OVERVIEW

Usually, one designs for a "regular adult" population in the age range of about 20 to 50 years. The anthropometry, biomechanics, physiology, psychology, attitudes, and behavior of these people are fairly well known. This is the group of most interest to industry and society as "movers" and contributors to the gross national product. Yet, other large population groups are of specific concern: pregnant women, children, the aging, and the disabled. These people need special ergonomic attention, but information about them is incomplete.

BACKGROUND

As discussed in Chapter 1, anthropometric information on military populations is fairly complete. By contrast, adult civilians are seldom measured as a large group, and their body dimensions must be inferred from those of soldiers. Fortunately, their body dimensions, as well as their physical and psychological capabilities and traits, are not very different; therefore, the ergonomic principles that help us design for the military also apply to civilian adults (and vice versa).

There are differences among adult men and women—for instance, in body sizes and in physical capabilities. Yet, in general, one can design nearly any workstation or any piece of equipment or tool so that it is usable by either women or men. In some cases, adjustability is needed, or one may have to provide objects in different ranges of dimensions. These adaptment ranges, however, are not gender specific but simply needed to fit different people.

Healthy adults between about 20 and 40 years of age are commonly considered "the norm." About them, most ergonomic information is available, and most ergonomic efforts are needed for them. One group among adults, however, needs special attention: pregnant women. Therefore, one section of the following text is devoted to them.

There is another group that draws everyone's attention: infants and children. The younger they are, the more different they are from our ergonomic norm, the adult. Specific ergonomic information is available for small children, but little is systematically known about teenagers.

Aging people are another large group who need special ergonomic attention. Body size and posture, physical capabilities, and psychological traits change, quickly for some, slowly for others. To accommodate the special traits of people during their later working years, their retirement, and their waning period poses challenging, yet ethically satisfying, tasks to ergonomists.

Another group of people, most but not all of them adults, also need special ergonomic consideration: the impaired and disabled. They differ from their peers in size, posture, and abilities. In many cases, proper ergonomic design of their environment and equipment can help them to overcome their handicaps and live a satisfactory life.

SPECIAL DESIGNS FOR WOMEN AND MEN?

Obviously, tight clothing must be cut differently for men and women in order to fit their bodies. But must one design special workstations and tools to fit each gender because, as group, males may do certain jobs better than females and vice versa?

The sections that follow are based on group averages and ranges taken from the literature—findings that may not hold for individuals. Furthermore, the capabilities and traits of males and females generally overlap—for example, in body size or muscle strength.

Size and Strength

Men are generally taller than women, who have relatively shorter legs, but wider hips. Women are weaker than men, usually attaining between 60 and 90 percent of the men's muscle strength and work output. (Kroemer 1999b, Kroemer et al. 1997). However, women's leg strength is only marginally lower, and their muscle tension developed per cross-sectional unit is equal to men's. Mobility in body joints is generally, but only marginally, greater in women. (See Chapter 1.)

Sensory Abilities

With regard to hearing, girls and adult females have lower absolute thresholds for pure tones than boys and men. Aging men have larger hearing losses than aging women. In vision, both static and dynamic, boys and men have better acuity than girls and women, and acuity declines earlier in females than in males. Females also have more vision deficiencies.

As regards taste, women detect sweet, sour, salty, and bitter stimuli at lower concentrations than men; however, some studies have not substantiated these findings. Women also can smell some substances more easily than men, but both taste and smell capabilities and preferences change within the female menstrual cycle and during pregnancy.

Little difference has been shown between the sexes in the threshold for temperature sensation. Females usually feel less warm (comfortable) than males in environments of 19 to 38°C initially, but adapt to the surrounding temperature more rapidly than males. Females often begin to sweat at higher temperatures than males and acclimatize to work in severely hot conditions, somewhat more slowly. Men may find it a bit easier to adjust to very hot and very cold conditions, but the differences are minor and conditional on numerous factors (Burse 1979; Shapiro et al. 1981).

Sensitivity to vibration is about the same in males and females; however, females are more sensitive to pressure stimuli on their body, except on the nose. Pain sensations (a complex and difficult topic for physiological and behavioral reasons) seem to be about the same in males and females (Baker 1987). In general, whatever differences there are in sensory functioning (see Chapter 4) between the sexes usually are small.

Motor Skills

Girls and women are more skillful than boys and men on perceptual and psychomotor tests, such as color perception, aiming and dotting, finger dexterity, inverted alphabet printing, and card sorting. Males perform better in speed-related tasks, on the rotary pursuit apparatus, and in other simple rhythmic eye–hand skills (Noble 1978).

Coping with Environmental Stress

Stress encompasses a wide variety of situations, including job pressures, marital and family tensions, and physical aspects of the environment such as climate and noise. (See Chapters 4 and 5.) Few gender-specific differences have been demonstrated. While females may be slightly more sensitive to sound and noise, the impact of these two potential stressors on performance, health, and social behavior seems to be similar for both sexes. Women appear to be better able to cope with having only a small amount of personal space than men: When forced into crowded quarters, men tend to maintain greater distances from others and react more negatively when people invade their personal space (Greene and Bell 1987). Redgrove (1976) suggested that women appear better able to cope with low arousal conditions of work (monotony and boredom), while men seem to cope better with higher arousal conditions (pressure).

Cyclical Variations

In their behavior under biological circadian variations, men and women react essentially the same. (See Chapter 5.) Of course, men do not have a menstrual cycle; hence, the question arises whether changes in women's physiological capabilities during the menstrual cycle are significantly different from a constant level, if that may be postulated of men.

During the menstrual cycle, women show distinct changes in basal temperature, in daily temperature variations, and, of course, in internal hormone production. While these changes may affect some nervous system activity and sensory functioning, which in turn might influence scores on specific sensitive tasks, there is no evidence of any significant variation in either overall performance or cognitive functioning between men and women. In comparison to men, women in general maintain a similar performance level throughout, in spite of premenstrual tension and pain, the occurrence of which is predictable and can be counteracted by medication and individual effort (Asso 1987; Nakatani et al. 1993; Patkai 1985). In a work task that required continual attention, Matthews and Ryan (1994) found some minor changes of little practical importance with the menstrual cycle, but after work, premenstrual women were lower in energy and mood than at other phases of their menstrual cycle. The mood changes suggest that, indeed, affective correlates of the menstrual cycle vary with situational factors. Matthews and Ryan state that women may be more vulnerable to work-induced fatigue and depression during the premenstrual phase.

Yet, in a study on steadiness of the hand while shooting a gun, women outperformed men consistently except during the week preceding their menses, while women taking oral contraception medication maintained a steady performance level, which was, however, below that of men (Hudgens et al. 1988).

Task Performance in General

The stereotypical question whether females or males "perform better" is, put in such a general way, unanswerable, because there are wide individual differences, as well as diverse performance measures. Yet, interest in gender-related differences in performance has apparently been around for a long time. The topic is laden with emotional, social, and political overtones, often biased by the point of view that the inferiority or superiority of one or the other gender should be established or dismissed.

One of the first unbiased collection of facts about the behavior of the sexes was Ellis's book *Man and Woman,* published by Black in London in six editions between 1894 and 1930.

The measure of a person's performance on a task depends on various attributes: the overall nature of the task, the specific conditions under which the task must be performed, the requirements for specific capabilities, the attitudes of the person performing the task, the subjective value of the task for the person performing it, and the social and task-related goals of the individual.

Regarding specific activities, only in those that require a large amount of muscular strength and power do men (on the average) develop a higher output than women. However, even here, individual capabilities, the selection of teams, training and skill, age, motivation often contradict statements based on "averages": Clearly, in many cases, strong women outperform weak men.

Hancock and coworkers (1988) reported that women appeared to be more strained by a repetitive, low-demand task than their male colleagues. In contrast, McCright (1988) found that females experienced less strain in low-demand tasks, but higher strain under high demand, compared with their male cohorts. In a series of tests in which performance and subjective workloads were assessed, women tended to perform slightly better than men on the majority of tasks, including grammatical reasoning, linguistic processing, mathematical processing, and recall (Schlegel et al. 1988).

Few statistics indicate differences in actual on-the-job performance. For example, Vail (1988) checked accidents in general aviation that could be tracked to pilot error for the years 1982 to 1985. Data adjusted for the proportions of male and female pilots indicated that female pilots had distinctly fewer accidents, and much less serious accidents, than their male colleagues. Female pilots had fewer accidents caused by errors in judgment and decision or in motor skills. The numbers of accidents due to weather and communication problems were approximately the same for males and females. McFadden (1997) analyzed the incidence of pilot error on U.S. airlines for the years 1986 to 1992 and found no significant difference in the performance of male and female pilots.

In sum, only when body size and body strength either favor or are necessary for the execution of a task do clear differences appear in the "average" task performance between the sexes. No other genetically based sex differences produce essential differences in the performance of males and females. While there are some other subtle variations, such as in sensory capabilities and in reactions to climatic stress, the differences between men and womem are by far overshadowed by interindividual differences. Even those few areas where small differences have been found, such as in the reaction to crowded conditions or in fine manual skills, the question is unanswered as to whether these reflect biologically or genetically based differences or whether they are merely the results of sex roles acquired through group traditions in a cultural context and under social pressure.

☛☛☛ ***No Special Designs for Women or Men***

As the literature makes clear, there no gender-specific traits that require or justify designing workstations or work tools specially for men or women. Properly adjusted ranges will fit all people, female and male. Nevertheless, certain jobs may be preferred by one sex or the other, though probably not exclusively so; for example, tasks that require brute force may be done mostly by strong men—but not all men are strong. ☚☚☚

DESIGNING FOR PREGNANT WOMEN

While pregnancy is one of the most common life events, surprisingly little systematic and scientific information is available in the literature for ergonomic purposes. Many brochures and books on anatomy, obstetrics, and gynecology contain "normative" tables about changes in body dimensions with pregnancy, but apparently, surveys of large population samples have been sparse in recent years.

☛☛☛ *In compiling anthropometric and ergonomic data on pregnant British women, Pheasant had to rely on body dimensions measured on pregnant Japanese women (Pheasant 1986, pp. 178–179).* ☚☚☚

Changes in size

The changes in a woman's body dimensions during pregnancy become apparent after two or three months, and, in general, increase throughout the course of pregnancy. The most obvious increases are in body weight, abdominal protrusion, and circumference. Figure 12–1 illustrates some of these changes. Fluegel and coworkers (1986) reported the changes from the fourth month of pregnancy measured on 198 German women: Waist circumference increased by 27 percent, weight by 17 percent, chest circumference by 6 percent, and hip circumference by 4 percent. Similar increases have been reported for American women with average increases of 17 percent in body weight, 8 percent in chest circumference, 4 percent in hip circumference, and 2 percent in abdominal circumference above the values measured 16 weeks after the onset of pregnancy in 105 white women with an average age of 26 years. Women who were heavy before pregnancy showed smaller percentage increases in the abdominal region than women who were not (Rutter et al. 1984).

Changes in distribution of mass

In designing for crash protection in automobiles, one must consider the biomechanically significant changes in the abdominal depth and circumference of pregnant women, accompanied by a shift in the center of mass of the body. Culver and Viano (1990) presented these changes in the form of ellipses. (See Figure 12–2.) Their geometries allow estimates of the contact area between the woman's body and restraining devices or the interior surfaces of automobiles in case of an impact.

Changes in mobility

Their increasing abdominal protrusion makes it increasingly difficult for pregnant women to get as close to work objects as they could before, and will again be able to after, pregnancy. The working area of the hands becomes smaller during pregnancy, and manipulating objects that are

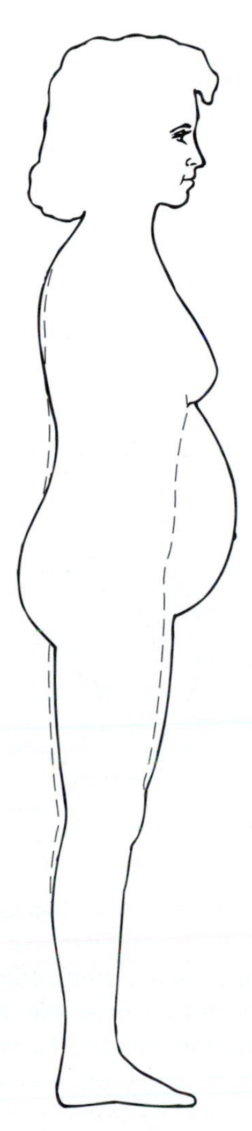

Figure 12–1. Changes in body dimensions and posture with pregnancy.

now further ahead of the spinal column generates an increased compression and bending strain on the spine and on ligaments and muscles in the back. This loading is, of course, also due to the increasing mass of the abdomen and its increasing moment arm with respect to the spinal column. The variations in abdominal shape also change the body posture, which, in the course of the pregnancy, assumes a backward pelvic rotation accompanied by forward movement of the trochanterion and brings about a flattening of the lumbar lordosis. These events explain, at least partly, the complaints of back strain and pain common with pregnancy.

Changes in mood and cognition

Morris et al. (1998) reported on the cognition and mood of 38 working women who were nearing the last trimester of pregnancy. Comparing their test results with those of a matched control group of nonpregnant female workers showed no significant differences, with the lone exception that the pregnant women felt less alert, vigorous, and energetic, probably because they were more easily fatigued.

Changes in performance

Physical performance capabilities also change during pregnancy. Although individuals vary greatly, with advancing pregnancy in general, there is a sharp decline in the ability to perform

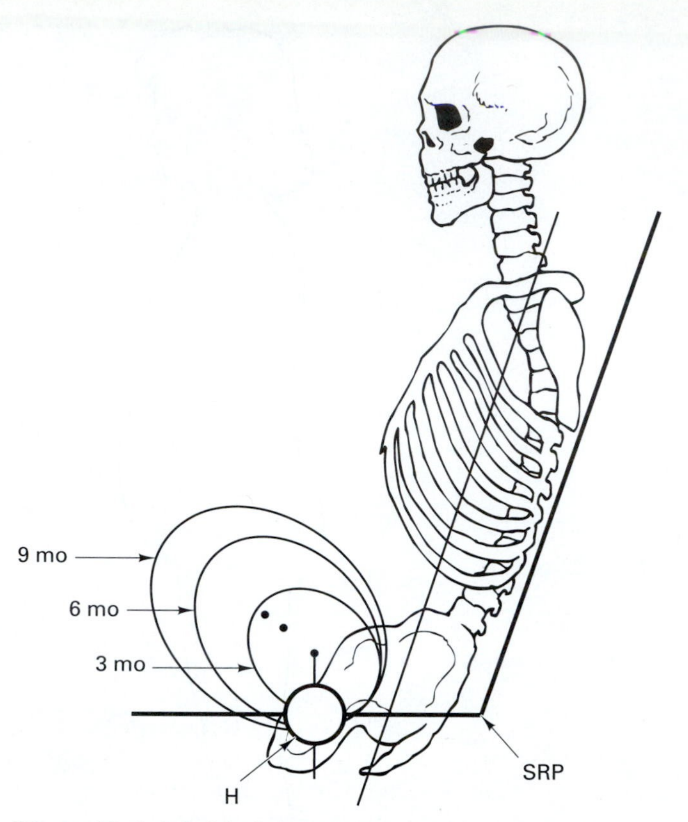

Figure 12–2. Model of changes in body dimensions with pregnancy. H; hip joint; SRP: seat reference point. Adapted from Culver and Viano, 1990. Reprinted with permission from *Human Factors,* Vol. 32. No. 6, 1990. Copyright 1990 by the Human Factors and Ergonomics Society, Inc. All rights reserved.

work that requires the exertion of large energies, much mobility, or far reaches, particularly if repeated or over long periods. Nicholls and Grieve (1992) a surveyed 200 residents of London, in the United Kingdom, who were between 29 and 33 weeks pregnant. They were asked about their current performance on certain tasks compared with their performance before becoming pregnant. The interviews used a five-point ordinal scale and concerned 46 activities that were preselected for inclusion in the study. Of these 46 tasks, 32 were found to be significantly more difficult to perform during pregnancy than before. Among the 32, the following were considered the hardest: picking an object up from the floor; working at a desk; walking upstairs; driving a car, getting in and out of a car, and using seat belts; ironing; reaching high shelves; using public toilets; and getting in and out of bed. The reasons reported for the difficulties were related to back pain, reduced reach and clearance, feeling unstable, being fatigued, having reduced mobility, and having difficulties in seeing objects near the body. In general, (suitable) sitting was found less straining than standing.

Many everyday tasks become more difficult with pregnancy. Yet, differences among individuals are striking, with some women finding only a few tasks more difficult and others finding nearly all activities harder to do.

To accommodate pregnant women, either at the workplace, in vehicles, or at home, the following ergonomic measures should be taken:

- Changing body dimensions—particularly increased abdominal protrusion—make it difficult to reach far objects. Thus, areas for manipulation should be kept close to the body and possibly somewhat elevated from their regular height.
- Work tasks should require as little force as possible, particularly in the vertical direction. Lifting of objects should be avoided.
- Suitable easily adjustable seats should be provided.
- Frequent freely selected rest periods should be allowed.
- More space than usual should be allowed for moving around, and obstacles should be avoided—particularly low objects that might be difficult to see.

DESIGNING FOR CHILDREN

Huge changes

The time between birth and early adulthood, about 18 years, is characterized by very large changes in body dimensions, body strength, skill, and other physical and psychological variables. At birth, we weigh about 3½ kilograms and are some 50 centimeters in length, of which the trunk represents about 70 percent. In the two decades that follow, body length increases three- to fourfold, weight increases about twentyfold, and body proportions change drastically. The trunk accounts for just over 50 percent of our stature when we are grown. (See Figure 12–3.) But these changes over time are individually quite different and appear to be related not only to genetic factors, but

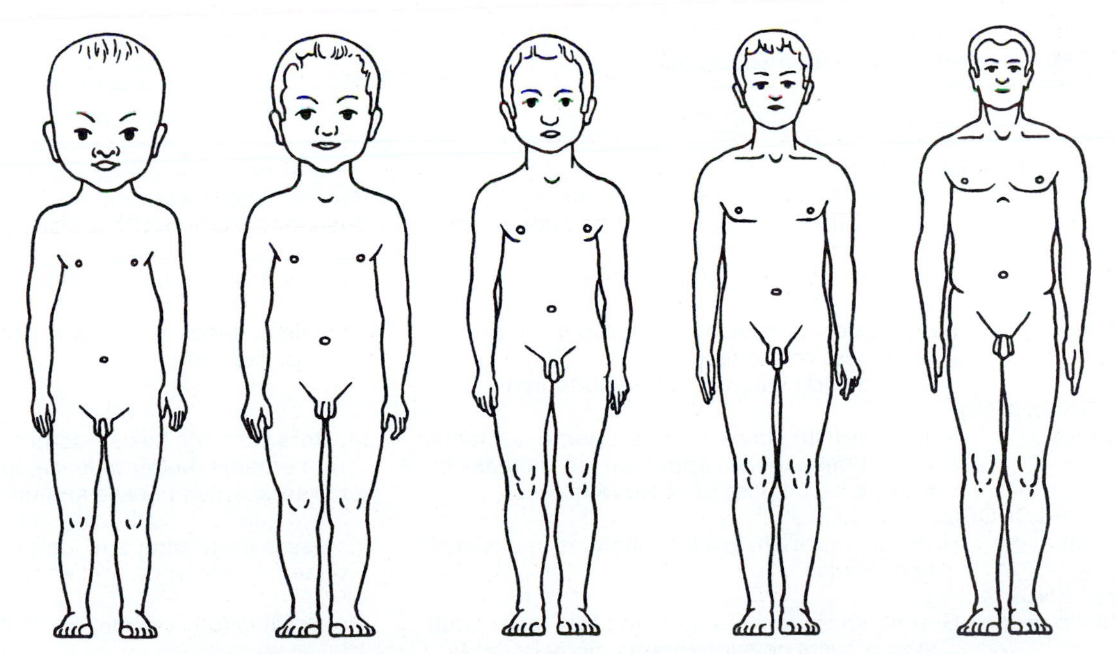

Figure 12–3. Changes in body proportions from birth to adulthood. (Modified from Fluegel et al. 1986.)

also to environmental variables. Thus, there is no "typical" boy or girl. In general, the rate of growth in boys is rapid during infancy (up to 2 years) and then slows until the onset of puberty (at about 11 years), when growth becomes strong again, reaching its peak at about 14 years and then slowing again. Final stature is attained in the early to middle twenties; yet, at 14 or 15 years, some boys have almost completed their growth, while others are just beginning a strong growth phase. In girls, the puberty growth spurt begins earlier, at about 9 years of age, and is fastest at about 12 years; full adult stature is often complete at age 16. Hence, at about 11 to 13 years, many girls are taller than boys of the same age. (See Table 12–1.)

Secular trends

Several secular trends have been observed in recent decades:

- The growth rate of children during the earlier years appears to have increased.
- Puberty is occurring earlier, as indicated by an earlier menarche in girls and adolescent growth spurt in boys and girls.
- The individual's final adult stature appears to have increased.

Such information is often more anecdotal than based on large scientific surveys. For example, Konz (1991) reported that Japanese boys at the age of 13 years in 1990 were an average of 159 cm tall, an increase of 18 centimeters since 1950; similarly, the average stature of girls 11 years of age was 146 cm in 1990, compared to 132 cm in 1950. And in the Netherlands, a survey of more than 600 children indicated significant differences among the body dimensions of children growing up in different provinces. Thus, information about children is specific not only to age groups, but also to areas of origin (Steenbekkers and Molenbroek 1990).

Anthropometry of children

Measurements on children are difficult to take, particularly when they are very young, with a short attention span, and do not understand or follow instructions. For example, the body length (stature) of babies and infants up to the age of two or three years is customarily measured with

TABLE 12–1. Children's Developmental Stages

Stage	*Physical characteristics*	*Motor skills*
INFANCY AND TODDLERHOOD		
0 to 6 months	Oversized head, short stubby limbs	Reaches and grasps, sits with support
6 to 9 months	Increases in weight, hence plump appearance	Sits alone, stands with assistance
9 to 15 months	Flexible limbs	Crawls, walks, stands alone
after 18 months	Gradual appearance of neck, protruding abdomen	Wobbly, stiff, flat gait; climbs
2 to 3 years	Head has become smaller in proportion to body; less roundness; lean muscles; curved back; still protruding abdomen	Flexible at knees and ankles; can run, jump, kick, hop
EARLY CHILDHOOD		
3 to 5 years	Rate of growth slows, body proportions change; loss of the babylike appearances, increase in muscle tissue; less back curvature	Masters walking; has smoother movements, better balance; turns corners; holds pencils and utensils
MIDDLE CHILDHOOD		
6 to 12 years	Horizontal growth, gradual changes in physical appearance	Increases in running and jumping distance, accuracy, and endurance
ADOLESCENCE		
12 to 18 years	Growth spurt peaks, hands and feet reach adult size, breasts develop in girls; body breadth increases; elongated trunks and legs for boys	Motor skills fully developed

the child lying on its back; later, the child's stature is taken when she or he stands. But the recumbent length is several centimeters larger than the stature.

Considering the importance given by pediatricians and parents to anthropometric information about the growth of children, it is utterly surprising how little information is available in the United States and the United Kingdom and how outdated it is. Snyder et al. (1975) measured American children more than a quarter of a century ago; the often-used growth charts by Hamill et al. (1979) rely on data that were measured in the period from 1963 to 1979 and then statistically smoothed and stratified. Of the two sets of stature data from the United States, the values reported by Snyder and coauthors are consistently lower than those by Hamill and coworkers. The difference is striking, but unexplained. The average statures of 18-year-olds reported by Hamill and coauthors also are larger than those given in Chapter 1 for American adults.

Sufficiently large, representative groups of children were measured fairly recently in France (Ignazi et al. 1996), Japan (Tsunawake et al. 1995) and (the former East) Germany (Fluegel et al. 1986); for British children, only estimates are at hand (Pheasant 1986). Excerpts from these data are listed in Table 12–2.

Given the many sources of variability, the data compiled in Tables 12–2 through 12–6 must be seen and used only in the context of the particular time and place to which they apply.

Designing For Body Size

In the United States, a rather large number of children are injured every year by head, neck, and hand entrapment. Thus, the U.S. Consumer Product Safety Commission sponsored an anthropometric study of American children, performed and published by Schneider and coinvestigators (1986). Their report not only provides related body dimensions, but also describes, in an exemplary fashion, the sampling and measuring strategies that are suitable for gathering such information. Based on these data, various recommendations and regulations specify the largest openings that still prevent children from moving through them, such as the distance between stakes of railings. Particularly critical dimensions are the head breadth, chest depth, and hand clearance diameter. These are compiled for American children in Table 12–3. Taking into account averages and standard deviations, the listing shows that up to the age of 12 years, girls have narrower heads, shallower chests, and smaller hand clearance diameters than boys. Thus, if openings are kept small enough not to let girls' bodies pass, boys should certainly not be able to squeeze through either.

Designing For Body Mass

For biomechanical design purposes, such as for restraint devices in automobiles, certain information is important: changes in body mass with age and changes in the location of the center of mass of the body, standing or sitting. Such information is compiled in Table 12–4 and shows that the location of the center of mass of the standing body, expressed in percent of body height above the floor, does not change much with increasing age and is quite similar for girls and boys. For seated children, the relative height of the center of mass above the seat decreases with age if expressed in percent of stature, but again, in a fairly similar fashion for girls and boys. This information suggests that no distinction need be made between boys and girls for the design of body restraint systems. Body dimensions of children are important for the design of furniture, particularly that used in schools. This poses problems because children of different body sizes may be combined in the same rooms from kindergarten on up. Thus, tables and chairs of very different sizes should be made available to fit all the children. Yet, this is often difficult to do for a variety of organizational reasons. The provision of adjustable chairs and tables, for example, might appear to be a suitable solution, but very young children might have great difficulties in adjusting the furniture to their size and liking.

TABLE 12–2. Average Stature (and Standard Deviation), in cm, of Children:
FRANCE (F), measured by Ignazi et al. (1996);
GERMANY (G), measured by Fluegel et al. (1986);
JAPAN (J), measured by Tsunawake et al. (1995);
UNITED KINGDOM (UK), calculated by Pheasant (1986);
UNITED STATES (US1), measured by Snyder et al. (1975);
UNITED STATE (US2), calculated by Hamill et al. (1979);

GIRLS							BOYS						
Age (years)	*F*	*G*	*J*	*UK*	*US1*	*US2*	*Age (years)*	*F*	*G*	*J*	*UK*	*US1*	*US2*
0	—	51.8	—		54.8 (3.6)	—	0	—	52.4	—	—	55.4 (4.0)	—
0.5	—	68.3	—		68.6 (2.3)	—	0.5	—	69.6	—	—	70.4 (2.4)	—
1	—	75.6	—		72.4 (2.9)	—	1	—	76.4	—	—	73.5 (3.2)	—
2	—	85.9	—	89	84.0 (3.4)	86.8	2	—	86.9	—	93	85.3 (3.4)	86.8
3	—	94.1	—	97	92.9 (4.4)	94.1	3	—	95.0	—	99	93.4 (3.9)	94.9
4	107.2 (5.2)	101.3	—	105	99.5 (4.3)	101.6	4	107.1 (4.6)	102.2	—	105	99.9 (3.8)	102.9
5	110.8 (4.7)	107.2	—	110	106.5 (4.7)	108.4	5	111.3 (4.6)	108.1	—	111	107.6 (5.0)	109.9
6	116.8 (4.9)	115.1	—	116	112.8 (5.0)	114.6	6	118.2 (5.8)	116.1	—	117	113.7 (4.8)	116.1
7	125.0 (5.4)	121.0	—	122	118.8 (5.0)	120.6	7	126.6 (5.7)	119.6	—	123	120.5 (4.7)	121.7
8	134.0 (5.0)	126.1	—	128	123.4 (5.3)	126.4	8	130.3 (4.8)	127.2	—	128	125.3 (5.8)	127.0
9	136.8 (5.7)	130.2	—	133	130.2 (5.9)	132.2	9	135.7 (4.6)	131.1	—	133	130.0 (5.8)	132.2
10	142.7 (7.7)	137.2	—	139	134.4 (6.1)	138.3	10	140.3 (5.4)	137.7	—	139	135.1 (6.3)	137.5
11	147.8 (6.9)	142.7	145.8 (7.3)	144	141.1 (6.8)	144.8	11	145.0 (7.6)	144.0	144.4 (6.5)	143	141.9 (5.3)	143.3
12	153.7 (6.2)	148.3	150.2 (5.8)	150	145.5 (6.5)	151.5	12	153.1 (8.1)	145.9	148.8 (6.2)	149	146.8 (7.1)	149.7
13	158.0 (4.9)	154.6	155.1 (4.5)	155	155.1 (6.2)	157.1	13	158.4 (9.0)	153.3	150.0 (6.9)	155	149.5 (7.8)	156.5
14	160.7 (6.7)	160.0	156.2 (5.3)	159	—	160.4	14	164.5 (8.9)	161.5	164.1 (6.8)	163	—	163.1
15	161.8 (6.7)	162.2	157.2 (5.2)	161	—	161.8	15	168.6 (6.9)	166.5	167.5 (5.6)	169	—	169.0
16	162.4 (6.0)	162.9	158.4 (6.1)	162	—	162.4	16	173.0 (7.1)	171.5	167.5 (6.0)	173	—	173.5
17	162.9 (5.2)	163.5	157.2 (5.0)	162	—	163.1	17	174.3 (5.7)	173.6	169.7 (4.0)	175	—	176.2
18	162.8 (5.5)	163.9	157.8 (5.7)	162	—	163.7	18	175.9 (7.6)	175.8	170.0 (5.7)	176	—	176.8

TABLE 12–3. Average Values (and Standard Deviations) in cm for Head Breadth, Chest Depth, and Hand Clearance Diameter in U.S. Children

	Head breadth		*Chest depth*		*Hand clearance diameter*	
Age (years)	*Girls*	*Boys*	*Girls*	*Boys*	*Girls*	*Boys*
0	10.3 (0.6)	10.4 (0.7)	9.0 (0.9)	9.3 (0.9)	3.21 (0.29)	3.33 (0.30)
0.5	11.4 (0.6)	11.7 (0.6)	9.9 (0.9)	9.9 (0.9)	3.55 (0.28)	3.72 (0.26)
1	12.3 (0.4)	12.6 (0.6)	10.4 (1.1)	11.0 (0.6)	3.86 (0.25)	4.14 (0.29)
2	13.0 (0.5)	13.3 (0.4)	11.3 (1.0)	11.6 (1.0)	4.10 (0.27)	4.24 (0.32)
3	13.3 (0.5)	13.5 (0.4)	11.8 (0.8)	12.0 (1.2)	4.30 (0.32)	4.51 (0.24)
4	13.5 (0.4)	13.8 (0.4)	12.2 (0.8)	12.5 (0.9)	4.50 (0.27)	4.57 (0.28)
5	13.6 (0.4)	14.0 (0.5)	12.7 (1.1)	13.0 (1.0)	4.66 (0.30)	4.82 (0.32)
6	13.7 (0.4)	14.0 (0.4)	13.2 (1.0)	13.3 (1.1)	4.79 (0.28)	4.99 (0.30)
7	13.9 (0.4)	14.2 (0.5)	13.5 (1.0)	14.1 (1.1)	5.01 (0.31)	5.16 (0.30)
8	14.0 (0.4)	14.2 (0.5)	13.7 (1.4)	14.3 (1.3)	5.08 (0.34)	5.28 (0.38)
9	14.1 (0.5)	14.3 (0.4)	14.4 (1.4)	14.8 (1.3)	5.22 (0.33)	5.42 (0.35)
10	14.1 (0.5)	14.4 (0.5)	14.7 (1.5)	15.2 (1.3)	5.42 (0.33)	5.56 (0.33)
11	14.2 (0.4)	14.6 (0.4)	15.7 (2.0)	16.2 (1.6)	5.60 (0.40)	5.85 (0.38)
12	14.5 (0.6)	14.5 (0.5)	16.2 (1.7)	16.8 (1.6)	5.82 (0.34)	6.03 (0.37)
13	14.6 (0.5)	14.5 (0.4)	17.9 (2.2)	17.2 (1.7)	6.16 (0.37)	6.06 (0.40)

Source: Data excerpted from Snyder et al. 1975.

TABLE 12–4. Body Mass and Location of the Center of Mass for U.S. Girls and Boys; Average and Standard Deviations

			Height (in percent of stature) of the center of mass of the body			
	Body mass (kg)		*Standing (above floor)*		*Seated (above seat)*	
Age (years)	*Girls*	*Boys*	*Girls*	*Boys*	*Girls*	*Boys*
0	4.6 (1.1)	4.8 (1.2)	59.4 (1.9)	58.5 (2.1)	50.2 (3.3)	48.0 (4.4)
0.5	6.7 (0.9)	7.4 (0.9)	58.1 (2.5)	59.1 (2.3)	47.1 (2.8)	46.6 (3.7)
1	8.9 (1.3)	9.5 (0.8)	58.1 (1.8)	58.5 (2.4)	44.6 (2.3)	45.6 (2.7)
2	11.2 (1.1)	12.2 (1.2)	57.5 (1.0)	57.5 (1.1)	41.3 (2.4)	39.3 (2.6)
3	12.8 (1.1)	14.2 (1.5)	59.3 (2.5)	58.9 (1.0)	39.1 (1.9)	37.6 (1.2)
4	15.4 (1.8)	15.8 (1.8)	58.8 (2.0)	59.7 (1.7)	37.9 (3.3)	37.2 (2.2)
5	17.7 (2.3)	18.3 (2.1)	59.3 (2.0)	58.9 (1.7)	35.3 (2.6)	36.6 (2.6)
6	19.3 (2.7)	20.8 (3.0)	59.3 (1.5)	59.1 (1.3)	34.0 (2.3)	35.0 (1.9)
7	21.8 (2.7)	23.2 (3.1)	58.6 (1.1)	58.7 (1.3)	33.3 (1.8)	33.1 (2.1)
8	24.2 (4.0)	25.3 (4.4)	58.0 (1.7)	58.6 (1.1)	32.2 (2.4)	32.3 (2.2)
9	27.7 (5.2)	27.7 (4.6)	58.0 (1.4)	57.9 (1.1)	30.7 (1.6)	32.1 (1.8)
10	30.6 (5.8)	30.4 (5.2)	57.5 (0.9)	58.0 (1.1)	30.2 (2.0)	31.1 (1.9)
11	34.4 (7.2)	35.4 (5.8)	57.4 (0.7)	57.7 (1.0)	29.5 (1.6)	30.0 (1.6)
12	38.1 (7.3)	38.8 (6.4)	57.4 (1.1)	57.8 (1.0)	29.4 (1.4)	30.1 (2.1)
13	48.0 (8.1)	40.7 (7.0)	57.4 (1.3)	58.0 (1.5)	29.2 (1.3)	29.7 (1.5)

Source: Data excerpted from Snyder et al. 1975.

TABLE 12–5. Average Torques (and Standard Deviations), in Ncm, Around Wrist, Elbow, and Knee Exerted by U.S. Children (Girls and Boys Combined)

Age (years)	*Wrist*		*Elbow*		*Knee*	
	Flexion	*Extension*	*Flexion*	*Extension*	*Flexion*	*Extension*
3	84 (47)	63 (22)	606 (156)	616 (111)	500 (197)	1673 (616)
4	122 (61)	61 (28)	731 (233)	724 (259)	468 (194)	1866 (710)
5	152 (79)	69 (30)	932 (319)	901 (285)	706 (351)	2301 (738)
6	224 (85)	90 (40)	1192 (299)	1034 (373)	956 (386)	2717 (961)
7	268 (105)	113 (47)	1687 (415)	1332 (441)	1175 (334)	3788 (1165)
8	352 (128)	122 (44)	2114 (506)	1612 (437)	1371 (564)	4762 (1391)
9	453 (188)	167 (74)	2248 (674)	1676 (527)	1986 (638)	5648 (1386)
10	434 (166)	164 (41)	2362 (603)	1596 (446)	2084 (842)	5553 (1826)
	n = 211	*n = 205*	*n = 495*	*n = 496*	*n = 267*	*n = 496*

Source: Data excerpted from Owings et al. 1975.

TABLE 12–6. Average Side Grip and Grasp Forces (and Standard Deviations), in *N*, *Exerted by U.S. Children (Boys and Girls Combined, n = 227)*

Age (years)	*Thumb-forefinger side grip* ("side pinch," see Fig. 8–26)*	*Power grasp ("grip strength")*
3	18.6 (4.9)	45.1 (14.7)
4	26.5 (5.9)	57.9 (17.7)
5	31.4 (7.8)	71.9 (18.6)
6	38.3 (5.9)	89.3 (22.6)
7	41.2 (6.9)	105.0 (32.4)
8	47.1 (9.8)	124.6 (33.4)
9	52.0 (9.8)	145.2 (35.3)
10	51.0 (8.8)	163.8 (37.3)

Source: Data excerpted from Owings et al. 1975.
*Pinch surfaces 20 mm apart.

Designing For Body Strength

While infants are still fairly uncoordinated and do not show great body strength (although their uncontrolled movements can inflict damage to others and themselves), body strength develops quickly during early and middle childhood. Hand strength shows a strong positive correlation with age, while, at least in the early years, there is little relation between strength, hand dominance, and gender. Tables 12–5 and 12–6 indicate torque and force capabilities measured on American children between 3 and 10 years of age. The tables are shown here for two reasons: First, they reflect how (average) strength increases with increasing age; second, they show the very large interindividual differences; with coefficients of variation ranging from 30 to 60 percent, a wide variation indeed.

The tables combine the measurements taken on girls and boys, largely because at these age groups no systematic differences between the genders can be observed. The slight decrease in strength from 9 to

10 years of age visible in some of the data sets is probably an artifact of the relatively few subjects in the ten-year-old age group who were measured. Between 10 and 12 years of age, boys are usually, but not consistently, slightly stronger than girls. Hand preference, as determined both by handwriting and throwing a ball, is not associated with strength up to the age of 12 years.

DESIGNING FOR THE AGING

In the United States, there is a curious use of terms: a "middle-aged" person becomes "older" at 45 years of age, "elderly" at 65 years, "old" as one reaches 75 years, and "very old" or "old old" if one lives beyond 85 years of age.

Age categories

Anthropometric and much demographic and capability-related information is usually collected in five-year intervals until the age of 65. Then it is just lumped together for the remaining years, with only occasionally a time marker set at 75 years of age.

Life expectancy

The number of years that humans can expect to live has increased dramatically. The average life span was less than 20 years in "prehistory" until about 1000 BC, and increased into the low 20s in Ancient Greece around the year 0. In the Middle Ages, about 1000 AD, life expectancy increased in Western Europe to the thirties, reaching the low forties in the 19th century. In Colonial America, until about 1700, the average life span was 35 years. It increased to about 50 years by 1900 and to about 75 around 1990 in the United States. Now, it approaches 80 years. Life expectancy depends on heredity, gender, climate, hygiene, nutrition, diseases, wars, and accidents.

Roles of the aging

The position of an older person in society has been quite variable in different eras and in different regions (Blaikie 1993; Shephard 1995). The aged person might be considered wise and experienced—a leader or an advisor—or a useless and expensive "appendix" that is removed from societal life [as was vividly portrayed by Margaret Mead (1901–1978)]. Intermediate positions also exist: Consider, for example, the forced early cessation of flying duties by airline pilots, the common "going into retirement" at about age 65 in most occupations, and the late-life activities of some professors and politicians in the United States.

What does a person owe to society, or society to a person? How much care can be expected from relatives and friends, how much from society, and what can the aging person return? These general, and many specific, concerns have been addressed in the United States by the Committee on an Aging Society (1988) and by Czaja (1990), both under the aegis of the National Research Council, and more recently by Fisk and Rogers (1997).

Changes in Anthropometry

Age brackets

Body dimensions in the aging population are measured as in younger adults, usually via a cross-sectional approach: One measures all available people and then lumps their measurements together within certain age brackets. This does not create a big problem in the "young adult" population, because dimensions do not change very much in the 20-to-40-year age span. (See Chapter 1.) However, it is a major problem in the description of the aging (as well as of children), for a couple of reasons:

- Among the aging, some persons change dimensions rapidly within a few years. For example, stature may change because of posture and a shrinking of the thickness of the spinal discs, weight may change because of changes in nutrition, metabolism, and health, and muscles and strength may change because of changes in activity levels, habits, and health. Other people, in contrast, show little change over long periods.
- The age brackets used for surveys are rather wide, usually encompassing decades or even longer time spans, as opposed to the common five years in younger cohorts. Hence, people with very different dimensions are contained in each sample.

Thus, chronological age is not a good criterion for classifying the aging (or children); they would be better described by a longitudinal procedure in which changes in body dimensions and capacities are observed within one individual over many years; yet, data of this sort are sparse (Annis et al. 1991, 1997; Hamill et al. 1997).

As a rule, anthropometric information reported in the literature on sufficiently large samples has been obtained from cross-sectional surveys. Available data are compiled in Table 12–7. The table exemplifies the current problems with anthropometric information: Most of the samples are exceedingly small, surveys are done for a few age ranges only, and there are no distinctions among ethnic origin, region, socioeconomic status, health, or other attributes that are co-determinants of anthropometry.

Given the limitations—in fact, the paucity—of these data from the United States, one anticipates even poorer and less complete data from other regions of the earth. Thorough discussions of American demography, with respect to both current and future aging cohorts, have been provided by Annis et al. (1991); Serow and Sly (1988); and Soldo and Longino (1988).

Height loss, weight gain

The apparent height loss with age (of about 1 cm per decade), starting in the thirties, may be a result of (a) a flattening of the cartilaginous disks between the vertebrae; (b) a flattening or thinning of the bodies of the vertebrae; (c) a general thinning of weight-carrying cartilage; (d) a change in the S-shape of the spinal column in the side view—particularly an increased kyphosis in the thoracic area (called humpback); (e) in some cases, scoliosis, a lateral deviation from the straight line displayed by the spinal column in the frontal view; and (f) possibly, a bowing of the legs and flattening of the feet. As groups, American men usually have their largest body weights in their thirties and then lose weight with aging; American women are relatively light in their twenties, but then increase their weight with age, becoming heaviest, on average, at about 60 years (Annis et al. 1991; Barlow et al. 1990; Stoudt 1981).

Old age is not necessarily a condition of disability, but rather an increase in the probability of a number of small changes in performance parameters (Rabbitt 1991, p. 776).

Changes in Biomechanics

In addition to the anthropometric changes that may occur with increasing age, there are numerous alterations in biomechanical features (discussed in Chapter 1; see also the 1997 book edited by Fisk and Rogers). These alterations take place in the bones, joints, and muscles, as well as in various skills.

Osteoporosis

Bones. The long bones in particular become larger in both their outer diameter and their inner diameter (hollower), and larger pores appear in the remaining bone. Total bone mass decreases.

In age-related osteoporosis, the bones become stiffer and more brittle (Ostlere and Gold 1991). Women and persons who exercise little are more likely to suffer from osteoporsis than men and active people.

Broken bones

The changes in bone structure are associated with an increased likelihood of breakage as a result of falls or other accidents in which sudden forces and impulses are exerted. Injuries to the pelvic girdle, hip joint, or femur are particularly frequent in older women, followed by injuries to the bones of the shoulder and arms.

Joints. The lining of joints, the bony surfaces in joints, the supply of synovial fluids, and the elasticity and resilience of joint capsules and ligaments are all reduced, leading to reduced mobility in the joints, often associated with pain.

Muscles. While well-used muscles can retain their capabilities into advanced age, the reduced use of muscles, often accompanied by decreased circulatory supply, generally leads to a loss of musculature and an ensuing diminution in strength capabilities.

Skills

Skills. While many or some skills can be retained by the aging with practice and good health, others deteriorate for the reasons already discussed and, possibly, because of reductions in the performance of the central and peripheral nervous systems, as well as diminished circulatory and metabolic capabilities. The reductions are most likely in activities that require the exertion of large energies or forces, often combined with endurance requirements and controlled through perceptual information, particularly via the visual and vestibular modes.

Manipulation skills require strength, mobility, and sensory control and are important for many tasks, on the job, at home, and during leisure. Despite their importance, no unified and standardized measuring techniques are at hand, although several attempts at formulating such techniques have been made (Scott and Marcus 1991). Thus, little reliable information on hand capabilities exists for adults in general, and next to nothing is known for aging persons.

Changes in Respiration and Circulation

Respiratory capabilities diminish with increasing age, mostly because the alveoli in the lungs are less able to perform the requisite exchanges of gases, oxygen, and carbon dioxide. (See Chapter 2.) Furthermore, the intercostal muscles and the chest diaphragm lose some of their ability to generate breathing space in the chest; hence, vital capacity decreases. This diminution is coupled with reduced blood flow and, possibly, emphysema that often results from smoking.

The elasticity of blood vessels seems to decrease as well. Resistance to the passage of blood in vessels may be increased due to deposits along their walls. Blood-cell production in the bone marrow is decreased. Thin aging people may have reduced volumes of body fluids.

The heart functions also change. The size of the heart may be reduced, and cardiac output is lower. The heart rate takes longer to return to its resting level after it has been increased. Neural control of the heart may be impaired (Spence 1989).

Changes in Nervous Functions

Sensations and perception

The ability to cope with the environment depends in large part on detecting, interpreting, and responding appropriately to sensory information. Both sensation (the reception of stimuli at sensors and the resulting neural impulses in the afferent part of the nervous system) and perception (the interpretation of the stimuli; see Chapter 3) change with age. Together with diminished arterial and venous flow in the blood vessels, a reduction in the number of cells changes the stimulation and

TABLE 12–7. Anthropometric Data on the Elderly: Means (and Standard Deviations), all in cm unless otherwise stated.

Age range: Sample size:	50–100[a] 822	60–69[b] 43	60-69[c] 72	65–69[d] 24	65–74[e] 72	65–90[f] 184	66–70[a] 169	70+[b] 12	70+[d] 20	70+[c] 28	72–91[e] 130	75–94[e] 40
Stature, against wall		172.8 (6.6)		171.9 (6.6)				171.5 (9.0)	170.4 (7.5)			
Stature, freestanding	175.1 (8.9)		172.6 (6.4)	171.2 (6.6)		169.0			169.6 (7.6)	171.9 (8.4)	168.4 (5.3)	
Sitting height	79.9 (5.3)	90.8 (3.0)	90.8 (2.9)	90.0 (2.9)				89.5 (3.5)	89.0 (3.4)	89.8 (3.9)	88.3 (3.1)	
Knee height		53.9 (2.5)	53.6 (2.5)					53. 5 (3.4)	53.2 (2.9)	53.7 (3.2)	53.8 (2.1)	
Popliteal height	42.1 (3.5)		42.1 (2.3)							42.1 (3.0)	44.0 (2.1)	
Thigh clearance height			19.7 (1.4)							14.8 (1.2)		
Hip breadth	37.4 (3.9)			36.0 (2.3)					35.8 (1.7)	37.8 (2.4)		
Bideltoid breadth			45.3 (2.4)	45.1 (2.1)					44.7 (1.6)	45.0 (1.7)	43.4 (2.3)	
Biacromial breadth			38.9 (1.7)							39.2 (1.8)	37.8 (1.6)	
Hand breadth	7.7 (0.6)		8.5 (0.4)	8.5 (0.4)					8.5 (0.4)	8.6 (0.4)	8.4 (0.4)	
Head breadth			15.5 (0.5)	15.5 (0.5)					15.5 (0.5)	15.5 (0.4)	15.4 (0.5)	
Foot breadth			9.8 (0.6)							9.9 (0.5)	10. 0 (0.5)	
Head circumference			57.1 (1.4)	57.1 (1.3)					58.0 (1.4)	57.4 (1.6)	56.9 (1.8)	
Calf circumference			35.9 (2.5)	36.0 (2.9)					34.7 (2.1)	35.3 (2.2)	34.3 (2.7)	
Chest circumference, relaxed			99.6 (7.1)	99.9 (6.3)					99.6 (5.5)	99.7 (5.9)	96.2 (7.6)	
Chest circumference, maximum			101.8 (6.9)	101.7 (6.1)					101.5 (5.4)	101.7 (5.7)	98.7 (7.4)	
Chest circumference, minimum			97.6 (7.2)	97.5 (6.5)					97.8 (5.6)	97.9 (6.0)	94.5 (7.6)	
Upper arm circumference			30.9 (2.7)	30.5 (2.6)					30.0 (2.4)	28.7 (2.8)		

Age range:	50–100[a]	60–69[b]	60-69[c]	65–69[d]	65–74[e]	65–90[f]	66–70[a]	70+[b]	70+[d]	70+[c]	72–91[e]	75–94[e]
Sample size:	822	43	72	24	72	184	169	12	20	28	130	40
Waist circumference			95.5 (9.3)	97.4 (3.9)					97.1 (8.0)	97.0 (7.6)		
Head length			19.6 (0.6)	19.6 (0.6)					19.5 (0.6)	19.7 (0.7)	19.7 (0.6)	
Hand length	17.5 (1.2)		18.9 (0.9)	18.9 (0.9)					18.8 (0.9)	19.0 (1.0)	18.8 (0.8)	
Buttock–knee length			58.6 (3.0)							58.4 (3.2)	59.1 (2.4)	
Buttock–popliteal length	46.3 (3.6)		48.2 (2.8)							48.1 (3.1)	47.2 (2.5)	
Elbow to middle finger length	44.2 (2.8)		46.8 (2.0)	46.8 (1.9)					46.6 (2.5)	46.9(2.8)	46.4 (1.8)	
Shoulder to elbow length			37.3 (1.8)	37.4 (1.7)					37.0 (2.1)	37.4 (2.2)	36.9 (1.7)	
Forward reach			84.2 (3.7)							85.0 (5.4)	86.9 (3.8)	
Span			178.7 (7.5)	178.8 (7.5)					177.6 (9.0)	179.2 (9.9)	174.0 (7.0)	
Skinfold (triceps) (right)				1.1 (0.4)		1.2 (0.3)				0.9 (0.4)	1.1 (0.4)	
Skinfold (subscapular) (right)				1.7 (0.8)						1.5 (0.7)	1.6 (0.7)	
Foot height				26.3 (1.2)	26.4 (1.2)				26.5 (1.3)	26.8 (1.4)	26.0 (1.0)	
Weight (kg)	63.7		76.6 (1.1)	76.4 (1.0)	65.6 (11.6)	63.7			74.3 (0.9)	75.3 (9.0)	69.0 (10.5)	63.7 (11.7)
Grip strength (left) (N)			432 (88)				323 (58)			352 (88)	262 (80)	
Grip strength (right) (N)			461 (88)				370 (68)			412 (88)	283 (78)	

Source: Excerpted with permission from "Anthropometry of the Elderly: Status and Recommendations," by P. L. Kelly and K. H. E. Kroemer, Human Factors, Vol. 32, No. 5, 1990.

References:

[a]Molenbroek (1987); Netherlands; average of males and females.

[b]Borkan, Hults, and Glynn (1983); United States; males only.

[c]Damon et al. (1972); United States; males only.

[d]Friedlander et al. (1977); United States; males only.

[e]Dwyer et al. (1987); United States; average of males and females.

[f]Pearson, Bassey, and Bendall (1985); United States; average of males and females.

conduction activities in the nervous system. This may lead to increased variability in reception and integration of, and hence response to, external and internal stimuli.

In the somesthetic system, the numbers of cells in the skin (dermis and epidermis) decrease, together with collagen and elastic fibers. Receptors such as Meissner's and Pacinian corpuscules also decrease in number. The reduction in the number of sensors and afferent fibers may be combined with reduced nerve conduction velocity.

Changes in Taste and Smell

The number of taste buds and the production of saliva in the mouth are often reduced. The tongue may fissue. The sense of smell is often said to be diminished, although this finding is not uniform (Belsky 1990; Hayslip and Panek 1989; Kermis 1984).

Changes in Visual Functions

"Loss of precision"

Changes in visual functions are tied to many concomitant anatomical, physiological, and psychological processes that develop with age (Committee on Vision 1987; Cowen 1988; Kline and Scialfa 1997). By analogy with the camera, the human eye as a photographic device loses precision. Structures that bend, guide, and transform light (see Chapter 4) change, reducing the amount of light reaching the retina and defocusing the image projected to the retina. It becomes more difficult to focus on near objects, particularly if they are elevated or move fast.

Watery eyes

The eyes water and tear, due to an accumulation of fluid on the outside. Water can affect the properties of the eye as a lens and can simply be annoying.

Changes in the Cornea

Droopy eyelids

Baggy eyelids may reduce the amount of light reaching the cornea.

Difficult to Focus

The cornea flattens, which limits the ability to focus. Fatty deposits can reduce the transmission of light and scatter arriving light.

Changes in the Pupil

Dimming

The opening of the pupil gets smaller, which reduces the amount of light entering the eyeball. This disorder, called senile miosis, has the most serious effect in dim light. There is a possible benefit from the smaller diameter, however, similar to having a smaller aperture in a camera lens: The depth of field may be enhanced, meaning that objects both near and far are in better focus, although they appear dimmer.

Changes in the Lens

Presbyopia

A common problem in people over 40 years is hardening of the lens, which reduces its ability to become thicker and more rounded for focusing on near objects. The condition is called presbyopia.

Changed perceived colors

The young eye has a slightly yellow-tinted lens, which acts as an ultraviolet filter for the retina. The aging eye becomes more yellow because of the development of fluorescent chromophores

of yellow color. A yellower lens is a stronger light filter, absorbing energy and hence both raising the threshold for the detection of light in general and specifically absorbing some of the blue and violet wavelengths. These changes alter the aging person's perception of colors: White objects appear yellow, blue is hard to detect, and blue and green are difficult to distinguish.

Dimmed light and veiling

With increasing age, water-insoluble dry protein becomes more prevalent, and macromolecules appear. These molecules decrease the transparency of the lens and thus the amount of light transmitted to the retina. The dispersion of light rays at the macromolecules may act like a veil through which one tries to see.

Cataracts

A large increase in insoluble proteins can form a cataract and cloud the lens of the eye. Cataracts can occur at any age, but are most common in the aging. Increased opacity distorts and decreases the available light. The effects of a cataract on vision depend on its size, location, and density. A small cataract in the center of the lens is likely to affect vision far more than even a large cataract at the periphery. Cataracts can cause blurred or double vision, spots, difficulty in seeing under too little or too much illumination, a change in the color of the pupil, and the sensation of having a film over the eye or looking through a waterfall. (The term *cataract* is derived from the Latin word for waterfall.)

Changes in the Vitreous Humor

The vitreous humor may yellow, increasing the problems already engendered by a yellowing lens. Liquid and gel portions may clump together, causing spots, or "floaters," to appear in the field of vision. Also, pockets of liquids may form. Together, clumping and liquefaction change the refractory properties of the eye, making the image formed on the retina less coherent. A sudden jolt or vibration may detach the posterior vitreous from the retina (probably as a result of macular edema), bringing about a severe vision impairment.

Changes at the Retina

The cones and rods of the retina—the sensors of color and light—are reduced in number in the aging person—particularly the cones at the fovea. Clarity of vision is diminished accordingly. Retinal pigment is reduced, a degenerative pigment (lipofuscin) begins to appear, and the retina becomes thinner, particularly at the periphery.

Glaucoma

The natural fluids produced in the eye may not drain well but collect inside the eyeball, and the ensuing pressure (glaucoma) may eventually destroy fibers of the optic nerve. Thus, regular checkups for glaucoma are recommended after the age of 40 to detect its occurrence before damage occurs.

Designing for Aging Vision

Individuals develop visual impairments of different types and magnitudes and at varying ages, but common experiences are difficulties with near vision (especially in dim light), reading small print, and distinguishing similar colors. In addition to these basic deficits, complex functions commonly deteriorate with age: visual perception, search, and processing, as well as depth perception and coping with glare.

Ocular motility, the ability to move the eyeball through muscular actions, becomes impaired, both in performing quick movements and in turning the eye to an extreme angle. Pursuit movements are impaired in most aging persons, which limits both the ability to follow a target smoothly and to move the gaze to a target and fixate thereon. Furthermore, the field of vision is reduced at the edges, particularly in the upward direction.

Simple Everyday Aids

Most of the visual problems of the aging can be dealt with fairly simply—for example, by providing proper corrective lenses, a higher intensity of lighting, and increased color contrast; using large characters with high contrast against the background; repositioning a computer screen; or shielding bright lights in the field of view. (See Chapter 4.)

Red and yellow should be preferred to indicate color accents, because green and blue become difficult to distinguish with increasing age. In fact, all kinds of contrast and sensitivities are reduced, which diminishes one's ability to perceive details of an object or a scene, particularly at twilight. Recognizing details from a cluttered background becomes difficult: Picking out individual faces in a crowd in a dim light is nearly impossible, and even reading a book with large print may be difficult if inadequate white space exists between the black letters.

Computer Work

At computer workstations, separate optical lenses may have to be used either for viewing the display or for reading and writing, particularly if the display is beyond about 50 cm, the common reading distance. The display should be placed low rather than high with respect to the eyes. (See Chapter 9.) On the screen, bright reds and yellows are relatively easy to distinguish and should be used instead of blues and greens. Since low-contrast images are difficult to see, aging workers may have difficulty viewing the green-on-dark lettering of older displays.

More Light?

Extra lighting (to increase luminance) can improve the visual ability of aging persons. But since glare is a problem for many, spotlights and other bright light sources must be placed carefully. Aging people often find it more difficult to adapt to sudden changes in lighting, particularly from bright to dim conditions. The detection of targets on a cluttered background is facilitated if the light-on-dark contrast of the target is enhanced.

Driving Automobiles

Many aging people drive automobiles, although they may avoid driving at night, on unfamiliar roadways, in congested areas, and during congested times. To help them, one should avoid light and color "clutter," such as that experienced in many commercial streets, where illumination and advertising lights compete with traffic lights. (In some European countries, no red or green lamps may be used near traffic lights.) Regarding the design of the "visual interior" of the automobile, the luminance of instruments should be higher than the common 300 cd m^{-2}, to avoid "washout" in direct sunlight; 1,200 cd m^{-2} are required. The common minimal contrast ratio of 5:1 is sufficient for passive displays, but the ratio should be about 20:1 for moving displays. Twenty-five minutes of arc should be the minimum character size. Critical visual information should be displayed in the central area of view. Red is most easily distinguished, followed by yellow and white. These colors should be pure in saturation. Such design recommendations (see Chapter 10) not only help older drivers, but also are advantageous for younger eyes.

Changes in Hearing

The ability to hear decreases quite dramatically in the course of one's life, first at the high frequencies between 10 and 20 kHz and then down to about 8,000 Hz, an impairment called presbycusis or age-related hearing loss. Yet, difficulties in hearing may extend into the lower frequencies, and they are often coupled with noise-induced hearing loss. (See Chapter 4.)

Biologic changes

The changes start at the pinna—the outer ear—which becomes harder and less flexible and may change in size and shape. Wax frequently builds up in the ear canal. Often, the Eustachian tube becomes obstructed, leading to an accumulation of fluid in the middle ear. There may also be arthrosics in the joints of the bones (anvil, hammer, and stirrup) of the middle ear, which fortunately, does not usually impair the transmission of sound to the oval window of the inner ear. There, the hair cells in the basilar membrane of the cochlea often atrophy and degenerate. Deficiencies in the bioelectric and biomechanical properties of the inner ear fluid set in, and of the cochlear partition degenerates mechanically, often together with a loss of auditory neurons. The deterioration causes either frequency-specific or more general deficiencies in hearing.

Hearing sensitivity affected

While it is estimated that 70 percent of all individuals over 50 years of age have some kind of hearing loss, the changes are individually quite different. Typically, in populations that do not suffer from industry or civilization-related noises, the hearing sensitivity in the higher frequencies is less reduced than in people from so-called developed countries. Changes related to the environment overlap with, and in some cases mask, age-dependent changes.

Speech hard to understand

Loss of hearing in the higher frequencies of the speech range reduces the person's understanding of consonants that have such high-frequency components. This explains why older persons often are unable to discriminate between phonetically similar words, making it difficult for them to follow conversations in noisy environments. Severe hearing disorders may lead to speech disorders, which, to some extent, may be psychologically founded. For example, if others have to speak loudly for you to hear, they may not want to interact with you because they may feel embarrassed. They may hesitate to speak if it is uncertain what level of loudness is required and whether one understands what is said.

Hearing aids

Technical hearing aids can improve a person's ability to hear by amplifying sounds for which hearing deficiencies exist. This, however, is difficult to do if the deficient areas are not known exactly—for example, because a person has not taken a hearing test recently. The simple amplification of sounds usually also enforces unwanted background noise. Selective suppression and amplification of sound is technically possible now, as mentioned in Chapter 4. Improvement is very difficult if the hearing loss is due to the destruction of structures of the middle and inner ear and, with current technology, is nearly impossible if the auditory nerves have been damaged. Further technical development may bring better help in the future; replacement of the cochlea, for example, by an electronic implant is already feasible.

Ergonomic interventions The preferred or comfortable sound intensity level for speech increases with age: For prose text, it is on average, from about 55 dB at 20 years of age to more than 80 dB at 80 years (Coren 1994). Furthermore, one can improve the clarity of the message, as discussed in Chapter 4, by providing sound signals that are easily distinguishable and of sufficient intensity, and by avoiding masking background sounds (noise). Another solution is to provide—at least some—information through other sensory channels, such as vision (eg, one might present illustrations and written text to accompany an auditory message), or to employ the sense of touch as in the use of Braille.

Changes in Somesthetic Sensitivity

The somesthetic senses include those related to touch, pain, vibration, temperature, and motion. (See Chapter 4.)

Touch In spite of decades of research, there is deplorably little reliable quantitative information about the changes, if any, in tactile sensitivity, including pain, with aging (Boff et al. 1986). Apparently, absolute thresholds increase, which may be associated with a loss of touch receptors, but that phenomenon and its explanation are still rather unresearched. There is a well-observed change in vibratory sensitivity, particularly in the lower extremities. This effect is used in the diagnosis of disorders of the nervous system. Among the possible explanations may be a reduction in blood supply to the spinal cord with ensuing damage to the nerve tracks, possibly a decline in the number of myelinated fibers in the spinal roots, or diminished blood flow to the peripheral structures of the body in general. Dietary deficiencies might also play a role.

Sensing Temperature Sensitivity to temperature also seems to decline with aging, but this may be partially offset by the oft-noticed desire of elderly persons to be in warmer temperatures, outside or within a building. The observed differences in temperature behavior may be associated with a decline in the body's temperature regulation system.

Perceiving Motion and Pain A decrease in information from kinesthetic receptors or a decrease in the use of that information in the central nervous system may contribute to aging persons' higher incidence of falls: With increasing age, one seems less able to perceive that one is being moved (such as in an automobile or airplane) or that one is moving certain body parts. Again, surprisingly little is known in a systematic fashion. The same is true for pain sensitivity, which appears much reduced as one gets older. Possible explanations are reductions in the number of Meissner's corpuscles and other receptor organs in the skin, in the number of myelinated fibers in the peripheral nervous system, and in the blood supply.

Changes in Psychometric Performance

With aging, reaction and response times to stimuli (see Chapter 4) typically increase. This phenomenon may be partially explained by deficiencies in the sensory peripheral parts of the nervous system, delays in the provision of information to the central nervous system by the afferent nerves, and a reduction in efficiency in the efferent part of the peripheral nervous system. Yet, successful psychometric performance involves perception, sensation, attention, short-term memory, decision making, intelligence, and personality, as well as motor behavior. Thus, while poor performance may be attributed to some of the more physiological factors, much depends on other processes of the mind that go beyond the scope of this text. Among the theories that purport to explain changes in central functions are those of neural noise, expectancy or set theory, complexity and information overload theories, and rigidity hypotheses. (See, for example, publications by the Committee on an Aging Society 1988; Belsky 1990; Berg 1998; Czaja 1990; Hayslip and Panek 1989; and Kermis 1984; and literature in geriatric psychology, physiology, and sociology.)

The aging of friends and relatives, as well as one's own aging, is of perpetual interest and concern. The discussion in this book of knowledge about changes with age and related research is by no means complete. Anecdotal observations and case studies abound in the literature, but systematic research and the compilation of its results are lacking. The recent phenomenon of greater numbers of older people, being active in politics and accumultating economic wealth, is prompting much research.

Research Needs

Available research findings do not provide a complete picture of the changes that occur with aging. This is largely due to the fact that cross-sectional research is nearly meaningless: Comparing the anthropometry or performance of persons of similar age gives little insight into the nature and effects of the aging process, and how to counteract them.

Chronological age is not a meaningful classifier; better classification systems need to be established. The so-called biological age is one attempt in that direction, but its definition is rather abstruse. Thus, a basic research task is to establish a suitable reference system, or reference systems, with proper scales and anchoring points.

It is obvious that many or almost all physical, perceptual, cognitive, and decision-making capabilities decline with age. Yet, some of those losses are slow and are not easily observed. Other faculties decline fast, or else they may deteriorate slowly at first and then quickly at some point in time, perhaps again stabilizing for a while. Some of these changes are independent of each other, but many are linked, directly or indirectly. Failing physical health may have effects on one's attitude and intentness, or failing eyesight might lead to a fall and serious injury, with an ensuing illness.

In addition, people have different coping strategies. One's failing ability to recall names, for example, may be overcome to some extent and for some time by developing mnemonic strategies. An aging person may reduce the boundaries of his or her activities by maintaining only those with which one is comfortable and that one can competently handle. Performing challenging physical or mental activities, perhaps purposefully so, can counteract reductions in capability significantly and for long periods of time.

Thus, in our current framework scaled by chronological age, one finds an immense variety of faculties that are maintained and perhaps in some respect even increased, of decreasing capabilities, and of either slow or fast changes with time.

There is much evidence that skilled performance, both mental and physical, can be maintained by many persons into advanced age (Rabbitt 1997), for several reasons:

- In most of our daily tasks, at or away from work, only a portion of our abilities is required. Thus, even with decreasing capacity, enough remains available to do what needs to be done.
- Certain tasks require a certain kind of experience and patience that may come with age.
- Continuing to use, and even to train, one's skills and functions tends to maintain them to an advanced age.

More research needs to be done.

Designing for the Older Worker

The U.S. Age Discrimination Acts have defined the "older" person as either age 40 or 45 and above. While one cannot set a particular year as the beginning of "aging," there is no doubt that certain work tasks become more difficult as one approaches retirement (Panek 1997). Among these tasks are strenuous physical exertions, such as moving heavy loads; tasks that require high mobility, particularly of the trunk and back; and work that requires high visual acuity and close focusing. On the other hand, anecdotal evidence suggests that tasks which require patience and experience-generated skills may be performed better by at least some older workers than by young persons. Nearly all age-related difficulties can be overcome or ameliorated by

- workplace and tool design and selection,
- the arrangement of work procedures and tasks,
- the provision of working aids, such as power-assisted tools or magnifying lenses, and
- managerial measures, such as assigning proper work tasks and providing work breaks.

Specific ergonomic measures are the logical results of the intent to counteract known deficiencies:

- Manipulation tasks are facilitated by providing proper work height, supports for the arms and elbows, and special hand tools, which may be power aided. (See Chapter 8.)
- Postural aids include the just-mentioned proper work height, coupled with the provision, in the shop or office, of an appropriately designed, easily adjustable seat. (See Chapter 9.)
- Vision deficiencies can be counteracted through the provision of corrective lenses, by intense and well-directed illumination, and by avoiding direct or indirect glare. (See Chapter 4.) If text must be read, on paper, on or a dial, or on a computer screen, the proper character size and contrast, as well as carefully selected color schemes, should be used. (For instance, avoid difficult visual discriminations of similar hues, particularly in the blue-green range.)

The auditory environment should be controlled to keep background noise at a minimum and to provide sufficient penetration of auditory signals that must be heard by selecting appropriate intensities and frequencies. (Avoid frequencies of about 4,000 Hz or greater; see Chapter 4.)

> The use of proper ergonomic measures, carefully selected and applied, is just "good human engineering," which would help all persons of all age groups, but is of particular importance for the older individual.

Designing for the Aging Driver and Passenger

Depending on whose statistics one reads, aging drivers may have fewer or more traffic accidents, but the pattern of accidents with age indicates increasingly missed or misinterpreted perceptual cues, slow or false reactions to cues, and wrong motor actions, such as accelerating instead of braking. There is a vast literature on the many traits of the aged driver. (See, for example, Eberhard and Barr 1992). Recommendations for the design of the interior of automobiles were discussed earlier.

Information Especially in public transportation (planes, trains, subways, buses, elevators, moving walkways, etc.), aging persons often report problems related to information (cues and signs) indicating direction, location, and use, such as the following:

- Where does the bus go that I see coming?
- Where does the train stop?
- Where is the exit to Brown Street?
- Where are we?
- How do I buy a ticket?

Many of these problems can be overcome by applying common "human-engineering principles," such as using signs with good lettering (of the proper contrast and size and with the proper symbols), appropriate illuminance and avoidance of glare, auditory announcements that are timely and understandable, and redundant information, such as showing the floor number in an elevator in large numerals and announcing it early over a public-address system.

Ingress and egress In buildings and transportation systems, entrances and exits, and passages and steps, are often difficult for the aging to negotiate, particularly in trams, trains, buses, or moving walkways. Uneven floors and damaged or misplaced floor coverings often pose problems in vehicles or hallways. Handholds used to be a problem in buses and trains, but most now provide a variety of loops, columns, and hand grips.

Use of walking aids and wheelchairs Many aged people use canes or walkers, and some need wheelchairs. While a cane usually does not pose a problem in public transportation (in fact, it may alert other passengers to be considerate and helpful), walking aids and particularly wheelchairs are often not easily accommodated and may be a serious deterrent to the use of public transportation.

☛☛☛ *"Human aging encompasses much more than physiologic change over the life course. Age-related changes are manifest across all aspects of life including physical, environmental, economic, and social aspects and mental well-being. Yet change along any dimension is not simply, or irrevocably, correlated with chronological age. The serious loss or compromise of capacity in one area, however, can accelerate the rate of decline in others. Such interactions are often complex. Poor health, for example, can require increased medical expenditures that divert income from other essential areas such as home upkeep or the purchase of food. Over time, such interactions can result in further erosion of functional capacity in the aging. Alternatively, a supportive social or physical environment may retard the rate of functional loss to some degree" (Soldo and Longino 1988, p. 103).* ☚☚☚

DESIGNING THE HOME FOR THE AGING

A 50-year-old who purchases a home (or a durable product) will be a rather different individual still using it 10, or 20, or more years later. Whether a person is able to live at home alone (perhaps with some outside help) or whether the person has to be cared for, either at home or in a supportive environment, is a complex function of various abilities and or disabilities. Figure 12–4 presents an overview of common disorders among the aged and how these problems can be alleviated, at least to some degree, by proper ergonomic measures (Czaja 1997; Peloqin 1994).

How much dependency?

Functional ability, or its opposite, dependency, is commonly assessed in terms of clusters of abilities: "instrumental activities of daily living" (IADL) or the less specific "activities of daily living" (ADL), shown in Table 12–8. Numerous surveys of elderly persons have been conducted using these two activity listings, in accordance with which persons have been classified into groups that require help and specifically designed environments. But there is a major problem with the IADLs and ADLs: While they are rather practical and self-explanatory, they lack specificity and objectivity and are difficult to measure. Clark et al. (1990) and Rogers et al. (1998) started to im-

Problems/ Manifestations	Senescence	Arteriosclerosis	Hypertension	Parkinson's disease	Peripheral neuropathy	Drowsiness	Cataracts/glaucoma	Arthritis	Paget's disease	Osteoporosis	Low back pain	Bronchitis/emphysema	Pneumonia	Diabetes	Senile dementia
General debility	☆		☆	☆	☆			☆	☆	☆	☆	☆	☆	☆	
Mobility	☆	☆		☆	☆		☆	☆	☆		☆				
Posture	☆			☆			☆	☆	☆	☆	☆				
Pain		☆		☆	☆			☆	☆		☆				
Incoordination		☆		☆	☆		☆	☆							
Reduced sensory input	☆	☆			☆	☆	☆								
Loss of balance	☆	☆		☆			☆								
Reduced joint mobility								☆	☆	☆	☆				
Weakness in muscles	☆			☆	☆			☆							
Auditory disorders	☆					☆		☆						☆	☆
Locating body in space		☆					☆	☆							
Shortening of breath		☆		☆								☆	☆		
Deformity								☆	☆	☆					
Memory impairment		☆													☆
Visual problems	☆	☆					☆							☆	
Disorientation		☆		☆											☆
Loss of sensation	☆				☆	☆									
Cognition disturbance		☆		☆											
Incontinence		☆													☆
Speech disorders				☆		☆	☆								
Touch disabilities	☆				☆			☆						☆	

Figure 12–4. Problems arising from common age-related disorders. (Adapted from Kemmerling 1991.)

TABLE 12–8. Measures of Daily Living

Instrumental activities of daily living (IADL)	*Activities of daily living (ADL)*
Managing money	Living
Shopping	Transfering between bed and chair
Light housework	Indoor and outdoor mobility
Laundry	Dressing
Preparing a meal	Bathing
Making a phone call	Toileting
Taking medication	

prove the situation by subdividing ADLs into more specific tasks such as lifting–lowering, pushing–pulling, bending–stooping, and reaching. Further work in this direction could result in a list of basic demands and activities similar to the elements of motion.

Since the 1970s, ergonomists and architects have found increasing attention to the needs and desires of aging persons, to design their living quarters to accommodate their diminishing faculties. It is helpful, for example, not to have steps or stairs that would be negociated, nor to store items in high or low cabinets for which one must reach up or bend down, not to have window or appliance controls that are difficult to manipulate.

Kitchen

One area in which ergonomics can have a great impact is the kitchen, where one stores, prepares, and serves food. In many households, the kitchen is the de facto living room and a phone-in message center. In the past, the kitchen was the woman's territory, but this is no longer true. The first "scientific study" of kitchens was completed by Lillian Gilbreth in the 1920s. Her classical investigation concerned mainly the flow of work, on which she did time and motion studies leading to methods that she pioneered together with her husband. Her redesigned kitchen reduced motions by nearly 50 percent.

"Seven principles" derived from time and motion studies, augmented by ergonomic findings, apply to the kitchen:

1. One should design for a small "work triangle," the corners of which are the refrigerator, sink, and stove or range.
2. The "work flow" for food preparation is to remove food from the refrigerator or cabinet, mix or otherwise prepare ingredients near the sink, cook on the range or in the oven, and serve. Kitchen components should facilitate that flow.
3. If there is a traffic flow caused by others, it should not cut through the patterns of the work triangle and work flow.
4. Items should be stored at the "point of first use," as determined by the work triangle and work flow.
5. The workspace for the hands should be at about elbow height or slightly below. Having things at this height facilitates manipulation and visual control. Counter and sink heights are derived from the elbow height. Note that it might be advisable to consider walking aids and that stools or chairs might be used.
6. The reaches to items stored in the kitchen should be at or slightly below eye height, to allow for visual control and easy arm and shoulder motions.
7. The motion and workspace should not be reduced or interrupted by doors of appliances and cabinets, which, therefore, should open "outward" from the working person.

Bathroom

Another area of major ergonomic concern is the bathroom, one of the busiest and, unfortunately, most dangerous rooms in the house. Basic fixtures in the bathroom include a bathtub or shower (or both), toilet, and a lavatory sink. Also, there are usually storage facilities for toiletries, towels, etc.

The bathtub and the shower are the two usual means for cleansing the whole body and are common sites of accidents. Their major danger stems from slipperiness between bare skin and the floor or walls. The more dangerous of the two is the bathtub, because of its slanted surfaces and the high sides above which one has to step, a procedure that is difficult for most people and particularly so for the elderly, who may have balance and mobility deficiencies. Kira (1976) described several techniques employed by most users getting in and out of a tub. They involve shifts in body weight: from the legs, to the arms and legs used together, and, finally, to the buttocks while entering the tub; in leaving, these shifts occur in the reverse order. There is a high potential for loss of balance and for slips and ensuing falls. The angle of the backrest and its slipperiness constitute the most critical design aspect regarding the person's resting in the tub. Proper handrails and grab bars, within easy reach both for sitting and for getting in and out of the tub, are important. The shower stall may also have a slippery floor to step on, but its lower enclosure rim makes it easier to move in and out.

Using the control handles for hot and cold water is quite often difficult for aged persons (Meindl and Freivalds 1992), particularly when they are not in their familiar home and have to cope with different handle designs, directions of movement, and resistances. Both better design principles and standardization would be helpful—for example, in the mode and direction of movement to regulate the temperature of the water.

The wash basin may be difficult to use if it is too far away, as it may be if inserted in a cabinet, so that one cannot step close to it. The faucet often reduces the usable area of the wash basin. Of course, the height of the sink is important, as are the water controls.

Kira's classical study has provided much information about the proper design, sizing, shaping, and location of the toilet bowl and seat. New information has come from several more recent studies listed by McClelland (1982). Asking the elderly about their problems and requesting suggestions from them has become a well-practiced and successful approach in applied gerontology (Nayak 1995, Mullick 1997).

A number of publications—for example, by Singer and Graeff (1988) and Peloquin (1994)—contain valuable ergonomic design recommendations for Western-style facilities. In other civilizations and other parts of the earth, quite different customs and conditions exist, for which at present only limited ergonomic information is available (Cai and You 1998; Ogawa and Arai 1995).

The Design of Nursing Homes

Severe functional disabilities are what we dread most when we think of old age, because they are at the core of our fears about growing old. In the United States, currently only one of four people age 65 has any problems negotiating life, but impairments are likely progressing. As more health deficiencies, mental problems, or functional disabilities occur with further aging, the elderly person first needs more help in his or her own home. Initially, that care may be privately secured, through relatives and friends or a hired person. For many, this is the beginning of a path that leads to a nursing home.

"Therapeutic nihilism"

Many aged persons suffer from (often self-diagnosed) treatable diseases and conditions, but have changed their attitude to accept aches, pains, and physical distress as something "normal" at an advanced age; and they may negelect or even refuse to take prescribed medication (Park and

Jones 1997). They may not want to appear weak and discouraged and hence do not want to "bother" the caregiver, even if they are generally ill and in need of help. This, of course, makes it particularly difficult to provide them with the help and care that they need and deserve.

Nevertheless, aged people are the most frequent consumers of physicians' services. Basically, it is the physician's job to diagnose an illness, prescribe a medical intervention, and effect a cure. But this is not the usual path with the elderly: The older one gets, the more likely one is to suffer from a chronic illness that cannot be cured, although occasionally it can be alleviated or covered up, at least for a while. Fighting disability requires diagnostic strategies different from those employed in battling illness. Among these strategies are techniques outside the physician's traditional realm of expertise; thus, the physician must collaborate with specialists in geriatric medicine, nurses, physical therapists, dietitians, psychologists, and ergonomists.

Room, board, care

There are different kinds of institutions for the elderly who cannot stay at home. Some simply offer room, board, and personal care to the residents. Others are more like a hospital, offering intensive medical services to seriously ill people. Some cater to certain religious or ethnic groups, some freely accept people with Alzheimer's disease or those who are bedridden, and others want only occupants who are not severely impaired, either physically or mentally. In the midst of this diversity, Belsky classified these institutions by the intensity of care they provide.

Nursing homes

In the United States, nursing homes must be classified as offering either "intermediate" or "skilled" care in order to be reimbursed by Medicaid or Medicare.

Residents who need assistance in functioning, but not intensive care, are in intermediate-care facilities. Here, the architectural and other ergonomic design recommendations for the private home also apply, as they facilitate the residents' efforts to look after themselves. In addition, design and organizational means must be considered that facilitate the caregivers' activities, such as easy cleaning, awareness of requirements for getting help, and emergency access.

While it is important that aged people have as much personal freedom as possible, being in an institution limits their choices in the most basic aspects of life, such as where to live, when to get up or lie down, what to do, and what meals to eat. Home management should carefully provide various choices for the residents, keeping their interests in mind above organizational considerations.

☛☛☛ *The quality of nursing facilities varies widely from "home" to "snake pit" (Belsky 1990, p. 107).* ☚☚☚

Patients in "skilled nursing care facilities" (as these institutions are called in the United States) usually have little ability to control themselves or their environment. Residents tend to be very old and in continual need of help and care. Owing to gender differences in longevity and types of illnesses and injuries, women are more likely than men to be disabled but not to have a life-threatening illness. Since nursing home care is, in the United States, largely financed by Medicaid, most residents are poor; many are single, divorced, or widowed.

Some of the earlier mentioned recommendations for ergonomic designs apply to the architecture and interior design of nursing homes for people who need intensive care, but now the aspects of providing 24-hour supervision and care, and possibly intensive medical treatment, prevail. Unfortunately, ergonomic information on architecture and interior design for nursing care facilities is piecemeal and incomplete, as is human-engineering information on the design of medical devices and equipment for health care and rehabilitation. These areas still require systematic research before general design guidelines can be expected (Committee on an Aging Society 1988; Gardner-Bonneau and Gosbee 1997; Hutchinson 1995; Vredenburgh et al. 1995).

With increased longevity, ergonomic design for the aging person, either still at work, retired from the job but still active, or needing care, has acquired great importance.

- *Changes among aging persons are not correlated with chronological age, but exhibit considerable variation.*
- *Changes within an aging individual or among aging people are not manifest as a simple linear decline, but show a variety of rates; change may even be arrested.*
- *Changes in function may produce different effects in the same person, and the variation among aged persons tends to increase with age.*
- *Different faculties may change at different rates, even independently from each other, but the serious loss or compromise of functional capacity in one area can accelerate the rate of decline in others.*
- *A supportive social or physical environment can retard the rate of functional loss to some degree (Committee on the Aging Society 1988, p. 7).*

> Fisk (1999) estimated that about half the problems reported by active adults aged 65 to 88 years had the potential to be improved by human factors intervention.

ERGONOMIC DESIGN FOR DISABLED PERSONS

Many of the design considerations discussed in the foregoing section regarding aging people are also relevant to disabled persons. Particularly appropriate are those design recommendations that help to alleviate, overcome, or circumvent impairments so as to attain the largest possible independence and everyday functioning capability.

At pedestrian street crossings, it is advantageous to combine traffic lights with acoustic signals that indicate to persons with impaired hearing and sight that they may or shall not cross the street. In addition to the familiar red and green lights, selected sounds help to indicate when it is safe to proceed. For example, a "cuckoo" sound may signal crossing in one direction and an electronic "chirp" indicate the orthogonal direction. This combination of light and sound is beneficial not only for impaired and aging persons, but also for all other pedestrians.

"Disability," Not "Handicap"

Disability derives from impairment

We are all only temporarily able bodied: As children, we lack the strength and skill that we hope to acquire during adulthood, and while aging, we lose some of the faculties that we previously enjoyed. Many suffer from injuries or illnesses that deprive them of certain capabilities, often only for some period of time, but possibly for the rest of their life. Estimates of the percentage of the working-age population with disabilities range from 8 to 17 percent in the United States, and the estimates for children and the elderly are also quite variable. Some of the difference stems from different definitions of disability (Levine et al. 1990; Gardner-Bonneau 1990).

Disability

The 1990 Americans with Disabilities Act (ADA) defines *disability* as "a physical or mental impairment which substantially limits one or more of an individual's major activities of daily living such as walking, hearing, speaking, learning, and performing manual tasks." The Committee on National Statistics of the U.S. National Research Council, as well as the World Health Organization, defines disability as "any restriction or lack [resulting from an impairment] of ability to perform an activity in the manner, or in the range, considered normal." Since human activities are varied, there are many different kinds of disabilities.

Impairment

In this context, an *impairment* is a chronic physiological, psychological, or anatomical abnormality of body structure or function caused by disease or injury, and work disability is a dysfunction in the vocation for which a person is trained, which is often selected early in adulthood and the choice of which may be influenced by impairment(s) existing during youth.

Handicap

The disadvantage that results from impairment and disability (and not the impairment or disability itself) properly is called a *handicap,* because it may entail a loss of income, social status, or social contacts. Some handicaps can be alleviated—for example, by providing a space to park a car next to a the entrance of a building used by persons with disabilities. Of course, it is much better to avoid a disability (and thus a handicap) by human-engineering means, such as replacing a natural eye lens impaired by cataracts with a clear artificial lens or by replacing a damaged body joint with a manufactured one.

Design for One

There is a wide range for judgment regarding the various impairments, disabilities, and handicaps, both with respect to their extent in relation to specific activities and regarding specific age groups. For example, judged against "normal adults," children, pregnant women and many elderly persons could be described as disabled.

In spite of individual differences, one often considers juveniles, adults, and pregnant women groups of people who have features that are sufficiently similar so as to allow ergonomic design recommendations that are generic to each group. This is not the case for many persons with disabilities, which are often very specific to the individual, and therefore, the person's special abilities (or lack of them) must be assessed to determine whether devices or other arrangements are needed for assistance.

Ergonomic Means to Enable the Disabled

Faste (1977) sorted and classified impairments as shown in Figure 12–5. The impairments (and ensuing disabilities) presented in the figure may result from very dissimilar conditions. For example, poor balance may result from conditions related to blood pressure, hemiplegia, paraplegia, amputation, multiple sclerosis, muscular dystrophy, cerebral palsy, Parkinson's disease, brain tumor, and other causes. The design matrix shown in Table 12–9 relates display design issues to impairments. Such charting identifies ergonomic challenges and indicates possible solutions.

In most impaired persons, a specific disability or a combination of disabilities can be compensated for by ergonomic means. For example, blind customers may be served at a sales or information counter to aid them in making transactions. Computers can be particularly beneficial for a paraplegic who is unable to do most physical tasks. With the computer, one does not have to manipulate papers, files, or other printed information, and one need not be able to write, draw, or handle a phone. Yet, the computer workstation itself, particularly the current QWERTY keyboard and the mouse, may be difficult or impossible for the impaired person to use. Thus, adaptive hardware (such as a stick attached to the head or hand or held in the mouth) is often needed to press the keys; or voice recognition software can replace many manual

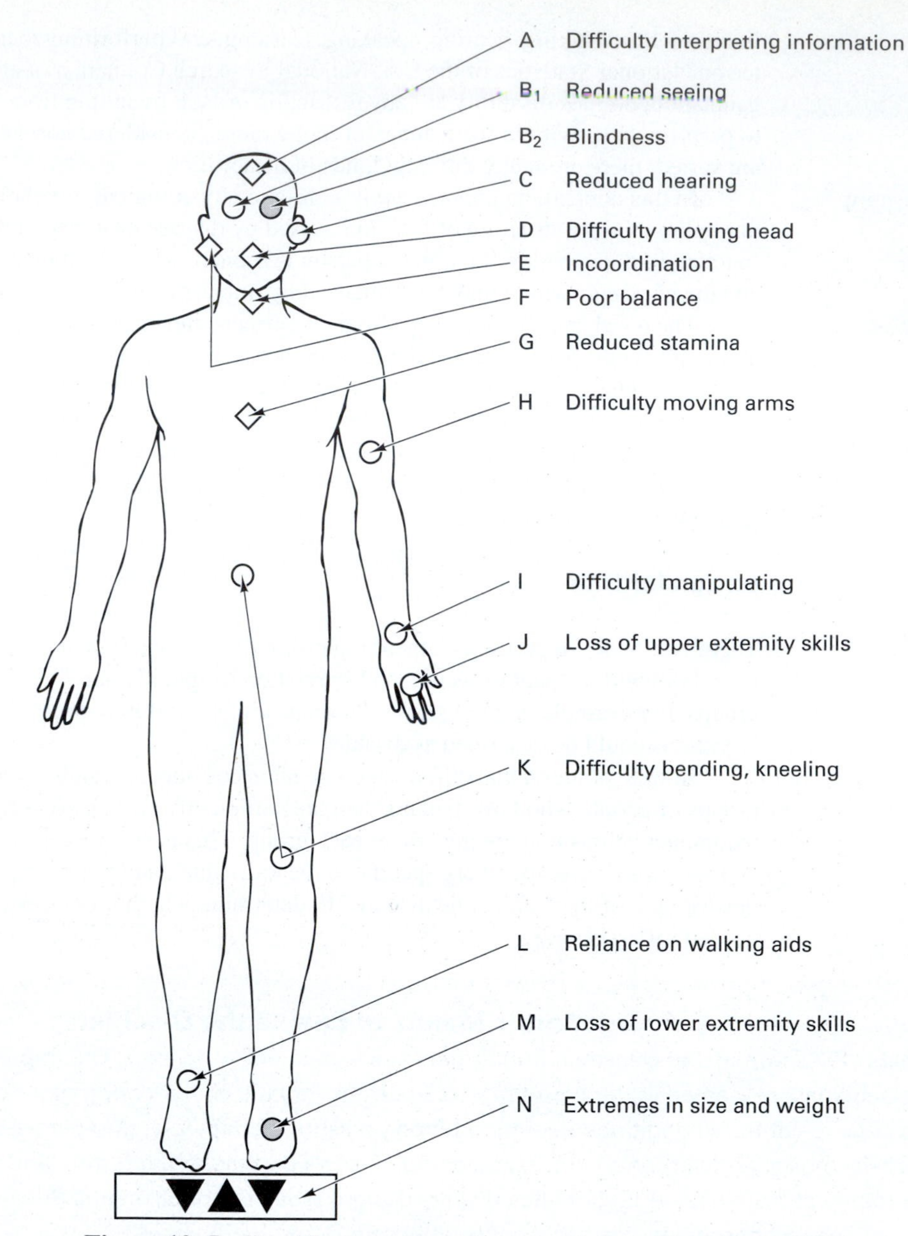

Figure 12–5. Classification of impairments. (Adapted from Faste 1977.)

inputs. Other software programs allow disabled individuals to control their environment in the home or office (Newell and Gregor 1997).

Casali and Williges (1990) described a systematic procedure for determining the most appropriate aid for a disabled computer user. As shown in Figure 12–6, the approach relies on a database containing information about available hardware and software. The residual abilities of the client regarding her or his

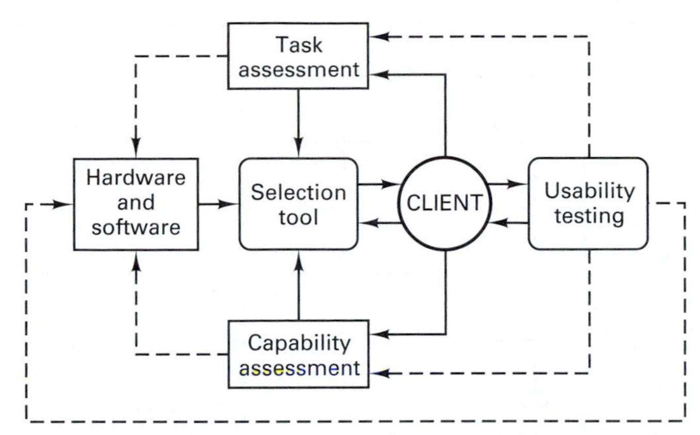

Figure 12–6. Systematic process for determining proper computer aids for an impaired client. (With permission from Casali and Williges 1990.)

use of the computer are assessed and compared with the task. After choosing a candidate solution, its usability is checked, and changes and improvements are introduced as needed.

Among the testing procedures and hardware that have been developed to quantitatively measure the residual capabilities of impaired persons are the Basic Elements of Performance (Kondraske 1988) and the Available Motions Inventory (AMI; see Dryden and Kemmerling 1990; and Smith and Leslie 1990). The AMI, for example, has a number of panels that contain switches, knobs, and other devices to be reached, turned, operated, or otherwise activated and manipulated. These panels are put into standardized positions within the work area, and each person's ability or inability to operate the controls as required is recorded, together with the time needed and the strength exerted. A large database is at hand that allows one to compare the individual's performance with that of other persons, able or disabled. Thus, the individual's capability or limitations regarding the performance of certain tasks that are of particular interest can be determined. The information obtained thereby facilitates a systematic approach to finding work tasks and work conditions suitable for a given individual's capabilities, or to modifying them to match the person's abilities.

Locomotion and transportation are serious problems for many disabled persons. Providing technical aids (eg, prostheses, crutches, and walkers) has been an age-old concern. Wheelchairs have been described for decades; a comprehensive review covering the time prior to the mid-1980s was undertaken by Zacharkow in 1988. In addition to affording movement, the wheelchair should provide stable, yet relaxed, support, particularly avoiding pressure points and sores (Troy et al. 1995). Accordingly, Zacharkow presented specific recommendations regarding the seat, backrest, armrest, leg support, etc. More recently, a large number of reports on wheelchair developments and uses have been published, describing client-propelled, attendant-pushed, and motor-driven chairs for indoor and outdoor use and even for racing. (See, e.g., the publications edited by Berg 1998 and Kumar 1997 and the journals *Biomechanical Engineering, Prosthetics and Orthotics International, Rehabilitation and Progress Reports,* and *Rehabilitation and Research Development.*) Much effort is given over to developing specialized engine-driven covered vehicles or the adaptation of existing automobiles to disabled drivers or passengers.

TABLE 12–9. Effects of Display on Users With Impairments. (Modified from Faste 1997, Cushman and Rosenberg 1991.)

		Impairments (see Fig. 12–5)														
	Display characteristics	*A*	B_1	B_2	*C*	*D*	*E*	*F*	*G*	*H*	*I*	*J*	*K*	*L*	*M*	*N*
Vertical location	high overhead	3	4	4		4	2	4						3	3	1
	requires looking up	2	2	4		3	1	3						1		
	requires looking straight ahead			4												
	requires looking down	2	2	4		3							1			1
Horizontal Location	directly in front			3												
	off to left or right side	3	2	4		3	3							2		
Viewing distance	about 0.5 m		1	4												
	about 1 m		2	4												
	farther than 1 m		4	4												
Orientation	horizontal			4												
	other	2	2	4		2										
Vertical size	small subtended angle		4	4												
	medium subtended angle		3	4												
	requires head movement			4		2										
Horizontal size	small subtended angle		4	4												
	medium subtended angle		2	4												
	requires head movement			4		3										
Content	shape code	3			1						1	4				
	color code	3		4												
	picture		1	4												
	map	3	2	4												
	pictogram	2	1	4												
	symbol	4		4												

		A	B_1	B_2	C_2	D	E	F	G	H	I	J	K	L	M	N
	identification label	3	2	4												
	dichotomous information	2	2	4												
	quantitative information	4	3	4												
	brief text	3	3	4												
	long text	4	4	4					1							
	audio cue supplement	2			2											
	audio cue only	3			3											
Exposure variables	used frequently	2														
	used occasionally	3	3	2												
	short viewing time	4	4	4												
	observer or display moving	4	4	4												
	dynamic display	3	4	4												
	interactive display	4	3	4	4						3	4				
Illumination	high contrast on display		1	4												
	low contrast on display	4	4	4												
	high contrast to surround			4												
	low contrast to surround	4	4	4												
	front lighted			4												
	translucent or back lighted			4												
	daylight		1	4												
	artificial light			4												
	glare present	4	4													
Other variables	legibility	4	4	4												
	readability	4	4	4												
	logic of location	4	2	4		1									1	
	logic of message content	4	3	4												

Legend: 1—Potential problem; 2—Problem; 3—Severe problem; 4—Impossibility.

The use of ergonomic knowledge in rehabilitation engineering is widespread, ranging from wrist splints to artificial limbs and from walking aids to special automobiles. Technology for people with disabilities has advanced to the stage where sophisticated devices are being systematically developed and implemented. These include devices for controlling actions by EMG signals, by movements of unimpaired body parts (such as the jaw), and by voice control or the direction of one's gaze. Such devices need to be selected according to a variety of criteria (Bogner 1998, Fernie 1997), including the following, listed by Batavia and Hammer in 1990:

- **Affordability.** The extent to which the purchase, maintenance, and repair of the device does not cause financial hardship to the consumer.
- **Dependability and durability.** The extent to which the device operates with repeatable and predictable levels of accuracy for extended periods of time.
- **Physical security.** The probability that the device will not cause physical harm to the user or other people.
- **Portability.** The extent to which the device can be readily transported to and operated in different locations.
- **Learnability and usability.** The extent to which the consumer can easily learn to use a newly received device and to which she or he can use it easily, safely, and dependably for the intended purpose.
- **Physical comfort and personal acceptability.** The degree to which the device provides comfort, or at least avoids pain or discomfort to the user, so that the person is motivated to use it in public or private.
- **Flexibility and compatibility.** The extent to which the device can be augmented by options and to which it will interface with other devices used currently or in the future.
- **Effectiveness.** The extent to which the device improves the user's capabilities, independence, and general objective and subjective situation.

Table 12–10 provides a ranking of several assist devices according to these criteria by a panel of disabled consumers.

TABLE 12–10. Ranking of Assistance Devices—1 is best

	For the motion impaired					*For the blind*			*For the deaf*			
	Wheel-chair	*Typing system*	*Robotic arm*	*Environmental control*	*Phone system*	*Type reader*	*Recording system*	*Orientation system*	*Alert System*	*Phone system*	*Speech recognition*	*Average ranking*
Effectiveness	1	2	1	1	1	2	2	2	1	1	1	1.36
Affordability	4	5	4	6	6	1	1	1	2	2	2	3.09
Operability	2	1	2	2	2	5	7	10	5	3	3	3.82
Dependability	3	4	3	3	3	3	3	6	4	5	5	3.82
Portability	*	*	*	*	*	7	4	3	6	4	8	5.33
Durability	8	8	8	7	7	4	5	11	3	6	6	6.64
Compatibility	13	3	6	4	5	6	6	13	10	7	4	7.00
Flexibility	7	6	5	5	4	11	9	15	8	9	7	7.82
Ease of maintenance	6	9	11	9	11	8	8	8	7	10	9	8.73
Securability	*	*	*	*	*	13	11	7	11	12	14	11.33
Learnability	14	7	10	11	10	9	16	14	14	11	10	11.45
Personal acceptance	5	11	7	13	9	15	15	12	13	14	13	11.55
Physical comfort	10	13	13	14	8	12	13	9	17	8	11	11.64
Supplier repair	9	10	12	8	12	10	10	16	16	16	12	11.91
Physical security	11	15	9	10	15	17	17	4	12	13	16	12.64
Consumer repair	12	12	14	12	13	14	12	5	15	17	15	12.82
Ease of assembly	15	14	15	15	14	16	14	17	9	15	17	14.64

Source: Adapted from Batavia and Hammer 1990.

SUMMARY

In body sizes and in physical capabilities, there are systematic differences between adult men and women. Yet, in general, one can design nearly any workstation and any piece of equipment or tool so that it can be used by either women or men. In some cases, one may have to adjust the device to the subject or provide objects of different dimensions than those in common use; yet, these adjustments and dimensions are generally not gender specific, but are simply needed to fit different people.

Commonly, one directs ergonomic efforts toward the effectiveness and well-being of the working-age population. However, there are large population groups that especially need ergonomic design. Among them are pregnant women, who, during their pregnancy, experience significant changes in body dimensions and in the capacity to perform demanding physical work. It is disappointing to note that fairly little systematic information on ergonomic design that would be useful for pregnant women has been published in the scientific literature.

Infants, children, and adolescents are another very large group of "ergonomic customers." From birth to adulthood, a person's body dimensions, physical characteristics, skill, intelligence, and attitude change enormously. Yet, descriptors of these developments are usually ordered by age (often stopping in the early teens), which is a rather meaningless classification because, at the same age, very large differences are likely to exist among individuals. This lack of information runs counter to the strong economic interest in providing goods to the young population, be such goods clothing, furniture, or toys. There also is a need for safeguarding and protecting children—for example, by properly designing cribs, railings, and toys or by providing restraints to protect them in case of an automobile accident.

Like children, aging people are also usually grouped in chronological age brackets, which is a rather meaningless classification scheme because physical characteristics, capabilities, and attitudes vary widely among individuals. As with children, one's capabilities can change (with the aging, usually decline) rapidly or slowly over time, or they may remain at about the same level for a while. Society and ourselves, as individuals have a great interest in providing an ergonomically properly designed and maintained environment to the older population, be it at the workplace, in the home, or in a health-care facility.

Helping impaired people to overcome their disabilities through technology is another major concern of ergonomics and rehabilitation engineering. Systematic procedures are at hand to determine the kind and extent of disabilities people have regarding whether they can carry out specific activities, and often help can be provided with mechanical devices or computers. Electronic devices require relatively little in the way of physical capabilities from the user and can control and operate rather complex machinery, store information, and transfer it to others and to the impaired person.

Thus, ergonomic knowledge and ergonomic procedures can be of help from childhood through adulthood and on into old age. The ergonomic principles and techniques used for architectural design and interior layout, for the workplace, and for equipment and tools are the same for everybody; they are just more important—and even critical—for the less able.

CHALLENGES

Most design is for "normal adults" in the age range of 20 to 40 years. Yet, is this a homogeneous group? Are there developments within a person during this age span that render the group heterogeneous?

Are there physiological reasons to presume that the musculature of women is, in and of itself, weaker than that of men?

Should one expect that a summation of small differences in sensory functions between the sexes make a difference in the performance of a professional task?

Is it true that females show better finger dexterity than males?

Are there specific occupations that appear to be better suited for women than men and vice versa?

Why would pregnant women have difficulty seeing objects on the ground in front of them?

What kind of seats would accommodate pregnant women in particular?

Would an air bag in an automobile promise to provide better forward protection to a pregnant women or a small child, compared with a shoulder seat belt?

What are the biomechanic problems related to side air bags for small persons (including children) compared with average-sized adults?

How would the proper design of restraints in automobiles be influenced by the age-related development of children?

Why is it so difficult to take anthropometric and biomechanical measurements on small children?

What kind of adjustments might be suitable to make for furniture used in kindergarten and elementary school?

Which professional activities appear to be particularly affected by changes related to aging?

Are there any professions and activities to which aging people appear particularly suited?

Why are there so few longitudinal studies of people?

What would it take to do such longitudinal studies?

Which physiological functions are likely to be maintained into old age by physical exercise?

What is the problem associated with sound amplification in most hearing devices?

Describe some approaches to assess changes in somesthetic sensitivities with age.

What tests might one use to qualify aging people, and possibly all adults, for a driver's license?

To what extent would the architectural layout of a house for aging persons differ from a layout suitable for younger people?

Discuss some schemes for classifying disabilities with an eye toward developing aids and devices that will assist the disabled.

Why are computers so helpful for disabled persons?

What can be done to make computers more acceptable to older persons?

Discuss some possible replacements for IADLs and ADLs.

Discuss the various uses of wheelchairs, either temporary or longterm, by persons with different impairments and for different purposes.

Can certain classification schemes be developed with regard to specific ergonomic design solutions?

Chapter 13

Why and How to Do Ergonomics

INTRODUCTION: THE EVOLUTION OF ERGONOMICS

When the first prehuman selected a stone blade, was that an pre-ergonomic act? Was affixing a handle to the blade early human engineering? Was the micro- and macroorganization mastered by the Romans to train, supply, equip, and lead their legions "systems ergonomics?" Or did the discipline start in the 19th century with the recognition that, for physiologic, psychologic, and sociologic reasons, laborers needed rest and recuperation to restore their capabilities for work? How about the rules of "scientific management" and of "work efficiency" of the early 20th century? Or the role of organized labor—of unions? Or aircraft and automobile design and manufacturing, computerization, and space exploration?

Evidently, human beings have always shaped tools and practices to their liking and for efficacy, on the individual and system levels and with no one event or time as the starting point. And what we generally began to call ergonomics in 1950 or human factors engineering in 1956 (see the introduction to this book) rests on millennia of experience and on intuitive and intentional development. During the second half of the 20th century, the wisdom derived from that experience became more organized, and more scientific, until it was formally incorporated into the design of work, of implements, and of human technology systems, as evident from the pioneering publications by McFarland in 1946; Lehmann 1953; Floyd and Welford 1954; McCormick 1957; Grandjean 1963. (Kroemer 1993b)

REASONS FOR USING ERGONOMICS

There are three major reasons for applying ergonomics:

1. **Moral Imperative.** To improve the human condition and quality of life, especially at work and in regard to health, safety, comfort, outcome, and enjoyment.

Certainly, work must be safe and healthy and should be comfortable. It can and should be enjoyable to achieve results at work and through work that we value personally.

2. **Progress in Knowledge and Technology.** To join the human quest to learn more about people and their desires, capabilities, and limitations, and to develop and apply new theories and practices.

Historically, thinkers and tinkerers, scientists and engineers, have lead the innate urge for progress. A standstill seems unacceptable: Objects and conditions should be better in the future.

3. **Economic Advantages.** To reduce the effort and cost expended in work systems that include humans as doers, users, and beneficiaries.

In many new designs of things and systems, the "human factor" already has been incorporated during the concept stage. Some examples are meat cutters and computer mice, hip joint replacements and portable phones, spacecraft and modern automobiles, and contemporary manufacturing and assembly plants. Some of these systems, such as high-performance aircraft or air traffic control networks, would be excessively dangerous or could not function at all without preplanned, designed-in human engineering.

Yet, many older objects and work systems, such as the typewriter keyboard and the traditional beverage delivery industry, were designed with little consideration of the human being as user or worker. In these cases, replacement, retrofitting and re-design are necessary to avoid injury or strain to humans and to remedy the inefficiency and inability of those systems to compete.

MICRO- AND MACROERGONOMICS

Ergonomics can make simple designs (such as toothbrushes, plier handles, and light switches) userfriendly and can render complex systems (subways, for example) safe, friendly, and easy to use. Human factors are important for nearly any item that we use, simple or complex, at work, at home, anytime, and anywhere. Of course, designing a public transportation system or a manufacturing plant is a much more complicated enterprise than designing a warning label.

Transportation and manufacturing systems comprise many people at different levels, all interacting with each other and with different technologies and all having different tasks and responsibilities. To design such sociotechnical systems in terms of "human-organization interface technology" requires *macroergonomics,* as defined by Hendrick and Kleiner (2000). Macroergonomics goes beyond the traditional microergonomic interface technologies that concern the specifics of hardware, software, the job, and the environment. Of course, even designing a toothbrush or a warning involves more than just considering the size of the hand or a person's reading acuity; so to label tasks micro- or macroergonomic can be a matter of judgment. The design of a work system through a consideration of relevant technical, environmental, and social variables and their interactions is a common ergonomic task; if organizational design and management aspects—and even economical and regional–political implications—are prominently involved, the task is indeed in the realm of macroergonomics (Kleiner and Drury 1999).

HOW DO WE DO ERGONOMICS?

Dynamics in New Designs

Obviously, it is best to design in an ergonomically correct fashion from the beginning, in the conceptual stage. This principle of "do it right from the start" applies to simple new items as well as to complex new human-technology systems.

Use and User

The concept team must be "use and user centered,": which implies a goal (use for what?), an instrument (use what?), a process (use how?) and a user (use by whom?). As Flach and Dominguez (1995) stated, the concept of use-centeredness has been pivotal to human engineering, but too often was not applied thoroughly. They present surgery as an example of a system, all of whose aspects must be integrated into a coherent whole.

Ergonomic Expertise

The concept and design team must have expertise in all related aspects of the system—including, in today's industry, marketing and sales. Human-factors design guidance is amply available, both in terms of experience with past similar products and processes and in the form of new information and data. Since 1994, in the United States alone, on average, each year about two major books on human factors or ergonomics came on the market. Besides the volume you are reading right now, there were books by Bailey (1996), Bhattacharya and McGlothlin (1996), Bridger (1995), Chapanis (1996), Karwowsky and Marras (1999), Konz (1995), Konz and Johnston (2000), Phillips (2000), Proctor and Van Zandt (1994), Salvendy (1998), Tayyari and Smith (1997), and Wickens, Gordon, and Liu (1998). In addition, several more specialized books appeared, for example on biomechanics and material handling. Burns and coauthors (1997) stated again what others had found earlier: The designer will utilize whatever is viable, useful, and usable, but stay away from information that is perceived as difficult.

☛☛☛ *If the engineer or manager believes that research data are of value, then that information will be sought, considered, and probably used. If the data are hard to find, understand, and apply, they will probably not be used.* ☚☚☚

Using ergonomic information

The use of new information depends on its how easy it is to access, interpret, and apply.

- **Accessibility** refers to the ease or difficulty of searching for information and obtaining it for use. Where does one look for the needed information (with the need often not yet clearly defined)? How does one obtain such information, especially if it comes from an obscure thesis or industry or military report? How does one use the information if it is declared confidential?
- **Interpretability** refers to the ability of the nonspecialist to understand, evaluate, and decide whether and how to apply the information to the design project. Making the judgment about whether to apply the data often requires a considerable understanding of the underlying theory and of complicated language on the part of the applier. How does one understand information if it is written in a foreign language or presented in a difficult format?
- **Applicability** involves risky judgment by the designer regarding the relevance, reliability, and applicability of (often very general, or very specific) theoretical information to the case at hand. The problem is threefold, at least. First, research may address problems that have only some overlap with the practical question. Second, most experimental data are obtained under laboratory conditions that rigorously control experimental variables and exclude confounding influences in such a manner that the laboratory work becomes "unrealistic." Third, theoretical data may just be too complex or too difficult to use.

CAD

Computer models can greatly facilitate the use of new information in the design of new products and processes or in the retrofitting of older designs. Yet, as discussed in Chapter 7, using an inadequate model, inputting misleading information, or using the output incorrectly all result in bad CAD.

9 Steps in product development

Major steps in the development of a new product (meaning anything that we can design, from the shovel to the automated teller machine to the subway system) are as follows:

1. Recognition of a need.
2. Statement of the need.
3. Concept of the product's utilization.
4. Exploration of candidate product designs.
5. Selection of the specific design of the product.
6. Technical development.
7. Production and distribution.
8. Use and maintenance.
9. Retirement and recycling.

Human factors

Of these nine steps, six involve human factors as major determiners of either the design of the product (steps 1, 3, 4, 5, and 8) or its production (step 7). The phases of selecting the product design features and the details of how the product will be used (steps 3, 4 and 5) are iterative and involve extensive and intensive ergonomic considerations (Chapanis 1995, 1996). These phases are threefold:

- The human-factors problems that are expected are to be eliminated.
- Skills that are required from the user and the maintainer shall be kept low.
- Training that is required of the user and the maintainer shall be minimal.

These ergonomic concerns and problems must be addressed early, comprehensively, and thoroughly: They "make or break" the usability and usefulness of the product (and its chances of succeeding in the market). Skill and training requirements should be as low as humanly possible to avoid use and maintenance problems. Their recognition (and subsequent elimination) can be made easier by comparisons with similar existing products, but usually, the realistic use of prototypes or the real-life employment of a preproduction specimen is necessary. Figure 13–1 helps in preempting ergonomic problems in either new or existing systems.

Ergonomic Interventions in Existing Designs

Most of the products and systems that we use daily have already been around for a while, and we may have good reasons to keep them; but some are in urgent need of improvement because of injuries associated with their use, safety concerns, the difficulty of using them, costly maintenance, insufficient productivity, or an inability to compete in the marketplace.

Figure 13–1 presents a stepwise sequence for improving a product or system ergonomically. The procedure not only applies to existing conditions, but can also be used for new designs.

What is Wrong?

We become aware of a problem either because the "ergonomic expert" (often the user) knows that something is wrong, or because of deficiencies in performance (concerning either quality or quantity), or because people simply may not like what is going on.

Identifying the true underlying problem (step 3) is the most important step: Missing the problem invariably leads to selecting a candidate solution that misses the goal (step 5).

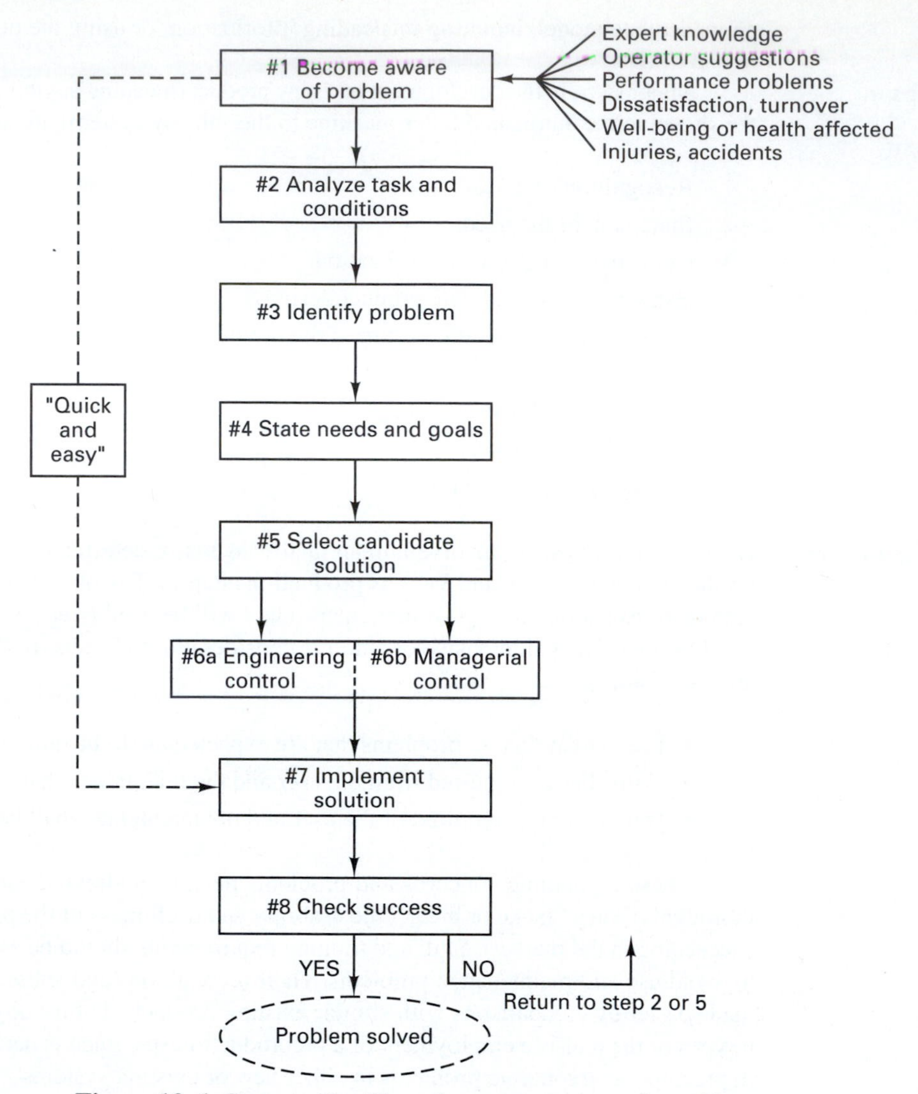

Figure 13–1 Steps to identify and preempt ergonomic problems.

Selecting the solution

Selecting the solution (either a technical or a managerial one, and often both) can be a difficult task, because several approaches with different consequences may be feasible. For example, should one load packages onto a truck by a forklift using pallets? That would require having pallets already available and might not fill the vehicle to capacity. So should the job be done by hand? If so, by how many people, and should they use a conveyer? (Or, in the case of a new system, should one send astronauts or robots on a first visit to the moon or to Mars?)

A large variety of ergonomic procedures and techniques exist for improving working conditions. Take computer workstations as an example. (See Chapter 9.) One can improve the layout of the room and workstation, improve the furniture (the chair, support for the display, support for the input device, and support for the source document), improve the lighting conditions and the physical climate of the workspace, reduce unwanted sound, and take into account the need for privacy. For "experimental cleanliness," one would like to introduce each intervention separately and then observe how it changes the person's feeling of well-being and performance. Yet, in reality, a comprehensive ergonomic approach combines these single measures. Therefore, a resulting improvement in attitude and work output is not easily traceable to any single intervention separately.

The Hawthorne Effect

Improvements that are observed immediately after the intervention may be due to the so-called Hawthorne effect. This was observed in the 1920s during experiments in the Hawthorne Works near Chicago: The ergonomic treatment consisted of improving the lighting conditions in a manufacturing and assembly task. Each rise in the lighting level was followed by improved output; but when the illumination level was finally lowered, performance still improved. In somewhat simplified terms, one may conclude that paying attention to the workers, taking their comments and activities seriously into account—in short, treating the workers as important persons—led to improved output regardless of the magnitude—and even the direction—of the overt ergonomic measure taken (Roethlisberger and Dickson 1939; Parsons 1984).

The Hawthorne effect is the delight of the ergonomic consultant, who can be fairly well assured that almost any reasonable measure taken in the client's facility will have positive effects. Even if these effects wear out after the intervention has been completed, one can come back a while later and do, in essence, the same thing again, with similar positive effects. And, of course, to have positive effects, even if they are only indirectly related to the actual measures taken, "is positive."

JUDGING THE EFFICACY OF ERGONOMIC INTERVENTIONS

Besides the subjective estimates, guesstimates, and gut feelings often relied on in making many ergonomic decisions, there are systematic ways to elicit choices from panels of experts or users. Doing so involves

- Defining the candidate solutions,
- Establishing criteria by which to judge the candidates,
- Laying out a score sheet according to the criteria,
- Obtaining paired-comparison scores from the participants,
- Determining weights for the various criteria, and
- Aggregating the data and accepting the candidate solution(s) with the best overall score.

The technique is straightforward, fast, and transparent, gives every participant the same voice, identifies the criteria, and shows how a change in their importance (weighting) influences the outcome in terms of the solution that is selected. The procedure works well for a small number of candidate solutions and criteria, but it can be used even in complex cases as discussed by Mitchell (1992).

Example: COOLING A TRAILER, PARKED IN THE SUN, THAT MUST BE LOADED BY HAND WITH PACKAGES

Candidate solutions: *Cool trailer by spraying water on roof and sides (WATER); cool trailer by installing air-conditioning (AC); cool trailer by forced air flow (FAN); no cooling of trailer, but person who is loading takes a 2-minute rest break in a cool room after every 5 minutes of loading (BREAKS).*

Criteria: *Effectiveness, availability, maintenance effort.*

Scoring: 1 to 10, with 10 best.

Raw scores given by raters:

	WATER	AC	FAN	BREAKS
EFFECTIVENESS	*8*	*10*	*5*	*1*
AVAILABILITY	*8*	*1*	*8*	*10*
MAINTENANCE	*4*	*2*	*5*	*10*
Total	**20**	**13**	**18**	**21**

With all criteria weighted the same, according to the raw scores, simply providing rest breaks would be the chosen solution.

With weights (1 to 10, with 10 highest) applied to the criteria, the scores are as follows:

	WATER	AC	FAN	BREAKS
EFFECTIVENESS × **10**	*80*	*100*	*50*	*10*
AVAILABILITY × **2**	*16*	*2*	*16*	*20*
MAINTENANCE × **5**	*20*	*10*	*25*	*50*
Total Scores	**116**	**112**	**91**	**80**

With weighted criteria, cooling the trailer by spraying water on its roof and sides is the preferred solution.

Optimal versus Good Solution

Seldom is one able to truly optimize a design: As Sheridan (on page 4 of his letter in the June 1988 issue of the *Human Factors Society Bulletin)* stated, "optimizing" literally means to obtain the most favorable solution—the greatest degree of perfection. But in the real world, there are constraints, often in terms of available time, resources, and money, that allow the ergonomist to find a good solution—the best one within the practical limitations—but not often "the optimal" solution. With respect to performance, this usually means a compromise, such as between a system's cost and quality. In the example just presented cooling a trailer may be the chosen solution to prevent the worker's overheating, although loading and unloading by mechanical means would avoid human labor altogether, but appears too costly.

A totally different kind of case is the necessity to design for the extreme emergency: The proverbial airline pilot flies for hours in boring monotony which may be interrupted by seconds or minutes of highly exciting demands. We must "optimize" the cockpit design for those seconds of panicky terror and make it "as good as possible" for routine operation.

At the beginning of this chapter, we gave three reasons for using ergonomics, associated with moral obligation, implementing progress, and gaining competitive and economic advantage. However, the decision still must be made regarding which of the possible human-factors approaches to take. This is often determined by considering the gains one might make and the efforts needed.

Measuring the Results of Ergonomics

There are many criteria by which to measure outcomes and efforts. A result can be judged "objectively," by the success or failure of the associated mission (often done by the military) or the system performance in terms of quantity (productivity) or quality, or by "subjective" attitudes such as loyalty, feelings of well-being and satisfaction, or a show of cooperative spirit and of social interactions on the part of the persons involved.

It is Good and It Feels Good

In the past, many work systems had a strong societal character (recall the oft-cited "Once an IBMer, forever an IBMer"), and the employees worked happily and diligently for the company, knowing that, under all foreseeable circumstances, their employment there would last for the rest of their working life. This kind of an unwritten social contract is rather seldom entered into now, because it was largely abandoned by the "lean and mean businesses" in the 1990s.

> Headlines on July 22, 1998:
> **"Best firms don't need employee loyalty"**—*USA TODAY,* page 11A.
> **"Companies Are Finding It Really Pays to Be Nice to Employees"**—The Wall Street Journal, page B1.

The "Almighty Dollar"

In Western industry and business, the effects of ergonomic interventions have been commonly measured by money values, with dollar signs affixed both to the human factors inputs and to the system outputs, such as performance, productivity, accidents, and injuries. This can be a fairly simple task: Take the cost of providing the workers with human-engineered tools that enhanced production, and divide it by the money made by producing (and selling) more gadgets. But such a cost–benefit calculation is difficult to carry out when the output depends not only on the workers, but also on circumstances over which they have little or no control, such as working at a speed that is determined by a machine, by the movement of an assembly line, or by the number of customers waiting in line. Under these conditions, ergonomic improvements are not likely to result in increased performance; however, they may result in improved quality of work, even if the tempo is machine paced, or in less strain on the person and in a better attitude and sense of well-being.

Cost–benefit analyses may be beset by serious methodological difficulties if they require the analyst to make value-laden assumptions. The costs of interventions are easier to count in dollars than are such benefits as a better work attitude, loyalty, and goodwill towards the employer or improved health, quality of life, and other positive economic side effects that may defy accurate estimation. For example, the art of estimating the number of cancers or occupational diseases prevented, or of work injuries avoided, is in its infancy. Health benefits may accrue far into the future, and a human life or a lost limb do not have an established market value. Nevertheless, cost–benefit, economic efficiency, and cost-effectiveness analyses are attempts at rational decision making, even if they are much more complex than is often naively suggested.

Recent publications by Alexander (1995), Burrows et al. (1998), Hendrick (1996), Jenkins and Rickards (1997), Kleiner and Drury (1999), McLeod and Morris (1996), Moore and Garg (1997), and Oxenburgh (1997), as well as the general literature, brim with case studies of the monetary and other successes of applying human engineering to new products and of ergonomic interventions to improve existing work systems.

"IMPROVING HUMAN AND SYSTEM PERFORMANCE, HEALTH, SAFETY, COMFORT, AND THE QUALITY OF LIFE"

As Rouse succinctly stated and demonstrated in a tongue-in-cheek fashion in 1991, the human-centered design of simple objects, as well of complex systems is, as a rule, better than what marketing and sales departments request and what engineering and production departments tend to deliver—see Figure 13–2.

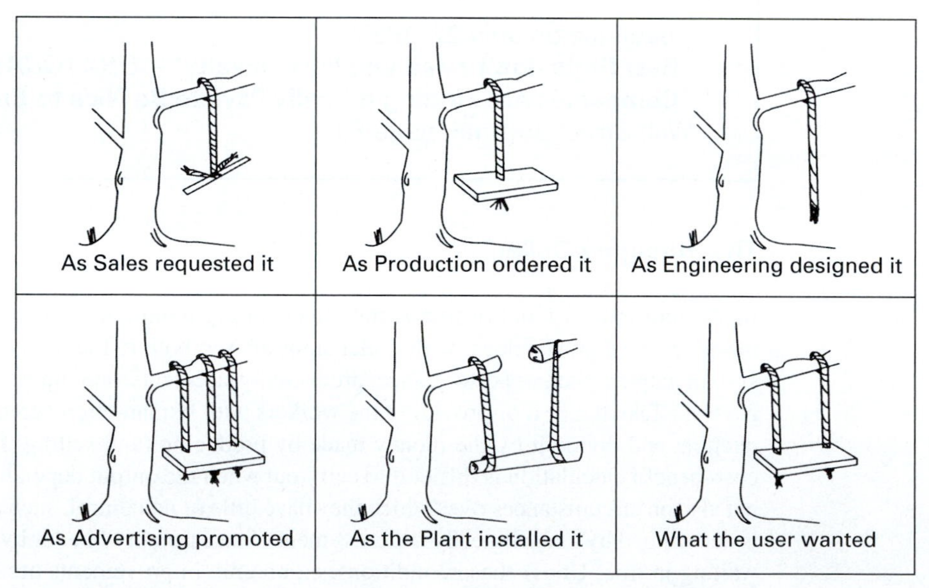

Figure 13–2 Problems in design (*CSERIAC Gateway 1991* 2(4), p. 1).

We close this chapter and the book with Pew's "Top-Ten List of Principles of Human Factors Practice" (reprinted with his kind permission):

- **Get into the design process early and often.**
- **The best way to gain an influence on a design is to join the design team.**
- **Pick problems to work on that you can solve and that will make a difference.**
- **Simple is better: If it takes too much to explain it, it won't be used.**
- **Bury complexity** [inside the system: make its everyday use simple].
- **Good design is transparent** [it appeals to common sense and is intuitively usable].
- **Don't design to eliminate human error; design to be forgiving of error.**
- **Design so that an operator can manage overload without performance consequences.**
- **Design to provide multiple ways of accomplishing the same thing.**
- **Design to promote human dignity.**

part 3

FURTHER INFORMATION

References

AARAS, A., HORGEN, G., BJORSET, H. H., RO, O., and THORESEN, M. (1998). Musculoskeletal, Visual and Psychosocial Stress in VDU Operators Before and After Multidisciplinary Ergonomic Interventions. *Applied Ergonomics 29,* 335–354.

AARAS, A. and RO, O. (1998). Supporting the Forearms at the Table Top Doing VDU Work. Pages 549–552 in S. Kumar (ED.), *Advances in Occupational Ergonomics and Safety.* Amsterdam, The Netherlands: IOS Press.

AARAS, A., RO, O. and HORGEN, G. (1998). Back Pain. A Prospective Three Years Epidemiological Study of VDU Workers. Pages 553–555 in S. Kumar (ED.), *Advances in Occupational Ergonomics and Safety,* Amsterdam, The Netherlands: IOS Press.

ACGIH (ED.). (1989). *Threshold Limit Values for 1980–90.* Cincinnati, OH: American Conference of Governmental Industrial Hygienists.

ACKERMAN, D. (1990). *A Natural History of the Senses.* New York, NY: Random House.

ADAMS, E, KANEKO, S. T., and RUTTER, B. (1993). Making Sure It's Right: Three Phases of Ergonomics Research in the Design of a Pointing Device. In *Proceedings of the Human Factors and Ergonomics Society 37th Annual Meeting* (pp. 443–447). Santa Monica, CA: Human Factors and Ergonomics Society.

ADAMS, M. A. and HUTTON, W. C. (1986). Has the Lumbar Spine a Margin of Safety in Forward Bending? *Clinical Biomechanics 1,* 3–6.

AKERBLOM, B. (1948). *Standing and Sitting Posture.* Stockholm, Sweden: Nordiska Bokhandeln.

ALBERY, W. B. and WOOLFORD, B. (1997). Design for Macrogravity and Microgravity Environments. Chapter 29 (pp. 935–963) in G. Salvendy (ED.), *Handbook of Human Factors and Ergonomics,* 2d ed. New York, NY: Wiley.

ALDEN, D. G., DANIELS, R. W., and KANARICK, A. F. (1972). Keyboard Design and Operation: A Review of the Major Issues. *Human Factors 14,* 275–293.

ALEXANDER, D. C. (1995). The Economics of Ergonomics: Part II. In *Proceedings of the Human Factors and Ergonomics Society 39th Annual Meeting* (pp. 1025–1027). Santa Monica, CA: Human Factors and Ergonomics Society.

ALFREDSSON, L., AKERSTEDT, T., MATTSSON, M., and WILBORG, B. (1991). Self-Reported Health and Well-Being Amongst Night Security Guards: A Comparison with the Working Population. *Ergonomics 34,* 525–530.

ALLREAD, W. G., MARRAS, W. S., and PARNIANPOUR, M. (1996). Trunk Kinematics of One-Handed Lifting, and the Effects of Asymmetry and Load Weight. *Ergonomics 39,* 322–334.

AMOORE, J. (1970). *Molecular Basis of Odor.* Springfield, IL: Thomas.

ANDERSON, C. K. and CATTERALL, M. J. (1987). The Impact of Physical Ability Testing on Incidence Rate, Severity Rate, and Productivity. Pages 577–584 in S. S. Asfour (ED.), *Trends in Ergonomics/Human Factors IV.* Amsterdam, The Netherlands: Elsevier.

ANDERSON, N. S. (1987). Cognition, Learning and Memory, Chapter 3 (pp. 37–54) in M. A. Baker (ED.), *Sex Differences in Human Performance.* Chichester, UK: Wiley.

ANDERSSON, G. B. J. (1991). Low Back Pain. In *Proceedings, Occupational Ergonomics: Work Related Upper Limb and Back Disorders* (not paginated). San Diego, CA: American Industrial Hygiene Association, San Diego Section.

ANDERSSON, G. B. J. (1999). Epidemiology of Back Pain in Industry. Chapter 51 (pp. 913–932) in W. Karwowski and W. S. Marras (EDS.), *The Occupational Ergonomics Handbook.* Boca Raton, FL: CRC Press.

ANDERSSON, G. B. J. and OERTENGREN, R. (1974). Lumbar Disc Pressure and Myoelectric Back Muscle Activity During Sitting—II. Studies on an Office Chair. *Scandinavian Journal of Rehabilitation Medicine 6,* 115–121.

ANDERSSON, B. J. G., OERTENGREN, R., NACHEMSON, A., and ELFSTROEM, G. (1974). Lumbar Disc Pressure and Myoelectric Back Muscle Activity During Sitting—I. Studies on an Experimental Chair. *Scandinavian Journal of Rehabilitation Medicine 6,* 104–114.

ANDERSSON, B. J. G., SCHULTZ, A. B. and OERTENGREN, R. (1986). Trunk Muscle Forces During Desk Work. *Ergonomics 29,* 1113–1127.

ANDERSSON, K., KARLEHAGEN, S., and JONSSON, B. (1987). The Importance of Variations in Questionnaire Administration. *Applied Ergonomics 18,* 229–232.

ANDRES, R. (1985). Impact of Age on Weight Goals. In *Proceedings, NIH Consensus Development Conference* (pp. 77–81). Bethesda, MD: National Institutes of Health.

ANKRUM, D. R. and NEMETH, K. J. (1995). Posture, Comfort, and Monitor Placement. *Ergonomics in Design April Issue,* 7–9.

ANNIS, J. F., CASE, H. W., CLAUSER, C. E., and BRADTMILLER, B. (1991). Anthropometry of an Aging Work Force. *Experimental Aging Research 17,* 157–176.

ANNIS, J. F. and MCCONVILLE, J. T. (1996). Anthropometry, Chapter 1 (pp. 1–46) in A. Bhattacharya and J.D. McGlothlin (Eds.), *Occupational Ergonomics.* New York, NY: Dekker.

ANSI/HFES 100–(2000, in preparation). *U.S. National Standard for Human Factors Engineering of Computer Workstations.* Santa Monica, CA: Human Factors and Ergonomics Society.

ANTONOVSKY, A. (1979). *Health, Stress, and Coping.* San Francisco, CA: Jossey-Bass.

ARNAUT, L. Y. and GREENSTEIN, J. S. (1990). Is Display/Control Gain a Useful Metric for Optimizing an Interface? *Human Factors 32,* 651–663.

ASCHOFF, H. (1981). *Handbook of Behavioral Neurobiology,* Vol. 4. New York, NY: Plenum.

ASFOUR, S. S., GENAIDY, A. M., and MITAL, A. (1988). Physiological Guidelines for the Design of Manual Lifting and Lowering Tasks: The State of the Art. *American Industrial Hygiene Association Journal 49,* 150–160.

ASHRAE (ED.) (1985). Physiological Principles for Comfort and Health. Chapter 8 in *1985 Fundamentals Handbook.* Atlanta, GA: American Society of Heating, Refrigerating, and Air-Conditioning Engineers.

ASIMOV, I. (1963). *The Human Body. Its Structure and Operation.* New York, NY: Signet.

ASIMOV, I. (1989). *Asimov's Chronology of Science and Discovery.* New York, NY: Harper & Row.

ASPDEN, R. M. (1988). A New Mathematical Model of the Spine and its Relationship to Spinal Loading in the Workplace. *Applied Ergonomics 19,* 319–323.

ASSO, D. (1987). Cyclical Variations. Chapter 4 (pp. 37–54) in M. A. Baker (ED.), *Sex Differences in Human Performance.* Chichester, UK: Wiley.

ASTIN, A. D. (1999). *Finger Force Capability: Measurement and Prediction Using Anthropometric and Myoelectric Measures.* Unpublished MS thesis, Dept. of Systems Engineering, Blacksburg, VA: Virginia Tech.

ASTRAND, P. O. and RODAHL, K. (1977). *Textbook of Work Physiology,* 2d ed. New York, NY: McGraw-Hill.

ASTRAND, P. O. and RODAHL, L. (1986). *Textbook of Work Physiology,* 3d ed. New York, NY: McGraw-Hill.

AUTHIER, M., LORTIE, M., and GAGNON, M. (1994). Handling Techniques: Impact of the Context on the Choice of Grip and Box Movement in Experts and Novices. Pages 687–693 in F. Aghazadeh (ED.), *Advances in Industrial Ergonomics and Safety VI.* London, UK: Taylor & Francis.

AYOUB, M. M. (1982). The Manual Lifting Problem: The Illusive Solution. *Journal Occupational Accidents 4,* 1–23.

AYOUB, M. M. and DEMPSEY, P. G. (1999). The Psychophysical Approach to Manual Materials Handling Task Design. *Ergonomics 42,* 17–31.

AYOUB M. M. and MITAL, A. (1989). *Manual Materials Handling.* London, UK: Taylor & Francis.

AYOUB, M. M., SELAN, J. L., and JIANG, B. C. (1984). *A Mini-Guide for Manual Materials Handling.* Institute of Ergonomics Research. Lubbock, TX: Texas Tech University.

AYOUB, M. M., SELAN, J. L., and JIANG, B. C. (1986). *Manual Materials Handling.* Chapter 7.2 (pp. 790–818) in G. Salvendy (ED.), Handbook of Human Factors. New York: Wiley.

AYOUB, M. M. and SMITH, J. L. (1988). Manual Materials Handling in Unusual Postures: Carrying of Loads. In *Proceedings of the Human Factors Society 32d Annual Meeting* (pp. 675–679). Santa Monica, CA: Human Factors Society.

BAILEY, R. W. (1996). *Human Performance Engineering.* Upper Saddle River, NJ: Prentice Hall.

BAKER, M. A. (ED.). (1987). *Sex Differences in Human Performance.* Chichester, UK: Wiley.

BALCI, R., AGHAZADEH, F., and WALY, S. M. (1998). Work-Rest Schedules for Data Entry Operators. Pages 155–158 in S. Kumar (ED.), *Advances in Occupational Ergonomics and Safety.* Amsterdam, The Netherlands: IOS Press.

BALLARD, B. (1995). How Odor Affects Performance: A Review. In *Proceedings, ErgoCon '95, Silicon Valley Ergonomics Conference and Exposition* (pp. 191–200). San Jose, CA: San Jose State University.

BALOGUN, J. A. (1986). Optimal Rate of Work During Load Transportation on the Head and by Yoke. *Industrial Health 24,* 75–86.

BARLOW, A. M. and BRAID, S. J. (1990). Foot Problems in the Elderly. *Clinical Rehabilitation 4,* 217–222.

BARKER, K. L. and ATHA, J. (1994). Reducing the Biomechanical Stress of Lifting by Training. *Applied Ergonomics 25,* 373–378.

BARNES, M.E. and WELLS, W. (1994). If Hearing Aids Work, Why Don't People Use Them? *Ergonomics in Design April Issue,* 19–24.

BASMAJIAN, J. V. and DELUCA, C. J. (1985). *Muscles Alive* (5th Ed.). Baltimore, MD: Williams & Wilkins.

BATAVIA, A. I. and HAMMER, G. S. (1990). Toward the Development of Comsumer-Based Criteria for the Evaluation of Assistive Devices. *Journal of Rehabilitation Research and Development 27,* 425–436.

BATTIE, M. C., BIGOS, S. J., FISHER, L., HANSSON, T. H., JONES, M. E., and WORTLEY, M. D. (1989). Isometric Lifting Strength as a Predictor of Industrial Back Pain Reports. *Spine 14,* 851–856.

BAUER, W. and WITTIG, T. (1998). Influence of Screen and Copy Holder Positions on Head Posture Muscle Activity and User Judgment. *Applied Ergonomics 29,* 185–192.

BEDNY, G.Z. and ZEGLIN, M.H. (1997). The Use of Pulse Rate to Evaluate Physical Work Load in Russian Ergonomics. *American Industrial Hygiene Association Journal 58,* 375–379.

BELSKY, J. K. (1990). *The Psychology of Aging. Theory, Research, and Interventions.* Pacific Grove, CA: Brooks/Cole.

BENDIX, T., POULSEN, V., KLAUSEN, K., and JESNEN, C. V. (1996). What Does a Backrest Actually Do to the Lumbar Spine? *Ergonomics 39,* 533–542.

BENNETT, P. B. and ELLIOTT, D. H. (EDS.). (1993). *The Physiology and Medicine of Diving,* 4th ed. Philadelphia, PA: Saunders.

BENSEL, C. K. and SANTEE, W. R. (1997). Climate and Clothing, Chapter 28 (pp. 909–934) in G. Salvendy (ED.), *Handbook of Human Factors and Ergonomics,* 2d ed. New York, NY: Wiley.

BEN-SHAKHAR, G. and FUREDY, J. J. (1990). *Theories and Applications in the Detection of Deception.* New York, NY: Springer.

BERANEK, L., BLAZIER, W., and FIGWER, J. (1971). Preferred Noise Criteria (PNC) Curves and Their Application to Rooms. *Journal Acoustical Society of America 50,* 1223–1228.

BERGER, E. H. and CASALI, J. G. (1992). Hearing Protection Devices. Chapter 8 in M. J. Crocker (ED.), *Handbook of Acoustics.* New York, NY: Wiley.

BERGQUIST-ULLMAN, M. and LARSSON, U. (1977). Acute Low Back Pain in Industry. *Acta Orthopaedica Scandinavica,* Supplement No. 170.

BERNARD, B. P. (ED.). (1997). Musculoskeletal Disorders and Workplace Factors. (DHHS NIOSH Publication No. 97-141). Washington, DC: U.S. Department of Health and Human Services.

BERNHARD, T. E. and JOSEPH, B. S. (1994). Estimation of Metabolic Rate Using Qualitative Job Descriptors. *American Industrial Hygiene Association Journal 55,* 1021–1029.

BERNS, T. A. R. and MILNER, N. P. (1980). *TRAM: A Technique for the Recording and Analysis of Moving Work Posture* (Report 80:23, pp. 22–26). Stockholm, Sweden: Ergolab.

BEST, J. B. (1997). *Cognitive Psychology,* 4th ed. St. Paul, MN: West Publishing.

BHATTACHARYA, A. and MCGLOTHLIN, J. D. (EDS.) (1996). *Occupational Ergonomics—Theory and Applications.* New York, NY: Dekker.

BIGOS, S. J., SPENGLER, D. M., MARTIN, N. A., ZEH, J., FISHER, L., and NACHEMSON, A. (1986). Back Injuries in Industry: A Retrospective Study, III. Employee-Related Factors. *Spine 11,* 252–256.

BISHU, R. R., WANG, W. HALLBECK, M. S., and COCHRAN, D. J. (1992). Force Distribution at Hand/Handle Coupling: the Effect of Handle Type. In *Proceedings of the Human Factors Society 36th Annual Meeting* (pp. 816–820). Santa Monica, CA: Human Factors Society.

BJOERING, G., JOHANSSON, L., and HAEGG, G. M (1998). The Pressure Distribution in the Hand When Holding Powered Drills, Pages 436–438 in S. Kumar (ED.), *Advances in Occupational Ergonomics and Safety.* Amsterdam, The Netherlands: IOS Press.

BLAIKIE, A. (1993). Images of Age: a Reflexive Process. *Applied Ergonomics 24,* 51–57.

BLANCHARD, B. S. and FABRYCKY, W. J. (1996). *Systems Engineering and Analysis,* 3rd ed. Englewood Cliffs, NJ: Prentice Hall.

BLINKHORN, S. (1988). Lie Detection as a Psychometric Procedure, Chapter 3 (pp. 29–39) in A. Gale (ED.), *The Polygraph Test. Lies, Truth, and Science.* London, UK: Sage.

BOBICK, T. G. and GUTMAN, S. H. (1989). Reducing Musculo-Skeletal Injuries by Using Mechanical Handling Equipment. Pages 87–96 in K. H. E. Kroemer, J. D. McGlothlin, and T. G. Bobick (EDS.), *Manual Material Handling: Understanding and Preventing Back Trauma.* Akron, OH: American Industrial Hygiene Association.

BOBJER, O., JOHANSSON, S., and PIGUET, S. (1993). Friction Between Hand and Handle. Effects of Oil and Lard on Textured and Non-Textured Surfaces; Perception of Discomfort. *Applied Ergonomics 24,* 190–202.

BOFF, K. R., KAUFMAN, L., and THOMAS, J. P. (EDS.). (1986). *Handbook of Perception and Human Performance.* New York, NY: Wiley.

BOFF, K. R. and LINCOLN, J. E. (EDS.). (1988). *Engineering Data Compendium: Human Perception and Performance.* Wright-Patterson AFB, OH: Armstrong Aerospace Medical Research Laboratory.

BOGNER, M. S. (1998). An Introduction to Design, Evaluation, and Usability Testing. Chapter 13 (pp. 231–247) in V. J. B. Rice, (ED.), *Ergonomics in Health Care and Rehabilitation.* Boston, MA: Butterworth-Heinemann.

BOGNER, M. S., WEINGER, M. B., LAUGHEREY, K. R., HAAS, E. C., and ADMUNSON, D. O. (1998). Warnings in Health Care. *CSERIAC Gateway 9*:3, 1–3.

BOHAN, M, CHAPARRO, A., FERNANDEZ, J. E., KATTEL, B. P., and CHOI, S. D. (1998). Cursor-Control Performance of Older Adults Using Two Computer Input Devices. Pages 541–544 in S. Kumar (ED.), *Advances in Occupational Ergonomics and Safety.* Amsterdam, The Netherlands: IOS Press.

BOHLE, P. and TILLEY, A. J. (1989). The Impact of Night Work on Psychological Well-Being. *Ergonomics 34,* 1089–1099.

BONNE, A. J. (1969). On the Shape of the Human Vertebral Column. *Acta Orthopaedica Belgica 35:* Fasc. 3–4, 567–583.

BOOTH-JONES, A. D., LEMASTERS, G. K., SUCCOP, P., ATTERBURY, M. R., and BHATTACHARYA, A. (1998). Reliability of Questionnaire Information Measuring Musculoskeletal Symptoms and Work Histories. *American Industrial Hygiene Association Journal 59,* 20–24.

BORG, G. A. V. (1962). *Physical Performance and Perceived Exertion.* Lund, Sweden: Gleerups.

BORG, G. A. V. (1982). Psychophysical Bases of Perceived Exertion. *Medicine and Science in Sports and Exercise 14,* 377–381.

BORG, G. A. V. (1990). Psychophysical Scaling with Applications in Physical Work and the Perception of Exertion. *Scandinavian Journal of Work, Environment and Health 16:Suppl. 1,* 55–58.

BORRELLI, G. A. (1680). *De motu animalium.* LUGDUNI BATAVORUM.

BOUDRIFA, H. and DAVIES, B. T. (1984). The Effect of Backrest Inclination, Lumbar Support and Thoracic Support on the Intra-Abdominal Pressure While Lifting. *Ergonomics 27,* 379–387.

BOWMAN, R. L. and DELUCIA J. L. (1992). Accuracy of Self-Reported Weight: A Meta-Analysis. *Behavior Therapy 23,* 637–655.

BOYD, P. R. (1982). *Human Factors in Lighting.* New York, NY: MacMillan.

BRAUNE, W. and FISCHER, O. (1889). *The Center of Gravity of the Human Body as Related to the Equipment of the German Infantryman.* (In German.) (Abh. d. Math. Phys. Cl. d. k. Saechs. Gesell. d. Wissenschaften Leipzig, 1889, 15. Translation in Human Mechanics (AMRL-TDR-63-123, pp. 1–56). Wright-Patterson AFB, OH: Aerospace Medical Research Laboratory, 1963.

BRENNAN, R. B. (1987). Trencher Operator Seating Positions. *Applied Ergonomics 18,* 95–102.

BRIDGER, R. S. (1995). *Introduction to Ergonomics.* New York, NY: McGraw-Hill.

BRINGELSON, L. S., YEOH, C. H., and WONG, C. B. (1998). An Empirical Investigation of Pointing Devices for Notebook Computers. Pages 545–548 in S. Kumar (ED.), *Advances in Occupational Ergonomics and Safety.* Amsterdam, The Netherlands: IOS Press.

BROUHA, L. (1967). *Physiology in Industry,* 2d ed. Riverside, NJ: Pergamon.

BROWN, D. A., COYLE, I. R., and BEAUMONT, P. E. (1985). The Automated Hettinger Test in the Diagnosis and Prevention of Repetition Strain Injuries. *Applied Ergonomics 16,* 113–118.

BROWN, I. D. (1994). Driver Fatigue. *Human Factors 36,* 298–314.

BROWN, J. R. (1972). *Manual Lifting and Related Fields: An Annotated Bibliography.* Toronto, ON: Labour Safety Council of Ontario, Ontario Ministry of Labour.

BROWN, J. R. (1975). Factors Contributing to the Development of Low-Back Pain in Industrial Workers. *American Industrial Hygiene Association Journal 36,* 26–31.

BROWNELL, K. D. (1995). Definition and Classification of Obesity. In K. D. Brownell and C. G. Fairburn (EDS.), *Eating Disorders and Obesity.* New York, NY: Guilford.

BRUNSWICK, E. (1956). *Perception and the Representative Design of Experiments,* 2d ed. Berkley, CA: University of California Press.

BUCHHOLZ, B., PAQUET, V., PUNNETT, L., LEE, D., and MOIR, S. (1996). PATH: a Work-Sampling Based Approach to Ergonomic Work Analysis for Construction and Other Non-Repetitive Work. *Applied Ergonomics 27,* 177–187.

BUCKLE, P. W., DAVID, G. C., and KIMBER, A. C. (1990). Flight Deck Design and Pilot Selection: Anthropometric Considerations. *Aviation, Space, and Environmental Medicine 61,* 1079–1084.

BUIS, N. (1990), Ergonomics, Legislation and Productivity in Manual Materials Handling. *Ergonomics 33,* 353–359.

BULLINGER, H., KERN, P., and BRAUN. M. (1997). Controls. Chapter 21 (pp. 697–728) in G. Salvendy (ED.), *Handbook of Human Factors and Ergonomics,* 2d. ed. New York, NY: Wiley.

BURGDORF, A., VAN RIEL, M., VAN WINGERDEN, J. P., VAN WINGERDEN, S., and SNIJDERS, C. (1995). Isodynamic Evaluation of Trunk Muscles and Low-Back Pain Among Workers in a Steel Factory. *Ergonomics 38,* 2107–2117.

BURGESS-LIMERICK, R. and ABERNETHY, B. (1997). Toward a Quantitative Definition of Manual Lifting Postures. *Human Factors 39,* 141–148.

BURNS, C. M., VICENTE, K. J., CHRISTOFFERSEN, K., and PAWLAK, W. S. (1997). Towards Viable, Useful and Usable Human Factors Design Guidance. *Applied Ergonomics 28,* 311–322.

BURROWS, E., THOMAS, G., and RICKARDS, J. (1998). A Pre-Intervention Benefit/Cost Methodology—Refining the Cost Audit Process. In *Proceedings of the 30th Annual Conference of the Human Factors Association of Canada* (pp. 131–136). Windsor, ON: Human Factors Association of Canada.

BURSE, R. L. (1979). Sex Differences in Human Thermoregulatory Responses to Heat and Cold Stress. *Human Factors 21,* 687–699.

BURT, S. and PUNNETT, L. (1999). Evaluation of Interrater Reliability for Posture Observations in a Field Study. *Applied Ergonomics 30,* 121–135.

BURTON, K. (1991). Measuring Flexibility. *Applied Ergonomics 22,* 303–307.

BUSH-JOSEPH, C., SCHIPPLEIN, O., ANDERSSON, G. B. J., and ANDRIACCHI, T. P. (1988). Influence of Dynamic Factors on the Lumbar Spine Moment in Lifting. *Ergonomics 31,* 211–216.

BUTLER, D., ANDERSSON, G. B. J, TRAFIMOW, J., SCHIPPLEIN, O. D., and ANDRIACCHI, T. P. (1993). The Influence of Load Knowledge on Lifting Technique. *Ergonomics 36,* 1489–1493.

CADY, L. D., BISCHOFF, D. P., O'CONNELL, E., THOMAS, P. C., and ALLAN, J. (1979a). Strength and Fitness and Subsequent Back Injuries in Fire Fighters. *Journal of Occupational Medicine 21,* 269–272.

CADY, L. D., BISCHOFF, D. P., O'CONNELL, E. R., THOMAS, P. C., and ALLAN, J. H. (1979b). Letters to the Editor: Authors' Response. *Journal of Occupational Medicine 21,* 720–725.

CAI, D. and YOU, M. (1998). An Ergonomic Approach to Public Squatting-Type Toilet Design. *Applied Ergonomics 29,* 147–153.

CAILLET, R. (1981). *Low Back Pain,* 3d ed. London, UK: Davis.

CAIN, W. S., LEADERER, B. P., CANNON, L., TOSUN, T., and ISMAIL, H. (1987). Odorization of Inert Gas for Occupational Safety: Psychophysical Considerations. *American Industrial Hygiene Association Journal 48,* 47–55.

CALDWELL, L. S., CHAFFIN, D. B., DUKES-DOBOS, F. N., KROEMER, K. H. E., LAUBACH, L. L., SNOOK, S. H., and WASSERMAN, D. E. (1974). A Proposed Standard Procedure for Static Muscle Strength Testing. *American Industrial Hygiene Association Journal 35,* 201–206.

CANNON, W. B. (1939). *The Wisdom of the Body.* New York, NY: Norton.

CAPLE, D. C. and BETTS, N. J. (1991). RSI—Its Rise and Fall in Telecom Australia 1981–1990. In *Proceedings of the 11th Congress of the International Ergonomics Association* (pp. 1037–1039). London, UK: Taylor & Francis.

CAPODAGLIO, P., CAPODAGLIO, E. M., and BAZZINI, G. (1995). Tolerability to Prolonged Lifting Tasks Assessed by Subjective Perception and Physiological Responses. *Ergonomics 38,* 2118–2128.

CAPODAGLIO, P., CAPODAGLIO, E. M., and BAZZINI, G. (1997). A Field Methodology for Ergonomic Analysis in Occupational Manual Materials Handling. *Applied Ergonomics 28,* 203–208.

CARLTON, R. S. (1987). The Effects of Body Mechanics Instruction on Work Performance. *The American Journal of Occupational Therapy 41,* 16–20.

CARVALHAIS, A. B., TEPAS, D. I., and MAHAN, R. P. (1988). Sleep Duration in Shift Workers. *Sleep Research 17,* 109–124.

CASALI J. G. and BERGER, E. H. (1996). Technology Advancements in Hearing Protection Circa 1995: Active Noise Reduction, Frequency-Amplitude-Sensitivity, and Uniform Attenuation. *American Industrial Hygiene Association Journal 57,* 175–185.

CASALI, J. G. and WRIGHT, W. H. (1995). Do Amplitude-Sensitive Hearing Protectors Improve Detectability of Vehicle Backup Alarms in Noise? In *Proceedings of the Human Factors and Ergonomics Society 39th Annual Meeting.* (pp. 994–998). Santa Monica, CA: Human Factors and Ergonomics Society.

CASALI, S. P. and WILLIGES, R. C. (1990). Databases of Accommodative Aids for Computer Users with Disabilities. *Human Factors 32,* 407–422.

CASEY, S. M. (1989). Anthropometry of Farm Equipment Operators. *Human Factors Society Bulletin 32,* 1–16.

CHAFFIN, D. B. (1981). Functional Assessment for Heavy Physical Labor. *Occupational Health and Safety 50,* 24, 27, 32, 64.

CHAFFIN, D. B. (1991). Occupational Ergonomics. In *Proceedings, Occupational Ergonomics: Work-Related Upper Limb and Back Disorders* (not paginated). San Diego, CA: American Industrial Hygiene Association, San Diego Section.

Chaffin, D. B. and Andersson, G. B. J. (1984). *Occupational Biomechanics.* New York, NY: Wiley.

Chaffin, D. B. and Andersson, G. B. J. (1991). *Occupational Biomechanics,* 2d ed. New York, NY: Wiley.

Chaffin, D. B., Andersson, G. B. J., and Martin, B. J. (1999). *Occupational Biomechanics,* 3rd. ed.New York, NY: Wiley.

Chaffin, D. B., Herrin, G. D., and Keyserling, W. M. (1978). Preemployment Strength Testing: An Updated Position. *Journal of Occupational Medicine 20,* 403–408.

Chandler, R. F., Clauser, C. E., McConville, J. R., Reynolds, H. M., and Young, J. W. (1975). *Investigation of Inertial Properties of the Human Body* (AMRL-TR-74-137). Wright-Patterson AFB, OH: Aerospace Medical Research Laboratory.

Chapanis, A. (Ed.). (1975). *Ethnic Variables in Human Factors Engineering.* Baltimore, MD: Johns Hopkins University Press.

Chapanis, A. (1995). Ergonomics in Product Development: A Personal View. *Ergonomics 38,* 1625–1638.

Chapanis, A. (1996). *Human Factors in Systems Engineering.* New York, NY: Wiley.

Chenoweth, D. (1983a). Fitness Program Evaluation: Results with Muscle. *Occupational Health and Safety 52,* 14–17 and 40–42.

Chenoweth, D. (1983b). Health Promotion: Benefit vs. Cost. *Occupational Health and Safety 52,* 37–41.

Cherry, N. (1987). Physical Demands of Work and Health Complaints Among Women Working Late in Pregnancy. *Ergonomics 30,* 689–701.

Cheverud, J., Gordon, C. C., Walker, R. A., Jacquish, C., Kohn, L., Moore, A., and Yamashita, N. (1990). *1988 Anthropometric Survey of U.S. Army Personnel* (Technical Reports 90/031 through 036). Natick, MA: U.S. Army Natick Research, Development, and Engineering Center.

Chi, C. F. and Lin, F. T. (1998). A Comparison of Seven Visual Fatigue Assessment Techniques in Three Data-Acquisition Tasks. *Human Factors 40,* 577–590.

Chiou, S., Bhattacharya, A., and Succop, P. A. (1996). Effect of Workers' Shoe Wear on Objective and Subjective Assessment of Slipperiness. *American Industrial Hygiene Association Journal 57,* 825–831.

Christensen, J. M. (1993). Forensic Human Factors Psychology—Part 2: a Model for the Development of Safer Products. *CSERIAC Gateway 4*:3, 1–5.

Christensen, J. M. and Talbot, J. M. (1986). Psychological Aspects of Space Flight. *Aviation, Space, and Environmental Medicine 57,* 203–212.

Christensen, J. M., Topmiller, D. A. and Gill, R. T. (1988). Human Factors Definitions Revisited. *Human Factors Society Bulletin 31,* 7–8.

CIE (1951). *CIE Proceedings, Volume 3.* Paris: Bureau Central de la Commission Internationale de l'Éclairage.

Ciriello, V. M. and Snook, S. H. (1983). A Study of Size, Distance, Height, and Frequency Effects on Manual Handling Tasks. *Human Factors 25,* 473–483.

Ciriello, V. M., Snook, S. H., and Hughes, G. J. (1993). Further Studies of Psychophysically Determined Maximum Acceptable Weights and Forces. *Human Factors 35,* 175–186.

Clark, M. C., Czaja, S. J., and Weber, R. A. (1990). Older Adults and Daily Living Task Profiles. *Human Factors 32,* 537–549.

Cohen, S., Frank, E., Doyle, W. J., Skoner, D. P., Rabin, B. S., and Gwaltney, J. M. (1998). Types of Stressors That Increase Susceptibility to the Common Cold in Healthy Adults. *Health Psychology 17,* 214–223.

Colle, H. A. and Reid, G. B. (1998). Context Effects in Subjective Mental Workload Ratings. *Human Factors 40,* 591–600.

Colligan, M. J. and Tepas, D. I. (1986). The Stress of Hours of Work. *American Industrial Hygiene Association Journal 47,* 686–695.

Collins, M., Brown, B., Bowman, K., and Carkeet, A. (1990). Workstation Variables and Visual Discomfort Associated with VDTs. *Applied Ergonomics 21,* 157–161.

Colquhoun, W. P. (1985). Hours of Work at Sea: Watch-keeping Schedules, Circadian Rhythms, and Efficiency. *Ergonomics 28,* 637–653.

COMMITTEE ON AN AGING SOCIETY (ED.). (1988). *The Social and Built Environment in an Older Society.* Washington, DC: National Academy Press.

COMMITTEE ON VISION (ED.). (1987). *Work, Aging, and Vision.* Washington, DC: National Academy Press.

CONGLETON, J. J., AYOUB, M. M., and SMITH, J. L. (1985). The Design and Evaluation of the Neutral Posture Chair for Surgeons. *Human Factors 27,* 589–600.

CONIGLIO, I., FUBINI, E., MASALI, M., MASIERO, C., PIERLORENZI, G., and SAGONE, G. (1991). Anthropometric Survey of Italian Population for Standardization in Ergonomics. In *Proceedings, 11th Congress of the International Ergonomics Association* (pp. 894–896). London, UK: Taylor & Francis.

CONRAD, R. and HULL, A. J. (1968). The Preferred Layout for Numerical Data Entry Key Sets. *Ergonomics 11,* 165–173.

CONWAY, K. and UNGER, R. (1991). Ergonomic Guidelines for Designing and Maintaining Underground Coal Mining Equipment. Pages 279–302 in A. Mital and W. Karwowski (EDS.), *Workspace, Equipment, and Tool Design.* Amsterdam, The Netherlands: Elsevier.

COOK, T. D. and CAMPBELL, D. T. (1979). *Quasi-Experimentation: Design and Analysis Issues for Field Settings.* Chicago: Rand McNally.

COOPER, R. AND SAWAF, A. (1998) *Executive EQ: Emotional Intelligence in Leadership and Organization.* New York, NY: Perigree.

COREN, S. (1994). Most Comfortable Listening Level as a Function of Age. *Ergonomics 37,* 1269–1274.

CORLETT, E. N. and BISHOP, R. P. (1976). A Technique for Accessing Postural Discomfort. *Ergonomics 19,* 175–182.

CORLETT, E. N., MADELEY, S. J., and MANENICA, I. (1979). Postural Targetting: A Technique for Recording Working Postures. *Ergonomics 22,* 357–366.

COSTA, G. (1996). The Impact of Shift and Night Work on Health. *Applied Ergonomics 27,* 9–16.

COURTNEY, A. J. (1994). The Effect of Scale-Side, Indicator Type, and Control Plane on Direction-of-Turn Stereotypes for Hong Kong Chinese Subjects. *Ergonomics 37,* 865–877.

COWEN, R. (1988). *Eyes on the Workplace.* Washington, DC: National Academy Press.

CRAIG, B. N., CONGLETON, J. J., KERK, C. J., LAWLER, J. M. and MCSWEENEY, K. P. (1998). Correlation of Injury Occurrence Data with Estimated Maximal Aerobic Capacity and Body Composition in a High-Frequency Manual Materials Handling Task. *American Industrial Hygiene Association Journal 59,* 25–33.

CRONK C. E. and ROCHE, A. F. (1982). Race- and Sex-Specific Reference Data for Triceps and Subscapular Skinfolds and Weight/Stature. *American Journal of Clinical Nutrition 35,* 347–354.

CROWELL, H. P. (1995). Human Engineering Design Guidelines for a Powered, Full Body Exoskeleton. (Report ARL-TN-60). Aberdeen Proving Ground, MD: U.S. Army Research Laboratory.

CSIKSZENTMIHALYI, M. (1990). *Flow: The Psychology of Optimal Experience.* New York, NY: Harper & Row.

CULVER, C. C. and VIANO, D. C. (1990). Anthropometry of Seated Women During Pregnancy: Defining a Fetal Region for Crash Protection Research. *Human Factors 32,* 625–636.

CUSHMAN, W. H. and ROSENBERG, D. J. (1991). *Human Factors in Product Design.* Amsterdam, The Netherlands: Elsevier.

CZAJA, S. J. (ED.). (1990). Aging. Special Issue, *Human Factors 32,* 505–622.

CZAJA, S. J. (ED.). (1990). *Human Factors Research Needs for an Aging Population.* Washington, DC: National Academy Press.

CZAJA, S. J. (1991). Work Design for Older Adults. Pages 345–369 in A. Mital and W. Karwowski (EDS.), *Workspace, Equipment, and Tool Design.* Amsterdam, The Netherlands: Elsevier.

CZAJA, S. J. (1997). Using Technology to Aid in the Performance of Home Tasks. Chapter 13 (pp. 311–334) in A. D. Fisk and W. A. Rogers (EDS.), *Handbook of Human Factors and the Older Adult.* San Diego, CA: Academic Press.

CZEISLER, C. A., DUMONT, M., and RICHARDS, G. S. (1990a). *Disorders of Circadian Function: Clinical Consequences and Treatment.* Paper presented at the NIH Consensus Development Conference on the Treatment of Sleep Disorders of Older People. Bethesda, MD: National Institutes of Health.

CZEISLER, C. A., JOHNSON, M. P., and DUFFY, J. F. (1990b). Exposure to Bright Light and Darkness to Treat Physiologic Maladaptation to Night Work. *The New England Journal of Medicine 322,* 1253–1259.

CZEISLER C. A., KRONAUER, R. E., and ALLAN, J. S. (1989). Bright Light Induction of Strong (Type 0) Resetting of the Human Circadian Pacemaker. *Science 244,* 1328–1332.

DAAMS, B. J. (1993). Static Force Exertion in Postures with Different Degrees of Freedom. *Ergonomics 36,* 397–406.

DALTROV, L. H., IVERSEN, M. D., LARSON, M. G., LEW, R., WRIGHT, E., RYAN, J., ZWERLING, C., FOSSEL, A. H., and LIANG, M. H. (1997). A Controlled Trial of an Educational Program to Prevent Low Back Injuries. *New England Journal of Medicine 337,* 322–328.

DANIELS, G. S. (1952). *The "Average" Man?* (Technical Note WCRD 53-7). Wright-Patterson AFB, OH: Wright Air Development Center.

DATTA, S. R. and RAMANATHAN, N. L. (1971). Ergonomic Comparison of Seven Modes of Carrying Loads on the Horizontal Plane. *Ergonomics 14,* 269–278.

DAVIES, B. T. (1978). Training in Manual Handling and Lifting. Pages 175–178 in C. G. Drury (Ed.), *Safety in Manual Materials Handling.* (DHEW (NIOSH). Publication No. 78–185). Cincinnati, OH: NIOSH.

DAVIS, J. R. and MIRKA, G. A. (1997). A Transverse Contour Model of Distributed Muscle Forces and Spinal Loads During Lifting and Twisting. In *Proceedings of the Human Factors and Ergonomics Society 41th Annual Meeting* (pp. 675–679). Santa Monica, CA: Human Factors and Ergonomics Society.

DE BRUIJN, I., ENGELS, J. A., and VAN DER GULDEN, J. W. J. (1998). A Simple Method to Evaluate the Reliability of OWAS Observations. *Applied Ergonomics 29,* 281–283.

DEEB, J. M., DRURY, C. G., and PIZATELLA, P. (1987). Handle Placement on Containers in Manual Materials Handling. In *Proceedings, 9th International Conference on Production Research* (pp. 417–423). Amsterdam, The Netherlands: Elsevier.

DEMPSEY, P. G. (1999). Prevention of Musculoskeletal Disorders: Psychophysical Basis. Chapter 60 (pp. 1101–1126) in W. Karwowski and W. S. Marras (EDS.), *The Occupational Ergonomics Handbook.* Boca Raton, FL: CRC Press.

DENNERLEIN, J. T., MOTE, C. D., and REMPEL, D. M. (1998). Control Strategies for Finger Movement During Touch-Typing. The Role of the Extrinsic Muscles During a Keystroke. *Experimental Brain Research 121,* 1–16.

DICKINSON, C. E., CAMPION, K., FOSTER, A. F., NEWMAN, S. J., O'ROURKE, A. M. T., and THOMAS, P. G. (1992). Questionnaire Development: An Examination of the Nordic Musculoskeletal Questionnaire. *Applied Ergonomics 23,* 197–201.

DINARDI, S. R. (ED.) (1997). *The Occupational Environment—Its Evaluation and Control.* Fairfax, VA: American Industrial Hygiene Association.

DOHERTY, E. T. (1991). Speech Analysis Techniques for Detecting Stress. In *Proceedings of the Human Factors Society 35th Annual Meeting* (pp. 689–693). Santa Monica, CA: Human Factors Society.

DOHRENWEND, B. P., RAPHAEL K. G., SCHWARTZ S., STUEVE, A., and SKODOL, F. (1993). The Structured Event Probe and Narrative Rating Method for Measuring Stressful Life Events. Pages 174–196 in *Handbook of Stress: Theoretical and Clinical Aspects,* 2d ed. New York, NY: Free Press.

DOWELL, W. R., PRICE, J. M., and GSCHEIDLE, G. M. (1997). The Effect of VDT Screen Distance on Seated Posture. In *Proceedings of the Human Factors and Ergonomics Society 41st Annual Meeting* (pp. 505–508). Santa Monica, CA: Human Factors and Ergonomics Society.

DRILLIS, R. and CONTINI, R. (1966). *Body Segment Parameters* (Report 1166-03). Office of Vocational Rehabilitation, Department of Health, Education and Welfare. New York, NY: New York University School of Engineering and Science.

DRUCKMAN, D. and BJORK, R. A. (EDS.) (1990). *In the Mind's Eye: Enhancing Human Performance.* Washington, DC: National Academy Press.

DRUCKMAN, D. and SWETS, J. A. (EDS.) (1988). *Enhancing Human Performance: Issues, Theories, and Techniques.* Washington, DC: National Academy Press.

DRURY, C. G., DEEB, J. M., HARTMAN, B., WOOLLEY, S., DRURY, C. E., and GALLAGHER, S. (1989). Symmetric and Asymmetric Manual Materials Handling—Part 1: Physiology and Psychophysics. *Ergonomics 32,* 467–489.

DRYDEN, R. D. and KEMMERLING, P. T. (1990). Engineering Assessment. Pages 107–129 in S. P. Sheer (Ed.), *Vocational Assessment of Impaired Workers.* Aspen, CO: Aspen Press.

DUCHON, J., WAGNER, J., and KERAN, C. (1989). Forward Versus Backward Shift Rotation. In *Proceedings of the Human Factors Society 33rd Annual Meeting* (pp. 806–810). Santa Monica, CA.: Human Factors Society.

DUECKER, J. A., RITCHIE, S. M., KNOX, T. J. and ROSE, S. J. (1994). Isokinetic Trunk Testing and Employment. *Journal of Occupational Medicine 36,* 42–48.

DUNHAM, R. B., PIERCE, J. L., and CASTANEDA, M. B. (1987). Alternative Work Schedules: Two Field Quasi Experiments. *Personnel Psychology 40,* 215–242.

DUPUIS, H. and ZERLETT, G. (1986). Whole-Body Vibration and Disorders of the Spine. *International Archives of Occupational and Environmental Health 59,* 323–336.

DUQUETTE, J, LORTIE, M., and ROSSIGNOL, M. (1997). Perception of Difficulties Related to Assembly Work: General Findings and Impact of Back Health. *Applied Ergonomics 28,* 386–396.

EASTMAN KODAK COMPANY. (Vol. 1, 1983; Vol. 2, 1986). *Ergonomic Design for People at Work.* New York NY: Van Nostrand Reinhold.

EBERHARD, J. W. and BARR, R. A. (EDS.). (1992). Safety and Mobility of Elderly Drivers, Part 2. Special Issue of *Human Factors 34,* 1–65.

EDHOLM, O. G. and MURRELL, K. H. F. (1974). *The Ergonomics Research Society. A History 1949 to 1970.* (No location given.): The Council of the Ergonomics Research Society.

EDMONDS, C., LOWRY, C., and PENNEFATHER, J. (1994). *Diving and Subaquatic Medicine.* Boston, MA: Butterworth-Heinemann.

EDWORTHY, J. and ADAMS, A. (1996). *Warning Design.* London, UK: Taylor & Francis.

EDWORTHY, J. and STANTON, N. (1995). A User-Centred Approach to the Design and Evaluation of Auditory Warning Signal: 1. Methodology. *Ergonomics 38,* 2262–2280.

ELBERT, K. E. K. (1991). *Analysis of Polyethylene in Total Joint Replacement.* Unpublished Doctoral Dissertion, Cornell University, Ithaca, NY.

ESTILL, C. F. and KROEMER, K. H. E. (1998). Evaluation of Supermarket Bagging Using A Wrist Motion Monitor. *Human Factors 40,* 624–632.

FAHRINI, W. H. (1975). Conservative Treatment of Lumbar Disc Degeneration, Our Primary Responsibility. *Orthopedic Clinics of North America 6,* 93–103.

FALLON, A. E. and ROZIN, P. (1985). Sex Differences in Perceptions of Desirable Body Shape. *Journal of Abnormal Psychology 94,* 102–105.

FALLON, E. F., DILLON, A., SWEENEY, M., and HERRING, V. (1991). An Investigation of the Concept of Designer Style and its Implications for the Design of CAD Man-Machine Interfaces. Pages 873–880 in W. Karwowski and J. W. Yates (EDS.), *Advances in Industrial Ergonomics and Safety III.* London, UK: Taylor & Francis.

FARRIS, B. A., LANDWEHR, H. R., FERNANDEZ, J. E., and AGARWAL, R. K. (1998). Physiological Evaluation of Mouse Pad Placement. Pages 487–490 in S. Kumar (ED.), *Advances in Occupational Ergonomics and Safety.* Amsterdam, The Netherlands: IOS Press.

FAST, A., SHAPIRO, M. D., and EDMOND, J. (1987). Low Back Pain in Pregnancy. *Spine 12,* 368–371.

FASTE, R. A. (1977). New System Propels Design for the Handicapped. *Industrial Design 2,* 51–55.

FERNANDEZ, J. E., AYOUB, M. M., and SMITH, J. L. (1991). Psychophysical Lifting Capacity Over Extended Periods. *Ergonomics 34,* 23–32.

FERNIE, G. (1997). Assistive Devices. Chapter 12 (pp 289–310) in A. D. Fisk and W. A. Rogers (EDS.), *Handbook of Human Factors and the Older Adult.* San Diego, CA: Academic Press.

FERNSTROEM, E. AND ERICSON, M. O. (1997). Computer Mouse or Trackpoint—Effects on Muscular Load and Operator Experience. *Applied Ergonomics 28,* 347–354.

FISHER, W. and TARBUTT, V. (1988). Some Issues in Collecting Data on Working Postures. In *Proceedings of the Human Factors Society 32nd Annual Meeting* (pp. 627–631). Santa Monica, CA: Human Factors Society.

FISK, A. D. (1999). Human Factors and the Older Adult. *Ergonomics in Design, January Issue,* 8–13.

FISK, A. D. and ROGERS, W. A. (EDS.) (1997). *Handbook of Human Factors and the Older Adult.* San Diego, CA: Academic Press.

FLACH, J. M. (1989). An Ecological Alternative to Egg-Sucking. *Human Factors Society Bulletin 32,* 4–6.

FLACH, J. M. and DOMINGUEZ, C. O. (1995). Use-Centered Design. *Ergonomics in Design, July Issue,* 19–24.

FLUEGEL, B., GREIL, H., and SOMMER, K. (1986). *Anthropologischer Atlas.* Berlin, Germany: Tribuene.

FOLKARD, S., and MONK, T. H. (EDS.). (1985). *Hours of Work.* Chichester, UK: Wiley.

FOLKMAN, S. and LAZARUS, R. S. (1988). The Relationship Between Coping and Emotion: Implications for Theory and Research. *Social Science and Medicine 26,* 309–317.

FOSTER, G. D., WADDEN, T. A., and VOGT, R. A. (1997). Body Image in Obese Women Before, During, and After Weight Loss Treatment. *Health Psychology 16,* 226–229.

FOX, J. G. (1983). Industrial Music. Pages 221–226 in D. J. Oborne and M. M. Gruneberg (EDS.), *The Physical Environment at Work.* New York, NY: Wiley.

FRANSSON-HALL, C., GLORIA, R., KILBOM, A., WINKEL, J., KARLQVIST, L., and WIKTORIN, C. (1995). A Portable Ergonomic Observation Method (PEO) for Computerized On-Line Recording of Postures and Manual Handling. *Applied Ergonomics 26,* 93–100.

FRANSSON-HALL, C., and KILBOM, A. (1993). Sensitivity of the Hand to Surface Pressure. *Applied Ergonomics 24,* 181–189.

FRASER, T. M. (1980). *Ergonomic Principles in the Design of Hand Tools.*(Occupational Safety and Health Series, No. 44). Geneva, Switzerland: International Labour Office.

FRAZER, L. (1991). Sex in Space. *Ad Astra 3,* 42–45.

FREDERICKS, T. K., GUNN, E., ROZEK, G., and BEERT, T. (1998). Maximum Acceptable Weight of Lift for an Asymmetrical Lowering Task Commonly Observed in the Parcel Delivery Industry. Pages 361–364 in S. Kumar (ED.), *Advances in Industrial Ergonomics and Safety.* Amsterdam, The Netherlands: IOS Press.

FREIVALDS, A. (1989). Understanding and Preventing Back Trauma: Comparison of U.S. AND European Approaches, Pages 55–63 in K. H. E. Kroemer, J. D. McGlothlin, and T. G. Bobick (EDS.), *Manual Material Handling: Understanding and Preventing Back Trauma* Akron, OH: American Industrial Hygiene Association.

FREIVALDS, A. (1999). Ergonomics of Hand Controls. Chapter 27 (pp. 461–478) in W. Karwowski and W. S. Marras (EDS.), *The Occupational Ergonomics Handbook.* Boca Raton, FL: CRC Press.

FROEBERG, J. E. (1985). Sleep Deprivation and Prolonged Working Hours. Chapter 6 (pp. 67–76) in S. Folkard and T. H. Monk (EDS.), *Hours of Work.* Chichester, UK: Wiley.

FRUITIGER, A. (1989). *Signs and Symbols.* New York, NY: Van Nostrand Reinhold.

FRY, H. J. H. (1986). Overuse Syndrome in Musicians 100 Years Ago. *Lancet September Issue,* 728–731 and *Medical Journal of Australia December Issue,* 620–625.

GAGNON, M. (1997). Box Tilt and Knee Motions in Manual Lifting: Two Differential Factors in Expert and Novice Workers. *Clinical Biomechanics 12,* 418–428.

GAGNON, M., PLAMONDON, A., and GRAVEL, D. (1995). Effects of Symmetry and Load Absorption of a Falling Load on 3D Trunk Muscular Movements. *Ergonomics 38,* 1156–1171.

GAGNON, M. and SMYTH, G. (1992). Biomechanical Exploration of Dynamic Modes of Lifting. *Ergonomics 35,* 329–345.

GALE, A. (ED.). (1988). *The Polygraph Test. Lies, Truth, and Science.* London, UK: Sage.

GALLAGHER, S. (1999). Ergonomics Issues in Mining, Chapter 106 (pp. 1893–1915) in W. Karwowski and W. S. Marras (EDS.), *The Occupational Ergonomics Handbook.* Boca Raton, FL: CRC Press.

GALLAGHER, S., MARRAS, W. S., and BOBICK, T. G. (1988). Lifting in Stooped and Kneeling Postures: Effects on Lifting Capacity, Metabolic Costs, and Electromyography of Eight Trunk Muscles. *International Journal of Industrial Ergonomics 3,* 65–76.

GALLAGHER, S. and MOORE, J. S. (1999). Worker Strength Evaluation: Job Design and Worker Selection. Chapter 21 (pp. 371–386) in W. Karwowski and W. S. Marras (EDS.), *The Occupational Ergonomics Handbook.* Boca Raton, FL: CRC Press.

GARCIA, D. T., WONG, S. L., FERNANDEZ, J. E., and AGARWAL, R. K. (1998). The Effect of Arm Supports on Muscle Activity in Shoulder and Neck Muscles. Pages 483–486 in S. Kumar (ED.), *Advances in Occupational Ergonomics and Safety.* Amsterdam, The Netherlands: IOS Press.

GARDNER-BONNEAU, D. J. (ED.). (1990). Assisting People with Functional Impairments. Special Issue of *Human Factors 32,* 379–475.

GARDNER-BONNEAU, D. J. and GOSBEE, J. (1997). Health Care and Rehabilitation, Chapter 10 (pp. 231–255) in A. D. Fisk and W. A. Rogers (EDS.), *Handbook of Human Factors and the Older Adult.* San Diego, CA: Academic Press.

GARG, A. and AYOUB, M. M. (1980). What Criteria Exist for Determining How Much Load Can be Safely Lifted? *Human Factors 22,* 475–486.

GARRETT, J. W. and KENNEDY, K. W. (1971). *A Collation of Anthropometry* (AMRL-TR-68-1). Wright-Patterson Air Force Base, OH: Aerospace Medical Research Laboratories.

GAVER, W. W. (1997). Auditory Interfaces. Chapter 42 (pp. 1003–1041) in M. Helander, T. K.Landauer, and P. Prabhu (EDS.) *Handbook of Human-Computer Interaction* 2d ed.Amsterdam, The Netherlands: Elsevier.

GAVRON, V. J. (1997). High-g Environments and the Pilot. *Ergonomics in Design, April 1997 Issue,* 18–23.

GAWANDE, A. (1998). *The Pain Perplex. The New Yorker September 21 1998,* 86–94.

GENAIDY, A. M., AL-SHEDI, A. A., and KARWOWSKI, W. (1994). Postural Stress in Industry. *Applied Ergonomics 25,* 77–87.

GENAIDY, A. M., ASFOUR, S. S., MITAL, A., and WALY, S. M. (1990). Psychophysical Models for Manual Lifting Tasks. *Applied Ergonomics 21,* 295–303.

GENAIDY, A. M., BARKAWI, H., and CHRISTENSEN, D. (1995). Ranking of Static Non-Neutral Postures Around the Joints of the Upper Extremity and the Spine. *Ergonomics 38,* 1851–1858.

GENAIDY, A. M., GUPTA, T., and ALSHEDI, A. (1990). Improving Human Capabilities for Combined Manual Handling Tasks Through a Short and Intensive Physical Training Program. *American Industrial Hygiene Association Journal 51,* 610–614.

GERARD, M. J., ARMSTRONG, T. J., FOULKE, J. A., and MARTIN, B. J. (1996). Effects of Key Stiffness on Force and the Development of Fatigue While Typing. *American Industrial Hygiene Association Journal 57,* 849–854.

GIACOMIN, J. and QUATTROCOLO, S. (1997), An Analysis of Human Comfort When Entering and Exeting the Rear Seat of an Automobile. *Applied Ergonomics 28,* 397–4–6.

GIL, H. J. C. and TUNES, D. B. (1977). Posture Recording: A Model for Sitting Posture. *Applied Ergonomics 20,* 53–57.

GILAD, I. and POLLATSCHEK, M. A. (1986). Layout Simulation for Keyboards. *Behaviour and Information Technology 5,* 273–281.

GILLILAND, K. and SCHLEGEL, R. E. (1994). Tactile Stimulation of the Human Head for Information Display. *Human Factors 36,* 700–717.

GODIN, G. and GIONET, N. (1991). Determinants of an Intention to Exercise of Electric Power Commission's Employees. *Ergonomics 34,* 1221–1230.

GOLEMAN, D. (1995). *Emotional Intelligence: Why It Can Matter More than IQ.* New York, NY: Bantam Books.

GOODMAN, L. S., DE YANG, L., KELSO, B., and LIU, P. (1995). Cardiovascular Effects of Varying g-suit Pressure and Coverage During $+g_z$ Positive Pressure Breathing. *Aviation, Space and Environmental Medicine 66,* 829–836.

GOPHER, D. and RAIJ, D. (1988). Typing With a Two-Hand Chord Keyboard: Will the QWERTY Become Obsolete? *IEEE Transactions on Systems, Man, and Cybernetics 18,* 601–609.

GORDON, C. C., CHURCHILL, T., CLAUSER, C. E., BRADTMILLER, B., MCCONVILLE, J. T., TEBBETTS, I., and WALKER, R. A. (1989). *1988 Anthropometric Survey of U.S. Army Personnel: Summary Statistics Interim*

Report (Technical Report NATICK/TR-89-027). Natick, MA: United States Army Natick Research, Development and Engineering Center.

GOULD, S. J. (1981). *The Mismeasure of Man.* New York, NY: Norton.

GOULD, S. J. (1988). A Novel Notion of Neanderthal. *Natural History 97,* 16–21.

GRAEBER, R. C. (1988). Aircrew Fatigue and Circadian Rhythmicity. Chapter 10 (pp. 305–344) in E. L. Wiener and D. C. Nagel (Eds.), *Human Factors in Aviation.* San Diego, CA: Academic Press.

GRANDJEAN, E. (1963). *Physiological Design of Work* (in German). Thun, Switzerland: Ott.

GRANDJEAN, E. (ED.). (1969). Sitting Posture. London, UK: Taylor & Francis.

GRANDJEAN, E., HUENTING, W., and NISHIYAMA, K. (1984). Preferred VDT Workstation Settings, Body Postures and Physical Impairments. *Applied Ergonomics 15,* 99–104.

GRANT, A. (1990). Homo-Quintadus, Computers and Rooms (Repetitive Ocular Orthopedic Motion Stress). *Optometry and Vision Science 67,* 297–305.

GREEN, B. F. and ANDERSON, L. K. (1955). The Tactual Identification of Shapes for Coding Switch Handles. *Journal of Applied Psychology 39,* 219–226.

GREENE, T. C. and BELL, P. A. (1987). Environment Stress. Chapter 5 (pp. 81–106) in M. A. Baker (ED.), *Sex Differences in Human Performance.* Chichester, UK: Wiley.

GREENSTEIN, J. S. (1997). Pointing Devices. Chapter 55 (pp. 1317–1348) in M. Helander, T. K. Landauer and P. Prabhu (EDS.) *Handbook of Human-Computer Interaction* (2nd ed.). Amsterdam, NL: Elsevier.

GREENWOOD, K. M. (1991). *Psychometric Properties of the Diurnal Type Scale of Torsvall and Akerstedt (1980). Ergonomics 34,* 435–443.

GREINER, T. M. (1991). *Hand Anthropometry of U.S. Army Personnel* (Technical Report TR-92/011). Natick, MA: U.S. Army Natick Research, Development and Engineering Center.

GREINER, T. M. and GORDON, C. C. (1990). *An Assessment of Long-Term Changes in Anthropometric Dimensions: Secular Trends of U.S. Army Males* (Natick/TR-91/006). Natick, MA: U.S. Army Natick Research, Development and Engineering Center.

GRIECO, A. (1986). Sitting Posture: An Old Problem and a New One. *Ergonomics 29,* 345–362.

GRIFFIN, M. D. and FRENCH, J. R. (1991). *Space Vehicle Design.* Washington, DC: American Institute of Aeronautics and Astronautics, Inc.

GRIFFIN, M. J. (1990). *Handbook of Human Vibration.* San Diego, CA: Academic Press.

GRIFFIN, M. J. (1997). Vibration and Motion. Chapter 25 (pp. 828–857) in G. Salvendy (ED.), *Handbook of Human Factors and Ergonomics,* 2d ed. New York, NY: Wiley.

GROENQUIST, R. and HIRVONEN, M. (1994). Pedestrian Safety on Icy Surfaces: Anti-Slip Properties of Footwear. Pages 315–322 in F. Aghazadeh (ED.), *Advances in Industrial Ergonomics and Safety VI.* London, UK: Taylor & Francis.

GUDJONSSON, G. H. (1988). How to Defeat the Polygraph Tests, Chapter 10 (pp. 126–136) in A. Gale (ED.), *The Polygraph Test. Lies, Truth, and Science.* London, UK: Sage.

GUIGNARD, J. C. (1985). Vibration. Chapter 15 (pp. 635–724) in L. V. Cralley and L. J. Cralley (EDS.), *Patty's Industrial Hygiene and Toxicology.* New York, NY: Wiley.

GUYLL, M. and CONTRADA, R. J. (1998). Trait Hostility and Ambulatory Cardiovascular Activity: Responses to Social Interaction. *Health Psychology 17,* 30–39.

GUYTON, A. C. (1979). *Physiology of the Human Body,* 5th ed. Philadelphia, PA: Saunders.

GYI, D. E. and PORTER, J. M. (1999). Interface Pressure and the Prediction of Car Seat Comfort. *Applied Ergonomics 30,* 99–107.

HAAS, E. C. and CASALI, J. G. (1995). Perceived Urgency of and Response Time to Multi-Tone and Frequency-Modulated Warning Signals in Broadband Noise. *Ergonomics 38,* 2313–2326.

HACKETT, T. P., ROSENBAUM, J. F., and TESAR, G. E. (1988). Emotion, Psychiatric Disorders, and the Heart. Pages 1883–1900 in E. Braunwald (ED.), *Heart Disease—A Textbook of Cardiovascular Medicine.* Philadelphia, PA: Saunders.

HADLER, N. M. (1997). Back Pain in the Workplace. *Spine 22,* 935–940.

HAGBERG, M. and REMPEL, D. (1997). Work-Related Disorders and the Operation of Computer VDT's, Chapter 58 (pp. 1415–1429) in M. G. Helander, T. K. Landauer, and P. V. Prabhu (EDS.), *Handbook of Human-Computer Interaction,* 2d ed. Amsterdam, NL: Elsevier.

HAHN, H. A. and PRICE, D. L. (1994). Assessment of the Relative Effects of Alcohol on Different Types of Work Behavior. *Ergonomics 37,* 435–448.

HAISMAN, M. F. (1988). Determinants of Load Carrying Ability. *Applied Ergonomics 19,* 111–121.

HALL, H. W. (1973). "Clean" Versus "Dirty" Lifting, An Academic Subject for Youth. *American Society of Safety Engineers Journal 18,* 20–25.

HAMILL, J. and HARDIN, E. C. (1997). Biomechanics. Page 699 in Chapter 26 in S. R. DiNardi (ED.), *The Occupational Environment—Its Evaluation and Control.* Fairfax, VA: American Industrial Hygiene Association.

HAMILL, P. V. V., DRIZD, T. A., JOHNSON, C. L., REED, R. B., ROCHE, A. F., and MOORE, W. M. (1979). Physical Growth: National Center for Health Statistics Percentiles. *American Journal of Clinical Nutrition 32,* 607–629.

HAMMER, W. and PRICE, D. (2001). *Occupational Safety Management and Engineering,* 5th ed. Upper Saddle River, NJ: Prentice Hall.

HAN, S. H., WILLIGES, B. H., and WILLIGES, R. C. (1997). A Paradigm for Sequential Experimentation. *Ergonomics 40,* 737–760.

HANCOCK, P. A. and MESHKATI, N. (EDS.). (1988). *Human Mental Workload.* Amsterdam, The Netherlands: Elsevier.

HANGARTNER, M. (1987). *Standardization in Olfactometry with Respect to Odor Pollution Control; Assessment of Odor Annoyance in the Community.* Presentations 87-75A.1 and 87-75B.3 at the 80th Annual Meeting of the APCA. New York, June 21–26.

HANSEN, L., WINKEL, J., and JORGENSEN, K. (1998). Significance of Mat and Shoe Softness During Prolonged Work in Upright Position: Based on Measurement of Low Back Muscle EMG, Foot Volume Changes, Discomfort and Ground Force Reactions. *Applied Ergonomics 29,* 217–224.

HARRISON, A. A., CLEARWATER, Y. A., and MCKAY, C. P. (1991). *From Antarctica to Outer Space: Life in Isolation and Confinement.* New York, NY: Springer.

HART, S. G. and STAVELAND, I. E. (1988). Development of NASA-TLX (Task Load Index): Results of Experimental and Theoretical Research. Pages 185–218 in P. A. Hancock and N. Meshkati (EDS.), *Human Mental Workload.* Amsterdam, The Netherlands. Elsevier.

HARVEY, R. and PEPER, E. (1997). Surface Electromyography and Mouse Use Position. *Ergonomics 40,* 781–789.

HASHEMI, L. and DEMPSEY, P. G. (1997). Body Parts and Nature of Injuries Associated with Manual Materials Handling Workers' Compensation Claims. In *Proceedings of the Human Factors and Ergonomics Society 41st Annual Meeting* (pp. 619–623). Santa Monica, CA: Human Factors and Ergonomics Society.

HAY, J. G. (1973). *The Center of Gravity of the Human Body.* Page 2044 in *Kinesiology III.* Washington, DC: American Association for Health, Physical Education, and Recreation.

HAYNE, C. R. (1981). Lifting and Handling. *Health and Safety at Work 3,* 18–21.

HAYSLIP, B. and PANEK, P. (1989). *Adult Development and Aging.* New York, NY: Harper & Row.

HEACOCK, H., KOEHOORN, M. and TAN, J. (1997). Applying Epidemiological Principles to Ergonomics: A Checklist for Incorporation Sound Design and Interpretation of Studies. *Applied Ergonomics 26,* 165–172

HEIDNER, F. (1915). *Type-Writing Machine.* Letter's Patent 1,138, 474, dated May 4, 1915; application filed March 18, 1914. United States Patent Office.

HELANDER, M. G. (1981). *Human Factors/Ergonomics for Building and Construction.* New York, NY: Wiley.

HELANDER, M. G. (1982). *Ergonomic Design of Office Environments for Visual Display Terminals* (Report for NIOSH). Blacksburg, VA: Virginia Tech (VPI & SU).

HELANDER, M. G. (ED.). (1988). Handbook of Human-Computer Interaction. Amsterdam: North-Holland.

HELANDER, M. G. and NAGAMACHI, N. (EDS.). (1992). *Design for Manufacturabilty. A Systems Approach to Concurrent Engineering and Ergonomics.* London, UK: Taylor & Francis.

HELANDER, M. G. and ZHANG, L. (1997). Field Studies of Comfort and Discomfort in Sitting. *Ergonomics 40I,* 895–915.

HELMERS, K. F., POSLUSZNY, D. M., and KRANTZ, D. S. (1994). Associations of Hostility and Coronary Artery Disease: A Review of Studies. Pages 67–96 in A. W. Siegman and T. W. Smith (EDS.), *Anger, Hostility and the Heart.* Hillsdale, NJ: Erlbaum.

HENDRICK, H. W. (1996). The Ergonomics of Economics Is the Economics of Ergonomics. In *Proceedings of the Human Factors and Ergonomics Society 40st Annual Meeting* (pp. 1–10). Santa Monica, CA: Human Factors and Ergonomics Society.

HENDRICK, H. W. and KLEINER. B. M. (1999). *Macroergonomics: An Introduction to Work System Analysis and Design.* Santa Monica, CA: Human Factors and Ergonomics Society.

HERKIMER COUNTY HISTORICAL SOCIETY (ED.). (1923). *The Story of the Typewriter 1873–1923. Published in Commemoration of the Fiftieth Anniversary of the Invention of the Writing Machine.* Herkimer, NY: Author.

HEUER, H. and OWENS, D. A. (1989). Vertical Gaze Direction and the Resting Posture of the Eyes. *Perception 18,* 353–377.

HEYMSFIELD, S. B., ALLISON, D. B., HESHKA, S., and PIERSON, R. N. (1995). Assessment of Human Body Composition. Chapter 14 (pp. 515–560) in D. B. Allison (Ed.), *Handbook of Assessment Methods for Eating Behaviors and Weight-Related Problems.* Thousand Oaks, CA: Sage Publications.

HIDALGO, J., GENAIDY, A., KARWOWSKI, W., CHRISTENSEN, D., HUSTON, R., and STAMBOUGH, J. (1995). A Cross-Validation of the NIOSH Limits for Manual Lifting. *Ergonomics 38,* 2455–2464.

HIDALGO, J., GENAIDY, A., KARWOWSKI, W., CHRISTENSEN, D., HUSTON, R., and STAMBOUGH, J. (1997). A Comprehensive Lifting Model: Beyond the NIOSH Lifting Equation. *Ergonomics 40,* 916–927.

HILL, S. G., IAVECCHIA, H. P., BYERS, J. C., BITTNER, A. C., ZAKLAD, A. L., and CHRIST, R. E. (1992). Comparison of Four Subjective Workload Rating Scales. *Human Factors 34,* 429–439.

HILL, S. G. and KROEMER, K. H. E. (1986). Preferred Declination of the Line of Sight. *Human Factors 28,* 127–134.

HIMMELSKIN, J. S. and ANDERSSON, G. B. J. (1988). Low Back Pain: Risk Evaluation and Preplacement Screening. *Journal of Occupational Medicine 3,* 255–269.

HIRSCH, R. S. (1970). Effect of Standard vs. Alphabetical Formats in Typing Performance. *Journal of Applied Psychology 54,* 484–490.

HOCKING, B. (1987). Epidemiological Aspects of "Repetition Strain Injury" in Telecom Australia. *The Medical Journal of Australia 147 (Sept.),* 218–222.

HOFFMAN, R. G. and POZOS, R. S. (1989). Experimental Hypothermia and Cold Perception. *Aviation, Space and Environmental Medicine 60,* 964–969.

HOLEWIJN, M. and LOTENS, W. A. (1993). The Influence of Backpack Design on Physical Performance. *Ergonomics 35,* 149–157.

HOLLAND, D. A. (1991). *Systems and Human Factors Concerns for Long-Duration Space Flight.* Unpublished Master's Thesis, Department of Systems Engineering, Blacksburg, VA: Virginia Tech.

HOLMES, T. H. and RAHE, R. H. (1967). The Social Readjustment Rating Scale. *Journal of Psychosomatic Research 11,* 213–218.

HOLZMANN, P. (1981). ARBAN—A New Method of Analysis of Ergonomic Effort. *Applied Ergonomics 13,* 82–86.

HOOD, D. C. and FINKELSTEIN, M. A. (1986). Sensitivity to Light. Chapter 5 (pp. 5.1–5.66) in K. R. Boff, L. Kaufman, and J. P. Thomas (EDS.), *Handbook of Perception and Human Performance.* New York. NY: Wiley.

HORNE, J. A. (1985). Sleep Loss: Underlying Mechanisms and Tiredness. Chapter 5, in S. Folkard and T. H. Monk (EDS.), *Hours of Work.* Chichester, UK: Wiley, 53–65.

HORNE, J. A. (1988). *Why We Sleep—The Functions of Sleep in Humans and Other Mammals.* Oxford, UK: Oxford University Press.

HOUSE, L. H. and PANSKY, B. (1967). *A Functional Approach to Neuroanatomy,* 2d ed. New York, NY: McGraw-Hill.

HUDGENS, G. A., FATKIN, L. T., BILLINGSLEY, P. A., and MAZURACZAK, J. (1988). Hand Steadiness: Effects of Sex, Menstrual Phase, Oral Contraceptives, Practice, and Handgun Weight. *Human Factors 30,* 51–60.

HUFFMAN, J. A. and LEHMAN, K. R. (1997). Pointing Devices in the Retail Environment. In *Proceedings of the Human Factors and Ergonomics Society 41st Annual Meeting* (pp. 415–419). Santa Monica, CA: Human Factors and Ergonomics Society.

HUGHES, R. E. (1995). Choice of Optimization Models for Predicting Spinal Forces in a Three-Dimensional Analysis of Heavy Work. *Ergonomics 38,* 2476–2484.

HUMAN FACTORS AND ERGONOMICS SOCIETY (1998). Human Factors and Ergonomics Society Strategic Plan. Page 388 in *Human Factors and Ergonomics Society Directory and Yearbook 1998–1999.* Santa Monica, CA: Human Factors and Ergonomics Society.

HUNT, D. P. and CRAIG, D. R. (1954). *The Relative Discriminability of Thirty-one Differently Shaped Knobs* (WADC-TR-54-108). Wright-Patterson AFB, OH: Wright Air Development Center.

HUTCHINSON, G. (1995). Taking the Guesswork out of Medical Device Design. *Ergonomics in Design April Issue,* 21–26.

IGNAZI, G., MARTEL, A, MOLLARD, R., and COBLENTZ, A. (1996). Anthropometric Measurements Evolution of a French School Children and Adolescent Population Aged Four to Eighteen. In *Proceedings of the 4th Pan Pacific Conference on Occupational Ergonomics* (pp. 111–114). Hsinchu, ROC: Ergonomics Society of Taiwan, 111–114.

ILO (ED.). (1974). *Introduction to Work Study.* Geneva, Switzerland: International Labour Office.

ILO (ED.). (1988). *Maximum Weights in Load Lifting and Carrying.* (Occupational Safety and Health Series, No. 59). Geneva, Switzerland: International Labour Office.

IMRHAN, S. N. (1998). Manual Torquing: A Review of Empirical Studies. Pages 439–442 in S. Kumar (ED.), Advances in Occupational Ergonomics and Safety. Amsterdam, The Netherlands: IOS Press.

IMRHAN, S. N. (1999). Push-Pull Force Limits. Chapter 23 (pp. 407–420) in W. Karwowski and W. S. Marras (Eds.), *The Occupational Ergonomics Handbook.* Boca Raton, FL: CRC Press.

ISO (ED.). (1985). *Evaluation of Human Exposure to Whole-Body Vibration* (ISO Standard 2631). Geneva, Switzerland: International Organization for Standardization.

ISO (ED.). (1987). *Mechanical Vibration and Shock: Mechanical Transmissibility in the Human Body in the Z Direction* (ISO Standard 7962). Geneva, Switzerland: International Organization for Standardization.

ISO (ED.). (1989). *Hot Environments* (ISO Standard 7243). Geneva, Switzerland: International Organization for Standardization.

ISO (ED.). (1995). *Ergonomics of the Thermal Environment—Estimation of the Thermal Insulation and Evaporative Resistance of a Clothing Ensemble* (ISO Standard 9920). Geneva, Switzerland: International Organization for Standardization.

JAEGER, M. (1987). *Biomechanical Human Model for Analysis and Evaluation of the Strain in the Spinal Column While Manipulating Loads* (in German). (Biotechnik Series 17, No. 33.) Duesseldorf, Germany: VDI Verlag.

JAEGER, M. and LUTTMANN, A. (1986). Biomechanical Model Calculations of Spinal Stress for Different Working Postures in Various Workload Situations. Chapter 15 (pp. 144–423) in N. Corlett, J. Wilson, and I. Manenica (EDS.), *The Ergonomics of Working Postures: Models, Methods and Cases.* London, UK: Taylor & Francis.

JASCHINSKI-KRUZA, W. (1991). Eyestrain in VDU Users: Viewing Distance and the Resting Position of Ocular Muscles. *Human Factors 33,* 69–83.

JENKINS, J. and RICKARDS, J. (1997). Can I Benefit from the Cost of Ergonomics? Exploring a Pre-Intervention Methodology. In *Proceedings of the 29th Annual Conference of the Human Factors Association of Canada* (pp. 139–147). Windsor, ON: Human Factors Association of Canada.

JENKINS, J. P. (ED.) (1991). *Human Performance for Long Duration Space Missions.* (Final Report, NASA-SSTAC Ad Hoc Committee). Washington, DC: NASA.

JENKINS, W. L. (1953). *Design Factors in Knobs and Levers for Making Settings on Scales and Scopes* (WADC-TR-53-2). Wright-Patterson AFB, OH: Aero Medical Laboratory.

JENSEN, M. C., BRANT-ZAWADZKI, M. N., OBUCHOWSKI, N., MODIC, M. T., MALKASIAN, D., and ROSS, J. S. (1994). Magnetic Resonance Imaging of the Lumbar Spine in People Without Back Pain. *New England Journal of Medicine 331,* 69–73.

JENSEN, R. C. (1985). A Model of The Training Process Devised from Human Factors and Safety Literature. Pages 501–509 in R. E. Eberts and C. G. Eberts (EDS.), *Trends in Ergonomics/Human Factors II.* Amsterdam, The Netherlands: Elsevier.

JOHANSSON, A., JOHANSSON, G., LUNDQUIST, P., AKESSON, I., ODENRICK, P., and AKELSSON, R. (1998). Evaluation of a Workplace Redesign of a Grocery Checkout System. *Applied Ergonomics 29,* 261–266.

JOHNSON, D. A. (1998). New Stairway—Old Problems. *Ergonomics in Design October Issue,* 7–10.

JOHNSON, L. C., TEPAS, D. I., COLGUHOUN, W. P., and COLLIGAN, M. J. (EDS.). (1981). *Biological Rhythms, Sleep and Shift Work.* New York, NY: Spectrum.

JOHNSON, R. C., DOAN, J. B., STEVENSON, J. M., and BRYANT, J. T. (1998). An Analysis of Subjective Responses to Varying a Load Centre of Gravity in a Backpack. Pages 248–251 in S. Kumar (ED.), *Advances in Industrial Ergonomics and Safety.* Amsterdam, The Netherlands: IOS Press.

JOHNSON, S. L. and LEWIS, D. M. (1989). A Psychophysical Study of Two-Person Manual Material Handling Tasks. In *Proceedings of the Human Factors Society 33rd Annual Meeting* (pp. 651–653). Santa Monica, CA: Human Factors Society.

JONES, C., MANNING, D. P., and BRUCE, M. (1995). Detecting and Eliminating Slippery Footwear. *Ergonomics 38,* 242–249.

JONES, D. F. (1972). Back Injury Research. *American Industrial Hygiene Association Journal 33,* 596–602.

JONES, R. G. (1990). Worker Independence and Output: The Hawthorne Studies Reevaluated. *American Sociological Review 55,* 176–190.

JORNA, G. C., MOHAGEG, M. F., and SNYDER, H. L. (1989).

JORNA, P. G. A. M. (1993). Heart Rate and Workload Variations in Actual and Simulated Flight. *Ergonomics 36,* 1043–1054.

JORNA, et al. (1989) Performance, Perceived Safety and Comfort of the Alternating Tread Stair. *Applied Ergonomics 20,* 26–32.

JUERGENS, H. W., AUNE, I. A., AND PIEPER, U. (1990). *International Data on Anthropometry* (Occupational Safety and Health Series No. 65). Geneva, Switzerland: International Labour Office.

KAHN, J. F. and MONOD, H. (1989). Fatigue Induced by Static Work. *Ergonomics 32,* 839–846.

KAHN, R. L and ROWE, J. (1998). *Successful Aging.* New York, NY: Pantheon.

KALEPS, I., CLAUSER, C. E., YOUNG, J. W., CHANDLER, R. F., ZEHNER, G. F., and MCCONVILLE, J. (1984). Investigation Into the Mass Distribution Properties of the Human Body and Its Segments. *Ergonomics 27,* 1225–1237.

KAMARCK, T. and JENNINGS, J. R. (1991). Behavioral Factors in Sudden Cardiac Death. *Psychological Bulletin 109,* 42–75.

KAMARCK, T. W., SHIFFMAN S. M., SMITHLINE, L, GOODIE, J. L., PATY, J. A., GNYS M., and YI-KUAN JONG, J. (1998). Effects of Task Strain, Social Conflict, and Emotional Activation on Ambulatory Cardiovascular Activity: Daily Life Consequences of Recurring Stress in Multiethnic Sample. *Health Psychology 17,* 17–29.

KANNER, A. D., COYNE, J. C., SCHAEFER, C., and LAZARUS, R. S. (1981). Comparison of Two Modes of Stress Measurement: Daily Hassles and Uplifts vs. Major Life Events. *Journal of Behavioral Medicine 4,* 1–39.

KANTOWITZ, B. H. and SORKIN, R. D. (1983). *Human Factors: Understanding People-System Relationships.* New York, NY: Wiley.

KARASEK, R. A., BAKER, D., MARXER, F., AHLBORN, A., and THEORELL, T. (1981). Job Decision Latitude, Job Demands, and Cardiovascular Disease: A Prospective Study of Swedish Men. *American Journal of Public Health, 71,* 694–705.

KARASEK, R. A., THEORELL, T., SCHWARTZ, J. E., SCHNALL, P. L., PIEPER, C. F., and MICHELA, J. L. (1988). Job Characteristics in Relation to the Prevalence of Myocardial Infarction in the U.S. Health Examination

Survey (HES) and the Health and Nutrition Examination Survey (HANES). *American Journal of Public Health 78,* 910–919.

KARHU, O., KARKONEN, R., SORVALI, P., and VEPSALAINEN, P. (1981). Observing Working Postures in Industry: Examples of OWAS Application. *Applied Ergonomics 12,* 13–17.

KARWOWSKI, W. (1988). Maximum Load Lifting Capacities of Males and Females in Teamwork. In *Proceedings of the Human Factors Society 32nd Annual Meeting* (pp. 680–682). Santa Monica, CA: Human Factors Society.

KARWOWSKI, W. (1991). Psychophysical Acceptability and Perception of Load Heaviness by Females. *Ergonomics 34,* 487–496.

KARWOWSKI, W. and MARRAS, W. S. (EDS.). (1999). *The Occupational Ergonomics Handbook.* Boca Raton, FL: CRC Press.

KARWOWSKI, W. and PONGPATANASUEGSA, N. (1988). Testing of Isometric and Isokinetic Lifting Strengths of Untrained Females in Teamwork. *Ergonomics 31,* 291–301.

KAUFMAN, J. E. and HAYNES, H. (EDS.). (1981). *IES Lighting Handbook, 1981. Application Volume.* New York, NY: Illuminating Engineering Society of North America.

KAZARIAN, L. and GRAVES, G. A. (1977). Compressive Strength Characteristics of the Human Vertebral Centrum. *Spine 2,* 1–14.

KEEGAN, J. J. (1952). Alterations to the Lumbar Curve Related to Posture and Sitting. *Journal of Bone and Joint Surgery 35,* 589–603.

KEELE, S. W. (1986). Motor Control. Chapter 30, (pp. 30.1–30.60) in K. R. Boff, L. Kaufman, and J. P. Thomas (EDS.), *Handbook of Human Perception and Human Performance.* New York, NY: Wiley.

KEESEY, R. E. (1995). A Set-Point Model of Body Weight Regulation. Chapter 9 (pp. 46–50) in Brownell, K. D.and C. G., Fairburn (EDS.), *Eating Disorders and Obesity,* New York, NY: Gilford.

KEESEY, R. E. and POWLEY, T. L. (1986). The Regulation of Body Weight. *Annual Review of Psychology, 37,* 109–133.

KEIR, P. J., BACH, J. M., ENGSTROM, J. W., and REMPEL, D. M. (1996). Carpal Tunnel Pressure: Effects of Wrist Flexion/Extension. In *Proceedings, American Society of Biomechanics 20th Annual Meeting* (pp. 169–170). Atlanta, GA: American Society of Biomechanics.

KELLER, E., BECKER, E., and STRASSER, H. (1991). An Objective Assessment of Learning Behavior with a Single-Hand Chord Keyboard for Text Inputs (in German). *Z. Arbeitswissenschaft 45,* 1–10.

KELLY, P. L. and KROEMER, K. H. E. (1990). Anthropometry of the Elderly: Status and Recommendations. *Human Factors 32,* 571–595.

KEMMERLING, P. T. (1991). *Human Factors Engineering for the Disabled and Aging.* Course Information ISE 5654, Fall Semester 1991. Blacksburg, VA: Virginia Tech.

KERMIS, M. D. (1984). *Psychology of Human Aging.* Boston, MA: Allyn and Bacon.

KEYSERLING, W. M. (1986a). Postural Analysis of the Trunk and Shoulder in Simulated Real Time. *Ergonomics 29,* 569–583.

KEYSERLING, W. M. (1986b). A Computer-Aided System to Evaluate Postural Stress in the Workplace. *American Industrial Hygiene Association Journal 47,* 641–649.

KEYSERLING, W. M. (1990). Computer-Aided Posture Analysis of the Trunk, Neck, Shoulders, and Lower Extremities. Pages 261–272 in W. Karwowski, A. M. Genaidy, and S. S. Asfour (EDS.), *Computer-Aided Ergonomics.* London, UK: Taylor & Francis.

KEYSERLING, W. M. (1998). Methods for Evaluating Postural Work Load. Chapter 11 (pp. 167–187) in W. Karwowski and G. Salvendy (EDS.), *Ergonomics In Manufacturing.* Dearborn, MI: Society of Manufacturing Engineers.

KING, K. B. (1997). Psychologic and Social Aspects of Cardiovascular Disease. *Annals of Behavioral Medicine 19,* 264–270.

KINNEY, J. M. (ED.). (1980). *Assessment of Energy Metabolism in Health and Disease.* Columbus, OH: Ross Laboratories.

KINNEY, J. S. AND HUEY, B. M. (EDS.). (1990). *Application Principles for Multicolored Displays.* Washington, D.C.: National Academy Press.

KITZES, W. F. (1996). *Forensic Safety Analysis: Investigation and Evaluation.* Chapter 3 (pp. 85–119) in Wiley Editorial Staff (ED.), *1996 Wiley Expert Witness Update—New Developments in Personal Injury Litigation.* New York, NY: Wiley.

KIRA, A. (1976). *The Bathroom.* New York, NY: Viking.

KIRK, J. and SCHNEIDER, D. A. (1990). *Physiological and Perceptual Responses to Load Carrying in Female Subjects Using Internal and External Frame Backpacks.* (Technical Report TR-91/023). Natick, MA: United States Army Natick Research, Development, and Engineering Center.

KLEINER, B. M. and DRURY, C. G. (1999). Large-Scale Regional Economic Development: Macroergonomics in Theory and Practice. *Human Factors and Ergonomics in Manufacturing 9,* 151–163.

KLESGES, R.C. (1995). Cigarette Smoking and Body Weight. Chapter 12 (pp. 61–64) in K. D. Brownell and C. G. Fairburn (EDS.), *Eating Disorders and Obesity.* New York, NY: Gilford.

KLINE, D. W. and SCIALFA, C. T. (1997). Sensory and Perceptual Functioning: Basic Research and Human Factors Implications. Chapter 3 (pp. 27–54) in A. D. Fisk and W. A. Rogers (EDS.), *Handbook of Human Factors and the Older Adult.* San Diego, CA: Academic Press.

KLOCKENBERG, E. A. (1926). *Rationalization of the Typewriter and of Its Use* (in German). Berlin, Germany: Springer.

KNAPIK, J. (1989). *Loads Carried by Soldiers: A Review of Historical, Physiological, Biomechanical, and Medical Aspects* (Technical Report T19-89). Natick, MA: United States Army Natick Research, Development, and Engineering Center.

KNAPICK, J., HARMAN, E., and REYNOLDS, K. (1996). Load Carriage Using Packs: A Critical Review of Physiological, Biomechanical and Medical Aspects. *Applied Ergonomics 27,* 207–216.

KNAPIK, J., HICKEY, C., ORTEGA, S., NAGEL, J., and DE PONTBRIAND, R. (1997). Energy Cost of Walking in Four Types of Snowshoes. In *Proceedings of the Human Factors and Ergonomics Society 41st Annual Meeting* (pp. 702–706). Santa Monica, CA: Human Factors and Ergonomics Society.

KNAUTH, P. (1996). Designing Better Shift Systems. *Applied Ergonomics 27,* 39–44.

KOGI, K. (1985). Introduction to the Problems of Shiftwork. Chapter 14 (pp. 115–184) in S. Folkard and T. H. Monk (EDS.), *Hours of Work.* Chichester, UK: Wiley.

KOGI, K. (1991). Job Content and Working Time: The Scope for Joint Change. *Ergonomics 34,* 757–773.

KONDRASKE, G. (1988). Rehabilitation Engineering: Towards a Systematic Process. *IEEE Engineering in Medicine and Biology Magazine 10,* 11–15.

KONZ, S. (ED.). (1991). Japanese Children. *Ergonomics 34,* 971.

KONZ, S. (1995). *Work Design. Industrial Ergonomics,* 5th ed. Scottsdale, AZ: Publishing Horizons.

KONZ, S. and JOHNSTON, S. (2000). *Work Design: Industrial Ergonomics* (5th ed.) Scottsdale, AZ: Holcomb Hataway.

KRAGT, H. (ED.). (1992). *Enhancing Industrial Performance: Experiences of Integrating the Human Factor.* London, UK: Taylor & Francis.

KRAMER, A. F., COYNE, J. T. and STRAYER, D. L. (1993). Cognitive Function at High Altitude. *Human Factors 35,* 329–344.

KROEMER, K. H. E. (1964). On the Effect of the Spatial Position of Keyboards on Typing Performance (in German). *Int. Zeitschrift Angewandte Physiologie einschl. Arbeitsphysiol. 20,* 240–251.

KROEMER, K. H. E. (1971). Foot, Operation of Controls. *Ergonomics 14,* 333–361.

KROEMER, K. H. E. (1972a). *Pedal Operation by the Seated Operator* (SAE Paper 72004). New York: Society of Automotive Engineers.

KROEMER, K. H. E. (1972b). Human Engineering the Keyboard. *Human Factors 14,* 51–63.

KROEMER, K. H. E. (1981). Engineering Anthropometry: Designing the Work Place to Fit the Human. In *Proceedings of the Annual Conference of the American Institute of Industrial Engineers* (pp. 119–126). Norcross, GA: AIIE.

KROEMER, K. H. E. (1982). *Development of LIFTEST, A Dynamic Technique to Assess the Individual Capability to Lift Material* (Final Report, NIOSH Contract 210-79-0041). Blacksburg, VA: Ergonomics Laboratory, IEOR Department, Virginia Tech.

KROEMER, K. H. E. (1985). Office Ergonomics: Work Station Dimensions. Chapter 18 (pp. 187–201) in D. C. Alexander and B. M. Pulat (EDS.), *Industrial Ergonomics.* Norcross, GA: Institute of Industrial Engineers.

KROEMER, K. H. E. (1986). Coupling the Hand with the Handle. *Human Factors 28,* 337–339.

KROEMER, K. H. E. (1988a). VDT Workstation Design. Chapter 23 (pp. 521–539) in M. Helander (ED.), *Handbook of Human Computer Interaction.* Amsterdam, The Netherlands: Elsevier.

KROEMER, K. H. E. (1988b). Ergonomics. Chapter 13 (pp. 183–334) in A. Plog (ED.), Fundamentals of Industrial Hygiene. Chicago: National Safety Council.

KROEMER, K. H. E. (1989a). Engineering Anthropometry. *Ergonomics 32,* 767–784.

KROEMER, K. H. E. (1989b). Cumulative Trauma Disorders: Their Recognition and Ergonomic Measures to Avoid Them. *Applied Ergonomics 20,* 274–280.

KROEMER, K. H. E. (1991a). Sitting at Work: Recording and Assessing Body Postures, Designing Furniture for Computer Workstations. Pages 93–109 in A. Mital and W. Karwowski (EDS.), *Work Space, Equipment and Tool Design.* Amsterdam, The Netherlands: Elsevier.

KROEMER, K. H. E. (1991b). Experiments with the TCK—A Keyboard with Built-in Wrist Rest and Only Eight Keys. Pages 537–542 in W. Karwowski and J. W. Yates (EDS.), *Advances in Industrial Ergonomics and Safety III,* London, UK: Taylor & Francis.

KROEMER, K. H. E. (1992). Avoiding Cumulative Trauma Disorders in Shop and Office. *American Industrial Hygiene Association Journal 53,* 596–604.

KROEMER, K. H. E. (1993a). Operation of Ternary Chorded Keys. *International Journal of Human-Computer Interaction 5,* 267–288.

KROEMER, K. H. E. (1993b). Psychology Plus Physiology Plus Biomechanics Equal Ergonomics? In *Proceedings of the 1993 International Industrial Engineering Conference.* (pp. 278–284). Norcross, GA: Institute of Industrial Engineers.

KROEMER, K. H. E. (1994). Locating the Computer Screen: How High, How Far? *Ergonomics in Design January Issue,* 40; and *October 1993 Issue,* 7–8.

KROEMER, K. H. E. (1997a). Design of the Computer Workstation. Chapter 57 (pp. 1395–1414) in M. G. Helander, T. K. Landauer, and P. V. Prabhu (EDS.), *Handbook of Human-Computer Interaction,* 2d ed. Amsterdam, NL: Elsevier.

KROEMER, K.H.E. (1997b). *Ergonomic Design of Material Handling Systems.* Boca Raton, FL: CRC Press/Lewis Publishers.

KROEMER, K. H. E. (1998). *Reviews of Publications Related to Keyboarding—In Chronological Order.* (Report, 15 February 1998). Radford, VA: K.H.E. Kroemer Ergonomics Research Institute, Inc.

KROEMER, K. H. E. (1999a). Engineering Anthropometry. Chapter 9 (pp. 139–165) in W. Karwowski and W. S. Marras (EDS.), *The Industrial Ergonomics Handbook.* Boca Raton, FL: CRC Press.

KROEMER, K. H. E. (1999b). Human Strength Evaluation. Chapter 11 (pp. 205–227) in W. Karwowski and W. S. Marras (EDS.), *The Industrial Ergonomics Handbook.* Boca Raton, FL: CRC Press.

KROEMER, K. H. E., KROEMER, H. J., and KROEMER-ELBERT, K. E. (1990). *Engineering Physiology: Bases of Ergonomics,* 2d ed. New York, NY: Van Nostrand Reinhold.

KROEMER, K. H. E., KROEMER, H. J., and KROEMER-ELBERT, K. E. (1997). *Engineering Physiology: Bases of Human Factors/Ergonomics,* 3d ed. New York: Van Nostrand Reinhold; Wiley.

KROEMER, K. H. E., MARRAS, W. S., MCGLOTHLIN, J. D., MCINTYRE, D. R., and NORDIN, M. (1990). Assessing Human Dynamic Muscle Strength. *International Journal of Industrial Ergonomics 6,* 199–210.

KROEMER, K. H. E., MCGLOTHLIN, J. D., and BOBICK, T. J. (EDS.). (1989). *Manual Material Handling: Understanding and Preventing Back Trauma.* Akron, OH: American Industrial Hygiene Association.

KROEMER, K. H. E. and ROBINETTE, J. C. (1968). *Ergonomics in the Design of Office Furniture. A Review of European Literature* (AMRL-TR 68-90). Wright-Patterson AFB, OH. Also published with shortened list of references (1969) in *International Journal of Industrial Medicine and Surgery 38,* 115–125.

KROEMER, K. H. E., SNOOK, S. H., MEADOWS, S. K., and DEUTSCH, S. (EDS.). (1988). *Ergonomic Models of Anthropometry, Human Biomechanics, and Operator-Equipment Interfaces.* Washington, DC: National Academy Press.

KRUMWIEDE, D., KONZ, S., and HINNEN, P. (1998). Floor Mat Comfort. (pp 159–162) in S. Kumar (ED.), *Advances in Occupational Ergonomics and Safety.* Amsterdam, The Netherlands: IOS Press.

KUMAR, S. (1997). The Effect of Sustained Spinal Load on Intra-Abdominal Pressure and EMG Characteristics of Trunk Muscles. *Ergonomics 40,* 1312–1334.

KUMAR, S., CHAFFIN, D. B., and REDFERN, M. (1988). Isometric and Isokinetic Back and Arm Lifting Strengths: Device and Measurement. *Biomechanics 21,* 35–44.

KUMAR, S. and MITAL, A. (EDS.). (1996). *Electromyography in Ergonomics.* London, UK: Taylor & Francis.

KUORINKA, I. and FORCIER, L. (EDS.). (1995). Work Related *Musculoskeletal Disorders: A Reference Book for Prevention.* London, UK: Taylor & Francis.

KUORINKA, I., JONSSON, B., KILBOM, A., VINTERBERG, H., BIERING-SORENSEN, F., ANDERSSON, G., and JORGENSEN, K. (1987). Standardized Nordic Questionnaires for the Analysis of Musculoskeletal Symptoms. *Applied Ergonomics 18,* 233–237.

KWALLEK, N. and LEWIS, C. M. (1990). Effects of Environmental Colour on Males and Females: A Red or White or Green Office. *Applied Ergonomics 21,* 275–278.

LANKHORST, G. J., VAN DE STADT, R. J., VOGELAAR, T. W., VAN DER KORST, J. K., and PREVO, A. J. H. (1983). The Effect of the Swedish Back School in Chronic Ideopathic Low Back Pain. *Scandanavian Journal of Rehabilitation and Medicine 15,* 141–145.

LATECK, J. C. and FOSTER, L. W. (1985). Implementation of Compressed Work Schedules: Participation and Job Redesign as Critical Factors for Employee Acceptance. Personnel *Psychology 38,* 75–92.

LATKO, W.A., ARMSTRONG, T. J., FOULKE, J. A., HERRIN, G. D., RABOURN, R. A., and ULIN, S. S. (1997). Development and Evaluation of an Observational Method for Assessing Repetition in Hand Tasks. *American Industrial Hygiene Association Journal 58,* 278–285.

LAUGHERY, K. R., JACKSON, A. S., and FONTENELLE, G. A. (1988). Isometric Strength Tests: Predicting Performance in Physically Demanding Transport Tasks. *In Proceedings of the Human Factors Society 32nd Annual Meeting* (pp. 695–699). Santa Monica, CA: Human Factors Society.

LAVENDER, S. A., CHEN, S. H., LI, Y. C., and ANDERSSON, G. B. J. (1998). Trunk Muscle Use During Pulling Tasks: Effects of a Lifting Belt and Footing Conditions. *Human Factors 40,* 159–172.

LAVENDER, S. A., OLESKE, D. M., NICHOLSEN, L., ANDERSSON, G. B. J., and HAHN, J. (1997). A Comparison of Four Methods Commonly Used to Determine Low-Back Disorder Risk in a Manufacturing Environment. In *Proceedings of the Human Factors and Ergonomics Society 41st Annual Meeting* (pp 657–660). Santa Monica, CA: Human Factors and Ergonomics Society.

LAZARUS, R. S. and COHEN, J. B. (1977). Environmental Stress, in L. Altman and J. F. Wohlwill (EDS.), *Human Behavior and the Environment: Current Theory and Research, Vol. 2.* New York, NY: Plenum.

LAZARUS, R. S. and FOLKMAN, S. (1984). *Stress, Appraisal, and Coping.* New York, NY: Springer.

LEAMON, T. B. (1994). Research to Reality: a Critical Review of the Validity of Various Criteria for the Prevention of Occupationally Induced Low Back Pain Disability. *Ergonomics 37,* 1959–1974.

LE BON, C. and FORRESTER, C. (1997). An Ergonomic Evaluation of a Patient Handling Device: The Elevate and Transfer Vehicle. *Applied Ergonomics 28,* 365–374.

LECLERCQ, S., TISSERAND, M., and SAULNIER, H. (1995). Tribological Concepts Involved in Slipping Accident Analysis. *Ergonomics 38,* 197–208.

LEE, C. H., HOSNI, Y. A., GUTHRIE, L. L., BARTH, T., and HILL, C. (1991). Design and Evaluation of a Work Seat for Overhead Tasks. Pages 555–562 in W. Karwowski and J. W. Yates (EDS.), *Advances in Industrial Ergonomics and Safety III.* London, UK: Taylor & Francis.

LEE, N., SWANSON, N., SAUTER, S. WICKSTROM, R. WAIKAR, A., and MANGUM, M. (1992). A Review of Physical Exercises Recommended for VDT Operators. *Applied Ergonomics 23,* 387–408.

LEE, Y. H. and CHEN, Y. L. (1996). An Isoinertial Predictor for Maximal Acceptable Lifting Weights of Chinese Subjects. *American Industrial Hygiene Association Journal* 57, 456–463.

LEHMANN, G. (1962). Praktische Arbeitsphysiologie, 2d ed. Stuttgart, Germany: Thieme.

LEHTO, M. and SALVENDY, G. (1995). Warnings: A Supplement Not a Substitute for Other Approaches to Safety. *Ergonomics 38,* 2155–2163.

LEVINE, D. B., ZITTER, M., and INGRAM, L. (EDS.). (1990). *Disability Statistics: An Assessment.* Committee on National Statistics, National Research Council. Washington, DC: National Academy Press.

LEWIS, J. R. (1994). Sample Sizes for Usability Studies: Additional Considerations. *Human Factors 36,* 368–378.

LEWIS, J. R., POTOSNAK, K. M. and MAGYAR, R. L. (1997). Keys and Keyboards. Chapter 54 (pp. 1285–1315) in M. G. Helander, T. K. Landauer, and P. V. Prabhu (EDS.), *Handbook of Human-Computer Interaction,* 2d ed. Amsterdam, The Netherlands: Elsevier.

LI, C., HWANG, S., and WANG M. (1990). Static Anthropometry of Civilian Chinese in Taiwan Using Computer-Analyzed Photography. *Human Factors 32,* 359–370.

LIN, L. and COHEN, H. H. (1995). Fall Accident Patterns Involved in Litigation. In *Proceedings of the Silicon Valley Ergonomics Conference & Exposition, ErgoCon 1995.* San Jose, CA: San Jose State University.

LITTERICK, I. (1981). QWERTYUIOP-Dinosaur in a Computer Age. *New Scientist January Issue,* 66–68.

LOCKE, J. C. (1983). Stretching Away from Back Pain Injury. *Occupational Health and Safety 52,* 8–13.

LOCKWOOD, A. H. (1989). Medical Problems of Musicians. *New England Journal of Medicine 320,* 221–227.

LOUHEVAARA, V., ILMARINEN, J., and OJA, P. (1985). Comparison of Three Field Methods for Measuring Oxygen Consumption. *Ergonomics 28,* 463–470.

LOHMAN, T. G., ROCHE, A. F., and MARTOREL, R. (EDS.) (1988). *Anthropometric Standardization Reference Manual.* Champaign, IL: Human Kinetics.

LOVELESS, N. E. (1962). Direction-of-Motion Stereotypes: A Review. *Ergonomics 5,* 357–383.

LOWE, B. and FREIVALDS, A. (1998). Design and Evaluation of Prototype Tool Handles. (pp. 417–420) in S. Kumar (ED.), Advances in Occupational Ergonomics and Safety. Amsterdam, The Netherlands: IOS Press.

LU, H. and AGHAZADEH, F. (1998). Modeling of Risk Factors in the VDT Workstation Systems. *Occupational Ergonomics 1,* 189–210.

LUNDERVOLD, A. (1951). Electromyographic Investigations of Position and Manner of Working in Typewriting. *Acta Physiologica Scandinavica 24,* 84–104.

LYKKEN, D. T. (1988). The Case Against the Polygraph Test. Chapter 9 (pp. 111–125) in A. Gale (ED.), *The Polygraph Test. Lies, Truth, and Science*. London, UK: Sage.

MACK, K., HASLEGRAVE, C. M., and GRAY, M. I. (1995). Usability of Manual Handling Aids for Transporting Materials. *Applied Ergonomics 26,* 353–364.

MALONE, R. L. (1991). *Posture Taxonomy.* Unpublished Master's Thesis, Department of Industrial and Systems Engineering, Blacksburg, VA: Virginia Tech.

MANDAL, A. C. (1975). Work-Chair with Tilting Seat. *Lancet,* 642–643.

MANDAL, A. C. (1982). The Correct Height of School Furniture. *Human Factors 24,* 257–269.

MARKLIN, R. W., SIMONEAU, G. G., and MONROE, J. F. (1997). The Effect of Split and Vertically-Inclined Computer Keyboards on Wrist and Forearm Posture. In *Proceedings of the Human Factors and Ergonomics Society 41st Annual Meeting* (pp. 642–646). Santa Monica, CA: Human Factors and Ergonomics Society.

MARKLIN, R. W., SIMONEAU, G. G., and MONROE, J. F. (1999). Wrist and Forearm Posture from Typing on Split and Vertically-Inclined Computer Keyboards. *Human Factors 41,* 559–569.

MARTENIUK, R. G., IVENS, C. J., and BROWN, B. E. (1996). Are There Task Specific Performance Effects for Differently Configured Numeric Key Pads? *Applied Ergonomics 27,* 321–325.

MARRAS, W. S., LAVENDER, S. A., LEURGANS, S. E., RAJALU, S. L., ALLREAD, W. G., FATHALLAH, F. A., and FERGUSON, S. A. (1993). The Role of Dynamic Three-Dimensional Trunk Motion on Occupationally-Related Low Back Disorders. *Spine 18,* 617–628.

Marras, W. S., Lavender, S. A., Leurgans, S. E., Fathallah, F. A. Ferguson, S. A. Allread, W. G., and Rajalu, S. L. (1993). Biomechanical Risk Factors for Occupationally Related Low Back Disorders. *Ergonomics 38,* 377–410.

Marras, W. S. and Mirka, G. A. (1996). Intra-Abdominal Pressure During Trunk Extension Motions. *Clinical Biomechanics 11,* 267–274.

Marras, W. S. and Rangarajulu, S. L. (1987). Trunk Force Development During Static and Dynamic Lifts. *Human Factors 29,* 19–29.

Marras, W. S. and Reilly, C. H. (1988). Networks of Internal Trunk-Loading Activities Under Controlled Trunk-Motion Conditions. *Spine 13,* 661–667.

Marras, W. S. and Schoenmarklin, R. W. (1991). Wrist Motions and CTD Risk in Industrial and Service Environments. In *Proceedings of the 11th Congress of the International Ergonomics Association* (pp. 36–38). London, UK: Taylor & Francis.

Martin, B., Rempel, D., Sudarsan, P, Dennerlein, J., Jacobson, M., Gerard, M., and Armstrong, T. (1998). Reliability and Sensitivity of Methods to Quantify Muscle Load During Keyboard Work. Pages 485–489 in S. Kumar (Ed.), Advances in Occupational Ergonomics and Safety. Amsterdam, The Netherlands: IOS Press.

Martin, B. J., Roll, J. P., and Gauthier, G. M. (1986). Inhibitory Effects of Combined Agonist and Antagonist Muscle Vibration on H-Reflex in Men. *Aviation, Space, and Environmental Medicine 57,* 681–687.

Martin, E. (1949). The Typewriter and Its Historical Development (in German). Aachen, Germany: Basten.

Martin, R. (1914). Lehrbuch der Anthropologie, 1st ed. (in German) Jena, Germany: Fischer.

Matthews, G. and Ryan, H. (1994). The Expression of the "Pre-Menstrual Syndrome" in Measures of Mood and Sustained Attention. *Ergonomics 37,* 1407–1417.

Mattila, M., Karwowski, W., and Vilkki, M. (1993). Analysis of Working Postures on Building Construction Sites: The Computerized OWAS Method. *Applied Ergonomics 24,* 405–412.

Max-Planck-Gesellschaft zur Foerderung der Wissenschaften (1968). *Tastatur.* German Patent 1,255,117.

Mazure, C. M. (1998). Life Stressors as Risk Factors in Depression. *Clinical Psychology: Science and Practice 5,* 291–313.

McAtamney, L. and Corlett, E. N. (1993). RULA: A Survey Method for the Investigation of Work-Related Upper Limb Disorders. *Applied Ergonomics 24,* 91–99.

McClelland, I. L. (1982). The Ergonomics of Toilet Seats. *Human Factors 24,* 713–725.

McConville, J. T., Churchill, T., Kaleps, I., Clauser, C. E., and Cuzzi, J. (1980). *Anthropometric Relationships of Body and Body Segment Moments of Inertia* (AFAMRL-TR-80-119). Wright-Patterson AFB, OH: Aerospace Medical Research Laboratory.

McDaniel, J. W. (1998). *Human Modeling: Yesterday, Today, and Tomorrow.* Keynote Address, First Digital Human Modeling for Design and Engineering Conference and Exposition, Dayton, OH, 28 April 1998.

McDaniel, J. W., Skandis, R. J., and Madole, S. W. (1983). *Weight Lift Capabilities of Air Force Basic Trainees* (AFAMRL-TR-83-0001). Wright-Patterson Air Force Base, OH: Air Force Aerospace Medical Research Laboratory.

McFadden, K. L. (1997). Predicting Pilot-Error Incidents of US Airline Pilots Using Logistic Regression. *Applied Ergonomics 28,* 209–212.

McGill, S. M. (1999a). Dynamic Low Back Models: Theory and Relevance in Assisting the Ergonomist to Reduce the Risk of Low Back Injury. Chapter 53 (pp. 945–965) in W. Karwowski and W. S. Marras (Eds.), *The Occupational Ergonomics Handbook.* Boca Raton, FL: CRC Press.

McGill, S. M. (1999b). Update on the Use of Back Belts in Industry: More Data—Same Conclusion. Chapter 74 (pp.1353–1358) in W. Karwowski and W. S. Marras (Eds.), *The Occupational Ergonomics Handbook.* Boca Raton, FL: CRC Press.

McGill, S. M. and Norman, R. W. (1987). Reassessment of the Role of Intra-Abdominal Pressure in Spinal Compression. *Ergonomics 30,* 1565–1588.

McGill, S. M., Norman, R. W., and Cholewicki, J. (1996). A Simple Polynomial That Predicts Low Back Compression During Complex 3-D Tasks. *Ergonomics 39,* 1107–1118.

McGill, S. M., Sharratt, M. T., and Seguin, J. P. (1995). Loads on Spinal Tissues During Simultaneous Lifting and Ventilatory Challenge. *Ergonomics 38,* 1772–1792.

McLeod, D. and Morris, A. (1996). Ergonomics Cost Benefits Case Study in a Paper Manufacturing Company. In *Proceedings of the Human Factors and Ergonomics Society 40th Annual Meeting* (pp 698–701). Santa Monica, CA: Human Factors and Ergonomics Society.

McMulkin, M. L. and Sivasubramanian, T. (1998). Hand Forces in Weaker and Stronger Members During Two-Person Lifts. Pages 316–319 in S. Kumar (Ed.), *Advances in Industrial Ergonomics and Safety.* Amsterdam, The Netherlands: IOS Press.

McVay, E. J. and Redfern, M. S. (1994). Rampway Safety: Foot Forces as a Function of Rampway Angle. *American Industrial Hygiene Association Journal 55,* 626–634.

McWright, A. (1988). Gender Differences in the Strain Responses to Job Demand. In *Proceedings of the Human Factors Society 32nd Annual Meeting* (pp. 853–856). Santa Monica, CA: Human Factors Society.

Meindl, B. A. and Freivalds, A. (1992). Shape and Placement of Faucet Handles for the Elderly. In *Proceedings of the Human Factors Society 36th Annual Meeting* (pp. 811–815). Santa Monica, CA: Human Factors Society.

Merrill, B. A. (1995). Contributions of Poor Movement Strategies to CTD/Solutions or Faulty Movements in the Human House. In *Proceedings of the Silicon Valley Ergonomics Conference & Exposition, Ergo-Con 95* (pp. 222–228). San Jose, CA: San Jose State University.

Meyer, D. E. and Kieras, D. E. (1997). A Computational Theory of Executive Cognitive Processes and Multi-task Performance: Part 1. Basic Mechanisms. *Psychological Review 104,* 3–65.

Meyers, J. M., Miles, J. A., Faucett, J., Janowitz, I., Tejeda, D. G., and Kabashima, J. N. (1997). Ergonomics in Agriculture: Workplace Priority Setting in the Nursery Industry. *American Industrial Hygiene Association Journal 58,* 121–126.

Meister, K. J. (1990). A Few Implications of an Ecological Approach to Human Factors. *Human Factors Society Bulletin 33,* 1–4.

Mellerowicz, H. and Smodlaka, V. N. (1981). *Ergometry.* Baltimore, MD: Urban and Schwarzenberg.

Mello, R. P., Damokosh, A. I., Reynolds, K. L., Witt, C. E., and Vogel, J. A. (1988). *The Physiological Determinants of Load Bearing Performance at Different March Distances* (Technical Report No. T15-88). Natick, MA: U.S. Army Research Institute of Environmental Medicine.

Metropolitan Life Foundation (Ed.). (1983). Comparison of 1959 and 1983 Metropolitan Height and Weight Tables. *Statistical Bulletin 64,* 6–7.

Michaels, S. E. (1971). QWERTY vs. Alphabetic Keyboards as a Function of Typing Skills. *Human Factors 13,* 419–426.

MIL-HDB 759, U.S. Army Missile Command (1981). *Human Factors Engineering Design for Army Material (Metric).* Philadelphia, PA: Naval Publications and Forms Center.

Miller, R. J. (1990). Pitfalls in the Conception, Manipulation, and Measurement of Visual Accommodation. *Human Factors 32,* 27–44.

Miller, R. L. (1977). Bend Your Knees. *National Safety News 115,* 57–58.

Miller, T. Q., Smith T. W., Turner, C. W., Guijarro, M. L., and Hallet, A. J. (1996). A Meta-Analytic Review of Research on Hostility and Physical Health. *Psychological Bulletin 119,* 322–348.

MIL-STD 1472F, (1999). *Department of Defense Design Criteria Standard.* AMSAM-RD-SE-TD-ST. Redstone Arsenal, AL: U.S. Army Aviation and Missile Command.

Minors, D. S. and Waterhouse, J. M. (1987). The Role of Naps in Alleviating Sleepiness During an Irregular Sleep-Wake Schedule. *Ergonomics 30,* 1261–1273.

Mirka, G. A. and Baker, A. (1996). An Investigation of the Variability in Human Performance During Sagittally Symmetric Lifting Tasks. *IIE Transactions 28,* 745–752.

MIRKA, G. A., BAKER, A., HARRISON, A., and KELAHER, D. (1998). The Interaction between Load and Coupling During Dynamic Manual Materials Handling. *Occupational Ergonomics 1,* 3–11.

MIRKA, G. A., KELAHER, D., BAKER, A., HARRISON, A., and DAVIS, J. (1997). Selective Activation of the External Oblique Musculature During Axial Torque Production. *Clinical Biomechanics 12,* 172–180.

MITCHELL, N. B. (1992). A Simple Method for Hardware and Software Evaluation. *Applied Ergonomics 23,* 277–280.

MOFFETT, J. A. K., CHASE, S. M., PORTEK, I., and ENNIS, J. R. (1986). A Controlled Perspective Study to Evaluate the Effectiveness of a Back School in the Relief of Chronic Low Back Pain. *Spine 11,* 120–121.

MOMTAHAN, K., HETU, R., and TANSLEY, B. (1993). Audibility and Identification of Auditory Alarms in the Operating Room and Intensive Care Unit. *Ergonomics 36,* 1159–1176.

MONK, T. H. (1989). Shift Worker Safety: Issues and Solutions. Pages 887–893 in A. Mital (ED.), *Advances in Industrial Ergonomics and Safety I.* Philadelphia, PA: Taylor & Francis.

MONK, T. H., FOLKARD, S., and WEDDERBURN, A. I. (1996). Maintaining Safety and High Performance on Shiftwork. *Applied Ergonomics 27,* 17–23.

MONK, T. H. and TEPAS, D. I. (1985). Shiftwork. Chapter 5 (pp. 65–84) in Cooper and Smith, M. J. (EDS.), *Job Stress and Blue Collar Work.* New York, NY: Wiley.

MONK, T. H. and WAGNER, J. A. (1989). Social Factors Can Outweigh Biological Ones in Determining Night Shift Safety. *Human Factors 31,* 721–724.

MONOD, H. and VALENTIN, M. (1979). The Predecessors of Ergonomy (in French). *Ergonomics 22,* 673–680.

MOON, R. E., VANN, R. D., and BENNETT, P. B. (1995). The Physiology of Decompression Illness. *Scientific American 273*(2), 70–77.

MOON S. D. and SAUTER, S. L. (EDS.). (1996). *Beyond Biomechanics. Psychosocial Aspects of Musculoskeletal Disorders in Office Work.* London, UK: Taylor & Francis.

MOORE, A., WELLS, R., and RANNEY, D. (1991). Quantifying Exposure in Occupational Manual Tasks with Cumulative Trauma Potential. *Ergonomics 34,* 1433.

MOORE, J. S. and GARG, A. (1997). Participatory Ergonomics in a Meat Packing Plant, Part I: Evidence of Lnag-Term Effectiveness. *American Industrial Hygiene Association Journal 58,* 127–131.

MOORE, T. G. (1974). Tactile and Kinesthetic Aspects of Pushbuttons. *Applied Ergonomics 5,* 66–71.

MORAY, N. (1988). Mental Workload Since 1979. *International Reviews of Ergonomics 2,* 123–150.

MORGAN, S. (1991). Wrist Factors Contributing to CTS Can Be Minimized, if Not Eliminated. *Occupational Health and Safety 60,* 47–51.

MORRIS, A. (1984). Program Compliance Key to Preventing Low Back Injuries. *Occupational Health and Safety 53,* 44–47.

MORRIS, N., TOMS, M., EASTHOPE, Y., and BIDDULPH, J. (1998). Mood and Cognition in Pregnant Workers. *Applied Ergonomics 29,* 377–381. (See also the related 1999 Letter to the Editor by H. Gross and H. Pattison in *Applied Ergonomics 30,* 177.)

MORROW, D. Y. and JEROME, L. (1990). The Influence of Alcohol and Aging on Radio Communication During Flight. *Aviation, Space and Environmental Medicine 61,* 12–20.

MOTOWIDLO, S. J., PACKARD, J. S., and MANNING, M. R. (1986). Occupational Stress: Its Causes and Consequences for Job Performance. *Journal Applied Psychology 71,* 618–629.

MUCKLER, F. A. and SEVEN, S. A. (1992). Selecting Performance Measure: "Objective Versus "Subjective" Measurement. *Human Factors 34,* 441–455.

MULLICK, A. (1997). Listening to People: Views about the Bathroom. In *Proceedings of the Human Factors and Ergonomics Society 41st Annual Meeting* (pp. 500–504). Santa Monica, CA: Human Factors and Ergonomics Society.

MUNSEL, A. H. (1942). *Book of Color.* Baltimore, MD: Munsell Color Book Corp.

MYUNG, R. and SMITH, J. L. (1997). The Effect of Load Carrying and Floor Contaminants on Slip and Fall Parameters. *Ergonomics 40,* 235–246.

NACHEMSON, A. (1989, May 5). Individual Factors Contributing to Low Back Pain. Presented at the Occupational Orthopaedics conference of the American Academy of Orthopaedic Surgeons, New York, NY.

NAGAI, H., HARADA, M, NAKAGAWA, M., TANAKA, T., GUNADI, B., SETIABUDI, M. L. J., UKTOLSEJA, J. L. A., and MIYATA, Y. (1996). Effects of Chicken Extract on the Recovery from Fatigue Caused by Mental Workload. *Applied Human Sciences 15,* 281–286.

NAKANISHI, Y. and NETHERY, V. (1999). Anthropometric Comparison between Japanese and Caucasian American Male University Students. *Applied Human Sciences 18,* 9–11.

NAKATANI, C., SATO, N., MATSUI, M., MATSUNAMI, M. and KUMASHIRO, M. (1997). Menstrual Cycle Effects on a VDT-Based Simulation Task: Cognitive Indices and Subjective Ratings. *Ergonomics 36,* 311–339.

NASA (1989). Man-Systems Integration Standards (Revision A). (NASA-STD 3000). Houston, TX: L.B.J. Space Center, SP 34-89-230.

NASA/WEBB (EDS.). (1978). *Anthropometric Sourcebook* (3 Volumes). (NASA Reference Publication 1024.) Houston, TX: LBJ Space Center.

NATIONAL ACADEMY OF SCIENCES (1980). Recommended Standard Procedures for the Clinical Measurement and Specification of Visual Acuity (Report of Working Group 39, Committee on Vision). Archives of *Opthalmology 41,* 103–148.

NATIONAL INSTITUTES OF HEALTH (ED.). (1985). *Health Implications of Obesity.* Conference Statement, 5(9). Washington, DC: U.S. Government Printing Office.

NATIONAL INSTITUTES OF HEALTH (ED.). (1990). *Noise and Hearing Loss.* (NIH Consensus Development Conference Statement, Vol. 8, No. 1). Bethesda, MD: National Library of Medicine, Office of Medical Applications of Research.

NATIONAL RESEARCH COUNCIL (ED.). (1979). *Odors from Stationary and Mobile Sources.* Washington, DC: National Academy Press.

NATIONAL RESEARCH COUNCIL (ED.). (1998). *Work-Related Musculoskeletal Disorders: A Review of the Evidence.* Washington, DC: National Academy Press.

NATIONAL RESEARCH COUNCIL (ED.). (1999). *Work-Related Musculoskeletal Disorders: Report, Workshop Summary, and Workshop Papers.* Washington, DC: National Academy Press.

NATIONAL RESEARCH COUNCIL COMMITTEE ON VISION (ED.). (1983). *Video Displays, Work and Vision.* Washington, DC: National Academy Press.

NAYAK, U. S. L. (1995). Elders-Led Design. *Ergonomics in Design January Issue,* 8–13.

NELSON, N. A. and SILVERSTEIN, B. A. (1998). Workplace Changes Associated with a Reduction in Musculoskeletal Symptoms in Office Workers. *Human Factors 40,* 337–350.

NEUFFER, M. B., SCHULZE, L. J. H., and CHEN, J. (1997). Body Part Discomfort Reported by Legal Secretaries and Word Processors Before and After Implementation of Mandatory Typing Breaks. In *Proceedings of the Human Factors and Ergonomics Society 41st Annual Meeting* (pp. 624–628). Santa Monica, CA: Human Factors and Ergonomics Society.

NEWELL, A. F. and GREGOR, P. (1997). Human Computer Interfaces for People with Disabilities. Chapter 35 (pp. 813–824) in M. G. Helander, T. K. Landauer, and P. V. Prabhu (EDS.), *Handbook of Human-Computer Interaction.* Amsterdam, The Netherlands: Elsevier.

NICHOLLS, J. A. and GRIEVE, D. W. (1992). Performance of Physical Tasks in Pregnancy. *Ergonomics 35,* 301–311.

NICHOLSON, A. S. (1989). A Comparative Study of Methods for Establishing Load Handling Capabilities. *Ergonomics 32,* 1125–1144.

NICHOLSON, A. S. (1991). Anthropometry and Workspace Design. Pages 3–28 in A. Mital and W. Karwowski (EDS.), *Workspace, Equipment and Tool Design.* Amsterdam, The Netherlands: Elsevier.

NICOGOSSIAN, A. E., HUNTOON, C. L., and POOL, S. L. (1989). *Space Physiology and Medicine,* 2d ed. Philadelphia, PA: Lea and Febiger.

NIOSH (1981). *Work Practices Guide for Manual Lifting.* (DHHS NIOSH Publication No. 81-122). Washington, DC: U.S. Government Printing Office.

NIOSH (Ed.). (1997). Alternative Keyboards. (DHHS NIOSH Publication 97-148). Washington, DC: U.S. Government Printing Office.

NIOSH Back Belt Working Group (Ed.). (1994). *Workplace Use of Back Belts.* (DHHS NIOSH Report 94-122). Cincinnati, OH: NIOSH.

Noble, C. E. (1978). Age, Race, and Sex in the Learning and Performance of Psycho-Motor Skills. Pages 287–378 in R. T. Osborne, C. E. Noble, and N. Weyl (Eds.), *The Biopsychology of Age, Race, and Sex.* New York, NY: Academy Press.

Nordin, M. (1991). Worker Training and Conditioning. In *Proceedings, Occupational Ergonomics: Work Related Upper Limb and Back Disorders* (not paginated). San Diego, CA: American Industrial Hygiene Association, San Diego Section.

Nordin, M., Andersson, G. B. J., and Pope, M. H. (1997). Musculoskeletal Disorders in the Workplace: Principles and Practices. St. Louis, MO: Mosby.

Norman, D. A. (1991). Cognitive Science in the Cockpit. *CSERIAC Gateway 2,* 1–6.

Norman, D. A. and Fisher, D. (1982). Why Alphabetic Keyboards Are Not Easy to Use: Keyboard Layout Doesn't Much Matter. *Human Factors 24,* 509–519.

Noyes, J. (1983a). The Qwerty Keyboard: A Review. *International Journal of Man-Machine Studies 18,* 265–281.

Noyes, J. (1983b). Chord Keyboards. *Applied Ergonomics 14,* 55–59.

Nunnely, S. A., French, J., Vanderbeek, R. D., and Stranges, S. F. (1995). Thermal Study of Anti-g Ensembles Aboard F-16 Aircraft in Hot Weather. *Aviation, Space and Environmental Medicine 66,* 309–311.

Nussbaum, M.A. and Chaffin, D. B. (1996). Development and Evaluation of a Scaleable and Deformable Geometric Model of the Human Torso. *Clinical Biomechanics 11,* 25–34.

Nygren, T. W. (1991). Psychometric Properties of Subjective Workload Measurement Techniques. *Human Factors 33,* 17–33.

Oborne, D. J. (1983). Vibration at Work. Pages 143–177 in D. J. Oborne and M. M. Gruneberg (Eds.), *The Physical Environment at Work.* New York, NY: Wiley.

Occhipinti, E. and Colombini, D. (1998). Proposed Precise Index for Assessment of Exposure to Repetitive Movements of the Upper Limbs (OCRA Index). Pages 467–470 in S. Kumar (Ed.), *Advances in Occupational Ergonomics and Safety.* Amsterdam, The Netherlands: IOS Press.

Occhipinti, E., Colombini, D., Frigo, C., Pedotti, A., and Grieco, A. (1985). Sitting Posture: Analysis of Lumbar Stresses with Upper Limbs Supported. *Ergonomics 28,* 1333–1346.

Occhipinti, E., Colombini, D., and Grieco, A. (1991). A Procedure for the Formulation of Synthetic Risk Indices in the Assessment of Fixed Working Postures. In *Proceedings, 11th Congress of the International Ergonomics Association* (pp. 3–5). London, UK: Taylor & Francis.

O'Donnell, R. D. and Egglemeyer, F. T. (1986). Workload Assessment Methodology. Pages 42.1–42.49 in K. R. Boff, L. Kaufman, and J. P. Thomas (Eds.), *Handbook of Perception and Human Performance,* Vol. II. New York, NY: Wiley.

Ogawa, I. and Arai, K. (1995). Ergonomic Considerations for Western-Style Toilets Used in Japan. In *Proceedings of the International Ergonomics Association World Conference 1995, Rio de Janeiro* (pp. 745–748). Rio de Janeiro, Brazil: Associação Brasileira de Ergonomia.

Ong, C. N. and Kogi, K. (1990). Shiftwork in Developing Countries: Current Issues and Trends. Pages 417–428 in A. J. Scott (Ed.), *Shiftwork.* Philadelphia, PA: Hanley and Belfus.

Oriet, L. P., and Dutta, S. P. (1989). Investigations of Modelling Two-Worker Lifting Teams. Pages 679–683 in A. Mital (Ed.), *Advances in Industrial Ergonomics and Safety I.* London, UK: Taylor & Francis.

Ostlere, S. J. and Gold, R. H. (1991). Osteoporosis and Bone Density Measurement Methods. *Clinical Orthopaedics 271,* 149–163.

Owens, D. A. and Leibowitz, H. W. (1983). Perceptual and Motor Consequences of Tonic Vergence. Pages 25–74 in C. Schor and K. Ciuffreda (Eds.), *Vergence Eye Movements: Basic and Clinical Aspects.* Boston, MA: Butterworths.

Owings, C. L., Chaffin, D. B., Snyder, R. G., and Norcutt, R. (1975). *Strength Characteristics of U.S. Children for Product Safety Design* (011903-F). Ann Arbor, MI: The University of Michigan.

Oxenburgh, M. S. (1997). Cost-Benefit Analysis of Ergonomics Programs. *American Industrial Hygiene Association Journal 58,* 150–156.

Panek, P. E. (1997). The Older Worker. Chapter 15 (pp. 363–394) in A. D. Fisk and W. A. Rogers (Eds.), *Handbook of Human Factors and the Older Adult.* San Diego, CA: Academic Press.

Panjabi, M. M., Goel, V., Oxland, T., Takata, K., Duranceau, J., Krag, M., and Price, M. (1992). Human Lumbar Vertebrae Quantitative Three-Dimensional Anatomy. *Spine 17,* 299–306.

Park, D. C. and Jones, T. R. (1997). Medication Adherence and Aging. Chapter 11 (pp. 257–287) in A. D. Fisk and W. A. Rogers (Eds.), *Handbook of Human Factors and the Older Adult.* San Diego, CA: Academic Press.

Parsons, H. M. (1974). What Happened at Hawthorne? *Science 18,* 922–932.

Parsons, H. M. (1990). Assembly Ergonomics in the Hawthorne Studies. In *Proceedings of the International Ergonomics Association Conference on Human Factors in Design for Manufacturability and Process Planning* (pp. 299–305). Buffalo, NY: Helander, Department of Industrial Engineering SUNYAB.

Parsons, K. C. (1995). International Heat Stress Standards: A Review. *Ergonomics 32,* 6–22.

Patkai, P. (1985). The Menstrual Cycle. Chapter 8 (pp. 87–96) in S. Folkard and T. H. Monk (Eds.), *Hours of Work.* Chichester, UK: Wiley.

Paul, J. A. and Douwes, M. (1993). Two-dimensional Photographic Posture Recording and Description: A Validity Study. *Applied Ergonomics, 24,* 83–90.

Paykel, E. S. (1997). The Interview for Recent Life Events. *Psychological Medicine 27,* 301–310.

Peloquin, A. A. (1994). *Barrier-Free Residential Design.* New York, NY: McGraw-Hill.

Peres, N. J. V. (1961). Process Work Without Strain. *Australian Factory 1,* 1–12.

Pheasant, S. (1996). *Bodyspace: Anthropometry, Ergonomics and the Design of Work,* 2d ed. London, UK: Taylor & Francis.

Perotto, A. O. (1994). *Anatomical Guide for the Electomyographer: The Limbs and Trunk.* Springfield, IL: Thomas.

Pew, R. W. (1995). Pew's Principles of Human Factors Practice. In *Proceedings of the Silicon Valley Ergonomics Conference & Exposition, ErgoCon'95* (pp. 3–4). San Jose, CA: San Jose State University.

Phillips, (2000). *Human Factors Engineering.* New York, NY: Wiley.

Plog, B. A. (Ed.) (2001, in press). *Fundamentals of Industrial Hygiene,* 5th ed. Itasca, IL: National Safety Council.

Pokorny, J. and Smith, V. C. (1986). Colorimetry and Color Discrimination. Chapter 8 (pp. 8.1–8.51) in K. R. Boff, L. Kaufman, and J. P. Thomas (Eds.), *Handbook of Perception and Human Performance.* New York, NY: Wiley.

Porter, G. (1996). Organizational Impact of Workaholism: Suggestions for Researching the Negative Outcomes of Excessive Work. *Journal of Occupational Health Psychology 1,* 70–84.

Post, D. L. (1997). Color and Human-Computer Interaction. Chapter 25 (pp. 573–615) in M. Helander, T. K. Landauer and P. Prabhu (Eds.) *Handbook of Human-Computer Interaction* (2nd ed.). Amsterdam, NL: Elsevier.

Potvin, J. R. (1997). Use of NIOSH Equation Inputs to Calculate Lumbosacral Compression Forces. *Ergonomics 40,* 650–655.

Potvin, J. R. and Bent, L. R. (1997). NIOSH Equation Distances Associated with the Liberty Mutual (Snook) Lifting Table Box Width. *Ergonomics 40,* 691–707.

Price, D. L. (1988). Effects of Alcohol and Drugs. Pages 489–551 in G. A. Peters and B. J. Peters (Eds.), *Automotive Engineering and Litigation,* Vol. 2. New York, NY: Garland.

Price, D. L., Radwan, M. A. E. and Tergou, D. E. (1986). Gender, Alcohol, Pacing and Incentive Effects on an Electronics Assembly Task. *Ergonomics 29,* 393–406.

Priel, V. Z. (1974). A Numerical Definition of Posture. *Human Factors 16,* 576–584.

PROCTOR, R.W. and VAN ZANDT, T. (1994). *Human Factors in Simple and Complex Systems.* Boston, MA: Allyn and Bacon.

PUTZ-ANDERSON, V. (1988). *Cumulative Trauma Disorders: A Manual for Musculoskeletal Diseases of the Upper Limbs.* London, UK: Taylor & Francis.

PUTZ-ANDERSON, V. and WATERS, T. (1991). *Revisions in NIOSH Guide to Manual Lifting.* Paper presented at the Conference, "A National Strategy for Occupational Musculoskeletal Injury Prevention," Ann Arbor, MI, April 1991.

RABBITT, P. (1991). Management of the Working Population. *Ergonomics 34,* 775–790.

RABBITT, P. (1997). Aging and Human Skill: A 40th Anniversary. *Ergonomics 40,* 962–981.

RADWIN, R. G. (1997). Force Dynamometers and Accelerometers. Chapter 33 (pp. 565–581) in W. Karwowski and W. S. Marras (EDS.), *The Occupational Ergonomics Handbook.* Boca Raton, FL: CRC Press.

RADWIN, R. G., BEEBE, D. J., WEBSTER, J. G., and YEN, T. Y. (1996). Instrumentation for Occupational Ergonomics. Chapter 7 (pp. 165–193) in A. Bahttacharya and J. D. McGlothlin (EDS.), *Occupational Ergonomics.* New York, NY: Dekker.

RAMSEY, J. D. (1995). Task Performance in Heat: A Review. *Ergonomics 32,* 154–165.

RASKIN, D. C. (1988). Does Science Support Polygraph Testing? Chapter 8 (pp. 96–110) in A. Gale (ED.), *The Polygraph Test. Lies, Truth, and Science.* London, UK: Sage.

RASTAKIS, L. (1998). Human-Centered Design Project Revolutionizes Air Combat. *CSERIAC Gateway 9*(1), 1–6.

REDFERN, M. S. and BIDANDA, B. (1994). Slip Resistance of the Shoe-Floor Interface Under Biomechanically Relevant Conditions. *Ergonomics 37,* 511–524.

REDGROVE, J. A. (1976). Sex Differences in Information Processing: A Theory and its Consequences. *Journal of Occupational Psychology 49,* 29–37.

REID, G. B. and NYGREN, T. E. (1988). The Subjective Workload Assessment Techniques: A Scaling Procedure for Measuring Mental Workload. In P. A. Hancock and N. Meshkati (EDS.), *Human Mental Workload* (pp. 185–218). Amsterdam, The Netherlands: Elsevier, 185–218.

REMPEL, D., SERINA, E., and KLINENBERG, E. (1997). The Effects of Keyboard Keyswitch Make Force on Applied Force and Finger Flexor Muscle Activity. *Ergonomics 40,* 800–808.

REMPEL, D., TITTIRANONDA, P., BURASTERO, S., HUDES, M., and SO, Y. (1999). Effect of Keyboard Keyswitch Design on Hand Pain. *Journal of Occupational and Environmental Medicine 41,* 111–119.

RICE, V. J. B. (ED.) (1998). *Ergonomics in Health Care and Rehabilitation.* Boston, MA: Butterworth-Heinemann.

RICE, V. J. B. (1999). Preplacement Strength Screening. Chapter 71 (pp. 1299–1319) in W. Karwowski and W. S. Marras (EDS.), *The Occupational Ergonomics Handbook.* Boca Raton, FL: CRC Press.

RIDDER, C. A. (1959). *Basic Design Measurements for Sitting* (Bulletin 616, Agricultural Experiment Station). Fayetteville, AR: University of Arkansas.

RIPPLE, P. H. (1952). Accommodative Amplitude and Direction of Gaze. *American Journal of Ophthalmology 35,* 1630–1634.

ROBINSON, G. S. and CASALI, J. G. (1995). Audibility of Reverse Alarms Under Hearing Protectors for Normal and Hearing-Impaired Listeners. *Ergonomics 38,* 2281–2299.

ROBINSON, G. S., CASALI, J. G., and LEE, S. E. (1997). Role of Driver Hearing in Commercial Motor Vehicle Operations: An Evaluation of the FHWA Hearing Requirement—Final Report (Report DTFH 61-95-C-00172; NTIS PB98-114606). Washington, DC: Federal Highway Administration.

RODGERS, S. H. (1997). Work Physiology—Fatigue and Recovery. Chapter 10 (pp. 268–297) in G. Salvendy (ED.), *Handbook of Human Factors and Ergonomics,* 2d ed. New York, NY: Wiley.

RODIN, J. (1993). Cultural and Psychosocial Determinants of Weight Concerns. *Annals of Internal Medicine 119,* 643–645.

ROEBUCK, J. A. (1995). *Anthropometric Methods. Designing to Fit the Human Body.* Santa Monica, CA: Human Factors and Ergonomics Society.

ROEBUCK, J. A., KROEMER, K. H. E., and THOMSON, W. G. (1975). *Engineering Anthropometry Methods.* New York, NY: Wiley.

ROEBUCK, J., SMITH, K., and RAGGIO, L. (1988). *Forecasting Crew Anthropometry for Shuttle and Space Station* (STS 88-0717). Downey, CA: Rockwell International.

ROETHLISBERGER, F. J. and DICKSON, W. J. (1943). *Management and the Worker.* Cambridge, MA: Harvard University Press.

ROGERS, A. S., SPENCER, M. B., STONE, B. M., and NICHOLSON, A. N. (1989). The Influence of a 1-hr Nap on Performance Overnight. *Ergonomics 32,* 1193–1205.

ROGERS, W. A., MEYER, B., WALKER, N., and FISK, A. D. (1998). Functional Limitations to Daily Living Tasks in the Aged: A Focus Group Analysis. *Human Factors 40,* 111–125.

ROHMERT, W. and RUTENFRANZ, J. (EDS.). (1983). Practical Work Physiology (in German), 3d ed. Stuttgart, Germany: Thieme.

ROSA, R. R. and COLLIGAN M. J. (1988). Long Workdays Versus Rest Days: Assessing Fatigue and Alertness with a Portable Performance Battery. *Human Factors 30,* 305–317.

ROSCOE, A. H. (1993). Heart Rate as a Psychophysical Measure for In-flight Work Assessment. *Ergonomics 36,* 1055–1062.

ROSS, L. E. and MUNDT, J. C. (1988). Multi-Attribute Modeling Analyses of the Effects of a Low Blood Alcohol Level on Pilot Performance. *Human Factors 30,* 293–304.

ROUSE, W. B. (1991). Human-Centered Design: Creating Successful Products, Systems, and Organisations. *CSERIAC Gateway 2*(4), 1–3.

ROWE, M. L. (1983). *Backache at Work.* Fairport, NY: Perinton.

RUTTER, B. G., HAAGER, J. A., DAIGLE, G. C., SMITH, S., MCFARLAND, N., and KELSEY, N. (1984). Dimensional Changes Throughout Pregnancy: A Preliminary Report. *Carle Select Papers 36,* 44–52.

SACKS, O. (1990). *The Man Who Mistook His Wife for a Hat.* New York, NY: Harper Perennial.

SAE (1973). *Subjective Rating Scale for Evaluation of Noise and Ride Comfort Characteristics Related to Motor Vehicle Tires* (SAE J1060, Recommended Practice 29.11). Detroit, MI: Society of Automotive Engineers.

SANDERS, M. S. and MCCORMICK, E. J. (1987) *Human Factors in Engineering and Design,* 6th ed. New York, NY: McGraw-Hill.

SAUTER, S. L., SCHLEIFER, L. M., and KNUTSON, S. J. (1991). Work Posture, Workstation Design, and Musculoskeletal Discomfort in a VDT Data Entry Task. *Human Factors 32,* 151–167.

SAYER, J. R., SEBOK, A. L., and SNYDER, H. L. (1990). Color-Difference Metrics: Task Performance Prediction for Multichromatic CRT Applications as Determined by Color Legibility. *Society for Information Display Digest,* 265–268.

SCERBO, M. W. (1995). Usability Testing, Chapter 4 (pp. 72–111) in J. Weimer (ED.), *Research Techniques in Human Engineering.* Englewood Cliffs, NJ: Prentice Hall.

SCHEIER, M. G. and BRIDGES, M. W. (1995). Person Variables and Health: Personality Predispositions and Acute Psychological States as Shared Determinants for Disease. *Psychosomatic Medicine 57,* 255–268.

SCHLEGEL, B., SCHLEGEL, R. E., and GILLILAND, K. (1988). Gender Differences in Criterion Task Set Performance and Subjective Rating. In *Proceedings of the Human Factors Society 32nd Annual Meeting* (pp. 848–852). Santa Monica, CA: Human Factors Society.

SCHMIDT, R. A. (1988). *Motor Control and Learning,* 2d ed. Champaign, IL: Human Kinetics.

SCHNALL, P. L., SCHWARTZ, J. E., LANDSBERGIS, P. A., WARREN, K., and PICKERING, T. G. (1992). Relation between Job Strain, Alcohol, and Ambulatory Blood Pressure. *Hypertension 19,* 488–494.

SCHNECK, D. J. (1992). *Mechanics of Muscle,* 2d ed. New York, NY: New York University Press.

SCHNEIDER, L. W., LEHMAN, R. J., PFLUG, M. A., and OWINGS, C. L. (1986). *Size and Shape of the Head and Neck from Birth to Four Years* (Final Report CPSC-C-83-1250). Ann Arbor, MI: The University of Michigan, Transportation Research Institute.

SCHNEIDER, S. and SUSI, P. (1994). Ergonomics and Construction: A Review of Potential Hazards in New Construction. *American Industrial Hygiene Association Journal 55,* 635–649.

SCHOLEY, M. and HAIR, M. (1989). Back Pain in Physiotherapists Involved in Back Care Education. *Ergonomics 32,* 179–190.

SCHULTZ, A. B. and ANDERSSON, G. B. J. (1981). Analysis of Loads on the Lumbar Spine. *Spine 6,* 76–82.

SCOTT, D. and MARCUS, S. (1991). Hand Impairment Assessment: Some Suggestions. *Applied Ergonomics 22,* 263–269.

SCRIPTURE, E. W. (1899). *The New Psychology.* New York, NY: Charles Scribner's Sons.

SEDGWICK, A. W. and GORMLEY, J. T. (1998). Training for Lifting; an Unresolved Ergonomic Issue? *Applied Ergonomics 29,* 395–398.

SEIBEL, R. (1972). Data Entry Devices and Procedures, Chapter 7 (pp. 312–344) in H. P. Van Cott and R. G. Kinkade (EDS.), *Human Engineering Guide to Equipment Design.* Washington, DC: U.S. Government Printing Office.

SEIDEL, H. (1988). Myoelectric Reactions to Ultra-Low Frequency, Whole-Body Vibration. *European Journal of Applied Physiology 57,* 558–562.

SEIDEL, H. and HEIDE, R. (1986). Long-Term Effects of Whole-Body Vibration: A Critical Survey of the Literature. *International Archives of Occupational and Environmental Health 58,* 1–26.

SELYE, H. (1978). *The Stress of Life,* rev. ed. New York, NY: McGraw-Hill.

SERINA, E. R. and REMPEL, D. M. (1996). Fingertip Pulp Response During Keystrikes. In *Proceedings of the American Society of Biomechanics 20th Annual Meeting* (pp. 237–238). Atlanta, GA: American Society of Biomechanics.

SEROW, W. J. and SLY, D. F. (1988). *The Demography of Current and Future Aging Cohorts.* Pages 42–102 in Committee on an Aging Society (ED.), America's Aging—The Social and Built Environment in an Older Society. Washington, DC: National Academy Press.

SESHAGIRI, B. (1998). Occupational Noise Exposure of Operators of Heavy Trucks. *American Industrial Hygiene Association Journal 59,* 205–213.

SESHAGIRI, B. and STEWART, B. (1992). Investigation of the Audibility of Locomotive Horns. *American Industrial Hygiene Association Journal 53,* 726–735.

SEVEN, S. A. (1989). Workload Measurement Reconsidered. *Human Factors Society Bulletin 32,* 5–7.

SHACKEL, B., CHIDSEY, K. D., and SHIPLEY, P. (1969). The Assessment of Chair Comfort. *Ergonomics 12,* 169–306.

SHAH, R. K. (1993). A Pilot Survey of the Traditional Use of the Patuka Round the Waist for the Prevention of Back Pain in Nepal. *Applied Ergonomics 24,* 337–344.

SHAPIRO, Y., PANDOLF, K. B., AVELLINI, B. A., PIMENTAL, N. A., and GOLDMAN, R. F. (1981). Heat Balance and Transfer in Men and Women Exercising in Hot-Dry and Hot-Wet Conditions. *Ergonomics 24,* 375–386.

SHARP, M., RICE, V., NINDLE, B. and WILLIAMSON, T. (1993). Maximum Lifting Capacity in Single and Mixed Gender Three-Person Teams. In *Proceedings of the Human Factors and Ergonomics Society 37th Annual Meeting* (pp. 725–729). Santa Monica, CA: Human Factors and Ergonomics Society.

SHARP, M. A. and LEGG, S. J. (1988). Effects of Psychophysical Lifting Training on Maximal Repetitive Lifting Capacity. *American Industrial Hygiene Association Journal 49,* 639–644.

SHERRICK, C. E. and CHOLEWIAK, R. W. (1986). Cutaneous Sensitivity, Chapter 12 (pp. 12.1–12.58) in K. R. Boff, L. Kaufmann, and J. P. Thomas (EDS.), *Handbook of Perception and Human Performance.* New York, NY: Wiley.

SHOLES, C. L. (1878). *Improvement in Type-Writing Machines.* Letter's Patent No. 207, 559, dated August 27, 1878; application filed March 8, 1878. United States Patent Office.

SIEKMANN, H. (1990). Recommended Maximum Temperatures for Touchable Surfaces. *Ergonomics 21,* 69–73.

SIERVOGEL, R. M., ROCHE, A. F., GUO, S., MUKHERJEE, D., and CHUMLEA, W. C. (1991). Patterns of Change in Weight/Stature from 2 to 18 Years: Findings from Long-Term Serial Data for Children in the Fels Longitudinal Growth Study. *International Journal of Obesity 15,* 479–485.

SILVERSTEIN, B. A. (1985). *The Prevalence of Upper Extremity Cumulative Trauma Disorders in Industry.* Unpublished Doctoral Dissertation, University of Michigan, Ann Arbor, MI.

SILVERSTEIN, B. A., ARMSTRONG, T. J., LONGMATE, A., and WOODY, D. (1988). Can In-plant Exercise Control Musculoskeletal Symptoms? *Journal of Occupational Medicine 30,* 922–927.

SIMONEAU, G. G., MARKLIN, R. W. and MONROE, J. F. (1999). Wrist and Forearm Postures of Users of Conventional Computer Keyboards. *Human Factors 41,* 413–424.

SINCLAIR, D. C. (1973). Mapping of Spinal Nerve Roots, in A. Jarrett (ED.), *The Physiology and Pathophysiology of the Skin,* Vol. 2, p. 349. London, UK: Academic Press.

SINGER, L. D. and GRAEFF, R. F. (1988). *A Bathroom for the Elderly* (Report on Grants 230-11-110H-150-8903081 and CAE-86-005-01). College of Architecture and Urban Studies. Blacksburg, VA: Virginia Tech.

SMITH, L., MACDONALD, I., FOLKARD, S., and TUCKER, P. (1998). Industrial Shift Systems. *Applied Ergonomics 29,* 273–280.

SMITH, M. J., KARSH, B. T., CONWAY, F. T., COHEN, W. J., JAMES, C. A., MORGAN, J. J., SANDERS, K., and ZEHEL, D. J. (1998). Effects of a Split Keyboard Design and Wrist Rest on Performance, Posture, and Comfort. *Human Factors 40,* 324–336.

SMITH, P. C. (1976). Behaviors, Results and Organizational Effectiveness: The Problem of Criteria, in M. D. Dunnette (ED.), *Handbook of Industrial and Organizational Psychology* (pp. 745–775). Chicago, IL: Rand McNally.

SMITH, R. V. and LESLIE, J. H. (EDS.) (1990). *Rehabilitation Engineering.* Boca Raton, FL: CRC Press.

SNOOK, S. H. (1987). Approaches to Preplacement. Testing and Selection of Workers. *Ergonomics 30,* 241–247.

SNOOK, S. H. (1988a). *The Control of Low Back Disability: The Role of Management.* Presented at the American Industrial Hygiene Conference, San Francisco, CA.

SNOOK, S. H. (1988b). Low Back Pain.

SNOOK, S. H. (1991). Low Back Disorders in Industry. In *Proceedings of the Human Factors Society 35th Annual Meeting* (pp. 830–833). Santa Monica, CA: Human Factors Society.

SNOOK, S. H. and CIRIELLO, V. M. (1991). The Design of Manual Handling Tasks: Revised Tables of Maximum Acceptable Weights and Forces. *Ergonomics 34,* 1197–1213.

SNYDER, H. L. (1985a). The Visual System: Capabilities and Limitations. Chapter 3 (pp. 54–69) in L. E. Tannas (ED.), *Flat-Panel Displays and CRTs.* New York, NY: Van Nostrand Reinhold.

SNYDER, H. L. (1985b). Image Quality: Measures and Visual Performance. Chapter 4 (pp. 71–90) in L. E. Tannas (ED.), *Flat Panel Displays and CRTs.* (pp. 71–90). New York, NY: Van Nostrand Reinhold.

SNYDER, R. G. (1975). Impact. Chapter 6 (pp. 221–295) in J. F. Parker and V. R. West (EDS.), *Bioastronautics Data Book.* (NASA SP-3006). Washington, DC: U.S. Government Printing Office.

SNYDER, R. G., SCHNEIDER, L. W., OWINGS, C. L., REYNOLDS, H. M., GOLOMB, D. H., and SCHORK, M. A. (1977). *Anthropometry of Infants, Children, and Youths to Age 18 for Product Safety Design* (Final Report UM-HSRI-88-17). Ann Arbor, MI: The University of Michigan, Highway Safety Research Institute.

SNYDER, R. G., SPENCER, M. L., OWINGS, C. L., and SCHNEIDER, L. W. (1975). *Physical Characteristics of Children as Related to Death and Injury for Consumer Product Safety Design* (Final Report, UM-HSRI-BI-75-5). Ann Arbor, MI: The University of Michigan, Highway Safety Research Institute.

SODERBERG, G. L. (ED.). (1992). *Selected Topics in Surface Electromyography for Use in the Occupational Setting: Expert Perspectives* (DDHS-NIOSH Publication 91-100). Washington, DC: U.S. Department of Health and Human Service.

SOLDO, B. J., and LONGINO, C. F. (1988). Social and Physical Environments for the Vulnerable Aged. In Committee on an Aging Society (ED.), *America's Aging—The Social and Built Environment in an Older Society.* Washington, DC: National Academy Press, 103–133.

SOLOMON, S. S. and KING, J. G. (1997). Fire Truck Visibility. *Ergonomics in Design April Issue,* 4–10.

SOMMERICH, C. M. and MARRAS, W. S. (1992). Temporal Patterns of Trunk Muscle Activity Throughout a Dynamic, Asymmetric Lifting Motion. *Human Factors 34,* 215–230.

SPELT, P. F. (1991). Introduction to Artificial Neural Networks for Human Factors. *Human Factors Society Bulletin 34,* 1–4.

SPENCE, A. P. (1989). *Biology of Human Aging.* Englewood Cliffs, NJ: Prentice Hall.

STAFF, K. R. (1983). *A Comparison of Range of Joint Mobility in College Females and Males.* Unpublished Master's Thesis, Texas A&M University, College Station, TX.

STAFFEL, F. (1884). On the Hygiene of Sitting (in German). *Zbl. Allgemeine Gesundheitspflege 3,* 403–421.

STAFFEL, F. (1889). *The Types of Human Postures and Their Relations to Deformations of the Spine* (in German). Wiesbaden, Germany: Bergmann.

STEENBEKKERS, L. P. A. and MOLENBROEK, J. F. M. (1990). Anthropometric Data of Children for Non-Specialist Users. *Ergonomics 33,* 421–429.

STEGEMANN, J. (1984). Physiology of Performance (in German), 3d ed. Stuttgart, Germany: Thieme.

STEKELENBURG, M. (1982). Noise at Work: Tolerable Limits and Medical Control. *American Industrial Hygiene Association Journal 43,* 402–410.

STEVENSON, J. M., ANDREW, G. M., BRYANT, J. T., GREENHORN, D. R., and THOMSON, J. M. (1989). Isoinertial Tests to Predict Lifting Performance. *Ergonomics 32,* 157–166.

STEVENSON, J. M., GREENHORN, D. R., BRYANT, J. T., DEAKIN, J. M., and SMITH, J. T. (1996). Gender Differences in Performance of a Selection Test Using the Incremental Lifting Machine. *Applied Ergonomics 27,* 45–52. Rebuttal and Reply. *Applied Ergonomics 27,* 133–137.

STOBBE, T. J. and PLUMMER, R. W. (1988). Sudden-Movement/Unexpected Loading as a Factor in Back Injuries. Pages 713–720 in F. Aghazadeh (ED.), *Trends in Ergonomics/Human Factors V.* Amsterdam, The Netherlands: Elsevier.

STOCKBRIDGE, H. C. W. (1957). *Micro-shape-Coded Knobs for Post Office Keys* (Techn. Memo No. 67). London, UK: Ministry of Supply.

STOUDT, H. W. (1981). The Anthropometry of the Elderly. *Human Factors 23,* 29–37.

STRAKER, L. M., STEVENSON, M. G., and TWOMEY, L. T. (1996). A Comparison of Risk Assessment of Single and Combination Handling Tasks: 1. Maximum Acceptable Weight Measures. *Ergonomics 39,* 121–140.

STRAKER, L. M., STEVENSON, M. G., TWOMEY, L. T., and SMITH, L. M. (1997). A Comparison of Risk Assessment of Single and Combination Handling Tasks: 3. Biomechanical Measures. *Ergonomics 40,* 708–728.

STUART-BUTTLE, C., MARRAS, W. S., and KIM, J. Y. (1993). The Influence of Anti-Fatigue Mats on Back and Leg Fatigue. In *Proceedings of the Human Factors and Ergonomics Society 37th Annual Meeting* (pp. 769–773). Santa Barbara, CA: Human Factors and Ergonomics Society.

SWANSON, N. G., GALINSKY, T. L., COLE, L. L., PAN, C. S., and SAUTER, S. L. (1997). The Impact of Keyboard Design on Comfort and Productivity in a Text-Entry Task. *Applied Ergonomics 28,* 9–16.

SWETS, J. A. and BJORK, R. A. (1990). Enhancing Human Performance: An Evaluation of "New Age" Techniques Considered by the U.S. Army. *Psychological Science 1,* 85–96.

TACHE, J. and SELYE, H. (1986). On Stress and Coping Mechanisms. Pages 3–24 in C. D. Spielberger and I. G. Sarason (EDS.), *Stress and Anxiety: A Sourcebook of Theory and Research,* Vol. 10. Washington, DC: Hemisphere.

TATTERSALL, A. J. and FOORD, P. S. (1996). An Experimental Evaluation of Instantaneous Self-Assessment as a Measure of Workload. *Ergonomics 39;* 740–748.

TAYYARI, F. and SMITH, J. L. (1997). *Occupational Ergonomics—Principles and Applications.* New York, NY: Chapman and Hall.

TENNANT, C. and ANDREWS, G. (1976). A Scale to Measure the Stress of Life Events. *Australian and New Zealand Journal of Psychiatry 10,* 27–32.

TEPAS, D. I. (1985). Flextime, Compressed Work Weeks and Other Alternative Work Schedules. Chapter 13 (pp. 147–164) in S. Folkard and T. H. Monk (EDS.), *Hours of Work.* Chichester, UK: Wiley.

TEPAS, D. I., PALEY, M. J., and POPKIN, S. M. (1997). Work Schedules and Sustained Performance. Chapter 32 (pp. 1021–1058) in G. Salvendy (ED.), Handbook of Human Factors and Ergonomics, 2d ed. New York, NY: Wiley.

TEVES, M. A., WRIGHT, J. E., and VOGEL, J. A. (1985). *Performance on Selected Candidate Screening Test Procedures Before and After Army Basic and Advanced Individual Training* (Technical Report No. T13/85). Natick, MA: United States Army Research Institute of Environmental Medicine.

THOMPSON, J. K. (1990). *Body Image Disturbance: Assessment and Treatment.* Elmsford, NY. Pergamon.

THOMPSON, J. K. (1995). Assessment of Body Image. Chapter 4 (pp. 119–148) in Allison, D. B. (ED.), *Handbook of Assessment Methods for Eating Behaviors and Weight-Related Problems.* Thousand Oaks, CA: Sage.

THOUMIE, P., DRAPE, J. L., AYMARD, C., and BEDOISEAU, M. (1998). Effects of a Lumbar Support on Spine Posture and Motion Assessed by Electrogoniometer and Continuous Recording. *Clinical Biomechanics 13,* 18–26.

TICHAUER, E. R. (1973). *The Biomechanical Basis of Ergonomics.* New York, NY: Wiley.

TITTIRANONDA, P., BURASTERO, S., and REMPEL, D. (1999). Risk Factors for Musculoskeletal Disorders Among Keyboard Users. *Occupational Medicine: State of the Art Reviews 14,* 17–38.

TOUGAS, G. and NORDIN, M. C. (1987). Seat Features Recommendations for Workstations. *Applied Ergonomics 18,* 207–210.

TROY, B. S., COOPER, R. A., ROBERTSON, R. N., and GREY, T. L. (1995). Analysis of Work Postures of Manual Wheelchair Users. In *Proceedings of the Silicon Valley Ergonomics Conference & Exposition, Ergo-Con'95* (pp. 227–234). San Jose, CA: San Jose State University.

TSUNAWAKE, N., TAHARA, Y., YUKAWA, K., KATSUURA, T., HARADA A. IWANAGA, K., and KIKUCHI, Y. (1995). Changes in Body Shape of Young Individuals from the Aspect of Adult Physique Model by Factor Analysis. *Applied Human Science 14,* 227–234.

TUREK, F. W. (1989). Effects of Stimulated Physical Activity on the Circadian Pacemaker of Vertebrates. Pages 135–147 in S. Daan and E. Gwinner (EDS.), *Biological Clocks and Environmental Time.* New York, NY: Guilford.

TURVILLE, K. L., PSIHOGIOS, J. P., and MIRKA, G. A. (1998). The Effects of Video Display Terminal Height on the Operator: A Comparison of the 15° and 40° Recommendations. *Applied Ergonomics 29,* 239–246.

TYRRELL, R. A. and LEIBOVITZ, H. W. (1990). The Relation of Vergence Effort to Reports of Visual Fatigue Following Prolonged Near Work. *Human Factors 32,* 341–357.

U.S. ARMY (1981). MIL-HDBK 759. *Human Factors Engineering Design for Army Material (Metric).* Redstone Arsenal, AL: U.S. Army Missile Command.

VAIL, G. J. (1988). A Gender Profile: U.S. General Aviation Pilot-Error Accidents. In *Proceedings of the Human Factors 32nd Annual Meeting* (pp. 862–866). Santa Monica, CA: Human Factors Society.

VAN COTT, H. P. and KINKADE, R. G. (EDS.). (1972). Human Engineering Guide to Equipment Design, rev. ed. Washington, DC: U.S. Government Printing Office.

VAN DEN HEEVER, D. J. and ROETS, F. J. (1996). Noise Exposure of Truck Drivers: A Comparative Study. *American Industrial Hygiene Association Journal 57,* 564–566.

VAN DER GRINTEN, M. P. (1991). Test-Retest Reliability of a Practical Method for Measuring Body Part Discomfort. In *Proceedings, 11th Congress of the International Ergonomics Association* (pp. 54–56). London, UK: Taylor & Francis.

VAN DER GRINTEN, M. P. and SMITT, P. (1992). Development of a Practical Method for Measuring Body Part Discomfort. Pages 311–318 in S. Kumar (ED.), *Advances in Industrial Ergonomics and Safety IV.* London, UK: Taylor & Francis.

VIDULICH, M. A. and TSANG, P. S. (1985). Assessing Subjective Workload Assessment: A Comparison of SWAT and the NASA-Bipolar Methods. In *Proceedings of the Human Factors Society 29th Annual Meeting* (pp. 71–75). Santa Monica, CA: Human Factors Society.

VILLANUEVA, M. B. G., SOTOYAMA, M., JONAI, H., TAKEUCHI, Y., and SAITO, S. (1996). Adjustments of Posture and Viewing Parameters of the Eye to Changes in the Screen Height of the Visual Display Terminal. *Ergonomics 39,* 933–945.

VINCENTE, K. J. and HARWOOD, K. (1990). A Few Implications of an Ecological Approach to Human Factors. *Human Factors Society Bulletin 33,* 1–4.

VINK, P., DOUWES, M., and VAN WOENSEL, W. (1994). Evaluation of a Sitting Aid: The Back-Up. *Applied Ergonomics 25,* 170–176.

VINK, P. and KOMPIER, M. A. J. (1997). Improving Office Work: a Participatory Ergonomics Experiment in a Naturalistic Setting. *Ergonomics 40,* 435–449.

VIRZI, R. A. (1992). Refining the Test Phase of Usability Evaluation: How Many Subjects Is Enough ? *Human Factors 34,* 457–468.

VIVOLI, G., BERGOMI, M., ROVESTI, S., CARROZZI, G., and VEZZOSI, A. (1993). Biochemical and Haemodynamic Indicators of Stress in Truck Drivers. *Ergonomics 36,* 1089–1097.

VON NOORDEN, G. K. (1985). *Binocular Vision and Ocular Motility,* 3d ed. St. Louis, MO: Mosby.

VREDENBURGH, A. G., SAIFER, A. G. and COHEN, H. H. (1995). You're Doing to Do What with That Thing? *Ergonomics in Design April Issue, 16–20.*

WAGNER, C. (1974). Determination of Finger Flexibility. *European Journal of Applied Physiology 32,* 259–278.

WADDEN, T.A., FOSTER, G.D., and LETIZIA, K.A. (1994). One-Year Behavioral Treatment of Obesity: Comparison of Moderate and Severe Caloric Restriction and the Effects of Weight Maintenance Therapy. *Journal of Consulting and Clinical Psychology 62,* 165–171.

WAIKAR, A., LEE, K., AGHAZADEH, F., and PARKS, C. (1991). Evaluating Lifting Tasks Using Subjective and Biomechanical Estimates of Stress at the Lower Back. *Ergonomics 34,* 33–47.

WANG, J., PIERSON, R. N., JR., and HEYMSFIELD, S. B. (1992). The Five Level Model: A New Approach to Organizing Body Composition Research. *American Journal of Clinical Nutrition 56,* 19–28.

WATERS, T. R. and PUTZ-ANDERSON, V. (1998). Assessment of Manual Lifting—The NIOSH Approach. Chapter 13 (pp. 205–241) in W. Karwowski and G. Salvendy (EDS.), *Ergonomics in Manufactoring.* Dearborn MI: Society of Manufacturing Engineers.

WATERS, T. R. and PUTZ-ANDERSON, V. (1999). Revised NIOSH Lifting Equation. Chapter 57 (pp. 1037–1061) in W. Karwowski and W. S. Marras (EDS.), *The Occupational Ergonomics Handbook.* Boca Raton, FL: CRC Press.

WARGO, M. J. (1947). Human Operator Response Speed, Frequency and Flexibility: A Review and Analysis. *Human Factors 9,* 221–238.

WASSERMAN, D. E. (1987). *Human Aspects of Occupational Vibrations.* Amsterdam, The Netherlands: Elsevier.

WASSERMAN, D. E. (1998). Vibration-Induced Cumulative Trauma Disorders. Chapter 21 (pp. 369–379) in W. Karwowski and G. Salvendy (EDS.), *Ergonomics in Manufacturing.* Dearborn. MI: Society of Manufacturing Engineers.

WASSERMAN, D. E., PHILLIPS, C. A., and PETROFSKY, J. S. (1986). The Potential Therapeutic Effects of Segmental Vibration on Osteoporosis. In *Proceedings of the 12th International Congress on Acoustics.* Paper F2-1.

WEBB, P. (1985) *Human Calorimeters.* New York, NY: Praeger.

WEBB, R. D. G. and TACK, D. W. (1988). Ergonomics, Human Rights and Placement Tests for Physically Demanding Work. Pages 751–758 in F. Aghazadeh (ED.), *Trends in Ergonomics/Human Factors V.* (pp. 751–758). Amsterdam, The Netherlands: Elsevier.

WEIMER, J. (1995). Developing a Research Project. Chapter 2 (pp. 20–48) in J. Weimer (ED.), *Research Techniques in Human Engineering.* Englewood Cliffs, NJ: Prentice Hall.

WEINSIER, R. L. (1995). Clinical Assessment of Obese Patients. Chapter 82 (pp. 463–468) in K. D. Brownell and C. G. Fairburn (EDS.), *Eating Disorders and Obesity,* New York, NY: Gilford.

WHINNERY, J. E. and MURRAY, D. G. (1990). Enhancing Tolerance to Acceleration (g_z) Stress: the "Hook" Maneuver. (NADC-90088-60). Warminster, PA: Naval Air Development Center.

WICKENS, C. D. and CARSWELL, C. M. (1997). Information Processing. Pages 89–129 in G. Salvendy (ED.), *Handbook of Human Factors and Ergonomics,* 2d ed. New York, NY: Wiley.

WICKENS, C. D., GORDON, S. E., and LIU, Y. (1998). *An Introduction to Human Factors Engineering.* New York, NY: Longman.

WIERWILLE, W. W. (1992). Visual and Manual Demands of In-Car Controls and Displays. Pages 299–300 in J. B. Peacock and W. Karwowski (EDS.), *Automotive Ergonomics: Human Factors in the Design and Use of the Automobile.* London, UK: Taylor & Francis.

WIERWILLE, W. W. and CASALI, J. G. (1983). A Validated Rating Scale for Global Mental Workload. In *Proceedings of the Human Factors Society 27th Annual Meeting* (pp. 129–133). Santa Monica, CA: Human Factors Society.

WIERWILLE, W. W. and EGGEMEYER, F. T. (1993). Recommendations for Mental Workload Measurement in a Test and Evaluation Environment. *Human Factors 35,* 263–281.

WIERWILLE, W. W., RAHIMI, M., and CASALI, J. (1985). Evaluation of 16 Measures of Mental Workload Using a Simulated Flight Task Emphasizing Mediational Activity. *Human Factors 27,* 489–502.

WIKER, S. F., CHAFFIN, D. B., and LANGOLF, G. D. (1989). Shoulder Posture and Localized Muscle Fatigue and Discomfort. *Ergonomics 32,* 211–237.

WIKTORIN, C., SELIN, K., EKENVALL, L., KILBOM, A., and ALFREDSSON, L. (1996). Evaluation of Perceived and Self-Reported Manual Forces Exerted in Manual Materials Handling. *Applied Ergonomics 27,* 231–239.

WILLIAMS, R. B., BAREFOOT, J. C., and SHEKELLE, R. B. (1985). The Health Consequences of Hostility. Pages 173–185 in M. A. Chesney and R. H. Rosenman (EDS.), *Anger and Hostility in Cardiovascular and Behavioral Disorders.* Washington, DC: Hemisphere.

WILLIAMS, V. H. (1988). *Isometric Forces Transmitted by the Fingers: Data Collection Using a Standardized Protocol.* Unpublished Master's Thesis, IEOR Dept., Virginia Tech, Blacksburg, VA.

WILLIAMSON, D. F. (1995). Prevalence and Demographics of Obesity. Chapter 68 (pp. 391–395) in K. D. Brownell and C. G. Fairburn (EDS.), *Eating Disorders and Obesity.* New York, NY: Guilford.

WILLIGES, R. C. (1995). Review of Experimental Design. Chapter 3 (pp. 49–71) in J. Weimer (ED.), *Research Techniques in Human Engineering.* Englewood Cliffs, NJ: Prentice Hall.

WILSON, G. F. and EGGEMEYER, F. T. (1994). Mental Workload Assessment. *CSERIAC Gateway 5*(2), 1–3.

WILSON, J. R. and CORLETT, E. N. (1990). *Evaluation of Human Work.* London, UK: Taylor & Francis.

WILSON, J. R. and CORLETT, E. N. (EDS.). (1995). *Evaluation of Human Work. A Practical Ergonomics Methodology.* London, UK: Taylor & Francis.

WING, A. M. (1983). Crossman, E. AND Goodeve. M. (1963): Twenty Years On. *Quarterly Journal of Experimental Psychology 35,* 245–249.

WINKELMOLEN, G. H. M., LANDEWEERD, J. A., and DROST, M. R. (1994). An Evaluation of Patient Lifting Techniques. *Ergonomics 37,* 921–932.

WINTER, D. A. (1990). *Biomechanics and Motor Control of Human Behavior,* 2d ed. New York, NY: Wiley.

WOFFORD, J. C. and DALY, P. S. (1997). A Cognitive-Affective Approach to Understanding Individual Differences in Stress Propensity and Resultant Strain. *Journal of Occupational Health Psychology 2,* 134–147.

WOOD, E. H., CODE, C. F., and BALDES, E. J. (1990). Partial Supination Versus GZ Protection. *Aviation, Space, and Environmental Medicine 61,* 850–858.

WOODSON, W. E. (1981). *Human Factors Design Handbook.* New York, NY: McGraw-Hill.

WOODSON, W. E. and CONOVER, D. W. (1964). *Human Engineering Guide for Equipment Designers,* 2d ed. Berkeley, CA: University of California Press.

WOODSON, W. E., TILLMAN, B., and TILLMAN, P. (1991). *Human Factors Design Handbook,* 2d ed. New York. NY: McGraw-Hill.

WRIGHT, W. C. (1993). *Diseases of Workers. Translation of Bernadino Ramazzini's 1713 De Morbis Articum.* Thunder Bay, ON: OH&S Press.

WYSZECKI, G. (1986). Color Appearance. Pages 9.1–9.57 in K. R. Boff, L. Kaufman, and J. P. Thomas (EDS.), *Handbook of Perception and Human Performance.* New York, NY: Wiley.

XING, L. (1988). Furniture Took Ages to Grow Legs. *China Daily October 31 Issue,* 5.

YETTRAM, A. L. and JACKMAN, J. (1981). Equilibrium Analysis for the Forces in the Human Spinal Column and its Musculature. *Spine 5,* 402–411.

YOULE, A. (ED.). (1990). *The Thermal Environment* (Technical Guide No. 8, British Occupational Hygiene Association). Leeds, UK: Science Reviews, Ltd. and H and H Scientific Consultants, Ltd.

YOUNG, S. L., BROGMUS, G. E., and BEZVERKHNY, I. (1997). The Forces Required to Pull Loads Up Stairs with Different Handtrucks. In *Proceedings of the Human Factors and Ergonomics Society 41st Annual Meeting* (pp. 697–701). Santa Monica, CA: Human Factors and Ergonomics Society.

YU, T., ROHT, L. H., WISE, R. A., KILIAN, D. J., and WEIR, F. W. (1984). Low-Back Pain in Industry: An Old Problem Revisited. *Journal of Occupational Medicine 26,* 517–524.

YUN, M. H. and FREIVALDS, A. (1995). Analysis of Hand Tool Grips. In *Proceedings of the Human Factors and Ergonomics Society 39th Annual Meeting* (pp. 553–557). Santa Monica, CA: Human Factors and Ergonomics Society.

ZACHARKOW, D. (1988). *Posture: Sitting, Standing, Chair Design and Exercise.* Springfield, IL: Thomas.

ZELLERS, K. K. and HALLBECK, M. S. (1995). The Effects of Gender, Wrist and Forearm Position on Maximum Isometric Power Grasp Force, Wrist Force, and Their Interactions. In *Proceedings of the Human Factors and Ergonomics Society 39th Annual Meeting* (pp. 543–547). Santa Monica, CA: Human Factors and Ergonomics Society.

ZIOBRO, E. (1991). A Contactless Method for Measuring Postural Strain. Pages 421–425 in W. Karwowski and J. W. Yates (EDS.), *Advances in Industrial Ergonomics and Safety III.* London, UK: Taylor & Francis.

Glossary of Terms

The following list contains descriptions of terms used in this book. In other context, terms may have additional meanings—see, for example, admittance.

A

***abduct*:** to pivot away from the body or one of its parts; opposed to adduct.

***absolute threshold*:** the amount of energy necessary to just detect a stimulus. Also called the detection threshold or merely threshold. Often taken as the value at which some specified probability of detection exists, such as 0.50 or 0.75. See differential threshold.

***absorbed power*:** the power dissipated in a mechanical system as a result of an applied force.

***accelerance*:** the ratio of acceleration to force during simple harmonic motion. (Accelerance is the inverse of apparent mass or effective mass.)

***acceleration*:** a vector quantity that specifies the rate of change of velocity.

***accommodation*:** in vision, an adjustment of the curvature (thickness) of the lens of the eye (which changes the eye's focal length) to bring the image of an object into proper focus on the retina.

***achromatic*:** lacking chroma or color.

***acoustics*:** the science of sound.

***acromion*:** the most lateral point of the lateral edge of the scapula. Acromial height is usually equated with shoulder height.

***action, activation (of muscle)*:** see contraction.

***activator*:** control(ler)

***acuity*:** the visual ability to discriminate fine detail. (Visual acuity is often expressed in terms of the angle subtended, or physical size, of the smallest recognizable object.) See contrast sensitivity; fovea; Landolt C; Snellen chart.

***adaptation*:** a change in sensitivity to the intensity or quality of stimulation over time. May be an increase in sensitivity (such as in dark adaptation) or a decrease in sensitivity (such as with continued exposure to a constant stimulus).

***adduct*:** to pivot toward the body or one of its parts; opposed to abduct.

***ADL*:** activities of daily living.

***admittance*:** the ratio of displacement to force in a vibrating mechanical system. The displacement and force may be taken at the same point or at different points in the same system during simple harmonic motion.

***ADP*:** adeno diphosphate. See mitochondrion.

***afferent*:** conducting nerve impulses from the sense organs to the central nervous system. See efferent.

***agonist*:** a contracting muscle, usually counteracted by an opposing muscle; same as protagonist. See antagonist.

***amplitude*:** the maximum value of a quantity (often a sinusoidal quantity). (Also called peak amplitude or single amplitude.)

***anastomosis*:** a connection between two blood vessels.

***anatomy*:** the science of the structure of the body.

***angular frequency*:** the product of the frequency of a sinusoidal quantity with 2π (in radians per second, rad s^{-1}).

***ankylosis*:** stiffening or immobility of a joint as a result of disease, together with a fibrous or bony union across the joint.

***annulus fibrosus*:** a ring of fiber that forms the circumference of an intervertebral disc.

***antagonistic muscles*:** pairs of muscles that act in opposition to each other, such as extensor and flexor muscles.

***anterior*:** at the front of the body; opposed to posterior.

***anthropometric dummy*:** a physical model constructed to reproduce the dimensions and ranges of movement of the human body for a specified percentile of an identified population.

***anthropometry*:** the measurement of human physical form (eg, height or reach).

***apparent mass*:** the ratio of force to acceleration during simple harmonic motion.

***arithmetic mean*:** see mean.

***arousal*:** a general term indicating the extent of readiness of the body. See autonomic nervous system.

***arteriole*:** a minute artery with a muscular wall.

***artery*:** a blood vessel carrying blood away from the heart. See vein.

***arthritis*:** a degenerated condition of joints characterized by inflammation. See osteoarthritis.

***arthrosis*:** degeneration of a joint.

***articular*:** relating to a joint.

***ATP*:** adenosine triphosphate. See mitochondrion.

***atrophy*:** wasting of body tissue.

***audio frequency*:** any normally audible frequency of a sound wave (eg, 20–20,000 Hz).

***audio-frequency sound*:** sound with a spectrum lying mainly at audio frequencies.

***audiogravic illusion*:** the apparent tilt of the body related to an auditory stimulus when an observer is exposed either to linear translational acceleration or deceleration or to centripetal acceleration. See oculogravic illusion.

***audiogyral illusion*:** apparent movement of a source of sound that is stationary with respect to an observer when the observer is rotated. See oculogyral illusion.

***autonomic nervous system*:** a principal part of the nervous system that is mainly self-regulating. It includes the sympathetic nervous system (involved in arousal) and the parasympathetic nervous system (involved in digestion and the maintenance of functions that protect the body). See nervous system.

***average*:** see mean.

***axilla*:** the armpit.

***axis*:** one of three mutually perpendicular straight lines passing through the origin of a coordinate system.

***axon*:** a nerve fiber that normally conducts nervous impulses away from a nerve cell and its dendrites to another nerve cell or to effector cells. Axons vary from about 0.25 to more than 100 μm in width and can be many centimeters long. Axons more than 0.5 μm thick are usually covered by a myelin sheath. See neuron; synapse.

B

***ballistic*:** relating to the parabolic motion characteristic of a projectile; in this book, the initial phase of such a motion.

***bandpass filter*:** a filter that has a single transmission band extending from a (nonzero) lower cutoff frequency to a (finite) upper cutoff frequency. Used to remove unwanted low- and high-frequency oscillations.

***bandwidth*:** the (nominal) bandwidth of a filter is the difference between the nominal upper and lower cut-off frequencies. The difference may be expressed either in hertz, or as a percentage of the pass-band center frequency, or as the interval between the upper and lower nominal cutoff frequencies, in octaves.

***bel*:** a unit of level when the base of the logarithm is 10. Used with quantities proportional to power. See decibel.

***biceps brachii*:** the large muscle on the anterior surface of the upper arm, connecting the scapula with the radius.

***biceps femoris*:** a large posterior muscle of the thigh.

***biodynamics*:** the science of (human) body motions in terms of a mechanical system; a subdivision of biomechanics.

***biomechanics*:** The science of the (human) body in terms of a mechanical system.

***biopsy*:** removal of tissue from a living person for diagnostic purposes.

***bite bar*:** a bar, plate, or mount held between the teeth of the upper and lower jaws so that there is no relative motion between the bar and the skull (head).

***blood pressure*:** pressure exerted on the artery walls by the blood. Systolic blood pressure is the maximal pressure produced by contraction of the left ventricle; diastolic blood pressure is the minimal pressure when the heart muscle is relaxed. Typically, the pressures are 120 mm Hg (systolic) and 80 mm Hg (diastolic), but values vary within and between individuals and depend on body posture, exercise, emotion, etc.

***brachialis*:** the muscle connecting the shoulder and upper arm with the ulna, crossing both shoulder and elbow joints.

***brain*:** that part of the central nervous system enclosed within the skull. See nervous system; electroencephalogram.

***breakthrough*:** the error appearing in the output of a system controlled by a human operator as a result of vibration transmitted via the limb of the operator. (Also called vibration-related error or feed-through.)

***brightness*:** individual perception of the intensity of a given visual stimulus.

***bump*:** a mild form of mechanical shock.

***bursitis*:** inflammation of a bursa. Bursae provide slippery membranes for tendons and ligaments to slide over bones. Bursitis may arise from infection or repeated pressure, friction, or other trauma.

***buttock protrusion*:** the maximal posterior protrusion of the right buttock.

C

***CAD*:** computer-aided design.

***cadaver*:** a dead body.

***candela (cd)*:** the SI base unit of the luminous intensity of a light source.

***canthus*:** juncture of the eyelids.

***capillary*:** minute blood vessel. Capillaries connect the arterial system to the venous system.

***cardiovascular*:** relating to the heart and the blood vessels or the circulation of blood.

***carpal bones*:** the scaphoid, lunate, triquetral, pisiform, trapezium, trapezoid, capitate, and hamate bones located in the proximal section of the hand.

***carpal tunnel syndrome*:** pressure of swollen tissue on the median nerve at the point where it passes through the carpal tunnel formed by the carpal bones and the transverse ligament.

***cartilage*:** the tough, smooth, white tissue (ie, gristle) covering the moving surfaces of joints.

***case-control study*:** in epidemiology, a study in which individual cases of disease are matched with individuals from a control group. See cohort.

Celsius**:** scale of temperature on which the melting point of ice is 0° and the boiling point of water is 100°.

center frequency**:** the geometric mean of the nominal cutoff frequencies of a passband filter (ie, center frequency = $(f_1 f_2)^{-2}$, where f_1 and f_2 are the cutoff frequencies).

centigrade**:** former name for the Celsius scale of temperature.

central nervous system**:** part of the nervous system consisting of the brain and the spinal cord, often abbreviated as CNS. See cerebellum; cerebral cortex; cerebrum; nervous system.

central tendency**:** a measure of a distribution of values as given by a typical value, such as the mean, geometric mean, median, or mode.

cerebellum**:** the posterior part of the brain consisting of two connected lateral hemispheres. The cerebellum is involved in muscle coordination and the maintenance of body equilibrium. See central nervous system.

cerebral cortex**:** the outermost layer of the cerebrum, consisting of "gray matter." The cerebral cortex is involved in sensory functions, motor control, and some higher processes. See central nervous system.

cerebrum**:** the largest part of the brain, consisting of two interconnected hemispheres with an inner core of "white matter" (myelinated fibers) and an outer covering of "gray matter" (unmyelinated fibers—the cerebral cortex). The cerebrum is involved in higher mental activities, including the interpretation of sensory signals, reasoning, thinking, decision making, and the control of voluntary actions. See nervous system; central nervous system; myelin.

cervical**:** relating to the neck. See vertebra.

cervicale**:** the posterior protrusion of the spinal column at the base of the neck caused by the tip of the spinous process of the seventh cervical vertebra.

chroma**:** attribute of color perception that determines to what degree a chromatic color differs from an achromatic color of the same lightness.

chromatic or achromatic induction**:** a visual process that occurs when two color stimuli are viewed side by side and each stimulus alters the color perception of the other. The effect of chromatic or achromatic induction is usually called simultaneous contrast or spatial contrast.

chronic**:** of long duration with slow progress; not acute.

clavicle**:** the collarbone linking the scapular with the sternum.

clinical**:** related to the consideration of symptoms of a disease as opposed to the scientific observation of changes.

closed-loop system**:** a system whose output is fed back and used to manipulate the input quantity. Closed-loop systems are frequently called feedback control systems. See open-loop system.

closed system**:** a system isolated from all inputs from outside. See open system.

CNS**:** central nervous system. See there.

cochlea**:** coiled, snail-shaped structure in the inner ear. The cochlea has a spiral canal making two and a half turns around a central core. The canal consists of three parallel fluid-filled subcanals separated from each other by Reissner's membrane and the basilar membrane. See Corti organ.

cognitive task**:** a task involving mental processes.

cohort**:** a defined population group. The term is used in epidemiology to refer to a group followed prospectively in a cohort study.

collimation**:** The process of making rays of light parallel (ie, as if they came from an object at infinite distance). Collimation may be achieved with an optical device such as a lens or mirror.

color illuminant**:** color perceived as belonging to an area that emits light as primary source.

color object**:** color perceived as belonging to an object.

color, related**:** color seen in direct relation to other colors in the field of view.

color surface**:** color perceived as belonging to a surface from which the light appears to be reflected or radiated.

color, unrelated**:** color perceived to belong to an area in isolation from other colors.

***compliance*:** the reciprocal of stiffness.

***concentric (muscle effort)*:** shortening of a muscle against a resistance.

***conditioned response*:** a response that anticipates an event. May refer to the response to a normally neutral stimulus that has been elicited by "conditioning."

***condyle*:** articular prominence of a bone.

***cone*:** receptor of light in the retina. Cones, which mediate color vision, are the only receptors located in the fovea. The density of cones decreases toward the periphery of the retina. Cones function in daylight (photopic) viewing conditions. See rod.

***confidence interval*:** for a normal distribution of measured data points, the range within which one value will lie with a given degree of probability.

***confounded*:** said of the results of an experiment when the observed effect may be caused by more than one variable.

***contraction*:** literally, "pulling together" the Z lines (that delineate the length of a sarcomere), caused by the sliding action of actin and myosin filaments. Contraction develops muscle tension only if external resistance against the shortening exists. Note that during an "isometric contraction" no change in sarcomere length occurs (a contradiction in terms!) and that in an "eccentric contraction" the sarcomere is actually lengthened. Hence, it is often better to use the term action, effort, or exertion instead of contraction. See muscle.

***contralateral*:** relating to the opposite side. See ipsilateral.

***contrast*:** the difference in luminances I_{max} and I_{min} between two areas. Suitable expressions for contrast include $(I_{max} = I_{min})/(I_{max} + I_{min})$. Note that several nonequivalent equations are used to describe contrast in the literature.

***contrast sensitivity*:** a measure of the ability of the visual system to detect variations in contrast. Contrast sensitivity is dependent on the angular size subtended by test objects. It is the dependent variable used in measuring visual performance as a function of spatial frequency. See acuity.

***control group*:** a group of persons with characteristics similar to those of an experimental group that is not exposed to the conditions under investigation.

***coordinate system*:** orthogonal system of axes to indicate the directions of motions or forces. By convention, biodynamic coordinate systems follow the "right-hand rule": The positive directions of the *x*-, *y*-, and *z*-axes are designated by the directions of the first finger, the second finger, and the thumb, respectively, of the right hand.

***Coriolis force*:** force that arises when a body that is undergoing rotation also undergoes translation. The force arises from a cross-coupling of the motions; the resultant motion is called Coriolis acceleration.

***Coriolis oculogyral illusion*:** the visual and postural illusion that occurs within a rotating environment when one moves the head about an axis orthogonal to the axis of rotation. A small light at the center of rotation will appear to rise diagonally forward to the right if the head is tilted to the left while undergoing clockwise rotation.

***cornea*:** transparent outer layer of the anterior portion of the eye.

***coronal plane*:** same as frontal plane.

***corpuscle*:** an encapsulated nerve ending. See Golgi–Mazzoni corpuscle; Golgi (tendon) organ; Krause's end bulb; Meissner's corpuscle; Merkel's disc; Pacinian corpuscle; Ruffini ending.

***correlation*:** in statistics, a relationship between two (or more) variables such that increases in one variable are accompanied by systematic increases or decreases in the other variable. See correlation coefficient; multiple correlation; product–moment correlation; rank-order correlation.

***correlation coefficient*:** a number that expresses the degree and direction of a relationship between two (or more) variables. A correlation coefficient of −1.00 indicates perfect negative correlation; +1.00 indicates perfect positive correlation; 0 indicates no correlation. See product–moment correlation.

***cortex*:** see cerebral cortex.

***Corti organ*:** complex structure in the cochlea of the inner ear associated with hearing. Includes the basilar membrane and attached hair cells; sounds impinging on the tympanic membrane (ie, the eardrum) cause a vibration of the ossicles (three small bones), which is then transmitted to the basilar membrane, causing movement and firing of the hair cells. See cochlea; ear.

***Coulomb damping*:** the dissipation of energy that occurs when a particle in a vibrating system is resisted by a force, the magnitude of which is a constant that is independent of displacement and velocity and the direction of which is opposite to the direction of the velocity of the particle. Also called dry friction damping.

***covariance*:** a measure of the extent to which changes in one variable are accompanied by changes in a second variable.

***CPU*:** Central processing unit of a computer.

***crest factor*:** the ratio of the peak value to the root-mean-square value of a quantity over a specified time interval.

***criterion*:** a characteristic by which something may be judged (plural: criteria).

***cross-axis coupling*:** the motion occurring in one axis due to excitation in an orthogonal axis.

***cross-sectional study*:** an investigation in which a group of persons is studied at one point in time. (Also called prevalence study.) See longitudinal study.

***CRT*:** cathode-ray tube.

***cumulative injury*:** see repetitive strain injury.

***cutaneous sensory system*:** sensory system with receptors in or near the skin. (Receptors include those responsible for the senses of touch, pressure, warmth, cold, pain, taste, and smell.) See proprioception; corpuscle.

***cycle*:** the complete range of values through which a periodic function passes before repeating itself.

D

***dactylion*:** the tip of the middle finger.

***damper*:** a device for reducing the magnitude of a shock or vibration by dissipation of energy. Also called an absorber.

***damping*:** the dissipation of energy with time or distance. See Coulomb damping.

***danger*:** an unreasonable and unacceptable combination of hazard and risk. See hazard, risk.

***dB*:** see decibel.

***dB(A)*:** A-weighted sound pressure level.

***decay time*:** for a shock pulse, the interval of time required for the value of the pulse to drop from some specified large fraction of the maximal value to some specified small fraction thereof.

***decibel*:** one-tenth of a bel. A level in decibels is 10 times the logarithm, to the base 10, of the ratio of powerlike quantities. The power level in decibels is $L_p = 10 \log_{10} (p^2/p_0^2) = 20 \log_{10}(p/p_0)$.

***degeneration*:** deterioration of physical tissues in the condition or function. Degeneration may be caused by injury or disease processes. The function may be impaired or destroyed.

***degrees of freedom*:** the number of independent variables in an estimate of some quantity.

***dendrite*:** a branching treelike process involved in the reception of neural impulses from neurons and receptors. Dendrites are rarely more than 1.5 mm in length. See axon.

***dependent variable*:** any variable, the values of which are the result of changes in an independent variable. See also experimental design.

***dermis*:** layers of skin between the epidermis and subcutaneous tissue. The dermis contains blood and lymphatic vessels, nerves and nerve endings, and glands and hair follicles.

***deterministic function*:** a function whose value can be predicted from knowledge of its behavior at previous times.

***diastole*:** the dilation and relaxation of the heart during which its cavities fill with blood. See systole.

***differential threshold*:** the difference in value of two stimuli that is just sufficient to be detected. Also called difference threshold. See absolute threshold.

***digits*:** the fingers (not the thumb) and toes.

***Diopter*:** a unit of measurement of the refractive power of lenses, equal to the reciprocal of the focal depth in meters. See Snellen chart.

***disability*:** loss of function or ability.

***disease*:** an interruption, cessation, or disorder of a body function. A disease is identified by at least two of the following: an identifiable cause, a recognizable group of signs and symptoms, and consistent anatomical alterations. See sign, syndrome.

***displacement*:** a vector quantity that specifies the change of position of a body with respect to a reference position.

***displacement control(ler)*:** a control, such as a hand control, that is operated by moving the control. The output is proportional to the displacement of the control. See also force, isometric, isotonic control.

***distal*:** the end of a body segment farthest from the center of the body. Opposed to proximal.

***dominant frequency*:** a frequency at which a maximum value occurs in a spectral-density curve.

***dorsal*:** at the back; also, at the top of hand or foot; opposed to palmar, plantar; and ventral.

***dorsum*:** the back or posterior surface of a part.

***double amplitude*:** see peak-to-peak value.

***double-blind*:** said of an experimental procedure in which neither the subject nor the person administering the experiment knows its crucial aspects (often, whether a substance being administered is a placebo or a drug).

***driving-point impedance*:** the ratio of force to velocity taken at the same point in a mechanical system during simple harmonic motion.

***dummy*:** test device or physical model simulating one or more of the anthropometric or dynamic characteristics of the human or animal body for experimental or test purposes.

***duodenum*:** the first part of the small intestine following the stomach.

***dynamic*:** relating to the existence of motion due to forces; not static or at equilibrium. See mechanics.

***dynamics*:** a subdiscipline of mechanics that deals with forces and bodies in motion.

***dynamic stiffness*:** (i) the ratio of change of force to change of displacement under dynamic conditions; (ii) the ratio of force to displacement during simple harmonic motion.

E

***ear*:** the organ of hearing. The ear consists of: (i) the external ear (the pinna and ear canal to the tympanic membrane or eardrum); (ii) the middle ear (the cavity beyond the eardrum and including the ossicles—that is, the malleus, incus, and stapes); and (iii), the inner ear (the cochlea, semicircular canals, utricle, and saccule). See Corti organ; vestibular system.

***ear–eye plane*:** a standard plane for orientation of the head, especially for defining the angle of the line of sight. The ear–eye plane is established by a line passing through the right auditory meatus (ear hole) and the right external canthus (juncture of the eyelids), with both eyes on the same level. The ear–eye plane is about 11 degrees more inclined than the Frankfurt plane. See frontal plane, medial plane, transverse plane.

***eccentric (muscle effort)*:** lengthening of a resisting muscle by external force.

***ECG*:** see electrocardiogram.

***EEG*:** see electroencephalogram.

***effective mass*:** see apparent mass.

***effector*:** a muscle or gland, at the terminal end of an efferent nerve, that produces an intended response.

***efferent*:** conducting of nerve impulses from the central nervous system toward the peripheral nervous system (eg, to the muscles). See afferent.

***effort*:** see contraction.

***EKG*:** see electrocardiogram.

***electrocardiogram (ECG; or German, elektokardiogramm EKG)*:** the graphical presentation of the time-varying electrical potential produced by heart muscle.

***electroencephalogram (EEG)*:** the graphical presentation of the time-varying electrical potentials of the brain.

***electromyogram (EMG)*:** the graphical presentation of the time-varying electrical potentials of muscle(s).

***electrooculogram (EOG)*:** the graphical presentation of the time-varying electrical potentials of the eye muscle(s).

***embolism*:** obstruction or occlusion of a vessel.

***EMG*:** see electromyogram.

***empirical*:** based on observation and experiment rather than theory.

***endocrine*:** relating to internal secretion of hormones by glands. The hormones are usually distributed through the body via the bloodstream.

***endolymph*:** a clear fluid in the membranous semicircular canals of the vestibular system. The flow of endolymph relative to the canals causes movement of the cupula and the firing of hair cells consistent with rotation of the head.

***end organ*:** term used to refer to a sensory receptor. See corpuscle.

***EOG*:** see electrooculogram.

***epicondyle*:** the bony eminence at the distal end of the humerus, radius, and femur.

***epidemiology*:** the study of the prevalence and spread of disease in a community.

***epidermis*:** the outer layer of the skin.

***equivalent comfort contour*:** outline of the magnitudes of vibration, expressed as a function of frequency, that produce broadly similar degrees of discomfort.

***equivalent continuous A-weighted sound pressure level*:** value of the A-weighted sound pressure level of a continuous, steady sound that, within a specified time interval *T*, has the same mean-square sound pressure as a sound under consideration.

***ergonomics*:** the discipline that examines human characteristics for the appropriate design of the living and work environment. Also called human factors or human engineering, mostly in the United States.

***etiology*:** a part of medical science concerned with the causes of disease. Also spelled aetiology.

***exertion*:** see contraction.

***experimental design*:** the plan of an experimental investigation. Experimental designs are aimed at maximizing the sensitivity and ease of interpretation of experimental measures (taken on the "dependent" variable as a result of manipulating the "independent variable"), while minimizing the influence of unwanted effects. The design is usually associated with some statistical measures used to test for significant results.

***extend*:** to move adjacent body segments so that the angle between them is increased, as when the bent leg is straightened; opposed to flex.

***extensor*:** a muscle whose contraction tends to straighten a limb or other body part. An extensor muscle is the antagonist of the corresponding flexor muscle.

***external*:** away from the central long axis of the body; the outer portion of a body segment.

***exteroception*:** the perception of information about the world outside the body. Involves the cutaneous sensory system and the senses of vision, hearing, taste, and smell. See proprioception; interoception.

***exteroceptor*:** any sensory receptor mediating exteroception.

***extrinsic variable*:** a variable external to a subject, the properties of which are not directly under the control of the subject.

***ex vivo*:** outside the living body. See in vivo.

***eye*:** see cone; fovea; retina; rod.

***eye movements*:** see saccade.

F

***facet*:** a small, smooth area on a bone, especially the superior and inferior articular facets of a vertebra. These form two small posterior joints that connect adjoining vertebrae (in addition to the larger anterior disc joint) and restrict the relative movement between the vertebrae.

***false negative*:** a term applied to results at one extreme of a distribution that are beyond the level required for members of the group. The criterion falsely suggests that the members are not part of the group. See false positive.

***false positive*:** a term applied to results at one extreme of a distribution that are beyond the level required for inclusion in some other group. The criterion falsely suggests that the members are part of the other group. See false negative.

***farsightedness*:** an error of refraction, when accommodation is relaxed, in which the parallel rays of light from an object at infinity are brought to focus behind the retina. Also called hyperopia or hypermetropia.

***fatigue*:** weariness resulting from bodily (or mental) exertion, reducing the performance. The condition can be removed and performance restored by rest.

***Fechner's law*:** law stating that the psychological sensation P produced by a physical stimulus of magnitude I increases in proportion to the logarithm of the intensity of the stimulus. Mathematically, $P = k \log I$. See Stevens' (power) law; Weber's law.

***feedback*:** input of some information to a system about the output of the system.

***feedback control system*:** see closed-loop system.

***feed-through*:** see breakthrough.

***femur*:** the thigh bone.

***fiber*:** see muscle.

***fibril*:** see muscle.

***filament*:** see muscle.

***filter*:** a device for separating oscillations on the basis of their frequency. A filter, which may be mechanical, acoustical, electrical, analog, or digital, attenuates oscillations at some frequencies more than those at other frequencies.

***Fitts' law*:** the motion time MT to a target depends on the length D of the path and the size W of the target (ie, on the difficulty of the activity). Mathematically, $MT = a + b \log_2(2D/W)$.

***flex*:** to move a joint in such a direction as to bring together the two parts it connects, as when the elbow is being bent; opposed to extend.

***flexor*:** a muscle whose contraction tends to flex. A flexor muscle is the antagonist of the corresponding extensor muscle.

***force control(ler)*:** a control, such as a hand control, that is operated by applying force to the control which does not move. The output is proportional to the force applied to the control. See also displacement, isometric, isotonic control.

***fovea*:** the fovea (centralis) of the eye is a small pit in the center of the retina that contains cones, but no rods. When a person looks directly at a point, its image falls on the fovea, which covers an angle of about 2 degrees. Visual acuity is normally greatest for images on the fovea.

***Frankfurt plane (occasionally spelled, falsely, Frankfort)*:** a standard plane for orientation of the head. The plane is established by a line passing through the right tragion (approximately the ear hole) and the lowest point of the right orbit (eye socket), with both eyes on the same level. See ear–eye plane; frontal plane; medial plane; transverse plane.

***free dynamic*:** in the context of muscle strength, an experimental condition in which neither displacement, nor its time derivatives, nor force is manipulated as an independent variable.

***frequency*:** the reciprocal of the fundamental period. Frequency is expressed in hertz (Hz), one unit of which is equals to one cycle per second.

***frequency weighting*:** a transfer function used to modify a signal according to a required dependence on vibration frequency.

***frontal plane*:** any plane at a right angle to the medial (midsagittal) plane dividing the body into anterior and posterior portions. Also called the coronal plane. See ear–eye plane; Frankfurt plane; medial plane; transverse plane.

***function disorder*:** any disorder for which there is no known organic pathology.

G

g, *acceleration due to gravity*: the acceleration produced by the force of gravity at the surface of the earth, 9.80665 ms^{-2}. Denoted *G* if used as a unit.

***gain*:** the amplification provided by a system.

***galvanic skin response*:** a measure of the electrical characteristics of the skin, usually the electrical resistance, which varies with emotional tension and other factors. A polygraph, or lie detector, uses the galvanic skin response as an indicator of a person's emotional state.

***Gaussian distribution*:** see normal distribution.

***geometric mean*:** the geometric mean of two quantities is the square root of the product of the two quantities.

***glabella*:** the most anterior point of the forehead between the brow ridges in the midsagittal plane.

***glenoid cavity*:** the depression in the scapula below the acromion into which fits the head of the humerus, forming the "shoulder joint."

***gluteal furrow*:** the furrow at the juncture of the buttock and the thigh.

***Golgi–Mazzoni corpuscle*:** an encapsulated nerve ending found in the dermis of the skin.

***Golgi (tendon) organ*:** a proprioceptive sensory nerve whose ending is mainly embedded within fibers of tendons at their junction with muscles. The Golgi organ is a "stretch" receptor activated by a change in tension between muscles and bone. See corpuscle.

H

***habituation*:** reduction in human response to a stimulus as a result of cumulative exposure to the stimulus. Habituation is often assumed to involve activity of the central nervous system. See adaptation.

***haemoglobin*:** the red respiratory protein of erythrocytes. Haemoglobin consists of globin, which is a protein, and haem, which is an iron compound. Also spelled hemoglobin.

***hand-transmitted vibration (or shock)*:** mechanical vibration (or shock) applied or transmitted directly to the hand–arm system, usually through the hand or its digits. A common example is the vibration from the handles of power tools.

***haptic*:** relating to the combined feeling from sensors in the skin (tactile sensation) and motion and displacement sensors (kinesthetic, proprioceptive sensation).

***harmonic*:** a sinusoidal oscillation whose frequency is an integral multiple of the fundamental frequency. The second harmonic is twice the frequency of the fundamental, etc.

***hazard*:** condition or circumstance that presents a potential for injury. See danger, risk.

***head-up display*:** a fixed display presenting information in the normal line of sight of the observer, such as in a windshield of an automobile or aircraft.

***helmet-mounted display*:** a visual display mounted on a helmet and moving with the head.

***hemoglobin*:** see haemoglobin.

***hernia*:** protrusion of a part of the body through tissues normally containing it.

***herniated disc*:** see prolapsed disc.

***high-pass filter*:** a filter that has a single transmission band extending from some critical cutoff frequency of interest. Used to remove unwanted oscillations at low frequencies.

***H-point*:** the pivot point of the torso and thigh on two-dimensional and three-dimensional devices used to measure a vehicle's seating accommodations.

***hue*:** attribute of color perception that uses color names and combinations thereof, such as bluish purple, yellowish green. The four unique hues are red, green, yellow, and blue, none of which contains any of the others.

***human analog (or surrogate)*:** in biodynamics, a body that has biodynamic properties representative of those of the human body. See dummy.

***human engineering*:** see ergonomics.

***human-factors (engineering)*:** see ergonomics.

***humerus*:** the bone of the upper arm.

***hyper*:** over, above, exceeding, excessive.

***hypertrophy*:** increase in bulk of some tissue in the body without an increase in the number of cells.

***Hypo*:** under, below, diminished.

***hypothesis*:** a supposition made as a starting point for reasoning or investigation, possibly without an assumption as to its truth. Hypotheses should be formulated such that they are amenable to testing by, for example, empirical research. See theory.

I, J

***IADL*:** instrumental activities of daily living.

***iliac crest*:** the superior rim of the pelvic bone.

***ilium*:** see pelvis.

***impact*:** a single collision of one mass with a second mass.

***impedance*:** the ratio of a harmonic excitation of a system to its response (in consistent units), both of whose arguments increase linearly with time at the same rate. See mechanical impedance.

***impulse*:** the integral with respect to time of a force taken over a time during which the force is applied; often simply the product of the force and the time during which the force is applied.

***independent variable*:** any variable whose values are independent of changes in the values of other variables; in an experiment, the variable that is manipulated so that its effect on one or more dependent variables can be observed. See also experimental design.

***inferior*:** below, lower, in relation to another structure.

***infra*:** below, low, under, inferior, after.

***infrasonic frequency*:** any frequency lower than normally audible sound waves (eg, below 20 Hz).

***infrasound*:** sound with a spectrum lying mainly at infrasonic frequencies.

***injury*:** damage to body tissue due to trauma.

***inner ear*:** the innermost part of the ear, containing the vestibular system and cochlea. Also called the labyrinth, a name derived from the complex mazelike structure within the bone.

***innervation*:** (i) provision of an organ with nerves; (ii) nervous stimulation or activation of an organ.

***inseam*:** a term used in tailoring to indicate the inside length of a sleeve or trouser leg. The inseam is measured on the medial side of the arm or leg.

***intelligence quotient (IQ)*:** a measure of intelligence defined as 100 times the mental age divided by the chronological age. (Average intelligence is therefore 100.)

***inter*:** between, among.

***internal*:** near the central long axis of the body; inside a body segment.

***interoception*:** the perception of information about the interior functioning of the body. Interoception involves receptors in the viscera, glands, and blood vessels and the senses of hunger, thirst, nausea, etc. See proprioception, exteroception.

***interoceptor*:** any sensory receptor mediating interoception.

***intersubject variability*:** variability among subjects. See intrasubject variability.

***interval scale*:** a scale in which differences between intervals have quantitative significance; although the intervals between values are significant, the absolute values are not. Allowable arithmetic operations are addition and subtraction, but not multiplication or division; thus, one cannot use proportions. The valid statistical operations on an interval scale include calculation of the mean value and of the standard deviation. See nominal; ordinal; ratio scale.

***intervertebral disc*:** flexible pad between the main bodies of vertebrae. Intervertebral discs have a soft jellylike core, the nucleus pulposus, enclosed by hard fibrous tissue that is attached to the bodies of the adjacent vertebrae. Discs make up about 25 percent of the length of the vertebral column. See annulus fibrosus; prolapsed disc; facet; vertebra.

***in toto*:** as a whole.

***intra*:** within, on the inside.

***intrasubject variability*:** variability within a subject. See intersubject variability.

***in vitro*:** in an artificial environment, such as in a test tube.

***in vivo*:** in the living body.

***involuntary action*:** an action not under voluntary control. Either a reflex action, a very well-learned action, or a normal action that proceeds from the genetic makeup of the body.

***ipsilateral*:** relating to the same side. See contralateral.

***ischaemia*:** see ischemia.

***ischemia*:** a deficiency in blood supply to a part of the body, often as a result of the narrowing or complete blockage of an artery or arteriole.

***ischial tuberosity*:** bony projection at the lower and posterior section of the coxal bone (also called the hip-bone) that is part of the pelvic girdle. When a person is seated on a flat, rigid surface, the contact pressure is usually greatest beneath the ischial tuberosities.

***ischium*:** the dorsal and posterior of the three principal bones that compose either half of the pelvis.

***iso*:** equal, the same.

***isoacceleration*:** a condition in which the acceleration is kept constant.

***isoforce*:** a condition in which the muscular force (tension) is constant. The term is synonymous with isotonic.

***isoinertial*:** a condition in which muscle moves a constant mass.

***isojerk*:** a condition in which the time derivative of acceleration, or jerk, is kept constant.

***isokinematic*:** a condition in which the velocity of muscle shortening (or lengthening) is constant. Depending on the given biomechanical conditions, this may or may not coincide with a constant angular speed of a body segment about its articulation.

***isometric*:** a condition in which the length of the muscle remains constant.

***isometric contraction*:** a muscular effort that causes tension, but no movement.

***isometric control*:** a control, such as a hand control, that can be operated by the isometric contraction of muscles. The control does not move, but responds to the applied force or torque. See also displacement, isometric, isotonic control.

***isotonic*:** a condition in which muscle tension (force) is kept constant, see isoforce. (In the past, isotonic was occasionally erroneously applied to any condition other than isometric.)

***isotonic control*:** a control, such as a hand control, that can be operated by the isotonic contraction of muscles. The control moves, but offers the same resistance to the applied force or torque at all positions. See also isometric control.

***jerk*:** a vector quantity that specifies the rate of change of acceleration.

***JND*:** just-noticeable difference. See there.

***joule (J)*:** the work done by a force of 1 N acting over a distance of 1 m. The joule is the SI unit of work. It is equivalent to 10^7 ergs and is the energy dissipated by 1 W in 1 s.

***just noticeable difference (JND)*:** the difference between two stimuli that is just barely perceived under some defined condition. See Weber's law.

K

***kinematics*:** a subdivision of dynamics that deals with the motions of bodies, but not the forces that cause them.

***kinesthetic*:** the feeling of motion, especially from the muscles, tendons, and joints. Also called somatosensory. See haptic, tactile.

***kinetics*:** a subdivision of dynamics that deals with forces applied to masses.

***knuckle*:** the joint formed by the meeting of a finger bone (phalanx) with a palm bone (metacarpal).

***Krause's end bulb*:** encapsulated corpuscle in parts of the skin, mouth, and other locations generally believed to be sensitive to cold.

***kyphosis*:** backward (convex) curvature of the spine. See lordosis, scoliosis.

L

***labyrinth*:** see inner ear.

***Landolt C*:** the letter C presented at various orientations as a test of visual acuity. (Also called Landolt ring.) See acuity, Snellen chart.

***latency*:** the period of apparent inactivity between the time a stimulus is presented and the moment that a specified response occurs.

***lateral*:** near or toward the side of the body; opposed to medial.

***LBP*:** low-back pain.

***ligament*:** fibrous band between two bones at a joint. Ligaments are flexible, but inelastic.

***lightness*:** visual perception of how much more or less light a stimulus emits in comparison to a "white" stimulus also contained in the field of view.

***linear function*:** one variable is said to be a linear function of another if changes in the first variable are directly proportional to changes in the second.

***linear system*:** a system in which the response is proportional to the magnitude of the excitation.

***line of sight*:** the line connecting the point of fixation in the visual field with the center of the entrance pupil and the center of the fovea in the fixating eye.

***longitudinal study*:** an investigation in which a person, or a group of persons, is studied over the course of time. See cross-sectional study.

***longitudinal wave*:** a wave in which the direction of displacement caused by the wave motion is in the direction of propagation of the wave.

***lordosis*:** forward (concave) curvature of the lumbar spine. See kyphosis; scoliosis.

***low-pass filter*:** a filter that has a single transmission band extending from zero frequency up to a finite frequency. (Used to remove unwanted high-frequency oscillations from a signal.)

***lumbago*:** a term used to describe pain in the middle and lower back. May be associated with sciatica, some combination of pulled muscles and sprained ligaments, or an unknown cause. See vertebral column; sciatica; prolapsed disc.

***lumbar*:** part of the back and sides between the ribs and the pelvis. See vertebra.

***lumen*:** the opening (cross section) of a blood vessel.

***lumen (lm)*:** the luminous flux emitted from a point source of uniform intensity of 1 candela into a unit solid angle. The lumen is the SI unit by which luminous flux is evaluated in terms of its visual effect.

***lux*:** the SI unit of illumination equal to 1 lm m^{-2}.

M

***macula*:** the maculae acousticae are the two patches of sensory cells in the utricle and saccule of the vestibular system of the inner ear. See vestibular system; otoliths.

***macula lutea*:** a small yellowish area (near the center of the retina of the eye) that contains the fovea where visual perception is more acute. See eye, fovea, retina.

***magnitude*:** measure of largeness, size, or importance.

***malinger*:** to feign an illness, possibly for compensation, sympathy, or the avoidance of work.

***malleolus*:** a rounded bony projection in the ankle region. The tibia has such a protrusion on the medial side, the fibula one on the lateral side.

***masking*:** a phenomenon in which the perception of a normally detectable stimulus is impeded by a second stimulus. The second stimulus may be presented at a different point in the same sensory system, causing "lateral masking." Alternatively, it may be presented at a different time, either before, in "forward masking," or after, in "backward masking," of the test stimulus.

***matched groups*:** an experimental procedure in which groups of subjects are matched for variables that may affect results, but that are not studied in the experiment.

***maximal value*:** the value of a function when any small change in the independent variable causes a decrease in the value of the function.

***maximum*:** the maximal value.

***mean*:** (i) the mean value of a number of discrete quantities is the algebraic sum of the quantities, divided by the number of quantities; (ii) the mean value of a function $x(t)$ over an interval between t_1 and t_2 is given by $x = \int_{t_1}^{t_2} x(t)\,dt$. The mean is a measure of central tendency. Also called the arithmetic mean or average. See geometric mean; median; mode.

***meatus*:** the juncture of the eyelids.

***mechanical advantage*:** in a biomechanical context, the lever arm (moment arm, leverage) at which a muscle works around a bony articulation.

***mechanical impedance*:** the ratio of force to velocity, where the force and velocity, may be taken at the same or different points in the same system during simple harmonic motion.

***mechanical system*:** an aggregate of matter comprising a defined configuration of mass, stiffness, and damping.

***mechanics*:** the branch of physics that deals with forces applied to bodies and their ensuing motions.

***mechanoreceptor*:** a receptor that responds to mechanical pressures, such as a touch receptor in the skin. See cutaneous sensory system; corpuscle.

***medial*:** near, toward, or in the midline of the body; opposed to lateral.

***medial plane*:** a vertical plane through the middle of the body (in the anatomical position) that divides it into right and left halves. Also called the midsagittal plane. See ear–eye plane; Frankfurt plane; frontal plane; transverse plane.

***median*:** the value, in a series of observed values, that has exactly as many observed values above it as below it; that is, the middle score in a distribution of scores ordered according to their magnitude. The median is a primary measure of central tendency for skewed distributions. See mean; mode.

***Meissner's corpuscle*:** specialized encapsulated nerve ending found in the papillae of the skin of the hand and foot, the front of the forearm, the lips, and the tip of the tongue. Meissner's corpuscle responds to pressure and vibration within a small area and is adapts rapidly. It is often a principal means of sensing vibration of the skin in the range from 5 to 60 Hz, depending on various conditions. Meissner's corpuscles are located in glabrous skin and are orientated perpendicular to the skin surface. Also called a tactile corpuscle. See Pacinian corpuscle.

***Merkel's disc*:** specialized free nerve ending (corpuscle) found immediately below the epidermis and around the ends of some hair follicles. The corpuscle is believed to respond to pressure applied perpendicularly to the skin at frequencies below about 5 Hz. See Ruffini ending.

***metacarpal*:** one of the long bones of the hand between the carpus and the phalanges.

***metacarpus*:** collectively, the five bones of the hand between the carpus and the phalanges.

***midsagittal plane*:** same as medial plane. See Frankfurt plane; frontal plane; transverse plane.

***minute (of arc) (')*:** unit of angular measure equal to 1/60 of a degree.

***mitochondrion (plural: mitochondria)*:** a small granular body floating inside a cell's cytoplasm. Contains a maze of tightly folded membranes within which oxygen and nutrients (brought into the cell by the blood via the circulatory system) are processed by enzymes to reform adenosine triphosphate, or ATP, from adenosine diphosphate, or ADP.

***MMH*:** manual material handling.

***modality*:** in psychology and physiology, an avenue of sensation (eg, the visual modality).

***mode*:** in statistics, the value, of a series of values, that is the most frequently observed. A measure of central tendency. See mean; median.

***model*:** a representation of some aspect of an idea, an item, or a functioning.

***modulation*:** the variation in the value of some parameter that characterizes a periodic oscillation.

***monotonic*:** a relationship between two variables in which, for every value of each variable, there is only one corresponding value of the other variable. The graphical representation of a monotonic function is a steadily rising or steadily falling curve.

***morbidity*:** the prevalence of disease in a population.

***motion sickness*:** vomiting (emesis), nausea, or malaise provoked by actual or perceived motion of the body or its surroundings.

***motivation*:** a desire, or an incentive, to achieve.

***motor(ic)*:** in life sciences, a term used to refer to processes or anatomical areas associated with muscular action.

***motor unit*:** the set of all muscle filaments under the control of one efferent nerve axon.

***multiple correlation (R)*:** the relation between a dependent variable and two (or more) independent variables. See correlation.

***multivariate*:** any procedure in which more than one variable is considered simultaneously.

***muscle*:** a tissue bundle of fibers, able to contract or lengthen. Specifically, striated muscle (skeletal muscle) that tends to move body segments about each other under voluntary control. There are also smooth and cardiac muscles.

***muscle contraction*:** shortening of muscle tissue as a result of contractions of motor units. Usually, tension develops between the origin and insertion of the muscle as it contracts. See contraction.

***muscle fibers*:** elements of muscle, containing fibrils.

***muscle fibrils*:** elements of muscle fibrils, containing filaments.

***muscle filaments*:** elements of muscle fibrils (polymerized protein molecules) capable of sliding along each other, thus shortening the muscle and, if doing so against resistance, generating tension. See contraction.

***muscle spindle*:** an end organ in skeletal muscle that is sensitive to stretching of the muscle in which it is enclosed. Muscle spindles taper at both ends and lie parallel to regular muscle bundles.

***muscle strength*:** the capability of a muscle to generate force (tension) between its proximal (origin) and distal (insertion) ends.

***myelin*:** fatty substance that encloses some nerve fibers. Myelinated fibers propagate nerve impulses faster than unmyelinated fibers.

***myo*:** a prefix referring to muscle.

***myopia*:** see nearsightedness.

N

***narrowband filter*:** a bandpass filter for which the passband width is relatively narrow (eg, one-third octave or less).

***natural frequency*:** a frequency of free vibration resulting from only elastic and inertial forces of a mechanical system. See resonance.

***nearsightedness*:** inability to see distant objects distinctly, owing to an error of refraction, if accommodation is relaxed, in which parallel rays of light from an object at infinity are brought to focus in front of the retina. Also called myopia.

***nerve*:** a whitish cord made up of myelinated or unmyelinated neural fibers (or both) and held together by a connective tissue sheath. Nerves transmit stimuli to the central nervous system from sensors (extero- and interoceptors) (ie, afferent nerves) or in the reverse direction, to effectors (ie, efferent nerves) from the central nervous system. See myelin; axon; dendrite; synapse; effector.

***nerve cell*:** see neuron.

***nervous system*:** a system composed of neural tissue controlling the human body's structures and organs. May be subdivided in many ways (eg, central nervous system and peripheral nervous system, or autonomic nervous system and somatic nervous system).

***neuromuscular*:** relating to the relationship between nerve and muscle, especially the motor innervation of skeletal muscle.

***neuron*:** the functional unit of the nervous system, consisting of cell body, dendrite, and axon. See synapse.

***nociceptor*:** specialized nerve endings involved in the sensation of pain.

***noise*:** (i) any disagreeable or undesired sound; (ii) sound, generally of a random nature, the spectrum of which does not exhibit clearly defined frequency components; (iii) electrical (or mechanical) oscillations of an undesired or random nature.

***nominal scale*:** the simplest form of a scale, qualitative and consisting of a set of categories or labels. A nominal scale facilitates the identification of items, to determine whether they are equivalent and to count them. See interval; ordinal; ratio scale.

***normal distribution*:** distribution with a probability density function $P(x) = (1/\sigma\sqrt{2\pi})e^{-x^2_p/2\sigma}$, where σ is the standard deviation of the signal and x_p is the instantaneous magnitude. Also called the Gaussian distribution.

***nucleus pulposus*:** the soft fibrocartilage central portion of the intervertebral disc.

***null hyphothesis*:** the assumption, used to test statistical significance, that there is no difference between sets of data. See statistical significance.

***nystagmus*:** eye movements consisting of a slow drift of the eyeball, followed by a rapid return to the original position.

O

***objective*:** in life sciences, possessing the property of being real and measurable in physical units. Often used in contrast to subjective.

***occipital*:** the back of the head, skull, or brain.

***occular*:** relating to the eye.

***octave*:** the interval between two frequencies that have a frequency ratio of 2.

***octave bandwidth filter*:** a bandpass filter for which the passband is one octave, that is, the difference between the upper and lower cutoff frequencies is one octave.

***oculogravic illusion*:** the apparent tilt of the body relative to the visual scene when an observer is exposed either to linear translational acceleration or deceleration or to centripetal acceleration. See audiogravic illusion.

***oculogyral illusion*:** apparent movement of a point of light that is stationary with respect to an observer when the observer is rotated. See audiogyral illusion.

***oculomotor*:** related to eye movements and their muscular control.

***OD*:** overuse disorder. See repetitive strain injury.

***olecranon*:** the proximal end of the ulna.

***omphalion*:** the centerpoint of the navel.

***one-half octave*:** the interval between two frequencies that have a frequency ratio of $2^{1/2}$ (= 1.4142).

***one-third octave*:** the interval between two frequencies that have a frequency ratio of $2^{1/3}$ (= 1.2599).

***open-loop system*:** a system with no feedback. See closed-loop system.

***open system*:** a system that is influenced by inputs from the outside. See closed system.

***orbit*:** the eye socket.

***ordinal scale*:** a scale in which items are placed in order according to a characteristic. An ordinal scale is the simplest quantitative scale: Items can be ranked, and large values imply that there is more of the characteristic, but do not indicate how much more. Nonparametric statistics is based on ordinal scales. The only other valid statistical operations on ordinal data are the determination of medians, percentiles, interquartile ranges, and similar procedures. See interval; nominal; ratio scale.

***orthogonal*:** (i) at right angles; (ii) said of variables that are independent of each other.

***os*:** bone.

***oscillation*:** the variation, usually with time, of the magnitude of a quantity with respect to a specified reference when the magnitude is alternately greater and smaller than some mean value.

***osteoarthritis, osteoarthrosis*:** degeneration of the joints, with some loss of the low-friction cartilage linings and the formation of rough deposits of bone. The term osteoarthrosis is preferred when there is no inflammation.

***osteoporosis*:** the wasting, or atrophy, of bone. Bone becomes more porous, grows more brittle, and changes its geometry. Osteoporosis is often associated with aging, but also can arise from immobilization of the limbs and nutritional deficiencies.

***otoliths*:** crystalline particles of calcium carbonate and a protein adhering to the gelatinous membrane of the maculae of the uticle and saccule.

***overuse injury*:** see repetitive strain injury.

P, Q

***Pacinian corpuscle*:** specialized oval-shaped encapsulated nerve ending. The Pacinian corpuscle responds to pressure over a diffuse area and adapts rapidly. It is a principal means of sensing vibration (eg, 45–400 Hz, depending on various conditions). Pacinian corpuscles are located in the palmar skin of the hands, in the plantar skin of the feet, on some tendons and ligaments, and in other locations. They are the largest nerve endings found in the skin (up to about 2 mm in length) and consist of concentric layers that make them look like a minute onion. Also called lamellated corpuscle. See Meissner's corpuscle.

***palmar*:** relating to the palm of the hand or sole of the foot. Also called plantar or volar. Opposed to dorsal.

***palpation*:** examination by feeling with the hands.

***parasympathetic nervous system*:** see autonomic nervous system.

***patella*:** the kneecap.

***pathology*:** the science of diseases.

***peak-to-peak value*:** the algebraic difference between the extreme values of a quantity.

***peak value*:** the maximal value of a quantity during a given interval. The peak value is usually taken as the maximum deviation of the quantity from the mean value.

***Pearson product–moment correlation*:** see product–moment correlation.

***pelvis*:** the basin-shaped ring of bone consisting of the ilium, pubic arch, and ischium.

***perception*:** awareness of some event; process by which the mind refers its sensations to external objects as their cause.

***peripheral nervous system*:** network of afferent (see there) and effluent (see there) nerves. Often abbreviated as PNS.

***phalanges (singular: phalanx)*:** bones of the fingers or toes. There are 14 on each hand, three for each finger and two for the thumb.

***phase*:** the fractional part of a period (in radians or degrees) through which a sinusoidal motion has advanced from some reference time.

***phon*:** unit of perceived loudness level. The loudness level of a sound is the sound pressure level of a 1,000-Hz pure tone judged by the listener to be equally loud.

***physiology*:** the biological science of the normal functions and phenomena of living organisms.

***pink vibration*:** a random vibration that has an equal mean-square acceleration for any frequency band having a bandwidth proportioned to the center frequency of the band. The energy spectrum of pink random vibration, as determined by octave, one-third-octave, etc., filters has a constant value. See white vibration.

***pitch*:** (i) perceived sound, depending on the frequency; (ii) rotational motion about the y-axis. See coordinate system.

***pivot*:** to rotate a body joint; to abduct or adduct the joint.

***placebo*:** a substance with no intended or known medicinal effect.

***plane*:** see ear–eye plane; Frankfurt plane; frontal plane; medial plane; transverse plane.

***plantar*:** relating to the sole of the foot. See palmar.

***plasma*:** the fluid portion of blood or lymph.

***plethysmograph*:** a device for the measurement and graphical representation of the changes in volume (often blood volume), of some part of the body.

***PNS*:** peripheral nervous system. See there.

***popliteal*:** pertaining to the ligament behind the knee or to the part of the leg behind the knee.

***posterior*:** at the back of the body; opposed to anterior.

***predisposition*:** a condition of having special susceptibility to a condition or disease.

***probability* (p):** an expression of the likelihood of occurrence of an event; usually expressed as a ratio of the number of occurrences of a given event of interest to the number of occurrences of all types of events considered.

***product–moment correlation* (r):** the linear correlation between two variables, based on the calculation of the mean of the products of the deviations of each score from the mean value of each variable. Also called the Pearson product–moment correlation.

***pronation*:** twisting the forearm about its long axis counterclockwise when one looks down one's own right arm, clockwise down one's left arm. Opposite of supination.

***prone*:** the position of the body when lying face downward. Opposite of supine.

***proprioception*:** the perception of information about the position, orientation, and movement of the body and its parts. See interoception; exteroception.

***proprioceptor*:** any sensory receptor that mediates proprioception.

***protagonist*:** same as agonist.

***proximal*:** the end of a body segment nearest the center of the body; opposed to distal.

***psychology*:** the science concerned with the behavior and mental processes of humans and animals.

***psychophysics*:** science concerned with the quantitative relations between the perception of stimuli and their physical characteristics.

***pure tone*:** a sound whose characteristic (eg, pressure) varies sinusoidally with time.

R

***radian (rad)*:** unit of angular measure equal to the angle subtended at the center of a circle by an arc the length of which is equal to the radius of the circle. (There are 2π radians in 360 degrees; 1 rad $\approx$ 57.3 degrees.)

***radius*:** the bone of the forearm on its thumb side.

***random process*:** a set (ensemble) of time functions that have no specific pattern or purpose but that can be characterized through statistical properties. Also called a stochastic process.

***range*:** the interval between the highest and lowest scores in a distribution.

***rank*:** in statistics, the position of a score relative to all other scores ordered according to their value. When ranked, scores form an ordinal scale.

***rank-order correlation*:** a correlation between two variables based on the differences in the rank orders of the two variables. Also called Spearman rank-order correlation.

***rate coding*:** the time sequence in which efferent signals arrive at a motor unit to cause contractions.

***rating scale*:** a scale of words or numbers along some dimension that may be used to report subjective reactions. Many types of rating scale may be devised. Scales may be bipolar, giving the options "good" and "bad," or unipolar. The number of steps on the scale is often in the range from 5 to 9. See scaling.

***ratio scale*:** a scale in which the ratios of values on the scale have quantitative significance. Ratio scales are the most powerful scales and hence are used for as many physical measurements as possible. They have true zero values, and an item that has twice as much of the characteristic that is of interest as another item is represented by a value twice as large. All valid mathematical operations can be performed on data from a ratio scale. See interval; nominal; ordinal scale.

***reaction time (RT)*:** the time between the presentation of a stimulus and the beginning of an observer's response. Simple reaction time is the time to make a single simple response to a single stimulus. Choice reaction time is the time involved when there are two or more stimuli and two or more corresponding possible responses to every stimulus.

***reclining posture*:** a body posture at some angle between seated upright and being supine.

***recruitment coding*:** the time sequence in which efferent signals arrive at different motor units to cause them to contract.

***reflex*:** an involuntary reaction in response to a stimulus applied peripherally and transmitted to the nervous centers of the brain or spinal cord. A reflex involves a receptor that is sensitive to the stimulus, an effector that responds to the stimulus, and a reflex arc between receptor and effector.

***reliability*:** the reliability of a test or measurement method is given by the degree to which it produces similar values when applied repeatedly under the same test conditions. See sensitivity; specificity; validity.

***repeatability*:** the extent to which the outcome of an experiment would recur if the experiment were repeated.

***repetition*:** performance of the same activity more than once.

***repetitive strain injury*:** an injury that arises from repeated strains (traumata) when a single strain (a trauma) gives no observable signs of injury. Often abbreviated as RSI. Also called cumulative injury or overuse injury.

***residual*:** in statistics, the part of the variance that cannot be attributed to the factors which have been considered.

***resonance*:** the increase in amplitude of oscillation of an electrical or mechanical system exposed to a forced oscillation whose frequency of excitation is equal or very close to the natural undamped frequency of the system.

***retina*:** the light-sensitive membrane covering the inner rear surface of the eye.

***rhythmic*:** the same action repeated in equal intervals.

***risk*:** probability of injury. See danger, hazard.

***rms*:** see root-mean-square value.

***rod*:** receptor of light in the retina. Rods function under conditions of low illumination (scotopic) vision. They do not mediate color vision or render perception of fine detail. There are no rods in the fovea. See cone.

***roll*:** rotational motion about the *x*-axis. See coordinate system.

***root-mean-square (rms) value*:** the square root of the average of the squares of a set of numbers.

***RSI*:** repetitive strain injury. See there.

***Ruffini ending*:** specialized encapsulated nerve ending (corpuscle) found in the dermis of hairy skin. The Ruffini ending responds to lateral stretching of the skin (and possibly warmth) and adapts slowly. See Merkel's disc.

S

***saccade*:** a quick jump of the eyes from one fixation point to another.

***saccule*:** the smaller of the two vestibular sacs in the membranous labyrinth of the inner ear. Like the utricle, the saccule contains a layer of receptors that are sensitive to translational forces arising from translational acceleration of the head or rotation of the head in an acceleration field. See vestibular system; macula; utricle.

***sacrum*:** segment of the vertebral column forming part of the pelvis. The sacrum is formed from the fusion of five sacral vertebrae. The sacrum articulates with the last lumbar vertebra, the coccyx, and the hip bone on either side.

***sagittal*:** ("in the line of an arrow shot from the bow") pertaining to the medial (midsagittal) plane of the body or to a parallel plane.

***saturation*:** attribute of color perception that determines the degree to which a chromatic color differs from an achromatic color, regardless of lightness.

***scalar*:** any quantity that is completely defined by its magnitude. See vector.

***scale*:** a procedure for placing items, individuals, events, or sensations in some series. Four types of scale are possible: a nominal scale, an ordinal scale, an interval scale, and a ratio scale. See these terms.

***scaling*:** in psychology, the determination of (a point on) a scale, along a psychological dimension (eg, discomfort) that has a continuous mathematical relation to some physical dimension (eg, magnitude of vibration). See rating scale; scale.

***scapula*:** the shoulder blade.

***sciatica*:** pain in the area of the sciatic nerve.

***scoliosis*:** a sideways curvature of the spine. See kyphosis; lordosis.

***semantic scale*:** a set of words describing various degrees of a characteristic, formed in a nominal or interval scale. See interval scale, nominal scale, rating scale, scale.

***semicircular canals*:** three small membranous ducts (tubes) that form loops of about two-thirds of a circle within each inner ear.

***sensation*:** the reception of stimuli at the sensors and their translation into neural impulses; followed by perception. See perception.

***sensitivity*:** (i) as applied to a transducer, the ratio of a specified output quantity (eg, electrical charge) to a specified input quantity (eg, acceleration); (ii) in screening, the proportion of individuals with a positive test result for the effect that the test is intended to reveal. See reliability; specificity; validity.

***sensorimotor*:** relating to a signal transmitted via a neural circuit from a receptor to the central nervous system and back to a muscle.

***shock*:** mechanical shock exists when a force, a position, a velocity, or an acceleration is suddenly changed so as to excite transient disturbances in a system.

***sign*:** in medicine, any abnormality that is discovered by a physician during an examination of a patient. (A sign is an objective symptom of disease.) See symptom.

***skeleton*:** in humans, all or part of the 206 separate bones in the body.

***skin*:** the outer covering of the body, consisting of epidermis, dermis, and subcutaneous tissues.

***slipped disc*:** see prolapsed disc.

***Snellen acuity*:** visual acuity measured using a standard chart containing rows of letters of graduated sizes, expressed as the distance in which a given row of letters is correctly read, compared to the distance at which the letters can be read by a person with clinically normal eyesight. For example, an acuity score of 20/50 indicates that the individual who was tested can read letters at the distance of 20 feet that a normally sighted person can read at 50 feet.

***Snellen chart*:** chart used to obtain approximate measures of visual acuity. Letters of various sizes are read at a fixed distance D (often 20 ft or 6 m). Visual acuity is given by D/D', where D' is the distance at which the smallest letters should be read by a normal eye. Normal acuity may be expressed as 20/20 (in ft) or 6/6 (in m). 6/12 indicates that letters which should be identified at 12 m were recognized only at 6 m. See diopter.

***somatic*:** relating to the body.

***somatosensory*:** see kinesthetic.

***sound*:** (i) acoustic oscillation capable of exciting the sensation of hearing; (ii) the sensation of hearing excited by an acoustic oscillation.

***spasm*:** an involuntary muscular contraction.

***Spearman rank-order correlation*:** see rank-order correlation.

***specificity*:** in screening, the proportion of individuals with a negative test result for what the test is intended to reveal (ie, true negative results as a proportion of the total of true-negative and false-positive results). See reliability, sensitivity; validity.

***spectrum*:** a description of a quantity as a function of frequency or wavelength.

***sphyrion*:** the most distal extension of the tibia on the medial side under the malleolus.

***spinal cord*:** the column of neural tissue running the length of the vertebral column down to the second lumbar vertebra.

***spinal nerve*:** in humans, one of the 31 pairs of nerves coming from the spinal cord: eight cervical, 12 thoracic, five lumbar, five sacral, and one coccygeal.

***spine*:** the stack of vertebrae. See vertebral column.

***spinous (or spinal) process of a vertebra*:** the posterior prominence.

***spondylosis*:** degeneration of vertebrae and their joints. Vertebral ankylosis; osteoarthritis of vertebrae. Narrowing of joint spaces or bony outgrowths on vertebrae or along edges of degenerating discs. These outgrowths may press on nerves at the point where they join the spinal cord. Some signs of spondylosis are apparent in most persons who are past middle age; they may cause stiffness and intermittent pain.

***static*:** at rest or in equilibrium; not dynamic.

***statics*:** a subdivision of mechanics that deals with bodies at rest.

***statistical significance*:** the probability that an result would have occurred if only chance factors were operating; hence, the degree to which the result may be attributed to some systematic effect. A probability less likely than 5 cases out of 100 is often chosen as the significance level, expressed as a significance of 5 percent, or $p = 0.05$. A lower probability is then required (ie, $p < 0.05$) if a result is not assumed to be a chance finding. The significance of a set of test results is determined by statistical tests. See type-I error, type-II error

***sternum*:** the breastbone.

***Stevens' (power) law*:** the relationship between the magnitude P of a psychological sensation produced by a stimulus of magnitude I given by $P = kI^n$. The value of n is a characteristic of the physical stimulus—for example, noise or vibration—while k depends on the units of measurement. See Fechner's law; Weber's law.

***stiffness*:** the ratio of change of force (or torque) to the corresponding change in translational (or rotational) displacement of an elastic element.

***stochastic process*:** see random process. (Derived from the Greek word for "guess.")

***strain*:** (i) in engineering, a dimensionless value given by the ratio of the deformation caused by a stress to the size of the material stressed; (ii) in life sciences, the experienced result of stress generated by physical or mental demands. See stress; stressor.

***stress*:** (i) in engineering, the load applied to a material that causes a change in its dimensions (ie, strain); (ii) in psychology, either the cause of strain or the resulting strain itself. See stressor.

***stressor*:** a cause of strain or stress. The term helps to remove the confusing use of the word "stress": the cause is a stressor and the effect may be called either the stress (psychological use) or the strain (engineering use).

***stylion*:** the most distal point on the styloid process of the radius.

***styloid process*:** a long, spinelike projection of a bone.

***sub-*:** a prefix designating below or under.

***subjective*:** in life sciences, something that is dependent on an individual. See objective; scale; scaling.

***superior*:** above, in relation to another structure; higher.

***supination*:** twisting the forearm about its long axis clockwise when one looks down one's own right arm, counterclockwise down one's left arm. Opposite of pronation.

***supine*:** the position of the body when lying face upward. Opposite of prone.

***supra-*:** prefix denoting above, superior, or in very large quantities.

***sympathetic nervous system*:** see autonomic nervous system

***symptom*:** in medicine, an abnormality in function, appearance, or sensation.

***synapse*:** the junction of one neuron with another, where nerve impulses are transmitted (chemically or electrically) from the axon of one neuron to the dendrites of the other.

***syndrome*:** in medicine, a combination of signs and symptoms that collectively indicate a disease.

***synergistic effect*:** the effect of a combination of two or more stressors. A synergistic effect is greater than the arithmetic sum of the effects of the individual stressors.

***systole*:** the contraction of the heart by which blood is driven from the ventricles through the aorta and pulmonary artery. See diastole.

T

***tactile*:** relating to touch or the sense of touch perceived through the skin. See corpuscle; haptic; kinesthetic; Meissner's corpuscle; Merkel's disc; Pacinian corpuscle; Ruffini ending.

***tarsus*:** the collection of bones in the ankle joint.

***tendon*:** fibrous cord joining a muscle to a bone.

***tendonitis*:** inflammation of a tendon.

***tendon reflex*:** a reflex muscular contraction elicited by a sudden stretching of a tendon. Sensors in tendons sense stretching and trigger a reflexive contraction of the muscle to oppose movement and maintain posture. The "knee jerk" caused by percussion of the tendon at the knee is a tendon reflex.

***tenosynovitis*:** inflammation of a tendon and its sheath.

***theory*:** a reasoned explanation of how something occurs, or may occur, but without absolute proof. A theory is more securely established than a hypothesis.

***thoracic outlet syndrome*:** compression of fifth cervical and first thoracic nerves and the subclavian artery by muscles in the region of the first rib and clavicle.

***threshold*:** see absolute threshold; differential threshold.

***tibia*:** the main bone of the lower leg (shinbone).

***tibiale*:** the uppermost point of the medial margin of the tibia.

***tinnitus*:** a subjective sensation of noises in the ear (ringing, whistling, booming, etc.) experienced in the absence of an external acoustic stimulus.

***tissue*:** a collection of similar cells and their surrounding structures.

***tolerance*:** the ability to endure a stimulus without harm.

***tone*:** (i) in muscles, a state in which there is normal muscular tension with slight stretching of the muscles maintained by proprioceptive reflexes; (ii) in acoustics, a sound at one given frequency.

***torsion*:** the act of twisting by the application of forces at right angles to the axis of rotation.

***touch*:** the act of, or the sensation that arises from, contact between the body and an object.

***tracking task*:** a task that involves continuously following a target. The two principal types of tracking task are pursuit tracking, in which movements of the target are directly indicated, and compensatory tracking, in which the difference between the actual and the desired location of the target is indicated.

***tragion*:** the point located at the notch just above the tragus of the ear.

***tragus*:** the conical eminence of the auricle (pinna, or external ear) in front of the ear hole.

***trait*:** a characteristic of a person.

***transducer*:** a device designed to receive energy from one system and supply energy, of either the same or a different kind, to another in such a manner that the desired characteristics of the input energy appear at the output.

***transfer function*:** a mathematical relation between the output (response) and the input (excitation) of a system.

***transfer impedance*:** in a mechanical sense, the ratio of the force taken at one point in a mechanical system to the velocity taken at another point in the same system during simple harmonic motion.

***transfer mobility*:** the ratio of the velocity taken at one point in a mechanical system to the force taken at another point in the same system during simple harmonic motion.

***translation*:** the (linear, straight-line) movement of an object so that all its parts follow the same direction (ie movement without rotation.)

***transmissibility*:** the nondimensional ratio of the response amplitude of a system in steady-state forced vibration to the excitation amplitude, expressed as a function of the vibration frequency. The ratio may be of forces, displacements, velocities, or accelerations.

***transmission*:** the sending of muscle force across one or more body joints.

***transverse plane*:** a plane across the body at right angles to the frontal plane and the medial (midsagittal) plane. See ear–eye plane, Frankfurt plane.

***trauma*:** an injury caused by harsh contact with an object. (From the Greek word for "wound.")

***triceps*:** the muscle of the posterior upper arm crossing the elbow.

***trochanterion*:** the tip of the bony lateral protrusion of the proximal end of the femur.

***tuberosity*:** a (large) rounded prominence on a bone.

***Type-I error*:** the erroneous rejection of a true hypothesis. A Type-I error is an error that arises in statistical tests and that is more likely when a low level of significance is required, such as $p = 0.05$, before rejecting the null hypothesis. See statistical significance.

***Type-II error*:** the failure to reject a false hypothesis. A type-II error is an error that arises in statistical tests and that is more likely when a high level of significance is required, such as $p < 0.001$, before rejecting the null hypothesis. See statistical significance.

U–Z

***ulna*:** the bone of the forearm on the side of the little finger.

***ultra-*:** -prefix meaning beyond, extreme, or excessive.

***ultrasonic frequency*:** any frequency higher than normally audible sound waves (eg, above 20,000 Hz).

***umbilicus*:** depression in the abdominal wall where the umbilical cord was attached to the embryo.

***utricle*:** the larger of the two vestibular sacs in the membranous labyrinth of the inner ear.

***validity*:** in tests, a measure of how well the test assesses what it purports to assess. See reliability; sensitivity; specificity.

***Valsalva maneuver*:** forced expiratory effort with either closed glottis or closed nose and mouth. The Valsalva maneuver may be used to inflate the eustachian tube; it may also increase one's tolerance to high magnitudes of downward acceleration, by preventing blood from returning to the heart from the head.

***variable*:** see dependent variable; independent variable.

***variance*:** the square of the standard deviation.

***vascular*:** relating to, or containing, blood vessels.

***vaso-*:** combining form denoting blood vessel. Also vas- and vasculo-.

***vasoconstriction*:** narrowing of a blood vessel.

***vasodilation*:** enlargement of a blood vessel.

***vasomotor*:** relating to the nerves that control the muscular walls of blood vessels.

***vasospasm*:** contraction of the muscular walls of the blood vessels.

***vector*:** a quantity that is completely determined by its magnitude and direction. See scalar.

***vein*:** a blood vessel carrying blood toward the heart. See artery.

***ventral*:** pertaining to the anterior side of the trunk.

***venule*:** a minute vein.

***vertebra*:** one of the bones of the spinal column. In humans there are 33 vertebrae: seven cervical vertebrae, 12 thoracic vertebrae, five lumbar vertebrae, five sacral vertebrae (fused as one bone, the sacrum), and four coccygeal vertebrae (fused as one bone, the coccyx).

***vertebral column*:** the series of 33 vertebrae that extend from the coccyx to the cranium; the backbone or spine.

***vertex*:** the top of the head.

***vertigo*:** an inappropriate sensation of movement of the body or the visual field caused by a disturbance of the mechanisms responsible for equilibrium.

***vessel*:** any duct, tube, or canal conveying body liquid, such as blood.

***vestibular system*:** collective term for the three semicircular canals and the two vestibular sacs (utricle and saccule) within the labyrinth of the inner ear.

***vestibule*:** any small cavity at the front of a canal; especially the middle part of the inner ear containing the utricle and saccule.

***vibration*:** the variation with time of the magnitude of a quantity that is descriptive of the motion or position of a mechanical system when the magnitude is alternately greater and smaller than some average value.

***viscera*:** the internal organs of the body, including the digestive, respiratory, urogenital, and endocrine systems and the heart, spleen, etc.

***viscosity*:** the resistance of a fluid to shear forces and, therefore, to flow.

***volar*:** relating to the palm of the hand or sole of the foot. See palmar, plantar.

***volition*:** conscious, voluntary action. See reflex.

***Weber-Fechner law*:** the combination of Weber's law and Fechner's law.

***Weber's law*:** law stating that the just-noticeable difference JND in the stimulus magnitude I is proportional to the magnitude of the stimulus. The relation may be expressed as $\Delta I/I$ = constant. See Fechner's law; Stevens's law.

***white vibration*:** a random vibration that has equal mean-square acceleration for any frequency band of constant width over the spectrum of interest. See pink vibration.

***x-axis*:** see coordinate system.

***yaw*:** rotational motion about a z-axis. See coordinate system.

***y-axis*:** see coordinate system.

***z-axis*:** see coordinate system.

Index